Coalbed Methane in China

Yan Song • Xinmin Zhang • Shaobo Liu

Coalbed Methane in China

Geological Theory and Development

Science Press
Beijing

Springer

Yan Song
China University
of Petroleum (Beijing)
Beijing, China

Xinmin Zhang
Xi'an Research Institute
China Coal Research Institute
Xi'an, Shaanxi, China

Shaobo Liu
Research Institute of Petroleum
Exploration and Development, PetroChina
Beijing, China

ISBN 978-981-33-4724-3 ISBN 978-981-33-4725-0 (eBook)
https://doi.org/10.1007/978-981-33-4725-0

Jointly published with Science Press
The print edition is not for sale in China (Mainland). Customers from China (Mainland) please order the print book from: Science Press.

This Springer imprint is published by the registered company Springer Nature Singapore Pte Ltd.
The registered company address is: 152 Beach Road, #21-01/04 Gateway East, Singapore 189721, Singapore

Foreword by Yan Song

The CBM National Basic Research Program of China (973 Program) will publish the *Basic Research Series of CBM Reservoir Forming Mechanism and Economic Development* (11 volumes), involving basic theories and applications in status quo and research, generation and storage, reservoir and accumulation, resource potential, and seismic exploration, economical, and efficient development of coalbed methane. It covers a wide range of areas, and is a very meaningful systematic science project. Asked by the Project Chief Scientist, I am happy to give the following text to show my support and congratulations.

CBM is an important unconventional natural gas resource. In the 1980s, the United States realized the commercial exploitation and utilization of CBM and established a coalbed methane industry with considerable scale. China is a country having abundant coal resources with quite rich CBM. According to the latest forecast, in the coalfield shallower than 2000 m, the CBM resources are 31×10^{12} m^3 (excluding the CBM in lignite), which is roughly equivalent to the conventional onshore natural gas resources in China. If the coalbed methane in lignite is calculated, the amount will be even more impressive. In terms of occurrence conditions and economic and social development demand of fossil energy resources in China, CBM is the most realistic replacement energy in China in the new century after coal, oil, and natural gas. At the same time, the development and utilization of CBM can eliminate the hazard of coalmine gas explosion and protect the environment.

Since the 1980s, China has carried out modern CBM technology research, development, and application. As of the first half of 2004, nearly 250 CBM wells have been drilled, and more than 10 CBM test well groups have been built in Liulin, Panzhuang, Dacheng, and Huainan, etc. Among them, three well groups have produced commercial CBM in Liujia, Fuxin; Panzhuang, Jincheng; and Shizhuang, Qinshui. In basic researches on coal reservoir characteristics and resource evaluation, and anthracite CBM development, etc., achievements have been made. However, in general, China's CBM industrialization process is slow and cannot meet the needs of national economic and social development.

CBM is different from conventional natural gas in geochemistry, reservoir, accumulation mechanism, flow mechanism, gas well production performance, etc. It is necessary to proceed with exploration and development of CBM using different theories and methods from conventional oil and gas. In addition, as China is a combination of several major plates after repeatedly colliding, and it is still affected by the joint action of the Eurasian, India, and Pacific plates. The coalification periods are more and long, and late reconstruction is frequent and strong, which makes the complexity and diversity of CBM geological conditions in China. Therefore, the US CBM theory developed on the single North American continent plate is not fully adapted to China's situation.

It is an urgent task for Chinese scientists and technicians to establish the basic CBM theory in line with China's geological features and providing scientific and technological support for building China's CBM industry. Supported by the Ministry of Science and Technology of the People's Republic of China, the 973 Program "Basic Reasearch on China Coalbed Methane Reservoir Forming Mechanism and Economic Development" was proposed, which gathers experts, scholars, and elites from the oil industry, coal industry, Chinese Academy of Sciences and higher institutes in China. It follows a new scientific research concept of multidisciplinary cooperation, and the combination of production with research, and changed the passive situation where departments and disciplines work separately.

The project focuses on national goals and key scientific issues and organizes various forces to carry out extensive and in-depth basic research from a high starting point on main scientific issues of CBM industrialization in China, such as the cause, reservoir performance, dynamics, genesis, accumulation, geophysical response, gas flow, and output mechanism of CBM. These results have theoretical guidance for the formation and development of China's CBM industry and represent the overall level of present basic researches on CBM in China.

Timely collection and publication of research results can demonstrate the situation of China's basic research on CBM and is an important link to strengthen academic exchanges, disseminate knowledge, and accelerate the transformation of scientific research results into real productivity. New scientific theories and technical methods will accelerate the industrialization of CBM in China and contribute to the development of CBM in the world. Let us all work together to meet the growing demand for clean energy in China's economic and social development.

August 2004

Yan Song
Academician of the Chinese
Academy of Sciences
Beijing, China

Foreword by Yan Song

CBM, commonly known as methane gas, is a self-generated, self-contained, unconventional natural gas that is present in coal seams in an adsorbed state. The development and utilization of CBM is a two-pronged thing. It can not only be used as a supplementary resource for conventional oil and gas, but more importantly, greatly improve the safe production conditions of coalmines and reduce or even eliminate coalmine accidents.

As a kind of unconventional natural gas resource with huge resources, CBM has gradually evolved from research to development and utilization. The United States is the first country to develop and utilize CBM. The CBM industry started in the 1970s and achieved large-scale commercial development in the 1980s. The production of CBM is growing rapidly. The annual output raised from less than 1×10^8 m^3 in 1980 to 100×10^8 m^3 in 1990. In the early 1990s, stable production was 200×10^8 m^3. In 2002, the annual output was 450×10^8 m^3, accounting for 7.9% of the annual production of natural gas in the United States. The development of CBM is quite successful in the US, specifically the San Juan Basin in Colorado and New Mexico and the Black Warrior Basin in Alabama. CBM wells are generally considered to be low yield, but some are quite productive. For example, in 1996, I visited the San Juan Basin under the control of ARCO, where 110 CBM wells were with daily gas production of 660×10^4 m^3. Therefore, it is of great theoretical and practical significance to study the high-yield regulations in the low production of CBM. Drawing on the successful experience of the United States, Australia has also carried out exploration and testing of CBM and achieved certain results. In addition, Czech, Poland, Belgium, the United Kingdom, Russia, Canada, and other countries have carried out exploration and development tests for CBM. At present, the world's research on CBM is deepening, and the development area is expanding, and the position of CBM in energy is rising.

China is a large coal resource country with abundant CBM resources (according to the estimate from the "7th Five-Year Plan," the resource shallower than 2000 m is 31×10^{12} m^3). China's CBM exploration and development is obviously behind the United States. Since the 1980s, the United States' CBM production technology has been actively introduced to carry out exploration and development experiments. However, the overall results are not significant. The main cause is that China's CBM geological conditions are complex, and the forming mechanism of CBM reservoirs is still not clear. The exploration and development of CBM are quite different from conventional natural gas, and there is a lack of perfect and mature theory. Therefore, the basic theoretical research on the exploration and development of CBM in China will be the premise to promote faster development of the industry. As a national key scientific and technological project, the "Development Research of Coal-forming Gas" performed 20 years ago, is a good example of proceeding the development of natural gas industry in China. I have worked with other scientists, appealing to the Ministry of Science and Technology of the People's Republic of China for the project establishment of CBM research. Today, this wish has finally been realized. The "Basic Research on CBM Reservoir Forming Mechanism and Economic Development" has been officially implemented. This is a good news and a great event that should be celebrated. The project will solve some major problems in the exploration and development of CBM in China, deepen the understanding of the mechanism of CBM accumulation and reclamation, and give birth to a promising future of CBM development.

I am happy to be appointed one of the experts to track the project. From the initiation of the project to the start of research, I have been paying attention to its progress and research results. The early achievement is remarkable, including many new discoveries, new understandings, new ideas, and innovations. Song Yan and Zhang Xinmin, two chief scientists, plan to publish the series of the *Basic Research Series of CBM Reservoir Forming Mechanism and Economic Development* (hereinafter referred to as "Series") (11 volumes). The "Series" contains all aspects of CBM exploration and development, including the references collected and analyzed —the theory basis "CBM Reservoir Forming Mechanism and Economic Development," and the results of various research projects and topics. The "Series" studies many basic and critical issues related to technology, theory and methods of CMP exploration and development from CMB forming dynamics process and resource contribution, heterogeneity, and control mechanism of physical properties, adsorption characteristics and gas storage mechanism, dynamic and accumulation conditions and model, CMB recoverable resources, seismic response of CMB reservoirs, basic theory and technology of CMB development. It is a project that has not been reached before. The purpose of the series is to provide a systematic theory of CMB exploration and development with Chinese characteristics, which is lacking in China. In addition, the chief scientists intend to fully demonstrate the outstanding achievements of China's CMB research to scholars in geosciences and CMB for mutual learning and exchange. The "Series" is knowledge accumulation, regular summaries, and innovations in this field. The publication of this series will be of great benefit to scholars, professionals, and college students engaged in CMB work, and will inevitably have an important impact and promotion on the CBM industry.

The chiefs and authors of the "Series" are young or middle-aged mainstay. The project gave them an opportunity. They are energetic, knowledgeable, diligent, and innovative. The "Series" drafted by them is solid and knowledgeable.

I would like to wish that the *Basic Research Series of CBM Reservoir Forming Mechanism and Economic Development* will be successfully published in succession, and become a comprehensive literature of CBM theory and practice.

August 2004

Yan Song
Academician of the Chinese
Academy of Sciences
Beijing, China

Preface

With the rapid development of the national economy, China's demand for energy continues growing, and the contradiction between oil and gas supply and demand has become increasingly prominent. From a net exporter of crude oil in 1993 to an importer, China's demand for crude oil import has increased year by year as domestic demand continues growing. In 2007, China's net crude oil imports were 15928×10^4 t, a year-on-year increase of 14.7%. The external dependence of crude oil reached 46.05%. According to the forecast, like crude oil, natural gas will have a supply gap in recent years. CBM is an unconventional natural gas resource-rich and less developed in China. The total resource of CBM with a depth of less than 2000 m in China is 32.86×10^{12} m^3, of which the technically recoverable resource is 13.90×10^{12} m^3. The prospective resources of CBM are roughly equivalent to the conventional natural gas in China. CBM is a clean energy source that is second after conventional natural gas. The development and utilization of CBM have important practical significance for ensuring the long-term rapid development of China's natural gas industry.

Coalmine safety is very important for the development of coal industry. Since 2001, with the increase of CBM (gas) extraction, the government has attached great importance to and invested a lot of human and financial resources. The death toll caused by coalmine accidents has generally declined. However, the number of coalmine accidents and casualties remains high. In 2006, there were 2945 coalmine accidents and 4746 deaths, including 327 gas accidents and 1319 deaths, accounting for 11.1% and 27.8% of industrial accidents in the coal industry, respectively. Statistics show that the number of people killed in gas accidents in recent years accounted for about 25% to 40% of the deaths from industrial accidents in the coal industry in China, causing huge economic losses. By "first producing gas and then coal," a coordinated development model between CBM and coal can be established, and the common development of CBM industry and coal industry will be realized, which will provide an important guarantee for the harmonious development of the national economy and society.

Environmental protection is an important part of a sustainable development strategy in China and the world in the twenty-first century. The air pollution caused by coal-based primary energy consumption structure is very serious, which leads to extensive acid rain and deteriorated urban air. The distribution of acid rain accounts for 8.4% of China's land area. The main component of CBM is methane (CH_4), while the "greenhouse effect" of methane is 22 times that of carbon dioxide (CO_2). In China, a large amount of CH_4 is directly discharged into the atmosphere due to coal mining. The higher CH_4 emissions make China face increasing pressure from the international community. As an energy, the CO_2 released by CBM burning the same calorific value is 50% less than oil and 75% less than coal. The pollutants produced by CBM combustion are generally only 1/40 of oil and 1/800 of coal. Accelerating the development of the CBM industry and rapidly increasing the proportion of natural gas in the national primary energy consumption structure can effectively reduce CH_4 and CO_2 emissions, which is an important way to improve and protect the living environment in China.

Based on the needs of national energy development, in 2002, the Ministry of Science and Technology of the People's Republic of China established the National Basic Research Program of China, the "Basic Research Series of CBM Forming Mechanism and Economic Development," which mainly addresses the subject issues of CBM exploration and development. The project was completed in November 2008. This book reflects the comprehensive research results of the project.

In terms of basic theoretical research, this book reveals the laws and mechanisms of CBM genesis, occurrence, accumulation, and permeability in China, systematically establishes CBM geological theory system and enriches and perfects natural gas geological theory. It is fundamental to CBM resource evaluation, accumulation prediction, and economic development.

1. Established a CBM genetic-type classification and tracer index system. It's proposed that secondary biogas is one of the important genetic types of CBM in China. The evidence of the existence of secondary biogas was determined. The generation conditions and formation mechanism of secondary biogas have been investigated.
2. Based on the theory of adsorption potential, models of coal adsorbing methane and sectional coal adsorption isotherms of different coal ranks at different temperatures and pressures were established. They reveal the adsorption during the formation of CMB, and provide a theoretical basis for understanding CBM accumulation mechanism, conducting CBM resource evaluation and development technology.
3. Defined the concept of CBM reservoir. Structure, caprock, and hydrodynamic conditions are controlling factors. Structure and hydrodynamics are controlling mechanisms on CBM reservoirs. This is a scientific basis for predicting CBM enrichment.
4. Established the theory of elastic energy and controlling effect of coal reservoir. It is pointed out that the mechanism of permeability changes when producing CBM is the combined effect of closed fractures caused by increasing stress and shrinked coal seams caused by gasification. That is to say, the theory of elastic self-regulation effect of coal reservoirs guides how to select favorable targets and drainage system.

In terms of method and technology, this book has formed a series of technologies covering CBM recoverable resource prediction, comprehensive geological evaluation, geophysical exploration, and optimization of production design, and served the implementation of coalbed methane development. Field application has proved effective.

1. Established a new "CBM resource classification system," which is consistent with conventional natural gas and international codes. A concept was proposed for technically recoverable resources of CBM and the systematic, scientific, and operable method for predicting CBM recoverable resources was developed for the first time in China. It is predicted that the technically recoverable resources of CBM are 13.90×10^{12} m^3 in China.
2. Established a prediction and evaluation method of CBM exploration and development zones, and found high, medium, and low metamorphic and unmetamorphic (lignite) favorable coal zones which are targets for the exploration and development of CBM resources.
3. Developed three-dimensional three-component seismic and AVO response detection techniques, which made good results in the CBM experimental area in Huainan. The technologies are worth promoting.
4. Independently developed a numerical simulation system for CBM reservoir evaluation, hydraulic fracturing, and pinnate horizontal well optimization design, which filled the gap in this field in China, and made good results in southern Qinshui Basin.

The book consists of 11 chapters. The first chapter introduces the status quo and research of CBM exploration and development. Chapters from the second chapter to the ninth chapter introduce the theory and evaluation technology. The tenth chapter and the eleventh chapter discuss the production mechanism and technology. The book is integral and covers the basic research results from exploration to development of CBM.

The preface and the first chapter of the book were written by Song Yan, Zhang Xinmin, and Liu Shaobo; the second chapter by Tao Mingxin and Xie Guangxin, et al.; the third chapter by Tang Dazhen and Wang Shengwei; the fourth chapter by Zhang Qun, Sang Shuxun, et al.; the fifth chapter by Qin Yong, Hou Quanlin, Song Yan, et al.; the sixth chapter by Song Yan, Liu Honglin and Hong Feng, et al.; the seventh chapter by Zhang Xinmin, and Zhao Jingzhou, et al.; the eighth chapter by Peng Suping and Huo Quanming, et al.; the ninth chapter by Liu Honglin, Zhang Xinmin, Qin Yong, Tang Dazhen, et al.; the tenth chapter by Hu Aimei and Zhang Sui'an, et al.; the eleventh chapter by Wan Yujin and Zhang Shicheng, et al. The final version was completed by Song Yan, Zhang Xinmin and Liu Shaobo.

In preparation of the book, thanks to the Ministry of Science and Technology of the People's Republic of China; China National Petroleum Corporation; PetroChina Company Limited; Xi'an Research Institute, China Coal Technology Engineering Group (Xi'an Research Institute); all responsible units, departments, and expert group. With joint efforts of all researchers, the project has been successfully completed, achieved expected goals, effectively implemented basic research innovation, and served the industry. It is expected the publication of this book can promote the basic research and application of CBM and the development of CBM industry in China.

Beijing, China — Yan Song
Xi'an, China — Xinmin Zhang
Beijing, China — Shaobo Liu

Contents

1 Status Quo and Research of CBM Exploration and Development

The development and utilization of CBM (coalbed methane) is of great significance in alleviating the shortage of conventional oil and gas supply, improving safe production of coalmine, implementing the strategy of sustainable development of national economy and protecting the atmospheric environment. CBM development firstly succeeded in the United States and has developed rapidly in Australia and Canada in recent years. Early CBM development in China began with gas drainage from coalmine. In the past decade, commercial development has been successfully carried out in southern Qinshui Basin and Fuxin Basin, and large-scale exploration has been carried out in coal-bearing basins throughout China.

1.1 CBM Development Status Quo in Some Countries

CBM resources estimated are 2980×10^{12} to 9609×10^{12} ft^3 (85×10^{12} to 265×10^{12} m^3) all over the world. The discovery of the Fruitland CBM area in the San Juan Basin in the United States triggered a worldwide exploration wave in the 1980s and early 1990s. Although that wave of exploration failed to find CBM fields comparable to those in the Fruitland CBM area, many economically significant (even small) CBM projects were launched in the United States during that period, and worldwide exploration continued. Commercial CBM projects have been carried out in the Bowen Basin of Queensland, Australia, and pilot projects have been carried out in other basins. CBM exploration or pilot projects are also ongoing in other countries, including Canada, the United Kingdom, China, Colombia and India. Canada declared commercial CBM sales in 2002 and has developed rapidly in recent years.

1.1.1 CBM Industry Development in the United States

At present, the CBM industry in the United States ranks first in the world in terms of technology and industrialization. The development status of the CBM industry in the United States represents the degree of exploration and development of CBM abroad.

Among 14 coal-bearing basins in the United States, the geological CBM reserves with depth below 1200 m reach 11.3×10^{12} to 24×10^{12} m^3, which are mainly distributed in 16 coal-bearing basins. In the eastern United States, commercial exploitation of small amount of CBM has been going on for more than 70 years. In fact, as early as 1943, Price and Headlee described the potential of the commercial CBM industry in details. CBM production in the western United States began about 40 years ago in the San Juan Basin. In the eastern and western United States, shallow coal seams accidentally became targets when well tests failed or deeper pay zones depleted in the early stage. Most wells produced very little CBM because only a few or no stimulation measures were taken to gas reservoirs at that time.

In the United States, sound CBM industry development during the first 20 years was determined by the following technical and non-technical issues: (i) The United States Bureau of Mines required degassing before underground mining to prevent explosion. (ii) The oil embargo imposed by OPEC in the 1970s led to the promulgation of measures by the federal government (Section 29 of the 1980 *Unexpected Profits Revenue Law of Crude Oil*) to encourage the development of unconventional natural gas resources. (iii) Progress in public sector research and technology was made by the U.S. Department of Energy, the Natural Gas Research Institute (now GTI) and others. (iv) Research was

Y. Song et al., *Coalbed Methane in China*,
https://doi.org/10.1007/978-981-33-4725-0_1

done by operating companies, especially by Amoco (now BP-Amoco). Due to the implementation of these research and experimental schemes, CBM exploration really began in the late 1970s. The well-known well Amoco-1 Cahn was drilled in the San Juan Basin in 1977. In 1977, USX and the United States Bureau of Mines jointly started the degassing project in coalbeds in vertical wells in the Oak Grove Coalfield in the Black Warrior Basin. The project was successful. From the mid to late 1980s, several basins in the United States were explored, and then development started in the San Juan and Black Warrior Basins.

In the 1980s, with the development and application of open-hole cavity completion technology and air drilling technology, the San Juan Basin and the Black Warrior Basin took the lead in realizing large-scale commercial development, and the CBM industry in the United States was born. After the 1990s, the CBM industry in the United States has made unprecedented progress in the research and test of drilling and completion technologies such as coring while drilling, intraformational horizontal well, directional pinnate horizontal well and composite well completion.

Native natural gas produced in the United States accounts for 85% of total gas consumption in the United States. The rest of the gas supply is imported from Canada. In order to solve the problem of energy shortage, more and more attention has been paid to the exploration and development of CBM. The proportion of CBM in natural gas has increased rapidly in recent years. In 2001, 24% of the U.S. energy supply came from natural gas, while CBM accounted for 8% of the natural gas produced in the U.S. mainland. In 2002, the output of CBM reached 450 × 10^8 m^3, equivalent to the total output of conventional natural gas in China (440 × 10^8 m^3). In 2007, the output of CBM in the U.S. reached 540 × 10^8 m^3, with 32,000 wells drilled (Fig. 1.1).

The following focuses on the current situation of CBM development technology in San Juan Basin, Black Warrior Basin and Powder River Basin.

The San Juan Basin is located in the south of Colorado and north of New Mexico. It is one of the basins in the south of Rocky Mountains and covers an area of 1.94 × 10^4 km^2. It is an asymmetric synclinal basin. The depth of CBM wells in this basin is generally 167.6–1219 m, and the thickness of main coalbeds is from 6.1 m to more than 12.2 m. The daily output of a CBM well can reach 22,653 m^3 and the output mainly comes from the Fruitland Formation. The sandstone under the Fruitland Formation also contains coal-formed gas, so gas is exploited from these two formations through the wells.

The first CBM well in the San Juan Basin was put into operation in 1953. It was completed with an ordinary open hole. Its production was 0.2 × 10^4 to 1.2 × 10^4 m^3/d, and stabilized at 0.54 × 10^4 m^3/d in the later period. The production lasted for nearly 40 years. Until 1977, when the first CBM field, the Xidaer gas field, was put into production, the potential of CBM development began to attract public notice. At present, 19 development zones have been formed in the basin, and more than 4000 CBM wells have been drilled, 3036 of which are production wells. In the basin, high-yield enrichment areas with about 777 km^2 and proven CBM reserves of 3000 × 10^8 m^3 can be delineated using 3 × 10^4 m^3/d per well as a gas production isoline. Up to the beginning of 2004, 2831.7 × 10^8 m^3 CBM has been produced in the basin (the in-situ CBM resource in the Fruitland Formation is estimated to be 14,158 × 10^8 m^3). CBM wells in this basin are basically vertical wells. Clear water or water-based drilling fluid are selected. In Area I, open-hole cavity completion was adopted by most wells, while in other areas perforation completion and hydraulic fracturing

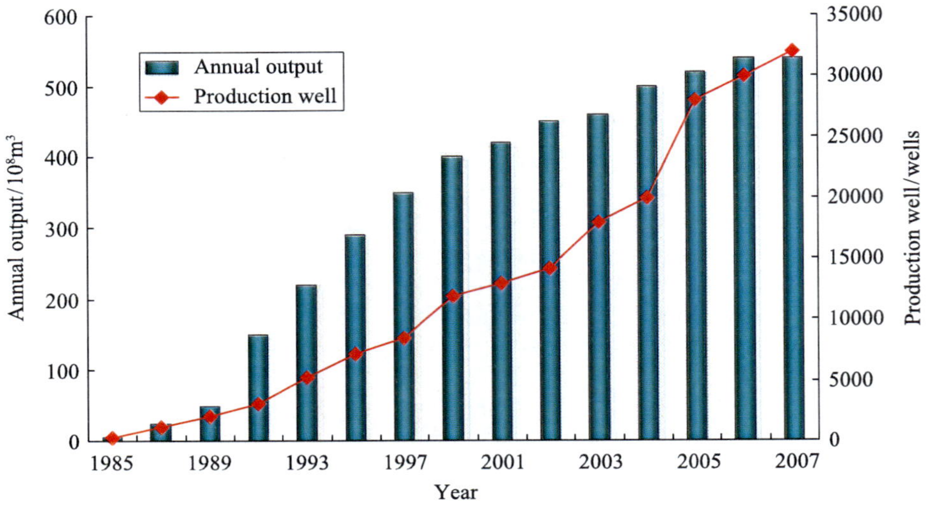

Fig. 1.1 CBM production growth chart in the United States (according to the U.S. Department of Energy, 2007)

stimulation were applied. In a well, fracturing fluid used is between 208.2 m^3 and 1135.7 m^3, with proppants between 100,000 lb[1] and 1200 lb, fracture length is more than 121.9 m and fracture height is less than 45.7 m. Although many difficulties rose out in the process of mining, the geological conditions and primary technologies ensured the success in the San Juan Basin.

The Black Warrior Basin is a relatively gentle, inclined, tectonic basin with a coal-bearing area of about 6950 km^2. The basin is about 370 km long from east to west, 302.5 km wide from south to north, and 240–1220 m deep. The total thickness of the coal seams is 6–12 m, with less than 1.3 m thickness of single layer, most of which are separated by sandstones. The coal rank is from high to low volatile bituminous coal. The CBM content is 6–20 m^3/t, 16 m^3/t on average, and the total CBM resources are 5663 × 10^8 m^3. In 1996, 2786 wells were drilled in the basin. The average output was 3000–6000 m^3/d per well. By 2002, 6000 CBM wells had been drilled in the basin, and the cumulative gas production was 396.44 × 10^8 m^3. Now 3474 wells are still in production, and the gas production is 9.34 × 10^6 m^3/d. CBM wells in the basin were completed by three methods: (i) "GOB wells" drilled in the mining area; (ii) horizontal well; and (iii) vertical well. 98% of the CBM wells are vertical wells, and the drilling fluid used is water-based or air. The volume of fracturing fluid used in one operation is 113.6–757 m^3, the pump rate is 210–2100 gal[2]/min, and the injection pressure is 3.45–15.9 MPa. 75% of the CBM wells adopted crosslinked gel fracturing fluid. The gel breaker is usually borate, sulfate or enzyme. The proppant is Alabama sand, injected 10,000 to 120,000 lb for one operation. The supported fracture width is from 1.27 cm to nearly closed. Depending on the distance from the wellbore and the placement of proppants in the fracture.

At present, the Black Warrior Basin mainly uses the following means to increase production: (i) drill infilling wells at well spacing of 550 m × 550 m in existing gas fields; (ii) improve drilling, completion and fracturing technologies, reduce development costs and improve operation efficiency, including multi-coal-seam recompletion, large volume of fracturing fluid, and multi-coal-seam hydraulic fracturing stimulation at the same time.

The Powder River Basin is located in southeastern Montana and northeastern Wyoming. It is a large sedimentary basin covering 6.7 × 10^4 km^2, and a large asymmetric syncline with a long axis of SE-NW direction. In the basin, the coal measures are Paleocene, 75% of the production areas are located in Wyoming, and 50% of the basin is considered to have CBM production potential. The coal seams are mostly subbituminous coal, and the methane in coal seams is biogenic. Compared with other coal basins, the methane content per unit volume in the basin is lower, about 0.8–1.13 m^3/t (generally 9.9 m^3/t in other coal basins). Commercial exploitation in the Powder River Basin depends on the high permeability, thick coal seams and good quality of produced water, which make up for the disadvantage of low gas content.

Industrial CBM development in the Powder River Basin began in 1986. Open hole completion and hydraulic fracturing stimulation were used in CBM wells to improve gas well production. The permeability is high and shallow sub-bituminous coal seams are easy to collapse after dehydration, the effect of hydraulic fracturing stimulation is not good, so most wells were completed with open holes. The well spacing is generally 400 m × 400 m to 550 m × 550 m. In the initial development stage, there were a few gas wells, specifically 18 wells in 1989, 1683 wells producing annual output of 21.68 × 10^8 m^3 in 1999, and 10,717 wells producing annual output of 89.71 × 10^8 m^3 by the end of 2002. At present, the Powder River Basin has become a main field for CBM development in the United States. As of February 2004, there were 12,155 production wells, but 3966 wells were shut down due to lack of water treatment equipment. The daily production of CBM is 0.246 × 10^8 m^3, and the daily water production is 228,002 m^3. The average daily gas production per well is 2025 m^3 and the daily water production per well is 1.287 m^3.

After many years of research and development practice in the United States, systematic theories and technologies for CBM exploration and development have been formed, including dewatering gas production, CBM stimulation (i.e. fracturing, caving, pinnate horizontal wells, injecting CO_2), and CBM reservoir numerical simulation. They are three "common technologies" of the CBM industry in the world, and have become the foundation and pillar of CBM exploration and development. At the same time, the United States has developed "Selection and Evaluation theory" and "Productivity Models" for mid-rank coal according to the geological characteristics of its own coalfields.

1.1.2 CBM Industry Development in Other Countries

The total CBM resources in Canada are about 6 × 10^2 to 76 × 10^{12} m^3, of which Alberta Province has about 11 × 10^{12} m^3. CBM development in Canada started relatively late, basically the same as that in China. Before 2001, CBM wells in Canada were about 70, but without production. In 2001, CBM wells increased to 100, and a breakthrough to single-well production was made. In 2002, the first commercial CBM project in Canada was established by EnCana and MGV. Since then, the exploration and

[1] 1 lb = 0.453592 kg.

[2] 1 gal (UK) = 4.54609 L, 1 gal (US) = 3.78543 L.

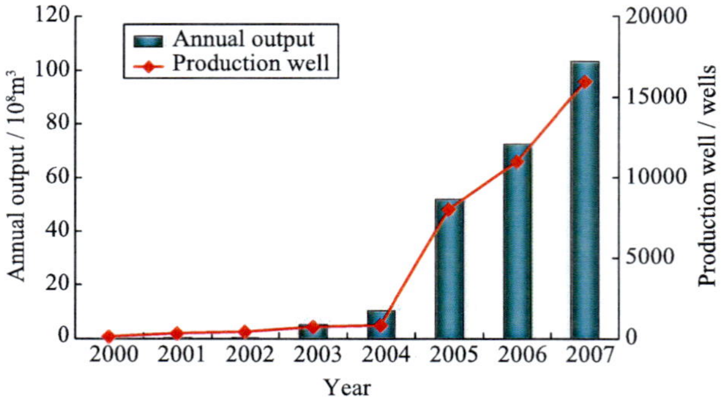

Fig. 1.2 CBM production growth chart of Canada (according to Canadian Unconventional Oil and Gas Association, 2008)

development of CBM in Canada has entered a period of rapid development. In 2005, the number of vertical CBM wells reached 6000, the output exceeded 30×10^8 m^3, accounting for about 1.8% of the total natural gas output. In 2007, Canada's CBM production reached 103×10^8 m^3, with a total of 16,000 wells drilled (Fig. 1.2).

Canada has made remarkable progresses in drilling CBM horizontal wells. CDX-Canada successfully drilled Canada's first single-branch CBM horizontal well in the Mannville coal seam in April, 2004. The horizontal section is 1000 m long. The cost from drilling to wellhead equipment installation is about 1.5 million Canadian dollars, and the daily output is about 7×10^4 m^3. By October 2005, CDX-Canada had completed five such horizontal wells in the Mannville coal seam (the maximum horizontal branch is 1200 m long) and 75 such horizontal wells in Alberta, of which 50 were completed by Trident. Trident announced the establishment of the first commercial CBM project in the Mannville coal seam in November 2005.

CBM has been exploited in Australia since 1976, mainly in the Bowen Basin of Queensland. CBM has been recovered from coal seams by surface drilling from 1987 to 1988. CBM production in Australia has increased year by year since 1996. From 2000 to 2001, the cost of CBM exploration in the Bowen Basin of Queensland alone amounted to 44.4 million US dollars, accounting for 37% of the total exploration cost of 120 million US dollars in the basin. In 2004, 277 CBM wells were drilled in Australia, and the CBM output reached 13.56×10^8 m^3, 78% of which was produced in Queensland, meeting 31% of the state's natural gas demand, and showing a sustainable growth momentum. The rapid development of CBM in Australia also benefits from policy encouragement and support. For example, the government requires that the gas content in the coal seam must be reduced to less than 3 m^3/t before mining coal. This ensures to "recover gas before mining coal", and promotes a large increase in the number of CBM drilling wells. In Queensland, the government requires power generated by natural gas must account for more than 13% of the state's total power. This policy has led to a substantial increase in natural gas demand and a large increase of investment in CBM exploration and development.

Production and utilization of gas from coalmines has a long history in Britain, Poland and countries of Commonwealth of Independent States (CIS). Gas drained from coal wells is mainly used for boiler gas or CBM power stations built in mining areas, only a small amount for civil use. At present, these countries are actively developing and applying new technologies of CBM power generation, but ground CBM development in these countries just began in recent years. In the United Kingdom, the Airth CBM field operated by British CBM Company was put into production in 1996, and the CBM produced is for generating power. Until recent, the British government realized that the proportion of CBM should be increased in the energy industry, and then began to encourage the exploitation of CBM—more CBM mining licenses will be issued, inapplicable regulations on the operation of CBM will be removed, and efforts will be made to support the fledgling CBM industry. The British government also provides tax incentives to CBM enterprises in accordance with the *Enterprise Investment Management Measures*. The investment can be recovered by reducing or exempting income tax or capital dividend tax. The Polish government has granted tax exemptions for 10 years to companies engaged in oil, gas and CBM exploration to attract a large number of domestic and foreign investors. At present, Russia and Ukraine are formulating preferential tax policies and regulations to encourage foreign companies to invest CBM development in their countries.

India has abundant CBM resources. CBM resources in lignite in the Gujarat Basin are about 3164×10^8 m^3. At present, the Indian government has delineated seven blocks in the Gondwanas coal basin for CBM exploration and development, and CBM exploration in Taniganj Coalfield and Jhjaria Coalfield has made rapid progress. The first CBM well was successfully drilled by National Petroleum Corporation of India in the Jhjaria Coalfield. The daily CBM output per well is 5000–6000 m^3, and the maximum daily output is 10,000 m^3. Similarly, CBM exploration results in the Bokaro Coalfield show that the daily CBM production per well can reach 5000–6000 m^3. It is worth noting that the potential CBM areas are far away from the natural gas production areas and the coal mining areas. This is conducive to the market development of CBM and does not affect the production of coalmines. However, the CBM industry in India has just started. Foreign investors in India are still facing challenges in terms of economic viability and other aspects of CBM projects. Besides the geological characteristics, the government's policy support and the marketing price of CBM are key factors to determine the success of CBM projects. These policies include: (i) CBM projects are exempted from taxes within seven years from the commencement of commercial production; (ii) the mining royalties should be low; (iii) exempt from import duties on materials and services necessary for CBM operations; (iv) market pricing principles are applicable for CBM; and (v) CBM is included in the definition and management of natural gas according to the *"Oil and Natural Gas Law"*, and the definition of natural gas should be extended in law to provide legal guarantee for the development of CBM industry.

1.2 CBM Industry Development Course in China—Theory and Technology Status

CBM exploration, development and utilization in China has gradually developed. In the initial stage, underground gas drainage was aimed at coal safety. In recent years, equal attention has been paid to both underground gas drainage and ground well group development of CBM.

1.2.1 Development Course

From the perspective of technology, CBM development in China has experienced three stages.

(1) Gas drainage from mining wells (1952–1988)

In 1952, a gas drainage station was established in the Longfeng Coalmine of Fushun Mining Bureau. From then on to 1989, the exploration and development of CBM in China mainly depended on draining gas from mining wells. Drainage and utilization of underground gas, determination of coal adsorption performance and CBM content were the main work in the stage. Important practical data have been accumulated for the prediction of CBM resources and the selection of favorable blocks.

(2) Introduction of modern CBM technology (1989–1995)

Modern CBM technologies were introduced to China from 1989 to 1995. In September 1989, the Ministry of Energy invited American experts on CBM to China to give a briefing. In November 1989, the Ministry of Energy held its first conference on CBM development in Shenyang. Subsequently, several research projects on CBM were funded by the National Eighth Five-Year Plan and local enterprises and the Global Environment Facility (GEF). CBM ground exploration and development experiments were carried out in Dacheng, Hebei Province and Liulin, Shanxi Province. In 1991, China's first academic monograph on CBM, *Coalbed Methane in China*, was published. At the same time, many foreign companies have invested to risk CBM exploration in China. During that period, China has introduced special testing equipment and application software for CBM. The introduction of equipment and personnel exchanges have made great progress in CBM resource evaluation, reservoir testing technology and mining technology.

(3) Formation and development of the CBM industry (1996–)

In order to accelerate the development of CBM in China, the State Council of the People's Republic of China approved the establishment of China United Coalbed Methane Co., Ltd. in early 1996. In the Ninth Five-Year Plan and the Tenth Five-Year Plan, the national research and test projects on CBM have been set up. In the same period, the National Planning Commission set up a first-class national geological exploration project—"China CBM Resource Evaluation" in order to promote the industrialization process of CBM. In 2002, the 973 Program set up the project "Basic Research Series of CBM Forming Mechanism and Economic Development", which systematically studies the key scientific problems that restrict the development of CBM in China from the basic theory and application. The results have been applied to the exploration and development of CBM.

In recent years, CBM industry has developed rapidly in China. By the end of 2007, more than 2000 ground CBM wells had been drilled. Drilling or well group production tests had been carried out in dozens of blocks, such as Qinshui, Hedong, Huaibei and Huainan, Liupanshui, Ningwu, Jixian, Hancheng, Enhong, Fuxin, Shenbei, Pingle, Fengcheng and Lengshuijiang in Shanxi, Shandong, Hebei, Guizhou, Anhui, Jiangxi, Hunan Provinces, etc. The proven geological CBM reserves are 1359×10^8 m^3, and the ground CBM production capacity is about 15×10^8 m^3.

1.2.2 Progress in Basic Research

As a new research field, CBM geology has been studied for only nearly 20 years in China (Song et al. 2005). In recent years, after fruitful researches, China has made breakthroughs to gas production experiments in Qinshui Basin and Ordos Basin (Zhang 1991; Li et al. 1996; Zhao 1999; Ye et al. 1999). These breakthroughs have created good conditions for basic research of CBM in China. On the basis of the 973 Program CBM project, innovative achievements have been made in some aspects, which basically represent the latest progress of CBM research in China.

1.2.2.1 Genesis Types and Identification of CBM

Like studies on conventional gas genesis, studies on CBM genesis are often based on geochemical methods to identify gas sources (Dai et al. 1986; Rice 1993; Scott 1993; Scott et al. 1994; Smith and Pallasser 1996). CBM has two primary origins: biogenic and thermal (Qin 2003). Biogenic CBM can be divided into primary biogenic gas from peat to soft lignite and secondary biogenic gas from bituminous coal. However, the stable carbon isotope composition of CBM in some areas shows the superposition of CBM of different origins.

Wang Yanlong et al. and Zhang et al. (2004) determined the origin of CBM in the Qinshui Basin in Shanxi Province, the Hongen Coalmine in Yunnan Province and the Xinji Coalmine in Huainan Province by studying CBM composition, carbon isotope and hydrogen isotope. For the first time, a large number of systematic hydrogen isotope analysis provided sufficient basis for determining the origin of CBM, and corresponding geochemical tracer indicators were established. The results show that the CBM in the southern Qinshui Basin is pyrolysis CBM, the CBM in the Liyazhuang Coalmine in the northwestern Qinshui Basin is secondary biogenetic CBM, and the CBM in the Xinji Coalmine in Huainan Coalfield is mixed CBM dominated by secondary biogenetic CBM. CBM is composed of secondary biogenetic methane, thermogenic methane, atmospheric nitrogen (N_2) residual CO_2 and heavy hydrocarbon compositions.

1.2.2.2 Characteristics and Evaluation of Coal Reservoirs

Coal reservoir characteristics, including fractures and pores, adsorbability, gas-bearing property, permeability and coal reservoir pressure, are important factors in CBM geological research. Traditional understanding of coal reservoirs is based on dry coal samples, and the “three-stage” evolution model of coal Langmuir volume was proposed. Through the study of the adsorbability and evolution law in balanced water, Chinese researchers have found the “two-stage” evolution model (Ye et al. 1999; Zhang and Yang 1999; Fu et al. 2002). Another important aspect of CBM geology is the non-linear characteristics and fractal description of coal reservoir fracture-pore system. Coal reservoir has been regarded as a “dual pore” system consisting of fractures and pores at home and abroad (Close 1993). Some researchers have tried to describe it by fractal theory and method (He 1995; Zhao et al. 1998). Fu Xuehai put forward that coal reservoir was a ternary fracture-pore medium composed of macrofractures, microfractures and pores, and established a fractal mathematical model and a classification scheme of the ternary fracture-pore system (Fu et al. 2001).

In view of problems and research hotspots in coal reservoir research in recent years, a lot of work has been done in the CBM project of the 973 Program. First, the parameters of high-rank coal reservoir were determined and the evaluation method of coal reservoir was put forward. Mercury intrusion experiment, low-temperature nitrogen adsorption and single-phase permeability measurement were proposed because they are easy to implement and facilitate parameter comparison. The parameters of coal reservoir physical properties were refined and optimized, and five parameters —porosity, average throat diameter, specific surface area, permeability and reservoir pressure gradient were selected to establish the evaluation index system of coal reservoir physical properties (Yao et al. 2005). Second, taking the Qinshui Basin as an example, geological models and fractal mathematical models of the pore systems of high-rank coal reservoirs were put forward, and the contribution of reservoir physical properties of each model was analyzed. According to the low permeability of the high-rank coal reservoir in China, large fracture systems were researched, and different development characteristics in low-permeability coal reservoirs of different coal ranks were found. Thermal evolution, sedimentary diagenesis and tectonic stress and strain are main causes for coal reservoir heterogeneity (Wang 2004).

1.2.2.3 Dynamic Field of CBM Reservoirs

Since the Ninth Five-Year Plan, China has carried out extensive work on controlling factors of CBM reservoir formation, such as tectonic stress field, thermal field and hydrodynamic field, and has achieved some results (Zhang et al. 1997; Wang et al. 1998; Yang and Tang 2000).

Macroscopically, tectonic settings control the geotectonic background for formation and evolution of a coal basin. From the perspective of the basin, the characteristics of tectonic stress field and the uneven distribution of internal stress are important factors on the occurrence, structure, physical properties, fracture development of coal reservoirs and cap rocks, and the difference of groundwater runoff conditions, thus affect the gas-bearing properties and recoverability of coal reservoirs. When studying how gas generates under the control of thermodynamics, researchers realized that the paleo-thermal field controlling late Paleozoic CBM enrichment in eastern China has the characteristics of “multi-stage and multi-heat-source superposition” (Ye et al. 2001). The relationship between thermodynamic conditions and CBM generation and preservation conditions can be divided into different types according to the coalification and gas generation characteristics in specific areas. Compared with other directions of CBM geological research, the preliminary studies on hydrodynamic conditions of CBM reservoir formation in China are relatively weak. But in recent years, remarkable progresses have been made in regional gas control, reservoir control, numerical simulation and evaluation theory of hydrogeological conditions (Luo and Yang 1997; Luo and Ye 2001; Ye et al. 2002).

In recent years, the influence of hydrochemical field on CBM reservoir forming process has been paid attention to. Through physical simulation, borehole draining test and

CBM well test, etc., Qin Yong et al. from CBM Project Group of the 973 Program have proposed that methane solubility increases logarithmically with the increase of pressure, and that methane solubility decreases exponentially with the increase of salinity. With the increase of salinity, methane solubility in coal seam water tends to converge at different depths (pressures).

In the past, researches on the dynamic field of CBM reservoir formation mainly focused on single factor and the summary of the laws, but less on the mechanism of gas control and the coupling of different factors. Through the study on the Qinshui Basin, Qin Yong et al. put forward a new idea to study the structural dynamic conditions of CBM reservoir formation by using coupled tectonic stress field, tectonic deformation and optical fabric of coal reservoir vitrinite. And the comprehensive effect of elastic self-regulation of coal base block was also put forward according to the opening and closing degree of natural fractures through simulation experiments. At constant coal rank, the comprehensive effect decreases gradually with the increase of reservoir fluid pressure. At constant fluid pressure, the comprehensive effect decreases gradually with the increase of coal rank. The areas with favorable CBM reservoir-forming dynamic conditions in the Qinshui Basin were predicted using the comprehensive effect (Fang et al. 2003; Fu et al. 2003, 2004a, b; Jiang et al. 2004; Sun et al. 2004; Wei et al. 2004; Jiang and Ju 2004).

1.2.2.4 Forming Process of CBM Reservoirs

CBM reservoir is different from conventional gas reservoir in terms of reservoir mechanism, forming process, boundary and fluid state (Song et al. 2005). The research on CBM reservoir forming process and model has been one of the fields that many researchers in China are concerned about. In view of the unconventionality of CBM reservoirs, many scholars have tried to give a certain meaning to CBM reservoirs in the geological research (Qian et al. 1997; Zhang et al. 2002). However, it is generally believed that CBM reservoirs are not only difficult to give an exact definition, but also difficult to define in geological space. Therefore, the definition and boundary of CBM reservoirs are still pending problems in the study of CBM reservoirs. The sealing conditions of CBM reservoirs can be divided into two types: water pressure and gas pressure. On this basis, many researchers classify gas reservoirs into different types according to the characteristics of the gas-bearing structure. However, there is no unified scheme to classify CBM reservoirs so far, and the study of CBM reservoir forming process tends to focus on geochemical methods, but less on comprehensive analysis.

In the past two years, Song Yan et al. have discussed the forming process and gas accumulation history of the CBM reservoirs in the Qinshui Basin based on the study of the geothermal gradient and burial history funded by the CBM project of the 973 Program. Comparative study has been carried out on the thermal evolution history, geochemical characteristics and gas accumulation history of the coalbeds in the Yangcheng area in the southern Qinshui Basin and the Huozhou area in the western Qinshui Basin. Then an idea was proposed that the forming of the present CBM reservoirs in the Qinshui Basin depended on a "critical geological moment", which determined the present CBM content, i.e. the forming period of the "effective overlying thickness" (the shallowest coal seams in the geological history). The temperature and pressure of the "critical geological moment" were discussed by combining forward modeling and inversion, and the gas content in the forming and evolving process was calculated. Based on relevant data at home and abroad, Su Xianbo et al. proposed four types of CBM reservoir boundary: fault boundary, hydrodynamic boundary, physical boundary and weathered and oxidize boundary, and discussed the mechanisms of the boundaries. The boundary of the CBM reservoir in the southern Qinshui Basin, the largest CBM reservoir found in China, was divided and determined (Su et al. 2004, 2005).

1.2.2.5 Adsorption and Desorption Mechanism

The Langmuir isothermal adsorption model and equation have been dominant in studying adsorption for a long time. In the 1970s abroad and the 1990s at home, the homogeneity of coal inner surface has been considered. Potential difference theory model, DR equation, GF equation and LF equation have been used to describe the adsorption characteristics of coal. Although the study of multi-component adsorption characteristics began in the late 1960s, it was mainly because of the need of exploration and development after the 1990s, more and more attentions were paid to the adsorption of N_2, CO_2 and mixed gases of different components (Schroeder et al. 2002; Ozdemir et al. 2003). The results show that the adsorption capacity of coal to different gases increases with the increase of coal rank, and the adsorption capacity of any coal to single-component gas has significant difference, the overall performance is $CO_2 > CH_4 > N_2$.

The influence of formation water has been less studied on CBM adsorption at higher temperature, pressure and to multi-component gas. Cui Yongjun et al. have made a great progress through simulation experiments: adsorption experiments on $CH_4 + N_2$ and $CH_4 + CO_2$ mixtures show that the adsorption characteristics of different coal to the same mixed gas are quite different, and the adsorption characteristics of the same coal sample to different mixed gas are different too. When the total adsorbed amount is between strong and weak adsorbents, the higher the proportion of strong adsorbents is, the larger the adsorbed amount is.

In previous studies, the effects of pressure and temperature on coal adsorption performance were discussed separately. According to the linear relationship between temperature and adsorption capacity, Zhang Qun et al. with the CBM Project Group of the 973 Program discussed the rule of coal adsorption performance under the jointing influences of pressure and temperature. It is concluded that the influence of pressure on coal adsorption capacity is greater than that of temperature at lower temperature and pressure zones, the amount of methane adsorbed by coal increases with the increase of temperature and pressure; while the influence of temperature on coal adsorption capacity is greater than that of pressure at higher temperatures and pressure zones, and the amount of methane adsorbed by coal decreases with the increase of temperature and pressure. At the same time, the study also reveals the changes of CBM adsorption parameters under gas–liquid-solid three-phase interaction through physical simulation experiments.

Desorption is a process opposite to adsorption, and it is very important to study desorption in CBM development. Compared with the study of adsorption, the study of desorption is relatively weak. In recent years, with the development and research of CBM in China, some new knowledge has been gained on the desorption of CBM. Zhang Sui'an et al. proposed that CBM adsorption and desorption are reversible processes and the desorption process lags through isothermal adsorption and desorption experiments on different samples, different single-component gases (CH_4, CO_2, N_2) and multi-component mixed gases with different proportions. At the same time, the generalized flow analysis of CBM in deformed dual media was carried out, and the pseudo-steady and unsteady seepage models of CBM in deformed dual media were established.

1.2.2.6 Evaluation of CBM Resources

CBM resource evaluation is a long-term basic work. Since the 1980s, many people have carried out national or regional CBM resource evaluation. Up to now, four times of CBM resources have been generated in China's systematic evaluation: 30×10^{12} to 35×10^{12} m^3 during the Seventh Five-Year Plan; 32.68×10^{12} m^3 during the Eighth Five-Year Plan period, calculated by the Xi'an Research Institute; 14.34×10^{12} m^3 from 1995 to 1998, calculated by China National Administration of Coal Geology; and 31.46×10^{12} m^3 from 1997 to 1999, calculated by China United Coalbed Methane Co., Ltd. and Xi'an Research Institute supported by National Planning Commission.

The outstanding problems in the evaluation of CBM resources are as follows: no attention has been paid to the recoverability of CBM and the corresponding amount of recoverable resources has not been obtained. The application of resource data and the comparison with foreign data are limited, and the CBM resources in lignite are neglected.

Based on these problems, Zhang Xinmin and Zhao Jingzhou put forward a new classification scheme for CBM resources (Lin and Chen 2004; Zhang et al. 2005) and the concept of technically recoverable CBM resources. Aiming at the particularity of CBM resources evaluation, two evaluating methods for technically recoverable CBM resources at different exploration and development levels were put forward for the first time, namely, numerical simulation and loss analysis. The new evaluation methods have achieved preliminary application results in the Qinshui Basin and the Fuxin Basin. The calculated CBM resources are not only comparable with conventional oil and gas resources, but also with international CBM resources. In addition, the research on CBM recoverability is highlighted, and the results have more practical values for CBM exploration and development.

1.2.3 Status and Progress of CBM Exploration and Development Technology in China

CBM occurs in underground coal seams in an adsorbed state, and coal reservoirs are deformed porous media with low permeability, which makes it difficult to identify and exploit CBM reservoirs. In recent years, technical research on CBM exploration and development have been strengthened at home and abroad, such as geophysical exploration, well drilling and completion, stimulation, etc. Especially for the mining technology of highly metamorphic anthracite CBM in China, substantial progress has been made through years of tackling key problems.

1.2.3.1 CBM Geophysical Exploration Technique

The role of geophysical data in CBM exploration is mainly manifested in the following aspects: to determine the thickness and burial depth of coalbeds by logging data; to determine the content of CBM by combining logging data with gas content (anhydrous and ash-free basis); to determine the structural development of coalbed by seismic data; to determine the fracture development of coalbed by multi-wave and multi-component and AVO data. In recent years, great progresses have been made to coal seams and CBM exploration by seismic exploration technology in China. Peng Suping et al. have carried out three-dimensional three-component seismic exploration in the Huainan Coalfield, and preliminarily established the pre-dictive methods of coal seam thickness, fracture development and CBM enrichment. According to objects, seismic attribute inversion is divided into four levels, i.e. P-wave post-stack inversion, P-wave prestack inversion, azimuth AVO inversion and multi-wave joint inversion. The lithologic parameters related to P-wave velocity are mainly inverted by P-wave post-stack inversion. AVO is the main means of P-wave pre-stack inversion (including azimuth AVO), which mainly inverts the parameters related to gas content. Azimuth AVO inversion is used to predict lithologic information of fractures and reservoir heterogeneity. Multi-wave joint inversion mainly detects fracture directional density, pressure and fluid properties, and determines the rock physical parameters precisely on the basis of P-wave inversion.

The existence of underground fractures leads to the azimuthal anisotropy of seismic wave field. The amplitude of seismic reflection varies with the change of offset and azimuth. Under the assumption of weak anisotropy, P-wave coefficient varying with offset and azimuth can be expressed as an analytical function of fracture parameters. In order to predict the distribution of underground fractures by using the law of seismic amplitude with offset and azimuth, fine analysis and processing were carried out on the full-azimuth 3D seismic data in the Huainan Coalfield. The seismic wave field near the primary coal seam presents obvious azimuthal anisotropy, and the law of seismic amplitude changing with azimuth is in good agreement with the result predicted with azimuthal anisotropy theory, showing that the P-wave azimuth AVO theory can be used to predict the fracture azimuth and density in the coalfield.

1.2.3.2 CBM Well Drilling and Completion Technique

There are two main types of ground CBM wells, vertical well and horizontal well, and the drilling methods are different. For coal seams at low pressure, rotary or percussive drilling is usually used, and air and foam are used as circulating medium. With low pressure, pore and permeability and easy to be polluted, underbalanced drilling is less harmful to coal seams. Conventional rotation is generally adopted for coal seams at high pressure. There are five kinds of CBM completion methods: openhole completion, casing completion, openhole-casing mixed completion, openhole cavity completion and horizontal draining liner completion.

In addition to conventional CBM drilling technology, new technology for directional pinnate horizontal wells has been developed in the United States. The daily gas production per well is from 3.4×10^4 to 5.7×10^4 m^3, and the recovery in five years is 85%. The advantages of directional pinnate horizontal well include increased effective supply range, improved conductivity, less damage to coal seam, high well production, high recovery, small wellsite, less environmental impact, less ground gathering and transportation facilities and good economic benefits. Langfang Branch of Research Institute of Petroleum Exploration & Development has put forward a scheme about the first directional pinnate horizontal CBM well in China after basic research for more than two years. Through numerical simulation and study of stimulation mechanism, it is recognized that formation pressure drop gradually spreads out from the pinnate horizontal well. On plane, the whole horizontal wellbore can be seen as the "source" of formation pressure decline and almost the whole coal seam area is utilized at the same time, so that the mining potential of coal seams has been fully exploited. The directional pinnate horizontal well technology is effective especially for the exploitation of highly metamorphic and low-permeability anthracite CBM in China.

1.2.3.3 CBM Stimulation Technique

(1) Hydraulic fracturing stimulation

Hydraulic fracturing stimulation is the measure most commonly used in CBM production. However, coal-bearing strata in China generally have undergone strong tectonic movement after coalification. As a result, the original structure of coal seam is often greatly damaged and the plasticity is greatly enhanced. During hydraulic fracturing stimulation, natural fractures and joints would not expand, and no long hydraulic cracks can be induced; plastic deformation may occur in the coal seams. Generally speaking, the stimulation effect is unsatisfactory. Therefore, it is very important to understand the fracture distribution law during hydraulic fracturing stimulation.

Zhang Shicheng et al. carried out triaxial mechanical test, static and dynamic test and hydraulic fracture inducing test on coal samples while conducting statistical analysis of domestic fractured CBM wells. Mechanical parameters such as Young's modulus, Poisson's ratio, volume compression coefficient, particle compression coefficient, pore elasticity coefficient and tensile strength of the coal rock in the Qinshui Basin were obtained. And a new understanding of the law of fracture distribution in the coal rock was established. When confining pressure is low, natural cleats are mostly open, and the seepage capacity is strong, so fractures will be induced along natural fractures. When confining pressure is high, natural cleats are mostly closed, so fractures will be induced in the direction perpendicular to the minimum principal stress. Different from conventional sandstone gas reservoirs, natural fractures (cleats) in coal seams are developed, and the elastic modulus of coal seam is small and the Poisson's ratio is large, so the propagation of fractures in coal seams is extremely complex—a large number of irregular fractures may be induced. On the basis of the above work and understanding, a mathematical model for hydraulic fracturing to coal rock was established (Shan and Zhang 2005).

In CBM exploration and development practices, PetroChina CBM Project Manager Department has established diagnostic methods for hydraulic fractures in coal seams: using well temperature to measure the height and location of fractures near wellbore zone; using ground SP to measure the orientation and length of fractures; using post-fracturing well test to measure the length and conductivity of fractures; using cross-well seismic "CT" method to determine the direction and height of fractures. In addition, China United Coalbed Methane Co., Ltd. used radioisotope and microseismic survey to measure the height, orientation and length of fractures (Li et al. 2002).

(2) Gas injection

Gas injection is a new method developed by Amoco Company in the United States to increase CBM production. It is deemed to be a bright future, and has attracted extensive attentions. N_2, CO_2 or flue gas is injected into coal seams to reduce partial methane pressure. It is conducive to the displacement and desorption of methane from coal seams, and can increase the single well production and recovery. This method can effectively improve CBM production potential, and can be used to develop CBM in deep and low-permeability coal seams. Because CO_2 injected helps maintain pore pressure, fractures and pores in deep coal seams that can be protected from closing. In order to improve CBM recovery technology, China United Coalbed Methane National Engineering Research Center Co., Ltd. carried out CO_2 injection and production test in Well TL-003, a well in the Zaoyuan well group in the southern Qinshui Basin, with the cooperation with Canadian experts. The test involved CO_2 injection, well shutdown (to achieve full displacement to methane by CO_2), gas drainage and gas components monitoring, and finally expected results were obtained.

1.3 Key Scientific and Technological Challenges in CBM Exploration and Development in China

Scientific and technological challenges in CBM exploration and development in China mainly include four aspects.

(1) Basic research

The basic research on CBM in China and abroad is still weak, and many aspects need to be strengthened and deepened, such as the geological environment and forming mechanism and evaluation system of coal reservoirs; adsorption and description models at high temperature and pressure, and control factors on coal adsorption; how porosity and permeability change during mining process, unstable seepage mechanism in deformed media and mathematical model of CBM desorption-seepage flow at multi-phase deformed media; how coupled dynamic fields and hydrochemistry influence CBM reservoir forming process, and numerical simulation of reservoir forming process; estimation and evaluation of new CBM resources and recoverable resources including lignite CBM.

(2) Serious shortage of CBM exploration blocks and targets

There are many coal-bearing basins with a long history of tectonic evolution and complicated geological conditions in China. Generally, coal wells are shallow, oil wells are deep, and the favorable targets of CBM wells are between the two, and almost no data from them. The exploration degree of CBM is still very low, the study of CBM reservoir-forming conditions and distribution laws is not deep enough, the prediction of high-yield areas is inaccurate, and the preparation of favorable target zones and blocks is seriously inadequate. It is an important task in the future to use basic researches to select favorable blocks, to use target evaluation method to evaluate coal-bearing basins and to provide more favorable targets for CBM exploration.

(3) Gap from foreign countries

The exploration and development of CBM requires special technology and equipment. At present, the technology of CBM production in China is not perfect in these aspects. Advanced production technology needs to be introduced and developed, including directional pinnate horizontal well, drilling inclined coal seams, low-cost air drilling operation, cavity completion, hydraulic fracturing stimulation, numerical simulation, etc.

(4) Incomplete produced water treatment system

Because of the large amount of water discharged with CBM mining, the treatment of produced water is a problem that must be faced. The United States has established a series of methods and measures, such as ion exchange, chemical treatment, reverse osmosis, artificial wetlands and so on. CBM exploration in China started late, and a perfect system for water treatment has not yet established. With the development of national economy, the strict requirement of environmental protection and the improvement of the awareness of environmental protection, the treatment of produced water will become an important chain in the development of CBM, similar to the United States and other developed countries. Technologies urgently needed to be developed include direct utilization, physical and chemical treatment, and reinjection of produced water.

References

Close JC (1993) Natural fracture in coal. In: Law BE, Rice DD (eds) Hydrocarbons from coal. AAPG Stud Geol 38:119–132

Dai JX, Qi HF, Song Y, Guan DS (1986) China's coalbed methane composition, carbon isotope types and their genesis and significance. Chin Sci (Series B) 12:1317–1326 (in Chinese)

Fang AM, Hou QL, Lei JJ et al (2003) Control of coalification on the characteristics of coalbed methane enrichment and occurrence. J Geol Univ 9(3):378–384 (in Chinese)

Fu XH et al (2002) Partial coal reservoir desorption and methane recovery in China. Coalfield Geol Explorat 28(2):19–22 (in Chinese)

Fu XH, Qin Y, Jiang B et al (2004a) Research on the "bottleneck" of CBM production of high-ranked coal reservoirs. Geolog Rev 50 (5):507–513 (in Chinese)

Fu XH, Qin Y, Li DH, Jian B, Wang WF (2004b) The effect analyses of hydraulic fracturing on anthracite reservoir of southern Qinshui Basin, Shanxi. In: Wang YH, Ge SR, Guo GL (eds) Mining science and technology. Balkema Publishers, Rotterdam, pp 321–324

Fu XH, Qin Y, Li GZ (2001) Influencing factors on coal reservoir permeability in the middle and south of Qinshui Basin, Shanxi. J Geomech 7(1):45–52 (in Chinese)

Fu XH, Qin Y, Zhang WH (2003) Coupling relationship between mechanical effects of high-ranked coal matrix and reservoir permeability. J Geologi Sci 9(3):373–377 (in Chinese)

He XQ (1995) Rheological dynamics of gas-bearing coals. China University of Mining and Technology Press, Xuzhou, pp 26–32 (in Chinese)

Jiang B, Ju YW (2004) Structural coal structure and its reservoir physical properties. Nat Gas Ind 24(6):27–29 (in Chinese)

Jiang B, Ju YW, Qin Y (2004) Textures of tectonic coals and their porosity. In: Wang YH, Ge SR, Guo GL (eds) Mining science and technology. Balkema Publishers, Rotterdam, pp 315–320

Li AQ (2002) Application of fracturing stimulation in coalbed methane exploration and development. In: Li WY, Ma XH, Zhao QB et al (eds) New progress in China's coalbed methane geological evaluation and exploration. China University of Mining and Technology Press, Jiangsu, pp 93–99 (in Chinese)

Li MC, Liang SZ, Zhao KJ (1996) Coalbed methane and exploration and development. Geological Publishing House, Beijing (in Chinese)

Lin DY, Chen CL (2004) Discussion on the calculation of recoverable resources of coalbed methane. China Coalfield Geol 3:15–17 (in Chinese)

Luo ZJ, Yang XL (1997) Study on hydrodynamic field of coalbed methane migration and accumulation. J Changchun Inst Geol 27 (3):356–358 (in Chinese)

Luo ZJ, Ye JH (2001) Three-phase coupling model of gas, water and solid produced with coalbed methane migration. J Changchun Univ Sci Technol 31(4):349–353 (in Chinese)

Ozdemir E, Morsi BI, Schroeder K (2003) Importance of volume effects to adsorption isotherms of carbon dioxide on coals. Langmuir 19:9764–9773

Qian K, Zhao QB, Wang ZC (1997) Theory and experimental technology for coalbed methane exploration and development. Petroleum Industry Press, Beijing (in Chinese)

Qin Y (2003) Progress and review of coalbed methane research in China. J Geologi Univ China 9(3):339–358 (in Chinese)

Rice DD (1993) Composition and origins of coalbed gas. In: Law BE, Rice DD (eds) Hydrocarbons from coal. AAPG Stud Geol 38:159–184

Shan XJ, Zhang SC (2005) Summary of research on hydraulic fracture propagation in coalbed methane wells. In: Song Y, Zhang XM et al (eds) Coalbed methane accumulation mechanism and theoretical basis of economic exploitation. Science Press, Beijing (in Chinese)

Schroeder K, Ozdemir E, Morsi BI (2002) Sequestration of carbon dioxide in coal seam. J Energy Environ Res 2:54–63

Scott AR (1993) Composition and origin of coalbed gases from selected basins in the United States. In: Proceedings of the 1993 international coalbed methane symposium, vol 1. The University of Alabama, School of Mines and Energy Development, Paper 9270, pp 174–186

Scott AR, Kaiser WR, Ayers WB et al (1994) Thermogenic and secondary biogenic gases, San Juan Basin. AAPG Bull 78(8):1186–1209

Smith JW, Pallasser R (1996) Microbial origin of Australia coalbed methane formation. AAPG Bull 80:891–897

Song Y, Zhang XM et al (2005) CBM reservoir accumulation mechanism and theoretical basis of economic exploitation. Science Press, Beijing (in Chinese)

Su XB, Lin XY, Song Y, Zhao MJ (2004) The classification and model of coalbed methane reservoir. Acta Geologica Sinica 78(3):662–666

Su XB, Lin XY, Zhao MJ, Song Y, Liu SB (2005) The upper paleozoic coalbed methane system in the Qinshui Basin, China. AAPG Bull 89(1):81–100

Sun ZX, Zhang W, Hu BQ (2004) Geochemical and geothermal factors controlling the origin of coal-bed methane in Qinshui Basin, China. Geochimica et Cosmochimica Acta 232

Wang MM, Lu XX, Jin H et al (1998) Hydrogeological characteristics of the Carboniferous-Permian coalbed methane enrichment area in North China. Petrol Experi Geol 20:385–393 (in Chinese)

Wang SW (2004) Evaluation of coal reservoirs in coalbed methane exploration and development. Nat Gas Ind 24(5):82–84 (in Chinese)

Wei C, Qin Y, Man L, Jian M (2004) Numerical simulation of geologic history evolution and quantitative prediction of CBM reservoir pressure. In: Wang YH, Ge SR, Guo GL (eds) Mining science and technology. Balkema Publishers, Rotterdam, pp 325–329

Yang Q, Tang DZ (2000) Effect of coal metamorphism on coalbed gas content and permeability in North China. Earth Sci J China Univ Geosci 25(3):273–278 (in Chinese)

Yao YB, Liu DM, Hu BL, Luo WL (2005) Application of geographic information system in comprehensive evaluation of coalbed methane resources. Coal Sci Technol 33(12):2–4 (in Chinese)

Ye JP, Shi BS, Zhang CC (1999) Coal reservoir permeability and influencing factors. J China Coal Soc 24(2):118–122 (in Chinese)

Ye JP, Wu Q, Wang ZH (2001) Control of hydrogeological conditions on the occurrence of coalbed methane. J Coal 26(5):459–462 (in Chinese)

Ye JP, Wu Q, Ye GJ et al (2002) Study on the dynamic mechanism of coalbed methane accumulation in southern Qinshui Basin. Geologi Rev 48(3):319–323 (in Chinese)

Zhang Q, Yang XL (1999) Isothermal adsorption characteristics of coal to methane under equilibrium water conditions. J China Coal Soc 24 (6):566–570 (in Chinese)

Zhang SA (1991) Tectonic thermal evolution and shallow coal-derived gas resources in major coalfields in China. Nat Gas Ind 11(4):12–17 (in Chinese)

Zhang SL, Tian SC, Chen JY (1997) Fault structure and reservoir dynamic system. Oil Gas Geol 18(4):261–266 (in Chinese)

Zhang XJ, Tao MX, Wang WC et al (2004) Generation and resource significance of biogenetic coalbed methane. Bull Mineral Geochem 23(2):166–171 (in Chinese)

Zhang XM, Han BS, Li JW et al (2005) Prediction methods of technically recoverable CBM resources. Nat Gas Ind 25(1):8–12 (in Chinese)

Zhang XM, Zheng YZ, et al (2002) Distribution of coalbed methane resources in China. In: Academic collection of the 80th anniversary of the Chinese Geological Society. Geological Publishing House, Beijing, pp 458–463 (in Chinese)

Zhao AH, Liao Y, Tang XY (1998) Quantitative study on fractal structure of coal pores. J China Coal Soc 23(4):439–442 (in Chinese)

Zhao QB (1999) Coalbed methane geology and exploration technology. Petroleum Industry Press, Beijing (in Chinese)

2 Genesis and Criteria of CBM

2.1 Geochemical Characteristics of CBM and Its Difference from Natural Gas

Driven by the demand on industrialization, CBM resources have become an emerging and global hotspot field. A series of achievements and progresses have been made in geological and geochemical research on CBM at home and abroad. Geochemistry of CBM is a new and important branch of the entire CBM system and gas geochemistry, not only because CBM is a direct target, but also because its geochemical composition contains rich information about the genesis, formation conditions, preservation, migration diffusion and development and utilization value of CBM. It also has important scientific significance and application value in enriching gas geochemical research.

2.1.1 Sample Testing Method

The carbon isotope in CBM samples tested and studied in this study was analyzed by a MAT-252 (stable isotope mass spectrometer) and a Delta Plus XP (stable isotope mass spectrometer), and PDB international standard with accuracy of $\leq \pm 0.25‰$. The hydrogen isotope was analyzed by a Delta Plus XP and SMOW international standard with accuracy of $\leq \pm 15‰$. Gas composition was tested using a MAT-271 (trace gas mass spectrometer). All the above test works were completed at the Key Laboratory of Gas Geochemistry of the Chinese Academy of Sciences. R^o and maceral analysis were conducted in the Xi'an Research Institute of CCTEG China Coal Research Institute.

2.1.2 Composition and Basic Characteristics

Data about the composition of CBM are very large and difficult to count. It is almost composed of CH_4, and less C_{2+} (heavy hydrocarbons), N_2 and CO_2. Tao Mingxin (2005) et al. tested the local CBM in China. The results show that, in addition to primary components, the CBM contains trace components Ar, H_2, He, H_2S, SO_2 and CO. For example, the desorbed CBM samples from the Sihe Coalmine in Jincheng City of Shaanxi Province were tested to have 85.53–99.85% of CH_4 and 4.63–30.87% of N_2, and different contents of C_2H_6, CO_2, Ar, H2S, SO_2 and He and other components.

Zhang Xinmin et al. (2002) collected CBM samples, about 6000 sets of data, from 358 mines of different geological ages and various coal ranks in China. The statistics of the composition show that CH_4 is dominant in the CBM, from 66.55 to 99.98%, and generally 85%–93%; CO_2 accounts for 0–35.58%, and generally less than 2%; the content of N_2 varies greatly, generally less than 10%; and the content of heavy hydrocarbon gas varies with the coal rank.

According to the analysis of 985 samples from American CBM wells by Scott et al. (1995), the composition of CBM and average content are CH_4 accounting for 93.2%, C_{2+} (heavy hydrocarbons) 2.6%, CO_2 3.1%, and N_2 1.1%.

Due to differences in thermal evolution and loss, the composition and content of CBM samples from different regions and even from the same region are different or variant. In addition, coalbeds in some coalfield or area mainly contains CO_2 or N_2, such as the Lower Silesian Coalfield in Poland and the Yaojie Coalfield in China. The composition of the coalbed gas in the Yaojie Coalfield is composed of CO_2 of 95.4%–96%, CH_4 of 0.138%, N_2 of 1.99%, Ar of 0.33%, H_2 of 0.005%, CO of 0.0065%, O_2 of 1.24%, and He of 0.0041% (Tao et al. 1994). The high concentration of CO_2 is not the result of normal coalification, but has special geneses or sources (Tao et al. 1991, 1992). It can be regarded as CBM in a broad sense.

Based on the above-mentioned data, CBM mainly contains more CH_4, and less N_2, CO_2 and C_{2+} (heavy hydrocarbons). Although seldom researches reported H_2S, SO_2 and CO, it does not mean that CBM does not contain these components. It is necessary to strengthen test and research on trace gas components in the future.

Y. Song et al., *Coalbed Methane in China*,
https://doi.org/10.1007/978-981-33-4725-0_2

2.1.3 Composition and Distribution of Isotope

Since taken as a new resource, CBM has been studied on isotope geochemistry, but only a dozen references were available before 1993, including the research results from Chinese scholar Dai Jinxing. Rice (1993) conducted a summary analysis of these documents. These results, mainly based on the study of carbon isotope of methane, indicate that the distribution of $\delta^{13}C$ ranges from −80 to −16.8‰ (PDB, the same below). Some documents also researched the carbon isotopes of ethane (C_2H_6) and CO_2 and the hydrogen isotope of methane, showing $\delta^{13}C_2$ from −32.9 to −22.8‰, $\delta^{13}C$ of CO_2 from −26 to + 18.6‰, $\delta^{13}D_{CH4}$ from −333 to −117‰ (SMOW, the same below). In the past decade, research results of CBM geochemistry have been reported. Representative references, such as Smith and Pallasser (1996) who studied the Permian CBM in the Sydney Basin and Bowen Basin, Australia, show that, for 307 gas samples, $\delta^{13}C$ is −78.9 to −18.0‰, and $\delta^{13}C$ of CO_2 is −15.5 to +16.7‰; for 88 gas samples, δD_{CH4} ranges from −255 to −152‰, with an average of −217‰; for 44 coal samples, δD is −93 to −162‰, with an average of −132‰. Kotarba (2001) studied the CBM in the Upper Silesian Basin and Lublin Basin in Poland, showing $\delta^{13}C_1$ is −67.3 to −52.5‰, and δD_{CH4} is −201‰. Hakan et al. (2002) reported the isotope characteristic of coal bed samples in the Zonguldak Basin in the western Black Sea of Turkey, showing for 13 CBM samples from the Carboniferous system, $\delta^{13}C_1$ is −51.1 to −48.3‰, and δD_{CH4} is −190 to −178‰; and for some samples, $\delta^{13}C_2$ is −37.9 to −25.3‰, $\delta^{13}C_3$ is −26.0 to −19.2‰, and $\delta^{13}C$ of CO_2 is −29.4 to −13.2‰. Aravena et al. (2003) studied the water soluble CBM in the Elk Valley Coalfield in southwestern Canada, and it is indicated by hydrocarbon test that the $\delta^{13}C$ is −65.4 to −51.8‰, δD is −415 to −303‰, which is related to the loss of hydrogen isotope of groundwater (δD is −163 to −148‰). The $\delta^{13}C_1$ of the Upper Carboniferous CBM in the Ruhr Basin, Germany, is −57.3 to −40.0‰, δD_{CH4} is −201 to −175‰; another sample has $\delta^{13}C_1$ of −85.9 to −85.1‰, δD_{CH4} is −260 to −257‰ (Thielemann et al. 2004).

In the middle 1980s, Chinese scholars studied the carbon isotope composition of CBM (CH_4 and CO_2) based on a national CBM (also known as coal-forming gas) project. For example, Dai Jinxing et al. (1986) tested and studied 42 CBM samples collected from 8 provinces and regions, and the distribution of $\delta^{13}C_1$ is −66.9 to −24.9‰. Tang Xiuyi et al. (1988) conducted studies on methane carbon isotope in the Kailuan Coalfield and Huainan Coalfield, and the $\delta^{13}C_2$ is −30.1 to −73.1‰. By studying the geochemical characteristics of CBM in the Xinji area of China, the $\delta^{13}C_1$ is −61.3 to −50.7‰, with an average of −56.6‰; the $\delta^{13}C_2$ is −26.7 to −15.9‰, and the $\delta^{13}C_3$ is −25.3 to −10.8‰; the $\delta^{13}CO_2$ is −39.0 to −6.0‰, with an average of −17.9‰; the $\delta^{15}N$ is −1 to + 1‰; $^3He/^4He$ is (1.13—3.20) × 10^{-7}, R/R_a is 0.08–0. 23.

Based on available tested and reported data of methane carbon isotope, the distribution of $\delta^{13}C_1$ and its correlation with R^o were plotted (Fig. 2.1). In order to reduce the impact of sample differences and facilitate data comparison, the data of primary desorbed gas from coal cores and drained gas from wells were selected, including some data of carbon isotope of coal-type methane. The $\delta^{13}C_1$ of Chinese CBM samples (Figure) is between −73.7 and −24.9‰, indicating that the distribution range is wide and the variation is complex, and on the whole, the $\delta^{13}C_1$ is poorly correlated with the thermal evolution degree (R^o) (specific features are described later).

There are few researches and reports on the hydrogen isotope composition of CBM in China, which is basically in a blank state. The δD_{CH4} of CBM in Xinji Anhui, Liyazhuang Shanxi and Enhong Yunnan is −244‰ to −196‰ (Tao et al. 2005a). In addition, the $^3He/^4He$ ratio of high concentration of CO_2 in the coal seam of Yaojie Gansu is at 10^{-8} scale, indicating that it is the gas sourced from crust (Tao et al. 1991, 2005b).

Based on the above-mentioned data, the composition of CBM is mainly composed of CH_4. The current distribution of $\delta^{13}C_1$ is about −80 to −16.8‰, δD_{CH4} is about −415 to −117‰, and $\delta^{13}C$ of CO_2 is −39 to + 18.6‰. Heavy hydrocarbons have fewer carbon isotope data which do not reflect their basic distribution. The isotope studies of oxygen and sulfur in CBM have not been reported. Nitrogen isotope and thorium isotope were only found in the above-mentioned Xinji and Yaojie. Overall, the research on the isotopic geochemistry of CBM is still less.

2.1.4 Difference and Specificity of Isotope Composition

According to statistics, the distribution of carbon isotope values of conventional natural gas in China ranges from −66.0 to −27.7‰ (Xu et al. 1994), and the highest reported value is about −25‰ (Zhao 2003). By contrast, the $\delta^{13}C_1$ distribution of CBM is much wider. In addition to the aforementioned data (about −80 to −16.8‰), there are some exceptions, such as the highest value of $\delta^{13}C_1$ in Germany CBM, which is −12.9‰, and the $\delta^{13}C_1$ of the Soviet CBM up to −10‰ (Dai et al. 1986). Tao Mingxin (2005) reported that the highest value of CBM obtained was −10‰. Ying Yupu et al. (1990) reported that the values of two CBM samples in the Nantong Coalfield were −13.3 and +14.2‰ (PDB). The second data is very anomalous, and

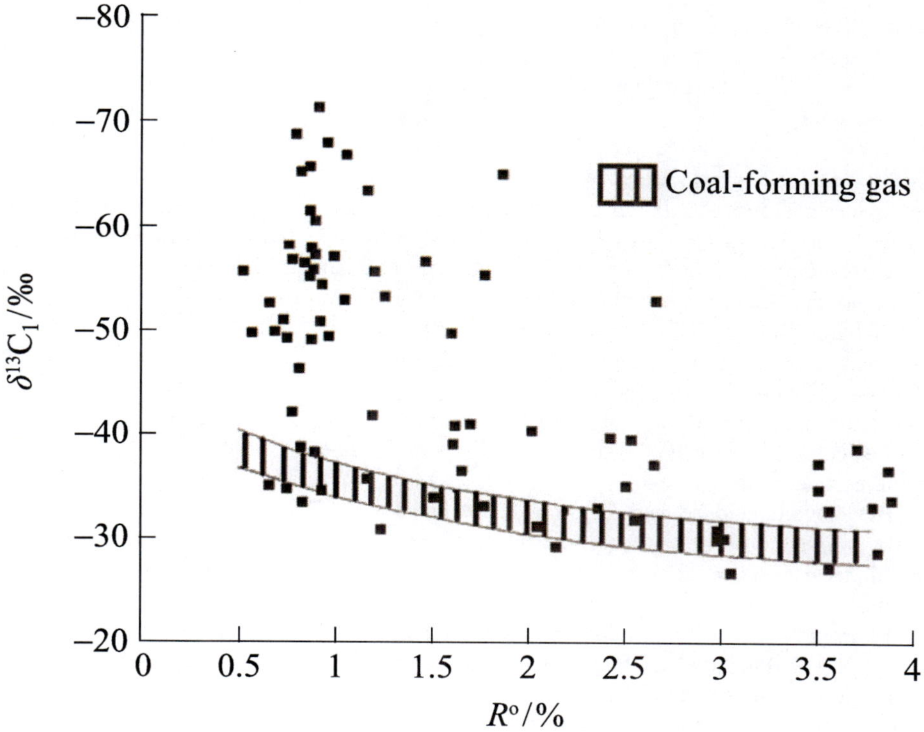

Fig. 2.1 Distribution of $\delta^{13}C_1$ and its relationship with R^o (Gao et al. 2002)

currently only one sample. More studies are needed. These data show that the carbon isotope is of very low value, and ore is abnormally high, which varies greatly.

The kerogen type of both CBM and coal-forming gas is Type III organic matter or humus. The difference is that coal-forming gas is conventional natural gas, but CBM is unconventional natural gas whose source rock and reservoir are the same. Therefore, theoretically, the carbon isotope composition of the two has certain comparability in the parental inheritance effect.

In order to visually compare the macroscopic distribution and basic differences of $\delta^{13}C_1$ in CBM and coal-forming gas, the relationship between $\delta^{13}C_1$ and R^o of coal-forming gas was proposed according to Dai Jinxing et al. (1992). $\delta^{13}C_1$(‰) = 14.12lgR^o −34.39, and within the range of ±2‰. The distribution of $\delta^{13}C_1$ in coal-forming gas was drawn. Figure 2.1 shows one of the most obvious features: the $\delta^{13}C_1$ of methane produced by the coal rock of the same rank or having similar R^o has a considerable variation, and the variation of $\delta^{13}C_1$ of the CBM in lowly mature coal rock is wide, while that of the CBM in highly mature coal rock is narrow. Although the $\delta^{13}C_1$ of CBM has a tendency to increase with the increase of the R^o value as a whole, the variation of $\delta^{13}C_1$ of the methane produced by coal rock with the same thermal evolution or similar R^o value is greatly different. The variation of $\delta^{13}C_1$ of the methane produced by coal rocks with different thermal evolution degrees is significantly different. Therefore, the correlation between $\delta^{13}C_1$ and R^o is very low. However, there is a significant positive correlation between R^o and $\delta^{13}C_1$ of conventional natural gas. Coal-forming gas has stable $\delta^{13}C_{1,}$ and is strongly correlated to its R^o, that is, the change is regular (Fig. 2.1). At the same thermal evolution degree, the $\delta^{13}C_1$ of CBM is generally lighter than that of coal-forming gas, except in a small amount of sample, they are equivalent. This is consistent with the phenomenon that the carbon isotope of coalbed methane is generally lighter noted by domestic and foreign scholars. For this conclusion, there are different understandings and interpretations (described later).

The above-mentioned characteristics reflect the significant differences in the carbon isotope composition and variation of CBM and conventional natural gas. Such composition and variation of CBM are likely results of secondary changes caused by various factors, but not their original characteristics.

The carbon isotope composition of CBM is complex, and the value of $\delta^{13}C_1$ varies greatly. Generally, there is a lack of understanding of its varying laws, which are difficult to be applied in practice (Tao 2005). Due to the complex forming mechanism of CBM, applicable natural gas classification standards cannot be directly used to explain the cause of CBM, and the fractionation of carbon isotope has some uncertainties in its data interpretation (Kotarba 1990). Although the carbon isotope of methane in low rank coal is light and the carbon isotope of methane in high coal ranks is heavy, the $\delta^{13}C_1$ of methane in the same coal rank is very wide (Rice 1993). Dai Jinxing et al. (1986) considered that the law of CBM and that of primary coal-forming gas ($\delta^{13}C_1$ vs. R^o) is not consistent, that is, the carbon isotope of methane in shallow coal seams is lighter, while the carbon isotope of coal-forming gas in medium and deep wells is basically the same. Tang Xiuyi et al. (1988) studied the carbon isotope of CBM in coalfields such as Kailuan Coalfield and Huainan Coalfield, and found that there is no inevitable relationship between $\delta^{13}C_1$ and R^o. The inherent connotation is that there are no obvious rules of the composition and variation of carbon isotope. But the results of this study indicate that the carbon isotope composition of various components of the CBM in Huainan has undergone significant secondary changes.

The above-mentioned data also show that not all $\delta^{13}C_1$ of CBM is lighter (lower) than those of coal-forming gas, there are higher ones, even the extremely highest $\delta^{13}C_1$, which is another characteristic or difference of CBM from coal-forming gas in carbon isotope composition. Although CBM with $\delta^{13}C_1$ higher than –20‰ is not universal, it exists objectively. For this special phenomenon, it has not attracted the general attention of academic circles, and it lacks in-depth research on its genesis. In addition to the traditional theory, the $\delta^{13}C_1$ of biological or organic methane is relatively low, and the highest $\delta^{13}C_1$ of conventional natural gas as described above is about –25‰. The $\delta^{13}C_1$ of non-biogenic methane is relatively high, such as the $\delta^{13}C$ of the mantle methane ejected from the eastern Pacific ridge, –17.6 to –15.0‰ (Welhan and Craig 1979). Therefore, this is an important scientific issue worthy of in-depth study on CBM and even carbon isotope geochemistry.

Internationally, the δD_{CH4} of CBM reported is –415 to –117‰, and most of them are in the range of –300 to –200‰. According to statistics, the distribution of δD_{CH4} in conventional natural gas in China is –312.8 to –83.9‰ (Dai 1990). From available data or more intuitively, the distribution of δD_{CH4} is not very different, but the δD_{CH4} of CBM is lower than the δD_{CH4} of conventional natural gas.

From the above-mentioned international references, although the specific δD_{CH4} of CBM is listed, the tracer significance or the range of tracer index is rarely discussed in detail. For example, Kotarba (1990) obtained δD_{CH4} of –256 to –211‰ from 10 CBM samples from the Nowa Ruda Coalmine in Poland, but it wasn't almost discussed. Rice (1993) reviewed related references, but only briefly discussed the hydrogen isotope composition and genetic type of CBM. In general, less δD_{CH4} data of CBM have been reported by now, and the specific characteristics of the distribution and change are still unclear. That is, the research level is relatively low. From the point of the composition characteristics, it may be more complicated than conventional natural gas. More data and researches are needed, particularly domestic research on the hydrogen isotope composition of CBM.

The $\delta^{13}C$ of CO_2 in CBM ranges from –39 to +18.6‰, while the $\delta^{13}C$ of biogenic CO_2, which is generally included in conventional natural gas, is lower than –10‰ (Dai et al. 1995; Tao et al. 1996; Xu et al. 1997). Obviously, the distribution of the former is not only much wider than the latter, but its highest value is also unusual.

Since there are fewer isotope data of other components, it is still difficult to compare CBM with conventional natural gas.

The above-mentioned data are only differences or characteristics of CBM and conventional natural gas in isotopic composition. They are results or appearances, and the causes or mechanisms may be different.

2.2 Control Factors on Carbon Isotope Indicator

Data show that the carbon isotope composition of CBM is much more complicated than conventional natural gas. Although it is generally believed that $\delta^{13}C_1$ and C_1/C_{1-5} are geochemical indicators for classifying the genesis of hydrocarbon gas, in terms of specific information on CBM, these two indicators vary greatly, and in fact it is very complicated to apply, especially for shallower CBM. The main or deeper reason lies in the failure to understand the factors that control or affect the carbon isotope composition and fractionation of CBM, as well as its specific performance and of impact.

2.2.1 Desorption and Fractionation of Methane Carbon Isotope

Coal is a highly porous medium with abundant organic matter. Coal has a large specific surface area, and organic matter has a strong adsorption effect on gaseous substances. As far as CBM is concerned, the pore surface of coal has a strong adsorption capacity; therefore CBM mainly occurs in adsorbed gas in coal seams. In addition to the specific properties and characteristics, such as large specific surface area, pressure and temperature are also important factors affecting the adsorption or storage of CBM in coal seams. As shown in Fig. 2.2, when the temperature is constant, the adsorption capacity of coal to methane (CH_4) increases with the pressure within a certain pressure range. Conversely, when the pressure drops, the adsorbed CBM (CH_4) gradually or partially becomes free, and the amount of adsorbed CBM increases as the temperature decreases.

It is generally believed that the composition and carbon isotopes of CBM change with fractionation during

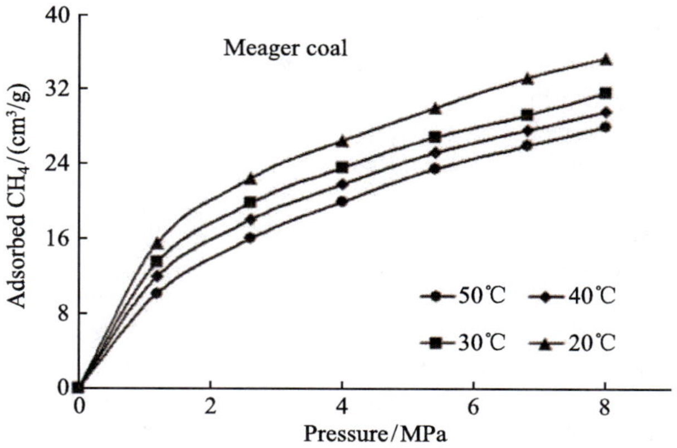

Fig. 2.2 Adsorption isotherms of coal to CH_4 at different temperatures (Cui et al. 2003)

desorption. In terms of CH_4, $^{12}CH_4$ is more easily desorbed than $^{13}CH_4$ (Dai et al. 1986; Tang et al. 1988; Tao et al. 2005a, b). In theory, the value of $\delta^{13}C_1$ in the CBM desorbed first is relatively low, while the value of $\delta^{13}C_1$ in the CBM desorbed late is relatively high. However, the intensity of isotope desorption and fractionation may be different under different conditions. At present, the exploration and development of CBM is developing rapidly, and the facilities for draining CBM in coalmines are also constantly developing. They provide more opportunities and conditions for collecting and testing desorbed gas from coal cores and gas samples from wells. In order to study the desorption and fractionation of carbon isotope in CBM, especially the desorption and fractionation of desorbed CBM and drained CBM samples, and to evaluate the effect of desorption and fractionation on the traceability of $\delta^{13}C_1$, fractionation experiments were conducted on the No.16 Permian coal cores taken in a well in Fuyuan, Yunnan Province. Sampling and desorption experi- ments were carried out at the wellhead. Nine samples of desorbed gas were collected in order, and the carbon isotope composition was tested. Table 2.1 lists the samples in the order of desorbed time. The $\delta^{13}C_1$ of the first three samples are –45.1‰, –46.4‰ and –46.9‰, respectively, with a slightly lighter trend. From the third sample, the $\delta^{13}C_1$ of the last 7 samples increases from –46.9 to –45.1‰, showing a heavier trend. For the first two samples, especially the first sample, it has a high $\delta^{13}C_1$, presumably due to the inclusion of free gas contained in the coal seam when it is underground. This result shows that the carbon isotope fractionation of the primary desorbed gas from the cores does not vary greatly, although $\delta^{13}C_1$ reflects the change, and generally becomes heavier, from –46.9 to –45.1‰, and the largest increment of 1.8‰. The results show that the carbon isotopic fractionation of the primary desorbed gas in the core has little change The desorption rate of CBM exhausted under formation conditions is relatively slower, and the variation of isotope fractionation should be smaller in some period. Therefore, it can be concluded that the variation of carbon isotope desorption and fractionation of the primary desorption of coal core and CBM discharged under formation conditions have no influence on its $\delta^{13}C_1$ tracer effect.

The distribution of the $\delta^{13}C_1$ of first desorption gas samples drained from seven coal cores taken from several coal seams in Well XS-01 in Xinji Coalmine in Huainan is –58.7 to –50.7‰, but the $\delta^{13}C_1$ of drained gas from the well is –51.7‰. It fell in the $\delta^{13}C_1$ range of desorbed gas, but it is slightly lower. That the $\delta^{13}C_1$ of the drained gas is slightly lower than that of the desorbed gas may be attributed to different coal seams which provide the coal cores and drained gas. In general, the $\delta^{13}C_1$ of desorbed gas and drained gas are relatively stable and are representative.

Tang Xiuyi et al. (1988) carried out desorption experiments on 20 coal samples (collected from several coalmines in Huainan, Huaibei, and Kailuan) before and after comminution. The $\delta^{13}C_1$ of desorbed gas of four samples are basically the same before and after comminution. The $\delta^{13}C_1$ of 11 coal samples before pulverized is significantly lower than that of the samples after comminution. The $\delta^{13}C_1$ of five coal samples before pulverized is higher than that of the samples after comminution. The maximum difference of the $\delta^{13}C_1$ of a sample reached 22.6‰. Three of four free (drained) gas samples from drilling boreholes have the $\delta^{13}C_1$ higher than the desorbed gas from the coal core taken at the same place, and another sample has the $\delta^{13}C_1$ 10‰ lower than the desorbed gas. This indicates that the carbon isotope compositions of methane desorbed from underground coal samples vary greatly, and are poorly representative. It has a great influence on $\delta^{13}C_1$ as a geochemical tracer. Therefore, the geochemical data of such samples is specific in application. We believe that sample type has a significant impact on the geochemical analysis results of CBM, and it causes evident variation in carbon isotope composition as well, thereby affecting the application of $\delta^{13}C_1$ as a geochemical tracer.

Table 2.1 Experiment results of methane carbon isotope of no.16 coal seam in Fuyuan, Yunnan

Sample no	1	2	3	4	5	6	7	8	9
$\delta^{13}C_1$/‰	–45.1	–46.4	–46.9	–46.5	–46.3	–46.2	–46.0	–46.0	–45.1

2.2.2 Impact of Secondary Biogenic Gas on Methane Carbon Isotope Composition

Secondary biogenic gas is produced with the action of bacteria after that coal is formed and uplifted with crustal movement. In fact, secondary biogenic gas is formed on the basis of thermogenic CBM. In recent years, secondary biogenic gas has been discovered in many coal-bearing basins around the world.

The adsorption and desorption of CBM are related to pressure and temperature. In terms of the forming background of secondary biogenic gas, the pressure of coal reservoir reduces and results in desorption of CBM in the process of uplifting or denudation of overlying strata, so that some thermogenic gas escapes. However, the reservoir not exposed is at some pressure which forces a certain amount of adsorbed CBM kept in the coal seam, even the free CBM can't escape. Scott et al. (1994) showed that the bituminous coal (R^o of 0.7%–1.0%) in the San Juan Basin should has a gas content of only 5–11 cm^3/g under normal conditions. But the actual content of CBM in this basin generally exceeds 16 cm^3/g, and some even exceeds 22 cm^3/g, reaching a saturation state. It is considered a result of secondary biogenic gas formed later. It is indicated by the result of Faiz et al. (2007) that the coal seams in the Sydney Basin adsorbed secondary biogenic gas at the late undersaturated state.

According to the desorption experiments on coal samples taken in multiple regions of China, the residual gas content after artificial desorption is about 0–3 m^3/t, roughly accounting for 1.5–30.0% (Zhang et al. 1999). Under formation conditions, the amount of gas remaining in coal seams should be greater. This further proves that a small amount of thermogenic gas remain in coal seams. Therefore, secondary biogenic gas generated later inevitably mixes with early thermogenic gas phase. We believe that it is rare for CBM consisting purely of secondary biogenic gas, and there must be thermogenic gas with varying amounts.

Scott et al. (1994) concluded that the geochemical composition of secondary biogenic gas is similar to that of primary biogenic gas. The main difference is that the thermal evolution of the coal rock exceeded that when primary biogenic gas was generated. In respect of the carbon isotope composition or tracer index of biogenic methane, Rightmire et al. (1984) believed that the $\delta^{13}C_1$ value (PDB, the same below) is generally –75 to –55‰, while the $\delta^{13}C_1$ of thermogenic methane is –40 to –25‰. By reviewing relevant data, Rice (1993) suggested that the $\delta^{13}C$ of biogenic methane is between –110 and –55‰. Smith and Pallasser (1996) proposed that the $\delta^{13}C_1$ of biogenic methane in the Sydney Basin and the Bowen Basin is –(60 ± 10)‰. The CBM with $\delta^{13}C_1$ of –80 to –60‰ in the Upper Silesian Basin is considered late (secondary) biogenic gas (Kotarba and Rice 2001).

According to the above-mentioned data, the academic circle believes that the high limit of the $\delta^{13}C$ of biogenetic methane is –60 to –50, and –55‰ is taken frequently. The $\delta^{13}C$ of thermogenic CBM is generally greater than –50‰ and it is mainly distributed in –45 to –30‰. There is a small amount of CBM with a higher $\delta^{13}C_1$. Due to the significant difference in carbon isotope composition of secondary biogenic gas and thermogenic gas, the generation of secondary biogenic gas and its mixture with thermogenic gas may produce new isotope effect. In order to provide a scientific basis for studying the carbon isotope geochemistry of CBM with secondary biogenetic methane and thermogenic methane, the variation of the carbon isotope was studied through experiments.

First, two groups of samples, which were very different in carbon isotope composition, were selected as original samples. One group (No.1) were taken in the Haila'er Basin, with methane content of 89‰ and a $\delta^{13}C_1$ of –73.2‰ which represents secondary biogenic gas. The other group (No.2 and No.3) were taken in the Qinshui Basin and the Jingyuan Coalfield, with methane contents of 98.96% and 97.5%, and $\delta^{13}C_1$ of –31.2‰ and –42.4‰, respectively, which represent thermogenic gas.

Then, mix No.2 with No.1 samples, No.3 with No.1 samples according to the volume ratio of 8:2, 6:4, 5:5, 4:6, 2:8, and prepare them into two series including 5 samples each, and representing methane mixed by secondary biogenic gas and thermogenic gas. Table 2.2 lists the experimental results of carbon isotope composition of mixed sample.

The experiment results have the following basic and common features:

(1) The $\delta^{13}C_1$ of any mixed sample is between the $\delta^{13}C_1$ of the two original samples. For example, the lowest $\delta^{13}C_1$ in the first series is 8‰ higher than that (–73.2‰) of the No.1 sample representing secondary biogenic gas. The maximum $\delta^{13}C_1$ is –9‰ lower than that (–31.2‰) of the No.2 sample representing thermogenic gas.
(2) As the content of biogenic gas decreases, the $\delta^{13}C_1$ of the mixed gas samples becomes higher in turn.

Table 2.2 $\delta^{13}C_1$ of mixed CBM samples (unit: ‰, PDB).

Mixing ratio		8:2	6:4	5:5	4:6	2:8
$\delta^{13}C_1$ of sample	No.2 sample/No.1 sample	–39.9	–51.2	–55.5	–57.7	–65.2
	No.3 sample/No.1 sample	–52.1	–57.9	–60.5	–63.3	–68.3

(3) Although the variation of $\delta^{13}C_1$ of the mixed samples does not strictly conform to the configuration or mixing ratio of the two original gases, it still shows obvious laws. That is, for every 10% increment of the ratio of secondary biogenic gas, the $\delta^{13}C_1$ decreases by about –4‰ in each sample in the first series and by –3‰ in each sample in the second series. This also seems to indicate that the greater the difference in the $\delta^{13}C_1$ of the two original gases, the greater the difference in the $\delta^{13}C_1$ values between mixed samples with different ratios.

The characteristics and laws of above-mentioned experimental results show that mixing two different genetic types of CBM can lead to significant variations in the carbon isotope composition of methane, and it is a regular variation. From the results of the experiment, it can be inferred that since the carbon isotope composition varies, the original correlation between the isotopes of components of thermogenic gas will vary as well.

2.2.3 Impact of Microscopic Composition of Coal Rock on Methane Carbon Isotope Composition

From the perspective of source rocks, there are also significant differences between coal and conventional natural gas. Coal rock is not only highly rich in organic matter, but also the macerals in different regions (even in the same region) and different types of coal rocks often vary to different degrees, or even there are large variations. The results of some studies have shown that the amount of gas generated by different macerals varies greatly (generally, exinite > vitrinite > inertinite), and the $\delta^{13}C_1$ of methane is different and complex (Li and Zhang 1990). Therefore, from the perspective of parental inheritance effect, it can be qualitatively considered that the variation of macerals in coal rock have some effects on the composition of carbon isotope since the amount of gas generated by different macerals in coal rock and the $\delta^{13}C_1$ of methane produced are different, especially for the coal rock with greatly varying macerals. In addition, considering catalytic action of inorganic component on hydrocarbon generation in coal rock, it is considered that the components of coal rock and their contents are factors that affect the carbon isotope composition of CBM. However, there has been still a lack of comprehensive and systematic research, and the impact has been still unclear (Tao 2005).

In addition to the above-mentioned CBM isotope desorption and fractionation, the mixing of secondary biogenic gas and the variation of coal maceral content have different effects on the carbon isotope composition of coalbed methane, there are also other aspects of research and understanding. Dai Jinxing et al. (1986) believe that, when coal seams become shallower, the CBM desorbs and diffuses. The ^{12}C-rich CH_4 desorbed in the lower primary coal seam will migrate upwards, and partially enter the upper desorption zone, thereby making the composition of CBM dry and carbon isotope composition light. In addition, the carbon isotope exchange and equilibrium reaction between original CO_2 and CH_4 makes the carbon isotope composition of CBM lighter.

2.3 Genetic Classification and Geochemical Tracer System

Classifying CBM by genetic types is an important task and goal in the study of CBM geochemistry, but the concepts or methods for conventional natural gas have been used currently. Rightmire et al. (1984) briefly mentioned two types of CBM, biogenic and thermogenic types. Rice (1993) divided CBM into biological (bacterial) gas and pyrolysis gas based on the isotopic composition of CBM and the R^o value of coal rock. Scott et al. (1994) found secondary biogenetic gas when studying CBM in the San Juan Basin in the United States and they were considered as the mixture of wet gas, *n*-alkanes and biogenic gas produced by bacteria after coalification.

Although the academic circle generally believed that $\delta^{13}C$ and C_1/C_{1-5} are main geochemical indicators for classifying genetic types of hydrocarbon gases. In terms of specific data above mentioned, these two indicators vary greatly and it is hard to be applied, especially for shallow CBM. Kotarba (1990) pointed out that, due to complex mechanism of CBM, the classification for natural gas can't be directly used to explain the genesis of CBM, and fractionation of carbon isotope makes uncertainties in data interpretation.

The carbon isotope composition of CBM is much more complicated than conventional natural gas, and deep and specific points or regular characteristics have not yet been clarified. There is still a lack of systematic classification criteria and understanding of the genetic types of CBM. Through the studies on the CBM samples taken in the Turpan-Hami Basin in Xinjiang, the Baojishan Coalmine in Gansu Province, the Zaoyuan (Shizhuang) and Panzhuang well group in Southern Qinshui Basin, and the Xinji Coalmine in Huainan Coalfield and Enhong Coalfield in Yunnan, five types of CBM were proposed, including primary biogenic CBM, thermogenic CBM, pyrolysis CBM, secondary biogenic CBM, and mixed CBM.

2.3.1 Primary Biogenic CBM

Primary biogenic CBM refers to primary biogenic gas formed with the action of microorganisms on organic matter during the diagenetic stage (lignite) of sediments (peat) in early burial stage, i.e. $R^o < 0.5\%$ or $R^o < 0.3\%$. Primary biogenic gas is from the fermentation of acetic acid by microorganisms and the reduction of CO_2 by methanogenic bacteria, and the latter is dominant. While primary biogenic gas appeared, peat was shallow, pressure was low, and pores were almost occupied by water, so gas adsorption was not significant, and primary biogenic gas usually escaped or dissolved in formation water, and then desorbed in the process of compaction and coalification. Therefore, it is generally believed that primary biogenic gas is difficult to retain in later coal seams (Scott et al. 1994).

Studies on the Shaerhu area in the Turpan-Hami Basin in Xinjiang indicate that the CBM is primary biogenic gas. The Shaerhu area is located in the middle of the southern slope of the Turpan-Hami Basin. Its prototype sedimentary structure is a small depression or fault depression in which Middle Jurassic coal seams were deposited. Late tectonic movements altered the depression into a nearly NE-trending asymmetry syncline with two groups of faults nearly trending EW and NS. The coal-bearing interval is the Middle Jurassic Xishanyao Formation (J_2x). The coal rock is less evolved, and composed of lignite, with R^o of 0.40–0.47%. Semibright coal and semidull coal are dominant, with a small amount of dull coal.

Desorbed gas samples taken in the Shaerhu area were tested. The result shows that the CBM contained CH_4 of 53.93%–64.76%, C_2H_6 of 0.1–0.14%, C_3H_8 of 0.08–0.13%, N_2 of 32.43–43.82%, CO_2 of 1.23–2.15%, and trace amount of Ar. The C_1/C_{1-n} ratios were 0.996–0.997 and C_1/C_2 ratios of 436.2–577.6, indicating extremely dry CBM.

The carbon and hydrogen isotopic compositions of seven CBM samples were tested, and the result shows the $\delta^{13}C_1$ ranges from –62.7 to –61.5‰, almost no change at all. Most scholars use –55 or –60‰ as a $\delta^{13}C_1$ threshold to divide biogenetic and thermogenic CBM. No matter which threshold is used, the carbon isotope composition of CBM in the Shaerhu area has the characteristics of biogenetic methane.

The δD_{CH4} is from –225 to –220‰, which, in terms of hydrogen isotope, is quite stable. By now studies on the hydrogen isotope of CBM are less, and those on the hydrogen isotope of biogenetic CBM are the least. The δD_{CH4} of the Shaerhu falls in the zone of secondary biogenic gas (–244 to –215‰) from Liyazhuang Coalmine, and it is also in the range of the δD_{CH4} from –242.5 to –219.4‰ of mixed CBM (more secondary biogenic gas and less thermogenic gas) from Xinji Coalmine. Comparatively, the δD_{CH4} of primary biogenetic gas varies little, while the δD_{CH4} of secondary biogenic gas changes obviously wider.

The coal rock is lignite with R^o values of 0.40–0.47%, reflecting that it is still in the stage of primary biogenic gas. The carbon and hydrogen isotopic compositions of CBM reflect the characteristics of biogenic gas. Therefore, the CBM in the Shaerhu area is primary biogenic gas. Biogenic gas has various sources, mainly CO_2 reduction, fermentation of acetic acid and methanol (such as methanol, methylamine, methyl sulfide, and the likes). The isotope composition of the biogenetic methane from different sources is different. Whiticar (1996) suggested that the biogenic gas from fermentation of acetic acid has relatively heavy carbon isotope and light hydrogen isotope, while the biogenic gas from CO_2 reduction is reversed—relatively light carbon isotope and heavy hydrogen isotope. Using the hydrocarbon isotope classification chart proposed by Whiticar et al. (1996), all CBM samples from the Shaerhu area fall in the CO_2 reduction segment (Fig. 2.3). This further proves that the primary biogenic gas is from reduction of microorganisms in the Shaerhu area.

2.3.2 Thermally Degraded CBM

Thermogenic gas mentioned by Rightmire et al. (1984) and Rice (1993) refers to CBM produced from organic matter in coal rock under the action of thermal (genetic) force. In fact, the organic matter in coal rock generating CBM under the action of heat involves two specific mechanisms—thermal degradation and thermal cracking, and two stages. Up to now, few reports have been released on the geochemical composition and identification indicators of pure thermogenic CBM in the world.

Systematic sampling and tests were carried out on the CBM in Baojishan Coalfield, Gansu Province, and then the geochemical composition, characteristics and interrelationship of the CBM were studied, as well as the genetic types and identification indicators, by considering the geology of the coalfield. As mentioned above, by now the international academic circle only classifies thermogenic CBM as a class, which has not yet been studied and subdivided. And no systematic research reports have been released on the geochemical composition and identification indicators of pure thermogenic CBM. Our study indicates that the CBM in the Baojishan Coalfield is typical or relatively pure thermogenic CBM. It is also a newly identified and classified genetic type, and belongs to generalized thermogenic CBM.

Administratively, Baojishan Coalfield is located in the Pingchuan District of Baiyin City (formerly Jingyuan County), Gansu Province, and geographically, it is in the eastern section of the North Qilian orogenic belt, where the main structures are the Baojishan complex syncline belt and reverse faults parallel to the axial of the syncline; and the

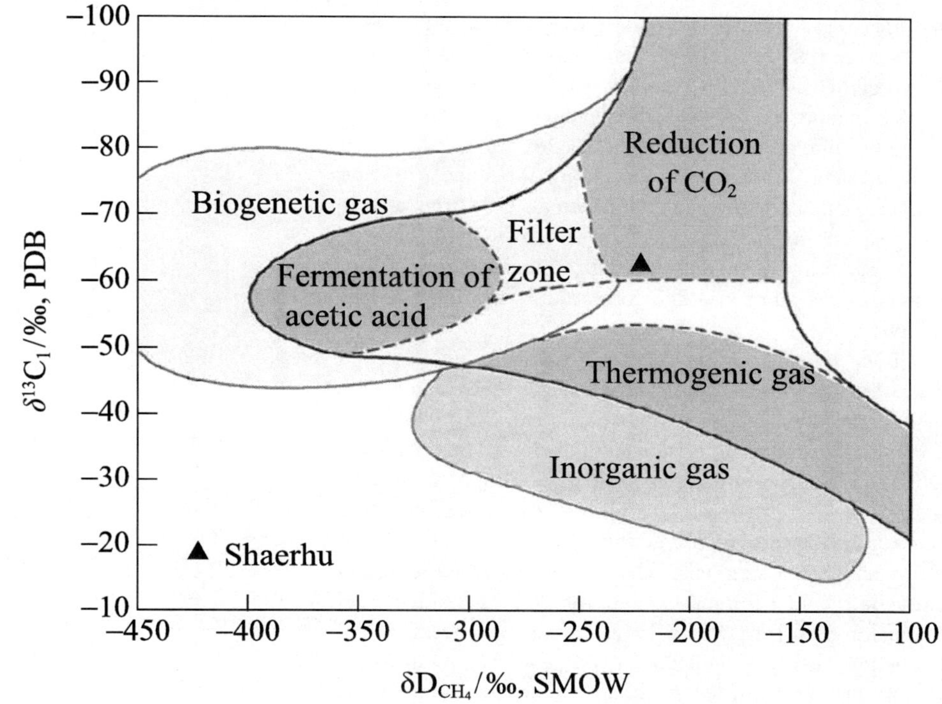

Fig. 2.3 Genesis and generation pathway identification chart of CBM in Shaerhu

main coal-bearing interval is Longfengshan Formation (J_2l) in lower Middle Jurassic. The Baojishan Coalfield is divided into three well fields where Weijiadi, Dashuitou and Baojishan wells were built, all of which produce high production of methane.

Gas samples were collected and tested, including drained gas samples from 9 underground holes in Weijiadi and Dashuitou, and desorbed gas samples from two coal cores from a aboveground hole in Baojishan Coalfield.

2.3.2.1 Composition of Drained CBM

In the Weijiadi gas sample, CH_4 is dominant, accounting for 89.45–99.73%; N_2 varies greatly, from 0 to 8.57%; CO_2 is 0.13–2.26%; heavy hydrocarbon is less and dominated by C_2H_6, 0.06–0.65%; and C_1/C_{1-n} is 0.994–0.999. In addition, there are traces of Ar, He, SO_2 and H2S. In the Dashuitou gas sample, CH_4 accounts for more, 99.20–99.84%; heavy hydrocarbon only contains C_2H_6, 0.06–0.65%; C_1/C_{1-n} is 0.993–0.999; CO_2 is 0.1–0.19% (Fig. 2.4); and some traces of SO_2 and He.

The above-mentioned data show that the composition and major characteristics of Weijiadi and Dashuitou CBM are basically the same, namely, mainly composed of methane and low-content heavy hydrocarbon, indicating the characteristics of dry gas. Additionally, the above-mentioned CBM has C_1/C_2 of 72.72–1664 (excluding 1664 in one sample, C_1/C_2 < 900 in other samples); the CDMI value is 0.100–2.267 (excluding 2.267 in one sample, CDMI < 1 in other samples). The range is narrow.

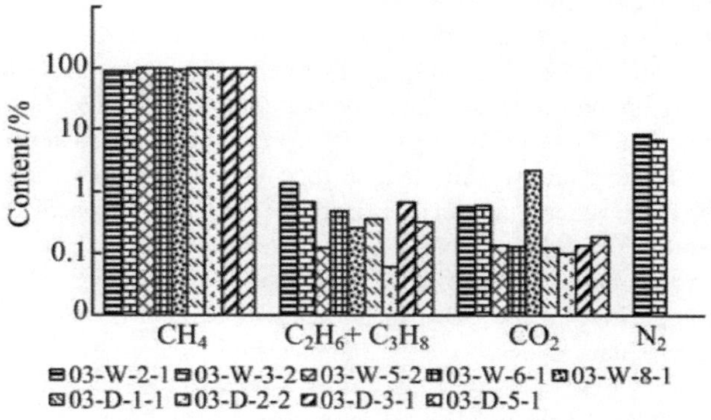

Fig. 2.4 Histogram (%) of major components of CBM drained from Baojishan Coalfield

2.3.2.2 Composition of Desorbed CBM

Two desorbed samples (WB-05 and WB-06) were collected from coal cores taken in an exploration well, WB-12, in southwestern Weijiadi Coalfield, constructed by Jingyuan Mining Bureau. The analysis result shows that the samples are composed of CH_4 of 76.35 and 82.92%, C_2H_6 of 9.86 and 6.11%, C_3H_8 of 1.29 and 0.50%, and traces of butane and pentane; and N_2 of 8.6%, 6.27%; CO_2 of 2.16% and 3. 28%

and traces of Ar and He; in addition, the drying coefficients C_1/C_{1-n} are 0.870 and 0.926, the C_1/C_2 values are 7.74 and 13.57, and CDMI values are 2.751 and 3.805, respectively.

The composition of desorbed methane samples is significantly different from that of the above-mentioned drained methane samples. The most outstanding difference is that the drained methane is dry gas, while the desorbed methane is wet gas. The mechanism and representative significance of the composition of the CBM produced under different conditions in the same coalfield or location are discussed as follows:

CBM is mainly present in coal seams in an adsorbed state. During the desorption process, methane is first desorbed, and then heavy hydrocarbons and CO_2, which are difficult to be desorbed. It is generally believed that methane, which has simpler structures, smaller molecule and smaller mass than heavy hydrocarbon, is more easily desorbed to a free phase. Due to such fractionation, the composition of CBM undergoes significant changes during desorption and extraction. Under formation conditions, CBM is slowly and gradually desorbed with a long extracting process. CBM being desorbed from coal cores is fast, even at a time, due to limited and broken coal cores, and at normal temperature and pressure, or even high temperature and low pressure. Therefore, under formation conditions, methane is first desorbed, and then heavy hydrocarbons, so that early CBM is relatively dry. By contrast, nearly all types of gas in coal cores are desorbed at a time, and the composition changes a little. In fact, coals in Baojishan Coalfield are mainly non-caking and weakly caking, their evolution is low, and they are generating wet gas. Both theoretical and practical works have shown that desorbed gas samples are more representative in terms of composition. Therefore, the composition of the CBM in Baojishan Coalfield is wet gas represented by desorbed gas samples from coal cores.

2.3.2.3 Composition of Isotopes

For the $\delta^{13}C_1$ values of nine extracted gas samples in Baojishan Coalfield, one is slightly lighter, –54.6‰, and the other are –46.2 to –35.1‰, and the average is –41.9‰ (Fig. 2.5, including a $\delta^{13}C_1$ value of a desorbed methane sample from coal cores). The δD_{CH4} value of the samples is –247.3 to –222.5‰, and the average is –235.6‰. It indicates that the carbon and hydrogen isotopic compositions of the methane samples extracted vary, but the variation is not great.

Theoretically, the $\delta^{13}C_1$ value of the extracted methane samples should gradually increase with the process of extraction (desorption). The $\delta^{13}C_1$ value of the methane desorbed at the second time from the WB-05 sample taken in the WB-12 exploration well is –40.4‰. It is slightly heavier than the average $\delta^{13}C_1$ of the extracted methane (–41.9‰). It is thus proved that the $\delta^{13}C_1$ value of the extracted methane is generally around the average value, and can represent the

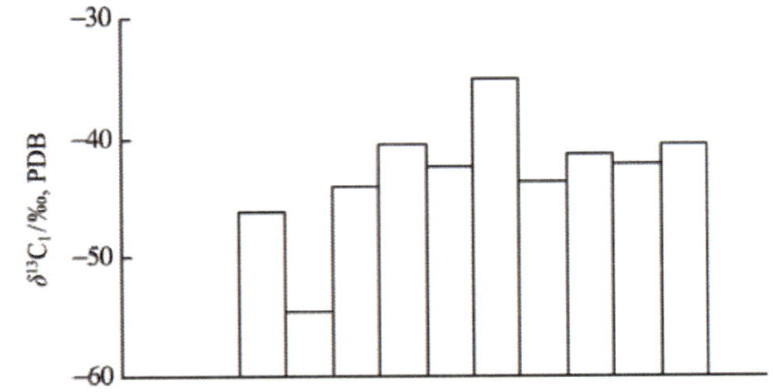

Fig. 2.5 Histogram of $\delta^{13}C_1$ of CBM in Baijishan coalfield

carbon isotope composition of the CBM in Baojishan Coalfield.

Scott et al. (1994) believed that thermogenic CBM was generated after the R^o value reached 0.5%. Caking coal and noncaking coal are dominant, and the R^o values of the above-mentioned two coalfields are 0.77–0.84%. As far as the thermal evolution and gas generation stage, Baojishan Coalfield is at early thermal genesis.

Judging from the hydrocarbon generation characteristics and the mechanism of thermogenic CBM, thermogenic CBM has two sources: thermal degradation and thermal cracking, which correspond to the early to middle (mature and highly mature) stage, and the late (over mature) stage of thermal evolution.

Based on the above-mentioned data, characteristics and other related analysis, this study concludes that the CBM in Baojishan Coalfield is thermogenic CBM and the coalfield is at the early stage of generating thermogenic CBM. At present, thermogenic CBM classified internationally includes two types—thermally degraded CBM and thermally cracked CBM, related to the early to middle (mature and highly mature) stage and the late (over mature) stage of thermal evolution of coal rock.

2.3.3 Thermally Cracked CBM

As mentioned above, CBM is generally divided into two types: biological (bacterial) gas and thermogenic gas. This book divides thermogenic CBM into two types: thermally degraded CBM and thermally cracked CBM.

We have tested and analyzed the geochemical composition of CBM in the southern part of the Qinshui Basin in Shanxi Province, and obtained new geochemical data; and in addition to $\delta^{13}C_1$ proposed by predecessors, additional geochemical tracer indicators, such as δD_{CH4}, C_1/C_{1+2}, C_1/C_2 and CDMI, were used to comprehensively study the CBM in the basin. It is concluded that the CBM in the basin is thermogenic CBM.

Located in the south and center of Shanxi region, the Qinshui Basin is a Carboniferous–Permian coal-bearing area in North China. The basin generally extends from north to south, 120 km wide from east to west and 330 km wide from north to south, and covers a total area of about 3 × 10^4 km^2. On the whole, the Qinshui Basin is a large-scale compound syncline. Its north and south ends are tilted, and the east and west wings are basically symmetrical. The structure in the basin is relatively simple. The marginal Lower Paleozoic (Cambrian and Ordovician) has a large dip, but the internal Upper Paleozoic (Carboniferous and Permian) and Mesozoic (Triassic) are relatively flat. Secondary folds are relatively developed, but their relieves and areas are small. Most faults are small structures.

Primary coal-bearing strata are Upper Carboniferous Taiyuan Formation and Lower Permian Shanxi Formation. The Upper Carboniferous Taiyuan Formation (C_3t) contains 4–14 coal layers. And the No.15 coal seam is the most stable in the whole basin, and very thick, generally 2.0–6.0 m. It is the primary coal seam of Taiyuan Formation. The Lower Permian Shanxi Formation (P_1s) contains 2–7 coal layers, 0.25–11.51 m thick and average 4.94 m. Among them, the No.3 coal seam is the thickest, 0.53–7.84 m, and widely distributed and laterally stable in the whole basin. It is the primary coal seam of Shanxi Formation. The metamorphism of the coal is generally high, and the coal types are from gas coal to anthracite. In the northern and southern parts of the basin, there are mainly anthracite and meager coal.

The Jincheng area in the southern part of the Qinshui Basin is a plot site for exploration and development of CBM in China, including Fanzhuang block (northeast), Panzhuang block (south) and Zhengzhuang block (northwest). The Carboniferous–Permian coal seams are generally shallower than 1000 m, and has high gas saturation, good preservation conditions and abundant CBM resources. The degree of metamorphism of the coal gradually decreases from southeast to northwest, and the R^o value decreases from 5.2 to 2.5, or even lower (Zhang and Tao 2000).

Samples were mainly taken in surface production wells in Fanzhuang block (Zaoyuan) and Panzhuang block (Fig. 2.6) in the southern part of the Qinshui Basin.

2.3.3.1 CBM Geochemical Composition Characteristics

(1) Compositional ratio

Compositional ratio is a basic index for classifying the genesis of hydrocarbon gas. Because the sampling conditions of the production wells are good and the gas samples are basically free from atmospheric pollution, Fanzhuang

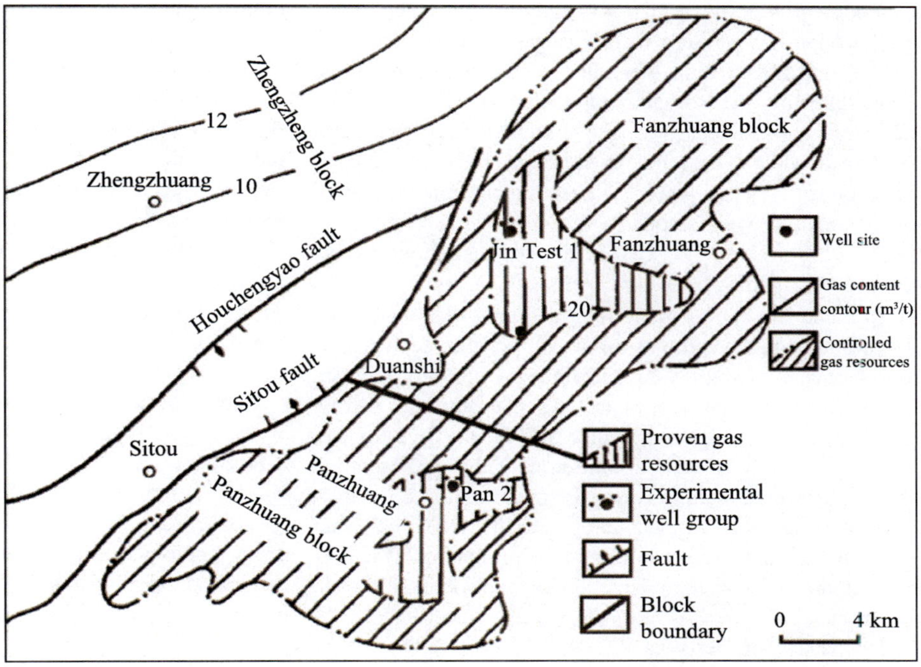

Fig. 2.6 Distribution of CBM blocks in southern Qinshui Basin (Zhang and Tao 2000)

block (Zaoyuan) and Panzhuang block were taken as representatives to discuss the compositional ratio and tracer indicator of the CBM in southern Qinshui Basin.

The CBM in the Fanzhuang block (Zaoyuan) is produced from the No.3 and No.15 coal seams. Samples from three gas wells (FZ002, FZ012 and FZ016) and a mixed gas sample from a gas gathering station were collected and analyzed, showing CH_4 of 98.16–98.99%, C_2H_6 of 0.012–0.029%, N_2 of 0.92–1.63%, CO_2 of 0.02–0.15%, Ar of 0.023–0.031%, and some trace SO_2 and H2S. The heavy hydrocarbons only contain C_2H_6 and the C_1/C_{1+2} ratio is 0.9997–0.9999, being extremely dry gas. The C_1/C_2 ratio is 3385–8249, also showing the characteristics of dry gas. The CDMI value is 0.02–0.2%, which is significantly lower than the CBM in the aforementioned Upper Silesian Basin and Lublin Basin.

The CBM in the Panzhuang block is produced from the No.3 coal seam. Its composition is similar to that of Fanzhuang CBM, but the methane content (98.96–99.55%) is slightly higher, ethane (0.007–0.01%) and nitrogen are slightly lower, C_1/C_{1+2} is up to 0.9999, C_1/C_2 12,390–14,137, and CDMI 0.13–0.29%, showing the characteristics of extremely dry gas.

(2) Composition and distribution of carbon and hydrogen isotopes

The $\delta^{13}C_1$ value of the CBM produced from single wells in the Zaoyuan zone is similar, –32.2 to –30.2‰. The $\delta^{13}C_1$ value of the gas mixture from gas gathering stations is –31.9‰, and it is within the range of the gas from single wells, reflecting the isotope composition of mixed gas. The δD_{CH4} value of the gas from single wells ranges from –193 to –145‰, while that of the mixed gas is –157‰ and falls in the range of –193 to –145‰, reflecting the mixed characteristics of hydrogen isotopes.

The $\delta^{13}C_1$ value of the CBM from three production wells (P001, P002 and P004 wells) in Panzhuang is –33.0 to –31.2‰, and the δD_{CH4} value ranges from –172 to –152‰. By comparison, the carbon isotope composition of the CBM from Zaoyuan is slightly lighter, but they are very close in the two blocks; the hydrogen isotope distribution is narrow in Panzhuang, but falls in the range of Zaoyuan. In Zaoyuan, the No.15 coal seam in Carboniferous Taiyuan Formation and the No.3 coal seam in Lower Permian Shanxi Formation are producing, while in Panzhuang, only the No.3 coal seam in Lower Permian Shanxi Formation is producing. This may result in the difference in the carbon isotope composition and variation.

The basic characteristics of CBM in the above two blocks are that the carbon isotope composition is fairly uniform (Fig. 2.7); the distribution of hydrogen isotope is relatively wide; and the carbon and hydrogen isotopic compositions are generally heavy, reflecting the characteristics of high evolution.

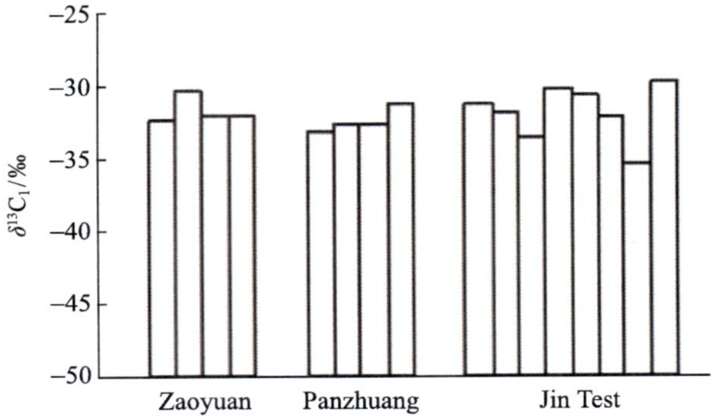

Fig. 2.7 Distribution of $\delta^{13}C_1$ (PDB) in southern Qinshui Basin (Jin test data from Hu et al. 2001)

In addition, according to Hu Guoyi et al. (2001), the distribution of $\delta^{13}C_1$ of the CBM from Jin Test wells, which shares the same Fanzhuang block with Zaoyuan wells, is –35.39 to –29.63‰ (expressed by "%" in the reference, such as –3.539%; "‰" in this book according to applicable provisions, Fig. 2.7). The overall change is not large, and the distribution of $\delta^{13}C_1$ in Zaoyuan and Panzhuang is very close (Fig. 2.7). This further shows that the factor causing the minor difference in the $\delta^{13}C_1$ value may be the difference in coal seams.

2.3.3.2 Genetic Types and Tracer Indicators

As mentioned earlier, most scholars take –55‰ as the boundary value to divide CBM into the biogenetic and the thermogenic. The distribution of $\delta^{13}C_1$ of various CBM samples taken in southern Qinshui Basin is –35.4 to –29.6‰, and the carbon isotope composition of thermogenic CBM is obviously heavier; the δD_{CH4} values of the samples are from –193 to –145‰, higher than the –250‰ threshold proposed by Rice (1993), and significantly heavier, reflecting the hydrogen isotopic composition of CBM in the high evolution stage.

From the analysis of the composition ratio and tracer index characteristics, the C_1/C_{1+2} ratio of the representative CBM from Zaoyuan and Panzhuang is 0.999–0.9999, and the C_1/C_2 ratio is 3385–14,137, belonging to extremely dry gas. There are two basic types of dry gas, one is generated from microbial action and the other is thermogenic gas generated in the late evolution stage, namely in the over-mature stage (generally $R^o > 2.0\%$). The heavy hydrocarbons (including liquid hydrocarbons) were further cracked into methane, i.e. dry gas (Xu et al. 1994). The above-mentioned characteristics and comparative analysis show that dry and even drier gas is the original feature of CBM in southern Qinshui Basin.

The R^o value of the coal seam in the study area is generally 2.8–4.8%, it has not only reached the overmature stage, but the degree of evolution has been quite high. In addition, the hydrocarbons in the CBM are almost composed of methane, indicating that the degree of evolution is high and the heavy hydrocarbons have been almost cracked. According to the gas generation stage proposed by Scott et al. (1994), when R^o reaches 3.0‰, it is the final stage of thermogenic methane generated at a large amount. Therefore, in combination with the geochemical characteristics of CBM and coal rock, the genetic type of the CBM in southern Qinshui Basin should be thermally cracked CBM.

2.3.4 Secondary Biogenic CBM

Rightmire et al. (1984) considered that biogenic methane would be formed in early coalification when the temperature was lower than 50 °C, the organic matter was decomposed by microorganisms, and the $\delta^{13}C$ of methane was generally –75 to –55‰. Rice (1993) believed that the distribution of $\delta^{13}C$ of bacterial methane produced by CO_2 reduction was in the range of –110 to –55‰, and the distribution of δD_{CH4} was –250 to –150‰. Scott et al. (1994) considered that the geochemical composition of secondary biogenic gas was similar to that of primary biogenic gas, but it differed mainly in that the thermal evolution of coal rock exceeded the forming stage of primary biogenic gas (R^o value below 0.30‰) and the coal seam was generally uplifted to shallow layers. In the Sydney and Bowen Basins in Australia (Smith and Pallasser 1996), the Upper Silesian Basin and Lublin Basin in Poland (Kotarba 2001), the Zonguldak Basin in Turkey (Hakan et al. 2002), the Elk Valley Coalfield in Canada (Aravena et al. 2003), and the Ruhr Basin in Germany (Thielemann et al. 2004), secondary biogenic gas was found.

In recent years, secondary biogenetic CBM has also been successively discovered in Xinji in Anhui Province, Liyazhuang Coalmine in Huozhou in Shanxi Province, and Enhong in Yunnan Province, and preliminary research was conducted[1] (Tao et al. 2005a). Among them, the CBM in Xinji in Anhui Province and in Enhong in Yunnan Province contains more thermogenic gas in addition to secondary biogenic gas. Therefore, the composition of CBM in the latter two regions should be a mixture of secondary biogenic gas and thermogenic gas (described later). The CBM in Liyazhuang in Shanxi Province is basically composed of secondary biogenic gas, which belongs to typical secondary biogenic CBM.

2.3.4.1 Geology of Liyazhuang Coalmine

The Huozhou Coalfield is located in the west of the Qinshui Basin and southern section of the Fen River, including five coalmines, namely Bailong, Xinji, Tuanbai, Caocun and Liyazhuang. Faults are developed, especially small faults. The lithology of the coalmines is very different in horizontal and vertical directions, as is the degree of structural development. Liyazhuang Coalmine is new in Huozhou Coalfield, and has complex geological conditions. The Permian contains coal, and its lithology is mainly siltstone, sandstone and mudstone. No.2 coal seam in the Shanxi Formation is primary, with a dip angle of 5°–15° (Wang 2002).

The coalmine has the following characteristics. The mining well has very high content of gas in the Huozhou Coalfield; the structure is deformed strongly, and the fracture is well developed; the modern tectonic stress field is extremely strong, shown by strong deformation and floor heave of roadway. In addition, the No.2 coal seam is shallow at 320–600 m and contains fat coal, with R^o of 0.87–0.96% and an average of 0.92%.

2.3.4.2 Geochemical Composition and Genetic Types

Four desorbed samples were taken from the No.2 coal seam in the Shanxi Formation of Liyazhuang Coalmine. The samples contain CH_4 of 68.35–99.35%, C_2H_6 of 0.01–0.022%, N_2 of 4.63–30.87%, a small amount of CO_2 (0.06–0.38%) and Ar, SO_2 and other components (Fig. 2.8). The dry index C_1/C_{1-5} of hydrocarbon component is greater than 0.999, indicating that the desorbed gas is dry.

The $\delta^{13}C_1$ of four desorbed gas samples is –59.1‰, –56.3‰, –61.7‰, and –61.5‰ respectively. From the point of view of desorption fractionation, the second sample is slightly heavier, while the other three samples are almost the same. Since the samples are from excavated faces, the gas is residual after most gas lost. From the available results and theories, the carbon isotope composition of the gas sample should be heavier than its original gas or its average, therefore it is estimated that the $\delta^{13}C_1$ of any gas sample is lower. The carbon isotope composition of Liyazhuang CBM has the typical characteristics of biogenic gas.

Liyazhuang CBM is evidently characterized by low $\delta^{13}C_1$. The coal rock contains gas or fat coal, with R^o of 0.87–0.96% and an average of 0.92%. By now there hasn't been mathematical expression between $\delta^{13}C_1$ and R^o. According to Xu Yongchang et al. (1997), when there is no subsidence in gas generation and the source rock has R^o <1.0% in a basin, the relationship between $\delta^{13}C_1$ and R^o of thermogenic CBM is

[1]Tao et al. 2007. The secondary biogenic methane found in the Xinji Coalfield, Anhui Province, China. AAAPG-2004, 6th International Conference, 228–229.

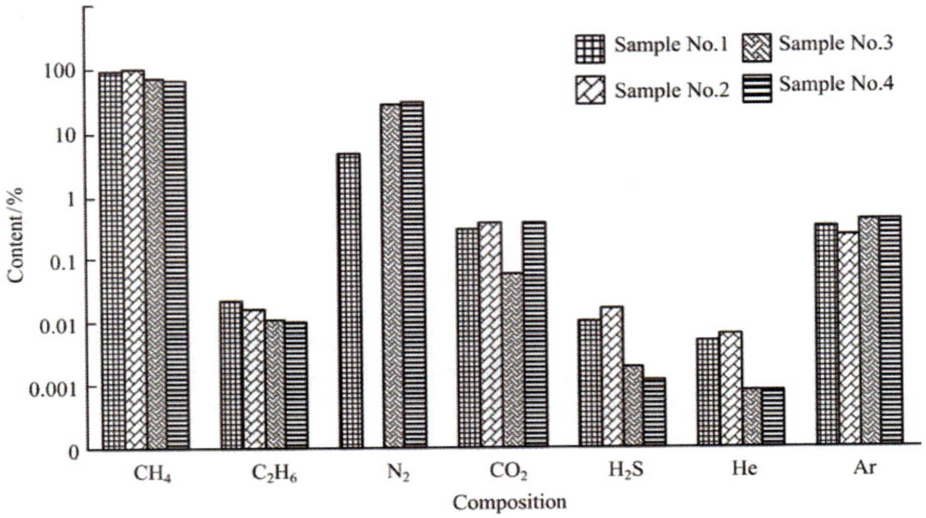

Fig. 2.8 Component content of CBM in Liyazhuang coalmine

$$\delta^{13}C_1 = 40.5\log R^o - 34.4(0.3\% R^o 1.0\%)$$

The geological characteristics of the Liyazhuang Coalmine meet the conditions above mentioned, namely R^o <1.0%, the thermogenic gas should be formed before the structure was uplifted, and secondary biogenic gas was mainly formed after the uplifting. In the case that the relationship between $\delta^{13}C_1$ and R^o has not yet been established, the above formula can be used to estimate the $\delta^{13}C_1$ of thermogenic methane. The average of $\delta^{13}C_1$ of the Liyazhuang CBM was calculated to be –35.9‰. Obviously, this estimate is much higher than the measured $\delta^{13}C_1$, and the carbon isotope composition is evidently contradictory to the thermal evolution stage of the gas fat coal, indicating that the methane is secondary biogenic gas.

The δD_{CH4} of the four desorbed gas samples is –244 to –215‰. It is significantly lower than that of Fanzhuang and Panzhuang CBM in the Qinshui Basin, but the value is just within the range of coaliferous biogenic gas (basically –280 to –210‰), when the upper limit of $\delta^{13}C$ of biogenetic methane is taken as –60‰. In the classification diagram of $\delta^{13}C_1$ and δD_{CH4}, the Liyazhuang CBM is located at the transition zone between biogenetic gas and mixed gas and is biased to the zone of biogenetic gas (Fig. 2.9). However, if the upper limit of $\delta^{13}C$ of biogenetic methane is taken as –55‰ which is universally used by most scholars, the sample points of Liyazhuang CBM in Fig. 2.9 will all fall in the zone of (secondary) biogenic gas, indicating that the CBM in this area is basically secondary biogenic gas.

Liyazhuang CBM contains $\delta^{13}C_2$ of –22.4 to –20.5‰ for ethane. It is consistent with the $\delta^{13}C_2$ of Xinji CBM (described later). It shows thermogenic characteristics and represents residual thermogenic CBM. This further proved that the methane was mainly formed in late evolution stage.

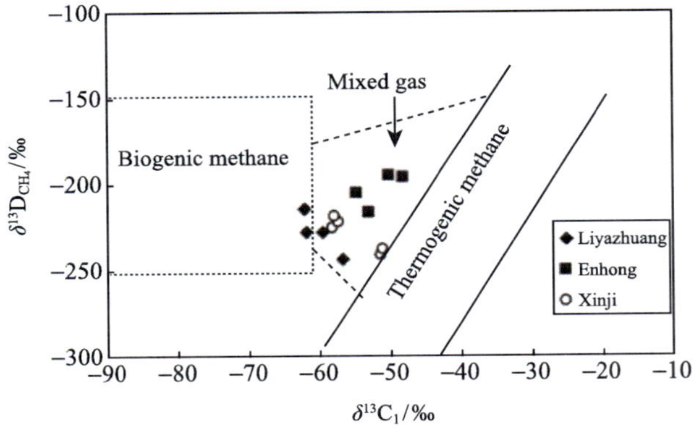

Fig. 2.9 $\delta^{13}C$ and δD_{CH4} tracer diagram for genetic types of Liyazhuang, Xinji, enhong CBM

In summary, the isotope composition and gas composition of Liyazhuang CBM have typical characteristics of biogenetic methane. In terms of the thermal evolution stage of the coal rock, and combined with the analysis of special geological structural phenomena in this area, there are conditions for secondary biogenic gas, so Liyazhuang CBM is secondary biogenic CBM. It should be noted that Liyazhuang CBM also contains trace amounts of thermogenic components such as ethane, and thermogenic methane.

2.3.5 Mixed CBM

Relevant theoretical analysis and experimental results show that thermogenic gas may be retained more or less because coal seams strongly adsorb gas, even if they were uplifted to near surface. Therefore, late secondary biogenic gas is inevitably mixed with early thermogenic gas. In other words, the authors believe that CBM which consists purely of secondary biogenic gas is rare, and such CBM generally contains varying amounts of thermogenic gas. Liyazhuang CBM is typical secondary biogenic gas, and is basically composed of biogenetic methane. But in most coalfields producing secondary biogenic gas, the produced gas is a mixture of secondary biogenic gas and thermogenic gas, such as the Xinji area in Anhui Province.

2.3.5.1 Xinji CBM

Geology

Xinji Coalmine belongs to the Huainan Coalfield. It is a Carboniferous-Permian coalfield in the southern margin of the North China Platform and located in the northern part of Anhui Province; its south is the Qinling-Dabieshan tectonic belt, east is the Tanlu fault zone, west is bounded by the Fuxin-Macheng fault, and north is the Mengbeng uplift, about 113 km long from east to west and 20–30 km wide from north to south, and has proven geological reserves of coal of over 147×10^8 t. Xinji Coalmine is structurally located at the southern flank of Xieqiao syncline and the middle section of Fufeng nappe. The tectonic line is NWW-EW. The Fufeng nappe pushed the Lower Proterozoic (Pt_1) metamorphic rocks and Cambrian strata onto the original system.

The coal-bearing strata in this area are Carboniferous Taiyuan Formation, Permian Shanxi Formation, Lower Shihezi Formation and Upper Shihezi Formation. The coal-bearing section consists of 45 layers, and the total thickness of the coal seams is up to about 40 m. Among them, the Carboniferous Taiyuan Formation contains thin coal seams. The most important recoverable coal seams are No.13-, No.11-2, No.8, No.6 and No.1 of the Permian, which are relatively stable and recoverable in most parts of the area (Fig. 2.10). The coal types are mainly gas coal, fat coal and coking coal. The average R^o is 0.85–0.97%.

Geochemical Composition and Genetic Types

(1) Geochemical composition

Results of the desorbed gas samples from Wells XS-02 and XS-01 show that the composition of the CBM is similar. The main characteristics are as the following:

The methane content is between 55.11 and 95.75%, mostly above 70%; the heavy hydrocarbon content is very low—ethane of 0.03–0.42% and propane of 0.04–0.18%. Propane was not detected in most samples. The C_1/C_{1-n} ratio is 0.993–1.0. The C_1/C_2 ratio is 188.6–2993.7, indicating extremely dry gas ($C_1/C_{1-5} < 0.99$). In terms of CBM, the

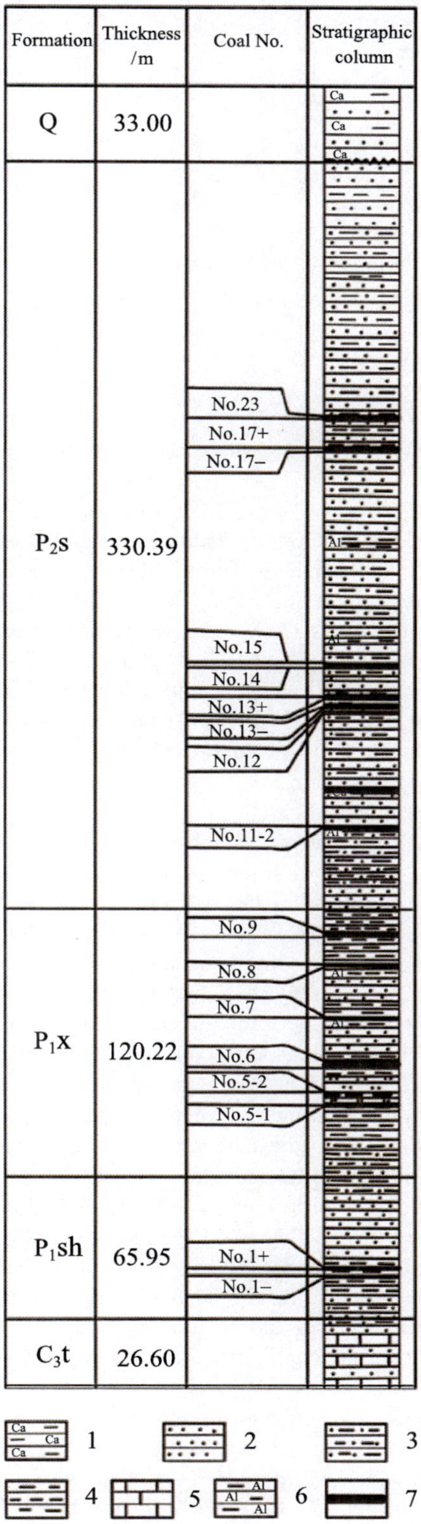

Fig. 2.10 Geological column of Xinji Coalmine 1. calcareous clay; 2. sandstone; 3. sandy mudstone; 4. mudstone; 5. limestone; 6. aluminous mudstone; 7. coal

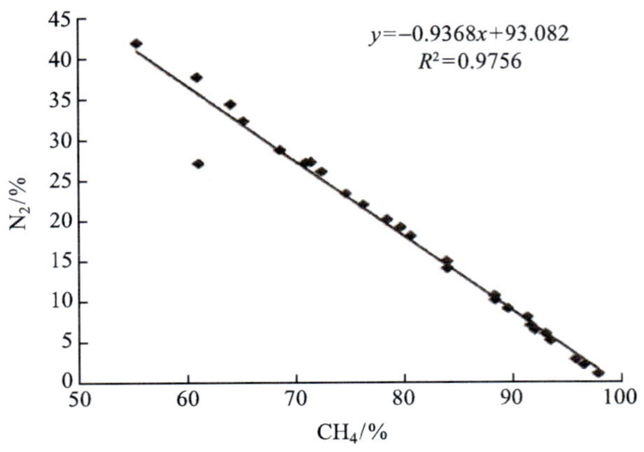

Fig. 2.11 Relationship between methane and N_2 in the CBM from Well XS-02

C_1/C_{1-5} ratio of the CBM containing secondary biogenic gas in the San Juan Basin in the United States is 0.77–1.00, but it is mainly dry gas and extremely dry gas (Scott et al. 1994). The biogenic gases in the Sydney Basin and Bowen Basin in Australia are also dry, and $C_1/C_2 > 1000$ is its characteristic index (Smith and Pallasser 1996). The C_1/C_2 ratio of Xinji CBM sample is < 1000, and some samples are close to 1000. The C_1/C_{1-n} ratio and C_1/C_2 indicators show that Xinji CBM has biogenic characteristics, but does not exclude thermogenic gas.

For gas components, non-hydrocarbon component is mainly N_2, 2.86–42.42%, and most less than 30%, and the content of N_2 and CH_4 has a good linear growth-decline relationship (Fig. 2.11). It is difficult to observe such excellent linear relationship in natural gas without secondary alteration, thereby indicating that the N_2 is mainly from atmosphere. The $\delta^{15}N$ of the gas sample is mainly at −1 to 1‰, which shows the isotope composition characteristics of atmospheric nitrogen. It also supports the above-mentioned conclusion. N_2 is a primary component of atmosphere, up to 78.08%. Sufficient atmospheric N_2 is soluble in surface water and infiltrates into underground coal seams with water, and its chemical properties are inactive and stable. It is exactly because the CBM in this area is dry gas, and the amount of atmospheric N_2 infiltration in different parts varies, which results in a good linear growth-decline relationship between CH_4 and N_2 in modern CBM.

The CO_2 in the Xinji CBM sample is only 0.51–1.93%. CBM with CO_2 below 5% has been found to be secondary biogenic gas in Sydney Basin and Bowen Basin, and the CO_2 content < 5% is a characteristic marker of biogenic CBM (Smith and Pallasser 1996; Ahmed and Smith 2001). The CO_2 content in Xinji CBM is much less than 5% compared to the CBM in Sydney Basin and Bowen Basin. A new simulation experimental study on thermogenic CBM by Kotarba and Lewan et al. (2004) found that, when their R^o changed from 0.3 to 1.5%, the CO_2 produced from two brown coal samples was 47.8 dm^3/kg and 71.4 dm^3/kg, the CH_4 16.3 dm^3/kg and 20.6 dm^3/kg, and the C_2/C_4 ratio 9.1 dm^3/kg and 10.2 dm^3/kg, respectively, and the amount of CO_2 is more than twice that of CH_4. The R^o of Xinji CBM is 0.88–0.91%, so a large amount of CO_2 would have been produced, but the actual content is extremely low, and significantly lower than that of thermogenic CBM. For example, the R^o of the coal from the Jingyuan Coalfield in Gansu Province is 0.6–1.18%, and the degree of evolution is similar to that of Xinji coal rock. However, the CO_2 content of the desorbed gas sample from the drilled coal core in Jingyuan Coalfield is 1.70–16.60%. In addition, the gas obtained from a sandstone core in Well XS-01 is mainly composed of N_2 (91.43‰), and the CO_2 content is as high as 7.18%, and the $\delta^{13}C_1$ is −21.4‰. The characteristics of organic genesis indicates the CO_2 may come from the coal seam. If the CO_2 in the sandstone comes from the coal seam and the N_2 is mainly from atmosphere, it reflects that the content of CO_2 in the early original CBM is quite low and the content of CH_4 is lower. The above comparative analysis shows that the CO_2 content in Xinji CBM has greatly reduced.

The $\delta^{13}C_1$ of the gas sample desorbed from the coal core in Xinji area is −61.3 to −50.7‰, with an average of −56.6‰. Among them, the $\delta^{13}C_1$ of the desorbed CBM in Well XS-02 is −61.3 to −54.7‰, with an average of 57.4‰; the $\delta^{13}C_1$ of the desorbed CBM in Well XS-01 is −58.7 to −50.7‰, with an average of 54.9, indicating that the $\delta^{13}C_1$ in these two wells doesn't vary greatly, but is very close.

The carbon isotope composition of methane is an important indicator for identifying the genetic type of natural gas. As described in this book, the limit of the $\delta^{13}C_1$ of microbial CBM is generally −55‰. According to Smith and Pallasser (1996), the value is −60 to −50‰. The $\delta^{13}C_1$ of the CBM containing secondary biogenic gas in Upper Silesian Basin is −79.9 to −44.5‰ (Kotarba 2001). Obviously, the $\delta^{13}C_1$ of Xinji CBM is basically in the range of microbial methane, but it biases to the heavier end. It can be deemed as the mixer of microbial methane and thermogenic methane.

The distribution of δD of Xinji CBM is −242.5 to −219.4‰, and it is basically the same to the δD_{CH4} (−244 to −215‰) of Liyazhuang CBM in Huozhou Coalfield.

The distribution of $\delta^{13}C_2$ of Xinji CBM is in the range of −26.7 to −15.9‰. The samples from Well XS-02 are uniformly distributed in the major coal seams of No.1 to No.11, and relatively complete. The $\delta^{13}C_2$ is −26.7 to −16.1‰, with an average of −22.1‰. The samples from Well XS-01 are only from the No.1–No.8 coal seams in the lower part, and the $\delta^{13}C_2$ is −23.6 to −15.9‰, with an average of −20.7‰. According to the summary by Rice (1993), the $\delta^{13}C_2$ of the CBM in the world is −32.9 to −22.8‰—the $\delta^{13}C_2$ in Upper Silesian Basin is −24.6 to −22.3‰ (Kotarba 2001). The $\delta^{13}C_2$ of Xinji CBM not only reflects the thermogenic characteristics of ethane, but also shows that the distribution of $\delta^{13}C_2$ is wider and heavier. This may reflect ethane decomposition by microbial in late stage, and the $\delta^{13}C$ is more enriched in the residual ethane. We also believes that Xinji CBM is the mixer of biogenic methane and thermogenic ethane.

The $\delta^{13}C_{CO2}$ of the Xinji CBM sample is distributed from –39.0 to –6.0‰, with an average of –17.9‰. The $\delta^{13}C_{CO2}$ (–39.0‰) of a sample is significantly lower, and that (–6.0 and –9.4‰) of another two samples is significantly heavier. For the rest 15 samples, the $\delta^{13}C_{CO2}$ is in the range of –29.2 to –12.2‰, with an average of –17.9‰. It is a very interesting phenomenon. The $\delta^{13}Cco_2$ of CBM reviewed by Rice (1993) is –26 to 18.6‰. The gas generated by pyrolysis at 250 °C, 350 °C and 400 °C is mainly CO_2, and the content is 97.3, 73.8 and 45.5‰, and the corresponding $\delta^{13}C$ of CO_2 is –19.3‰, –21.9‰, and –21.6‰ respectively (Smith and Pallasser 1996). From the comparative analysis of above-mentioned data, it can be confirmed that the CO_2 in Xinji CBM was generated by pyrolysis. But the $\delta^{13}C_{CO2}$ varies greatly, and seems heavier than the CO_2 from normal pyrolysis. It should be the residual after microbial reduction.

(2) Comprehensive discussion on genetic types of Xinji CBM

According to the gas composition and isotopic composition above mentioned, it is shown that Xinji coalbed methane shows the characteristics of microbial genesis, and thermogenic ethane and atmospheric N_2. They are the mixed gas with different genesis.

The R^o value of the coal seam where the Xinji CBM sample was taken is 0.88–0.91%. According to the generation stage proposed by Scott (1993), Xinji coal rock is in the evolution stage (R^o of 0.6–0.8%) generating the maximum amount of wet gas and thermogenic methane (R^o of 0.8–1.0%). It should have produced a large amount of thermogenic CO_2 and heavy hydrocarbons. However, as mentioned earlier, Xinji CBM is extremely dry. The CDMI value is 0.64–3.06% (mostly 0.96–2.08%), and it is in the range of the CDMI value (0.25–4.1%) of the CBM in the Upper Silesian Basin (Kotarba 2001), and relatively concentrated, indicating that the CBM in these two regions is similar in genesis. Dry gas is not only a feature of biogenic gas, but also a result of secondary alteration in the Xinji area.

The $\Delta\delta^{13}C_{C2-C1}$ of the Xinji CBM sample is 30.7–57.4, which shows that the carbon isotope compositions of CH_4 and C_2H_6 are very different. And the correlation is also extremely low, indicating that the two belong to different genetic types. On the genetic type tracer plot $-\delta^{13}C_1$ vs. δD_{CH4}, the sample points of Xinji CBM all fall onto the mixed zone of biogenetic gas with thermogenic gas

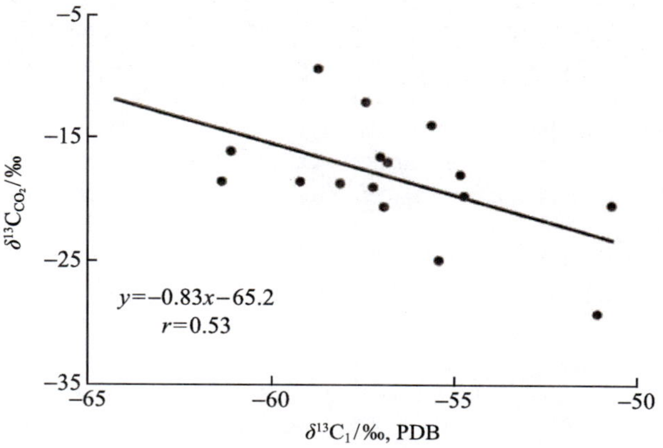

Fig. 2.12 Correlation between the $\delta^{13}C_1$ and $\delta^{13}C_{CO2}$ of Xinji CBM

(Fig. 2.9). This proves that the Xinji CBM is the mixture of two types of CBM. There is a negative correlation between the $\delta^{13}C_1$ and the $\delta^{13}C_{CO2}$ of Xinji CBM (Fig. 2.12), which not only shows that there is a genetic relationship between the two, but also reflects the isotope bifurcation characteristics of CO_2 reduced to methane by microorganisms. That is, the secondary biogenic methane in the Xinji area was produced by the reduction of CO_2.

The R^o of the Xinji CBM is 0.88–0.91%, with an average of 0.895%. By using the expression between $\delta^{13}C_1$ and R^o proposed by Xu Yongchang et al. (1997), an average R^o of 0.895% was taken in the expression, and then the estimated average $\delta^{13}C$ of the Xinji thermogenic methane is –56.57‰. It should be intermediate of biogenic methane and thermogenic methane.

The statistical analysis of a large number of $\delta^{13}C_1$ value data of conventional biogenic gas in relevant regions of the world shows the peak value is about –75 to –70‰ (Tao et al. 2007). In the same evolutionary stage, the $\delta^{13}C_1$ of Type III organic matter including coal is generally heavier than that formed by Type I organic matter. Taking into account the above-mentioned information and factors, three values (namely the end member $\delta^{13}C_1$ of biogenic methane, –70‰; the estimated $\delta^{13}C_1$ of thermogenic methane, –36.35‰; and the measured $\delta^{13}C_1$ average, –56.57‰) were taken, and the result shows that the secondary biogenic methane accounts for 60.1% and the thermogenic methane accounts for 39.9%. The result quantitatively testifies that Xinji CBM is the mixture of biogenetic gas with thermogenic gas, and dominated by secondary biogenetic gas. The experimental result of isotope effect of mixing microbial methane and thermogenic methane is the foundation to the estimating method.

The northward thrusting and extruding activities in the famous Dabie Mountain tectonic belt on the southern part of the Xinji area have caused strong uplift and denudation in the later period. The Carboniferous–Permian coal-bearing strata are directly covered by unconsolidated Quaternary sediments. Moreover, developed faults provide good conditions for surface water to penetrate and prosperous bacteria. The characteristic of atmospheric N_2 in CBM reflects strong penetration of surface water, and indicates favorable environment and material conditions for secondary biogenic gas.

2.3.5.2 Enhong CBM

Geology

Enhong Coalfield is located in Qujing and Fuyuan County in eastern Yunnan Province. The coal-bearing strata are Upper Permian Xuanwei Formation, and it contains up to 18–73 coal layers, of which 8–20 layers can be mined, totally 10–30 m thick. The coal types are dominated by coking coal, followed by meager coal, and the R^o value is 1.24–1.43%. Based on the calculation of CBM resources in 24 evaluated blocks of Enhong Coalmine, the CBM resources below 2000 m are 612.92 × 10^8 m^3, and the resources below 1000 m account for 82%, which are preferable to exploration and development in Yunnan Province (Deng et al. 2004).

Geochemical Composition and Genetic Types

Two gas samples from No.9 and No.19 coal seams in Enhong Coalmine contain methane of 67.65 and 93.92%, ethane of 0.574% and 0.733% respectively, and C_1/C_{1-5} ratio greater than 0.99, indicating dry gas. N_2 accounts for 30.28 and 4.70%, and CO_2 for 1.06% and 0.51% respectively.

The $\delta^{13}C_1$ of gas samples taken from four coal seams (No.9, No.10, No.15, and No.19) is –54.5‰, –52.8‰, –47.9‰, and –49.8‰, respectively, showing a tendency to become lighter upward. The δD_{CH4} is distributed from –206 to –196‰. The $\delta^{13}C_2$ is from –25.1 to –22.6‰, without great variation. The $\delta^{13}C_{CO2}$ is in –30.5 to –23.9‰, evidently lighter.

The thermal evolution of Enhong coal rock is higher than that in Xinji and Liyazhuang, which has reached the stage of coking coal and meager coal. Taking the R^o average of 1.35% and substituting it to the expression between $\delta^{13}C_1$ and R^o ($\delta^{13}C_1 = 40.5\log R^o–34.4$), the $\delta^{13}C_1$ of the thermogenic CBM in Enhong is −29‰, much higher than that of the actually measured values. In the identification diagram of $\delta^{13}C$ versus $\delta^{13}D$, the Enhong CBM samples also fall into the zone of the mixture of biogenic and thermogenic gases (Fig. 2.10). This indicates that Enhong CBM contains secondary biogenic gas. The carbon isotope of ethane in Enhong CBM is characterized by thermogenic genesis.

There is also a strong structural uplift in the Enhong area, and the overlying strata of the Upper Permian coal-bearing strata are almost missing except the Middle and Lower Triassic, and there is only sporadic Cenozoic ranging from less than 1 m to tens of meters. And multiple groups of faults are extremely developed. The coal seams vary greatly in depth due to folds, ranging from less than 300 m to over 1000 m. The above characteristics indicate that the Enhong area has geological structural conditions for the formation of secondary biogenic gas. And CBM is also a mixture of secondary biogenetic gas and thermogenic gas, especially the upper coal seam (such as coal seams No.9 and No.10) which has more prominent characteristic of secondary biogenic gas.

2.3.5.3 Resources Significance of Secondary biog- Enetic CBM

Studies have shown that primary biogenic methane generated in early stage is difficult to retain in later coal seams. Therefore, previous resources evaluation, zone selection and exploration and development of CBM were mainly considering the problems from the perspective of thermogenic gas, while the potential of secondary biogenic CBM resources has been rarely considered or underestimated.

Research and exploration practices in recent years have shown that secondary biogenetic CBM is not only a type of CBM, but also has a large resources potential. Secondary biogenic gas produced by microbial degradation in shallow coal seams can upgrade or increase the gas content to varying degrees, thereby making a compensation for the decrease caused by the loss of thermogenic CBM due to the uplift in late stage to some extent. For example, the San Juan Basin in the United States is the basin with the highest CBM yield in the world. In 1992, about 80% of CBM from the United States was produced in the San Juan Basin, and its secondary biological CBM accounted for more than 15% of CBM production. By 1992, the cumulative production of secondary biogenic CBM had reached 34 × 10^8 m^3. The CBM content in the San Juan Basin is higher than expected, and the generation of secondary biogenic gas is one of the reasons responsible for it (Scott et al. 1994). Judging from the current preliminary research, there is secondary biogenetic CBM in some coal-bearing basins in China. For example, the Liyazhuang Coalmine in Huozhou Coalfield has secondary biogenic gas, and it has become a high-yield mine in Huozhou Coalfield.

Evidently, the resources potential of secondary biogenic CBM can't be negligible due to shallow burial depth and low development cost—all within the exploration depth. Therefore, it should be an important resource which needs to be intensively studied and evaluated.

2.3.6 Tracer Indicator System

Based on the above-mentioned research results of genetic types of CBM, combined with relevant data, classification schemes and corresponding tracer index systems are proposed as follows:

Primary biogenic CBM: $\delta^{13}C_1 < -55‰$; $\delta D_{CH4} = -(225 \pm 25)‰$; $C_1/C_{1-n} \geq 0.99$; $C_2 \leq 0.2\%$; $R^o \leq 0.5\%$.

Thermogenic CBM: $\delta^{13}C_1 > -55‰$ (–50 to –35‰); $\delta D_{CH4} \geq -250‰$; positive correlation between $\delta^{13}C_1$ and δD_{CH4}; $C_1/C_{1-n} \leq 0.95$; average $CO_2 \approx 4\%$; CDMI 90%, $R^o = 0.5–2.0\%$.

Thermally cracking CBM: $\delta^{13}C_1 > -40‰$; $\delta D_{CH4} > -200‰$; positive correlation between $\delta^{13}C_1$ and δD_{CH4}; $C_1/C_{1+2} \geq 0.99$; CDMI $\leq 0.15\%$; $R^o > 2.0\%$.

Secondary biogenic CBM: $\delta^{13}C_1 < -55‰$; $\delta D_{CH4} = -(225 \pm 25)‰$; $C_1/C_{1+2} \geq 0.99$; $C_2 \leq 0.2\%$; $R^o > 0.5\%$.

Mixed CBM: $\delta^{13}C_1 \geq -55‰$; $\delta D_{CH4} = -250–150‰$; $\delta^{13}C_1 = -27$ to $-15‰$; $\delta^{13}C_{CO2} = -(22 \pm 18)‰$; $C_1/C_{1+2} \geq 0.95$; CDMI = 0.5–0.15%; $R^o > 0.5\%$.

Compared with international research results of CBM genetic types, the scheme involves thermally degraded CBM, thermally cracking CBM and mixed CBM, and a comprehensive tracer index system is proposed for the first time.

2.4 Characteristics and Formation Mechanism of Secondary Biogenic CBM

Biogenetic CBM is formed by microbial decomposition of organic matter at low temperature (generally below 50 °C) and shallow formation. Microorganisms produce simple compounds such as acetic acid and CO_2 by degrading soluble organic matter in coal seams, and anaerobic methanogenic bacteria use these simple compounds to generate methane. There are two different pathways for the formation of biogenic coalbed methane, namely the fermentation of acetic acid by microorganisms and the reduction of CO_2 by methanogenic bacteria (Rice 1993). According to current research data, the generation of biogenic gas is mainly based on the latter pathway.

Primary biogenic gas is formed in the early burial stage of coal-forming matters, that is when $R^o < 5\%$ or $R^o < 3\%$. Secondary biogenetic gas is formed after coal, that is, the thermal evolution of coal (source) rock exceeds the forming stage of primary biogenic gas, and the R^o value of coal rock is generally about 0.30–1.50. The formation conditions and mechanisms are that coal seams are generally uplifted to the shallow, temperature reduces, and an environment is established for microbial activities In the process of surface water pouring into the coal seam mainly through structural fissures, a large number of bacteria are brought into the coal seam, resulting in the prosperity of bacteria in the coal seam. Secondary biogenic gas is formed by the action of organisms on wet gas, *n*-alkanes and other organic matters produced during coalification process (Scott 1994).

2.4.1 Characteristics

Through comprehensive studies on CBM composition, carbon and hydrogen isotope composition, and geological conditions in the Liyazhuang Coalmine in Shanxi Province and the Xinji Coalmine in Anhui Province, main characteristics of secondary biogenic gas are as follows:

① The CBM contains mainly methane, and the dry index C_1/C_{1-5} is more than 0.99, which indicates dry gas.
② The $\delta^{13}C_1$ is generally less than –55‰, basically in the range of biogenetic methane, or much lower than the theoretical estimate of thermogenic methane, indicating non-thermogenic gas.
③ The δD_{CH4} of the biogenic gas is –250 to –200‰ according to available data.
④The carbon isotope composition of heavy hydrocarbons shows the characteristics of thermogenic methane, but the content of the heavy hydrocarbons is extremely low, indicating that the heavy hydrocarbons have been degraded, and only thermogenic gas left.
⑤ The content of CO_2 is extremely low, generally less than 2%. It may be residual CO_2 after reduction by microorganisms.
⑥ On the genetic identification chart of $\delta^{13}C_1$ and δD, the sample points fall in the zone of the mixture of biogenic gas and thermogenic gas, and biases to biogenic gas, indicating the CBM is biogenic and contains some residual thermogenic gas.
⑦ The R^o significantly exceeds the evolution stage of primary biogenic gas.
⑧ After uplifted, fractures are developed in the coal system and its overlying strata, and provide a good geological environment for bacteria carried by water into the coal seams, and finally secondary biogenic gas was produced.

2.4.2 Organic Geochemistry and Microbial Degradation

Based on the researches on biomarker compounds of soluble organic matter in coal rock samples, more researches were carried out on the forming mechanism and parental material of secondary biogenetic CBM.

2.4.2.1 Soluble Organic Matter in Coal Rock

Coal rock samples were collected from the Sihe Coalmine, Fucheng 71 Coalmine, and Liyazhuang Coalmine in the Qinshui Basin, Shanxi Province, the Xinji Coalmine, Zhangji Coalmine, Panji Coalmine in the Huainan Coalfield, Anhui Province, and the Enhong Coalmine in Qujing, Yunnan Province. The Sihe Coalmine, Fucheng 71 Coalmine and Liyazhuang Coalmine are highly thermally evolved, and the vitrinite reflectivity (R^o) of the coal rock is about 4%. It is the primary source rock for thermogenic CBM in the Qinshui Basin. The thermal evolution in the Liyazhuang Coalmine, Xinji Coalmine, Zhangji Coalmine and Panji Coalmine is relatively low, and the R^o ranges from 0.85 to 0.97. The thermal evolution in the Enhong Coalmine is relatively high, and the R^o is 1.24–1.43%.

All the coal rock samples were pulverized to below 80 meshes, and the soluble chloroform bitumen "A" was obtained by Soxhlet extraction for 48 h, with refined chloroform solvent. The chloroform bitumen "A" was precipitated with refined petroleum ether and infiltrated in a silica gel-alumina chromatographic column. Alkane, aromatic and non-hydrocarbon components were washed out by refined petroleum ether, dichloromethane and ethanol, respectively. GC-MS analysis was carried out on the saturated hydrocarbons. The instrument used is GC6890N/MSD5973N GC-MS spectrometer. The column is HP-5 elastic quartz capillary column (30 × 0.32 × 0.25 mm). The column was heated from 80–290 °C at 4 °C/min, and then kept for 30 min.

Table 2.3 shows the content of chloroform bitumen "A" and its family components in coal rock samples. The content of chloroform bitumen "A" varies greatly in different coal rock samples. The content of chloroform bitumen "A" in the Sihe Coalmine and the Fucheng 71 Coalmine is extremely low, only accounting for 0.009–0.011% of the sample weight. The total hydrocarbon content in these two samples is very low too, only 0.0018–0.0042%. In the family components of chloroform bitumen "A", the content of non-hydrocarbons is very high (67.4–70.51%), the content of aromatic hydrocarbons is very low (0.21–2.27%), showing the distribution characteristics—non-hydrocarbons > saturated hydrocarbons > asphaltene > aromatic hydrocarbons. And the ratio of saturated hydrocarbons to aromatic hydrocarbons is high, ranging from 16–99. The above-mentioned characteristics reflect the high thermal evolution degree of the coal rock and most soluble organic matters have cracked into dry gas.

The coal rock in the Liyazhuang Coalmine has the highest content of chloroform bitumen "A", accounting for 0.359%. And the content of total hydrocarbons is relatively high, accounting for 0.0693%. Asphalt has the highest content in chloroform bitumen "A" (68.25%), followed by aromatic hydrocarbons and non-hydrocarbons. The content of saturated hydro- carbon is the lowest 5.77%, and the ratio of saturated hydrocarbons to aromatic hydrocarbons is extremely low (0.42). The above-mentioned characteristics reflect low thermal evolution degree of the coal rock, and there are abundant soluble organic matters in the coal rock, which can provide sufficient nutrients to later microbial activities in coal seams. The low content of saturated hydrocarbons in chloroform bitumen "A" reflects possible degradation of saturated hydrocarbons by microorganisms.

The content of chloroform bitumen "A" in the Xinji, Zhangji and Panji coal samples is high, ranging from 0.669 to 2.996%. The total hydrocarbon content is also high, ranging from 0.1734 to 0.7789%. Among chloroform bitumen "A" in the Zhangji No.11 and Xinji No.13 coal

Table 2.3 Extracts and their family components in coal rock samples

Coalmine	Layer	R^o/%	Chloroform bitumen "A"/%	Saturated hydrocarbons /%	Aromatic hydrocarbons /%	Non-hydro-carbons/%	Asphaltene/ %	Saturated hydrocarbons/ Aromatic hydrocarbons	Total hydrocarbons /%
Sihe	P_1s	3.93	0.009	20.73	0.21	70.51	8.54	98.71	0.0018
Fucheng 71	C_3t	4.00	0.011	36.22	2.27	67.40	11.74	15.96	0.0042
Liyazhuang	P_1s	0.94	0.359	5.73	13.58	12.44	68.25	0.42	0.0693
Zhangji No.11	P_2	0.85–0.97	0.669	4.69	21.23	55.67	18.40	0.22	0.1734
Zhangji No.13	P_2	0.85–0.97	2.958	4.92	21.41	28.75	44.92	0.23	0.7789
Panji No.1	P_2	0.85–0.97	2.996	3.74	22.73	25.11	48.42	0.16	0.7931
Xinji No.13	P_2	0.85–0.97	0.874	3.34	22.88	58.66	15.12	0.14	0.2291
Enhong	P_2	0.85–0.97	0.008	39.12	7.56	50.27	3.05	1.86	0.0039

samples, the content of non-hydro- carbon is the highest, ranging from 55.67 to 58.66%, then aromatic hydrocarbons from 21.23 to 22.88%, asphalt from 15.12 to 18.40%, and saturated hydrocarbons from 3.34 to 4.69%. The chloroform bitumen "A" in the Zhangji No.13 and Panji No.1 coal samples contain the highest asphalt ranging from 44.92 to 48.42%, followed by non-hydrocarbons from 25.11 to 28.75%, aromatic hydrocarbons from 21.41 to 22.73%, and saturated hydrocarbons from 3.74 to 4.92%. The ratio of saturated hydrocarbons to aromatic hydrocarbons in coal rock samples from Xinji Coalmine, Zhangji Coalmine, and Panji Coalmine is very low, ranging from 0.14 to 0.23. The above-mentioned characteristics reflect that the thermal evolution of the coal rock had been mature. So a large amount of liquid hydrocarbons had been formed. Abundant soluble organic matters provided sufficient nutrients for microbial activities in middle and late evolution stages. The low content of saturated hydrocarbons in the chloroform bitumen "A" reflects possible degradation of saturated hydrocarbons by microorganisms.

The content of the chloroform bitumen "A" in coal rock samples from the Enhong Coalmine is very low, being 0.001%, and the total hydrocarbon content is extremely low, being 0.0039%. The result is similar to the chloroform bitumen "A" (0.009–0.011%) and total hydrocarbon content (0.0018–0.0042%) in the coal rock samples in the highly evolved Sihe Coalmine and Fucheng 71 Coalmine. In the chloroform bitumen "A" in the coal rock samples from the Enhong Coalmine, the content of non-hydrocarbons is relatively high, being 50.27‰, followed by saturated hydrocarbons of 39.12% which was the highest in the samples analyzed. The ratio of saturated hydrocarbons to aromatic hydrocarbons is relatively high, being 1.86. The above-mentioned characteristics reflect that the coal rocks have experienced relatively high evolution.

2.4.2.2 Thermal Evolution and Sedimentary Environment

Thermal Evolution

The total ion current maps of saturated hydrocarbons in three coal rock samples taken from the Qinshui Basin and Huoxi Basin in Shanxi Province all have a unimodal pre-peak distribution, with a dominant peak at nC_{17} or nC_{18}. Among them, the carbon number distribution of saturated hydrocarbons in coal rocks from the Sihe Coalmine and Fucheng 71 Coalmine is in C_{14} to C_{31}. The content of the alkane with low carbon number is significantly higher than that of the alkane with high carbon number, and even carbon number is dominant. The $\sum nC_{21}^{-}/\sum nC_{22}+$ is 3.26–6.52 and the OEP is less than 1.0, reflecting high evolution of the coal rock samples. In addition, it also indicates that soluble organic matter is subject to thermal cracking evidently, resulting in the cracking of long-chained alkane with high carbon number into short-chained alkane with low carbon number. As a result of thermal cracking of soluble organic matter, a large amount of methane was produced, and became the main CBM source in the Sihe Coalmine and Fucheng 71 Coalmine (Fig. 2.13).

The carbon number distribution of saturated hydrocarbons of the coal rock in the Liyazhuang Coalmine is C_{12} to C_{33}, and the main peak carbon is nC_{17}. The alkane with low carbon number is slightly higher than the alkane with high carbon number, and it also shows the dominance of even carbon number. The nC_{21}^{-}/nC_{22}^{+} value is 1.45, and the OEP value is less than 1.0. Most of alkanes with high carbon number still exist, reflecting that the thermal evolution of the coal rock is not very high, and the soluble organic matter hasn't yet been thermally cracked.

The saturated hydrocarbons in the coal rock samples from the Xinji Coalmine, Zhangji Coalmine and Panji Coalmine also show a unimodal post-peak distribution, with a dominant peak at nC_{27} or hopane. The carbon number distribution of saturated hydrocarbons is C_{17} to C_{33}. The content of the alkane with low carbon number is significantly lower than that of the alkane with high carbon number, showing not obvious odd or even predominance. The nC_{21-}/nC_{22+} value is 0.19–0.34, and the OEP is larger than 1.0, reflecting that the thermal evolution of the coal rock has been mature, and liquid hydrocarbons have been formed at a large quantity. The thermal evolution of the coal rock samples in these areas is lower than that of the coal rock samples in the Qinshui Basin.

The carbon number distribution of coal rock samples in the Enhong Coalmine is C_{17} to C_{33}, with a predominant peak at nC_{18}. The alkane with low carbon number is significantly lower than that of the alkane with high carbon number, showing obvious even predominance. The nC_{21-}/nC_{22+} value is 1.63, and the OEP nearly approaches 1.0. The alkane with high carbon number is mostly degraded, reflecting higher thermal evolution of the coal rock and thermal cracking of soluble organic matter.

In the coal rock samples from the Sihe Coalmine and Fucheng 71 Coalmine, tricyclic terpane is abundant, and C_{19} and C_{20} tricyclic terpane and C_{21} to C_{26} tricyclic terpane are included. They have relative abundance of 70–82% on the m/z = 191 mass chromatogram. However, the abundance of hopane compounds is relatively low, 13–26%. The C_{32} homo-hopane (Fig. 2.13) was not detected. The above-mentioned characteristics also reflect that the coal rock samples from the Sihe and Fucheng 71 Coalmines are highly thermally mature.

In the saturated hydrocarbon samples of the coal rock in the Liyazhuang Coalmine, hopane is dominated in the terpane series and a relative abundance of 84% in the m/z = 191 chromatogram; C_{32} homo-hopane was not detected;

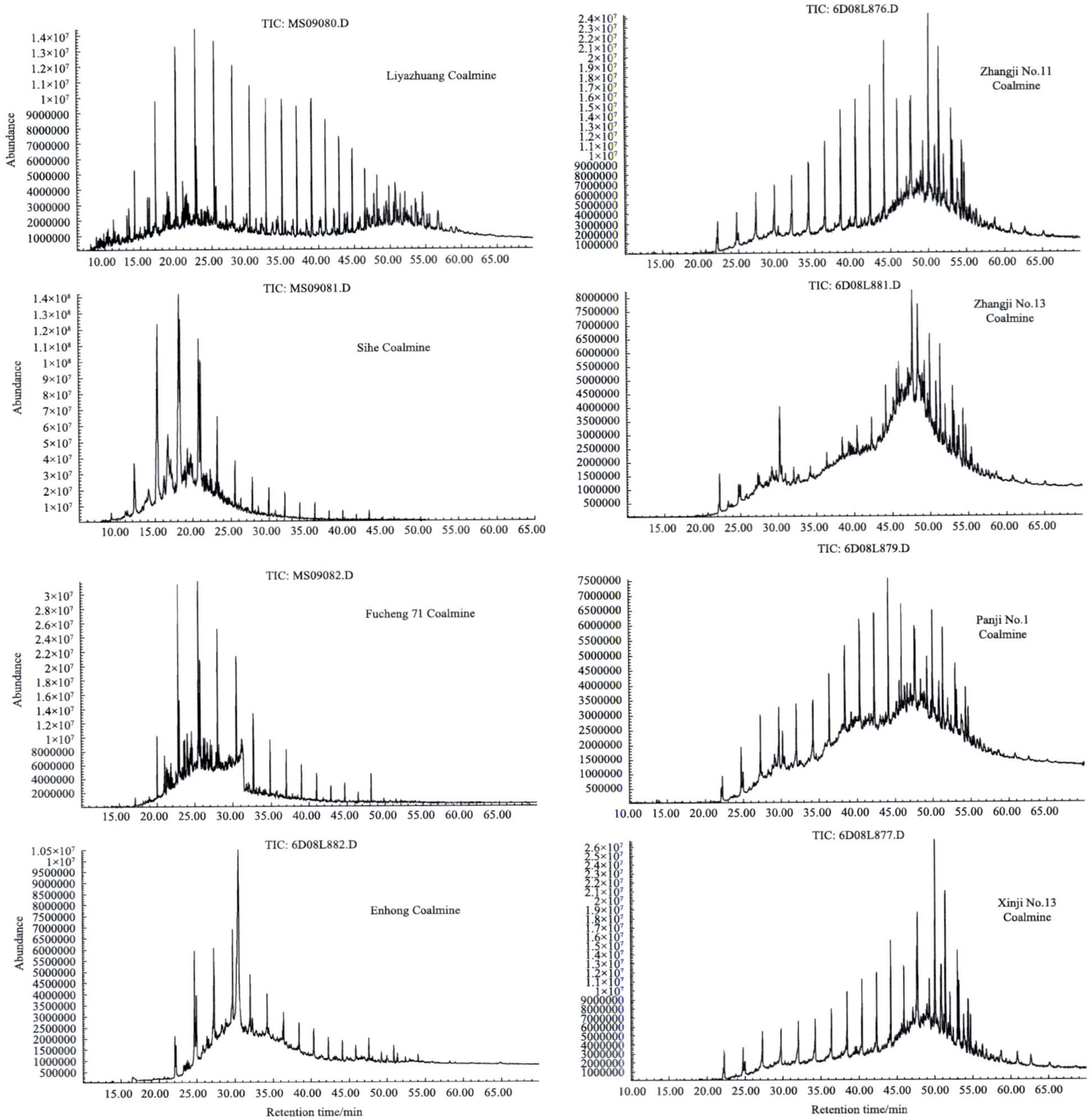

Fig. 2.13 Total ion flow diagram of saturated hydrocarbons in coal rock samples

the relative abundance of tricyclic terpane is very low (4.2%); and only C_{19} and C_{20} tricyclic terpane was detected (Fig. 2.14), reflecting relatively low thermal evolution of the coal rock from the Liyazhuang Coalmine.

In the saturated hydrocarbons of coal rock samples in Xinji Coalmine, Zhangji Coalmine and Panji Coalmine, hopane compounds are dominant; the relative abundance of hopane is more than 98% on the $m/z = 191$ mass chromatogram; C_{32} homo-hopane is abundant; and the abundance of tricyclic terpane is less than 2% (Fig. 2.14), reflecting low thermal evolution of the coal rock.

In the saturated hydrocarbons of coal rock samples in the Enhong Coalmine, hopane compounds are dominant; C_{19} and C_{20} tricyclic terpane, and C_{19} to C_{20} tricyclic terpane are

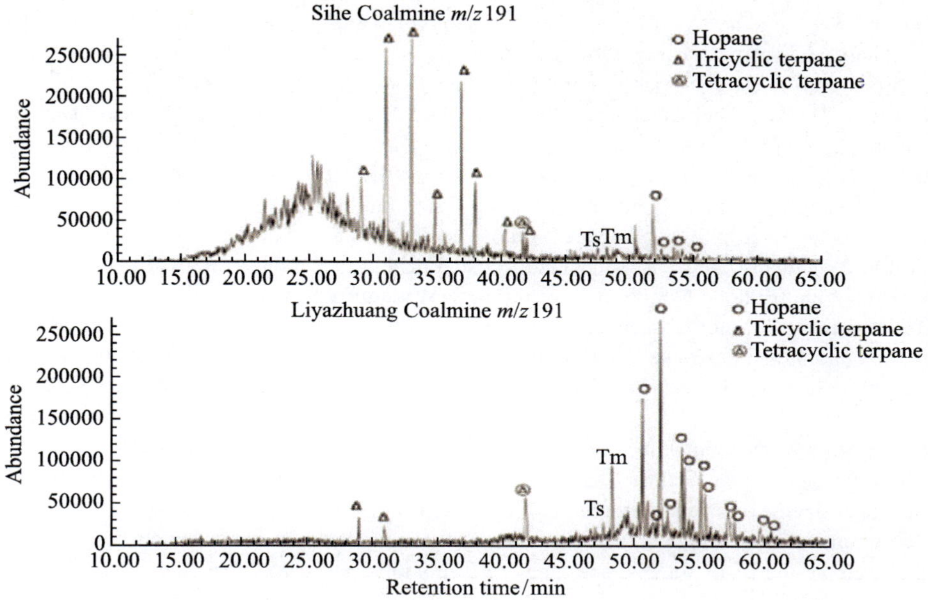

Fig. 2.14 m/z = 191 mass spectrogram

included; on the $m/z = 191$ chromatogram, the relative abundance of tricyclic terpane is more than 80%; but the abundance of hopane compounds is less than 20% (Fig. 2.14), indicating that the thermal evolution of the coal rock is relatively high.

Ts/(Ts + Tm) value, namely the $C_{27}18\alpha$(H)-22,29,30 trisnorhopane/[$C_{27}18\alpha$(H)-22,29,30 trisnorhopane + $C_{27}17\alpha$ (H)-22,29,30 trisnorhopane], is commonly used as a maturity parameter. In the epigenetic stage, the $C_{27}18\alpha$(H)-22, 29, 30 trisnorhopane was more stable than $C_{27}17\alpha$(H)-22, 29, 30 trisnorhopane (Seifert and Moldowan 1978). The Ts/(Ts + Tm) values of saturated hydrocarbons in the coal rock samples in the Sihe Coalmine and Fucheng 71 Coalmine range from 0.35 to 0.50. The Ts/(Ts + Tm) values of saturated hydrocarbons in the coal rock samples in the Liyazhuang Coalmine are 0.15, relatively high. The Ts/(Ts + Tm) values of saturated hydrocarbons of coal rock samples in Xinji Coalmine, Zhangji Coalmine and Panji Coalmine are between 0.02 and 0.05. It shows that the Sihe Coalmine and Fucheng 71 Coalmine in the Qinshui Basin and the Enhong Coalmine in Qujing have the highest maturity. The maturity of the Liyazhuang Coalmine is relatively higher, while the Xinji Coalmine, Zhangji Coalmine and Panji Coalmine have the lowest maturity (Table 2.4).

The $\beta\alpha$-C_{30} moretane/$\alpha\beta$-C_{30} hopane ratio reflects the evolution degree of the coal rock to a certain extent. The data show that during the diagenetic evolution of organic matter, $\beta\alpha$-C_{30} moretane transformed into more stable $\alpha\beta$-C_{30}

Table 2.4 Distribution of saturated hydrocarbons in coal rock samples

Coalmine	$\Sigma nC_{21-}/\Sigma nC_{22+}$	Pr/Ph	Pr/ nC_{17}	Ph/ nC_{18}	Gammacerane/ $\alpha\beta$-C_{30} hopane	OEP	Ts/ (Ts + Tm)	$\beta\alpha$-C_{30} moretane/ $\alpha\beta$-C_{30} hopane	$\alpha\beta\beta$/ ($\alpha\beta\beta$ + $\alpha\alpha\alpha$) C_{29}-sterane	20S/ (20S + 20R) $\alpha\alpha\alpha C_{29}$-sterane
Sihe	6.52	1.33	0.85	0.84	–	0.98	0.50	0.09	0.38	0.43
Fucheng 71	3.26	0.59	0.33	0.49	–	0.95	0.35	0.06	0.47	0.50
Liyazhuang	1.45	1.80	0.29	0.17	0.057	0.97	0.15	0.09	0.50	0.59
Zhangji No.11	0.19	2.33	2.00	0.37	0.034	1.28	0.03	0.33	0.34	0.39
Zhangji No.13	0.34	2.25	3.60	0.89	0.045	1.17	0.03	0.55	0.30	0.38
Panji No.1	0.24	1.38	2.75	0.50	0.036	1.27	0.05	0.34	0.38	0.44
Xinji No.13	0.24	2.00	2.40	0.43	0.028	1.37	0.02	0.36	0.31	0.39
Enhong	1.63	0.47	0.67	0.56	–	1.11	0.45	0.10	0.43	0.46

hopane. With the increase of maturity, $\beta\alpha$-C_{30} moretane/$\alpha\beta$-C_{30} hopane became smaller (Xiang 1989). For saturated hydrocarbons in the coal rock samples in the Sihe Coalmine, Fucheng 71 Coalmine, Liyazhuang Coalmine, and Enhong Coalmine, the $\beta\alpha$-C_{30} moretane/$\alpha\beta$-C_{30} hopane ratio is 0.06–0.10. For saturated hydrocarbons in coal rock samples in the Xinji Coalmine, Zhangji Coalmine and Panji Coalmine, the $\beta\alpha$-C_{30} moretane/$\alpha\beta$-C_{30} hopane ratio is 0.33–0.55 (Table 2.4). It also reflects that the Sihe and Fucheng 71 Coalmine and Enhong Coalmine have the highest maturity. The maturity of the Liyazhuang Coalmine is relatively higher, and the Xinji Coalmine, Zhangji Coalmine and Panji Coalmine have the lowest maturity.

Sedimentary Environment

The pristane/phytane (Pr/Ph) is a parameter for identifying the redox degree of paleoenvironment. This value is larger in the samples of coal and coal measure strate with terrigenous materials as the main source rock, and it has obvious predominance of pristane. Especially for continental coals, it is lower in samples derived from salt water environment (Pr/Ph < 1.0) (Peters and Moldowan 1993). For the samples in this study, the Pr/Ph ratio of coal rock samples from Fucheng 71 Coalmine and Enhong Coalmine is less than 0.6, indicating a partially reduced environment. All other samples have obvious predominance of pristane, and the Pr/Ph value is greater than 1.3 (Table 2.4), reflecting that the sedimentary environment might be brackish and fresh, and the ancient water was not very deep, but strongly oxidized.

2.4.2.3 Microbial Degradation

Organic matters are selectively degraded by microorganisms. In general, the resistance of biomarker compounds to biodegradation is subject to the following ascending order: normal paraffins, isoprenoids, sterane, hopane, rearranged hopane, aromatized sterane and porphyrin (Chosson et al. 1992). When organic matter and crude oil are degraded by microorganisms, *n*-alkanes may be seriously deficient, and isoprenoids are better preserved than normal paraffins; therefore the variation of relative abundance of isoprenoids versus normal paraffins can be used as a parameter to measure whether a sample is subject to microbial degradation.

Not only microorganisms preferentially degrade normal paraffins selectively, but also strong thermal degradation allows normal paraffins to be preferentially consumed over isoprenoids. The value of $\sum$isoprenoid alkanes/$\sum$normal paraffins of coal samples is 0.036–0.373 (Table 2.5). Among them, the $\sum$isoprenoid alkanes/$\sum$normal paraffins in the highly mature Sihe Coalmine and Fucheng 71 Coalmine in the Qinshui Basin and the Enhong Coalmine in Qujing, Yunnan Province is relatively high, being 0.139–0.373, while that of the less mature Liyazhuang Coalmine is relatively high (0.311) too. It is close to that in the Sihe Coalmine and Fucheng 71 Coalmine, indicating that isoprenoids are abundant in the Liyazhuang Coalmine, which may be related to the very developed microorganisms in the coal rock.

For coal rock samples from Xinji Coalmine, Zhangji Coalmine, and Panji Coalmine, the total ion current map shows that some or all of the *n*-alkanes have been degraded, but their $\sum$isoprenoid alkanes/$\sum$normal paraffins ratio is not very high, ranging from 0.033 to 0.190. This is probably due to the high degree of microbial degradation of coal rock samples, and the isoprenoid alkane is also partially depleted due to microbial degradation.

Compared to normal paraffins and isoprenoids, hopanes are strongly resistant to microbial degradation. The ratios of $\sum$hopanes/$\sum$normal paraffins (Table 2.5) are extremely different, ranging from 0.004 to 2.409. Among them, the ratio of $\sum$hopanes/$\sum$normal paraffins in the Sihe Coalmine and Fucheng 71 Coalmine in the Qinshui Basin is lower, 0.004–0.007. The ratio of $\sum$hopanes/$\sum$normal paraffins in

Table 2.5 Biomarker compounds in coal rock samples

Coalmine	Layer	$\sum$isoprenoid alkanes/$\sum$normal paraffins	$\sum$hopanes/$\sum$normal paraffins	$\sum$hopanes/$\sum$isoprenoid alkanes
Sihe	P_1s	0.373	0.004	0.034
Fucheng 71	C_3t	0.217	0.007	0.011
Liyazhuang	P_1s	0.311	0.095	0.306
Zhangji No.11	P_2	0.033	0.761	23.300
Zhangji No.13	P_2	0.190	2.409	12.692
Panji No.1	P_2	0.036	0.520	14.526
Xinji No.13	P_2	0.047	1.143	24.444
Enhong	P_2	0.139	0.051	0.364

the Liyazhuang Coalmine and the Enhong Coalmine is low, 0.051–0.095. The ratio of ∑hopanes/∑normal paraffins in the Xinji Coalmine, Zhangji Coalmine and Panji Coalmine is high, 0.520–2.409, reflecting that the coal rock samples from the Xinji Coalmine, Zhangji Coalmine and Panji Coalmine suffered strong microbial degradation. The coal rock samples from the Liyazhuang Coalmine and the Enhong Coalmine undergone minor microbial degradation. No signs of microbial degradation were left in the coal samples from the Sihe Coalmine and Fucheng 71 Coalmine, but they suffered strong thermal degradation, which converted hopane to tricyclic terpane.

Since hopane is more resistant to antimicrobial degradation than isoprenoids, the ∑hopanes/ ∑isoprenoid alkanes ratio also reflects the extent that the sample suffered microbial degradation. The ∑hopanes/∑isoprenoid alkanes ratios (Table 2.5) are between 0.011 and 24.444, representing evident differences. For coal samples without microbial degradation from the Sihe Coalmine and Fucheng 71 Coalmine, the values are less than 0.1. The Liyazhuang Coalmine and the Enhong Coalmine suffered from slight microbial degradation, so the values are less than 1.0. The coal samples with stronger microbial degradation from Xinji Coalmine, Zhangji Coalmine and Panji Coalmine have the ∑hopanes/∑isoprenoid alkanes ratios higher than 10.

Figure 2.15 is a correlation diagram between the ∑hopanes/∑normal paraffins ratio and the ∑hopanes/ ∑isoprenoid alkanes ratio of the coal rock samples analyzed. With the increase of the ∑hopanes/ ∑normal paraffins ratio, the ∑hopanes/∑isoprenoid alkanes ratio increases evidently. For coal samples subject to different microbial degradation, the ∑hopanes/ ∑isoprenoid alkanes ratio and the ∑hopanes/∑normal paraffins ratio vary greatly. For coal rock samples from the Xinji Coalmine, Zhangji Coalmine and Panji Coalmine suffered from slight microbial degradation, both the ∑hopanes/ ∑isoprenoid alkanes ratio and the ∑hopanes/∑normal paraffins ratio are higher. They are two magnitudes higher than those in the coal rock samples from the Liyazhuang Coalmine and Enhong Coalmine. The latter is one magnitude higher than those in the coal samples from the Sihe Coalmine and Fucheng 71 Coalmine.

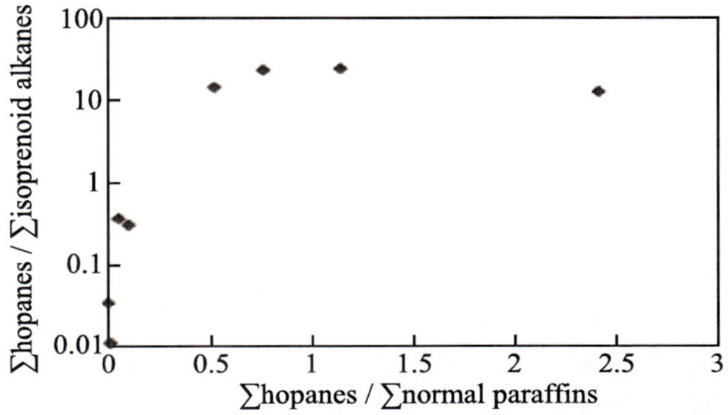

Fig. 2.15 Crossplot of ∑hopanes/∑normal paraffins ratio and ∑hopanes/∑isoprenoid alkanes ratio

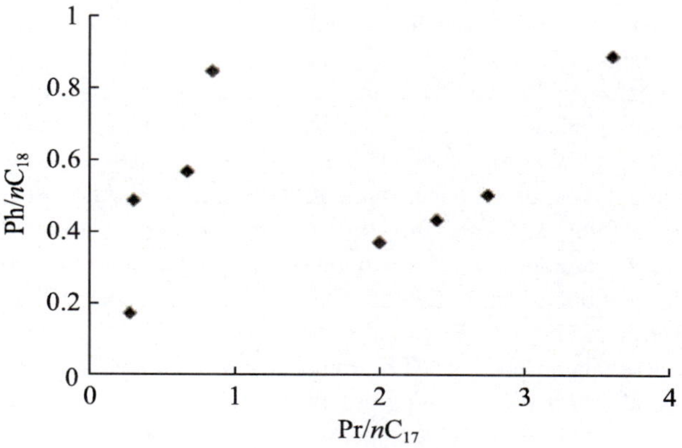

Fig. 2.16 Crossplot of Pr/nC_{17} and Ph/nC_{18} for coal samples analyzed

The ratios of Pr/nC_{17} and Ph/nC_{18} reflect the degradation of organic matter to some extent. Figure 2.16 is the crossplot of Pr/nC_{17} and Ph/nC_{18} for the coal samples analyzed. For coal samples suffered thermal degradation, from the Sihe Coalmine, Fucheng 71 Coalmine, Liyazhuang Coalmine and Enhong Coalmine, the Pr/nC_{17} ratios are generally less than 1.0, and the ratios of Pr/nC_{17} and Ph/nC_{18} have tendencies to increase with the increase of degradation degree. For coal samples suffered stronger microbial degradation, from the Xinji Coalmine, Zhangji Coalmine and Panji Coalmine, the Pr/nC_{17} ratios are generally greater than 2.0, and the Pr/nC_{17} and Ph/nC_{18} ratios have tendencies to increase with the increase of degradation degree. The increase of Pr/nC_{17} ratio caused by microbial degradation is greater than that resulted from thermal evolution.

2.4.2.4 Detection of iC_{25} and iC_{30} Acyclic Isoprenoid Alkanes and Their Geological Significance

Active microorganisms inevitably leave traces in coal seams, namely biomarker compounds related to the action of microorganisms. Ahmed and Smith discovered biomarker compounds from organic solvent extracts from the Permian coal rocks in the Sydney Basin and Bowen Basin in Australia. They reflect the degradation of organic matter in the coal rocks and provide a basis for identifying secondary biogenic CBM in the basins (Ahmed and Smith 2001). Noble and Henk (1998) studied the source rocks of the biogenic gas at commercial scale in Indonesia, and found biomarker compounds from methanogens, and used them to determine the active intensity of methanogens.

Most acyclic isoprenoids lower than C_{20} may be originated from the phytyl side chain of chlorophyll a, and secondary sources are chlorophyll b and bacteriochlorophyll a. Long-chained acyclic isoprene compounds more than C_{20} are generally derived from archaea, which are the diagenetic products of the predecessors of organisms C_{20}, C_{40}, or even higher carbon numbers (Thompson and Kenncutt 1992). Some types of archaea, such as halophilic, thermophilic acidophilus and methanogenic bacteria, contain saturated and unsaturated acyclic isoprenoid alkanes with carbon chain number from C_{14} to C_{30} (Tornabene et al. 1979; Holzer et al. 1979; Rowland et al. 1982).

Tornabene et al. (1979) found acyclic isoprenoid compounds with different degrees of saturation dominated by C_{30} squalene series, C_{25} pentame- thylprenyl series, and C_{20} phytyl series in the neutral lipids of nine methanogenic bacteria. Holzer et al. (1979) reported that C_{25} isoprenoid alkanes in methanogenic bacteria are mainly tail-to-tail irregular skeletons (2, 6, 10, 15, 19-pentamethyleicosane). Risatti et al. (1984) found a regular backbone of the head-to-tail linkage (2, 6, 10, 14, 18-pentamethylicosane) in methanogenic bacteria. The tail-to-tail irregular isoprenoid compounds are special biomarker compounds for archaea input in sediments (Volkman and Maxwell 1986). Petrov et al. (1990) found head-to-head irregular isoprenoid alkanes of C_{21} to C_{39} in crude oil, which is believed from the lipid of archaea wall. Stefanova (2000) found long-chained head-to-head irregular isoprenoid alkanes of C_{38} to C_{40} in coal rock samples, which are deemed biomarkers of totally anaerobic methanogenic bacteria in ancient sediments.

C_{20} or higher-chained acyclic isoprenoid series is generally deemed from archaea and diagenetic products of predecessors of C_{20}, C_{40}, or organism with higher carbon numbers. Studies on archaea lipids showed that methanogenic bacteria synthesized acyclic isoprenoid compounds as structural components of their biological cell membranes (Tornabene et al. 1979). In particular, C_{25} isoprenoid alkanes is used as a characteristic biomarker compound for methanogenic archaea in marine sediments (Brassell et al. 1981; Risatti et al. 1984). And C_{30} isoprenoid alkanes are also considered to be characteristic marker compounds of archaea (Schouten et al. 1997; Russell et al. 1997).

Distribution Characteristics of Acyclic Isoprenoid Alkanes in Coal Rock Samples

There are low-carbon-number acyclic isoprenoid alkanes, namely iC_{16}, iC_{18}, pristane (Pr) and phytane (Ph), but no high-carbon-number acyclic isoprenoid alkanes detected, in the coal samples from the Sihe Coalmine in the Qinshui Basin. In the coal rock samples from the Fucheng 71 Coalmine in the Qinshui Basin, both low-carbon-number acyclic isoprenoid alkanes, including iC_{18}, Pr and Ph, and high-carbon-number acyclic isoprenoid alkanes, were found, namely iC_{23} to iC_{26}, iC_{23} to iC_{30}. In the coal rock samples from the Liyazhuang Coalmine, acyclic isoprenoid alkane is very rich, including low-carbon-number iC_{15}, iC_{16}, iC_{18}, Pr, Ph and iC_{21}, and low- carbon-number iC_{23} to iC_{30}, and those higher than iC_{30} (Fig. 2.17). Low-carbon-number acyclic isoprenoid alkanes were detected in the coal rock samples from the Xinji Coalmine, Zhangji Coalmine and Panji Coalmine in the Huainan Coalfield and the Enhong Coalmine in Qujing, namely iC_{18}, iC_{21}, Pr and Ph. No high-carbon-number acyclic isoprenoids were found.

The coal samples from the Fucheng 71 Coalmine show dominant phytane, and the pristane/phytane (Pr/Ph) ratio is 0.59, reflecting a partially reduced sedimentary environment. The sulfur content is high in the organic solvent extracts of the samples, which also indicates a partially reduced environment. In general, bacterial microbial is developed in a sulfur-rich environment (Peters and Moldowan 1993; Hughes et al. 1995; Russell et al. 1997). Therefore, the long-chain acyclic isoprenoid alkanes detected in the samples are signs of bacteria input in the sedimentary environment. In the coal rock samples from the Liyazhuang Coalmine, acyclic isoprenoid alkanes are very rich, including iC_{23} to iC_{30} and those higher than iC_{30} (Fig. 2.18). This illustrates the development of bacteria, especially methanogens in the coal samples from the Liyazhuang Coalmine.

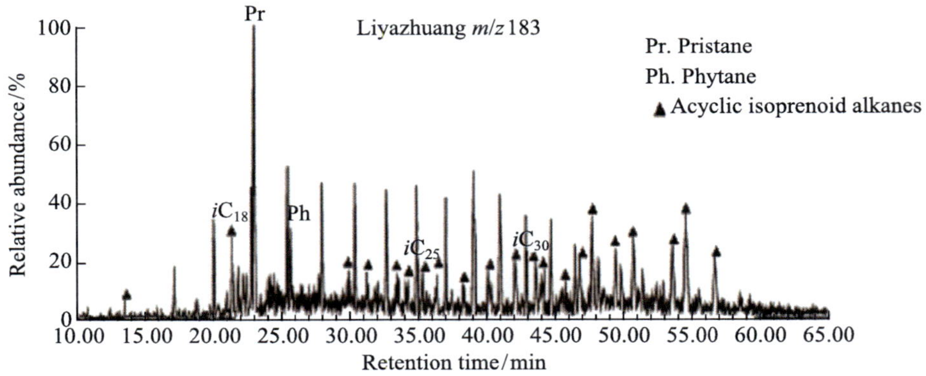

Fig. 2.17 Distribution characteristics of acyclic isoprenoid alkanes in the extracts from Liyazhuang Coalmine

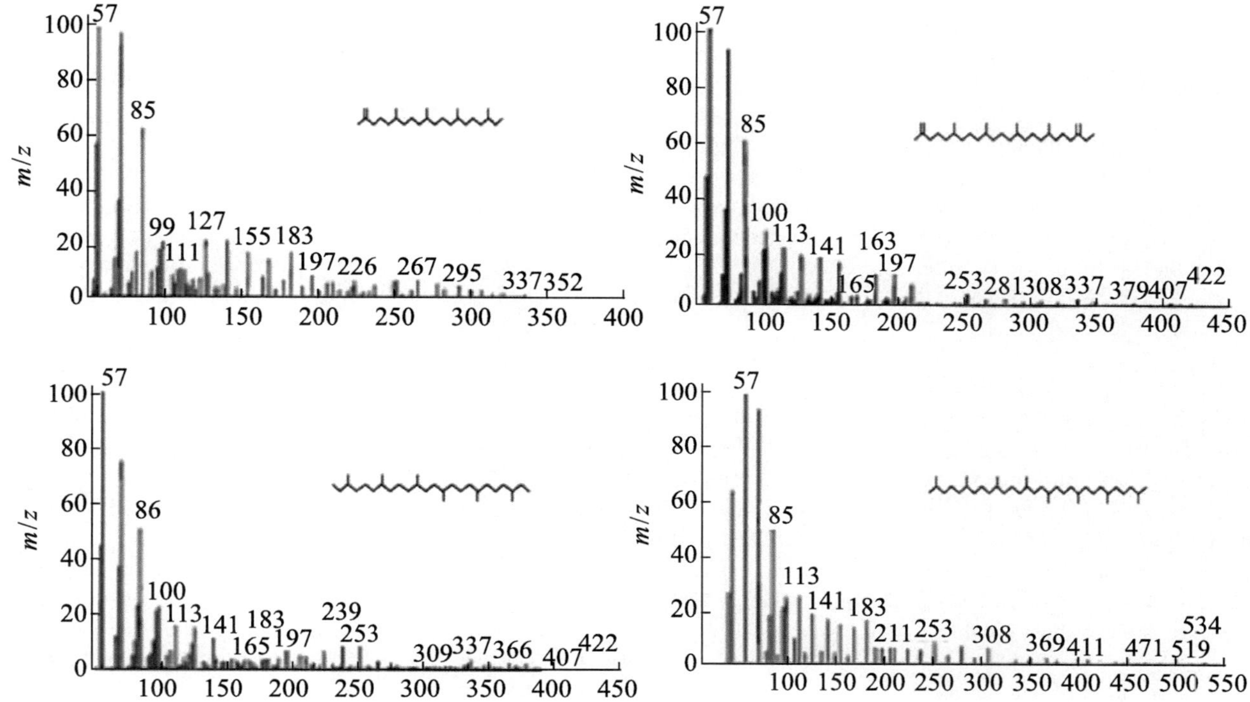

Fig. 2.18 Mass spectra of C_{25}, C_{30}, and C_{38} acyclic isoprenoid alkanes

The coal rock sample has a Pr/Ph of 1.80, which reflects an oxidized environment. Therefore, it can be concluded that the long-chained acyclic isoprene alkanes are not bacteria input during the deposition of the coal rock, but the traces of bacterial activities in the coal seams after coalification.

Among the coal samples from the Xinji Coalmine, Zhangji Coalmine and Panji Coalmine, only low-carbon-number acyclic isoprene alkanes were detected and the relative abundances is low. This is probably due to the high degree of microbial degradation of soluble organic matter in coal samples, and acyclic isoprenoid alkanes have been degraded.

Detection of iC_{25} and iC_{30} Acyclic Isoprenoid Alkanes and Their Geological Significance

The acyclic isoprenoid alkanes in coal rock samples of the Liyazhuang Coalmine are very rich, including low-carbon-number iC_{15}, iC_{16}, iC_{18}, Pr, Ph, iC_{21} and iC_{22}, and high-carbon-number iC_{23} to iC_{40}. They all have relatively high abundance.

In the coal rock samples in the Liyazhuang Coalmine, the isoprenoid alkanes detected are iC_{15} head-to-tail regular series and iC_{30} isoprenoid alkanes with both head-to-tail regular series and head-to-head regular series. The detected long-chained isoprenoid alkanes are all head-to-head regular series.

The detected isoprenoid alkanes iC_{15}, iC_{16} and head-to-head regular series (iC_{34} to iC_{39}) indicate the development of methanogenic bacteria in the coal rock samples. The pristane/phytane values (Pr/Ph) are 1.80, reflecting that the sedimentary environment of the coal rock was partially oxidized. Therefore, the detected high-carbon-number (>C_{25}) acyclic isoprenoid alkanes are not bacteria input during the deposition of the coal rock, but the traces of bacterial activities, especially the development of fully anaerobic hyperthyroidisms in the coal seams after coalification. The thermal evolution of the coal rock is relatively high-the R^{o} is up to 0.92%. Therefore, the CBM whose carbon isotope shows biogenetic characteristics is unlikely primary biogenic gas, but secondary biogenic CBM. The abundant long-chained acyclic isoprene alkanes detected in the coal rock samples are direct organic geochemical evidences for secondary biogenic CBM in the Liyazhuang Coalmine.

2.4.2.5 Experiments on Methanogenic Bacteria and Microbial Colonies in Coal Rocks

Experimental Study on Heterotrophic Bacteria

Methanogenic bacteria are archaea capable of converting organic or inorganic materials into methane under anaerobic conditions. There are generally two types of methanogenic bacteria, one uses H2 and CO_2 to produce methane, and the other uses formate to produce methane. So far, more than 200 types of methanogenic bacteria have been isolated and identified, and they are classified into 5 orders of 3 classes in the aquatic archaea phylum. Whether coal can be degraded by anaerobic microorganisms such as methanogens is a concern of geosciences, therefore study on methanogens in coal seams is of great significance for understanding the production and formation conditions of CBM.

(1) Samples and experiments

Coal rock samples from the Xinji Coalmine, Panji No.13 coal seam in No.1 Coalmine, and Zhangji Coalmine in Huainan Coalfield were taken as experimental samples. Coal powder samples from the Xinji Coalmine and Panji No.1 Coalmine were analyzed to the enrichment and isolation of methanogenic bacteria. The anaerobic tubes were kept at 30 °C and 60 °C, and meanwhile a control group was set. After culturing, the colors of cultures in the tubes were different from that in the control group.

(2) Identification of heterotrophic bacteria

Coal is composed of extremely complex organic and inorganic matters, and methane is only the last product of biogenic methane generated by methanogenic bacteria. First, some anaerobic bacteria and facultative anaerobic bacteria (such bacteria can survive under aerobic or anaerobic conditions, conduct aerobic respiration under aerobic conditions, and ferment under anaerobic conditions) decompose macromol-ecular organic matters into small molecular organic matters such as monosaccharides. Then the small molecule organic matters are decomposed by hydrogen-producing acetogenic bacteria population to produce acetic acid, H_2, and CO_2. Finally, methano-genic bacteria produce CH_4 using components such as CO_2, acetic acid and H_2. Therefore, studying heterotrophic bacteria (heterotrophic bacteria) involved in coal decomposition is also important for unders- tanding coal decomposition and gas production.

Culture media were prepared in laboratory, and heterotrophic bacteria in coal rock were isolated and classified.

The Xinji samples were cultured at 37 °C. There were mainly three types of heterotrophic bacteria isolated: (i) white, round, opaque, large and flat, smooth-edged and wet colony (Fig. 2.19a); (ii) light yellow, round, opaque, smooth-edged and wet colony (Fig. 2.19b); (iii) orange, round, smooth-edged and wet colony (Fig. 2.19c).

The Xinji samples were cultured at 4 °C for 1 month. There were mainly 4 types of colonies (low-tempera- ture bacteria): (i) white, round, translucent, neat-edged, wet and smooth colony; (ii) light yellow, round, opaque, wet, smooth, large and flat colony (Fig. 2.19d); (iii) yellow, round, opaque, wet, middle-uplifted, and neat-edged colony (Fig. 2.19e); (iv) orange, round, opaque and wet colony (Fig. 2.19f).

The Panji No.1 samples were cultured at 37 °C, and the above three types of heterotrophic bacteria were isolated too: (i) white, round, opaque, neat-edged and wet colony; (ii) light yellow, round, opaque, neat- edged and wet colony;

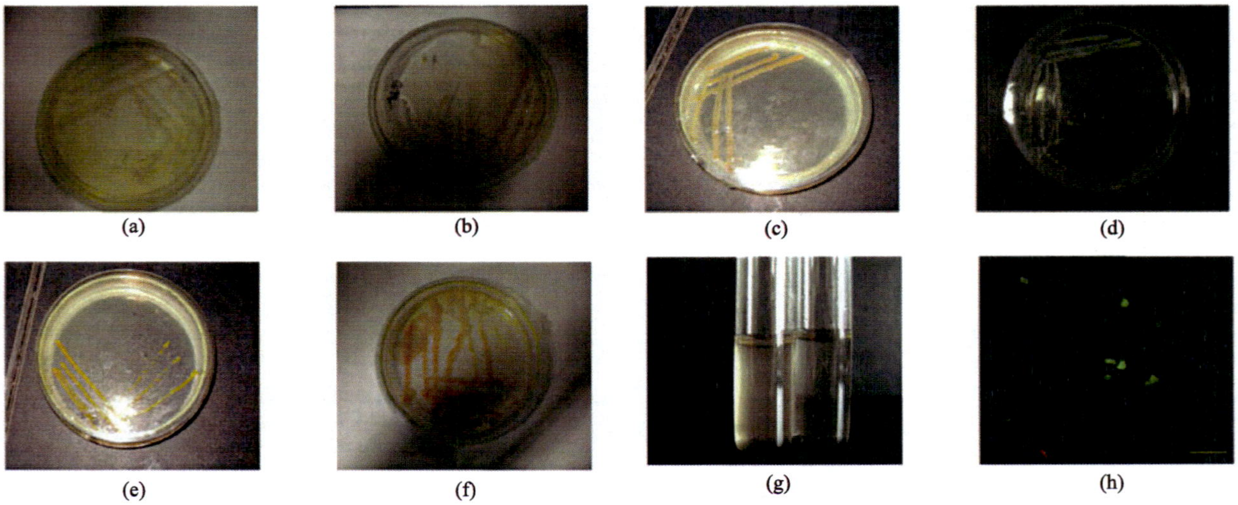

Fig. 2.19 Separation and identification of heterotrophic bacteria and enrichment of methanogens in coal rock **a** white, round, opaque, wet and smooth colonies; **b** light yellow, round, opaque, wet and smooth colonies; **c** orange, round, smooth-edged, wet colonies; **d** light yellow, round, opaque, wet and smooth colonies; **e** yellow, round, middle-uplifted and smooth-edged, wet colonies; **f** orange, round, opaque, wet and smooth colonies; **g** methanogenic bacteria cultured in liquid; **h** fluorescence from methanogenic bacteria

Table 2.6 Dominant strains, quantity and culturing temperature and characteristics of heterotrophic bacteria

Strain	Near-source genus	Quantity / (points/g)	Culturing temperature/°C	Characteristics	Cell morphology
J. anibacter	*J. anophelis*	5×10^6	4–37	Colored, small, wet, smooth, nearly opaque, and neat-edged	Sphere
Bacillus	*B. pseudofirmus*	1×10^5	37	Large, thin, light orange, wet, smooth, transparent, and neat-edged	Rod
	B. cereus	8×10^7	4–37	White, unsmooth with possible floccules, nearly opaque, unneat-edged, and growing rapidly	Stick, chain
	B. horikoshii	5×10^5	37	Thin, light brown, nearly transparent, irregular-edged	Rod
	B. thuringiensis	5×10^4	4–16	Milky white, thicker, smoother, nearly opaque, neat-edged	Stick, chain
Dietzia	*D. natronolimnaea* *D. daqingensis*	2×10^4	16–37	Red, some slightly orange, small, wet, smooth, nearly opaque, neat-edged, and growing slowly	Sphere
Brachy-bacterium	*B. sacelli*	4×10^4	4–16	Light yellow, milky white, small, nearly opaque, and neat-edged	Sphere

(iii) yellow, round, middle-uplifted, opaque, neat-edged and wet colony.

The Panji No.1 samples were cultured at 4 °C for 1 month. There were mainly 4 types of colonies: (i) white, round, translucent, neat-edged, small and wet colony; (ii) light yellow, round, translucent, neat-edged, small and wet colony; (iii) yellow, round, middle-uplifted, opaque, smooth and wet colony; (iv) orange, round, opaque, smooth and wet colony.

The heterotrophic bacteria in the Zhangji coal samples were isolated and identified, and the data of main population of heterotrophic bacteria were obtained. The number of heterotrophic bacteria was counted. About 20 strains were isolated, and 7 dominant bacteria (more than 10×10^3 points/g coal powder) were studied (Table 2.6).

Enrichment and Isolation of Methanogenic Bacteria

Simulating the growth conditions of methanogens, methanogenic bacteria were enriched in the Zhangji samples under strict anaerobic conditions (Fig. 2.19g). The culturing characteristics and spontaneous fluorescence under microscope can confirm the presence of methanogenic bacteria (Fig. 2.19g).

Gas Production Experiment on Methano-genic Bacteria from Coal

The coal samples for experiments were taken from the No.13-1 coal seam of the Zhangji Coalmine. According to experimental requirements, they were divided into three groups. The first group (4 samples) are inoculated samples, which were aseptically inoculated into the methanogen culture medium for gas production experiment. The second group (3 samples) are transferred samples, which were aseptically inoculated into a methanogen culture medium for anaerobic culture, and then the cultures were transferred for a second time into a fresh medium for continuous gas production. The third group is a control group including activated sludge samples. It was inoculated into an anaerobic culturing tube for gas production. Finally, gas samples produced by the three groups of experiments were collected and analyzed (Figs. 2.20, 2.21).

As shown in Fig. 2.20, the inoculated samples (the first group) and the control sample (the third group, No.1 sludge sample) were compared. The methane content of the gas produced by the inoculated samples is higher than that produced by the control samples, especially the inoculated Sample b, from which the methane content is four times

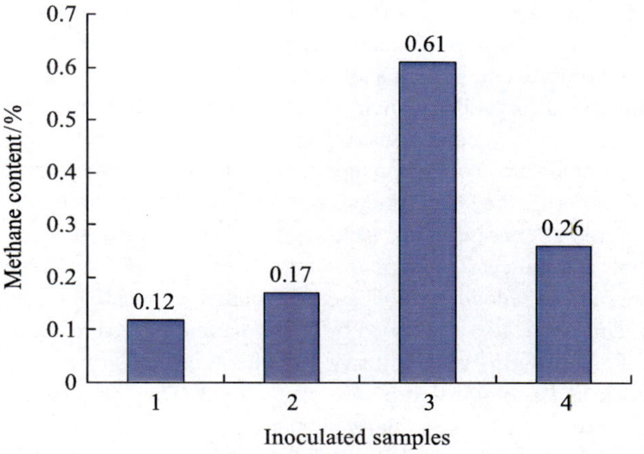

Fig. 2.20 Methane content of the gas produced by inoculated coal samples 1. No.1 sludge sample; 2. inoculated Sample a; 3.inoculated Sample b; 4.inoculated Sample c

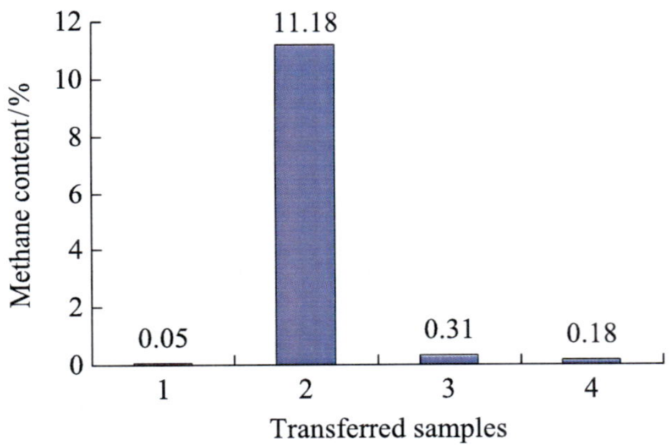

Fig. 2.21 Methane content of the gas produced by transferred coal samples 1. No.2 sludge sample; 2. transferred Sample a; 3. transferred Sample b; 4. transferred Sample c

more than that of the control samples. It confirmed the presence of methanogens in the inoculated samples.

The methane content of the gas produced by transferred Sample a (11.18‰) is significantly higher than the methane content of the gas produced by the control sample (0.05‰). This further proves the presence of methanogenic bacteria in the inoculated coal samples and obvious methanogenic activities. However, the methane contents of the gas produced by transferred Sample b and Sample c are extremely low. This may be attributed to the sealing capacity of the culture bottles. If it is not good, the samples might be polluted by air.

The above experimental results prove that methanogenic bacteria exist in the coal seams in the Zhangji Coalmine, and microbial gas can be produced. Moreover, the methane content of the gas produced by the transferred samples is higher than that of the inoculated samples, indicating that after transferred in laboratory, the amount of methanogens and the production of methanogenic bacteria were remarkably improved. In terms of reproduction and activities of microbial organisms such as methanogenic bacteria, the environmental conditions of the coal seams in Huainan Coalfield are certainly superior to those in laboratory. Therefore, the experiment results not only prove the microbial conditions for producing secondary biogenic gas, but also indicate geological microbial conditions for a large amount of secondary biogenic gas in Huainan Coalfield. The results were also confirmed by the above-mentioned research results of CBM geochemistry, organic geochemistry of coal rock and geological structure of coalfield. They all confirm the existence of secondary biogenic gas in Huainan Coalfield, and also shows that the unilateral research conclusions are scientific and reliable.

2.4.2.6 Organic Geochemistry of Coal Rock and Generating Mechanism of CBM

Coal is a kind of combustible organic rock with high organic matter and high organic carbon content. However, the formation of CBM is not only related to the abundance of organic matter in coal rock, but also closely related to the thermal evolution of coal rock, the hydrogeology and tectonic geological conditions of coal seam. Coal rock is both a source rock of CBM and a reservoir of CBM. Complex polymeric organic compounds in coal and soluble organic matter adsorbed in coal seams generate CBM under thermal action and microbial activity, namely thermogenic CBM and biogenic methane.

Thermogenic CBM is produced during coalification with rising temperature, and functional groups rich in hydrogen and oxygen in the complex polymeric organic compounds are thermally degraded or soluble organic matters in coal seams are cracked at high temperature. Thermogenic CBM is divided into pyrolyzed gas and thermal-cracking gas. Pyrolyzed gas is produced with pyrolysis of complex polymeric organic matters, while thermal-cracking gas is produced with thermal cracking of liquid hydrocarbons at high temperature.

Biogenetic CBM is the result of degradation of soluble organic matters in coal rock by microorganisms in coal seams, including primary and secondary CBMs. Primary biogenic CBM refers to the gas generated by the action of microorganisms on organic matter in the early burial stage of coal seams (in undegraded coal). Secondary biogenic CBM is the gas dominated by methane and generated by anaerobic methanogenic bacteria under the living and propagating conditions provided by microbial degradation of coaliferous wet gas and organic matters in coal seams, when microorganisms are brought by surface water into coal-bearing basins (Scott et al. 1994).

The content and composition of organic matter in coal rock and characteristics of biomarker compounds reflect the formation, thermal evolution and biological processes of coal rock, and also provide information on the genesis of CBM in coal seams.

In the samples from the Sihe Coalmine and Fucheng 71 Coalmine in the Qinshui Basin, the content of chloroform bitumen "A" is very low, accounting for only 0.009–0.011%; the total hydrocarbon content is very low, 0018–0.0042%; the vitrinite reflectivity (R^{o}) is about 4%; in the saturated hydrocarbons, the content of low-carbon-number alkane is significantly higher than that of high-carbon-number alkane, and even carbon number is dominant; the $\sum nC_{21-}/\sum nC_{22}+$ ratio is 3.26–6.52; the OEP is close to 1.0; the Ts/(Ts + Tm) ratio is high, 0.35–0.50; the $\beta\alpha$-C_{30} moretane/$\alpha\beta$-C_{30} hopane ratio is 0.06–0.09. All of these parameters indicate that the thermal evolution of the coal rock is very high. The content of soluble organic matter (liquid hydrocarbons) is extremely low, reflecting liquid hydrocarbons were cracked to dry gas at high temperature. As a sedimentary organic matter with high organic carbon content but poor parent material, gas generation was dominant during thermal evolution. Especially when the coal rock became mature, dry gas production is high; therefore, there are conditions for generating a large amount of thermally cracked CBM.

In the samples from the Liyazhuang Coalmine in the Huoxi Basin, the content of chloroform bitumen "A" is relatively high, accounting for 0.535%; the total hydrocarbon content is relatively high, accounting for 0.0693%; the saturated hydrocarbons/aromatic hydro- carbons ratio is very low (0.42); the amount of low-carbon-number alkanes is slightly higher than the amount of high-carbon-number alkanes, and even carbon number is dominant; in the saturated hydrocarbons, most high-carbon-number alkanes are existent, and the $\sum nC_{21-}/\sum nC_{22+}$ ratio is 1.45, the OEP is less than 1.0, the Ts/(Ts + Tm) ratio is higher, 0.15, and the $\beta\alpha$-C_{30} moretane/$\alpha\beta$-C_{30} hopane ratio is 0.09. These parameters reflect that the thermal maturity of the coal rock is not very high, and the soluble organic matter has not been thermally cracked. The low content of saturated hydrocarbons in the chloroform bitumen "A" reflects possible degradation of saturated hydrocarbons by microorganisms.

The content of soluble organic matter (liquid hydrocarbons) is high, but the thermal evolution degree is low, indicating the coal rock is not in the stage of generating a large amount of thermogenic gases. However, acyclic isoprenoid alkanes in the coal rock sample are very rich, including low-carbon-number iC_{15}, iC_{16}, iC_{18}, Pr, Ph, iC_{21}, and iC_{22}, and high-carbon-number iC_{23} to iC_{40}, and their abundance is relatively high. In addition, head-to-tail regular series iC_{15} and iC_{30} and head-to-head regular series were detected. The long-chained isoprenoid alkanes detected are all head-to-head regular series, indicating developed methanogenic bacteria in the coal rock samples.

The above-mentioned characteristics indicate that abundant soluble organic matters in coal rocks provide sufficient nutrients for microbial activities in middle and late evolution stages. It is the material basis for the formation of secondary biogenic CBM, and the development of methanogenic bacteria in the coal rock creates conditions for the formation of secondary biogenic CBM. At shallow coal seams and suitable temperature, bacteria activities continue to supply gas, so that the Liyazhuang coal seams are rich in secondary biogenetic CBM.

In the coal rock samples from the Xinji Coalmine, Zhangji Coalmine and Panji Coalmine, the contents of chloroform bitumen "A" are high, between 0.669 and 2.996%; the total hydrocarbon contents are high, between 0.1734 and 0.7789%; the saturated hydrocarbons/aromatic hydrocarbons ratios are relatively low, ranging from 0.14 to 0.23; the contents of low-carbon-number alkanes are significantly lower than those in high-carbon-number alkanes, and the predominance of odd or even carbon number is not obvious; the $\sum nC_{21-}/\sum nC_{22+}$ ratios are 0.19–0.34; the OEP values are less than 1.0; the Ts/(Ts + Tm) ratios are high, between 0.02 and 0.0; and the $\beta\alpha$-C_{30} moretane/$\alpha\beta$-C_{30} hopane ratios are 0.33–0.54.

The thermal evolution of the coal rocks in the Xinji Coalmine, Zhangji Coalmine, and Panji Coalmine is relatively low, with R^o between 0.85 and 0.97%. The low content of saturated hydrocarbons in chloroform bitumen "A" reflects possible degradation of the saturated hydrocarbons by microorganisms. High $\sum$hopanes/$\sum$normal paraffins and $\sum$hopanes/$\sum$isoprenoid alkanes ratios reflect strong microbial degradation of the saturated hydrocarbons.

The above-mentioned characteristics reflect that the coal rock in this area has entered a mature stage. At this stage, a large amount of liquid hydrocarbons were formed. Therefore, there is abundant soluble organic matter in the coal rock, which provides sufficient nutrients for the microbial activities in the middle and late evolution stages of coal seam. The microbial degradation characteristics of saturated hydrocarbons provide evidence of microbial development in coal rock. Therefore, it is believed that the CBM of Xinji Coalmine, Zhangji Coalmine and Panji Coalmine in Huainan in Anhui Province is mainly secondary biogenetic gas.

The content of chloroform bitumen "A" in the coal rock samples of the Enhong Coalmine in Qujing, Yunnan was very low, being 0.001%. The total hydrocarbon content was extremely low as well, being 0.0039%. The saturated rock/aromatic hydrocarbons value of coal rock sample was relatively high, being 1.86.

In the coal rock in the Enhong Coalmine, the R^o is between 1.24 and 1.43%; the amount of the low-carbon-number alkane is significantly higher than that of the high-carbon-number alkane in the saturated hydrocarbons, and the predominance of odd or even carbon number is obvious; the $\sum nC_{21-}/\sum nC_{22+}$ ratio is 1.63; the OEP is less than 1.0; the $\beta\alpha$-C_{30} moretane/ $\alpha\beta$-C_{30} hopane ratio is 0.06–0.10. The high-carbon- number alkanes in the samples have almost been degraded, reflecting that the thermal evolution is relatively high, and the soluble organic matter has been thermally cracked. The $\sum$hopanes/$\sum$normal paraffins and $\sum$hopanes/$\sum$isoprenoid alkanes ratios reflect slight microbial degradation of the saturated hydrocarbons, so secondary biogenetic gas contributes to the CBM in the Enhong Coalmine.

References

Ahmed M, Smith JW (2001) Biogenic methane generation in the degradation of eastern Australian Permian coals. Org Geochem 32:809–816

Aravena R, Harrison SM, Barker JF, Abercrombie H, Rudolph D (2003) Origin of methane in the Elk valley coalfield, southeastern British Columbia Canada. Chem Geol 195:219–227

Brassell SC, Wardroper AM, Thompson ID et al (1981) Specific acyclic isoprenoids as biological markers of methanogenic bacteria in marine sediments. Nature 290:693–696

Chosson P, Connan J, Dessort D, Lanau C (1992) In vitro biodegradation of steranes and trepans: a clue to understanding geological situations. In: Moldowan JM, Albrecht P, Philp RP (eds) Biological markers in sediments and petroleum. Prentice Hall, Englewood Cliffs, pp 320–394

Cui YJ, Zhang QL, Yang XL (2003) Adsorption properties and varying law of equal adsorption heat of different coals. Nat Gas Ind 23 (4):130–131 (in Chinese)

Dai JX (1990) Several characteristics of hydrogen isotope of organic alkane gas in China. Pet Explor Dev 5:27–32 (in Chinese)

Dai JX, Pei XG, Qi HF (1992) China natural gas geology, vol. 1. Petroleum Industry Press, Beijing, pp 6–75. (in Chinese)

Dai JX, Qi HF, Song Y, Guan DS (1986) China's coalbed methane composition, carbon isotope types and their genesis and significance. Chin Sci (Ser B) 12:1317–1326 (in Chinese)

Dai JX, Song Y, Dai CS, Chen AF, Sun ML, Liao YS (1995). Inorganic gas and conditions of gas reservoirs in eastern China. Science Publishing House, Beijing, pp 12–20, 211 (in Chinese)

Deng MG, Gui BL, Pu CJ, Luo QL (2004) Prospects for CBM exploration and development in Yunnan Enhong mining area and countermeasures. China Coal 30(1):48–50 (in Chinese)

Faiz M, Saghafi A, Sherwood N, Wang I (2007) The influence of petrological properties and burial history on coal seam methane reservoir characterization, Sydney Basin, Australia. Int J Coal Geol 70:193–208

Gao B, Tao MX, Zhang JB, Zhang XB (2002) Distribution characteristics and controlling factors of carbon isotope in coalbed methane. Coalf Geol Explor 30(3):14–17 (in Chinese)

Hakan H, Namik MY, Cramer B et al (2002) Isotopic and molecular composition of coal-bed gas in the Amasra region (Zonguldak Basin–western Black Sea). Org Geochem 33:1429–1439

Holzer G, Oro J, Tornabene TG (1979) Gas chromatographic-mass spectrometric analysis of neutral lipids from methanogenic and thermoacidophilic bacteria. J Chromatogr 186:795–809

Hu GY, Liu SS, Li JM et al (2001) Coalbed methane genesis in Jincheng area. Qinshui Basin Oil Gas Geol 22(4):319–321 (in Chinese)

Hughes WB, Holba AG, Dzuo LI (1995) The ratios of dibenzothiophene to phenanthrenes and pristine to phytane as indicators of depositional environment and lithology of petroleum source rocks. Geochem Et Cosmochem Acta 59:3581–3598

Kotarba MJ (1990) Isotopic geochemistry and habitat of the natural gases from the upper carboniferous zacler coal-bearing formation in the Nowa Ruda coal district (Lower Silesia, Poland). Org Geochem 16(1–3):549–560

Kotarba MJ (2001) Composition and origin of coalbed gases in the Upper Silesian and Lublin Basins, Poland. Org Geochem 32:163–180

Kotarba MJ, Rice DD (2001) Composition and origin of coalbed gases in the Lower Silesian, southwest Poland. Appl Geochem 16:895–910

Kotarba MJ et al (2004) Characterizing thermogenic coalbed gas from Polish coals of different ranks by hydrous pyrolysis. Org Geochem 35:615–646

Li M C, Zhang W C. 1990. Shallow CBM in Coal Fields in China. Beijing: Science Press. 1–225, 145–149 (in Chinese)

Noble RA, Henk FH Jr (1998) Hydrocarbon charge of a bacterial gas field by prolonged methanogenesis: an example from the East Java Sea, Indonesia. Org Geochem 29:301–314

Peters KE, Moldowan JM (1993) The biomarker guide. Prentice Hall, Englewood Cliffs, pp 1–50

Petrov AA, Vorobyova NS, Zemskova ZK (1990) Isoprenoid alkanes with irregular "head-to-head" linkages. Org Geochem 16:1001–1005

Rice DD (1993) Composition and origins of coalbed gas. In: Law BE, Rice DD (eds). Hydrocarbons from Coal. AAPG Studies in Geology, vol 38, pp 159–184

Rightmire CT, Eddy GE, Kirr JN (1984) Coalbed methane resources of the United States. AAPG Stud Geol 17(7–8):1–14

Risatti JB, Rowland SJ, Yon DA, Maxwell JR (1984) Stereochemical studies of acyclic isoprenoids XII. Lipids of methanogenic bacteria and possible contributions to sediments. Org Geochem 6:93–104

Rowland SJ, Lamb NA, Wilkinson CF, Maxwell JR (1982) Confirmation of 2, 6, 10, 15, 19-pentamethylleicosane in methanogenic bacteria and sediments. Tetrahedron Lett 23:101–104

Russell M, Grimalt JO, Hartgers WA et al (1997) Bacterial and algal markers in sedimentary organic matter deposited under natural sulphurization conditions (Lorca Basin, Murcia, Spain). Org Geochem 26:605–625

Schouten S, van der Maarel MJEC, Huber R et al (1997) 2, 6, 10, 15, 19-pentamethylicosenes in methanolobus bombayensis, a marine methanogenic archaeon, and in methanosarcina mazei. Org Geochem 26:409–414

Scott AR (1993) Composition and origin of coalbed gases from selected basins in the United States. In: Proceedings of the 1993 international coalbed methane symposium. The University of Alabama, School of Mines and Energy Development, Paper 9270, vol 1, pp 174–186

Scott AR, Kaiser WR, Ayers WB et al (1994) Thermogenic and secondary biogenic gases. S Juan Basin AAPG Bull 78(8):1186–1209

Scott AR, Zhou NJ, Levine JR (1995) A modified approach to estimating coal and coal gas resources: example from the Sand Wash Basin. Colorado AAPG Bull 79(9):1320–1336

Smith JW, Pallasser R (1996) Microbial origin of Australia coalbed methane formation. AAPG Bull 80:891–897

Stefanova M (2000) Head to head linked isoprenoids in Miocene coal lithotypes. Fuel 79:755–758

Tang XY, Yang YC, Liu DM, Shen P, Shen QX (1988) Several issues concerning coal-forming gas components and carbon isotope of methane. In: Annual report of the open research laboratory of biochemistry and gas geochemistry, lanzhou institute of geology, Chinese academy of sciences. Gansu Science and Technology Press, Lanzhou, pp 240–252 (in Chinese)

Tao MX (2005) Current status and development trend of coalbed methane geochemistry. Prog Nat Sci 15(6):648–652 (in Chinese)

Tao MX, Chen FY, Ma YX (1994) Carbon dioxide outburst and prevention in Yaojie Coalmine, Gansu Province. J Nat Disasters 3 (3):85–89 (in Chinese)

Tao MX, Shi BG, Li JY, Wang WC, Li XB, Gao B (2007) Secondary biological coalbed gas in the Xinji area, Anhui province, China: evidence from the geochemical features and secondary changes. Int J Coal Geol 71:358–370

Tao MX, Wang WC, Xie GX, Li JY, Wang YL, Zhang XJ, Zhang H, Shi BG, Gao B (2005a) Secondary biological CBM found in some coal fields in China. Sci Bull 50(Supplement I):14–18 (in Chinese)

Tao MX, Xu YC, Shi BG, Jiang ZT, Shen P, Li XB, Sun ML (2005b) Mantle degassing and deep geological structures of different types of fault zones in China. Chin Sci (D Ser) 35(5):441–451 (in Chinese)

Tao MX, Xu YC, Chen FY, Shen P, Sun ML (1991) Characteristics of helium isotope in CO_2 in Yaojie coalfield and its significance. Sci Bull 12:921–923 (in Chinese)

Tao MX, Xu YC, Ma YX (1992) Carbon dioxide outburst and research in coalfields. Prog Earth Sci 7(5):40–441 (in Chinese)

Tao MX, Xu YC, Shen P et al (1996) Geotectonics and geochemical characteristics and accumulation conditions of the mantle gas reservoirs in eastern China. Chin Sci (D Ser) 26(6):531–536 (in Chinese)

Thielemann T, Cramer B, Schippers A (2004) Coalbed methane in the Ruhr Basin, Germany: a renewable energy resource? Org Geochem 35:1537–1549

Thompson KFM, Kenncutt HMC (1992) Correlations of gulf coast petroleum, on the basis of branched acyclic alkanes. Org Geochem 18(1):103–119

Tornabene TG, Langworthy TA, Holzer G et al (1979) Squalenes, phytanes and other isoprenoids as major neutral lipids of methanogenic and thermoacidophilic "archaebacteria." J Mol Evol 13:73–83

Volkman JK, Maxwell JR (1986) Acyclic isoprenoids as biological marker. In: Johns RB (ed) Biological markers in the sedimentary record. Elsevier, New York, pp 1–42

Wang DH (2002) 3D seismic exploration of deep coal seams in Huozhou mining area. Shanxi Sci Technol 5:1–3 (in Chinese)

Welhan JA, Craig H (1979) Methane and hydrogen in East Pacific Rise hydrothermal fluids. Geophys Res Lett 6:829–831

Whiticar MJ (1996) Stable isotope geochemistry of coals, humic kerogens and related natural gases. Int J Coal Geol 32:191–215

Xiang LB (1989) Distribution of medium steroid and terpane biomarkers in coal-bearing strata. Mar Orig Pet Geol 3(1):76–83 (in Chinese)

Xu YC et al (1994) Theory and application of natural gas genesis. Science Press, Beijing, pp 49–357 (in Chinese)

Xu YC, Shen P, Tao MX et al (1997) Geochemistry on mantle-derived volatiles in natural gases from eastern China oil/gas provinces (II)–helium, argon and hydrocarbons in mantle volatiles. Sci China (Ser D) 40(3):315–321

Ying YP, Wu J, Li RW et al (1990) Discovery and genesis of abnormally heavy carbon isotope of coalbed methane in China. Sci Bull 35(19):1491–1493 (in Chinese)

Zhang JB, Tao MX (2000) Geological significance of methane carbon isotope in coalbed methane exploration: a case study on Qinshui Basin. Acta Sedimentol Sin 18(4):611–614 (in Chinese)

Zhang Q, Feng SL, Yang XL (1999) Residual gas content in coal and its influencing factors. Coalf Geol Explor 5:26–28 (in Chinese)

Zhang XM, Zhuang J, Zhang SA et al (2002) China Coalbed methane geology and resources evaluation. Science Press, Beijing (in Chinese)

Zhao JZ (2003) Theory and application of natural gas accumulation in foreland basin. Petroleum Industry Press, Beijing, pp 21–27 (in Chinese)

3 Characterization and Controlling Factors of CBM Reservoirs

3.1 CBM Reservoir Space

3.1.1 Pore System

Coal is a complex porous medium where the pore refers to the space not filled with solids (organic matter and minerals). According to the size, coal pores can be divided into macropores (>1000 nm), mesopores (100 to 1000 nm), small pores (10 to 100 nm) and micropores (<10 nm). Gas penetrates by laminar flow and turbulent flow in the macropores, and exists by means of capillary condensation, physical adsorption and diffusion in the micropores.

3.1.1.1 Pore System in Highly Metamorphic Coals in Qinshui Basin

Basic Structure

Microscopically, the common pores in the high-rank coal in the Qinshui Basin are primary pores which are dominated by vitrinite, fusinite and semi-fusinite structures with hollow cells, and followed by interstitial pores. The mineral pores found are intercrystalline pores which are mainly composed of clusters of pyrite crystals. Open pores are also visible in the samples with highly mosaic structures. It should be noted that gas pores are widely distributed in the high-rank coal reservoirs in the Qinshui Basin. They are pores formed by gas generating and accumulating during coal gasification, and related to the increase of coal rank and the rapid hydrocarbon generation of coal seams in the anomalous geothermal background of the Yanshanian. The pores even show expansive and bursting features, and locally connect with micro- fractures (Fig. 3.1).

Development Degree of Pore System

(1) Results of mercury intrusion experiments

The measured porosity of 40 coal samples taken in the Qinshui Basin is from 0.90 to 10.90%, with an average of 3.35%, which is generally low. From the perspective of hydrocarbon migration, when the porosity is above 10%, it is conducive to hydrocarbon migration; when the porosity is below 10% or lower, hydrocarbon migration is difficult. The low porosity of coal reservoirs in the Qinshui Basin is an inherent weak point of highly metamorphic coal reservoirs. Mercury intrusion experiments reflect the pore distribution ranging from 72 to 50,000 nm in diameter, including partial small pores, all mesopores, and partial macropores. The overall pore distribution is dominated by micro-intermediate pores (0–200 nm), accounting for more than 80%, and less macropores and mesopores (>200 nm) (Fig. 3.2).

From the perspective of coal seam and plane distribution, there is no significant difference in porosity. Therefore, in terms of porosity, the reservoir porosity development in the Qinshui Basin is not ideal for CBM production, which is also an inherent weak point in highly metamorphic coal reservoirs. There is no significant correlation between the total porosity and the vitrinite reflectance, which indicates that the porosity is controlled by many other factors.

The development degree of pores with different sizes in coal is called pore distribution (also known as pore structure). The pore distribution characteristics can better reflect the pore properties of coal.

The pore distribution of 40 coal samples taken in the Qinshui Basin is characterized by micro-intermediate pores

Y. Song et al., *Coalbed Methane in China*,
https://doi.org/10.1007/978-981-33-4725-0_3

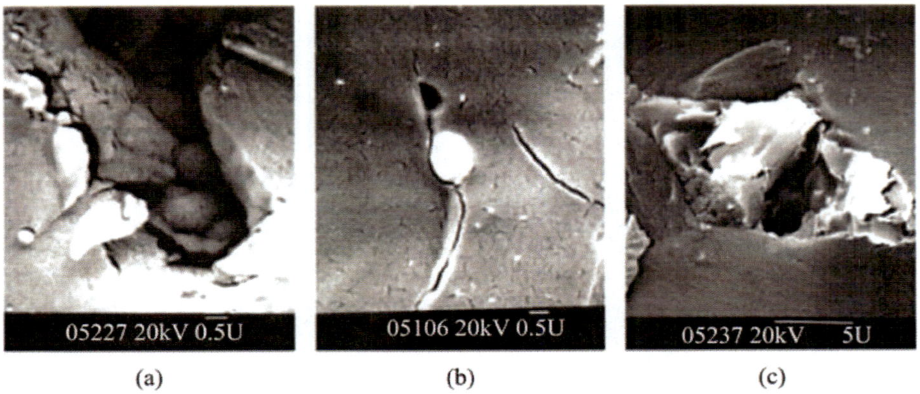

Fig. 3.1 Ultra-microscopic characteristics of pores in the coal in Qinshui Basin. **a** metamorphic pores; No. 15 coal sample from Yangquan No. 1 Coalmine in northern Qinshui Basin, yqyk15-3, R^{o} = 2.96%, 12000×; **b** pores connect with microfractures; No. 2 coal sample from Qianxin Coalmine, Qinyuan in central Qinshui Basin, qyqx2-3, R^{o} = 1.66%, 10,000×; **c** burst pore; Sihe Coalmine, Jincheng in southern Qinshui Basin, jcsh3-l, R^{o} = 3.14%, 4500×

Fig. 3.2 Pore distribution in the coal in Qinshui Basin from mercury intrusion experiments, northern samples: 1–17; central samples: 18–33; southern samples: 34–40

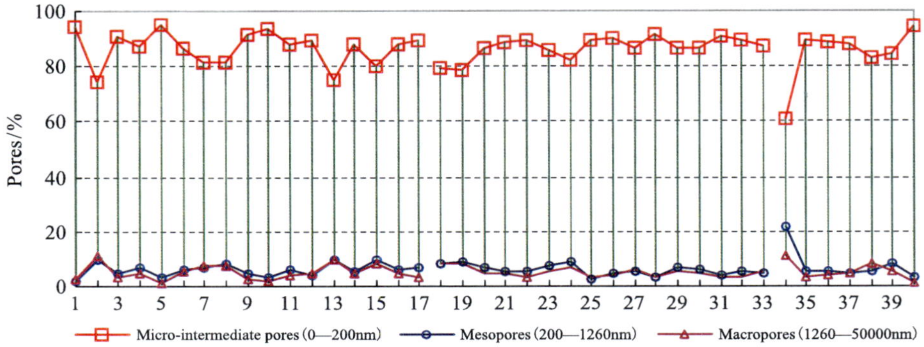

(0–200 nm), not less than 61.17%, and mostly higher than 80%; but macropores and mesopores (pore size >200 nm) are not developed, which is a major feature of the pore system in highly metamorphic coal reservoirs, according to mercury intrusion experiments. Further analysis of experimental data reveals that the pore structure also shows significant differences in the high percentage of micro-intermediate pores and the low percentage of mesopores and macropores. The difference was observed as the change of the percentage of small-intermediate pores in different samples and the change of the percentage of the pore interval with mesopores and above. The production of CBM is controlled by the pores at each section, so the effect of the change in the percentage of each pore interval measured by mercury intrusion experiments is significant on the penetration of CBM. Ideally, the percentage of each pore interval is more than a critical value and the internal configuration is reasonable. If the pores in a certain pore size range are not developed due to some factors, it will be a bottleneck of fluid penetration, making CBM difficult to produce. This phenomenon is the sensitivity of permeability to pore structure, and will be described later. In the 40 samples, the pores of 0 to 200 nm account for 61.17 to 94.74%, and 86.30% on average, and the coefficient of variation is 0.07. The distribution of pores of 0 to 200 nm was measured by low-temperature nitrogen adsorption. Other experimental parameters include mean throat diameter (μm), discharge pressure (MPa), mercury removal efficiency (%), sorting coefficient (pm) and homogenous coefficient (Table 3.1). The mean throat diameter varies from 0.01 to 0.56 μm, and 0.071 μm on average.

The average throat diameter of the middle- and low-rank coal pore system in the Ordos Basin is 5.92–14.69 μm. It is obvious that the average throat diameter of the high-rank coal reservoirs in the Qinshui Basin is lower than that of the middle- and low-rank coal reservoirs in the Ordos Basin by 2–3 orders of magnitude, which reflects the obvious characteristics of micro-intermediate pores of high-rank coal pore system. The variation of the discharge pressure is from 0.91 to 11.55 MPa. The change from small to large discharge pressure indicates that the percentage of macropores and mesopores decreases, while the percentage of

Table 3.1 Experimental parameters of coal reservoirs in Qinshui Basin from mercury intrusion experiments

Coalmine	Sample	Macro type of coal rock	Reflectance R^o/%	Porosity/ %	Average throat diameter/μm	Displacement pressure/MPa	Mercury removal efficiency/%	Sorting coefficient/μm	Homogenous coefficient
Yangquan No. 1	yqyk15-1	Semibright coal	3.01	3.4	0.06	4.10	54.36	0.36	0.62
	yqyk15-2	Bright coal	2.75	3.1	0.02	10.59	50.48	0.04	0.74
	yqyk15-3	Bright coal	2.96	4.1	0.08	5.08	73.44	0.12	0.49
	yqyk15-4	Semidull coal	2.89	3.6	0.04	8.91	64.00	0.10	0.73
	yqyk15-5	Semibright coal	2.86	1.2	0.10	4.28	79.52	0.33	0.45
	yqyk15-7	Semidull coal	3.02	4.8	0.08	4.28	47.51	0.05	0.58
	yqyk15-8	–	–	4.6	0.03	8.91	48.37	0.06	0.73
	yqyk15-9	Bright coal	2.92	5.8	0.01	11.55	49.16	0.05	0.77
Xinjing, Yangquan	yqxj3-i	Semibright coal	2.23	0.9	0.09	4.28	66.28	0.14	0.49
	yqxj3-2	Bright coal	2.48	1.0	0.09	3.76	68.01	0.25	0.48
	yqxj3-3	–	–	3.9	0.06	6.04	45.46	0.09	0.82
	yqxj3-4	Semibright coal	2.46	1.5	0.09	3.76	68.01	0.25	0.48
	yqxj3-5	Bright coal	2.34	2.9	0.05	7.18	52.04	0.01	0.72
Kaiyuan, Shouyang	syky3-l	Semibright coal	2.70	2.7	0.07	4.28	53.77	0.10	0.60
	syky9-l	Semibright coal	4.00	4.0	0.08	4.28	65.72	0.06	0.46
Shizhuang, Yangquan	yqsd15-l	Semibright coal	1.70	1.1	0.08	3.92	73.82	0.06	0.51
Duanwang, Shouyang	sydw15-l	Dull coal	2.14	1.0	0.08	5.08	75.79	0.08	0.59
Zuoquan	hm0302	Semibright coal	3.35	2.2	0.03	8.54	60.03	0.07	0.67
	hmzk0102	Bright coal	3.17	2.5	0.04	8.91	62.38	0.01	0.79
Qinxin, Qianyuan	qyqx2-l	Semibright coal	1.99	0.9	0.08	4.10	75.53	0.04	0.57
	qyqx2-2	Semibright coal	1.74	1.3	0.08	4.66	67.60	0.09	0.50
	qyqx2-3	Semibright coal	1.66	1.3	0.07	5.31	79.05	0.11	0.55
	qyqx2-4	Semibright coal	1.55	2.8	0.08	4.28	66.57	0.03	0.64
Changcun, Changzhi	czcc3-l	Semibright coal	1.96	5.1	0.05	6.88	56.16	0.05	0.79
	czcc3-2	Semibright coal	2.21	5.5	0.02	11.06	48.20	0.13	0.75
	czcc3-3	Semibright coal	2.12	3.9	0.06	8.18	49.31	0.15	0.77
	czcc3-4	Semibright coal	2.16	2.7	0.07	3.76	49.56	0.08	0.57
	czcc3-5	Semibright coal	2.07	4.3	0.03	10.59	65.01	0.16	0.75
	czcc3-6	Semibright coal	2.08	3.4	0.06	5.08	67.42	0.07	0.70

(continued)

Table 3.1 (continued)

Coalmine	Sample	Macro type of coal rock	Reflectance R^o/%	Porosity/%	Average throat diameter/μm	Displacement pressure/MPa	Mercury removal efficiency/%	Sorting coefficient/μm	Homogenous coefficient
Jingfang, Changzhi	czjf3-l	Semidull coal	1.79	5.9	0.04	7.18	56.99	0.10	0.78
	czjf3-3	Semibright coal	1.93	2.3	0.08	3.92	63.73	0.10	0.55
Wangzhuang, Changzhi	czwz3-2	Semibright coal	1.79	3.6	0.06	4.10	49.36	0.09	0.57
Sihe, Jincheng	jcsh3-l	Bright coal	3.43	7.6	0.56	0.99	52.37	1.37	1.12
	jcsh3-2	Semidull coal	3.41	1.1	0.06	7.18	77.01	0.10	0.70
	jcsh3-3	Bright coal	3.20	1.5	0.05	6.59	63.57	0.11	0.77
	jcsh3-4	Semibright coal	2.84	2.2	0.05	7.50	58.12	0.11	0.79
Xiaoxian, Gaoping	gpxx3-l	Bright coal	2.97	3.9	0.05	5.54	61.56	0.07	0.65
Guohe, Xihe	xhgh3-l	Bright coal	3.04	4.0	0.03	8.54	68.38	0.08	0.77
	xhgh3-2	–	–	5.3	0.04	10.59	66.95	0.21	0.82

intermediate pores increases. The smaller the discharge pressure, the more favorable the pore structure. The mercury removal efficiency ranges from 45.46 to 79.52%, with an average of 61.47% in the Qinshui Basin. In the Ordos Basin, the mercury removal efficiency of the middle- and low-rank coal reservoir is over 80%, and above 90% in some samples. This indicates that the coal reservoir in the Qinshui Basin has a significant decrease in mercury removal efficiency due to the significant increase in metamorphism. Generally, the higher the efficiency of mercury removal, the better the permeability of the pore system. Therefore, the overall low mercury removal efficiency of the high-rank coal reservoir indicates that the permeability is poor in the Qinshui Basin.

(2) Results of low-temperature nitrogen adsorption experiments

The relative percentage of mesopores, small pores and micropores calculated by low-temperature nitrogen adsorption method can not completely reflect the configuration of the pores in the sample, but can still be used as a reference to discuss the relative development degree of mesopores, small pores and micropores of the high-rank coal reservoirs in the Qinshui Basin.

Table 3.2 shows the main parameters and results of low-temperature nitrogen isotherm adsorption experiments. In the Qinshui Basin, the BET specific surface area of the coal samples is 0.001–0.98 m^2/g, with an average of 0.17 m^2/g; the BJH cumulative surface area is 0.0504–1.141 m^2/g, with an average of 0.345 m^2/g. Figure 3.3 shows the distribution of mesopores, pores and micropores of the samples determined by low-temperature nitrogen adsorption experiments. For most samples, small pores are most developed, mesopores are secondary, and micropores are least developed. In some samples, mesopores (in sample No. 1) or micropores (in sample No. 20) are most developed.

Combination of the results of ordinary microscope, scanning electron microscope, mercury intrusion experiments and low-temperature nitrogen adsorption experiments provides the following conclusions of the development characteristics of the pore system of high-rank coal reservoirs in the Qinshui Basin: (i) The pores include primary pores, metamorphic pores, exogenous pores, mineral pores and microfractures. The primary pores mainly have cell pores and interstitial pores; the metamorphic pores are mainly gas pores; the exogenous pores have breccia pores and particle pores; the mineral pores are mainly intercrystalline pores and intergranular pores. All these genetic types of pores have been observed, gas pores are most developed, and other types of pores are generally not very developed. (ii) Microfractures, macropores, mesopores, small pores and micropores are developed to some extent, and form a complete pore network connective at the microscopic level. It provides necessary channels for production of CBM.

Table 3.2 Specific surface areas and pore sizes of coals in Qinshui Basin from low-temperature nitrogen adsorption experiments

Coalmine	Sample	BET specific surface area/(m^2/g)	BJH total pore volume/ (mL/g)	BJH specific surface area/(m^2/g)	BJH average pore diameter/nm	Reflectance R^o/%
Yangquan No.1	yqyk15-8	0.0399	0.00146	0.2430	24.14	–
	yqyk15-9	0.1790	0.00107	0.3350	12.82	2.92
Xinjing, Yangquan	yqxj3-i	0.0905	0.00122	0.0660	74.06	2.23
	yqxj3-2	0.1030	0.00082	0.0504	64.78	2.48
Shidian, Yangquan	yqsd15-l	0.4040	0.00256	0.5140	19.95	1.70
Kaiyuan, Shouyang	syky3-l	0.5280	0.00234	0.2580	36.28	1.73
Duanwang, Shouyang	sydw15-l	0.2860	0.00253	0.2990	33.93	2.14
Zuoquan	hm0302	0.0490	0.00210	0.4030	20.53	3.35
Qinxin, Qinyuan	qyqx2-l	0.1440	0.00120	0.0970	50.62	1.99
	qyqx2-3	0.3670	0.00230	0.1930	47.01	1.66
	qyqx2-4	0.0730	0.00092	0.0570	65.03	1.55
Changcun, Changzhi	czcc3-l	0.1730	0.00180	0.1610	14.59	1.96
	czcc3-2	0.1520	0.00120	0.2060	23.47	2.21
	czcc3-3	0.1450	0.00120	0.2740	18.23	2.12
	czcc3-4	0.1270	0.00120	0.0940	50.52	2.16
	czcc3-5	0.1690	0.00130	0.1820	27.82	2.07
Jingfang, Changzhi	czjf3-l	0.2490	0.00290	0.6420	17.99	1.79
	czjf3-2	0.9800	0.00390	1.1410	13.52	1.89
	czjf3-3	0.6140	0.00260	0.6810	15.27	1.93
	czjf3-4	0.1340	0.00140	0.1260	44.26	1.95
Wangzhuang, Changzhi	czwz3-2	0.2710	0.00220	0.8090	10.72	1.79
Sihe, Jincheng	jcsh3-l	0.1920	0.00200	0.3300	24.45	3.43
	jcsh3-3	0.5070	0.00140	0.7740	7.18	3.20

(iii) The general characteristics of the pore system are that small pores are most developed, mesopores and micropores are almost similar or micropores are less developed than mesopores, and macropores are the least developed. But in some samples micropores or mesopores are most developed, followed by small pores, and macropores are least developed. (iv) The pore-throat distribution is quite different. The internal arrangement of the pores in some samples is reasonable, as developed pores in every section to some extent, which provides continuous channels for gas migration. In other samples, the internal arrangement of the pores is very unreasonable, showing as that the percentage of pores in some section is extremely low, which may cause discontinuous channels for gas production and a permeable bottleneck. (v) The results of mercury intrusion experiments shows that generally undeveloped macropores result in poor permeability.

3.1.1.2 Pore System of Medium- and Low-Rank Metamorphic Coal Reservoirs in Eastern Ordos Basin

The porosity measured by mercury intrusion experiments reflects the percentage of pores with diameter greater than 7.2 nm in coal reservoirs, which is the apparent porosity. The porosity of the coal samples taken in eastern Ordos Basin is between 2.5 and 8.5%. The mercury intrusion experiment data do not fully reflect the configuration of the various types of pores in the samples, but they can show the developed characteristics of any type of pores in the medium- and low-rank metamorphic coal reservoirs (Fig. 3.4).

The adsorption isotherms of the coal samples can be divided into the following types: Type a and Type b curves have been slowly rising; Type c and Type d begin to rise quickly when the relative pressure is greater than 0.8, and rise sharply at the end; Type e presents two sections—the

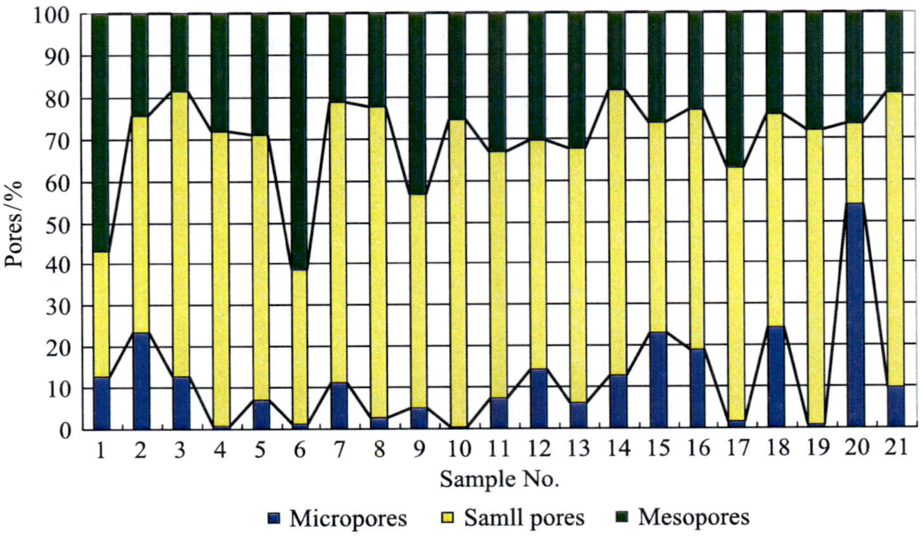

Fig. 3.3 Pores percentage measured by low-temperature nitrogen adsorption experiments

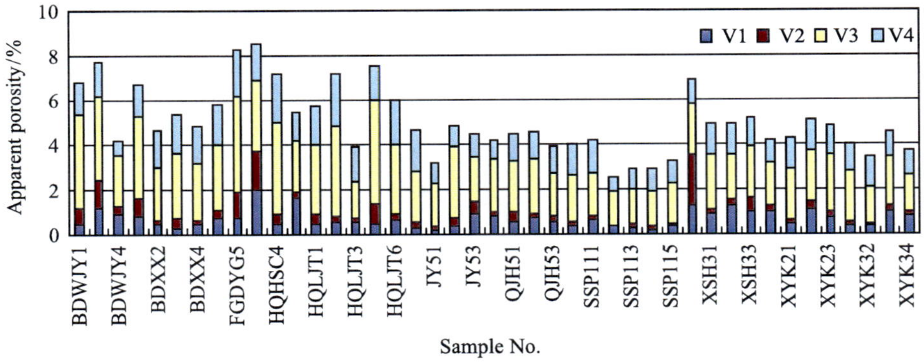

Fig. 3.4 Pore distribution of medium- and low-rank metamorphic coals in eastern Ordos Basin. V1, macropores; V2, mesopores; V3, intermediate pores; V4, micropores

first slowly reduces and then slowly rises, and the second quickly decreases and then sharply rises; Type f first slowly decreases, and then rises sharply at the end (Fig. 3.5). It is generally believed that the first section of the curve rises slowly, indicating a transition from monolayer to multi-layer adsorption; in the second section, the curve rises sharply, indicating that capillary condensation occurs in larger pores and results in the sharp increase of absorption. The phenomenon that the curve appears twisting and the adsorption volume in the first section decreases with the increase of pressure needs to be discussed in depth.

According to the theory of adsorption and agglomeration, during the adsorption experiment on the solids with capillary pores, capillary agglomeration will occur in the pore with a Kelvin radius when the relative pressure increases. If the pressure is reduced after increasing, the phenomenon that the adsorbate gradually desorbs and evaporates will occur. Due to the specific shape of the capillary, the relative pressure at which agglomeration and evaporation may be similar or different in the same pore. If the relative pressures during agglomeration and evaporation are the same, the adsorption branch overlaps with the desorption branch of the adsorption isotherm. On the contrary, if the two relative pressures are different, the two branches will separate to form a so-called adsorption loop. According to pore structures and whether an adsorption loop can generate, the pores in the coal can be divided into three types: Type I pores are open and gas-permeable, including cylindrical pores with open ends and parallel plate pores open at four sides. Such pores can produce an adsorption loop. Type II pores are gas-impermeable, and closed at one end. They include cylindrical pores, parallel plate pores, wedge-shaped pores, and tapered pores. Such pores do not create an adsorption loop. Type III are pores in a special form, i.e. pores with narrow bottlenecks (like ink bottles). Although these pores are closed at one end, it can produce an adsorption loop, and there is a clear sign on the loop caused by the pores, that is, the desorption branch has a sharply falling inflection point.

The sample adsorption loops are mostly Type a and Type b. The return branch of type a rapidly decreases and

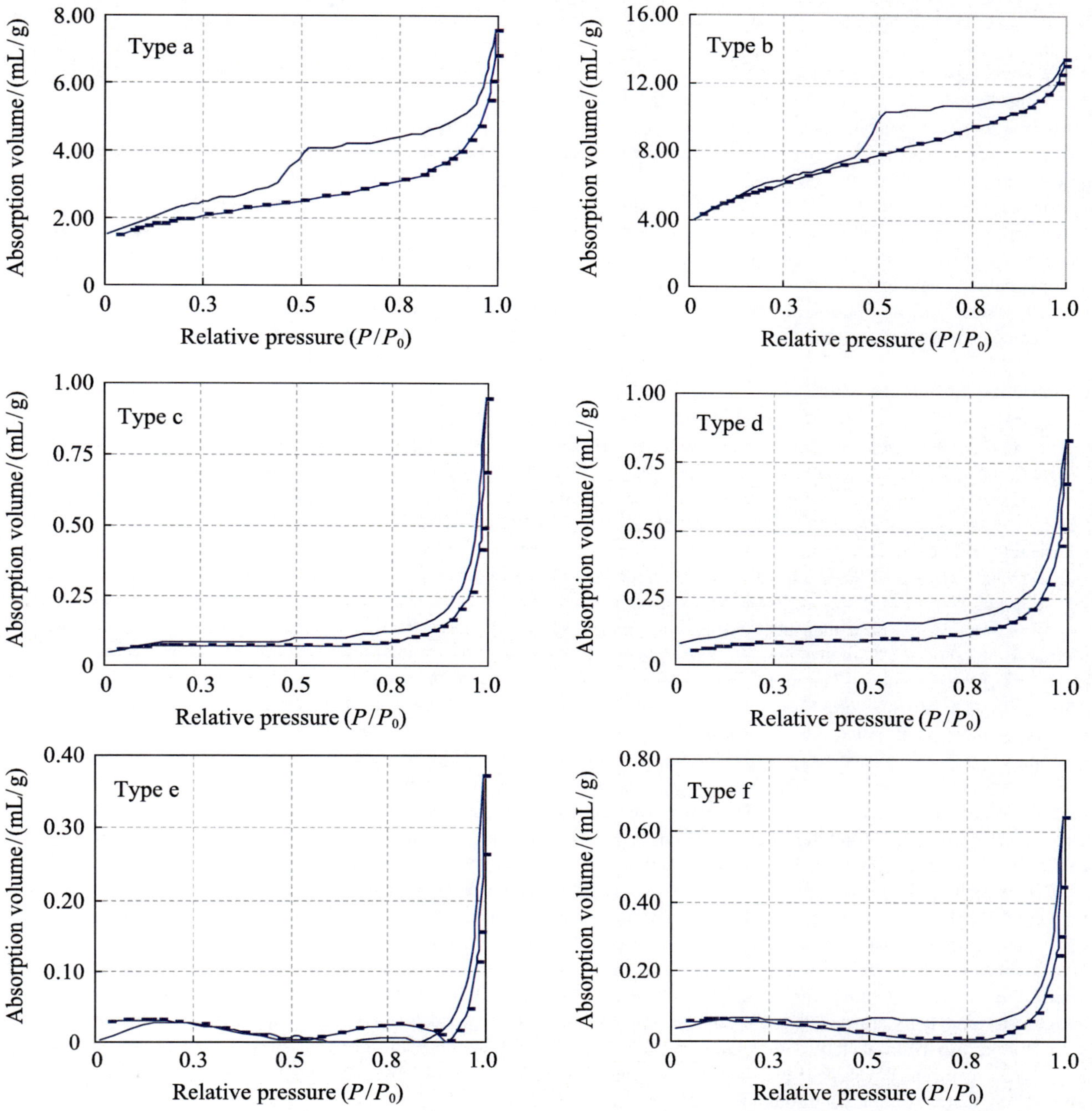

Fig. 3.5 Types of low-temperature nitrogen adsorption isotherms of medium- and low-rank metamorphic coals in eastern Ordos Basin

gradually slows down when the relative pressure is at 0.5–0.9, and an inflection point appears at around 0.5, indicating that the return line mainly appears at large relative pressure, and with an obvious inflection point, which means the pore system is more complicated. First, at lower relative pressure, the adsorption branch and the desorption branch are substantially coincident, indicating that the small-sized pores are mostly gas-impermeable and closed at one end, that is, Type II pores; at relatively high pressure, an adsorption loop appears, indicating that there are larger pores, open Type I pores are existent and Type II pores may be existent because Type II pores do not contribute to the return line. The shape of Type b adsorption loop slightly lags behind the adsorption branch. Actually, there is no loop or the loop is very small, reflecting that the pore system is mainly composed of gas-impermeable Type II pores closed at one end.

3.1.2 Fracture System

3.1.2.1 Large Fracture System

The large fracture system in coal reservoirs refers to the fracture system that is visible by the naked eye under natural conditions. It doesn't include faults. It consists of a cleat system, an expansion joint system and an exogenous joint system. The development of large fracture systems is the key to restricting the permeability of coal reservoirs. Taking the No. 3 coal reservoir in the Chengzhuang Coalmine as a case, next the development and genetic characteristics of the large fracture system and its control on the permeability of the coal reservoir will be described.

Development Characteristics

The Chengzhuang Coalmine is a monoclinic structure with a dip angle of 3°–5°. The faults in the coalmine are sparse. In the range of 120 km^2, 12 faults with a fault throw of less than 5 m each were found. The extension length is mostly less than 100 m, and there is only one fault with an extension length of 200 m. The thickness of the coal reservoir is stable, about 6 m, and contains a tonstein intercalation about 20 cm thick. The micro-components are more composed of vitrinite components, more than 80% on average, and less inertia group, about 15% on average, but the stable group is rarely or hard to distinguish due to the high rank of the coal. The ash yield is 8–18%, and about 13% on average. The pore types have plant cell residual pores, matrix pores and secondary gas pores; the porosity of the basement block is 6–11%, and 8% on average. The cleat porosity is 0.18–0.63%, and about 0.4% on average. The transformation to the coal reservoirs by structural stress resulted in significantly different permeability of the coal reservoirs.

The exogenous joints are divided into two types. One type cut through a coal seams into the top and bottom of the coal seam; the other type cut through the whole or most of a coal seam but do not enter the top and bottom of the coal seam, the latter accounting for more than 90%. Dense exogenous joint belts are developed at almost equal intervals. Within 5 to 10 m, usually 1 to 3 joints cut through an entire coal seam are developed. In some dense exogenous jointed belt, there is a joint cut through the coal seam into the top and bottom. The exogenous joints that cut through coal seams into their tops and bottoms have the characteristics of multiple active periods. There is structural coal of different thicknesses in the joint fractures, and multiple stages of calcite veins in the fractures, indicating that these joints are primary channels for fluid exchange between inside and outside. The orientations of the exogenous joints are northeast and northwest. The coal seams are broken at two joint groups. In the No. 3 coal reservoir of the Chengzhuang Coalmine, laterally, the exogenous joint that cuts through the coal seams into their tops and bottoms can extend more than 100 meters; that cuts through an entire coal seam can extend tens to hundreds of meters; and that doesn't cut through an entire coal seam can extend over a dozen meters, generally no more than 20 m.

Gas expansion joints are developed. The occurence of the expansion joints is consistent with that of the cleats. The expansion joints are almost equally spaced. The width of the expansion joint is usually about 2 times that of the cleat, and the height is usually 3 to 10 times that of the cleat. The surface of the expansion joint is smooth and straight, and has the characteristics of pure tension joints. Although the upper and lower limits of the expansion joints are not as regular and neat as the cleats, there are significant differences in the degree of development of the expansion joints in different coal seams. Usually they are most developed in bright coal, then semibright coal and its transitional type, and not developed in semidull coal and dull coal.

The development of the cleat system is mainly constrained by the components of coal rock, generally developed in vitrain and bright coal, with equal spacing or nearly equal spacing. The cleats in vitrain are most constrained by coal rock stratification, so do the cleats in bright coal.

The large fracture system has obvious directionality. The exogenous joints have two dominant directions, namely northeast to southwest and northwest to southeast. The dominant direction of the expansion joints and the cleats is northeast to southwest (Fig. 3.6).

The large fracture system is obviously heterogenous. The exogenous joints are closely grouped and banded locally. The gas expansion joint and cleat systems has obvious component selectivity. They are most developed in bright coal, but not developed in dull coal, resulting in the differential development of the large fracture system on section. The porosity of the coal seam with developed large fractures is high, and the difference between the seam with high porosity and that with low porosity is at least an order of magnitude. The directionality and spatial heterogeneity of the large fracture systems lead to changes in the permeability with azimuth and horizon, and can also lead to the permeability anisotropy of the coal seams.

Control Factors on Large Fracture Systems

(1) Influences of coal seam structure and the top and bottom of coal seam on large fracture systems
The thickness of the tonstein and the lithology of the top and bottom of the coal seam directly affect the development of exogenous joints. Usually, when the thickness of the tonstein is greater than 0.5 m, exogenous joints are not developed or their density

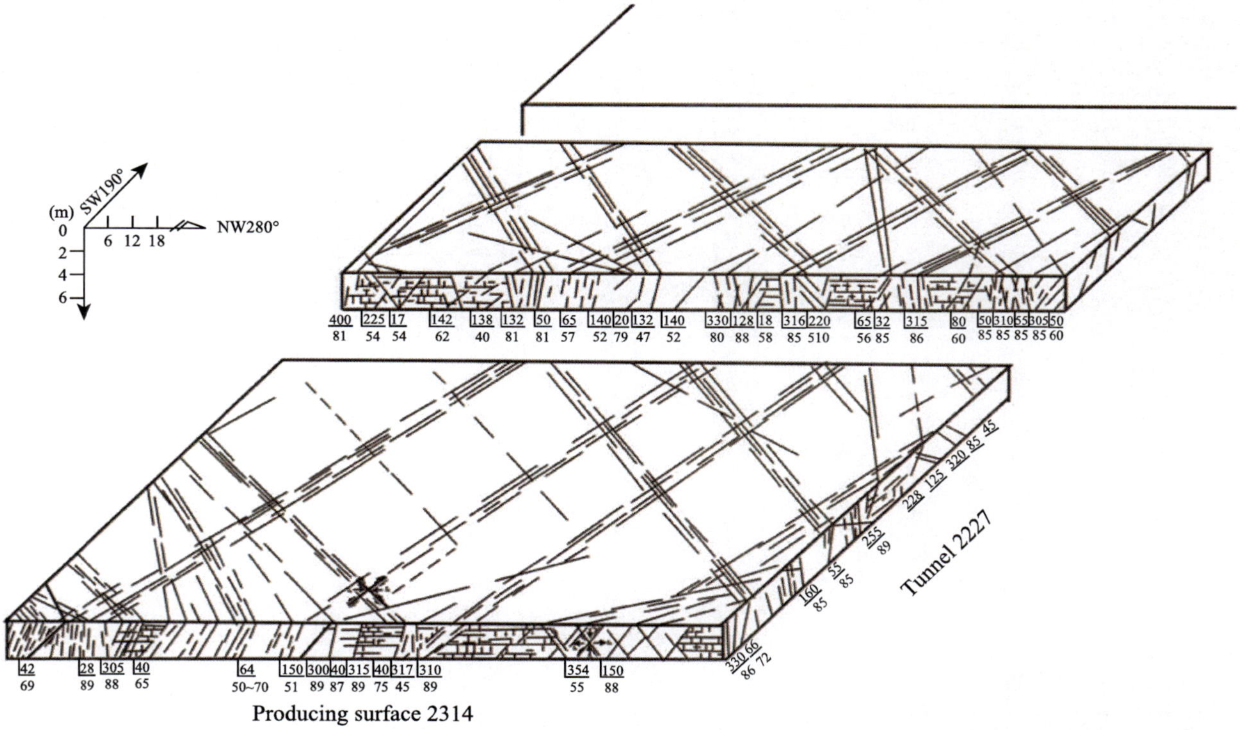

Fig. 3.6 Observed joint and cleat systems in Erpan block of Chengzhuang Coalmine

significantly reduces in the tonstein. If the direct top and bottom of the coal seam is sandstone, exogenous joints are not developed or only developed inside the coal seam; if the direct top and bottom of the coal seam is siltstone and mudstone, a few exogenous joints may extend into the direct top and bottom (Fig. 3.7). Expansion joints are generally not developed in the tonstein thick more than 0.2 m, but a few may appear in the tonstein thick less than 0.2 m.

(2) Influences of coal rock type and coal minerals on large fracture systems
Mineral content and coal rock type have obvious constraining effects on gas expansion joints and cleats in coal seams. In the bright coal, both cleats and expansion joints are well developed. On the contrary, cleats are sparse, and expansion joints are generally not developed in the dull coal (Table 3.3).

(3) Influences of fault and fold on large fracture systems
Faults and fold structures have important control effects on the development of exogenous joint systems in coal seams. Usually in the vicinity of faults, exogenous joints are developed abnormally, often in dense groups or belts, and there are many exogenous joints that cut through coal seams. In the area where the curvature of the fold is large, the density of exogenous joints significantly increases. The influences of fault and fold structures on the development of cleats and inflation joints are not obvious.

Genesis Model of Large Fracture Systems

According to the development law of the large fracture system in coal seams, and the inference of the order of cut-through, the cleats are products of fluid discharge during coalification (Fig. 3.8a); the expansion joints are products of further fluid discharge in coal during secondary superimposed metamorphism of coal (Fig. 3.8b), and the exogenous joints are structural products that utilized and modified the expansion joints and cleat systems (Fig. 3.8c).

According to the developmental cross-cutting relationship, the exogenous joints in the study area can be divided into four periods (Table 3.4).

3.1.2.2 Microfracture System

Microfractures are channels communicating pores with macrofractures. According to width (*W*) and length (*L*), microfractures can be divided into four types. In the Qinshui Basin, microfractures are more developed, the northern ones

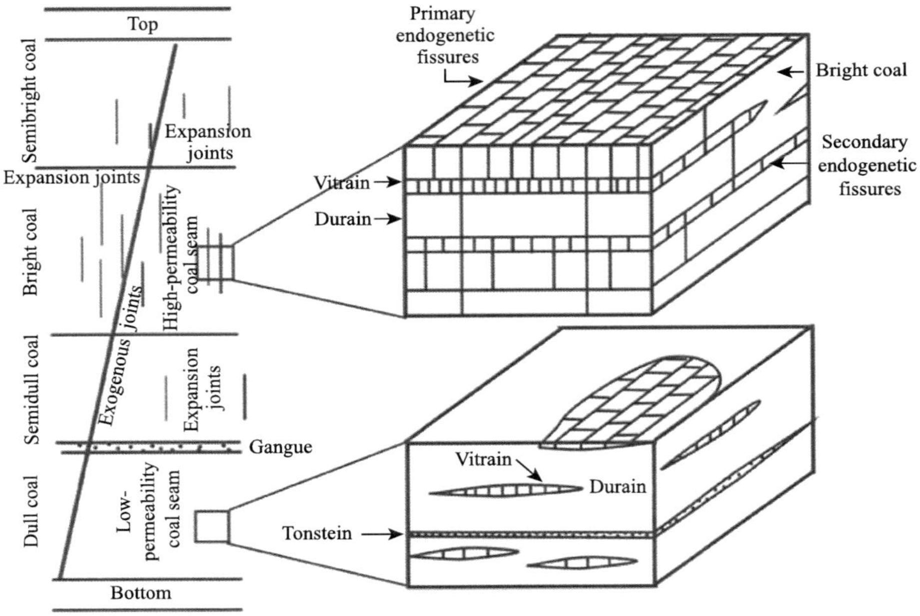

Fig. 3.7 Schematic fracture distribution in coal reservoirs in Chengzhuang Coalmine

are more developed than the central and southern ones (Table 3.5). Most of the microfractures were filled with minerals in later stages.

3.2 Characterization and Mathematical Model of Favorable Coal Reservoirs

3.2.1 Pore System Model

3.2.1.1 Modeling and Analysis of Pore System

Cluster Analysis of Pore System

Pore system cluster analysis is useful for the case that there is no prior classification, i.e. how to classify samples or indicators. Cluster analysis, also known as point group analysis, is a multivariate statistical analysis method that quantifies the objects according to the relationship between the objects in terms of nature or cause. This classification method considers all factors comprehensively. According to the objects, the cluster analysis method is further divided into Q-type and R-type cluster analysis. Q-type cluster analysis means to classify samples, and R-type cluster analysis means to classify variables. To measure the similarity (or correlation) between objects, we first need to define some metrics, namely cluster statistic. Q-type cluster statistic involves similarity coefficient, correlation coefficient and distance coefficient (Euclidean distance). This study used Q-type analysis to cluster samples. The Euclidean distance was chosen as the Q-type clustering statistic, which is defined as follows:

In the orthogonal coordinate system, the distance between the two sample points X_i and X_j is

$$d_{ij} = \sqrt{\sum_{k=1}^{m} (x_{ik} - x_{jk})^2}$$

The smaller the d_{ij}, the closer the properties of the samples X_i and X_j are. Therefore, samples with relatively small d_{ij} should be classified into one category. In order not to make d_{ij} too large (just divide the samples with relatively small d_{ij} in n samples into one class and reduce the d_{ij} by the same multiple, which does not affect the classification result), the above formula is rewritten as

$$d_{ij} = \sqrt{\frac{1}{m} \cdot \sum_{k=1}^{m} (x_{ik} - x_{jk})^2} \quad (i, j = 1, 2, \ldots, n)$$

$[d_{ij}]_{n \times n}$ is a real symmetric matrix, and $d_{11} = d_{22} = \cdots = d_{nn} = 0$.

This clustering method selects the Ward Method, which is obtained by Ward according to the principle of variance analysis. If the classification is reasonable, the dispersion sum of squares between similar samples is small, and the dispersion sum of squares between class and class is larger. Assuming that class G_p is merged with class G_q into a new G_r, the distance recursion formula of G_r and any class G_i is

$$D_{ir}^2 = \frac{n_i + n_p}{n_r + n_i} D_{ip}^2 + \frac{n_i + n_q}{n_r + n_i} D_{iq}^2 - \frac{n_i}{n_r + n_i} D_{pq}^2$$

The result of the Ward Method is better. It requires that the distance between samples must be Euclidean distance.

Table 3.3 Relationship between coal rock types and development characteristics of large fractures

Coal type	Cleat development	Expansion joint development
Bright coal	In the vitrain, the density of the primary cleat group is 9–10 lines/5 cm, and the density of the secondary is 5–6 lines/5 cm; the carrier size is (4–6) mm × (50–70) mm × (80–150) mm, and the percentage of the carrier is 12–17%. In the bright coal, the density of the primary cleat group is 4–5 lines/5 cm, and the density of the secondary is 2–3 lines/5 cm; the carrier size is (10–15) mm × (80–120) mm × (150–300) mm, the percentage of the carrier is 20–25%; the cleat porosity is 0.675–0.729%	The density of the primary expansion joints is 8–15 lines/20 cm, and the density of the secondary is 4–10 lines/20 cm; the carrier size is (100–150) mm × (200–1000) mm × (500–1500) mm, and the percentage of the carrier is 05–1.62%; the porosity of the expansion joints is 1.05–1.62%
Semibright coal	In the vitrain, the density of the primary cleat group is 9–10 lines/5 cm, and the density of the secondary is 5–6 lines/5 cm; the size of the carrier is (3–5) mm × (40–60) mm × (60–120) mm, the percentage of the carrier is 8–10%. In the bright coal, the density of the primary cleat group is 4–5 lines/5 cm, and the density of the secondary is 2–3 lines/5 cm; the carrier size is (8–12) mm × (60–100) mm × (120–200) mm, the percent of the carrier is 18–20%; the cleat porosity is 0.43–0.54%	The density of the primary expansion joints is 6–12 lines/20 cm, and the density of the secondary is 3–8 lines/20 cm. The carrier size is (80–120) mm × (150–800) mm × (300–1200) mm, and the percentage of the carrier is 85–1.22%. The porosity of the expansion joints is 0.85–1.22%
Semidull coal	In the vitrain, the density of the primary cleat group is 10–12 lines/5 cm, and the density of the secondary is 6–7 lines/5 cm; the size of the carrier is (2–4) mm × (10–20) mm × (50–80) mm, the percentage of the carrier is 3–5%. In the bright coal, the density of the primary cleat group is 3–4 lines/5 cm, and the density of the secondary is 1–2 lines/5 cm; the carrier size is (5–7) mm × (30–50) mm × (80–100) mm, the percent of the carrier is 3–5%; the cleat porosity is 0.24–0.365%	The density of the primary expansion joints is 2–3 lines/20 cm, and the density of the secondary is 1–2 lines/20 cm. The carrier size is (50–100) mm × (100–600) mm × (200–1000) mm, and the percentage of the carrier is 5–8%. The porosity of the expansion joints is 0.35–0.52%
Dull coal	In the vitrain, the density of the primary cleat group is 10–14 lines/5 cm, and the density of the secondary is 6–9 lines/5 cm; the size of the carrier is (1–3) mm × (5–10) mm × (20–30) mm, the percentage of the carrier is 1–2%. In the bright coal, the density of the primary cleat group is 4–6 lines/5 cm, and the density of the secondary is 2–4 lines/5 cm; the carrier size is (2–4) mm × (15–30) mm × (50–80) mm, the percent of the carrier is 2–4%; the cleat porosity is 0.11–0.24%	Not developed

Before cluster analysis, original quantitative data should be standardized first. There are many methods for data standardization. This study uses sum standardization. Sum normalization refers to the ratio of each observation of a variable to the sum of all observations of the variable. Therefore, after transformation, the element values of the data matrix are at [0, 1], and the sum of all observations of a variable is equal to 1. The specific transformation formula is

$$x'_{ij} = \frac{x_{ij}}{x_{\cdot j}} \quad (i = 1, 2, \ldots, nj = 1, 2, \ldots, m)$$

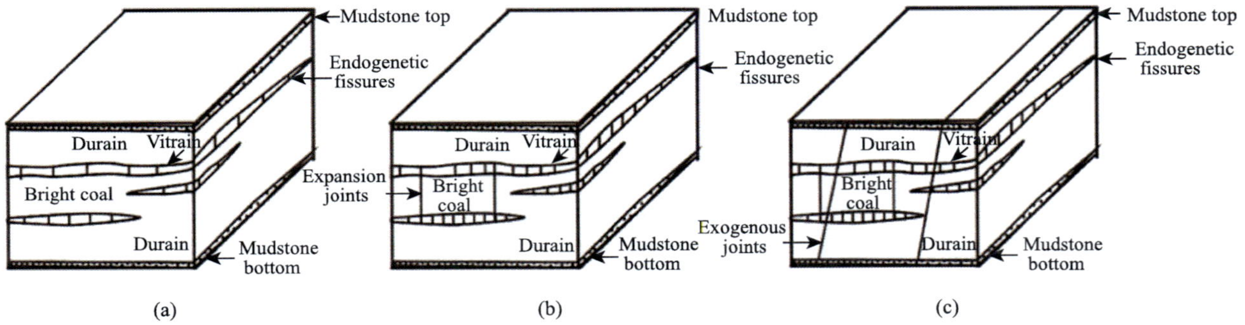

Fig. 3.8 Genesis models of large fractures in coal reservoirs. **a** Endogenetic fissures developed in coal seams; **b** endogenetic fissures fully developed into gas expansion joints; **c** exogenous joints developed in coal seams

Table 3.4 Joint features and principal stress

Period	Joint group	Occurrence	Principal stress
I	1	60°–75°/NW ∠50°–60°	Vertical σ_1, and horizontal σ_2 and σ_3
	2	60°–75°/SE ∠50°–60°	
II	3	310°–330°/NE ∠60°–70°	NS horizontal σ_1, vertical σ_2 and horizontal σ_3
	4	30°–40°/SE ∠88±	
III	5	35°–50°/NW ∠88±	WE horizontal σ_1, vertical σ_2 and horizontal σ_3
	6	310° ±/NE ∠70°–88°	
IV	7	350°/SW 260° ∠ 85°	σ_1 SE135° ∠ 10°, nearly vertical σ_2, and σ_3 NE45°

where x'_{ij} is transformed data; x_{ij} is the data before transformation; $x_{\cdot j}$ is the sum of the observations of j variables; and $x_{\cdot j} = \sum_{k=1}^{n} x_{kj}$, n is the total number of samples.

Nearly 40 samples taken in 13 mining areas, including northern Yangquan–Shouyang, central Qinyuan–Changzhi and southern Jincheng–Gaoping, were analyzed to investigate the general characteristics and differences in the development of the pore system of the high-rank coal reservoirs in the Qinshui Basin. Table 3.6 shows the results of mercury intrusion experiments on 39 samples after standardization.

Two kinds of parameters about the pore structure were obtained. One type of parameter is the pore percent of each pore interval. It can qualitatively determine gas desorption and permeation in the pore system. The other type of parameters includes porosity, average throat diameter, displacement pressure, mercury removal efficiency, sorting coefficient and homogenous coefficient. These parameters mainly reflect the permeation properties of the pore system. Two kinds of parameters were selected as variables to carry out Q-type cluster analysis on the samples (Fig. 3.9). 39 samples were divided into 4 categories. This classification reflects the samples from the perspective of adsorption, desorption and permeation. The properties of each type of samples were averaged in each cluster (Table 3.7). The average value is concentrated reflection of the characteristics of the samples in each cluster, which is the model or approximate model of the samples in the cluster, so it can be studied as the common feature of the samples in the cluster. However, this ideal result has a certain gap with actual sample properties. Therefore, this study selected representative measured samples as an example of the model, the properties of which are very close to the average value of each cluster of samples. After comparative analysis, four samples of jcsh3-l, yqyk15-9, czjf3-l and qyqx2-3 were selected as typical samples of clusters I, II, III and IV (Table 3.8) and model analysis was carried out.

Fractal Mathematical Model of Pore System

Mandebrot (1982) founded a fractal theory in the mid 1970s. The research object is a random, irregular, self-similar or statistically self-similar system that exists widely in nature and human social activities. The fractal theory provides insight into the fine structure hidden in chaos by means of the principle of self-similarity, and provides a simple quantitative description tool for discovering the regularity of disciplines. In recent years, the fractal theory has been widely used in geology, and has achieved a series of achievements in structural faults, petroleum-bearing structure distribution, metallogenic regularity and prediction,

Table 3.5 Statistical microfractures in coal reservoirs in Qinshui Basin

Coalmine	Sample	Fracture density/(fractures/9 cm^2)					Fracture/%				Reflectance R^o/%
		Type A	Type B	Type C	Type D	Total	Type A	Type B	Type C	Type D	
Yangquan No.1	yqyk15-l	0	1	11	43	55	0	1.8	20.0	78.2	3.01
	yqyk15-2	0	1	13	48	62	0	1.6	21.0	77.4	2.75
	yqyk15-3	0	0	7	32	39	0	0	17.9	82.1	2.96
	yqyk15-4	0	0	4	6	10	0	0	40.0	60.0	2.89
	yqyk15-5	0	2	5	13	20	0	10.0	25.0	65.0	2.86
	yqyk15-7	0	1	17	33	51	0	2.0	33.3	64.7	3.02
	yqyk15-9	0	1	8	49	58	0	1.7	13.8	84.5	2.92
Xinjing, Yangquan	yqxj3-l	1	4	8	13	26	3.8	15.4	30.8	50.0	2.23
	yqxj3-2	0	1	2	9	12	0	8.3	16.7	75.0	2.48
	yqxj3-4	0	1	22	26	49	0	2.1	44.9	53.0	2.46
	yqxj3-5	0	0	16	24	40	0	0	40.0	60.0	2.34
Shidian, Yangquan	yqsd15-l	0	1	9	20	30	0	3.3	30.0	66.7	1.70
Kaiyuan, Shouyang	syky3-l	1	2	7	23	33	3	6.1	21.2	69.7	1.73
	syky9-l	0	1	5	26	32	0	3.1	15.6	81.3	1.97
Duanwang, Shouyang	sydw15-l	0	1	22	21	44	0	2.3	50.0	47.7	2.14
Zuoquan	hm0302	0	0	1	5	6	0	0	16.7	83.3	3.35
	hmzk0102	0	3	5	0	8	0	37.5	62.5	0	3.17
Qinxin, Qinyuan	qyqx2-l	0	0	7	12	19	0	0	36.8	63.2	1.99
	qyqx2-2	2	3	9	16	30	6.7	10.0	30.0	53. 3	1.74
	qyqx2-3	0	0	9	17	26	0	0	34.6	65.4	1.66
	qyqx2-4	0	1	8	35	44	0	2.3	18.2	79.5	1.55
Changcun, Changzhi	czcc3-l	1	2	15	29	47	2.1	4.3	31. 9	61.7	1.96
	czcc3-2	0	0	1	2	3	0	0	33.3	66.7	2.21
	czcc3-3	0	2	3	8	13	0	15.4	23.1	61.5	2.12
	czcc3-4	2	1	10	16	29	6. 9	3.5	34.5	55.1	2.16
Changcun, Changzhi	czcc3-5	0	0	4	3	7	0	0	57.1	42.9	2.07
	czcc3-6	0	2	1	2	5	0	40.0	20.0	40.0	2.08
Jingfang, Changzhi	czjf3-l	0	10	14	103	127	0	7.9	11.0	81.1	1.79
	czjf3-2	0	1	3	5	9	0	11.1	33.3	55.6	1.89
	czjf3-3	0	1	2	30	33	0	3	6.1	90.9	1.93
	czjf3-4	0	0	2	5	7	0	0	28.6	71.4	1.95
Wangzhuang, Changzhi	czwz3-2	0	0	4	9	13	0	0	30.8	69.2	1.79
Sihe, Jincheng	jcsh3-l	0	5	9	10	24	0	20.8	37.5	41.7	3.43
	jcsh3-2	0	0	4	6	10	0	0	40.0	60.0	3.41
	jcsh3_3	0	2	3	16	21	0	9.5	14.3	76.2	3.20
	jcsh3-4	0	2	12	10	24	0	8.3	50.0	41.7	2.84
Xiaoxian, Gaoping	gpxx3-l	0	2	9	2	13	0	15.4	69.2	15.4	2.97
Guohe, Xihe	xhgh3-l	0	2	14	28	44	0	4.6	31.8	63.6	3.04

Table 3.6 Mercury intrusion pore structure parameters sum standardization

Area	Coalmine	Sample	No.	Porosity/%	Average throat diameter/μm	Displacement pressure/MPa	Mercury removal efficiency/%	Sorting coefficient/μm	Homogenous coefficient	Pore content/%		
										0–0.1 μm	0.1–0.63 μm	0.63–25 μm
Northern Qinshui Basin	Yangquan No.1	yqyk15-1	1	0.028	0.021	0.017	0.023	0.066	0.024	0.028	0.0085	0.0131
		yqyk15-2	2	0.025	0.007	0.043	0.021	0.007	0.029	0.022	0.0393	0.0534
		yqyk15-3	3	0.033	0.029	0.021	0.031	0.022	0.019	0.027	0.0182	0.0157
		yqyk15-4	4	0.029	0.014	0.037	0.027	0.018	0.028	0.026	0.0280	0.0227
		yqyk15-5	5	0.010	0.036	0.018	0.033	0.060	0.017	0.028	0.0131	0.0063
		yqyk15-7	6	0.039	0.029	0.018	0.020	0.009	0.022	0.026	0.0255	0.0257
		yqyk15-8	7	0.037	0.011	0.037	0.020	0.011	0.028	0.024	0.0271	0.0365
		yqyk15-9	8	0.047	0.004	0.047	0.020	0.009	0.030	0.024	0.0319	0.0383
	Xinjing, Yangquan	yqxj3-i	9	0.007	0.032	0.018	0.028	0.026	0.019	0.027	0.0188	0.0127
		yqxj3-2	10	0.008	0.032	0.015	0.028	0.046	0.019	0.028	0.0138	0.0101
		yqxj3-3	11	0.032	0.021	0.025	0.019	0.016	0.032	0.026	0.0248	0.0195
		yqxj3-4	12	0.012	0.032	0.015	0.028	0.046	0.019	0.027	0.0174	0.0252
		yqxj3-5	13	0.024	0.018	0.029	0.022	0.002	0.028	0.022	0.0391	0.0485
	Kaiyuan, Shouyang	syky3-l	14	0.022	0.025	0.018	0.022	0.018	0.023	0.026	0.0215	0.0237
		syky9-l	15	0.033	0.029	0.018	0.027	0.011	0.018	0.024	0.0383	0.0394
	Shidian, Yangquan	yqsd15-l	16	0.009	0.029	0.016	0.031	0.011	0.020	0.026	0.0235	0.0236
	Duanwang, Shouyang	sydw15-l	17	0.008	0.029	0.021	0.032	0.015	0.023	0.027	0.0270	0.0161
Northern Qinshui Basin	Zuoquan	hm0302	18	0.018	0.011	0.035	0.025	0.013	0.026	0.024	0.0332	0.0404
		hmzk0102	19	0.020	0.014	0.037	0.026	0.002	0.031	0.024	0.0369	0.0391
	Qinxin, Qinyuan	qyqx2-l	20	0.007	0.029	0.017	0.031	0.007	0.022	0.026	0.0278	0.0228
		qyqx2-2	21	0.011	0.029	0.019	0.028	0.016	0.019	0.026	0.0215	0.0228
		qyqx2-3	22	0.011	0.025	0.022	0.033	0.020	0.021	0.027	0.0225	0.0182
		qyqx2-4	23	0.023	0.029	0.018	0.028	0.005	0.025	0.025	0.0298	0.0276
	Changcun, Changzhi	czcc3-l	24	0.041	0.018	0.028	0.023	0.009	0.031	0.025	0.0357	0.0337
		czcc3-2	25	0.045	0.007	0.045	0.020	0.024	0.029	0.027	0.0115	0.0161
		czcc3-3	26	0.032	0.021	0.034	0.021	0.027	0.030	0.027	0.0182	0.0201
		czcc3-4	27	0.022	0.025	0.015	0.021	0.015	0.022	0.026	0.0209	0.0294
		czcc3-5	28	0.035	0.011	0.043	0.027	0.029	0.029	0.027	0.0152	0.0175
		czcc3-6	29	0.028	0.021	0.021	0.028	0.013	0.027	0.026	0.0278	0.0262
	Jingfang, Changzhi	czjf3-l	30	0.048	0.014	0.029	0.024	0.018	0.030	0.026	0.0242	0.0228
		czjf3-3	31	0.019	0.029	0.016	0.027	0.018	0.021	0.027	0.0222	0.0177
	Wangzhuang, Changzhi	czwz3-2	32	0.029	0.021	0.017	0.021	0.016	0.022	0.026	0.0182	0.0279
Southern Qinshui Basin	Sihe, Jincheng	jcsh3-l	33	0.062	0.200	0.004	0.022	0.250	0.043	0.018	0.0860	0.0549
		jcsh3-2	34	0.009	0.021	0.029	0.032	0.018	0.027	0.027	0.0224	0.0156
		jcsh3-3	35	0.012	0.018	0.027	0.026	0.020	0.030	0.026	0.0234	0.0192
		jcsh3-4	36	0.018	0.018	0.031	0.024	0.020	0.031	0.026	0.0200	0.0235
	Xiaoxian, Gaoping	gpxx3-l	37	0.032	0.018	0.023	0.026	0.013	0.025	0.025	0.0225	0.0390
	Guohe, Xihe	xhgh3-l	38	0.033	0.011	0.035	0.028	0.015	0.030	0.025	0.0318	0.0275
		xhgh3-2	39	0.043	0.014	0.043	0.028	0.038	0.032	0.028	0.0127	0.0076

Table 3.7 Average parameters of cluster sample analysis from mercury intrusion experiments

Cluster	Number of samples	Porosity percent/%			Porosity/%	Average throat diameter/μm	Displacement pressure/MPa	Mercury removal efficiency/%	Sorting coefficient/μm	Homogenous coefficient	Qualitative evaluation on desorption and permeation
		Micropores-small pores	Mesopores	Macropores							
I	1	61.17	22.08	11.49	7.60	0.560	0.99	52.37	1.370	1.12	Most favorable
II	8	79.33	8.61	8.76	3.63	0.039	8.19	56.21	0.046	0.69	Favorable
III	18	88.04	5.67	4.70	3.81	0.053	6.77	57.26	0.115	0.71	Unfavorable
IV	12	90.02	5.31	3.60	1.48	0.082	4.61	72.32	0.139	0.53	Unfavorable

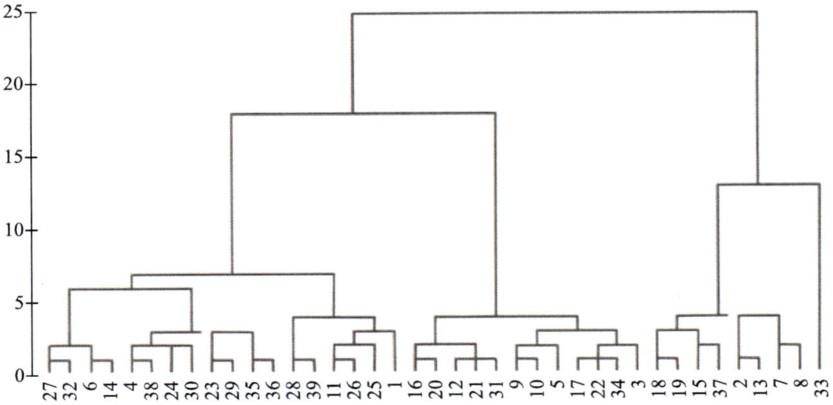

Fig. 3.9 Q-type clustering diagram with pore percent and mercury intrusion parameters as variables

quasi-crystal structure, seismic survey and porous media with permeation systems. Pfeifer and Avnir (1983) used molecular adsorption to study the pore characteristics of reservoir rocks, which are considered to have the properties of fractal structures. Later, Katz and Thompson (1985) and Kroch (1988) observed rock sections by scanning electron microscopy and found that the pore distribution in sandstone, shale and carbonate rocks has good fractal properties.

Fractal method is mathematical and used to more accurately quantify the pore system of coal reservoirs. To fractalize the pore system, the first step is to determine the distribution characteristics of some important parameters (such as pore volume, pore specific surface area, etc.). The distribution characteristics of the fractals of these parameters are then calculated, that is, their fractal dimensions are calculated. In particular, the fractal theory is still a theory of statistical self-similarity. The fractal model of the pore system is based on the mathematical model of statistical observations and experiments. Whether the fractal dimension is accurate and credible, and can reflect the real geological significance still depends on the observational scale and whether the observations are carefully taken.

Fractal dimension is an important parameter in the study of fractals in porous media with permeation systems. To facilitate mathematical and experimental measurements, a commonly used dimension definition is a box dimension, defined as follows:

Let F be any non-empty bounded subset of R^n. Let $N(A, \delta)$ denote the minimum number of the set with the maximum diameter δ and cover F, then the upper and lower box dimensions of F are defined as

$$\overline{\dim}BF = \overline{\lim_{\delta \to 0}} \frac{\ln N(F, \delta)}{\ln(1/\delta)}$$

$$\underline{\dim}BF = \underline{\lim_{\delta \to 0}} \frac{\ln N(F, \delta)}{\ln(1/\delta)}$$

If the upper and lower box dimensions are equal, the box dimension of F is defined as:

$$\dim BF = \lim_{\delta \to 0} \frac{\ln N(F, \delta)}{\ln(1/\delta)}$$

As mentioned above, the pore system in coal reservoirs is very complicated. We can use Menger sponge to simulate the pore characteristics and study its fractal structure. Imagining an initial cube, in the case of level 1, the unit cube is divided into 27 cubes with $r_1 = 1/3$, and 20 of them are left ($N_1 = 20$). In the second level, the remaining cubes are divided into 729 cubes with $r_2 = 1/9$, and 400 of them are left ($N_2 = 400$). Repeat the process, until the final image is called Menger sponge. If the length of the initial cube is L, then divide it into N equal small cubes, and then remove some of them according to some rules (for example, remove the 6 cubes at the center of any surface and the cube at the center of the volume), the number of remaining cubes is A_1. After k operations, the remaining cube has a side length of

$$r_k = \frac{L}{\left(\sqrt[3]{N}\right)^k} = \frac{L}{N^{k/3}},$$

The total number of cubes is

$$A_k = A_1 k, \quad or \ A_k = \left(\frac{L}{r_k}\right)^{D_b} = L^{D_b} r_k^{-D_b} \tag{3.1}$$

where $D_b = \frac{\ln(A_{k+1}/A_k)}{\ln(r_k/r_{k+1})}$, is called pore volume fractal dimension.

From the above formula, the pore volume can be derived as

$$V_h = L^3 - A_k r_k^3 \tag{3.2}$$

Take Eq. (3.1) into the Eq. (3.2), and

$$V_h = L^3 - L^{D_b} r_k^{3-D_b}$$

Table 3.8 Parameters of cluster samples analysis from mercury intrusion experiments

Cluster	Typical samples	Porosity percent/%			Porosity/%	Average throat diameter/μm	Displacement pressure/MPa	Mercury removalefficiency/%	Sorting coefficient/μm	Homogenous coefficient
		Micropores-small pores	Mesopores	Macropores						
I	jcsh3-1	61.17	22.08	11.49	7.60	0.560	0.99	52.37	1.370	1.12
II	yqyk15-9	81.57	8.18	8.03	5.80	0.010	11.55	49.16	0.050	0.77
III	czjf3-1	86.86	6.21	4.78	5.90	0.040	7.18	56.99	0.100	0.78
IV	qyqx2-3	89.64	5.78	3.82	1.30	0.070	5.31	79.05	0.110	0.55

Due to constant Land D_b,

$$V_h \propto r_k^{3-D_b}$$

Take a differential on both sides of the above formula, then

$$\frac{dVh}{dr} \propto r_k^{2-D_b} \quad (3.3)$$

Mercury intrusion is a technical means to determine the pore size and distribution in porous media. During mercury intrusion experiment, in order to overcome the internal surface tension between mercury and solids, pressure $P(r)$ must be applied before mercury fills in the pore with a radius r. For a cylindrical pore, $P(r)$ and r satisfy the Wash Burn equation, i.e.

$$P(r) = (-4\sigma\cos\theta)/(r \times 10^2) \quad (3.4)$$

where $P(r)$ is applied pressure, MPa; r is the pore radius of coal sample, nm; σ is the surface tension on mercury, σ = 480 mN/cm; θ is the contact angle of mercury with solid surface, θ = 140°.

Finish Eq. (3.4), then

$$P(r) \times r = 1.5 \times 10^7 \text{ or } r = 1.5 \times 10^7/P(r) \quad (3.5)$$

Derive both sides of Eq. (3.5), then

$$dr = -[r/P(r)]dP(r) \quad (3.6)$$

During the experiment, the total pore volume at a given pressure is equal to the volume of mercury injected, i.e.,

$$dVh = dV_{P(r)} \quad (3.7)$$

Take Eqs. (3.3), (3.5), and (3.6) into Eq. (3.7), then

$$dV_{P(r)}/dP(r) \propto r^{4-D_b}$$

Take logarithm on both sides of Eq. (3.7) and take $\text{In}r = -\text{In}P(r)$ into it, then,

$$\ln[dV_{P(r)}/dP(r)] \propto (D_b - 4)\ln P(r) \quad (3.8)$$

Equation (3.8) is the fractal mathematical model of the pore system in coal reservoirs. In practical applications, the slope k is obtained by plotting $\ln[dV_{P(r)}/dP(r)]$ and $\ln P(r)$, then $D_b = 4+k$.

Pore System Models

(1) Model 1

The representative sample of Model 1 is jcsh3-l taken from the No. 3 primary coal seam in the Sihe Coalmine. The vitrinite reflectance (R^o) is 3.43%. Figure 3.10 shows the mercury intrusion and removal curves. Figure 3.11a shows the isothermal adsorption curves. The porosity of the sample is relatively high, reaching 7.60%, and the mercury intrusion saturation is also the highest in all samples, up to 45%. In the sample, mesopores are most developed, then small pores, and the last micropores; macropores are also relatively developed. The fractal dimensions of the pore volume are D_b = 3.1816. The pore volume of the sample is sized and fractal, and the fractal dimensions are 2.9857 and 2.4367 (Fig. 3.12a). The average throat is 0.5 μm, an order of magnitude larger than the pore throats in other types of samples, indicating that the large throats account for more. The mercury intrusion curve has a wide pore hysteresis loop and a large volume difference (pressure difference) between injected and removed mercury amounts. This indicates that there are more open pores and better pore connectivity in the pore interval tested by mercury intrusion. The low-temperature nitrogen adsorption line has a distinct adsorption loop and has an inflection point, indicating that the pore system is open, including cylindrical pores open at both ends and parallel plate pores open on four sides. Such pore morphology is beneficial to desorption and diffusion of CBM. Therefore, the pore structure of such samples is most favorable for desorption, diffusion and permeation of CBM, and the reservoir represented by it is favorable for CBM exploration and development.

(2) Model 2

The representative sample of Model 2 is yqyk15-9 taken from the No. 15 primary coal seam in the Yangquan No. 1 Coalmine. The vitrinite reflectance (R^o) is 2.92%. Figure 3.10 shows the mercury intrusion and removal curves. Figure 3.11b shows the low-temperature nitrogen adsorption isotherms. The porosity of the sample is lower than that of Model 1. The mercury intrusion saturation is 25%, which is nearly 20% lower than that of the Model 1, indicating that the pore development is much poorer than that of the Model 1, and the pores with smaller diameters are more developed. The fractal dimensions are 3.2308, significantly higher than that of the Model 1, because the vitrinite reflectance is lower than that of the Model 1, the fractal dimension of the pore volume decreases as the coal rank increases (Fig. 3.12b). In terms of pore structure, the sample contains mainly small pores about 20% higher than the Model 1, and mesopores about 20% lower than the Model 1. Similar to the Model 1, the mercury intrusion curve of Model 2 has a macropore hysteresis loop and a large volume difference (pressure difference) between mercury intrusion and removal. This indicates that there are many open pores

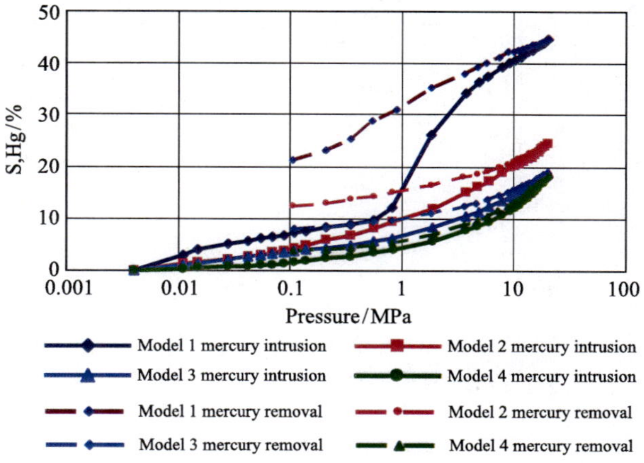

Fig. 3.10 Mercury intrusion and removal curves of typical coal samples taken in the Qinshui Basin

and good pore connectivity. The low-temperature nitrogen adsorption isotherm has an obvious adsorption loop and an inflection point, indicating that the pore system is open. Such pore morphology is conducive to desorption and diffusion of CBM. Comprehensive evaluation of the pore structure of Model 2 samples shows it is conducive to desorption, diffusion and permeation of CBM. The reservoir represented by it is favorable for CBM exploration and development.

(3) Model 3

The representative sample of the Model 3 is czjf3-l taken from the No. 3 primary coal seam in the Jingfang Coalmine. The vitrinite reflectance is 1.79%. Figure 3.10 shows the mercury intrusion and removal curves. Figure 3.11c shows the low-temperature nitrogen adsorption isotherms. The porosity of the sample is lower than that of the Model 1, and is basically equal to the Model 2. The mercury intrusion saturation is 18%, nearly 7% lower than that of the Model 2. Small pores are the most developed, 15% higher than the Model 2; mesopores and macropores are about 8% lower than that of Model 2. The fractal dimensions of the pore volume are 3.2656 (Fig. 3.12c). The mercury intrusion curve has a narrow pore hysteresis loop, and the difference between mercury intrusion and removal (pressure difference) is small, which indicates that there are many closed pores and medium pore connectivity. The low-temperature nitrogen adsorption isotherm has an adsorption loop and has an inflection point, but it is not as obvious as the adsorption loop of the Model 1 and Model 2, indicating that the pore system transitions from open to closed pores. Such pore morphology is not conducive to desorption and diffusion of CBM relative to Model 1 and Model 2. The pore structure of such samples is not conducive to desorption and diffusion of CBM, and the reservoir represented by it is common or poor for CBM exploration and development.

(4) Model 4

The representative sample of the Model 4 is qyqx2-3 taken from the No. 2 primary coal seam in the Qinxin Coalmine in Qinyuan. The vitrinite reflectance is 6.66%. Figure 3.10 shows the mercury intrusion and removal curves. Figure 3.11d shows the low-temperature nitrogen adsorption isotherms. The porosity of this sample is very low, only 1.3%. The mercury intrusion saturation is also very low, only 17%. In the sample, small pores are the most developed, accounting for more than 70%, then mesopores, micropores and macropores are the least developed. The fractal dimensions of the pore volume are 3.437 (Fig. 3.12d). The mercury intrusion curve has a small pore hysteresis loop, and the volume difference (pressure difference) between mercury intrusion and removal is small, which indicates that the semi-closed pores are absolutely advantageous and the pore connectivity is poor. The low-temperature nitrogen adsorption isotherm has almost no adsorption loop, which also indicates that the pore system is mainly composed of gas-tight pores closed at one end. Such pore morphology is extremely detrimental to desorption and diffusion of CBM. The pore structure of such samples is most detrimental to desorption, diffusion and permeation of CBM, and the reservoir represented by it is the poorest reservoir for CBM exploration and development.

3.2.1.2 Contribution of Pore System Structure to Reservoir Physical Properties

The pore system structure of high-rank coal reservoirs has significant effects on two reservoir physical properties: reservoir space and reservoir permeability.

Relationship Between Reservoir Pore System and Physical Properties

From lignite to highly metamorphic coal, the porosity continuously reduces, but the reduced are mainly biological pores and interparticle pores. These pores are mainly macropores and mesopores and have little effect on the adsorption capacity of CBM. The reason for the decrease in porosity is that overburden pressure causes the pores in coal seams to be compressed and liquid hydrocarbons in coalification process to block the pores with physical structures. With the degree of coalification to the stage of highly metamorphic coal, the condensation of aromatic hydrocarbons, the overflow of gas molecules and the appearance of coal molecular crystals, a large number of molecular structural pores with a diameter of less than 10 nm are formed,

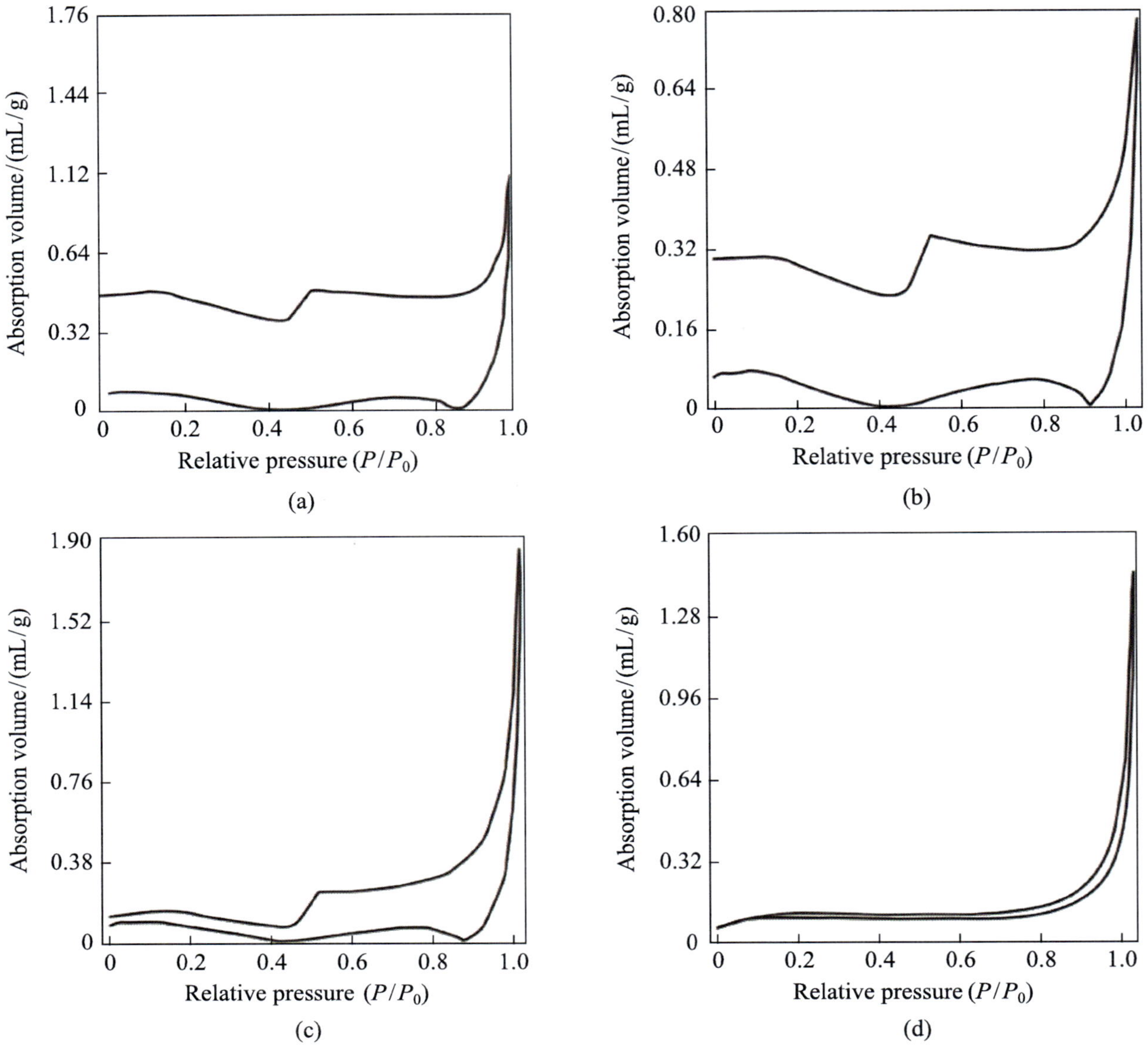

Fig. 3.11 Low-temperature nitrogen adsorption curves of pore Model 1 to Model 4

which are beneficial to the adsorption of methane. Although the total pore volume of highly metamorphic coal reduces, micropores and small pores are very developed, generally above 70%, which make the surface area in high-rank coal is mainly composed of micropores and small pores (Fig. 3.13). Under the same reservoir pressure conditions, its adsorption capacity is significantly enhanced. At the same time, due to coalification to highly metamorphic coal stage, the moisture in coal reduces and the content is very low, so that the displacement of methane molecules by water molecules is greatly reduced. This ensures that the internal surface area increases with the increase of small pores and micropores, and the ability to adsorb methane increases too, improving the reservoir performance.

In the geological history of high-rank coal, the gas production is high. Because the macropores and mesopores are not developed enough, and the endogenous fractures are closed, the permeability is low. In addition, structural changes and the resulting changes in reservoir pressure would not lead to dissipation of CBM to a large amount, which is conducive to the continuous generation and accumulation of CBM.

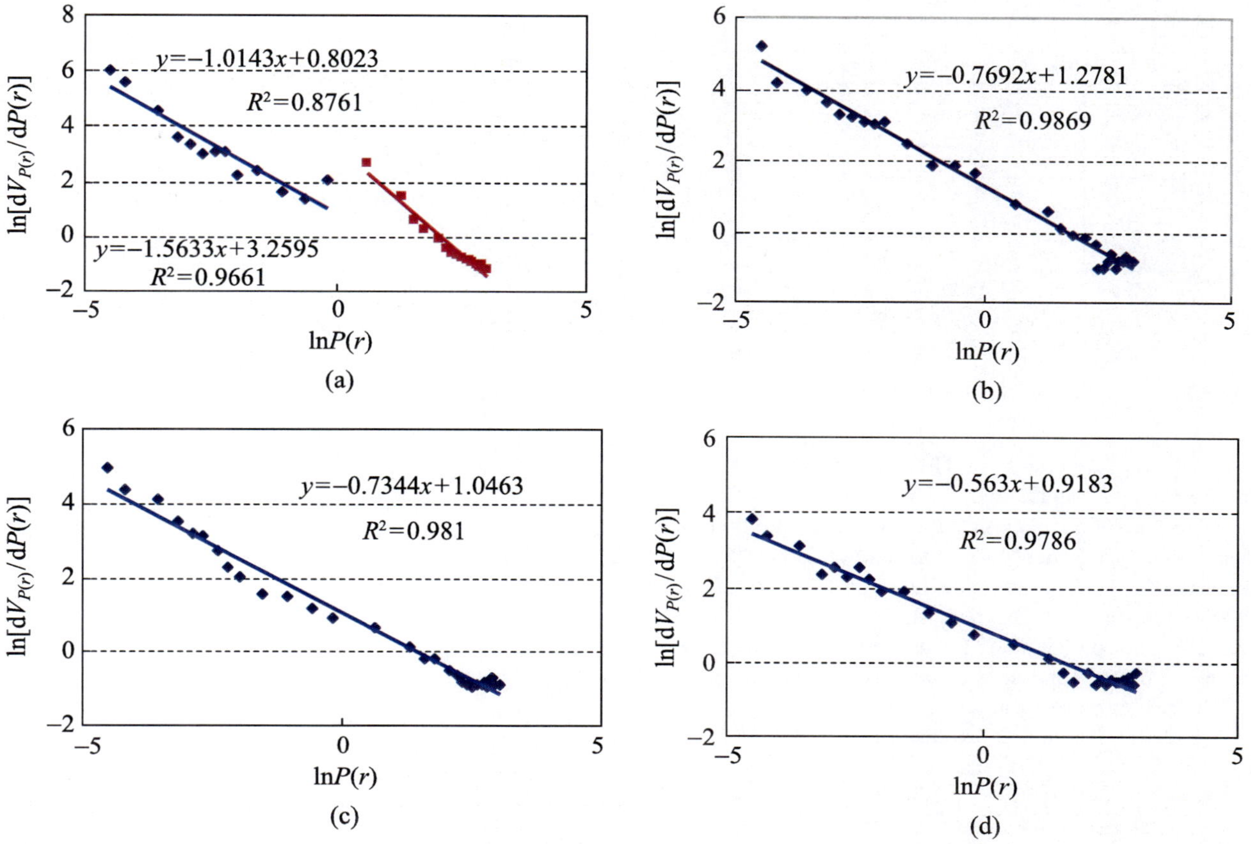

Fig. 3.12 Fractal characteristics of the pore volume of Model 1 to Model 4

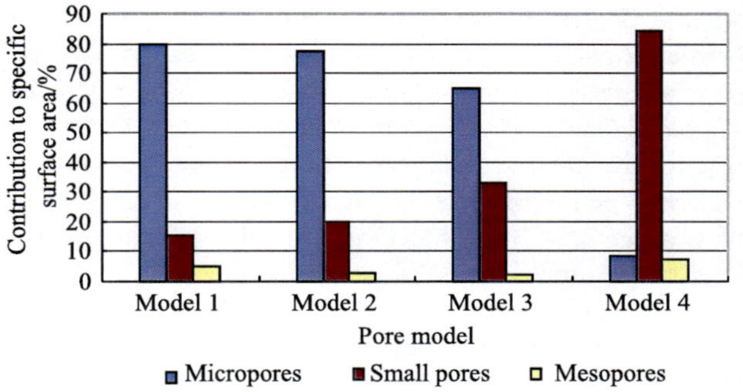

Fig. 3.13 The contribution of micropores and small pores in Model 1 to Model 4 to specific surface area

Influence of Pore Structure on Desorption, Diffusion and Permeation of CBM

The migration of CBM can be simplified into three continuous processes of desorption, diffusion and permeation. Each process and behavior of gas producing from coal seams are closely related to the pore size. The realization of each behavioral process requires the pores with corresponding pore sizes to develop to a certain extent and the pore morphology is favorable for gas migration. This requires a reasonable configuration of the pore structure. Only a well-configured pore structure can produce gas. If the pores with certain pore sizes are not developed, it will prevent infiltrating, called a "bottleneck" phenomenon in high-rank coal reservoirs.

In high-rank coal, generally micropores and small pores are more developed than macropores and mesopores. With complex pore morphology, large internal surface area, and strong adsorption capacity, it is difficult to desorb gas from the inner surface of the pores. However, it cannot be generalized. It is the fact that the difference in the pore structure of high-rank coal is very large (Fig. 3.14).

Pore structure determines whether gas can be desorbed and how much gas can be desorbed from the inner surface of coal and whether gas diffusion can proceed smoothly and thoroughly. The pore model 1 is dominated by mesopores, followed by small pores, and the pore morphology is open. Such pore structure is the most favorable for gas desorption, diffusion and laminar flow permeation. The Model 2, Model 3 and Model 4 have the most small pores, then mesopores

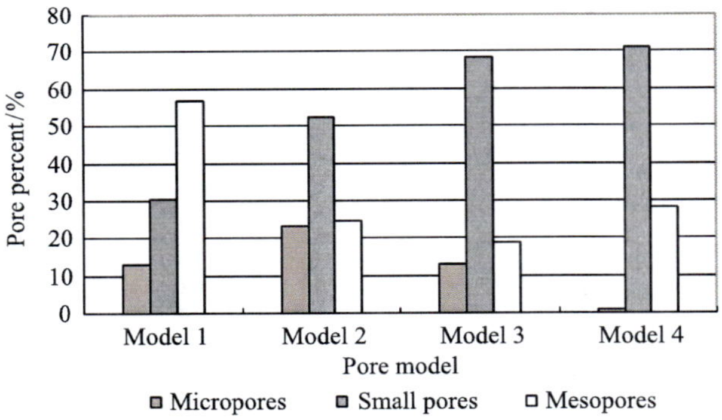

Fig. 3.14 Pore distribution in Model 1 to Model 4

and the least micropores. The percent of small pores increases significantly, and the contribution of the small pores to the surface area increases in order. The pore morphology is open in the Model 2 to semi-closed in the Model 3 and Model 4. Under the same reservoir conditions (similar storage pressure and gas content), from the Model 2 to the Model 4, the conditions of gas desorption and diffusion are better, so do the conditions of laminar flow permeation.

The measured permeability of 18 samples is from 0.021 mD[1] to 5.81 mD, with an average of 0.493 mD, the coefficient of variation is 2.71, indicating that the permeability of the high-rank coal is very variable. The tested permeability of the coal reservoirs collected in the Qinshui Basin is 0.042–45.8 mD. Analysis of the porosity and the permeability shows no significant correlation between them, but some samples also show low porosity and low permeability. Factors affecting permeability are very complex, including porosity, pore structure, ultra-microfractures, micro- fractures and macrofractures, fracture opening and mineral fillings, but their control may be different in different areas.

In the pore larger than 75 nm, gas permeates in several ways: molecular slipping and laminar flow, intense laminar flow, and turbulent flow. Wall (1965) proposed a method for calculating permeability based on capillary pressure curves. This method assumes that pore system is composed of capillary bundles of different sizes. He divides pore volume into several levels with equal volume according to the statistical point of view, and then calculates the permeability of each level, which is called interval permeability, and finally superimposes them to obtain total permeability. For a rock sample, the total permeability is calculated as

$$
\begin{aligned}
k &= \frac{\phi^{4/3}}{8n^2}\left[r_1^2 + 3r_2^2 + 5r_3^2 + \cdots + (2n-1)r_n^2\right] \\
&= \frac{\phi^{4/3}\left[\sum_{i=1}^{n}(2i-1)r_i^2\right]}{8n^2}
\end{aligned}
\tag{3.9}
$$

where k is the permeability of the rock sample, mD; ϕ is the porosity (decimal) of the rock sample; n is the levels of equivalent pore volumes, generally taking $n = 10$; r is the corresponding pore throat radius, cm, $r_1 > r_2 > \cdots > r_n$.

The contribution (%) to interval permeability is

$$
\left[(2i-1)r_n^2 \Big/ \sum_{i=1}^{n}(2i-1)r_i^2\right] \times 100 \tag{3.10}
$$

The contribution of the interval permeability of the representative samples of the pore models was calculated according to the above formula. The results show that the pores in different pore size sections have different contributions to the permeability due to the difference in pore size and content (Table 3.9). The pore with a pore diameter smaller than 1 μm has little contribution to the permeability. When the pore diameter is larger than 1 μm, the contribution to the permeability begins to rise significantly, which is basically consistent with sandstone. The percent of the pore greater than 1 μm is extremely low in high-rank coal, but contributes the most to the permeability, up to 99% (Fig. 3.15).

Therefore, in high-rank coal there is a bottleneck of gas permeation due to the low pore percent of some pore interval. Pore system is only one of the channels through which CBM is produced. Another factor that improves the permeability of coal seams is micro- fractures that connect pore system to macrofractures. When we take downhole samples in the Jingfang coalmine in Changzhi, a large fault was found, and a sample was collected near the fault. The measured permeability is the highest among all tested samples, which is 5.81 mD. Fracture observation also shows that D-type microfractures in the sample are the most developed in all 40 samples, reaching 103 fractures/9 cm^2. This indicates that microfracture development can greatly improve reservoir permeability. The genesis, occurrence and production mechanism and mining technology of the Antrim shale gas in the Michigan Basin are the same as those of CBM. The pore system of the Antrim shale gas is very close to the pore system of highly metamorphic coal reservoirs.

[1]Darcy, 1 D = 0.986923 × 10^{12} m^2, 1 mD = 1 × 10^3 D = 1 × 10^3 μm^2

Table 3.9 The contribution to the interval permeability of representative coal samples of pore models

Interval permeability/nm	Contribution to permeability/%			
	Model 1	Model 2	Model 3	Model 4
	jcsh3-l	yqyk15-9	czjf3-l	qyqx2-3
0–36.4	0.00803	0.02489	0.02846	0.07997
36.4–40.5	0.00002	0.00008	0.00007	0.0003
40.5–45.1	0.00003	0.00007	0.00009	0.00035
45.1–60.0	0.00009	0.00023	0.00022	0.00091
60.0–80.4	0.00016	0.00025	0.0003	0.0012
80.4–152.6	0.00068	0.00141	0.0012	0.00339
152.6–892.6	0.01539	0.01012	0.00788	0.01967
892.6–1339.7	0.04788	0.07412	0.03614	0.08857
1339.7–5368.6	0.11276	0.3473	0.19996	0.52061
5368.6–26267.8	1.81069	4.90951	3.51873	5.70493
26267.8–183874.9	98.00428	94.63201	96.20693	93.58009

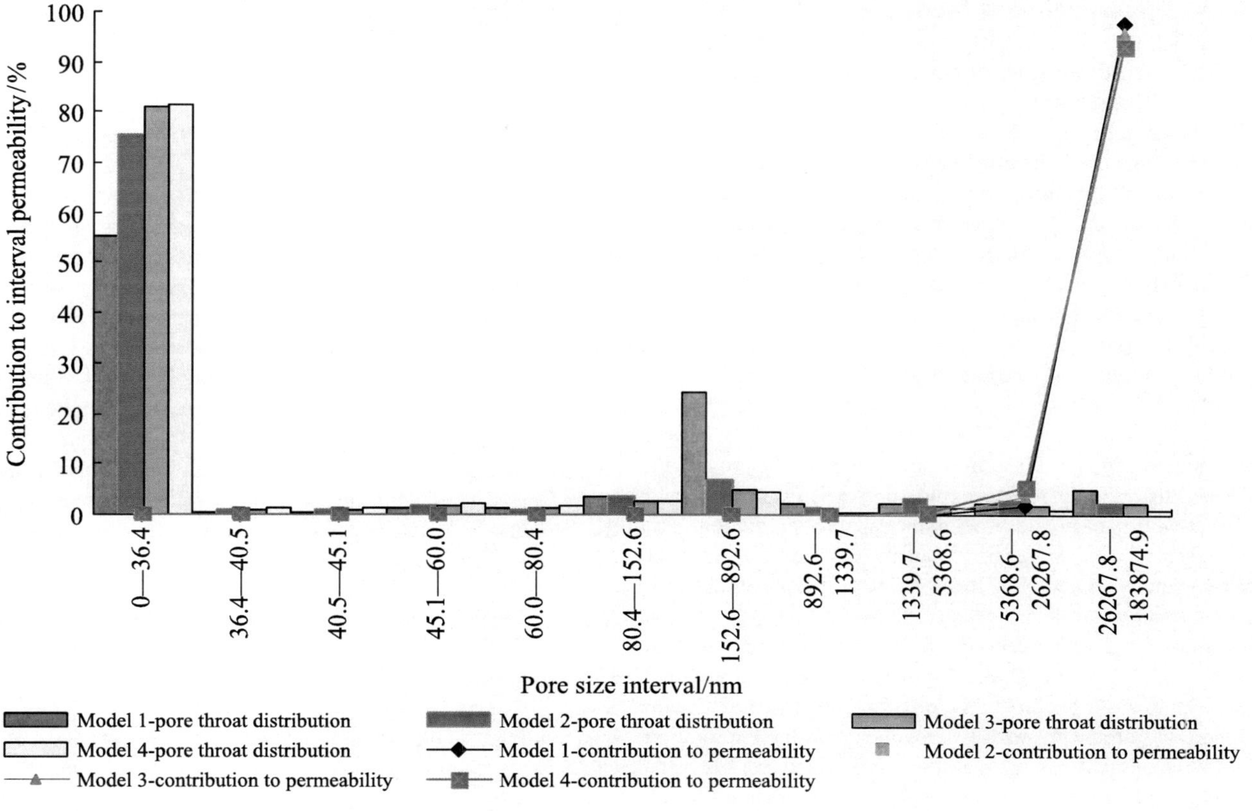

Fig. 3.15 Contribution of pore sizes to total permeability

The shale has been successfully developed because the exogenous microfractures and macrofractures are very developed, with high density and variable dispersion, which greatly improves the reservoir permeability. Australian scholar Paul particularly emphasized the influence of microfractures and mineral fillings on the recoverability of CBM in his proposed CBM production model. He pointed out that microfracture size, density, continuity and connectivity contribute more to total permeability, which is an important and useful inspiration for exploration and development of CBM in the highly metamorphic coal in the Qinshui Basin. The coal reservoirs in the Qinshui Basin have undergone strong structural changes in the later period, forming many secondary structural units, some of which should have the conditions to form a suitable fracture system. This study shows that the microfractures in the Shouyang Coalmine in northern Yangquan are generally more developed than those in the central and southern parts, but the microfracture development is strongly heterogeneous and may be subject to local tectonic activities, coal rock composition and mechanical properties. In addition, the opening and closing of fractures depend on in situ stress conditions. Therefore, in an highly metamorphic coal area, when evaluating a new block, tectonic development, microfracture development and in situ stress conditions will be studied in depth to make an objective evaluation.

3.2.2 Heterogeneous Models of CBM Reservoir

3.2.2.1 Modelling Method and Coalmine Selection

The heterogeneity of coal reservoir physical properties involves coal rock heterogeneity, permeability anisotropy and pore-cleat (fracture) heterogeneity of coal reservoirs, and their relationships. The heterogeneity severely restricts the accumulation, preservation and recoverability of CBM. In the Blackburn and San Juan Basins of the United States, the overall structural conditions are simple, but the heterogeneity of coal reservoir physical properties causes the gas production range of an adjacent gas well (in a 2–3 km^2 gas field) to differ by more than 6 times (Close 1993; Gash et al. 1993; Gayer and Harris 1996). The same problem is also encountered in the Qinshui Basin where the heterogeneity causes distortion of CBM exploration and evaluation, and affects well pattern and production capacity prediction.

Reservoir heterogeneity generally refers to planar heterogeneity, interlayer heterogeneity, intralayer heterogeneity and micro-heterogeneity. Intralayer heterogeneity is an important factor affecting the exploration and development of CBM. The so-called intralayer heterogeneity model of the pore system refers to the possible influences of several important parameters with heterogeneity on the exploration and development of CBM. In this book, coefficient of variation (v_k), coefficient of outburst (T_k) and coefficient of rank (J_k) are used to reflect the heterogeneity of the pore system of coal reservoirs. The coefficient of variation (v_k) is the ratio of the standard deviation of a property of a group of samples to its average value (Eq. 3.11). The coefficient of outburst (T_k) is the ratio of the maximum value of a property of a group of samples to its average value (Eq. 3.12). The coefficient of rank (J_k) is the ratio of the maximum value of a property of a group of samples to its minimum value (Eq. 3.13).

$$v_k = \frac{\sigma}{\bar{x}} \tag{3.11}$$

where σ is the standard deviation of a property of a group of samples; $\bar{x}$ is the average of the property of the group of samples.

$$T_k = \frac{x_{\max}}{\bar{x}} \tag{3.12}$$

where $x_{\max}$ is the maximum of a property of a group of samples; $\bar{x}$ is the average of the property of the group of samples.

$$J_k = \frac{x_{\max}}{x_{\min}} \tag{3.13}$$

where $x_{\max}$ is the maximum of a property of a group of samples; $x_{\min}$ is the minimum of the property of the group of samples.

Studies have shown that coefficient of variation (v_k), coefficient of outburst (T_k) and coefficient of rank (J_k) can well reflect the heterogeneity of reservoir. In general, the larger the coefficient of variation (v_k), the coefficient of outburst (T_k) and the coefficient of rank (J_k), the stronger the heterogeneity. The mercury intrusion parameters such as porosity, pore structure and microfracture can reflect the desorption and permeation of the pore system. In seven mining areas of the Qinshui Basin from the north to the central to the south, detailed coal stratification was carried out. Considering sample test, area and primary coal seams, etc., 4 mining areas were selected, i.e. the Yangquan No. 1 Coalmine in the north of the basin (No. 15 coal seam), the Qinxin Coalmine (No. 2 coal seam) in the northern and central parts, the Changcun Coalmine (No. 3 coal seam) in the central and southern parts, and the southern Sihe Coalmine (No. 3 coal seam).

3.2.2.2 Intralayer Heterogeneous Models

Intralayer Heterogenous Model of No. 15 Coal Seam in Yangquan No. 1 Coalmine

The total thickness of the coal seam in the underground observation section is about 7 m, and 11 coal layers and 5 gangue layers can be divided (Table 3.10). All the gangue layers are mudstone. The macroscope coal types are from semidull coal to bright coal. The coal rock is very heterogeneous—the porosity, mesopores, macropores, displacement pressure, average throat diameter, sorting coefficient, pore morphology, pore connectivity, microfractures (fractures/9 cm^2) and other parameters change very obviously, and most parameters are very different, indicating that the structure of the pore system is very different. In terms of single layer, some layer is better, another is poor in physical properties. As a result, vertical and horizontal permeability channels (relatively) and barriers appear alternately, which brings great difficulties to the production of coalbed methane. However, in terms of the whole coal reservoir with more channels, the permeability of impermeable coal layers or the layers with ultra-low permeability can be improved by reservoir stimulation such as hydraulic fracturing, so that the gas in these layers can be produced with the gas from permeable layers. Therefore, for such coal seam, although the physical heterogeneity is very strong, due to the large thickness and the existence of coal layers with better physical properties, the reservoir properties in terms of gas desorption and permeation can be improved by stimulation, so that the reservoir can be used for CBM development.

Intralayer Heterogeneous Model of No. 2 Coal Seam in the Qinxin Coalmine

The total thickness of the coal seam in the underground observation section is about 2 m, and four coal layers and one gangue layer can be divided (Table 3.11). The macroscope coal types are semibright coal, semidull to semibright coal and semidull coal. The minimum coal layer is about 0.4 m thick. The thickness of the gangue layer is about 0.1 m. The heterogeneous parameters of the sample indicate that the coal seam is not very heterogeneous and the pore system is relatively homogeneous. The disadvantage of the pore system is the low porosity of the layers, only 0.9–2.8%. The pores are all partially closed, and the connectivity is poor, which makes the physical properties of any layer very poor, and barriers in both vertical and horizontal directions. Reservoir stimulation may be ineffective, so the coal seam is difficult to be promising.

Intralayer Heterogeneous Model of No. 3 Coal Seam in Changcun Coalmine

The total thickness of the coal seam in the underground observation section is about 6.2 m, including three coal layers and two gangue layers (Table 3.12). The macroscopic types are all semibright coal. The minimum layer thickness is greater than 0.6 m. The parameters such as porosity, mesopores, macropores, displacement pressure, average throat diameter, sorting coefficient and number of microfractures change very obviously, and most parameters are very different, which indicates that the structure of the pore system is different. In terms of single layer, some layer is a bit better, another is common in physical properties, resulting in some differences in reservoir permeability vertically and laterally. In terms of the whole coal seam, the properties are moderate, and the permeability of the layers can be improved by reservoir stimulation. Therefore, for such coal seam, although the physical property heterogeneity is strong, in view of the large thickness through the reservoir stimulation to improve the reservoir performance, it can still be used as a favorable reservoir for CBM development.

Intralayer Heterogeneous Model of No. 3 Coal Seam in Sihe Coalmine

The total thickness of the coal seam in the underground observation section is about 6 m, including four coal layers and one gangue layer (Table 3.13). The macroscopic coal types are bright coal, semibright coal and semidull coal. The minimum layer thickness is greater than 0.6 m. The changes of porosity, mesopores, macropores, displacement pressure, average throat diameter and sorting coefficient are obvious, and most parameters are very different, which indicates that the structure of the pore system is very different. In terms of single layer, some layer is good, another is moderate or poor in physical properties. In terms of the whole coal seam, the physical properties are moderate, and the permeability of the layer with poor physical properties may be improved by reservoir stimulation. Therefore, for such coal seam, although the physical property heterogeneity is strong, due to the large thickness through the reservoir stimulation to improve the reservoir performance, it can still be used as a favorable reservoir for CBM development.

3.3 Forming Mechanism and Controlling Factors of Favorable Coal Reservoirs

3.3.1 Coal Diagenesis

3.3.1.1 Control of Sea Level on Heterogeneity

The physical properties of coal reservoirs are characterized by planar heterogeneity, interlayer heterogeneity and intralayer heterogeneity. Planar heterogeneity refers to the geometry and lateral continuity of a single coal reservoir, and the planar variation and directionality of permeability and porosity within the coal reservoir. Interlayer heterogeneity refers to the difference between vertical coal reservoirs, involving sequency cycle, permeability and porosity, and the distribution of isolators in the coal reservoir. Intralayer heterogeneity refers to vertical permeability difference and the distribution of gangues in coal reservoirs. Due to different sedimentary environments during transgression and regression, coal seams differ in terms of morphology, plane distribution, coal rock composition, ash and distribution of gangue interlayers, and these factors control the permeability of the coal seams.

Table 3.10 Intralayer heterogeneous model of No. 15 coal seam in Yangquan No. 1 coalmine

	Mercury intrusion parameters	Heterogeneous parameters									
		Coefficient of variation	Coefficient of outburst	Coefficient of rank	Sample no.	Porosity shape	Connectivity	Micro fracture	Desorption	Permeation	Evaluation result
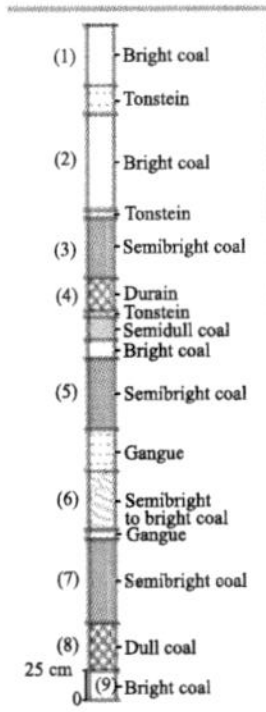	Porosity/%	0.36	1.52	4.83	(1)	Partial closed pore	Poor	55	Unfavorable	Poor	Poor layer
	Micro-small pores/%	0.08	1.10	1.27	(2)	Very open pore	Good	62	Very favorable	Very good	Good layer
	Mesopores/%	0.42	1.64	4.63	(3)	Partial closed pore	Poor	39	Unfavorable	Poor	Poor layer
	Macropores/%	0.59	2.02	8.47	(4)	Partial closed pore	Poor	10	Unfavorable	Poor	Poor layer
	Average throat diameter/μm	0.62	1.91	10.00	(5)	Partial closed pore	Poor	20	Unfavorable	Poor	Poor layer
	Displacement pressure/MPa	0.43	1.60	2.82	(6)	–	–	–	–	–	–
	Mercury removal efficiency/%	0.21	1.36	1.67	(7)	Open pore	Good	51	Favorable	Good	Common layer
	Sorting coefficient/μm	0.94	2. 60	9.00	(8)	Very open pore	Good		Very favorable	Very good	Good layer
	Homogenous coefficient	0.19	1.21	1.71	(9)	Very open pore	Good	58	Very favorable	Very good	Good layer

Table 3.11 Heterogeneous model of No.2 coal seam in Qinxin Coalmine

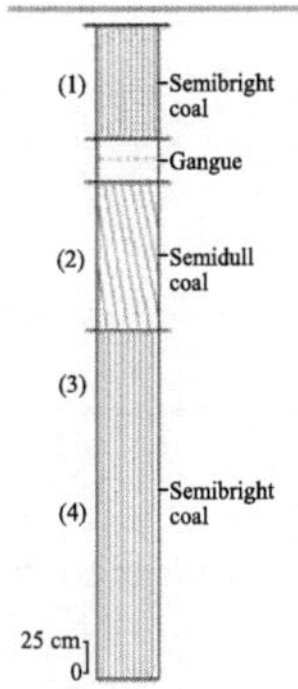

Mercury intrusion parameters	Heterogeneous parameters									
	Coefficient of variation	Coefficient of outburst	Coefficient of rank	Sample no.	Porosity shape	Connectivity	Micro fracture	Desorption	Permeation	Evaluation result
Porosity/%	0.53	1.78	3.11	(1)	Partial closed pore	Poor	19	Unfavorable	Unfavorable	Poor layer
Micro-small pores/%	0.02	1.02	1.05							
Mesopores/%	0.16	1.17	1.39							
Macropores/%	0.17	1.21	1.52	(2)	Partial closed pore	Poor	30	Unfavorable	Unfavorable	Poor layer
Average throat diameter/μm	0.01	1.03	1.14							
Displacement pressure/MPa	0.12	1.16	1.26	(3)	Partial closed pore	Poor	26	Unfavorable	Unfavorable	Poor layer
Mercury removal efficiency/%	0.08	1.10	1.19							
Sorting coefficient/μm	0.57	1.67	3.67	(4)	Partial closed pore	Poor	44	Unfavorable	Unfavorable	Poor layer
Homogenous coefficient	0.10	1.13	1.28							

Table 3.12 Heterogeneous model of No. 3 coal seam in Changcun Coalmine

	Mercury intrusion parameters	Heterogeneous parameters									
		Coefficient of variation	Coefficient of outburst	Coefficient of rank	Sample no.	Porosity shape	Connectivity	Micro-fracture	Desorption	Permeation	Evaluation result
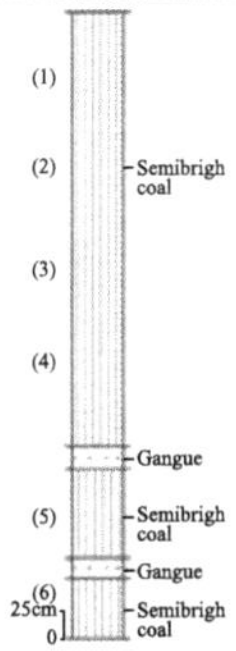	Porosity/%	0.25	1.33	2.04	(1)	Open pore	Better	47	Very favorable	Very favorable	Good layer
	Micro-small pores/%	0.04	1.04	1.11	(2)	Partial closed pore	Good	3	Unfavorable	Favorable	Common layer
	Mesopores/%	0.41	1.66	3.10	(3)	Partial closed pore	Good	13	Unfavorable	Favorable	Common layer
	Macropores/%	0.30	1.41	2.09	(4)	Partial closed pore	Good	29	Unfavorable	Unfavorable	Poor layer
	Average throat diameter/μm	0.40	1.49	3.50							
	Displacement pressure/MPa	0.39	1.46	2.94	(5)	Partial closed pore	Good	7	Unfavorable	Favorable	Common layer
	Mercury removal efficiency/%	0.15	1.21	1.40							
	Sorting coefficient/μm	0.43	1.50	3.20	(6)	Partial closed pore	Good	5	Unfavorable	Unfavorable	Poor layer
	Homogenous coefficient	0.11	1.10	1.39							

Table 3.13 Heterogeneous model of No.3 coal seam in Sihe Coalmine

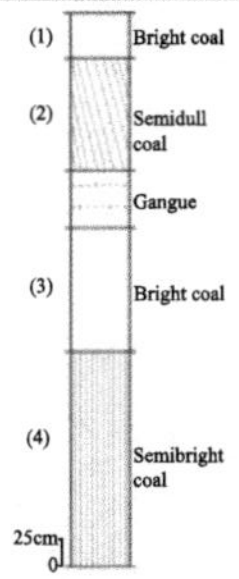

	Mercury intrusion parameters	Heterogeneous parameters									
		Coefficient of variation	Coefficient of outburst	Coefficient of rank	Sample no.	Porosity shape	Connectivity	Micro-fracture	Desorption	Permeation	Evaluation result
	Porosity/%	0.98	2.45	6.91	(1)	Very open pore	Good	24	Very favorable	Very favorable	Good layer
	Micro-small pores/%	0.17	1.10	1.47							
	Mesopores/%	0.85	2.33	4.30							
	Macropores/%	0.64	1.94	3.51	(2)	Partial closed pore	Common	10	Unfavorable	Unfavorable	Poor layer
	Average throat diameter/μm	1.40	3.11	11.20							
	Displacement pressure/MPa	0.55	1.35	7.58	(3)	Partial closed pore	Common	21	Unfavorable	Unfavorable	Poor layer
	Mercury removal efficiency/%	0.17	1.23	1.47							
	Sorting coefficient/μm	1.50	3.24	13.70	(4)	Partial closed pore	Common	24	Unfavorable	Unfavorable	Common layer
	Homogenous coefficient	0.22	1.33	1.60							

Control of Sea Level on Planar Heterogeneity of Coal Reservoirs

The control of sea level on the planar heterogeneity of coal reservoirs is mainly reflected in the control of the morphology, lateral continuity and connectivity.

The coal seam forming during the rapid rise of sea level is thin, the distribution range is small, and the lateral continuity and connectivity are poor. It is often "lenticular" and the planar heterogeneity is strong. The No. 7 and No. 11 coal seams distributed in the transgressive system are thick more than ten centimeters to several tens of centimeters, and the lateral distribution is unstable. In addition, due to the diversification of the coal-forming environment, the ash content and coal rock composition vary greatly.

During the slowly rising period of sea level, restricted bays, distributary bays, and distributary channels and lakes were filled with silt, and swamps were widely distributed, resulting in coal seams with large thickness, wide distribution, lateral stability and weak heterogeneity, such as No. 15, No. 2 and No. 3 coal seams in the late stage of the high-stand system tract. In addition, due to the simple coal-forming environment, the ash content and coal-rock composition changed little.

Control of Sea Level on Intralayer Heterogeneity

The control of sea level on intralayer heterogeneity is mainly reflected in the number of gangue layers and the ash content of coal reservoirs.

The coal seam forming in the transgressive system is thin, and it is dominated by delta front peat swamp, which was less affected by flooding. Therefore, the gangue interlayers are less, usually no or only one layer, such as No. 7, No. 10, No. 11, No. 12 coal seams (Table 3.14), and the vertical heterogeneity is relatively weak. The coal seam forming in the high-stand system tract is thick and mostly in the delta plain environment, which was greatly affected by flooding. Therefore, the gangue interlayers are more, usually above one layer, such as No. 2, No. 3, No. 8 (15) and No. 9 coal seams, and the vertical heterogeneity is relatively strong. Gangue interlayers, as barriers, directly affect the vertical permeability of coal seams. The more the gangue interlayers there are, the stronger the vertical heterogeneity is.

For a coal seam in the high-stand system tract, usually the ash content is high in the upper and lower parts and low in the middle part of the coal seam (Fig. 3.16). For example, the ash content is high and the resistivity is low in the lower and upper parts of the No. 9 coal seam, and the lower part of the No. 8 coal seam; the ash content is low and the resistivity is high in the middle part of the No. 9 coal seam, and the upper part of the No. 8 coal seam.

For the coal seam in the transgressive system tract, the ash content usually decreases gradually from the bottom to the top of the coal seam. This is mainly because the overlying formation is mostly limestone deposited in a clear water environment. In the late stage of the swamp development, less terrigenous materials were injected. For the coal seam in the high-stand system tract, the porosity of the middle part is higher than that of the top and bottom; while for the coal seam in the transgressive system tract, the permeability of the middle and upper parts is higher than that of the bottom.

Control of Sea Level on Interlayer Heterogeneity

The control of sea level on the interlayer heterogeneity of coal reservoirs is mainly reflected in the control of the ash and microscopic components. The coal seam forming in the transgressive system tract dominated by the rising sea level is few and thin (mostly more than ten centimeters to several tens of centimeters), mainly deposited in delta front and tidal flat environments, and less influenced by flooding and crevasse channels, so the ash content is low, with an average of 10–20%, which is conducive to the development of pores and fractures. In addition, the water covering the swamp is deep, the reduction is strong, so the gelled component is rich and more conducive to forming cleats. The coal seam forming in the high-stand system tract dominated by the declining sea level is more and thick (mostly several tens of centimeters to several meters), mainly deposited in the delta plain environment, and greatly affected by flooding and crevasse channels, so the ash content is high, with an average of 25–28%, and pores are filled to a large extent, which is not conducive to the preservation of pores and the formation of fractures. In addition, the water covering the swamp is shallow, the oxidization is relatively strong, so the gelled component is less and not conducive to the formation of cleats.

3.3.1.2 Relationship Between Ash and Porosity and Permeability

Ash directly affects the properties of coal reservoirs. According to the quantitative relationship between ash yield and porosity, the lower the ash yield, the higher the porosity of the coal reservoir (Fig. 3.17), and the better the reservoir properties. According to the relationship between the ash yield of the sample and the porosity, there is a linearly negative correlation between the two, i.e.

$$Y = -0.6538X + 14.764 \quad (3.14)$$

where Y is ash yield, %; X is porosity, %; the correlation coefficient is 0.5448.

Based on the relationship between sample ash and permeability, the correlation between the two is poor (Fig. 3.18). This is mainly because the permeation depends on the fracture development.

Table 3.14 Features of coal seams in Xishan Coalmine, Taiyuan Region

Coal seam No.	Stability	Thickness/m	Average gelation/%	Average ash/%	Average sulfur/%	Gangue	System tract
01	The most instable	<0.5	50	27	<1	0	High stand
02	Instable	0.5–1	–	30	<0.5	0	High stand
03	Instable	0.6–0.8	70	28	0.65	0	High stand
1	The most instable	0.4–0.6	–	30	2.19	0	High stand
2	Stable	2–5	73	21	0.83	>2	High stand
3	More stable	1–3	70.2	28	<0.5	1–2	High stand
4	Instable	0.5–1.2	50	28	<1	1–2	High stand
5	Instable	0.5–1	–	33	0.53	1–2	High stand
6	More stable	1–2	86	27	2.44	>2	High stand
7	More stable	0.7–1	79	17	2.63	0	Transgressive
8 (15)	Stable	2–5	79	20	2.63	>2	High stand
9	Stable	2–5	76	23	1.34	>2	High stand
10	The most instable	0.3–0.5	–	–	1.62	0	Transgressive
11	The most instable	0.2–0.5	85	–	5.3	0	Transgressive
12	The most instable	–	–	–	–	0	Transgressive

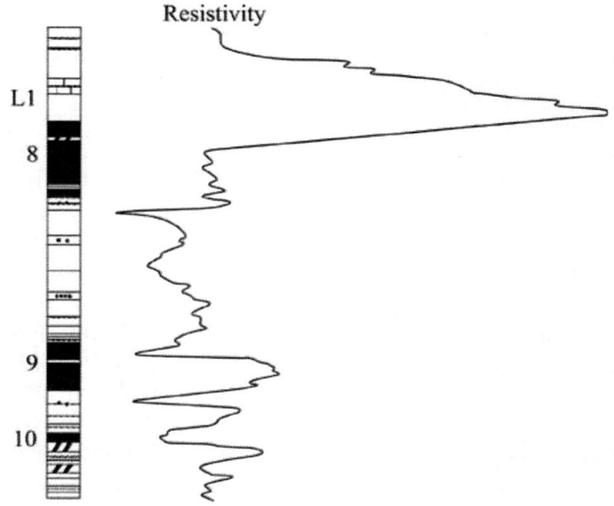

Fig. 3.16 Resistivity curve of 667 pores in Taiyuan region

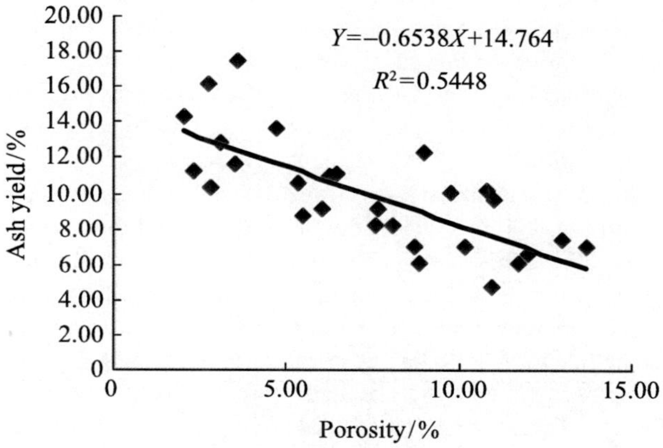

Fig. 3.17 Relationship between coal ash and porosity in Qinshui Basin

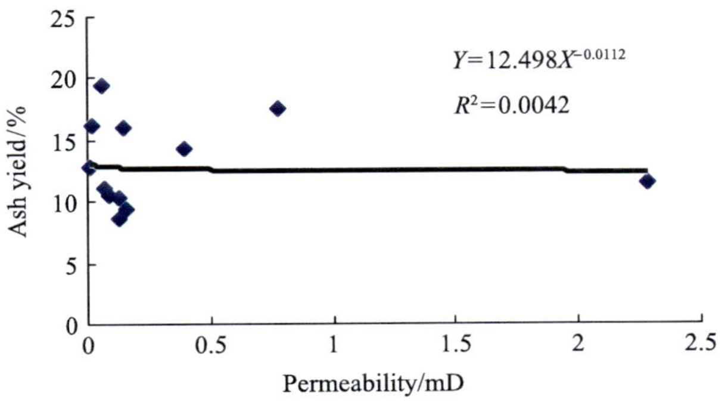

Fig. 3.18 Relationship between Carboniferous and Permian coal ash and permeability in Qinshui Basin

3.3.1.3 Control of Coal Facies

In order to clarify the main controlling factors causing the differential development of the pore system of the high-rank coal reservoirs in the Qinshui Basin, vitrinite reflectance (R^o, %), C (flammable composition) (%), ash (%), mineral (%), GI, TPI, V/I and WI were used to analyze 37 samples.

Factor analysis can be divided into R, Q and corresponding factor analysis according to research objects. R-factor analysis studies the relationship between variables. Q-factor analysis studies the relationship between samples based on variables. Both of them use a few common factors to extract most information of research objects. Since they identify the primary contradiction of controlling original observation data, it is easier to reveal the genetic or spatial connection of research objects and facilitates direct geological interpretation and logical inference by studying the characteristics of common factors. However, R-factor analysis and Q-factor analysis isolate variables from samples, so many useful geological information would miss. On this fact, a corresponding analysis is generated. It combines R-factor analysis with Q-factor analysis to analyze variables and samples uniformly, so it is more conducive to geological interpretation.

R-factor analysis was performed on 8 indicators of 37 samples. Table 3.15 shows the first, second and third factor loads. The cumulative percentage of variance of the first three factors is 79.48%, which indicates that the first three factors can reflect most of the original data. The variance of the first factor f_1 is 40.41%; the variance of the second factor f_2 is 24.57%; and the variance of the third factor f_3 is 14.77%. This indicates that the geological factor is dominant in the region, followed by f_2 and f_3. The primary factors controlled by f_1 are TPI and WI (Fig. 3.19), both of which reflect the sedimentary environment of coal seams, i.e., coal facies. The primary indicators controlled by f_2 are ash content and mineral content. The most important one is the ash content, which can reflect the redox (pH) of the swamp. The primary variables controlled by f_3 are GI and V/I, which also reflect the sedimentary environment. In addition, according to the results of the corresponding analysis, the vitrinite reflectance and the carbon content reflecting the degree of coal metamorphism contributed little to the three main factors and did not constitute an independent main factor. It can be inferred that for the high-rank coal reservoirs in the Qinshui Basin, the high rank is the primary cause for the overall unfavorable condition of the pore system, and the sedimentary environment or coal facies is likely to cause the structural difference of the pore system. The sedimentary environment or coal facies may indirectly control the physical properties by directly controlling the differential development of the pore system of the high-rank coal reservoirs. It is not difficult to understand that a specific sedimentary environment or coal facies provides a specific material basis, and for the sediments in different environments (geological backgrounds), even after the same post-metamorphism and reservoir reformation, the final reservoir physical properties may be very different. Therefore, the sedimentary environment or coal facies plays a dominant role in the differential development of the pore system of the high-rank coal reservoirs in the Qinshui Basin.

From the microscopic composition, microscopic type and facies characteristics of the representative samples for pore Model 1, Model 2, Model 3 and Model 4 (Tables 3.16, 3.17, 3.18), the Model 1 has very high GI and V/I, but very low TPI and WI, and the vitrinite reflectance of matrix is very high, which reflects that the sedimentary environment is lower delta and distributary bay, and the coal facies is typical livimg water peat swamp with strong water flow activity, strong microbial activity and strong cover water. The Model 2 has high TPI and WI, and the micro-coal type is mainly vitrite, reflecting that the forming environment is upper delta, and the coal facies is forest peat swamp which is extremely humid with deep water. The difference between the Model 3 and the Model 4 is small, and the relevant indicators reflect that they were formed in braided river or dry deltas, and the coal facies is dry peat swamp. As far as the vitrinite reflectance is concerned, the metamorphic degree of the Model 1 and Model 2 is much higher than that of the Model 3 and Model 4. If coal rank is a primary controlling factor, according to general understanding, as coal rank increases, reservoir properties such as porosity and permeability will become worse, but the pore structures in the Model 1 and the Model 2 are obviously superior to those in the Model 3 and the Model 4, indicating more favorable reservoirs. The coal metamorphism is not the dominant controlling factor on the differential development of the pore system in the high-rank coal reservoirs in the Qinshui Basin.

After we analyze the geological environment of the representative samples of the pore models, corresponding analysis was conducted on eight variables of all 37 samples. Table 3.15 shows the first three R-factor loads. Table 3.19

Table 3.15 Three R-factor loads (f_1, f_2 and f_3)

No.	Variable	f_1	f_2	f_3
1	R^o/%	0.0016	0.0025	−0.0015
2	C/%	−0.0318	−0.0675	−0.0265
3	Ash/%	−0.0853	0.1235	−0.0022
4	Mineral/%	−0.0338	0.0738	0.0217
5	GI	0.0635	−0.0019	0.0949
6	TPI	0.1303	0.0479	−0.0474
7	V/I	0.0744	0.0074	0.0628
8	WI	0.1342	0.0314	−0.0358
Cumulative/%		40.142	64.714	79.481

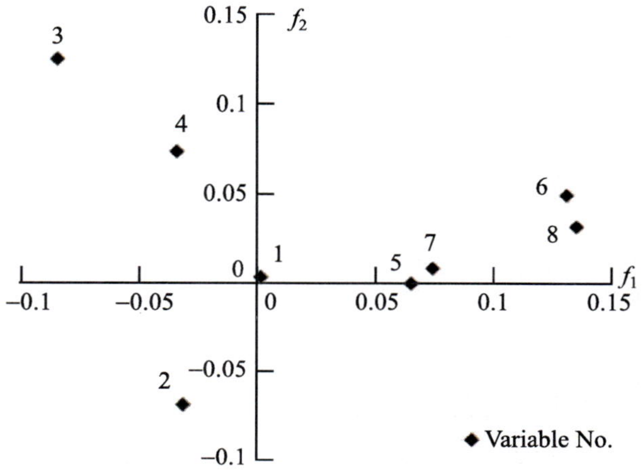

Fig. 3.19 f_1 and f_2 load plan of 8 variables

shows the first three Q-factor loads. Figure 3.20 shows the primary factor distribution.

f_1 and f_2 are factors reflecting the sedimentary environment, and the corresponding analysis makes the true geological significance more obvious. The different values of the indicators in different regions indicate the characteristics of different peat swamp sedimentary facies. The sample distribution in different regions indicates their respective causes and differences in causes. Further analysis found that the samples in Zone I were all caused by forest peat swamp, and some samples (No. 32, No. 34, No. 37) were the products of live peat swamp. The samples in Zone II were almost all caused by dry peat swamp, and the samples in Zone II are generally high in ash content and mineral content, while the ash content and mineral content of Zone II_2 are much lower than that of Zone II_1. Therefore, along the axis of f_2, the ash content and the mineral content increase from the bottom up, which reflects the oxidation was enhanced and the reducibility was reduced in the sedimentary environment. Now it is clear that the corresponding analysis of the eight variables of the 37 samples is actually good classification of the causes of the samples.

In the cluster analysis, classes I and II samples with relatively reasonable pore structures and relatively good physical properties are in good correspondence with the samples from active peat swamp and forest peat swamp in Zone I shown in Fig. 3.20. Class III samples with the same moderate physical properties and class IV samples with the worst physical properties correspond to the samples belonging to dry peat swamp. The good sample correspondence characterized by two classes (pore structure and genetic environment) further proves that the coal facies is the primary controlling factor on the differential development of the pore system and physical properties of the high-rank coal reservoirs in the Qinshui Basin.

The samples of live peat swamp facies in lower delta and delta bay and those of forest peat swamp facies in upper delta generally have relatively favorable pore systems, so the

Table 3.16 Quantitative statistical results of microscopic components of representative samples for porosity models

Sample no.	Macro coal rock	Vitrinite						Total	Inertinite						Total	Exinite				
		T	C1	C2	C3	C4	VD		SF	F	Mi	Ma	Scl	ID		Sp	Cu	Re	FI	ED
jcsh3-1	Bright coal	0.6	3.8	68.7	13.2	0	2.9	89.2	0	1.5	0	2.2	0	6.5	10.2	0	0	0	0	0
yqyk15-9	Bright coal	0	65	11.4	1.6	0	4.8	82.8	0.8	8.1	0	0	0	5.6	14.5	0	0	0	0	0
czjf3-1	Semidull coal	0.8	19	28.9	2.5	0	5.8	57.0	5.0	16.5	8.2	1.6	0	6.6	37.9	0	0	0	0	2
qyqx2-3	Semibright coal	1.4	7.5	41.5	6.1	1.4	2.9	60.8	15.0	4.8	2.0	7.4	0	7.6	36.8	0	0	0	0	2

Table 3.17 Quantitative statistical results of microscopic coal types of representative samples for porosity models

Sample no.	Vitrite	Microite	Exinite	Clarite		Vitrinertinite		Durite		Trima-cerite	Mineralised coal
				Vitrite-rich	Exinite-rich	Vitriner-tinite	Vitriner-toliptite	Microite-rich	Exinite-rich		
jcsh3-l	23.0	1.3	0	0	0	73.0	1.9	0	0	0	0.8
yqyk15-9	80.3	3.4	0	0	0	10.3	4.3	0	0	0	1.7
czjf3-l	16.9	8.4	0	0	0	54.2	17.8	0	0	0.6	2.1
qyqx2-3	5.4	7.7	0	0	0	32.3	50.7	0	0	3.9	0

Table 3.18 Tested and coal facies parameters of representative samples for porosity models

Sample No.	Porosity/%	Microfracture/(fractures/9 cm^2)	R^o/%	Moisture/%	Ash/%	C/%	H/%	Mineral/%	Coal facies parameters			
									GI	TPI	V/I	WI
jcsh3-1	7.6	24	3.43	1.31	13.22	91.985	3.136	0.6	11.43	0.08	8.75	0.06
yqyk15-9	5.8	58	2.92	1.11	14.80	90.427	3.365	2.7	5.71	4.35	5.71	4.01
czjf3-1	5.9	127	1.79	0.34	11.93	88.932	4.183	3.4	2.09	1.11	1.50	0.57
qyqx2-3	1.3	26	1.66	0.30	4.34	90.772	4.415	0.3	2.49	0.51	1.65	0.20

Table 3.19 The first three Q-factor loads of 37 samples

Sample no.	f_1	f_2	f_3	Sample no.	f_1	f_2	f_3
1	−0.0030	−0.0244	−0.0173	20	−0.0153	−0.0421	−0.0145
2	0.0935	0.0344	0.0007	21	0.0199	−0.0325	0.0258
3	0.0457	0.0075	−0.0210	22	−0.0183	−0.0173	−0.0094
4	−0.0291	−0.0141	−0.0192	23	−0.0177	−0.0209	−0.0091
5	−0.0019	−0.0230	−0.0173	24	−0.0300	−0.0145	−0.0158
6	0.0046	0.0027	−0.0362	25	−0.0332	0.0322	0.0258
7	0.0301	0.0268	−0.0077	26	−0.0130	−0.0254	0.0027
8	−0.0302	0.0087	−0.0096	27	−0.0119	−0.0386	0.0056
9	0.0205	−0.0004	0.0550	28	−0.0265	0.0033	−0.0151
10	0.0052	−0.0212	−0.0214	29	−0.0234	−0.0108	0.0004
11	0.1081	−0.0006	−0.0324	30	−0.0225	0.0270	−0.0022
12	0.0163	−0.0136	0.0122	31	−0.0218	−0.0099	0.0060
13	−0.0135	−0.0129	0.0152	32	0.0004	−0.0082	0.0540
14	−0.0013	−0.0183	−0.0072	33	−0.0544	0.0538	−0.0253
15	−0.0594	0.0809	−0.0256	34	−0.0117	−0.0009	0.0390
16	−0.0368	0.0500	0.0323	35	−0.0064	−0.0345	0.0022
17	0.0686	0.0337	0.0017	36	0.0903	0.0170	0.0055
18	−0.0283	0.0092	0.0059	37	−0.0040	−0.0015	0.0203
19	−0.0185	−0.0382	−0.0176				

above environments are favorable for forming coal reservoirs. The samples with poor pore structures are generally the product of dry peat swamp which is generally difficult to form favorable reservoirs. In addition, the interactive environment or transitional environment in which the two coal facies interact may lay the foundation for forming coal reservoirs with medium physical properties.

The change of coal facies (peat swamp) is much more complicated than that of conventional oil and gas reservoirs. In the Qinshui Basin, there are significant differences in the coal facies in different sections (vertically) and different regions (laterally), and the changes are regular. The vertical and horizontal changes of the coal facies lead to changes in the composition of the coal rock, which in turn affects the pore structure, gas volume and coalbed methane adsorption to some extent. In addition, the coalbed sedimentary environment determines the geometry of the coal seam, in turn the geometry macroscopically controls the lateral and vertical changes in the physical properties of the coal seam. The coal seams forming in offshore peat swamps are generally stable and distributed widely. The coal seams forming in continental peat swamps are narrowly distributed laterally and the phase changes largely. The coal seams forming in delta plain peat swamps are between the two. The quality of coal seams is also strongly influenced by the sedimentary environment. The presence of gangues in the coal seam increases the heterogeneity. In the Qinshui Basin, favorable sedimentary facies can not be always deposited permanently to form thick and continuous coalbed methane reservoirs; generally, the favorable coal layer in the coalmines does not exceed 3 m; and the frequent change of the sedimentary environment (coal facies) resulted in strong heterogeneity of the pore system and the reservoir properties in vertical and lateral directions.

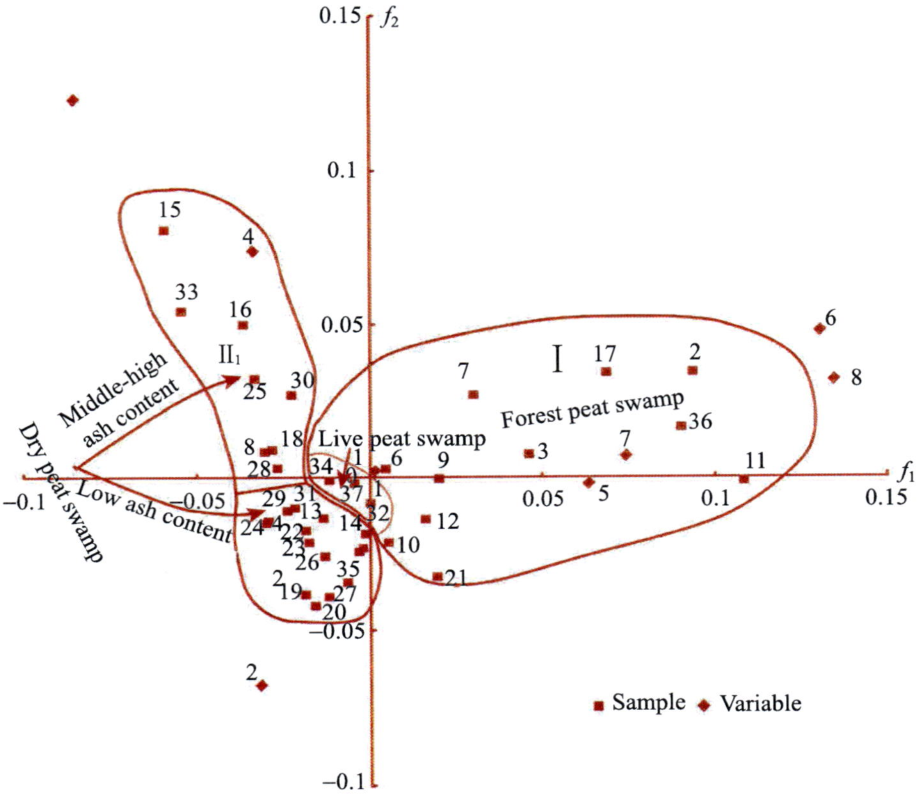

Fig. 3.20 Correspondence analysis of coal samples, metamorphic parameters and sedimentary environment parameters

3.3.1.4 Diagenetic Transformation

Cleats control the permeability of coal reservoirs. Once the cleats are filled with minerals, the permeability of the coal reservoir will drop sharply.

There are three groups of cleat systems in the primary coal seams of the Qinshui Basin. The first group is at 1°–80°, generally 35°–40°; the second group is at 275°–357°, most at 280°–295°; and the third group is 330°–360°, most at 340°–350°. The first group is the most developed, followed by the second group, and the third group is the least. In the Jincheng mining area, Yangquan mining area and Tunliu mining area, the cleats are dense and the most developed, while they are less dense in other coalmines. The fillings in the cleats are mainly calcite and locally quartz. The third group is filled to the largest extent (Table 3.20).

Cleat statistics show that, in the lower coal seams of the Qinshui Basin, the density of endogenous cleats is higher in deep coal seams, but relatively lower in shallow coal seams, and the cleat density is the highest in the Jincheng and Yangquan mining areas, the Lu'an and Huoxi mining areas. The fillings can be divided into two categories: one is clay minerals dominated by kaolinite and illite; the other is carbonate minerals dominated by carbonate cements. Authigenic pyrite is rare.

Authigenic illite is mica-type mineral forming during diagenesis. It not only records the temperature, pressure and medium conditions experienced by the sediment, but also records the age at which the sediment experienced the most ancient temperature. The illite K/Ar isotope dating results and coal seam burial and thermal evolution history analysis show that the earliest illite was developed at about 240 Ma, i.e. the Triassic at 1000–1500 m, and the vitrinite reflectance is 0.5–1.0%. At that time the coal seam was rapidly buried, and cleat systems with significant amount and scale appeared, and were filled with authigenic illite. The second large-scale illite formation occurred at about 150 Ma, and the forming temperature reached 150–230 °C, which was the most active period of the Yanshanian magma intrusion. The third illite formation was at 65–100 Ma when the paleo-thermal field recovered to normal, the structural uplift caused the coal seam to become shallower, the temperature decreased sharply, coalbed methane and formation water volume contracted, and some cleats were developed and filled with illite (Table 3.21).

Clay minerals in the form of epigenetic crystalline are more common in metamorphic anthracite. The anthracite in the Jincheng mining area contains flaky illite, finely fibrous palygorskite-sepiolite mineral, pompon-like chlorite, and

Table 3.20 Basic characteristics of tectonic fractures and their fillings in Qinshui Basin

Location	Coal seam	Occurrence	Fillings	Density/ (fractures/m)
Gaoyang in Finxi	No. 3	35°∠83°, 37°∠82°	Closely filled but not obvious	59
Liangdu in Jiexiu	No. 2	5°∠82°, 65°∠76°, 75°∠78°	Closely filled but not obvious	60
Bailong in Huozhou	No. 2	l°∠82°, 80°∠81°, 75°∠80°	Calcite	77
Liuchagou in Lingshi	No. 2	24°∠83°	Calcite	57
Yicheng	No. 2	331°∠42°, 304°∠55°	Calcite	17
Ning'aogou in Qinshui	No. 2	280°∠83°, 285°∠77°	Closed cleats not filled obviously	53
Xigou in Yangcheng	No. 3	325°∠83°, 328°∠89°, 330°∠70°	Closed cleats filled with calcite	28
Qinyuan	No. 3	285°∠83°, 307°∠71°	Calcite	73
Lu'an	No. 3	276°∠88°, 304°∠67°, 324°∠80°	Calcite	72
Chengzhuang in Jincheng	No. 3	66°∠82°, 33°∠86°	Calcite	35
Fanzhuang in Jincheng	No. 3	318°∠80°, 304°∠83°	Calcite	55
Xinzhi in Huozhou	No. 9	275°∠87°, 285°∠87°, 22°∠85°	Calcite	124
Beizhuang in Qinyuan	No. 10	287°∠86°, 22°∠79°, 357°∠62°	Calcite	93
Xiaohepu in Lu'an	No. 15	20°∠51°, 315°∠73°, 345°∠76°	Calcite	48

Table 3.21 K/Ar dating results of illite in cleat

Mining area	Sample	Rank	Depth/m	Illite polytype	K/Ar dating/Ma
Jincheng	Jin-1	WY-III	550.0	1 M	245 ± 3
Jincheng	Jin-2	WY-III	560.0	1 M	211 ± 3
Jincheng	Jin-3	WY-III	450.5	1 M	210 ± 3
Jincheng	Jin-4	WY-III	476.8	1 M	78 ± 3
Yangquan	Y-l	WY-III	459.8	1 M	238 ± 3
Yangquan	Y-2	WY-III	446.7	1 M	190 ± 3
Huoxi	H-l	JM	348.2	1 M	205 ± 3
Yangcheng	Yang-1	PM	457.0	1 M	239 ± 3
Yangcheng	Yang-2	PM	463.6	1 M	201 ± 3
Lu'an	L-l	SM	570.3	1 M	237 ± 3
Lu'an	L-2	SM	570.3	1 M	60 ± 3
Shouyang	S-l	PM	648.1	1 M	242 ± 3
Shouyang	S-2	PM	576.0	1 M	197 ± 3
Shouyang	S-3	PM	543.0	1 M	100 ± 3

bamboo-leaf-like chlorite, which are products of hydrothermal activities at some temperature. The minerals in the lowly metamorphic coal series in the Huoxi Coalfield are dominated by kaolinite with primary sedimentary morphology, but the illite crystallinity is generally low. In the argillaceous rocks in the moderately to highly metamorphic coal series, the minerals are dominated by illite which is flaky with sharp edge and straight side, and the crystallinity is high. Some low- temperature hydrothermally altered minerals such as palygorskite and chlorite was found in the moderately to highly metamorphic coal.

Yanshanian hydrothermal veins, such as calcite and quartz, are widely developed in the coal-bearing strata of the Qinshui Basin. Among them, calcite veins are the most common. For comparison with illite, samples from the coal-bearing strata were taken and investigated to inclusions. The results show that the minimum temperature of the inclusions in the Jincheng-Yangcheng-Qinshui area is greater than 220 °C; the uniform temperature of the inclusions in the Yangquan mining area on the northern margin is greater than 220 °C; the uniform temperature of the inclusions in the central part of the Qinshui Basin is lower than that in the north and south sides; and the measured temperature of the inclusions corresponds well with the illite forming temperature. Under the microscope, the inclusions are usually 10–40 μm, the gas-liquid ratio is mostly 5–20%, and the shape is relatively regular, indicating original inclusions (Table 3.22). In general, the inclusions in magmatically hydrothermal deposits are large (20–90 μm), with low gas-liquid ratio and regular shape; while the inclusions in non-magmatically hydrothermal deposits are small (less than 3 μm), with low gas-liquid ratio (3–8%) and irregular shape. The shape and size indicate that the inclusions in the Qinshui Basin are products of non-magmatically and magmatically hydrothermal activities. The observation under the microscope heating/colding stage shows that the double-phase inclusions are of NaCl–H_2O type.

The chemical compositions of early and late illite are similar, reflecting that the illitization event took place in a wide range, the formation of illite was closely related to the rapid burial of coal seams, magmatism and buried depth, as well as to groundwater activities.

3.3.2 Changes in Porosity and Permeability During Coalification

3.3.2.1 Relationship Between Metamorphism and Porosity

The statistics (Fig. 3.21) of the mercury intrusion results of 103 samples with different coal ranks and vitrinite reflectance of 0.5–4.0% indicates that the mercury intrusion porosity changes with coal rank are almost consistent with the results of previous studies, that is, a trend of high–low–high. In general, the compaction and shrinkage of less metamorphic coal is not strong enough, so the structure is loose, the moisture content is high, and the porosity is large. With further compaction and shrinkage, the degree of metamorphism increases, the moisture gradually decreases and the porosity decreases. When the vitrinite reflectance reaches 2.5%, the porosity begins to increase due to the increase of secondary fractures. Figure 3.22 is the previous result of the relationship between total pore volume and carbon content in coal. It shows that lignite has the most developed pore structure; as the degree of coalification increases, the porosity gradually becomes lower; and in the metamorphic coal stage ($C_{\text{daf}} > 86\%$), the porosity begins to increase again, mainly due to the increase of the rank of the aromatic coal layers.

The above-mentioned law of coal seam porosity has a clear corresponding relationship with macromolecular structure changing with coal rank. According to the theory of macromolecular structure, the process of coalification is actually a process of thermal decomposition and thermal polycondensation of organic components in coal. A series of chemical changes shown as low-molecular compounds and aromatization and polycondensation of aromatic systems are accompanied by a series of physical changes such as changes in coal pores and fracture systems.

The aromaticity of coal can be quantitatively measured by ^{13}C NMR (Yang 1996). Studies have found that the apparent aromatic carbon ratio in bituminous coal increases rapidly with the increase of coal rank; in anthracite, the increasing rate of the aromatic carbon ratio decreases significantly. The intersection of bituminous coal and anthracite coal ($R^{\circ} = 2.5\%$) is the transitional point at which condensing reaction replaces aromatization reaction and becomes the dominant geochemical reaction during coalification (Fig. 3.23).

3.3.2.2 Relationship Between Coal Metamorphism and Pore Size Distribution

Pore size distribution has a close relationship with the degree of coalification. According to the results of Chen (2001), the distribution of pores in different ranks of lignite is relatively uniform; in long-flame coal, micropores increase significantly, while macropores and mesopores decrease significantly; to the bituminous coal stage of moderate coalification, macropores and micropores are dominant, while mesopores are less; in highly metamorphic coal, such as lean coal and anthracite, micropores account for the majority, while mesopores and macropores with pore diameters larger than 100 nm only account for about 10%.

Statistics show that the average pore volume of coal seams in China accounts for 40.97%, the average of transition pores is 26.61%, and that of mesopores and macropores is 12.84% and 19.855%, respectively. The average median radius of the pores is concentrated between 4.0 and 8.0 nm, and the proportion of micropores is the highest.

The average pore throat diameter obtained by the mercury intrusion method can represent the overall condition of the

Table 3.22 Mineralogy of the inclusions in the Paleozoic coal series in Qinshui Basin

Location	Vein	Inclusion/μm	Decrepitation temperature/°C	Uniform temperature/°C	Shape	Gas-liquid ratio	Phase		Others
							One-phase	Double-phase	
Jin Test 1	Calcite	10–20	299	230	Negative crystal	10–20	–	NaCl–H_2O	Abundant inclusions
Yangquan	Calcite	3–15	289	227	Negative crystal	10–20	–	NaCl–H_2O	Less inclusions
Lu'an	Quartz	3–10	278	201	Negative crystal	10–15	–	NaCl–H_2O	Small and less inclusions
Liangdu, Jiexiu	Calcite	20–35	301	172	Regular	5–25	–	NaCl–H_2O	Mostly double-phase inclusions
Yangcheng	Quartz	<10	294	238	Regular	20–35	–	NaCl–H_2O	Abundant inclusions
Yicheng	Calcite	5–15	313	243	Negative crystal	<20	–	NaCl–H_2O	Abundant inclusions
Qinshui	Calcite	4–15	–	251	Negative crystal	5–15	CO_2–CH_4	NaCl–H_2O	Abundant inclusions
Qinyuan	Quartz	5–30	287	228	Negative crystal	8–20	CO_2–CH_4	NaCl–H_2O	Abundant inclusions

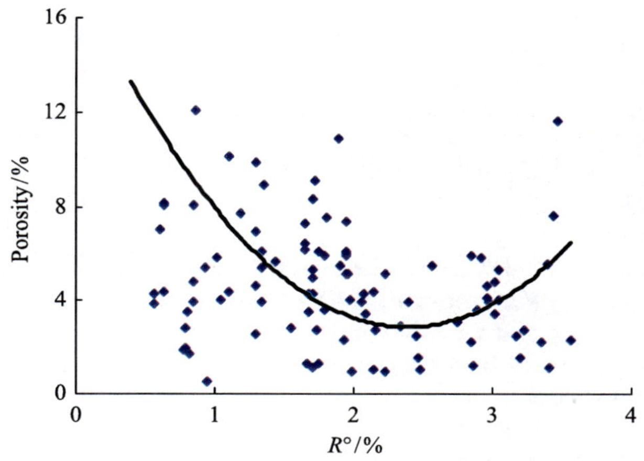

Fig. 3.21 Mercury intrusion porosity versus coal rank

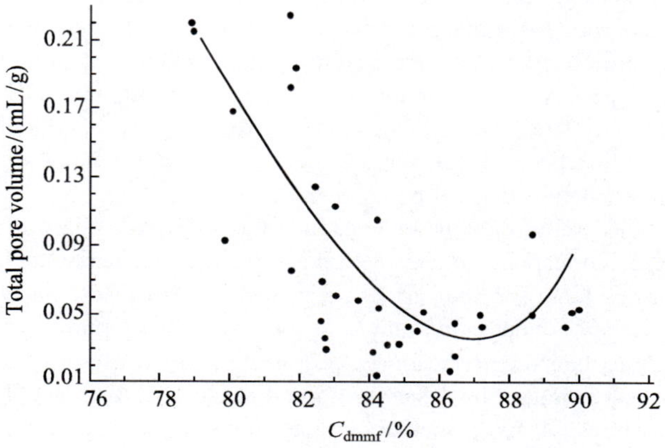

Fig. 3.22 Total pore volume versus carbon content (Chen 2001)

pore structure of samples. Figure 3.24 shows the crossplot of the average pore diameter with the coal rank of 103 coal samples in China. It can be seen that the average pore diameter has a specific correspondence with the coal rank. The pore structure with an average pore diameter of less than 1 μm is widely distributed in the samples of various coal ranks. However, the pore structure with an average pore diameter of more than 1 μm is only distributed in the coal samples at medium to low rank, and rarely seen in the anthracite samples at high rank (R° > 2.5%).

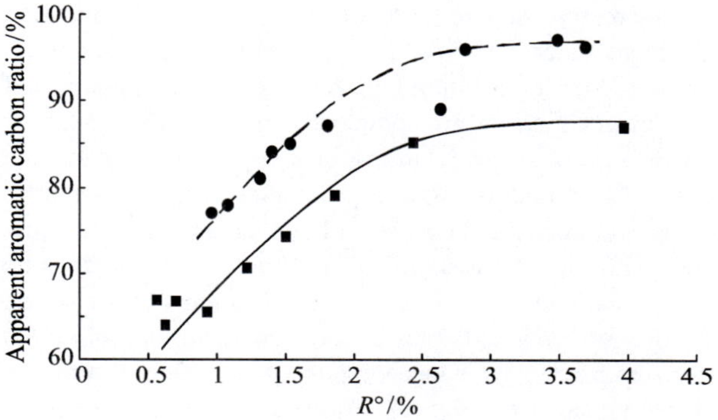

Fig. 3.23 Apparent aromatic carbon ratio versus coal rank (adapted from Yang 1996)

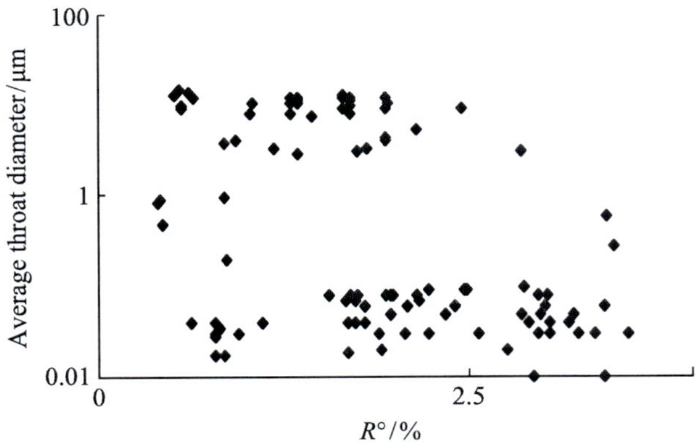

Fig. 3.24 Crossplot of average throat diameter with coal rank

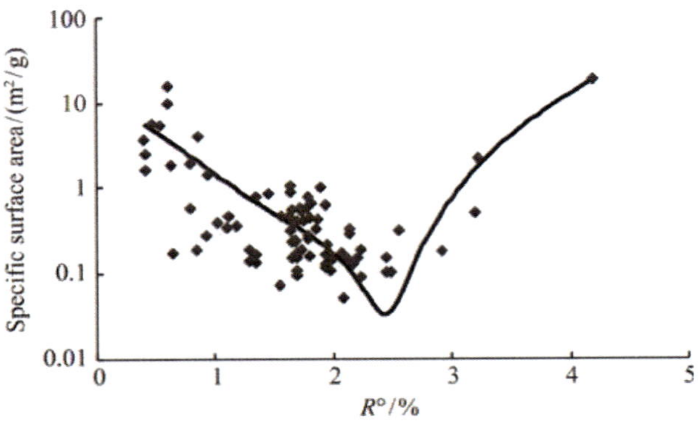

Fig. 3.25 Relationship between specific surface area and coal rank

3.3.2.3 Relationship Between Coal Metamorphism and Specific Surface Area of Pores

The specific surface area of pores is an important indicator for characterizing the pore structure of coal, which can be determined by low-temperature liquid nitrogen adsorption. The adsorption isotherms were obtained by measuring the adsorption at different relative pressures, and then the specific surface was calculated using the Langmuir equation. The pore size distribution was calculated using the BJH method. The results of low-temperature nitrogen adsorption mainly reflect the specific surface area and pore volume of micropores and small pores in coal.

Generally, micropores constitute the adsorption space of coal, correspond to the microporosity in coal matrix, and have a large specific surface area; small pores constitute the capillary condensation and diffusion space; mesopores constitute the slow permeation space; macropores constitute the strong laminar flow space composed of cleats and structural fractures. A large specific surface area indicates a strong ability to adsorb coalbed methane, and the main contributor to the specific surface area is micropores. It is generally believed that adsorption capacity to gas increases as coal rank increases. According to this rule, specific surface area should also increase with the increase of coal grade, but the results of low-temperature nitrogen absorption on some coal samples in China are not completely the same (Fig. 3.25). It can be seen that the relationship between specific surface area and coal rank of some coal samples in China is similar to that of porosity and coal rank. In the coal at middle or low rank, with the increase of coal metamorphism, the specific surface area gradually decreases; to the anthracite stage, the specific surface area begins to increase again; the minimum specific surface area is located at the intersection of bitumite and anthracite ($R^\circ = 2.5\%$). The CO_2 isotherm adsorption experiments conducted by Bustin and Clarkson (1998) showed that the micropore volume and surface area of the coal samples decreased first and then increased with the increase of the coal rank, and the minimum value appeared in the bitumite stage.

For example, the porosity of the anthracite in the Jincheng mining area is 4.2%, and the specific surface area of the pores is as high as 18.47 m^2/g, of which the specific surface area of the micropores is dominant, up to 94.5% (Fig. 3.26a); the porosity of the lean coal in the Lu'an mining area is 1.71%, and the specific surface area of the pores is only as high as 0.228 m^2/g, of which small, micropores and mesopores is 52.2%, 40.3%, 7.5%, respectively (Fig. 3.26b). Both the amount and the distribution of the total surface area of the pores are very different, which reflects that the absorption capacity of the anthracite in the Jincheng mining area is much larger than that of the lean coal in the Lu'an mining area.

3.3.2.4 Relationship Between Coal Metamorphism and Cleat Development and Coal Seam Permeability

Cleats are results of the joint action of tension, fluid pressure and tectonic stress in coal. Metamorphism is an important factor on promoting the development of endogenous fractures in coal. Therefore, different metamorphic stages provide different characteristics of cleat development. Judging from the cleat development in the coals at different coal ranks in different mining areas, the general cleat density of the coal at medium rank is higher, and the cleat density of the coal at low or high rank is lower. In medium-rank coal, the cleats in fat coal and coking coal are most developed, which is related to the large amount of fluid generated and released concentratedly during the second coalification. After fat coal, fluid production decreased rapidly, and fluid pressure dropped drastically, which may cause the environmental pressure to be greater than the fluid pressure, and close original cleats. Downhole observation also found that the cleats were filled or partially filled with cements in anthracite. The cements are almost calcite film, followed by

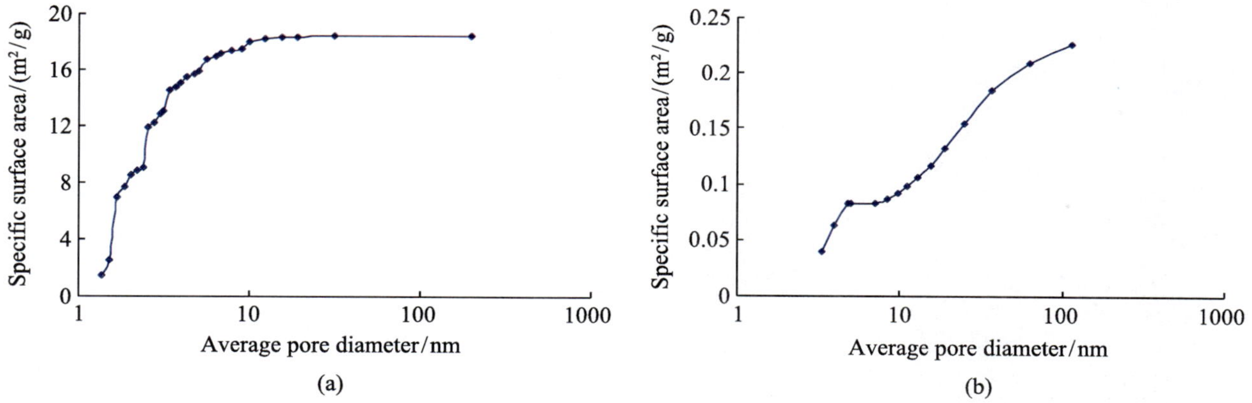

Fig. 3.26 Distribution of specific surface area of Jincheng anthracite and Lu'an lean coal, **a** Jincheng anthracite; **b** Lu'an lean coal

clay or pyrite. The presence of cements indicates that the cleats were once open and fluid migration occurred. The fluid was volatile carried by underground hydrothermal fluid. It penetrated into coal seams through fractures and then precipitated as calcite vein or membrane after cooling. Some researchers believe that the closure of cleats was caused by the filling and scuffing of secondary components in the high-rank coal stage.

The understanding of Ammosove and Eremin (1963) on the relationship between cleat density and coal rank is generally accepted. That is, cleat density gradually increases from low-rank coal, and to the maximum in medium-rank coal (i.e. coking coal); after that, cleat density decreases gradually as coal rank increases. The overall distribution of cleats is normal. Levine (1996) concluded that cleat density increases with increasing coal rank, and after reaching the maximum, the density remains unchanged with the increase of coal rank, that is, from lignite to intermediate-volatile bitumite, the cleat spacing reduces gradually, but keeps substantially unchanged when the reflectivity reaches 1.5% or more.

In addition, cleat density is affected by the composition of coal. Generally, cleats are only developed in bright coal layers and rarely extend to dull coal layers. The higher the vitrinite content, the higher the cleat density. Minerals in coal seams hinder the development of cleats. Downhole observations often finds cleats ending at the boundary with higher ash content. Cleat density is also controlled by the thickness of bright coal layers. For example, when the thickness of a bright coal layer is less than 1.5 cm, the cleat density is 38 cleats/5 cm; and when the thickness is more than 3 cm, the density is only 19 cleats/5 cm.

3.3.3 Structural Stress and Strain Response

3.3.3.1 Mechanical Properties of Medium- and High-Rank Metamorphic Coal

The increase of temperature leads to the decrease of the strength of coal, while the increase of confining pressure leads to the increase of the strength. The influences of temperature and confining pressure are very different on different ranks of coal.

Experimental studies conducted under conditions of simultaneous increase in temperature and confining pressure have shown that samples deformed at higher temperature and higher pressure tend to have lower strength. The strength of coal rock samples under low temperature and pressure conditions (200 °C and 200 MPa) is not affected by changes in the composition and vitrinite reflectance of the sample, which is basically 380 to 400 MPa. However, as temperature and pressure increase, the strength evolution gradually shows different. For rock samples with different compositions and vitrinite reflectance, the magnitude, rate and regularity of their strength changes are different. The samples with high vitrinite reflectance (R° of 3.0% ±), or those with low vitrinite reflectance (R° = 1.7% to 1.8%) change very different. In the experimental range of full strain <10%, the samples with low vitrinite reflectance generally show the regularity of the strength gradually increases with the uniform gradient, i.e. the increase of strain. For the samples with high vitrinite reflectance, a significant strain hardening tendency appears after reaching a certain strain (Fig. 3.27).

On the one hand, the macroscopic mechanical behavior of coal rock deformation is that under the condition of simultaneous increase of temperature and pressure, the effect of

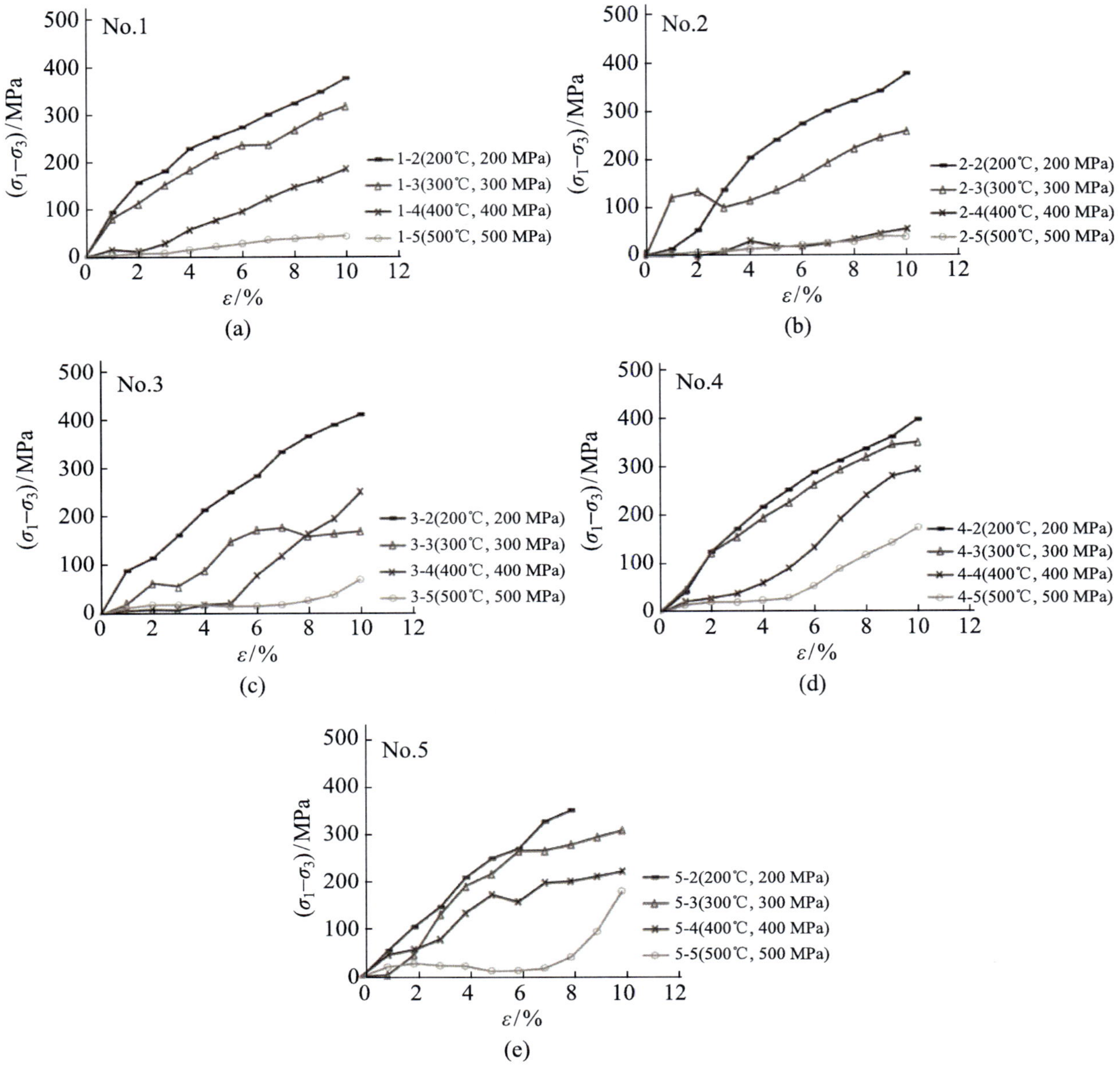

Fig. 3.27 Macroscopic mechanical performance of coal samples during experiment. **a** Sample No. 1 (Shidian, Yangquan); **b** sample No. 2 (Wangzhuang, Changzhi); **c** Sample No. 3 (Xiaoxian, Gaoping); **d** Sample No. 4 (Guohe, Xihe); **e** Sample No. 5 (Qinxin, Qinyuan)

temperature is far greater than the effect of pressure, that is, as the temperature and pressure increase, the strength of the sample gradually decreases. The full strain experienced by the sample also directly affects the rheology of the coal rock. When the strain is low, the deformation of all samples has the same trend; but when the strain is more than 5%, the strength of the coal sample is significantly different, especially at high temperatures (400 or 500 °C). At 400 °C and 400 MPa, a strain hardening phenomenon occurs when the strain reaches 5%, and the sample strength increases rapidly; but at 500 °C and 500 MPa, when the strain reaches over 8%, a weak strain hardening phenomenon occurs.

The macroscopic mechanics of coal rock deformation has the same performance as its microscopic and micromechanical characteristics, but there are also some differences. The original rock of all deformed samples contains only a few microfractures (Fig. 3.28a, b), and the experimental deformation shows brittle deformation at low temperature and pressure (200–300 °C), and ductile deformation at relative high temperature and pressure (>300 °C, Fig. 3.29),

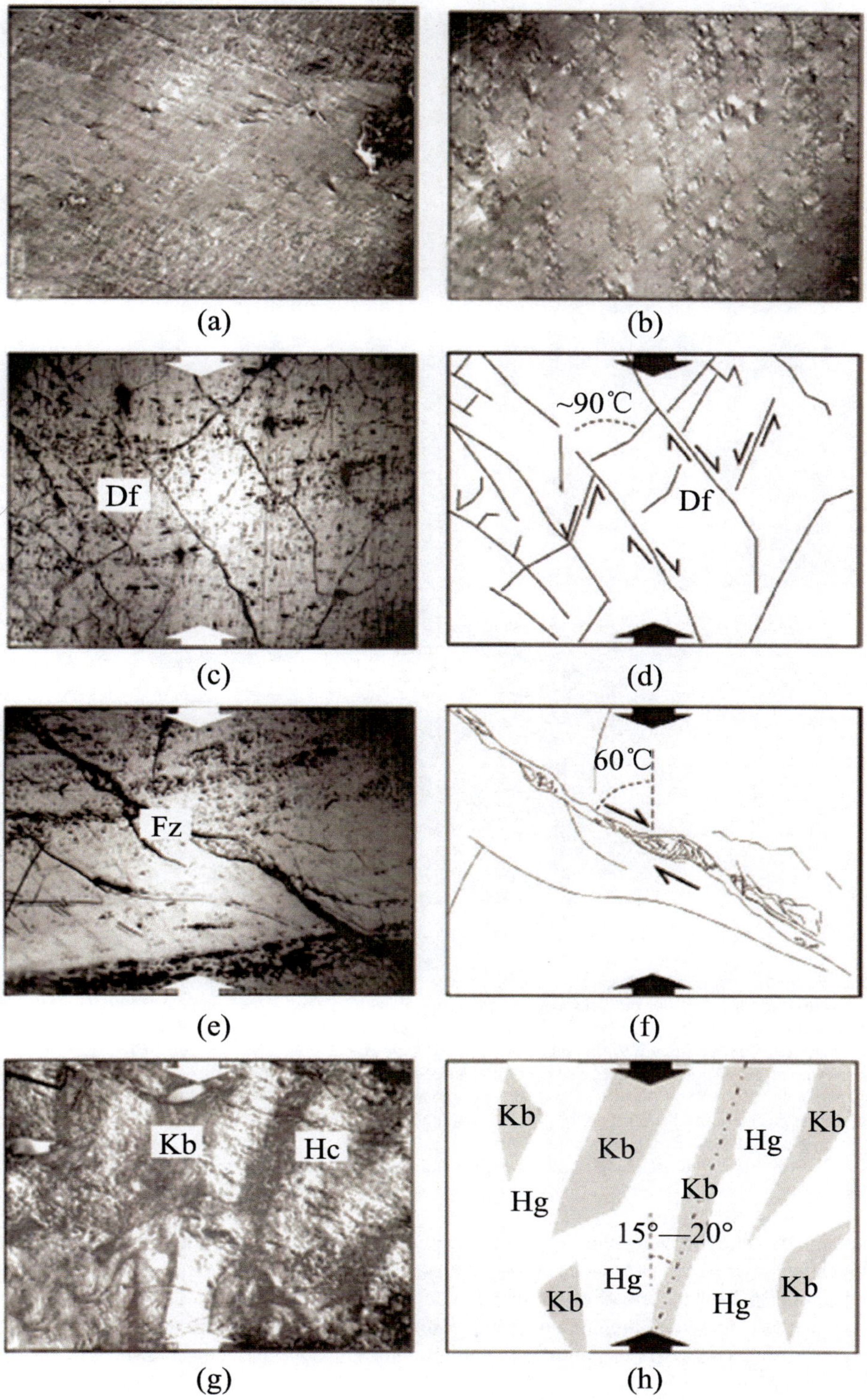

Fig. 3.28 Brittleness and brittle-ductile transition of undeformed and low-temperature deformed coal rocks, and their microstructures and mechanical analysis. **a** Undeformed coal (taken in Wangzhuang, Changzhi); **b** undeformed coal (taken in Guohe, Xihe); **c** and **d** samples at 200 °C and 200 MPa (taken in Xiaoxian, Gaoping); **e** and **f** samples at 300 °C and 300 MPa (taken in Xiaoxian, Gaoping); **g** and **h** samples at 300 °C and 300 MPa (taken in Shidian, Yangquan); the σ_1 of any sample is shown on a upright photo which bottom side is 1.5 mm long

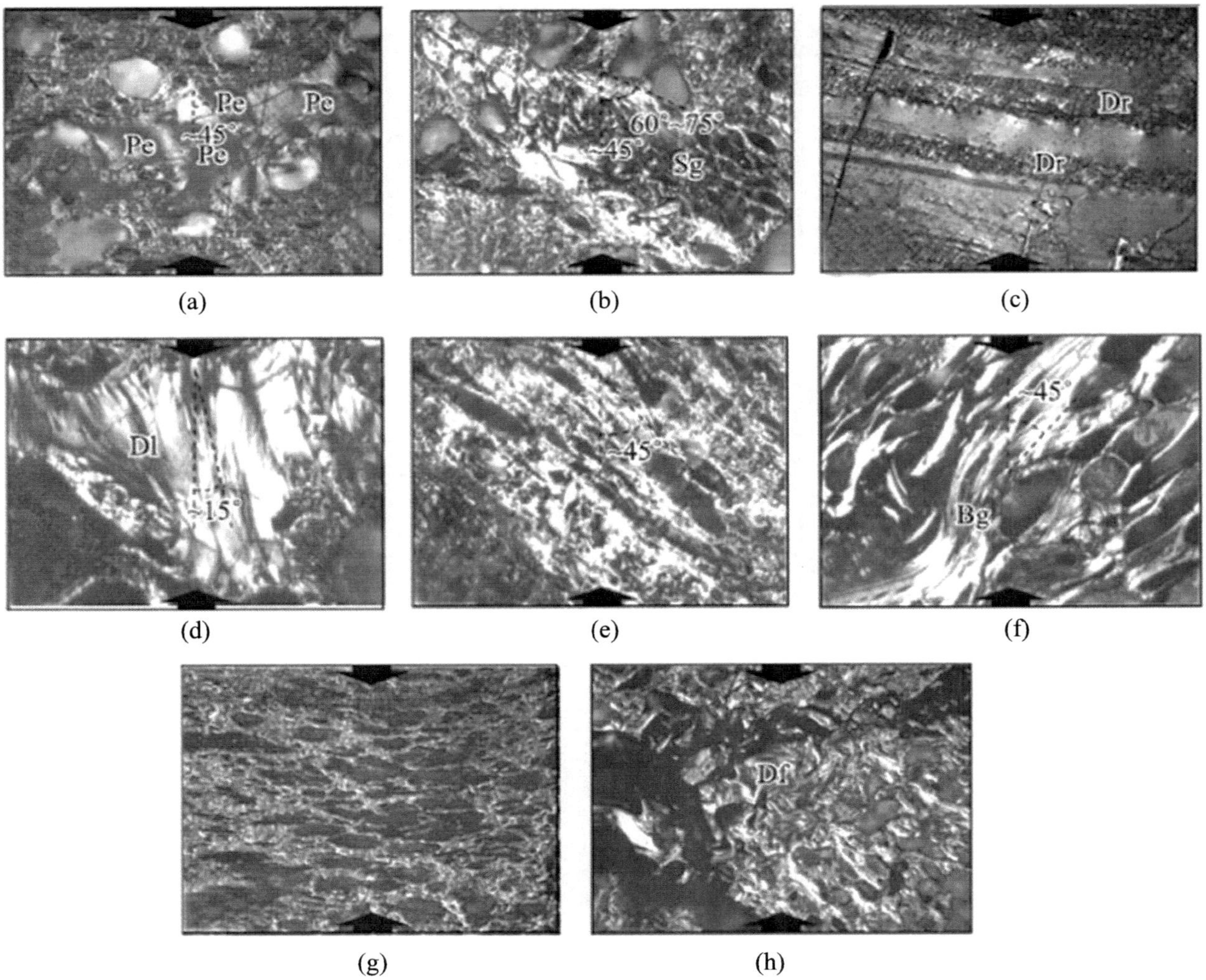

Fig. 3.29 Microstructural characteristics of ductile coal. **a** Flaky extinction; the long axis of the flake is at an angle of 45° with σ_1 (Pe flaky extinction, 300 °C, 300 MPa; Wangzhuang, Changzhi); **b** recrystallization of inert particles into fine particles (Sg), the long axis of the monomer and the aggregate is at an angle of 45° and 60°–75° with σ_1, respectively (400 °C, 400 MPa; Shidian, Yangquan); **c** stripped, dynamic, recrystallized grains (Dr, 400 °C, 400 MPa; Xiaoxian, Gaoping); **d** deformed texture in the inert body (Dl, $\sigma_1 = 15°$, 400 °C, 400 MPa; Shidian, Yangquan); **e** flattened and elongated inert particles, unidirectional orientation $\sigma_1 = 45°$, surrounded by recrystallized particles (400 °C, 400 MPa, Shidian, Yangquan); **f** wavy inert particles (Bg); plastic flow deformation ($\sigma_1 = 15°$, 400 °C, 400 MPa; Wangzhuang, Changzhi); **g** flattened and elongated vitrinite group; **h** the shape of inert body with flow deformation (Df, 400 °C, 400 MPa; Wangzhuang, Changzhi); **(a)**, **(b)**, **(c)**, **(e)**, **(f)** and **(h)** show the photo with a bottom side of 0.75 mm; **(d)** and **(g)** show the photo with a bottom side of 1.5 mm

and the transition between the two. Correspondingly, various types of microstructures with brittle deformation and/or crystalline plastic deformation are developed, such as ruptures, micro- fractures, unidirectional elongation and orientation of crystals (inert plastids), intragranular wavy extinction, deformed texture, kinks, fine granulation, dynamic recrystallized grains and crystalline plastic flow pleats.

Composition, temperature and pressure are the most important factors restricting the fluidity of coal rock. Comparison of the mechanical properties under the same temperature and pressure conditions shows that, on the one hand, the sample with higher vitrinite reflectance has higher strength, while the sample with lower vitrinite reflectance has lower strength; on the other hand, the mechanical strength changes different at different temperature and pressure. At higher temperature and pressure (>200 °C and 200 MPa), a negative correlation appears between strength and temperature and pressure. With the increase of temperature and pressure, strength reduces significantly. The mechanical behavior at lower temperature and pressure is significantly different. The sample deformed at 100 °C and 100 MPa has relatively low strength regardless of the vitrinite reflectance. The dependence of coal rock on temperature and pressure is directly reflected in the change of the microstructure after deformation. At low temperature and low pressure, the deformation shows brittle fractures, usually irregular and jagged, with rough surface and poor directionality. In addition, it is important that the fracture surface is microscopically open, which fully indicates that such deformation is very favorable for the opening of macroscopic and microscopic fractures. With the increase of temperature and pressure, the deformation of coal rock obviously tends towards plastic deformation.

Traditional tectonic physics describes rock with ductility in the case that stress doesn't drop significantly when strain reaches 5% (penetrating fractures appear). The microstructure of coal rock may be either typical intragranular plastic deformation (crystalline plastic deformation) or penetrating micro-fractures. However, in the definition of structural geology, the ductile deformation of rock emphasizes the appearance of the plastic deformation mechanism (crystal plastic deformation) of the crystal in the deformed rock. Although the macroscopic mechanical properties of the deformed samples under the experimental conditions in this book are completely consistent with the ductile characteristics defined by structural physics, their microstructural performance has changed a lot. Samples deformed at 200 °C and 200 MPa are generally characterized by typical signs of brittle deformation. Penetrating fractures are uniformly distributed in Sample No. 1 and No. 2; typical microfracture zones were found in Sample No. 5 and No. 3, which have a angle of 30° to 45° with the principal compressive stress σ_1. 200 °C and 200 MPa, and 300 °C and 300 MPa are the temperature and pressure for the brittle-ductile transition of the coal rock in the Qinshui Basin during experiments. Under that transition condition, a characteristic microstructural combination took place in the deformed coal rock. The typical microstructural type is the appearance of flaky extinction phenomenon. It is an uneven extinction phenomenon induced by microfractures, which not only retains the characteristics of microfractures, but also develops a wavy extinction phenomenon. From 300 °C and 300 MPa, crystal ductile deformation gradually dominates, and the microstructure evolves from elongated inertia particles to wavy extinction, deformed texture, kinks, fine granulation, dynamic recrystallized particles and microscopic crystalline plastic flow pleats, etc. In addition, with the gradual increase of temperature and pressure, the deformation characteristics of coal rock samples also have regular changes. At lower temperature and pressure, the flattening and elongation of inert particles, unidirectional orientation, the wavy extinction and deformed texture inside crystal particles are dominant, then toward high temperature, dynamic recrystallized particles are gradually increasing, and the shape of the particle boundary becomes more complicated.

Under lower temperature and pressure conditions, fractured coal shows more characteristics of low- pressure fractures and exhibits a trend toward typical tensile fractures. Under higher temperature and confining pressure conditions, the fractures developed in coal rock are more similar to those induced at high pressure. The development of the fracture system has an important influence on the porosity and permeability of coal rock. After fractured at low pressure, typical loose or weakly consolidated tectonic rocks, such as broken coal and super-crushed coal, are formed, and various properties and directions of fractures are developed and interpenetrated, resulting high porosity and permeability. Fractured coal at high pressure is quite different. Flat fracture surfaces closely arranged, regular and closed are typical representatives. Although they have certain influences on the porosity and permeability of coal rock, it is not significant.

Low-Pressure Fractures

Microfracture zone. The narrow fracture zone with an angle of about 30°–60° with the principal compressive stress is generally characterized by extremely irregular overall shape, boundary shape, internal microstructure and related structural forms. It usually cuts through coal rock components with different vitrinite reflectances, and its orientation is independent of the composition and reflectance of the coal rock. The fractured characteristics in the zone are very significant, including: (i) irregular boundary and shape; (ii) angular debris; (iii) particle size from micron to tens of microns (mainly 3–5 μm), and the debris with variable shapes is randomly distributed within the fracture zone and forms loose coal; 4 large debris is cemented by fine debris. The debris is generally nearly equiaxed; and in some cases, especially for larger debris, the long axis is significantly longer than the other two axes and has a certain directionality (Fig. 3.30).

High-Pressure Fractures

Microlens strips. Intergranular microlens strips are quite developed in some deformed samples. Similar to microfracture zones, they are typically characterized by straight and regular boundaries and spaced throughout the sample. They are distributed along the external compressive stress at about 30° to 60° (Figs. 3.31, 3.32). Fine lenticular fragments having long parallel axes are widely developed, which are generally several micrometers to several tens of micrometers in length. The fine lenticular fragments have straight and regular boundaries and do not have any transition zones, but exhibit significant linearity.

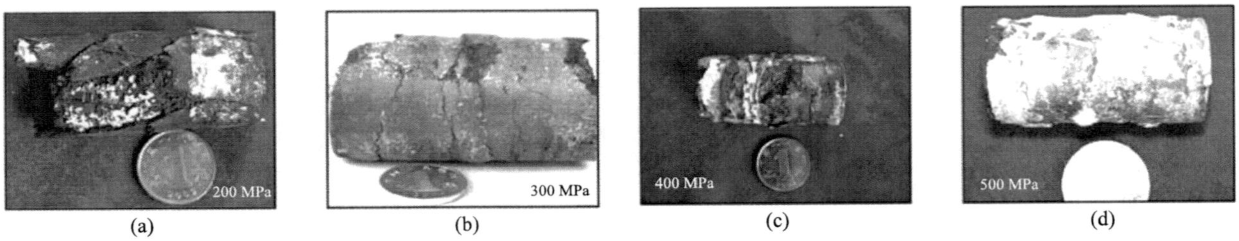

Fig. 3.30 Fractured deformation of coal rock samples during experiment. **a** Low-pressure fractures (200 MPa); **b**, **c** transitional fractures (300 MPa, 400 MPa); **d** high-pressure fractures (500 MPa)

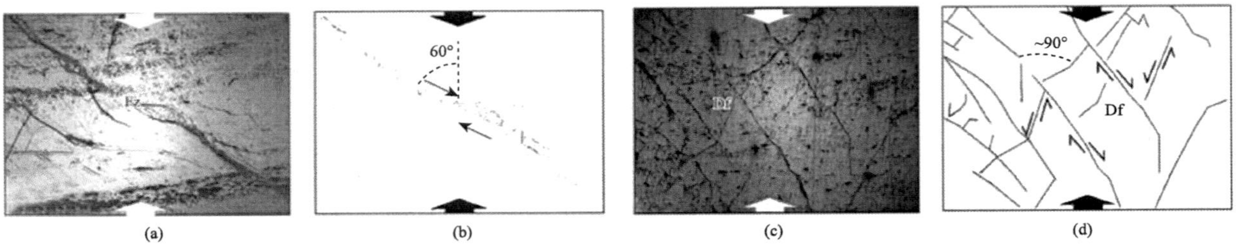

Fig. 3.31 Microfractures in deformed samples during experiment at a pure shear condition. **a**, **b** Low-pressure fractures; **c**, **d** high-pressure fractures

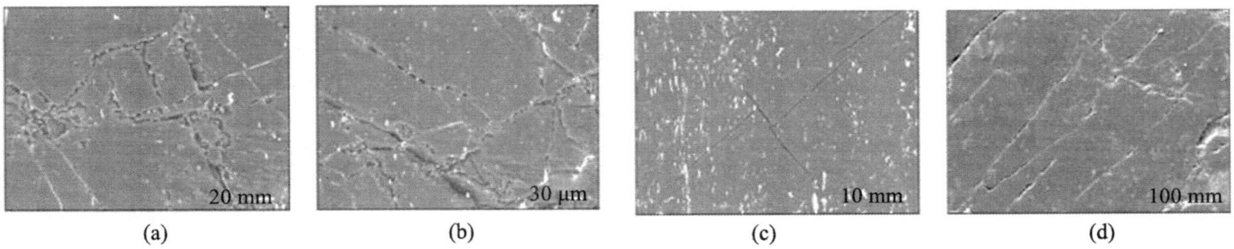

Fig. 3.32 Submicrostructural patterns in deformed samples during experiment at a pure shear condition. **a**, **b** low-pressure fractures; **c**, **d** high-pressure fractures

Penetrating microfracture system. Similar to macro-joints, conjugate microfracture systems are often developed in experimentally deformed coal rocks and evenly distributed in deformed samples. The fracture is often straight and regular with a weak (micron) opening. The length, direction and density of the microfractures are relatively uniform throughout the deformed sample. Two groups of conjugate shear-fracture structures appear in most samples, which may cross each other. But in general, one group is more developed, while the other group is often limited. In addition, their direction with respect to the principal compressive stress is generally about 45°, but there is a large change. With progressive deformation, the developmental direction of the penetrating fracture changes significantly, and σ_1 gradually changes from small to large, from 30° to 60°. Factors that affect its directionality include temperature, pressure, and progressive strain process.

Boundary fractures on kink and kink belt. Kink belt is one of the most common microstructures. Single or conjugated kinetic belts have different degrees of development. A single kink belt has an angle of 15°–30° with the principal stress, while a conjugated kink belt has an acute bisector parallel with the minimum principal stress. The development of the two groups of kink belts in a conjugate kink belts is not equal. The boundary of the kink belt is often inharmonious, so that microfractures often occur along such boundary, forming a fractured boundary. Where fractures are very developed, they may accumulate into groups, and the misplaced part is in the shape of micro-cleaved rock.

3.3.3.2 Mechanical Properties of Medium- and Low-Rank Metamorphic Coals

Coal rock samples taken from different parts of the Ordos Basin have significant compositional differences and

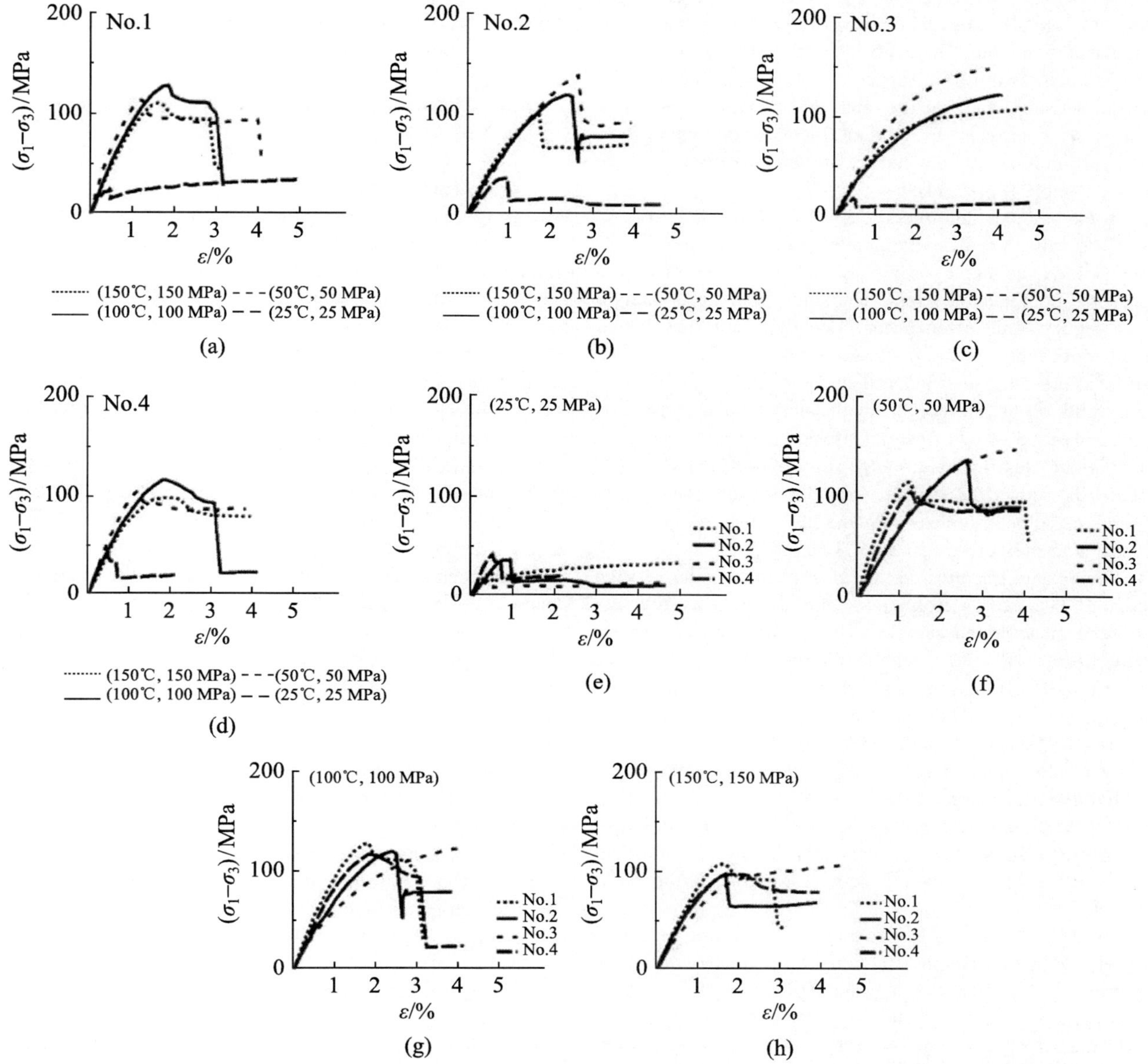

Fig. 3.33 Macroscopic mechanical performance of experimentally deformed coal in the Ordos Basin. Sampling location: Sample No. 1 taken in Xixian Coalmine; Sample No. 2 taken in Dongyungou; Sample No. 3 taken in Wangjiayan, Baode; Sample No. 4 taken in Huoshan Village and Hequ Country; **a**–**d** show the stress-strain curves of Sample No. 1—Sample No. 4; **e**–**h** show the mechanical performances of the samples at the same temperature and pressure

structural anisotropy. The obvious manifestation of the influence of coal rock components on the structure is: Sample No. 3 has the highest content of inert components; and at low temperature and low pressure (50 °C and 50 MPa to 150 °C and 150 MPa) and in the strain range (<4%), no fractures were found in the samples during the experiment (Fig. 3.33).

Under the conditions of temperature below 150 °C and pressure below 150 MPa, the deformation performance of the medium-rank and low-rank coal samples in the Ordos Basin is dominated by fracturing and microfracturing, indicating the important role of these two effects in coal rock deformation. Of course, the deformation characteristics under different temperature and pressure conditions show certain differences. As temperature and pressure decrease, the fracture extension decreases, but the fracture penetration increases gradually.

In coal rock samples deformed at 150 °C and 150 MPa, the strain is generally shown by one or several high-pressure fractures parallel to each other. The fractures are straight, regular and smooth. Generally, the extension is small, and single fractures mostly disappear inside the sample, and the angle between the fracture and the maximum principal compressive stress is 45°. Several fractures combine to form a larger, uniform fracture surface and end the primary strain of the sample (Fig. 3.34).

In the samples deformed at 50 °C and 50 MPa, 100 °C and 100 MPa, a few nearly parallel fractures are developed, which have an angle of mostly 60° with the maximum principal compressive stress. The single fracture is straight and regular and extends far. Locally, there are lateral feathery fissures indicating shearing direction. The fractured surfaces are often distributed densely, and there are angular fragments on the surface, lenticular and angular, especially on the overlap of two fractures. Beside the primary fracture zone, there are often small-range short fractures parallel to it and dispersed in the sample. Irregular fracture zones were found in some samples, showing the effect of gradually weakening pressure (Figs. 3.35, 3.36).

Short-range fractures are the most characteristics of such kind of coal rock deformation, and they are evenly distributed in deformed samples. Unlike the primary fracture zone aforementioned (i.e., the micro- fracture zone), the short-range fractures are more straight and regular, generally a few microns to hundreds of microns, with fissures below micron level, and no fault clay or fragments within them. They can penetrate a component layer into another layer with different component. They are relatively stable in length and direction in deformed samples. Relative to the principal compressive stress, they are generally 30° to 45°, and their orientations change even in different parts of the same sample.

At 25 °C and 25 MPa, fracture belts are generally developed in coal samples, showing the characteristics of the fractures induced at low pressure. The characteristic structural type is a narrow fracture belt at an angle of about 30° with the principal compressive stress. Generally, such belts are extremely irregular in whole body, boundary, internal microstructures, and related tectonic patterns. The body generally extends along the boundaries with different mineral particles, including particles of different mineral species and the same mineral with different lattice orientations. Also, they often cut inside particles, especially in the particles having more developed internal cleavage (Fig. 3.37).

Secondary microfractures and microfracture belts in different directions are densely distributed near the primary fracture belt and controlled by the latter. Their occurrence, orientation and scale are mainly controlled by local shear stress. The secondary micro-fracture belt is 30°–45° from the primary fracture belt in the secondary microstructure. They generally appear in groups, but are not parallel to each other, and their straight and clear boundaries indicate the significance of shearing during formation.

3.3.3.3 Microscopic Deformation Mechanism of Coal Rock

Fracturing Nucleation

Fracturing nucleation during the deformation of coal rock, that is, the occurrence of fracture surface or microfracture surface caused by intracrystal dislocation and entanglement and then hardening, is controlled by many factors. Such nucleation may begin at particle boundaries, causing the boundary of the friction particle to slip, or begin at some end of the adjacent particle or intragranular inclusion, or begin at the boundary with heterogeneous vitrinite. Fracturing nucleation is mainly caused by the impact on particle contact, local stress concentration induced by internal defects, elastic, plastic and directive disharmony between different components. Their appearance can also be caused by kink action. The boundary of a kink belt is an important place for the accumulation of intragranular strain and the uncoincidence of crystal structures.

It is worth noting that, unlike the rock with multi-component crystals as the main composition, the fractures in deformed coal samples tend to be more oriented and relatively more uniform.

Fracture Expansion and Association

During deformation, fractures expand to a larger scale with the increase of strain, by several ways such as extension, composition, style transition, and association.

The factors controlling the growth of micro-fractures or subcritical fractures include: (i) The most common is the accumulation of elastic strain energy, which is the main mechanism related to rapid fracture propagation. When the stress is concentrated at the end of the fracture, the elastic strain energy accumulates, and controls fracture expansion; (ii) with tight dislocation and entanglement or the development of dense kinks, crystal plasticity further limits deformation and may cause rapid hardening of fractures, where crystal plasticity may contribute to fracturing; (iii) at the open particle boundary or 3 nodes, due to point defects and vacancy concentration, vacancy diffusion can lead to the development of pores, causing damage; (iv) lattice structure changes cause stress concentration by producing products with different volumes, and may cause pores, resulting in damage. In addition, hydraulic fracturing and stress corrosion also cause rapid fracture propagation.

3.3.3.4 Stress and Strain Response

Coals at different ranks are different in composition, structure, gas content, cleat and fracture development. These

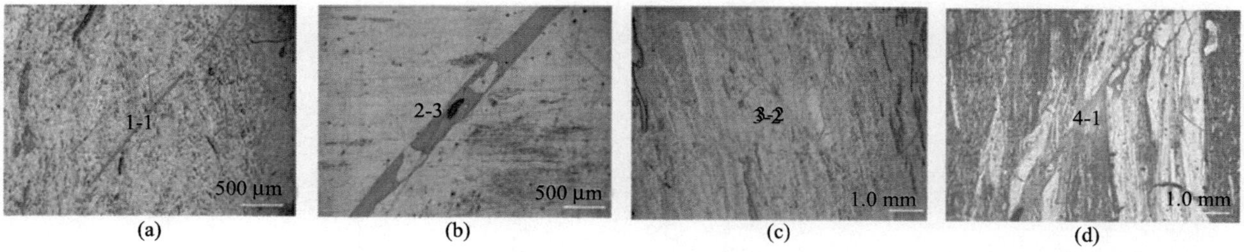

Fig. 3.34 Microscopic deformation characteristics of coal rock samples at 150 °C and 150 MPa. **a**–**d** Show Sample No. 1—Sample No. 4

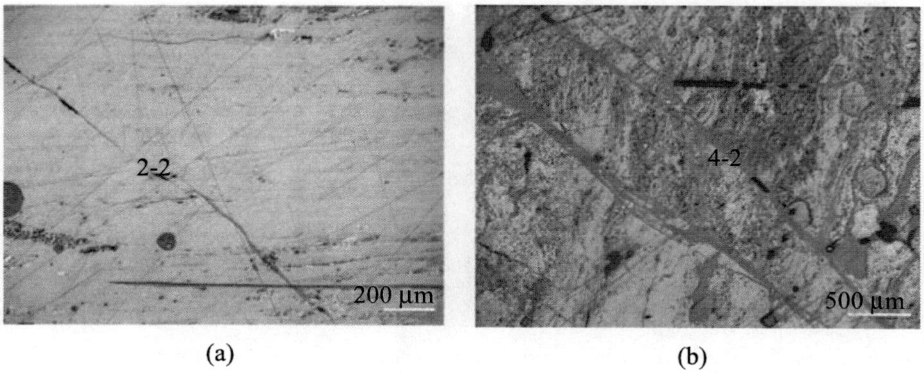

Fig. 3.35 Microscopic deformation of coal rock samples at 100 °C and 100 MPa. **a** Sample No. 2; **b** sample No. 4

Fig. 3.36 Microscopic deformation of coal rock samples at 50 °C and 50 MPa. **a** and **b** Sample No. 2; **c** and **d** sample No. 4

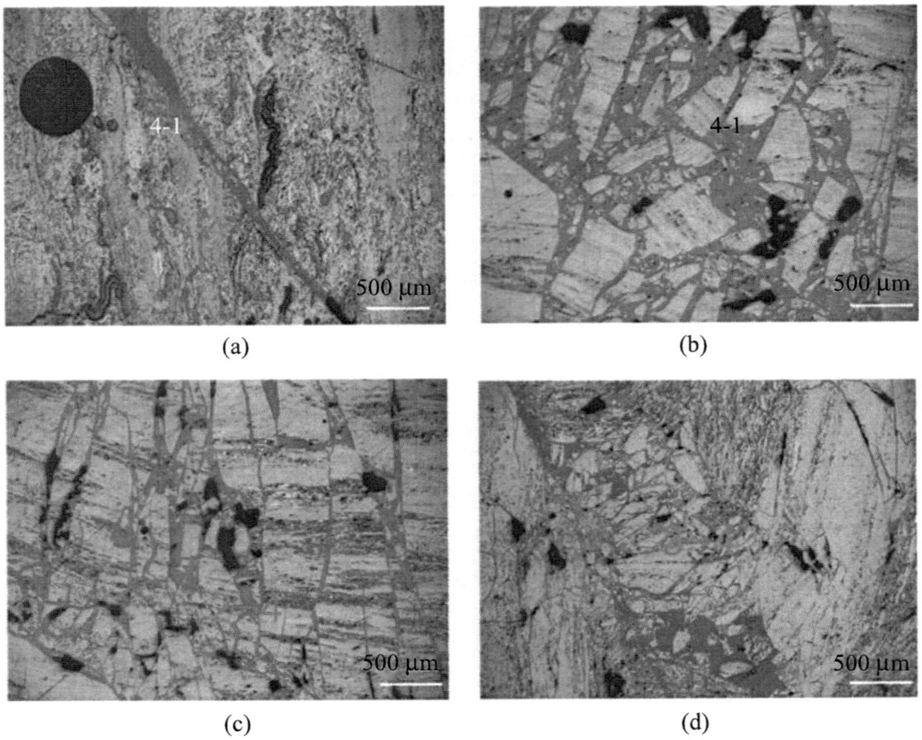

Fig. 3.37 Microscopic deformation of coal rock samples at 25 °C and 25 MPa. **a** Sample No. 1; **b, c** Sample No. 2; **d** Sample No. 4

factors affect and control the permeability of coal reservoirs, of which fractures are direct. It is more effectively to predict favorable coal reservoirs by exploring the relationship between exogenous fractures and permeability and combining with other methods of fracture research.

The stress and strain strength of the coal rock in the Qinshui Basin is negatively correlated with temperature at high temperature and high pressure, and is positively correlated with confining pressure. For samples of different ranks, their strength tends to be consistent at low and high temperature and pressure. At 300 °C and 300 MPa, 400 °C and 400 MPa, the strength of the coal rock samples is significantly different, that is, the effect of confining pressure on high-rank coal is higher than that of temperature; for low-rank anthracite samples, strong strain hardening occurs when strain reaches a certain value (ε is about 5%). At higher temperature and pressure (500 °C, 500 MPa), the strength of the coal samples tends to be consistent, showing a distinct docile state, which indicates that the influence of confining pressure is greater than temperature.

The cleats are developed in the coals with high vitrinite content in the Qinshui Basin, while the cleats in the coals with high inertinite content are not developed, that is, the components affect the development of the cleats, and the degree of the cleat development affects the degree of the strain. Therefore, at certain temperature and pressure, the coal with low inertinite content has the lowest strain (300 °C, 300 MPa).

The coal rock in the Ordos Basin is at medium- low rank. At low temperature and low pressure (25 °C, 25 MPa), the sample reaches its strength limit at lower strain and is destructed. Obviously, low-rank coal rocks are susceptible to brittle deformation due to their low thermal evolution and relatively high ash content, so their resistance to strain is very weak. If inertinite content is high, the coal rock will not break easily when deformed at low strain. It can be concluded that the coal rock with high vitrinite content is easily broken and the performance of storing coalbed methane is relatively good.

3.3.4 Control of Basin Evolution on Coal Reservoir Physical Properties

3.3.4.1 Physical Properties and Geological Factors

The permeability of coal seams is very heterogeneous. The permeability of different coal seams is different, even in the same coal seam, the permeability is different in vertical and lateral directions. In general, there are many internal and external factors controlling the permeability of coal seams. Internal factors are fractures, coal ranks, coal rock types, components and structures. External factors include in situ stress, effective stress, burial depth, Klinkenberg effect and geological structural effect. A large amount of production data and permeability sensitive experiments show that internal and external factors that play the controlling role are fractures in coal reservoirs and effective stress, respectively.

Coupling Relationship Among Controlling Factors

A coal reservoir is a three-dimensional geological body that exists at a certain depth underground. The opening of fracture, reservoir pressure, mechanics and permeability are closely related to geostress.

The permeability (K) is functional with reservoir physical properties (f), geostress field (σ), pressure field (P), temperature field (T), electric field (e) and magnetic field (m).

$$K = F(f, \sigma, P, T, e, m) \quad (3.15)$$

The physical properties of coal reservoirs can be determined by fracture dimension (D_f), pore dimension (D_p), engineering modulus (E), Poisson's ratio (v), matrix maximum shrinkage (ε_{max}), and Pressure (P_{50}) at 1/2 ε_{max}.

$$f = f_1(D_f, D_p, E, v, \varepsilon_{max}, P_{50}) \quad (3.16)$$

The geostress field is coupled by gravity stress (σ_v), tectonic stress (σ_t), pore pressure (P), thermal stress (σ_r), and matrix shrinkage stress (σ_s).

$$\sigma = f_2(\sigma_v, \sigma_t, P, \sigma_r, \sigma_s) \quad (3.17)$$

In addition, the geostress field, pressure field, temperature field and electromagnetic field are functions of time and space.

$$\sigma/P/T/e/m = f_3(x, y, z, t) \quad (3.18)$$

Since coal reservoirs are discontinuous and anisotropic, their physical properties are also function of space.

Control of Fractures on Permeability

To some extent, the fracture system in coal reservoirs is an important factor controlling the permeability. In theory, developed fractures are conducive to increasing permeability.

McKee et al. (1988) concluded that permeability is proportional to the square of fracture width. According to Levine (1996), permeability is proportional to the cube of fracture width and inversely proportional to the spacing of fracture.

$$K = \frac{(1.013 \times 10^9)W_x^3}{12S} \quad (3.19)$$

where K is permeability, mD; S is fracture spacing, mm; W_x is fracture width, mm. The direction in which the fracture is the most open means the direction in which the permeability is the largest.

Control of Stress on Permeability

The internal stress existing in the earth's crust is called geostress and is mainly composed of gravity stress, tectonic stress, pore pressure, thermal stress and residual stress. In laboratory, coal core columns are used to measure the permeability at confining pressure, which reflects that coal permeability is extremely sensitive to stress. The absolute permeability of coal reservoir fractures is a function of effective stress acting on coal reservoir, and it decreases with the increases of effective stress.

Effective stress is the difference between vertical stress and pore pressure, i.e.,

$$\sigma_e = \sigma_v - \beta P$$

where σ_e is effective stress, MPa; σ_v is vertical stress, MPa; β is a constant, dimensionless; P is pore pressure, MPa.

Vertical pressure is caused by the weight of overburden. Pore pressure is normally equal to hydrostatic pressure. Obviously, effective stress increases with the burial depth of coal reservoirs, and accordingly, the stress acting on the fracture system increases, and leads to the closure of the fracture system and the decrease of the absolute permeability of the fracture system. Somerton et al. (1975) found the following relationship between effective stress and permeability.

$$K = 1.03 \times 10^{-0.31}\sigma$$

where K is permeability, mD; σ is effective stress, MPa.

It can be seen that it is a power function between effective stress and permeability, and permeability decreases as effective stress increases.

3.3.4.2 Control of Burial History on Permeability

During the basin evolution process, the increase of the burial depth of coal reservoir is not conducive to the improvement of coal reservoir permeability. This apparent trend can be seen from the trend of the absolute permeability of coal reservoirs in the San Juan Basin, the Black Warrior Basin, and the Piceance Basin in the United States (Fig. 3.38).

The burial history of the Upper Paleozoic coal seams in the Qinshui Basin has undergone the following four stages. The first stage started from the Late Triassic when coal seams formed, and then were buried first slowly and later rapidly through the Hercynian and Indosinian periods. From the Late Carboniferous to the end of the Early Permian, the study area was in a stable and balanced settlement stage as a platform, at the settlement rate of 7 m/Ma, during which filling period, accumulation period and buried period alternated. From the Late Permian to the Late Triassic, it was a

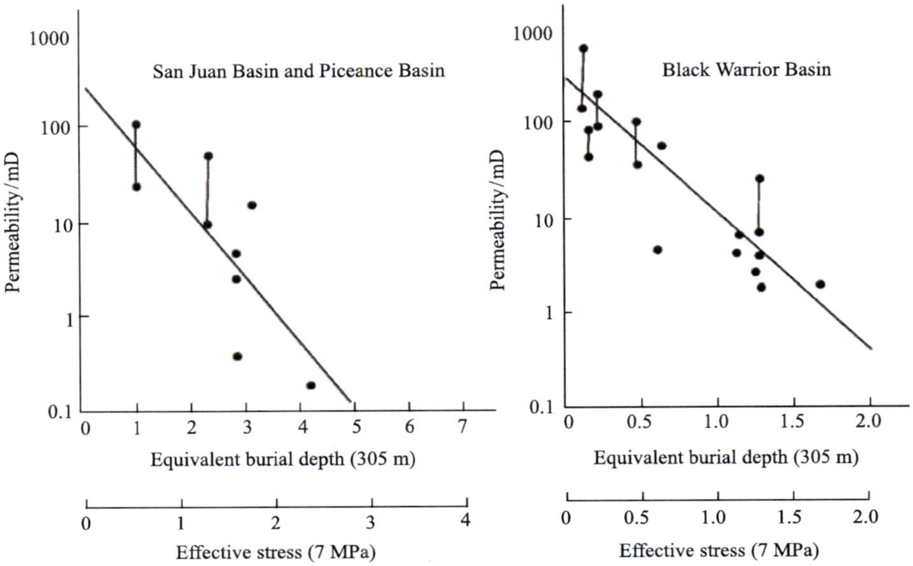

Fig. 3.38 Permeability versus burial depth of coal reservoirs in some regions of the United States (according to Mckee et al. 1988)

quick settlement stage at the maximum rate of 95 m/Ma, and the coal seams were buried at 2200–4200 m as the crust settled deeply. The second stage started from the Early Jurassic to the Middle Jurassic during which it was the early Yanshanian, and a stablely fluctuating stage. Influenced by the Indosinian movement during the Early Jurassic, the study area was uplifted and denuded, so that the Early Jurassic formation was missing in the whole area, the Triassic formation was locally denuded, and the coal seams are shallower. The maximum uplift was greater than 700 m. During the Middle Jurassic, fault block basins took place under the control of the Yanshanian movement. The crust slowly settled, sedimentation began again, and the coal seams were buried deeply again. But the increased depth was small, only about 200 m. The average sedimentation rate was about 9.5 m/Ma. The third stage was from the late Jurassic to the Early Cretaceous during which it was the middle Yanshanian, and the depth of the coal seams reduced significantly. The earth's crust was in a state of structural uplift, the depth of the coal seam was reduced by about 800 m, and the average rate of uplift was about 13.6 m/Ma. The fourth stage started from the Late Cretaceous to today, the coal seams are shallower and shallower. From the early Himalayan to the late Yanshanian, the crust was slowly uplifted, the sedimentation almost stopped, and the Triassic strata were almost denuded, so that the overlying formation including coal seams were severely denuded, and shallow coal seams were weathered. During the late Himalayan period, structural differentiation re-emerged and the earth's crust was basically stable or slightly subsided.

From the perspective of basin evolution, in the first and second stages of sedimentary burial, tectonic activities were

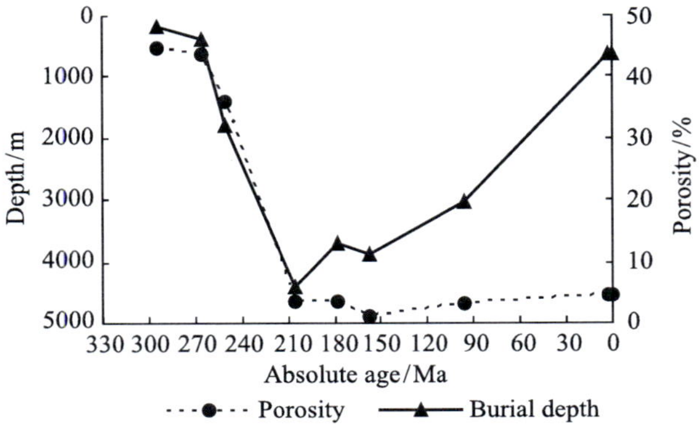

Fig. 3.39 Depth versus porosity during the sediment evolution of coal reservoirs

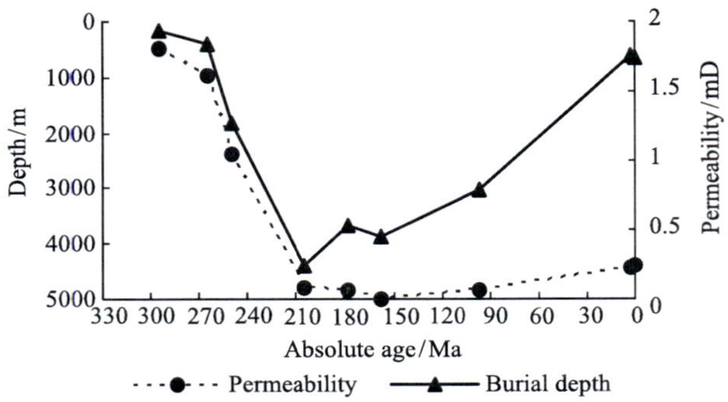

Fig. 3.40 Depth versus permeability during the sediment evolution of coal reservoirs

basically stable; then with the gradual increase of burial depth, the compaction of the overlying rock was enhanced, and the porosity and permeability of coal seams were reduced (Figs. 3.39, 3.40). During the third and fourth stages, tectonic and volcanic activities were frequent, and a series of fractures developed under the control of tectonic stress field and geothermal field, resulting in improved permeability.

3.3.4.3 Thermal and Maturity Histories on Permeability

Thermal and maturity histories indicate that the coal reservoir in the Qinshui Basin is the result of deep metamorphism and regional magma thermal metamorphism. In the first stage of basin evolution, i.e. before the Yanshanian, the coal metamorphism was mainly controlled by deep metamorphism at normal paleotemperature. At the end of the Triassic, the R° was 1–1.4%, belonging to fat and coking coals, and the permeability of the coal reservoir didn't change much. Then during the third stage, i.e. the Yanshanian, the strong tectonic movement uplifted and folded the area, and the Carboniferous–Permian coal seams and overlying strata were denuded, while the Moho depth uplifted and local magma intruded, forming a regionally high geothermal field which accelerated the evolution of the coal quickly from fat and coking coals to lean and anthracite coals. The increase in reservoir temperature and coal rank led to a significant increase in coal reservoir permeability (Figs. 3.41, 3.42).

Thermal magma metamorphism in the area caused the deterioration of the coal in a short period under conditions of small pressure change, and high and rapidly increasing temperature. Therefore, the rapid regional metamorphism did not change the depth of the coal seam because the confining pressure to generate the same degree of the coal is less than the confining pressure caused by deep metamorphism, which is conducive to the preservation of the fracture system. Rapid dehydration and defunctionalization makes it easier to retain the fracture system during rapid shrinkage and reconstruction of polycyclic aromatic hydrocarbon. At stable pressure and stress, thermal baking by magma causes a large amount of volatile matter in the coal to precipitate, leaving many round or tubular pores in dense groups. That not only led to a larger free space for coal molecules, the relatively poor ordering and directionality of coal molecules arrangement, the increase of molecular spacing and pore size, but also formed a large amount of hydrocarbons in a short time due to pyrolysis or cracking of organic matter. The hydrocarbon gas may leave a large number of pore structures in communication with the fracture system when it releases from coal.

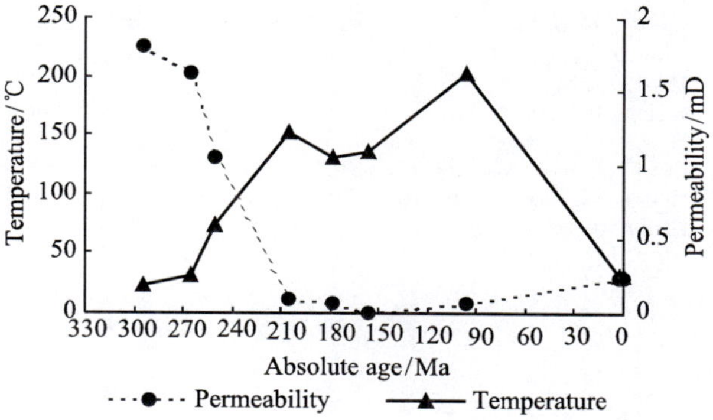

Fig. 3.41 Temperature versus permeability during the sediment evolution of coal reservoirs

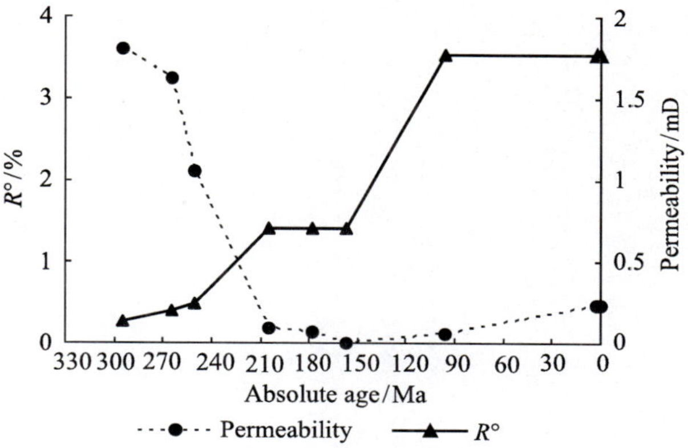

Fig. 3.42 Maturity versus permeability during the sediment evolution of coal reservoirs

References

Ammosove II, Eremin IV (1963) Fracturing in coal. IIZDAT Publishers, Office of Technical Services, Moscow

Bustin RM, Clarkson CR (1998) Geological controls on coalbed methane reservoir capacity and gas content. Int J Coal Geol 38(1–2):3–26

Chen P (2001) Nature, classification and utilization of coal in China. Chemical Industry Press, Beijing, pp 81–83 (in Chinese)

Close JC (1993) Hydrocarbons from coal. AAPG Stud Geol 38:119–130

Gash BW, Volz RF, Potter G et al. (1993) The effects of cleat orientation and confining pressure on cleat porosity, permeability and relative permeability in coal. The 1993 International Coalbed Methane Symposium. University of Alabama, Tuscaloosa

Gayer R, Harris I (1996) Coalbed methane and coal geology. Geol Soc 1–338

Katz AJ, Thompson AH (1985) Fractal stone pores: implications for conductivity and formation. Phys Rev Lett 54(3):1325–1328

Kroch CE (1988) Sandstone fractal and Euclidean pore volume distributions. J Geophys Res 93(B4):3286–3296

Levine JR (1996) Model study of the influence of matrix shrinkage on absolute permeability of coal bed reservoirs. Geol Soc Publ 199:197–212

Mandebrot BB (1982) The fractual geometry nature. W H Fremen, New York

McKee CR, Bumb AC, Koening RA (1988) Stress-dependent permeability and porosity of coal. Rocky Mountain Association of Geologist, pp 143–153

Pfeifer P, Avnir D (1983) Chemistry in nonintegral dimensions between two and three. J Chem Phys 79(7):3369–3558

Somerton WH, Soylemezoglu IM, Dudley RC (1975) Effect of stress on permeability of coal. Int J Rock Mech Mining Sci Geol Abst 12:129–145

Yang Q (1996) Coal metamorphism in China. Coal Industry Press, Beijing, pp 163–164 (in Chinese)

4 Coal Absorption Characteristics and Model Under Reservoir Conditions

Under reservoir conditions, CBM is mainly absorbed on the pore surface of coal. The capacity of CBM storage is mainly related to factors such as metamorphic environment, composition, ash yield, moisture, reservoir pressure and temperature. Under similar metamorphism and sedimentary conditions, coal rock composition, ash yield, moisture and pore surface area are different, which can be reflected by isothermal absorption experiments on coal samples. After thermal evolution, coal seams with same metamorphism may be different in burial depth due to late structural deformation, so that the reservoir pressure and temperature change greatly. The reservoir temperature and pressure change with the burial depth of the coal seam. In the coal seam shallower than 2000 m, the normal reservoir pressure varies from 0.1 to 20 MPa, and the temperature varies from 15 to 75 °C. For such change, it is impossible to do several isothermal absorption experiments at different temperatures on the same metamorphic coal sample. It is necessary to build a model.

In the CBM field, the existing adsorption model represented by Langmuir equation is often used to describe the adsorption behavior of coal, which is mainly used to quantitatively describe the relationship between absorption capacity and pressure, but cannot reflect the comprehensive influence of temperature and pressure change on absorption capacity with the change of formation depth. Even the existing formula describing the relationship between absorption, temperature and pressure is only an empirical relation established on the basis of experimental data. How to establish a absorption model under reservoir conditions, or to establish a quantitative relationship between absorption, temperature and pressure, is an urgent problem to be solved in the CBM field. In order to study the absorption characteristics and description model of coal at various temperature and pressure, we conducted, for the first time at home and abroad, temperature-variable and pressure-variable absorption experiments on coal samples, and isothermal absorption experiments on same coal samples, and studied and compared the adsorption characteristics of coal under the combined action of temperature and pressure. According to the theory of absorption potential, by the use of absorption characteristic curves, a model describing the relationship between pressure, temperature and absorption was established, which can be used to calculate the adsorption quantity of coal under any combined action of temperature and pressure with in the experimental pressure range by using the single isothermal adsorption experimental results of coal sample. More importantly, accurate calculation of gas absorption under formation conditions, and accurate prediction of coalbed methane content and resources can provide more reliable resource forecast, and be conducive to new understanding of absorption capacity, improvement of absorption theory and study of coalbed methane accumulation mechanism.

We can only obtain the absorption amount on the experimental pressure point through absorption experiments. To obtain the absorption capacity at given pressure, we must fit experimental points by a description model to establish an absorption isotherm. Using the description model, we can calculate the absorption amount at any pressure, so experimental results can be fully utilized. The Langmuir model was often used to describe coal isothermal absorption experiments and made good results. During the study, the absorption constants of some lignite and anthracite No. 1 samples were found abnormal, and the Langmuir volume and pressure were negative or large ($V_L = -25.47 - 254.15\ cm^3/g$), which are not consistent with actual data. The experimental results and the absorption constants are meaningless, and it indicates that the Langmuir model is not suitable for describing isotherm absorption curves of lignite and high metamorphosed anthracite. Then the whole coalification series is divided into three stages according to the description by the Langmuir model. By referring to the theory of absorption potential, the theoretical model of absorption potential was extended and modified, and the isothermal absorption model quantitatively describing the

Y. Song et al., *Coalbed Methane in China*,
https://doi.org/10.1007/978-981-33-4725-0_4

coal samples with vitrinite reflectance less than 0.4% and more than 7.0% was built in the highest experimental pressure range. The prediction is close to the measured value. The method to obtain the maximum absorption amount was developed, and the maximum absorption amount calculated is comparable with the Langmuir volume. The model can be used to process experimental data and calculate the absorption amount at any pressure within the experimental pressure range. It provide a platform for accurately describing the isothermal absorption curve of dark lignite and high metamorphic anthracite, applying experimental results, comparing absorption capacity, and evaluating underground absorption capacity and CBM resources.

4.1 Experimental Study on Methane Absorption Characteristics Under the Combined Influences of Temperature and Pressure

In the history of coal sample absorption experiments, references reported mainly on isothermal absorption experiments, (that is, changing the pressure under the same conditions of experimental temperature and others, and measuring the volume of gas adsorbed when the absorption equilibrium is reached at different pressure).

However less references reported on isobaric and temperature-variable absorption experiments, nor on temperature-variable and pressure-variable absorption experiments. In order to simulate formation temperature and pressure, isothermal absorption experiments generally use supercritical temperature and pressure. The highest experimental pressure is from low pressure absorption (less than 5 MPa), medium pressure absorption (5–12 MPa) to high pressure absorption (12–35 MPa), and the experimental temperature is from normal temperature (less than 50 °C) to high temperature (50–100 °C). In order to study the absorption characteristics and absorption calculation model under the combined influence of underground temperature and pressure, two experimental methods were used for the same coal samples, one for isothermal absorption experiments at different temperatures, and the other for variable temperature and pressure. Experimental temperature and pressure depend on the performance of experiment equipment, so medium pressure absorption and high pressure absorption were conducted, and the temperature was normal and high.

4.1.1 Isothermal Absorption Experiments

The experiment method is static capacity. The instrument used is a high-pressure isothermal adsorber manufactured by Raven Ridge of the United States. The highest experimental temperature is 50 °C and the highest experimental pressure is 15 MPa.

The coal samples include gas coal, coking coal, meager coal and anthracite (Table 4.1).

The coal sample was broken to 60–80 mesh. Before the isothermal absorption experiment, the equilibrium moisture treatment was carried out on the coal sample, and the moisture content was tested to restore to the water content state under the formation conditions. The conditions for the equilibrium moisture content experiment were conducted at 30 °C, and 96–97% moisture for about 5 days.

Isothermal absorption experiments were carried out on four ranks of treated samples at 20 °C, 30 °C, 40 °C, 50 °C, respectively. Two samples from Huajiahu and Yangquan were tested at 25 °C, 35 °C, 45 °C, and 50 °C respectively (Fig. 4.1).

The following real gas state equation was used to calculate the absorption amount (V) at each pressure point (P) of each sample.

$$PV = nZRT \tag{4.1}$$

where n is the mole fraction of gas, mol; Z is a gas compression factor, dimensionless; R is a molar gas constant, J/(mol K); T is thermodynamic temperature, K.

According to the Langmuir equation,

$$P/V = P/V_L + P_L/V_L \tag{4.2}$$

If, $A = 1/V_L$ and $B = P_L/V_L$, Eq. (4.2) is the function of P/V and P.

$$P/V = P/V_L + P_L/V_L = AP + B \tag{4.3}$$

Using the least squares method to find the slope (A) and intercept (B) of the linear regression equation of P/V and P, and the Langmuir volume (V_L) and Langmuir pressure (P_L) are obtained, i.e.,

$$V_L = 1/A \tag{4.4}$$

$$P_L = BV_L \tag{4.5}$$

4.1.2 Temperature-Variable and Pressure-Variable Absorption Experiments

In order to understand the absorption characteristics under the joint influence of temperature and pressure, we decided to simulate the temperature and pressure of the coal seam shallower than 2000 m, and carry out the experiments at variable temperature and pressure, high temperature and high pressure.

Table 4.1 Coal sample parameters for isothermal absorption experiment

Sample	Coal	Location	Coal seam no	M_{ad}/%	A_d/%	V_{daf}/%	R_{max}/%
HNXJ	Gas coal	Xinji, Huainan	1	1.19	10.34	33.30	0.88
HJH	Gas coal	Huajiahu	8	1.94	13.40	–	0.88
LLMW	Coking coal	Miaowan, Liulin	4	0.64	4.72	20.26	1.32
TLQZ	Meager coal	Quzhuang	3	1.06	11.78	13.56	2.06
YQ	Anthracite	Yangquan	15	0.91	11.28	–	2.69
QSZZ	Anthracite	Zhengzhuang	3	1.11	13.89	7.09	3.56

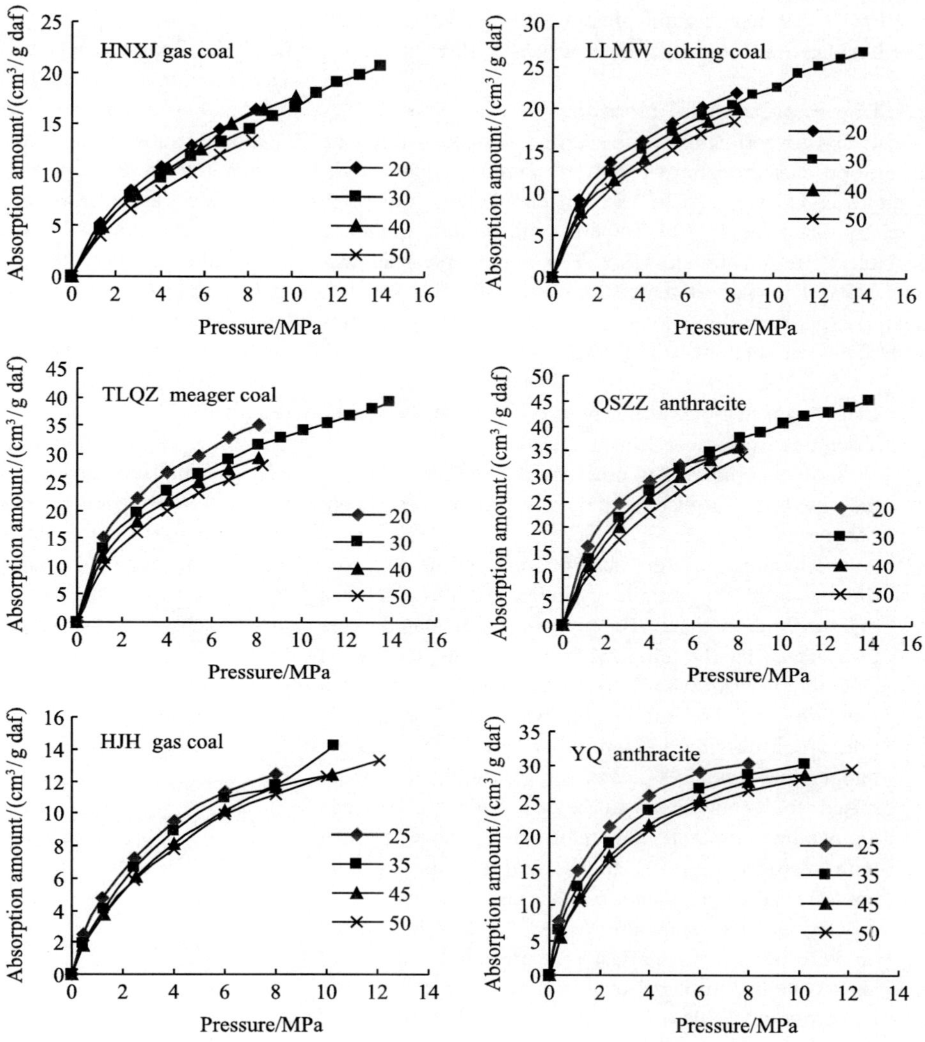

Fig. 4.1 Isothermal absorption curves and absorption amounts

4.1.2.1 Experimental Samples

Four representative coal samples of different coal ranks were selected for absorption experiments at variable temperature pressure (Table 4.2), and they were subjected to isothermal absorption experiments at 30 °C to compare the changes in absorption capacity.

4.1.2.2 Experimental Design

The changes in temperature and pressure are related to buried depth, and there is a certain law. In general, normal surface temperature is 15–20 °C, the temperature increases by 3 °C, and the pressure increases by 1 MPa when increasing every 100 m starting from the surface. According

Table 4.2 Coal samples parameters for variable temperature and pressure

Sample	Rank	Coal seam no	M_{ad}/%	A_d/%	V_{daf}/%	R_{max}/%
BRDG6	Long flame coal	6	5.92	6.02	30.83	0.55
LLMW	Coking coal	4	0.64	4.72	20.26	1.32
HZ02	Meager coal	15	0.72	12.01	10.36	2.10
ZZZX(ZK2-1)	Anthracite	3	1.80	11.64	6.89	3.75

to this rule, absorption experiments were carried out on the coal samples of the above four ranks at different depths. The experimental temperature was 18–72 °C. The pressure was 1–19 MPa. 12 temperature–pressure points was selected. The simulated underground depth was 100–1900 m.

4.1.2.3 Experimental Method

Absorption experiments were carried out to study the absorption characteristics of the coal samples at high temperature and pressure, and variable temperature and pressure, by using the IS-100 isothermal absorption apparatus introduced from Terratek, USA. The highest pressure was 34.5 MPa. The highest temperature was 100 °C. The following experimental methods were used for the experiments at variable temperature and pressure:

① Equilibrium moisture treatment to coal samples before the absorption experiment (Zhang et al. 1999).
② Volume experiment and pressure experiment following the method for isothermal absorption experiment (Zhang and Cao 2003).
③ At the beginning of the absorption experiment, closed the valve of the sample cylinder, fill the reference cylinder with methane. The pressure is calculated target pressure. After the temperature was stable, started the isothermal absorption experiment and recorded the pressure and temperature at different times. The balance pressure at the first pressure point was 1 MPa, and the temperature was 18 °C. After completing the first point, raised the temperature to 21 °C. After the balance temperature is stable, the reference cylinder is filled with methane to reach the second calculated target pressure. The calculated second target pressure was 2 MPa, and the temperature was 21 °C. Repeated the above process until the last pressure point. Figure 4.2 shows the experimental results at variable temperature and pressure. Table 4.2 lists the measured absorption of the coal samples at different depths. Figure 4.3 shows the absorption capacity curve.

The period of the absorption experiment at variable temperature and pressure is long. Generally, it takes more than one month to complete a group of experiments.

4.1.2.4 Processing of Experimental Data

The processing of the experiment data at variable temperature and pressure is more complicated than those of isothermal absorption experiment, because the balance temperature at each pressure point is different, a part of absorption gas is desorbed after the balance is destroyed when the temperature rises; at the same time, a part of gas is adsorbed again when the pressure rises with the increase of temperature. Considering the above problems, first calculated the total mole fraction of free gas before and after each equilibrium at different temperatures according to the real gas state Eq. (4.1); then calculated the mole fraction of desorbed or adsorbed gas in the sample cylinder after increasing the temperature. The difference is the mole fraction of actual adsorbed gas at that point after the temperature increases. The specific method is as follows:

Similar to isothermal absorption, used Eq. (4.1) to determine the mole fraction before equilibrium (n_1) and that after balance (n_2), and then the mole fraction of desorbed or adsorbed gas (n_3) in the sample cylinder after the temperature increased. The mole fraction of adsorbed gas (n) of the coal sample is

$$n = n_1 - n_2 - n_3 \tag{4.6}$$

where n is the mole fraction of adsorbed gas; n_1 is the mole fraction before balance; n_2 is the mole fraction after balance; n_3 is the mole fraction of desorbed gas after increasing temperature.

Total volume of adsorbed gas (V):

$$V = n \times 22.4 \times 1000 \tag{4.7}$$

Gas absorption amount (V'):

$$V' = V'/M \tag{4.8}$$

where M is the weight of coal sample, g; V' is the total volume of adsorbed gas, cm^3.

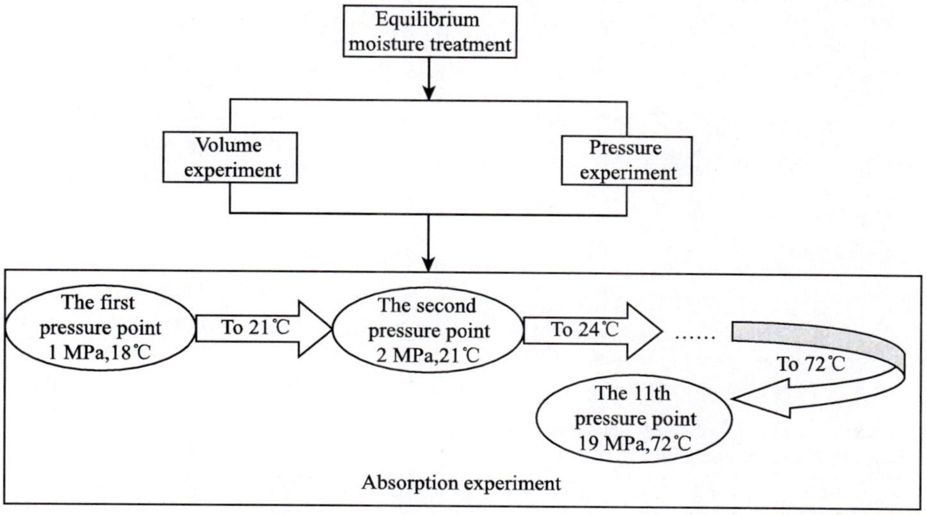

Fig. 4.2 Flow chart of temperature- and pressure-variable absorption experiment

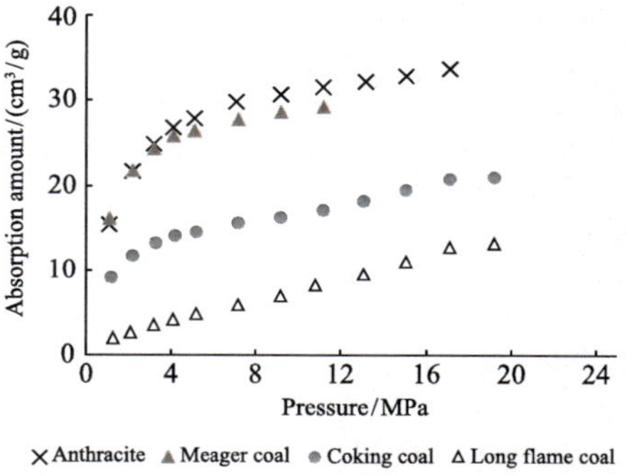

Fig. 4.3 Absorption curves at different temperature and pressure. X, ▲, ●, △, from bottom to top, represent 18 °C, 21 °C, 24 °C, 27 °C, 30 °C, 36 °C, 42 °C, 48 °C, 54 °C, 60 °C, 66 °C and 72 °C, respectively

4.2 Absorption Characteristics of Coal Under Formation Conditions

Absorption experiments were conducted on coal samples at constant, variable and high temperature and pressure above. Then we investigated how temperature affects coal absorption capacity, what are isobaric absorption characteristics, and absorption amount under the combined effects of temperature and pressure.

4.2.1 Effect of Temperature on Coal Absorption Capacity

In the experimental results, the Langmuir volumes are different, and no obvious relationship with temperature (Fig. 4.4). The Langmuir pressure is related to temperature. It increases with temperature (Fig. 4.5), indicating that temperature enhances the activity of gas molecules and reduces the absorption capacity. If the absorption amount reaches one-half of the Langmuir volume, high pressure is needed. Isobaric absorption can be obtained by the Langmuir constant. The isobaric absorption diagram (Fig. 4.6) shows that the absorption amount decreases with increasing temperature; and the higher the coal metamorphism, and the greater the slope, the larger the decreased amount (Fig. 4.7). For example at 8 MPa, when temperature increases from 20 to 50 °C, the absorption capacity of the HNXJ-1 coal sample reduces from 13.37 cm^3/g daf to 11.24 cm^3/g daf, and that of the QSZZ-3 coal sample reduces from 35.23 cm^3/g daf to 28.90 cm^3/g daf. It can be seen that the influence of temperature on coal absorption capacity cannot be ignored.

According to the Langmuir volume and Langmuir pressure obtained at different temperatures, absorption capacities at different pressures were calculated. It was found that the

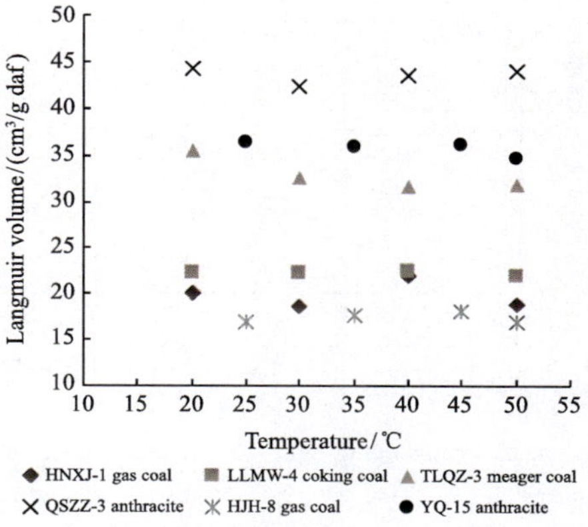

Fig. 4.4 Langmuir volume of isothermal absorption of the same sample at different temperatures

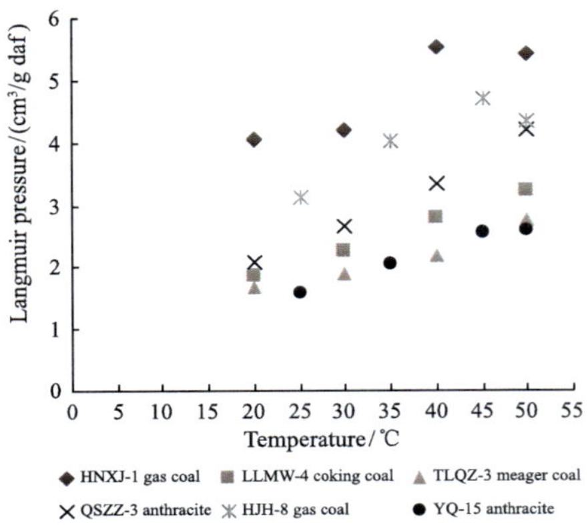

Fig. 4.5 Langmuir pressure of isothermal absorption of the same sample at different temperatures

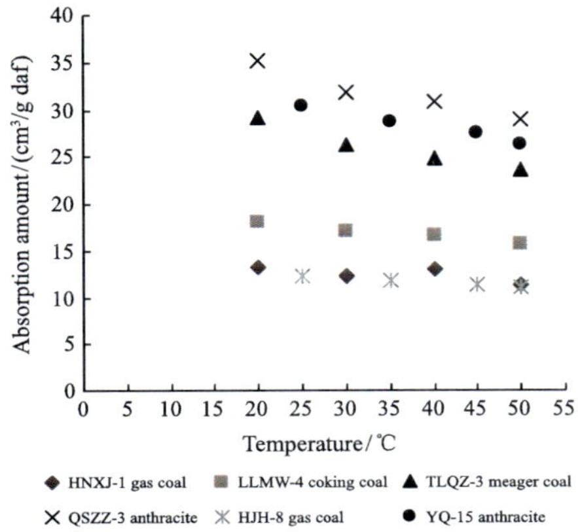

Fig. 4.6 Isobaric absorption amount of the same sample

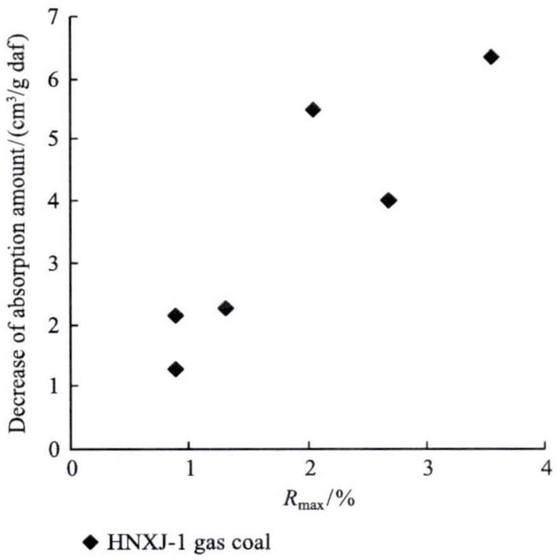

Fig. 4.7 Decrease of absorption amount of metamorphic coal

temperature of the same sample was linearly negatively correlated with the gas content at the same pressure (Figs. 4.8 and 4.9). The lower the pressure, the better the correlation. The coefficients in different linear formulas are different at different pressures and show regular changes. The amount of absorption at any temperature and the same pressure can be calculated by using a linear equation.

4.2.2 Effects of Temperature and Pressure on Coal Absorption Capacity

4.2.2.1 Coal Absorption Capacity Investigated by Isothermal Experiments

According to the Langmuir constant from isothermal absorption experiments on the same samples at 25 °C, 35 °C, 45 °C, 50 °C or 20 °C, 30 °C, 40 °C, 50 °C, and the Langmuir model, we built the linear relation between absorption amount and temperature. Using the relation, we calculated the absorption amount of each sample at any pressure and temperature, that is, the absorption amount under the combined influences of temperature and pressure (Fig. 4.10). The temperature on Fig. 4.10 is shown in Table 4.3.

The above is the absorption amount under the combined influence of normal geothermal gradient and pressure gradient, calculated based on the absorption constants at different temperatures. This method can be used to calculate the absorption capacity under the influences of different geothermal gradients and pressure gradients.

At the same temperature and pressure, when the isothermal absorption curve under the combined influences of temperature and pressure intersects with the isotherm absorption curve, their absorption amounts are close. When

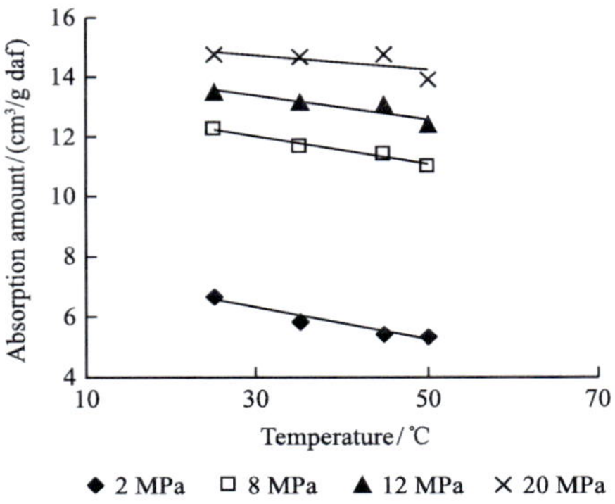

Fig. 4.8 Isobaric absorption amount of HJH-8 coal sample

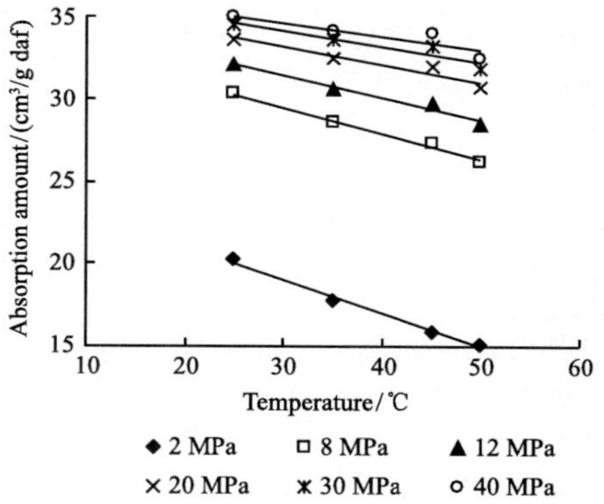

Fig. 4.9 Isobaric absorption amount of YQ-15 coal sample

the temperature and pressure in the former case is smaller than that in the latter case, the absorption amount in the former case is greater than that in the latter case, and the smaller the temperature and pressure, the larger the difference of the absorption amount. When the temperature and pressure in the former case is larger than that in the latter case, the absorption amount in the former case is smaller than that in the latter case, and the larger the temperature and pressure, the larger the difference of the absorption amount. The larger difference is caused by the change in temperature (Fig. 4.10).

For the coal different in metamorphism, the trend of the absorption curve is different under the combined influence of

Fig. 4.10 Absorption curves under the combined influence of temperature and pressure and isothermal absorption curves

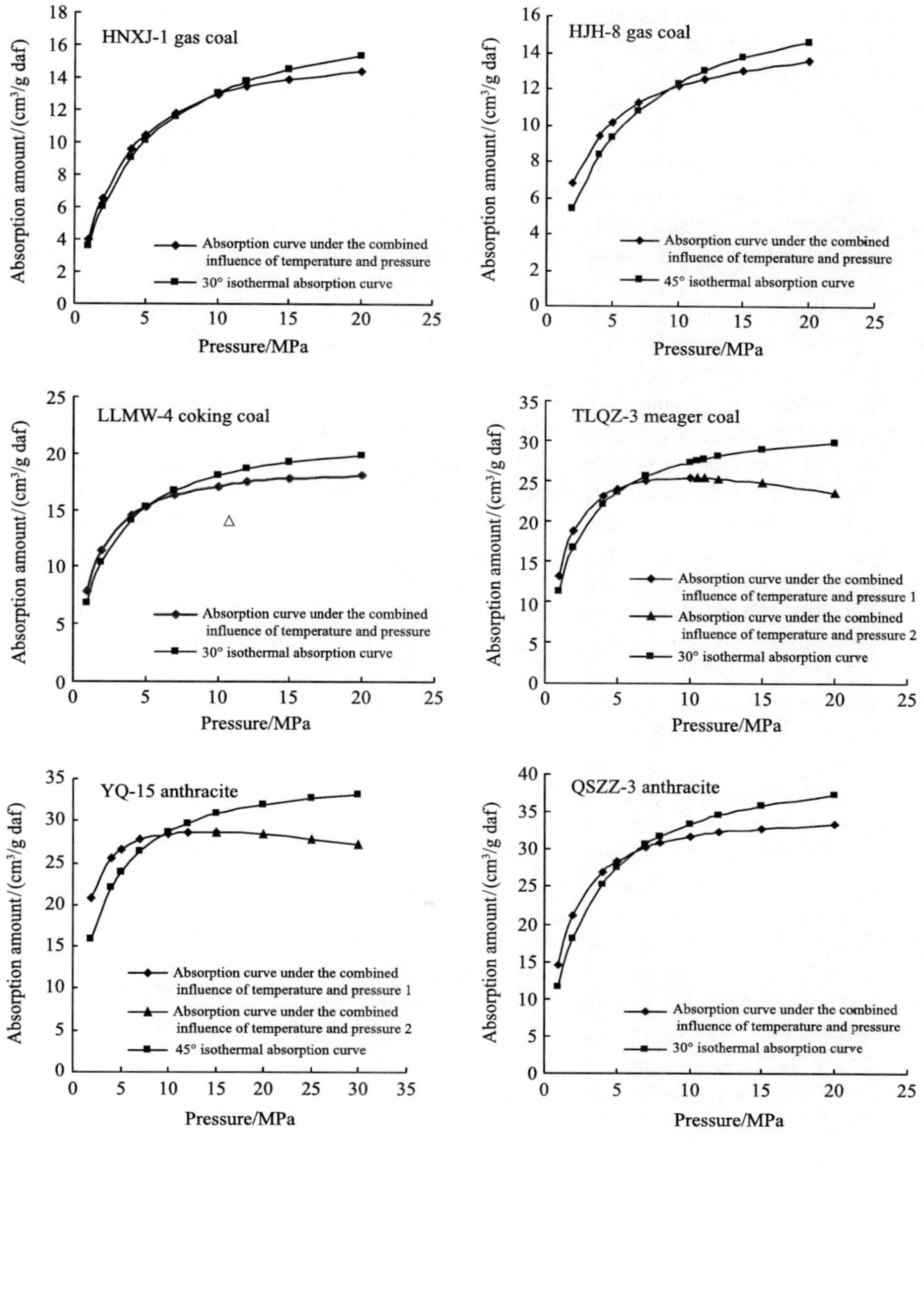

Table 4.3 Temperature on Fig. 4.10

Pressure/MPa	1	2	4	5	7	10	12	15	20	25	30	35
Temperature/°C	18	21	27	30	36	45	51	60	75	90	105	120

temperature and pressure. For example, samples HNXJ-1, HJH-8, LLMW-4 and QSZZ-3 show that the absorption capacity increases slowly with the increase of temperature and pressure, indicating that the influence of pressure on the absorption capacity plays a leading role, and the influence of temperature on the absorption capacity is less than pressure. When sample TLQZ-3 at temperature less than 45 °C and pressure less than 10 MPa, and sample YQ-15 at temperature less than 60 °C and pressure less than 15 MPa, their absorption amounts increase with the increase of temperature and pressure, indicating that pressure plays an leading role and the influence of temperature on the absorption amount is less than the influence of pressure. When sample TLQZ-3 at temperature greater than 45 °C and pressure greater than 10 MPa, and sample YQ-15 at temperature greater than 60 °C and pressure more than 15 MPa, their absorption amounts decrease with the increase of temperature and pressure, indicating that the influence of temperature is greater than the influence of pressure on the absorption amount (Fig. 4.10).

From the experimental results of six coal samples (Fig. 4.5), the absorption curves under the combined action of temperature and pressure have two distinct trends: one is that the absorption amount increases with increasing temperature and pressure; the other that the absorption amount increases first, and then decreases after the maximum value appears. This depends on the Langmuir pressure. The temperature of isothermal absorption of samples HNXJ-l, HJH-8, LLMW-4 and QSZZ-3 is 30 °C, and the Langmuir pressure is greater than 2.27 MPa. When the pressure is less than 40 MPa and the temperature is less than 135 °C, the absorption amount keeps increasing. For samples TLQZ-3 and YQ-15, when the temperature is greater than 45 °C and the pressure is greater than 10 MPa, the absorption amount decreases. For sample TLQZ-3, the isotherm absorption temperature is 30 °C, and the Langmuir pressure is 1.92 MPa; for YQ-15, the isothermal absorption temperature is 25 °C, and the Langmuir pressure is 1.59 MPa; for YQ-15, the isothermal absorption temperature is 35 °C, and the Langmuir pressure is 2.04 MPa. Research results show the Langmuir pressure increases with the increase of temperature. It is estimated that the Langmuir pressure is less than 2 MPa when the isothermal absorption temperature of sample YQ-15 is 30 °C. According to the Langmuir pressure of 6 coal samples and other data, when the temperature is 30 °C, and the Langmuir pressure is more than 2 MPa, the absorption amount of the coal samples increases under the combined influence of temperature and pressure; when the Langmuir pressure is less than 2 MPa, the absorption amount increases first and then decreases; different coals have different maximum inflection points on the absorption curves. It is derived that the smaller the Langmuir pressure, the earlier the inflection point appears; the pressure range of the inflection point is 10–16 MPa, and the temperature range is 45–60 °C. Laboratory statistics on the Langmuir pressure of coal samples different in metamorphism indicates that the Langmuir pressure varies widely from 1.11 to 13.97 MPa (Fig. 4.11), even the Langmuir pressure of the coal samples similar in metamorphism is quite different. The Langmuir pressure is mostly greater than 2 MPa. The vitrinite reflectance of the coal sample with the Langmuir pressure less than 2 MPa ranges from 0.87 to 3.68%. With limited experimental samples, now it is not very clear whether the absorption amount of any coal sample in this metamorphic stage with the Langmuir pressure less than 2 MPa will appear the maximum inflection point.

The conditions of the maximum inflection point of the absorption amount influenced by both temperature and pressure are that the absorption amount at low pressure is large, and that at high pressure is small; the absorption amount increased by increasing a unit pressure is less than that decreased by increasing temperature.

4.2.2.2 Coal Absorption Capacity Investigated by Temperature- and Pressure-Variable Experiments

It is an indirect method to study the absorption capacity and absorption characteristics by isothermal absorption experiments on the same coal samples at different temperature is an indirect method rather than a direct measurement under the

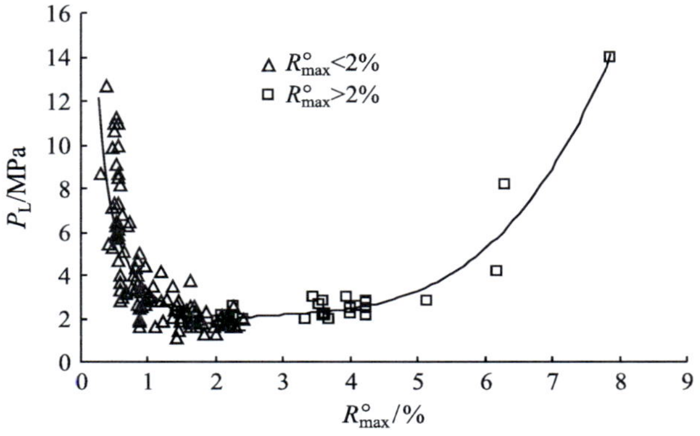

Fig. 4.11 Langmuir pressure versus R^o_{max}

simultaneous change of temperature and pressure. However, to learn real absorption characteristics, it is necessary to carry out absorption experiments at variable temperature and pressure. Absorption experiments at variable temperature and pressure were conducted on four coal samples at different ranks by simulating normal temperature and pressure. Figure 4.12 shows the absorption characteristics of the coal samples under the combined influence of temperature and pressure.

① As temperature and pressure increased, the absorption amount of the four coal samples increased, but the maximum value didn't appear. The results of isothermal absorption experiments on the four samples at 30 °C show that the Langmuir pressure is greater than 2.7 MPa. The Langmuir pressure at different temperatures is greater than 2 MPa and the absorption capacity increases, indicating the result of the increase in absorption amount at variable temperature and pressure is correct. Moreover, the experiment has initially proved the correct conclusion.

② When the temperature is less than 30 °C, the absorption capacity of anthracite, meager coal and coking coal at variable temperature and pressure is greater than the isothermal absorption capacity at the same pressure; and the former curve is above the 30 °C isotherm absorption curve. For long flame coal, the absorption capacity at variable temperature and pressure is always greater than the isothermal absorption capacity at the same pressure; and the former absorption curve is always above the 30 °C isothermal absorption curve. In theory, this phenomenon should not occur.

③ When the experimental conditions of two methods are the same, i.e. at 30 °C and 5 MPa (the fifth pressure point), the two curves of anthracite, meager coal and coking coal cross each other, and the absorption amount is the same. For long flame coal, the absorption amount at variable temperature and pressure is larger than the isothermal absorption amount, and the two curves are the closest.

④ When the temperature of the experiment at variable temperature and pressure is higher than that of the experiment at constant temperature, i.e. 30 °C, the absorption amount of anthracite/meager coal is significantly lower than the isothermal absorption at the same pressure, indicating that the equilibrium moisture in the coal decreases with increasing temperature. However, the influence of temperature on the absorption amount plays a major role. The absorption capacity of coking coal at variable temperature and pressure is lower than

Fig. 4.12 Comparison of absorption at variable temperature and pressure with 30 °C isothermal absorption of coal samples at different metamorphic stages. From low to high, • represents the absorption at 18 °C, 21 °C, 24 °C, 27 °C, 30 °C, 42 °C, 48 °C, 54 °C, 60 °C, 66 °C, 72 °C, respectively

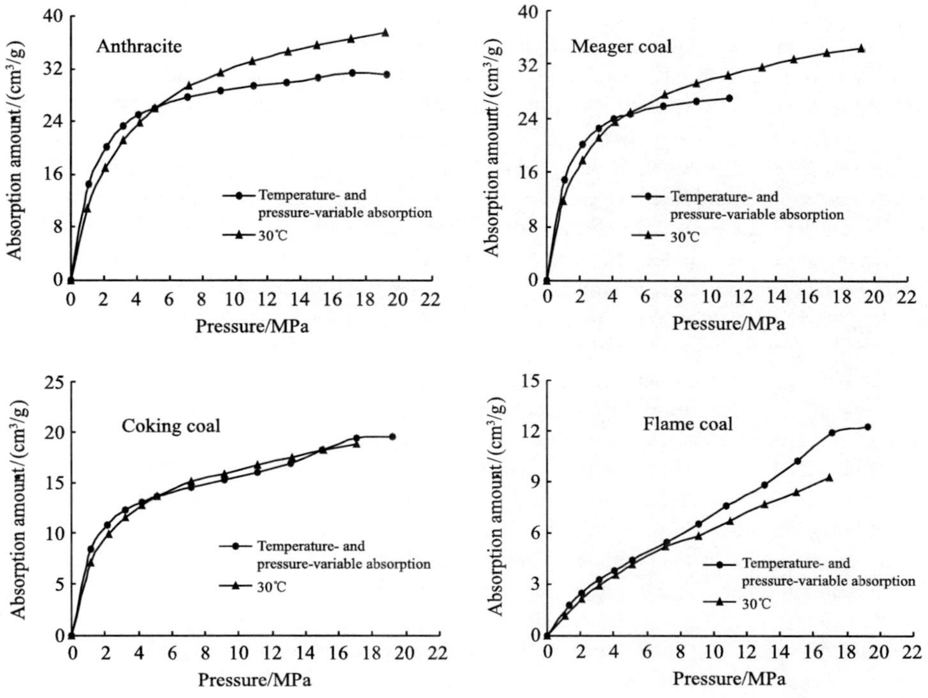

the isothermal absorption amount at the same pressure. But when the temperature is 60 °C and the pressure is 15 MPa (the tenth pressure point), the two curves cross again, and then the absorption amount at variable temperature and pressure starts to be larger than the isothermal absorption amount at the same pressure. This indicates that the absorption amount of coking coal decreases with increasing temperature, but when the equilibrium moisture decreases, the absorption amount begins to increase. The absorption capacity of long flame coal at variable temperature and pressure is always larger than the isothermal absorption capacity at the same pressure. The reason is that the equilibrium moisture in long flame coal is high, up to 15.11%, so that during the experiment at variable temperature and pressure, the equilibrium moisture controls the absorption. As the temperature increases, the equilibrium moisture decreases, and the absorption amount is larger than the isothermal absorption at the same pressure.

According to the comparison of absorption characteristics of different coal ranks in three temperature ranges during the above experiments at variable temperature and pressure and 30 °C isotherm experiments, when the temperature increased from 18 °C to 21 °C, 24 °C, ..., 72 °C, and the pressure increased from 1 to 2 MPa and 3 to 19 MPa, the equilibrium moisture in the the coal samples changed too. The equilibrium moisture is defined as the water content when the coal sample reaches a state of water saturation at 30 °C and relative humidity of 96%. It is similar to the definition of the highest intrinsic moisture in coal quality tests, with the same test principle but different methods. The main purpose of testing equilibrium water is to restore the state of the coal under reservoir conditions. Joubert and Grein (1973) found that when the coal seam did not reach the critical moisture, the increase of water led to the decrease of methane absorption. After the critical moisture, methane absorption no longer decreased with the increase of water. We also proved the conclusion by experiments. Therefore, we conducted absorption experiments under the condition equal to or greater than the equilibrium moisture to represent underground conditions. Since the equilibrium moisture of the coal was tested at 30 °C and relative humidity of 96–97%, it can be considered that, during the experiment at variable temperature and pressure, the equilibrium moisture does not change with temperature before 30 °C. However, when the experimental temperature increases from 30 °C to 36 °C, 42 °C, ..., 72 °C, the equilibrium moisture may decrease, the pores and surface originally adsorbed by water begin to absorb methane, so methane adsorption ability increases, especially in the coal at low rank, this phenomenon is more obvious.

4.2.3 Change Mechanism of Adsorption Capacity Under Reservoir Conditions

According to the true gas state equation $PV = nZRT$, at constant volume (V), the mole fraction of gas (n) is proportional to pressure (P) and inversely proportional to temperature (T); the mole fraction increases by only increasing pressure, but it decreases by only increasing temperature; when both temperature and pressure increase, the increase of the mole fraction depends on the extent of respective increase of the two. When pressure increases largely, the increase of the mole fraction caused by pressure is greater than the decrease of the mole fraction caused by temperature, so final mole fraction increases. When temperature increases greatly, the decrease of the mole fraction caused by temperature is greater than the increase of the mole fraction caused by pressure, so final mole fraction decreases. Following normal pressure gradient and temperature gradient, the mole fraction increases at 19 MPa and 72 °C, indicating that the influence of pressure is greater than the influence of temperature.

4.3 Adsorption Model Under Reservoir Conditions

The adsorption capacity and description model under the combined influence of temperature and pressure are guarantee to the accurate calculation of coal adsorption amount, accurate prediction of CBM content and resources, understanding of adsorption capacity under formation conditions, improvement of adsorption theory and research of coalbed methane mechanism. On the basis of adsorption experiments on coal samples at constant and variable temperatures and pressures, we built the model relating temperature, pressure and adsorption capacity under formation conditions by introducing the theory of adsorption potential.

4.3.1 Characteristics Curve of Methane Adsorption

The theory of adsorption potential holds that there is an adsorption potential ε everywhere in the adsorption space, and ε is equivalent to the work required to attract 1 mol of gas from external space to a point. The main force of physical adsorption is dispersive force, and the dispersive force is the only temperature-independent force. It can be concluded that the relationship of ε-V_{ad} (adsorbed phase volume) in the same adsorption system will not change with temperature, namely, the ε-V_{ad} curve of an adsorbent to the same type of gas is unique. In adsorption study, the ε-V_{ad} is called a characteristic curve. Many solid–gas physical adsorption studies have proved the correctness of this idea, such as adsorption of benzene, CO_2, etc. by activated carbon. Methane adsorption by coal belongs to physical adsorption. Since the force is mainly dispersive force, the characteristic curve of methane adsorption by the same coal should be unique.

To obtain the characteristic curve of the coal-methane adsorption system, two parameters are needed: adsorption potential ε and adsorption phase volume V_{ad}. The calculation idea of ε is that methane molecules are adsorbed on coal surface, so ε is equivalent to the work required to compress methane at equilibrium pressure P to the saturated vapor pressure at temperature T, i.e.

$$\varepsilon = \int_{P}^{P_0} V_{\text{free}} \mathrm{d}P = \int_{P}^{P_0} \frac{RT}{P} \mathrm{d}P = RT \ln \frac{P_0}{P} \tag{4.9}$$

where ε represents adsorption potential; V_{free} is the volume of the gas in the free state; P represents equilibrium pressure; P_0 is the saturated vapor pressure at temperature T; T represents equilibrium temperature; and R is a gas constant.

To obtain the characteristic curve, another parameter V_{ad} (adsorption phase volume) is needed, which can be calculated by the following formula.

$$V_{ad} = \frac{m}{\rho_{ad}} = \frac{V \times 16}{22400 \times \rho_{ad}} \tag{4.10}$$

where V_{ad} represents the volume of equilibrium adsorbed phase; m is the mass of adsorbed gas; V is the amount of equilibrium adsorbed gas; ρ_{ad} represents the density of the adsorbed phase. In this Book, the density of the adsorbed phase of methane is 0.375 g/cm^3.

When calculating ε under adsorption equilibrium conditions using Eq. (4.9), the saturated vapor pressure P_0 is a parameter that needs to be determined first. When $T < T_C$ (critical methane temperature), P_0 is existent and can be obtained from experiments. When $T > T_C$, since gas at above the critical temperature cannot be liquefied, there is no saturated vapor pressure, so it can only be calculated by estimating or empirical formula. The most common method is the empirical formula proposed by Dubinin (1960):

$$P_0 = P_C \left(\frac{T}{T_C} \right)^2 \tag{4.11}$$

where P_C represents critical pressure; T_C represents critical temperature.

Amankwah and Schwarz (1995) found that the modified Dubinin formula would be better when studying the characteristic curves of activated carbon adsorbing CH_4 and H_2. The formula for calculating the saturated vapor pressure is shown in Eq. (4.12), where k is related to the adsorption system.

$$P_0 = P_C \left(\frac{T}{T_C} \right)^k \tag{4.12}$$

In this Book, the improved Dubinin formula proposed by Amankwah and Schwarz (1995) was used to calculate P_0. By changing the k value, the characteristic curve at the highest correlation coefficient is obtained, and the calculation formula of the next step is determined thereby. Since the coal reservoir temperature is much higher than the critical methane temperature (−82.6 °C), the saturated methane vapor pressure P_0 calculated from the k value has no physical meaning and can only be regarded as a parameter in the characteristic curve equation. It is found that the correlation coefficient of the characteristic curve of the experimental coal sample gradually increases with the increase of the k value, and the results show that when $k \geq 2.7$, the correlation coefficient exceeds 0.99, so in the following discussion, data analysis is based on the P_0 calculated by $k = 2.7$. As for the relationship between the k value and the correlation coefficient, it is not possible to give a reasonable explanation based on present researches on high-pressure gas or supercritical gas adsorption.

Taking $k = 2.7$, six characteristic curves were plotted (Fig. 4.12). The results show that the ε-V_{ad} curves of each coal sample at four temperatures almost fall on the same curve, and the ε-V_{ad} is logarithmic, indicating that the ε-V_{ad} is independent of temperature. This further proves the interaction between coal and methane molecules is the dispersion force, and the adsorption process is physical.

According to the theory of adsorption potential, when ε = 0, the space between the adsorption potential surface and the adsorbent is the limit adsorption space, which is equivalent to the volume of the adsorption phase when saturation is reached. According to the relationship between ε and V_{ad}, the volume of the adsorption phase at ε = 0 is the limit adsorption phase volume. The ultimate adsorption amount V can be obtained by converting V_{ad} to the standard state. The ultimate adsorption capacities of the six kinds of coals were calculated in Table 4.4. It can be seen that the ultimate adsorption amount calculated by the theory of adsorption potential is very close to the average saturated adsorption amount obtained by the Langmuir equation. The ultimate adsorption amount calculated by the theory of adsorption potential is only one, but the V_L value obtained by the Langmuir equation is slightly different with temperature.

Table 4.4 Langmuir volume (V_L) and data of characteristic curve

Coal		HNXJ-1 gas coal	LLMW-4 coking coal	TLQZ-3 meager coal	QSZZ-3 anthracite	HJH-8 gas coal	YQ-15 anthracite
V_L calculated by Langmuir equation/(cm^3/g)	Max	25.47	28.34	40.04	54.93	21.97	45.04
	Min	−27.95	−29.06	−45.27	−57.93	−27.76	−48.82
	Ave	26.24	28.79	41.79	56.81	25.24	47.13
Ultimate adsorption amount calculated by characteristic curve/ (cm^3/g)		24.38	29.58	45.32	59.40	25.05	52.91

4.3.2 High-Pressure Isothermal Adsorption Curve and Its Correction to k Value

In order to verify the objectivity of the logarithmic relationship between the adsorption potential and adsorption phase volume and the applicability of the adsorption models of different ranks of coal at high pressure, high-pressure isotherm adsorption experiments were carried out respectively on dry coal samples tt22x-026, tt23x-030, tt23x-032 and tt98x-022. The vitrinite reflectances of the coal samples are 0.55%, 4.25%, 0.7%, 1.32%, respectively. The experimental temperature is 30 °C, the final pressures of the coal samples are 23.96, 23.95, 22.03, 23.97 MPa. In addition, high-pressure isothermal adsorption experiments were conducted on samples tt22x-026 and tt98x-022 for equilibrium moisture at 16.96 MPa and 17.09 MPa, respectively, in order to analyze the applicability of the models under different experimental conditions.

The characteristic curves of methane adsorption of the coal samples (Fig. 4.13) show that each rank of the coal sample has a good logarithmic ε-V_{ad} relationship under different experimental conditions, and the correlation coefficient is above 99%. This fully demonstrates the objectivity of the logarithmic relationship between the adsorption potential and the volume of the adsorbed phase under high pressure conditions.

However, as can be seen from Fig. 4.14, when the experimental pressure is higher than 16 MPa, although the adsorption potential and the adsorption phase volume are still logarithmic, according to the theory of adsorption potential and the relationship between the adsorption potential and the adsorption phase volume, $\varepsilon = 0$ reached within the experimental pressure range, that is, the saturated adsorption amount has been reached. However, the experimental data shows that within this range, the adsorption amount continues to increase, which contradicts the result of the model inference. And it can be seen from Table 4.5 that the ultimate adsorption amount obtained from the characteristic curve differs greatly from the V_L value obtained by the Langmuir equation. Therefore, we initially believe that the adsorption potential-adsorption phase volume model needs further correction and improvement when the final experimental pressure is high.

Analysis shows that when the final experiment pressure is greater than 16 MPa, it is higher than the saturated vapor pressure (16.08 MPa) at 30 °C. In this case, according to the adsorption potential formula $\varepsilon = RT \ln P_0/P$, if $\varepsilon = 0$, it must be $P = P_0$, that is, the equilibrium pressure is equal to the saturated vapor pressure at 30 °C. In other words, it can be used to estimate the ultimate adsorption volume only when the final pressure is less than the saturated vapor pressure (16.08 MPa) at 30 °C. When the experimental pressure is outside this range, the ultimate adsorption capacity at 30 °C is obtained by model, and the model must be corrected.

First increased the pressure point at $\varepsilon = 0$, that is, increased the saturated vapor pressure by delaying the occurrence of saturated vapor pressure. According to the improved Dubinin formula $P_0 = P_C[T/T_C]^k$, the k value in the model needs to be adjusted. Through continuous experiment, it is found that when $k = 4.5$, that is $P_0 = P_C(T/T_C)^{4.5}$, not only the correlation of the logarithmic relationship between the adsorption potential and the adsorption phase volume increases, but also they agree well with the experimental data, and the saturated absorption amount obtained by the logarithmic relationship is also very close to the maximum adsorption amount obtained by the Langmuir equation, as shown in Fig. 4.15 and Table 4.6.

4.3.3 Absorption Model

After fitting the ε-V_{ad} relationship through isothermal adsorption experiments on nearly 100 coal samples, it's found that the characteristic curves of all coal samples shows a logarithmic relationship between the adsorption phase

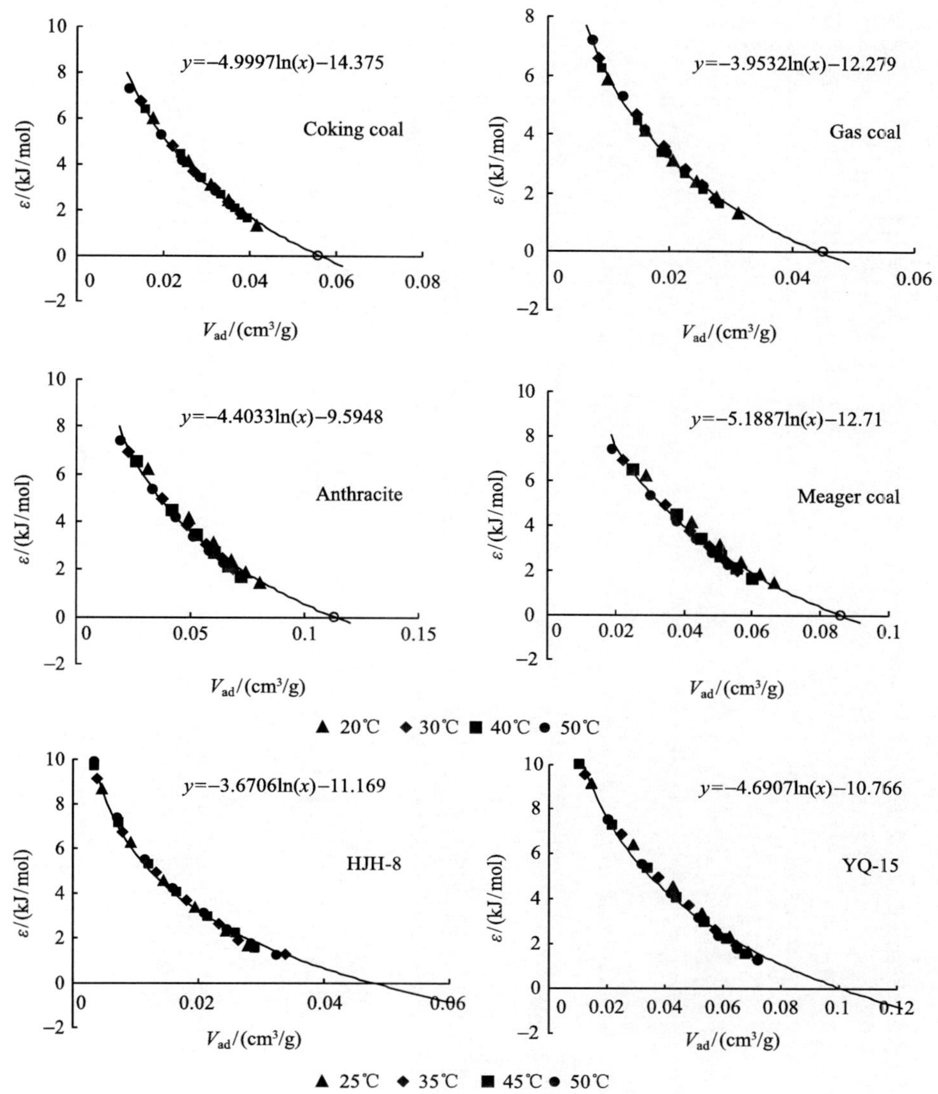

Fig. 4.13 Characteristic curves of methane adsorption

volume and the adsorption potential, and the correlation coefficient is very high. If the relationship between the adsorbed phase volume and the gas volume in the standard state is linear, the adsorbed phase volume of methane can be replaced by the amount of methane adsorbed in the standard state, that is, the adsorption amount and the adsorption potential are also logarithmic.

From ε-V_{ad}, $\varepsilon = RT \ln \frac{P_0}{P} = a \ln V + b$ is possible, i.e.,

$$\ln V = -\frac{RT}{a}\ln P + \left(\frac{RT}{a}\ln P_0 - \frac{b}{a}\right) \tag{4.13}$$

Taking $P_0 = P_C\left(\frac{T}{T_C}\right)^k$ into Eq. (4.13),

$$\ln V = -\frac{RT}{a}\ln P + \frac{RT}{a}[\ln P_C + k(\ln T - \ln T_C)] - \frac{b}{a} \tag{4.14}$$

Let $A' = \frac{R}{a}$, $B' = -\frac{b}{a}$,

$$\ln V = -A'T\ln P + A'T \times [\ln P_C + k(\ln T - \ln T_C)] + B' \tag{4.15}$$

At the constant temperature T, let

$$A = -A'T \tag{4.16}$$

$$= A'T \times [\ln P_C + k(\ln T - \ln T_C)] + B' \tag{4.17}$$

Equation (4.15) becomes

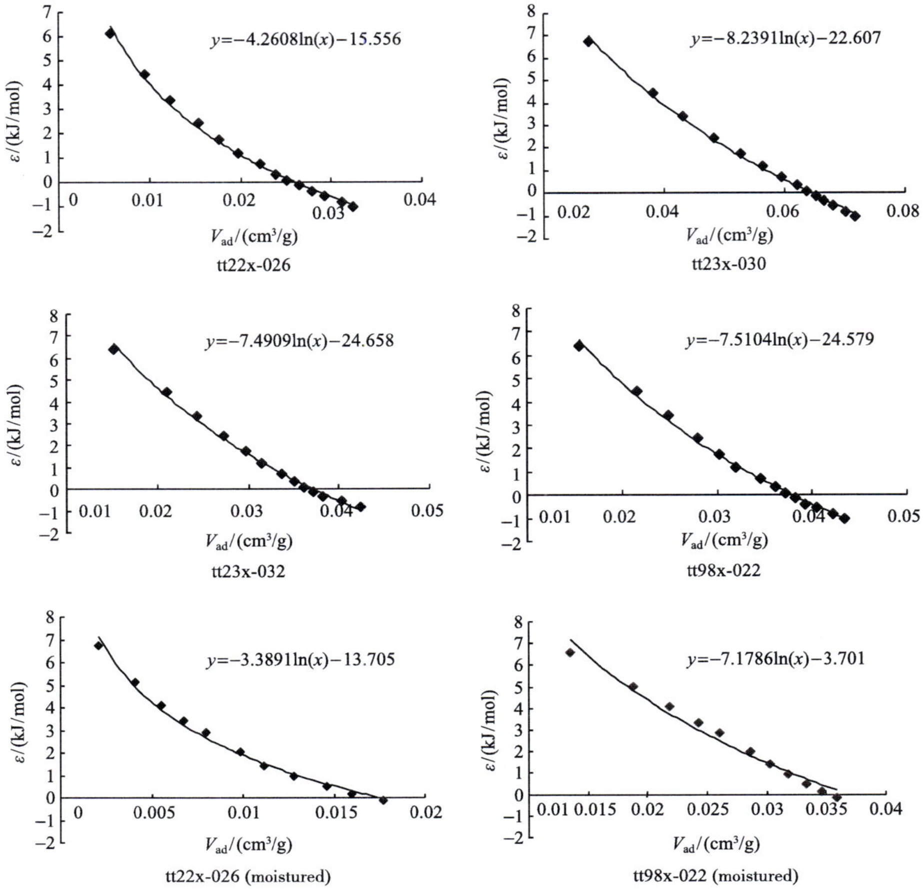

Fig. 4.14 High-pressure characteristic curves based on uncorrected models

Table 4.5 Langmuir volume and data of characteristic curves

Method	Dry coal samples				Moistured coal samples	
	tt22x-026	tt23x-030	tt23x-032	tt98x-022	tt22x-026	tt98x-022
V_L calculated by Langmuir equation/ (cm^3/g)	24.22	41.66	24.58	25.56	17.11	21.34
Ultimate adsorption amount calculated by characteristic curve/ (cm^3/g)	13.63	33.77	19.52	19.90	9.20	19.33

$$\ln V = A \ln P + B \tag{4.18}$$

From the logarithmic relationship between the adsorption potential and the adsorption amount, a logarithmically linear relationship is existent between the adsorption amount and the pressure at the same temperature. In order to verify the conclusion, the experimental results of the dozens of coal samples were studied. It's found that there is a logarithmically linear relationship between the adsorption amount and the pressure for all the coal samples, and the correlation is very high, basically at 0.98 or more (Fig. 4.16). It is known from the theory of adsorption potential that the ε-V_{ad} characteristic curve of the same adsorption system is unique (verified by previous studies), so the *A* and *B* values are unique at any temperature, and so are the *A*′ and *B*′ values. If the *A*′ and *B*′ values of a coal sample adsorbing methane are known, the adsorption amount at any different temperature and pressure can be obtained according to Eq. (4.15). According to the relationship of *A*′ and *B*′ and *A* and *B* in Eq. (4.16) and Eq. (4.17), it can be seen that, using the relationship between the adsorption amount and the pressure from the isothermal adsorption experiment at constant

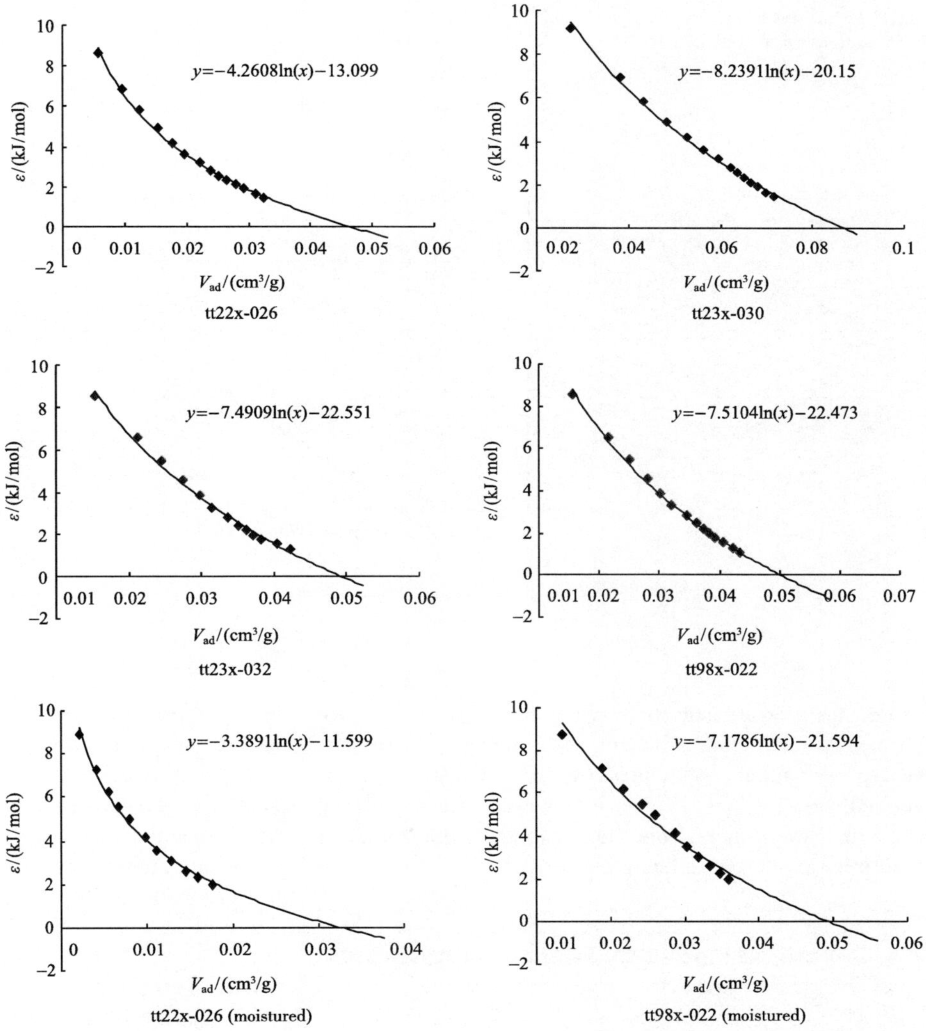

Fig. 4.15 High-pressure characteristic curves based on corrected models

Table 4.6 Langmuir volume and data of characteristic curves

Method	Dry coal samples				Moistured coal samples	
	tt22x-026	tt23x-030	tt23x-032	tt98x-022	tt22x-026	tt98x-022
V_L calculated by Langmuir equation/ (cm^3/g)	24.22	41.66	24.58	25.56	17.11	21.34
Ultimate adsorption amount calculated by characteristic curve/ (cm^3/g)	22.35	43.60	25.87	26.34	17.13	25.92

temperature, A and B are available, and then A' and B' dependent of temperature are available, and finally the amount of adsorption at any other temperature according to Eq. (4.16). Isothermal adsorption experiments were performed at different temperatures on different coal samples to verify the model proposed.

4.4 Verification and Application of Models

According to the theory of adsorption potential, the relationship between the adsorption potential and the adsorption phase volume of methane was preliminarily established,

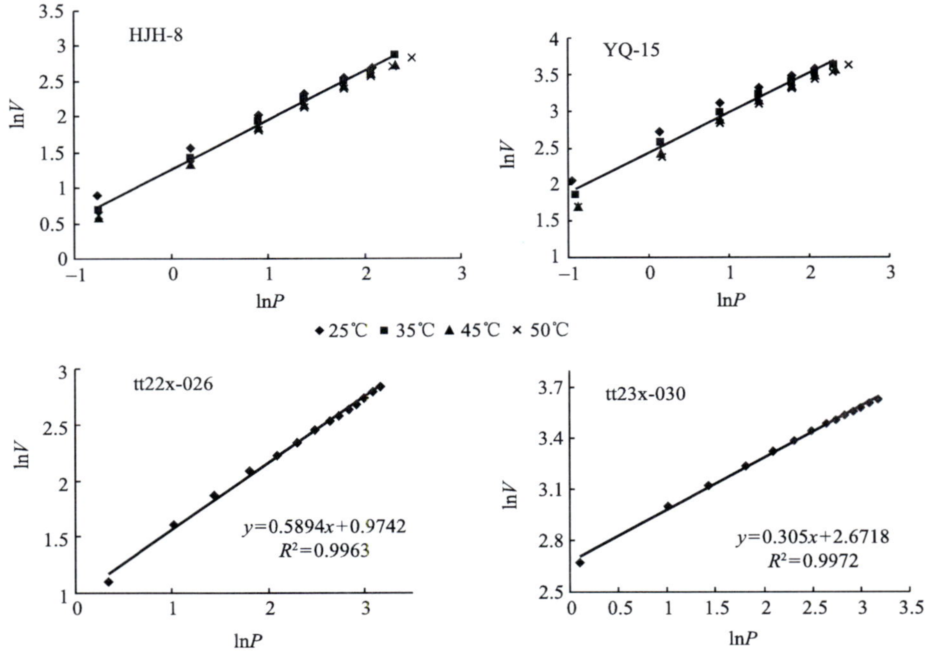

Fig. 4.16 Linear ln *V*-ln *P* relationship of coal samples

namely the characteristic curve equation. Based on the above physical model, the mathematical model of methane adsorption amount and pressure and temperature was derived, that is, $V = f(P, T)$. Next we use the model and isothermal experimental data to verify the applicability of adsorption models at different temperatures.

4.4.1 Verified Models by Isothermal Adsorption Experiments

In this study, gas coal samples from Huainan mining area, coking coal samples from Hedong mining area, meager coal samples from Lu'an mining area and anthracite samples from Jincheng mining area were used to conduct adsorption experiments at 30 °C and build the relationship between ln*V* and ln*P*. After obtaining *A* and *B*, *A*′ and *B*′ of the four kinds of coal samples are available. Taking the experimental pressures at four temperatures into Eq. (4.15), and the adsorption amount calculated by the model can be obtained (Since the highest experimental pressure is about 10 MPa, $k = 2.7$). The isothermal adsorption curves at 20, 30, 40, and 50 °C were calculated by the method above. The theoretical and measured results (Fig. 4.17) show that the predicted curves are very close to the measured, except gas coal at 40 and 50 °C and meager coal and anthracite at 20 °C.

The adsorption isotherms at 25, 35, 45, and 50 °C were predicted using the adsorption experiments of the No. 8 coal seam in Huajiahu mining area and the No. 15 coal seam in Yangquan mining area (Fig. 4.18). The predicted curves of the two types of coals at four temperatures are very close to the measured points. This further proves the correctness of the model and indicates that, within given temperature and pressure ranges, when the measured conditions and the predicted conditions are very close, the model can be used to predict the amount of adsorption at other temperatures and pressures based on the experimental data at some temperature.

4.4.2 Temperature- and Pressure-Variable Experiment Results and Models

In this study, coal samples of four coal ranks, including long flame coal, coking coal, meager coal and anthracite, were made experiments at variable temperature and pressure, following the experimental conditions and methods mentioned above. The experimental temperature is 18–72 °C, and the pressure is 1–19 MPa. A total of 12 temperature and pressure points were taken, involving low, medium and high temperature, constant and high pressure conditions. The isothermal adsorption experimental data at 30 °C (the highest pressure at about 19 MPa), were used to establish models to predict the experimental results of 12 temperature and pressure points at variable temperature and pressure.

Figures 4.19 and 4.20 show the results of 30 °C isothermal adsorption experiments and those at variable temperature and pressure. The two kinds of curves intersects each other at nearly the fifth experimental point for long flame coal, coking coal, meager coal and anthracite. Before this point, the adsorption capacity at variable temperature

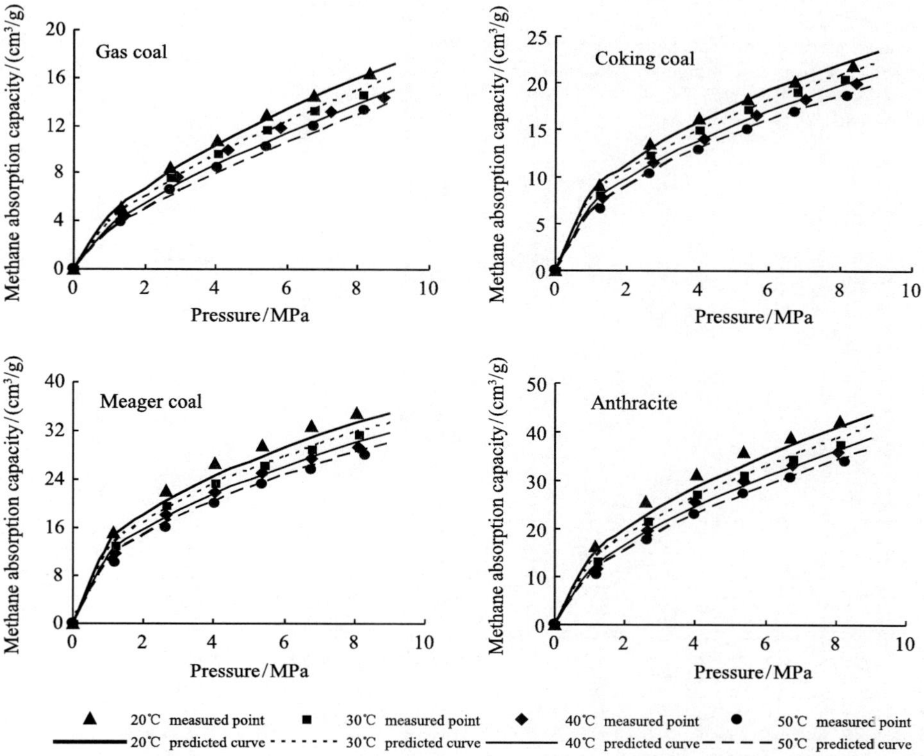

Fig. 4.17 Adsorption isotherms based on 30 °C characteristic curve

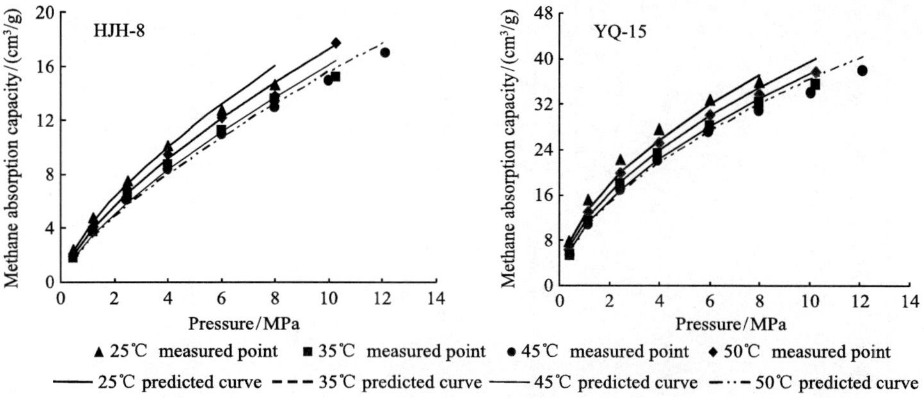

Fig. 4.18 Adsorption isotherm curves based on 35 °C characteristic curve

and pressure is higher than that at constant 30 °C. This is mainly because the former temperature is lower than 30 °C, but the pressure is similar, so that the adsorption at low temperature is higher than that at high temperature at the same pressure. This is consistent with previous studies. Starting from the sixth experimental point, the two kinds of experiments show different performances for four ranks of coal samples. With the increase of pressure, the adsorption amount of meager coal and anthracite at variable temperature and pressure is gradually lower than that at 30 °C; the adsorption amount of coking coal at variable temperature and pressure is also lower than the isothermal adsorption amount, but the trend is not obvious; the adsorption amount of long flame coal at variable temperature and pressure is gradually higher than the isothermal adsorption amount.

The highest pressure of temperature- and pressure-variable and 30 °C isotherm adsorption experiments is about 16 MPa. First used the model formula at low to medium pressure, that is, $k = 2.7$, to predict the adsorption amount at 18 °C, . . ., 72 °C, and 1–19 MPa, based on 30 °C isothermal adsorption experimental data, and then compared with the experimental results (Fig. 4.19). Before 30 °C, the predicted adsorption amounts of long flame coal and coking coal are close to the experimental results. However, as the

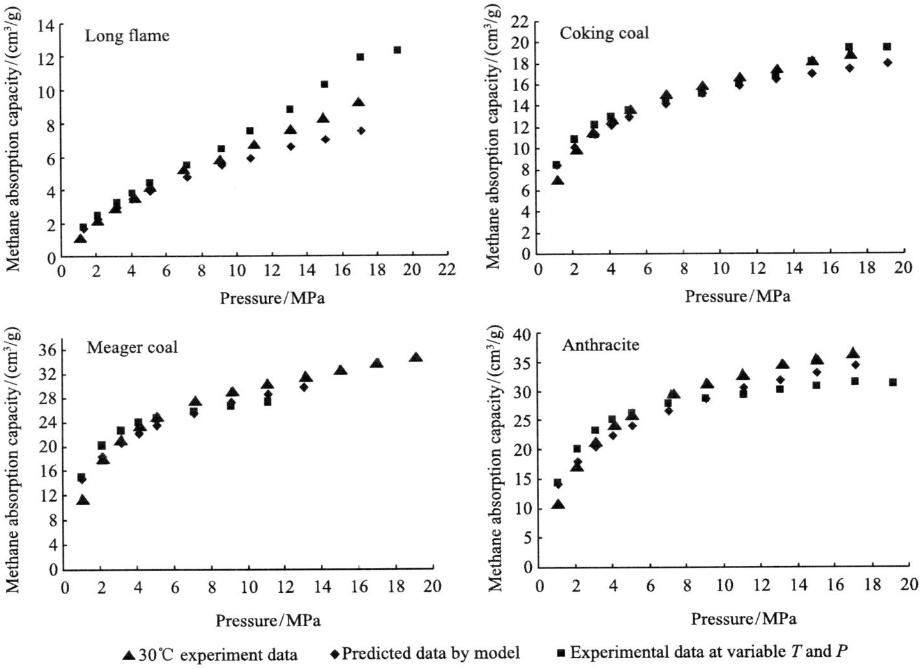

Fig. 4.19 Comparison of predicted and experimental results (k = 2.7). ■ from low to high at 18, 21, 24, 27, 30, 36, 42, 48, 54, 60, 66 and 72 °C

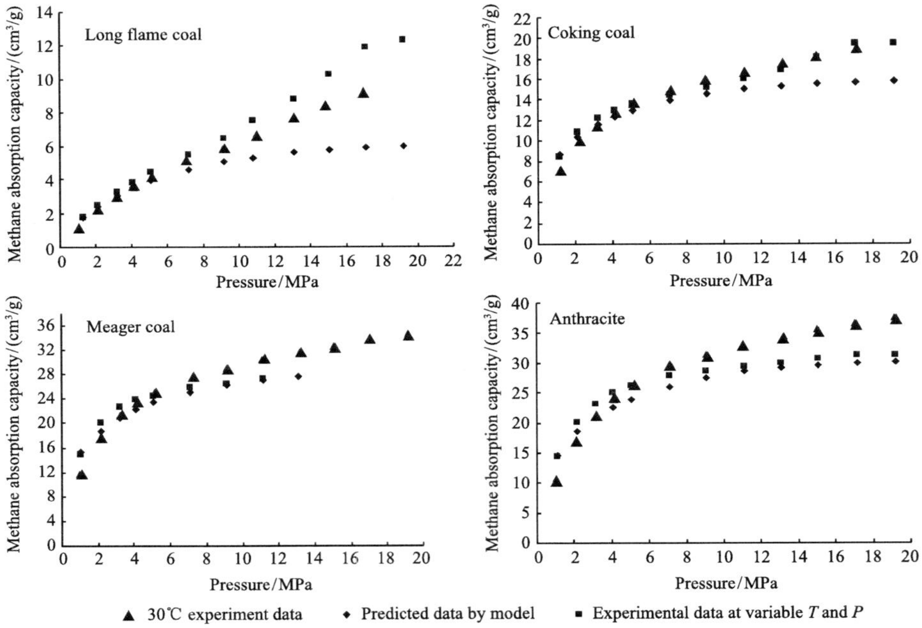

Fig. 4.20 Comparison of predicted and experimental results (k = 4.5). ■ from low to high at 18, 21, 24, 27, 30, 36, 42, 48, 54, 60, 66 and 72 °C

temperature and pressure increase, the difference between the two is getting larger and larger. One cause is that the model formula of $k = 2.7$ is not accurate because the experimental pressure is higher than 16 MPa; the other is the model predicts the adsorption amount at other temperatures based on the isothermal adsorption at 30 °C, without considering the effect of changes in moisture and other substances on the adsorption results at higher temperatures. For meager coal and anthracite, the difference between the predicted adsorption amount and the experiment results keeps very obviously, because the model formula taking $k = 2.7$ is not accurate.

Then taking $k = 4.5$, the adsorption amounts at 18 °C, …, 72 °C and 1–19 MPa based on the 30 °C isothermal adsorption experimental data were predicted and compared with the experimental results at variable temperature and pressure (Fig. 4.20). The behaviors of long flame coal and coking coal are similar to those when $k = 2.1$, that is, the predicted adsorption amounts are very close to the experiment results. As the temperature increases, the difference between the two becomes larger and larger. The predicted adsorption capacities of meager coal and anthracite are very close to the experiment results. The causes are those described above. In addition, it also provides a reliable basis for the rationality of the model.

The adsorption amounts of meager coal and anthracite predicted by models are very close to experiment results, but those of long flame coal and coking coal predicted by models are quite different from experiment results. The reason may be, during the temperature-and pressure-variable experiment, with the increase of experimental temperature, the moisture in the pores of the coal sample evaporates (the coal sample was found obviously drier after the experiment), and accordingly the increased pore surface area absorbs more methane. Such influence is greater than the increase of temperature, so the adsorption amount is larger than 30 °C isothermal adsorption amount. According to the isothermal adsorption experiment results and adsorption mechanism, as temperature increases, adsorption should decrease. So the amount of adsorption calculated by the model should be credible, which is good correction to the influence of other factors in the process of variable temperature and pressure.

From the above research results, it is clear that the prediction model established in this project can be used to predict the adsorption capacity at any reservoir temperature and pressure based on the results of isothermal adsorption experiments of a coal sample at any temperature (Fig. 4.20). The relative error is small, and the predicted result is very close to the experimental result. Although there is a certain error, it is enough to solve problems in macroscopic geological research.

4.4.3 Coal Adsorption Capacity in Qinshui Basin

In a basin or a coalfield, the burial depth is different due to the late tectonic deformation, and the burial depth causes different adsorption amounts due to temperature and pressure changes. It is impossible to accurately understand the distribution law of the adsorption amount under the coal reservoir conditions of the basin or coalfield by sampling at any depth and carrying out experiments at any temperature. The best method is to use limited experimental data to find the distribution law. Next, taking the Qinshui Basin as an example, we describe how to estimate the adsorption capacity at any buried depth and coal metamorphic condition.

4.4.3.1 Characteristics of Coal Adsorption

The Qinshui Basin is the key research area in this study. In order to study the coal adsorption characteristics and controlling factors in the Basin, researchers went to coalmines to collect coal samples and CBM samples and other data, and then carried out isothermal adsorption experiments on the coal samples and related analysis.

Many isothermal adsorption experiments show that the maximum Langmuir volume of the No. 3 coal seam of the Shanxi Formation changes largely, from 19.70 to 55.36 m^3/g daf, and the Langmuir volume in the basin has a certain regularity. The adsorption capacity of dry ashless coal is the strongest in the southern part, 46 m^3/g daf, then becomes weak from the southern edge toward northwest, and finally to the lowest near the Huozhou, 20 m^3/g daf. From the southern edge toward northeast, the adsorption capacity is also weakening, lower near the Zhangzi area, 34 m^3/g daf. From the Anze area toward north, the Langmuir volume contours are distributed in the north–south direction, the adsorption capacity in the east is stronger than that in the west, and the east to the Heshun-Songta area is a strong adsorption area with the Langmuir volume of 42 m^3/g daf. From the east toward west, the Langmuir volume gradually becomes smaller and smaller, to 28 m^3/g daf in the Gujiao mining area (Fig. 4.21).

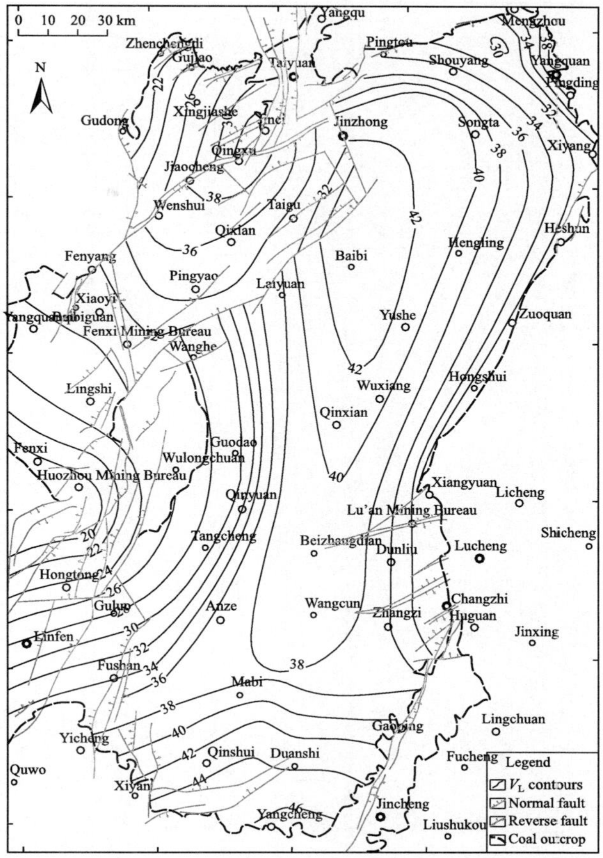

Fig. 4.21 V_L distribution map (isothermal adsorption) in Qinshui Basin (unit: cm^3/g)

4.4.3.2 Relationship Between Langmuir Volume and Coal Metamorphism

In the Qinshui Basin, the range of coal metamorphism is wide, from fat coal, coking coal and lean coal to meager coal and anthracite, and the vitrinite reflectance is 0.90%–5.13%. From the Langmuir volume contours and the reflectance contours of the No. 3 coal seam, the Langmuir volume is closely related to the metamorphic distribution law (Fig. 4.22). The laws of the two are basically the same: in the area with high coal metamorphism, the Langmuir volume is large; in the area with low coal metamorphism, the Langmuir volume is small. The coal metamorphism in the Jincheng mining area is the largest, the coal is anthracite; the coal metamorphism in Huozhou is the lowest, the coal is fat coal, and the Langmuir volume is the smallest. These phenomena are sufficient to show that coal metamorphism has a controlling effect on the adsorption capacity.

Regression analysis shows that the Langmuir volume and R_{max} of the coal in the basin is logarithmic and well correlated (Fig. 4.23). The Langmuir volume is available when R_{max} is 1.08–4.96%.

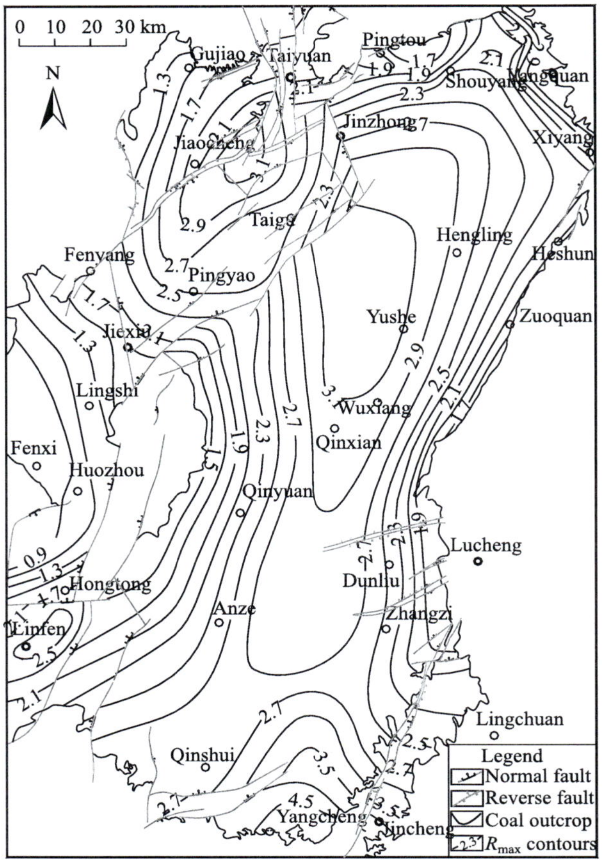

Fig. 4.22 Vitrinite reflectance contours of the No. 3 coal seam in Qinshui Basin (unit: %)

Regression analysis shows that the Langmuir pressure and R_{max} of the coal is polynomial. The Langmuir pressure is available when the R_{max} is 1.08–4.96%. The Langmuir pressure and the Langmuir pressure can be used to predict the adsorption capacity.

4.4.3.3 Coal Adsorption Capacity

Depth, Temperature and Pressure

With available data of coalfield exploration, coalbed methane exploration, coalfield prediction, the buried depth map of No. 3 coal seam was plotted (Fig. 4.24), and then reservoir temperature and pressure were estimated. The current geothermal gradient is low in the Qinshui Basin. The geothermal gradient measured in Well QC 1 is averaged 2.62 °C/hm between 500 and 1100 m, which was used in the Project. 31 coal seams in the Qinshui Basin were measured. The reservoir pressure is lower, generally at 3.68–9.72 kPa/hm. The average pressure gradient in the south of the Tunliu area is 6.73 kPa/hm, and 6.51 kPa/hm in the Yangquan area. The average value of the basin is 6.68 kPa/hm. Temperatures and pressures at different depths were calculated based on the reservoir temperature and pressure gradients (Table 4.7).

R_{max}

The coal metamorphism in the Qinshui Basin is distributed as strips in a wide range. Coal metamorphism is the main factor controlling the amount of coal adsorption. To know the adsorption amount at different buried depths, it is necessary to know the metamorphism of the coal. By overlapping burial contours with R_{max} contours, and selecting the intersection of the two contours, the R_{max} is available at different buried depths.

Coal Absorption Capacity

From the above studies, the relationship between coal metamorphism and adsorption constant is available. And using the reservoir adsorption model (Eq. 4.11), the adsorption amount is available at different temperatures, pressures, depths and coal metamorphism (Figs. 4.25 and 4.26). According to the adsorption amount, the distribution map of coal adsorption capacity of the reservoirs shallower than 2000 m (under the reservoir conditions) was plotted under the influence of depth, pressure, temperature, coal metamorphism, ash and equilibrium moisture in the Qinshui Basin (Fig. 4.27), as well as the distribution map of coal adsorption under the combined influence of depth, pressure, temperature and coal metamorphism (Fig. 4.28). The adsorption amount increases from the edge to the basin

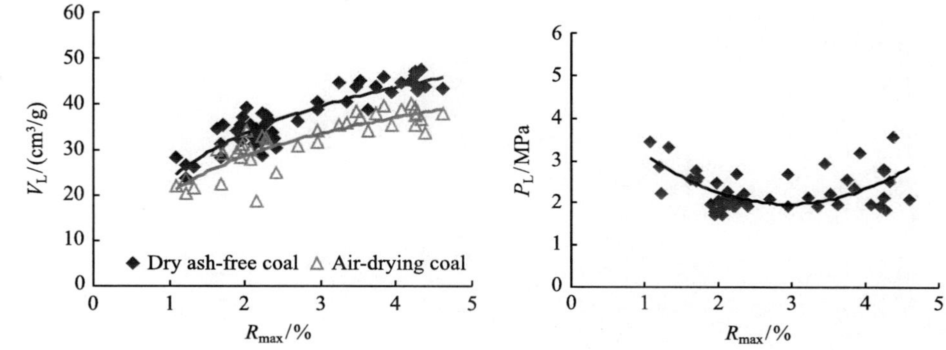

Fig. 4.23 V_L and P_L versus R_{max}

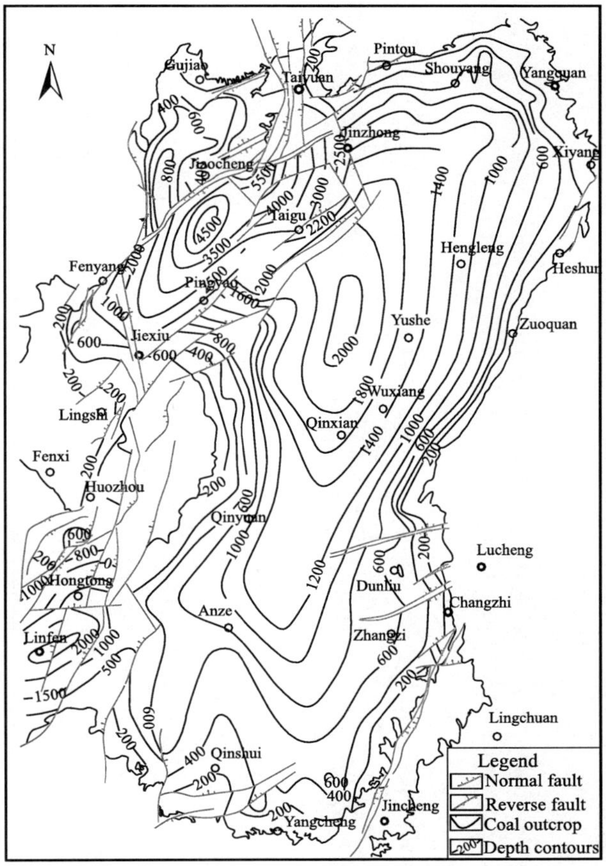

Fig. 4.24 Buried depth map of no. 3 coal seam in Qinshui Basin (unit: m)

center, and the distribution of adsorption contours are not exactly the same as the Langmuir volume and the vitrinite reflectance contours, especially in the south, but similar in the north and central parts. They are roughly the same as the burial contours, indicating that coal metamorphism has an effect on adsorption amount, but the effect of burial depth is greater than that of coal metamorphism, that is, the combined effect of pressure and temperature is greater than the effect of coal metamorphism.

4.4.4 Scientific and Applied Values

The temperature and pressure in the coal reservoir change with the depth. The temperature and pressure change in different basins. Even in the the same seam with the same metamorphic degree, they may be different at different depths. Such changes in temperature and pressure cause the difference in coal adsorption capacity. The deeper the burial, the higher the pressure and the higher the temperature. Pressure is a very important condition for the increase of coal adsorption, while temperature increase reduces the amount of coal adsorption. Such an increase and a decrease bring a major problem in predicting coal adsorption. Inaccurate prediction of adsorption also means inaccurate prediction of gas content. In turn such inaccuracies can affect a series of problems such as reliable CBM resources and gas production capacity. Using the established coal adsorption model under reservoir conditions, and isothermal adsorption experimental data or empirical formula of Langmuir volume and pressure, the adsorption amount at any reservoir temperature and pressure can be more accurately predicted. This method makes the application of isothermal adsorption experimental data and the prediction of gas content possible, and is very applicable.

The model for predicting coal adsorption under reservoir conditions is based on the theory of adsorption potential. A large amount of experimental data have proved the model reliable. Application in the Qinshui Basin has proved it feasible.

Table 4.7 Coal reservoir pressure and temperature

Depth/m	Pressure/MPa	Temperature/°C
200	1.34	15.93
400	2.67	21.17
600	4.01	26.41
800	5.34	31.65
1000	6.68	36.89
1200	8.02	42.13
1400	9.35	47.37
1600	10.69	52.61
1800	12.02	57.85
2000	13.36	63.09

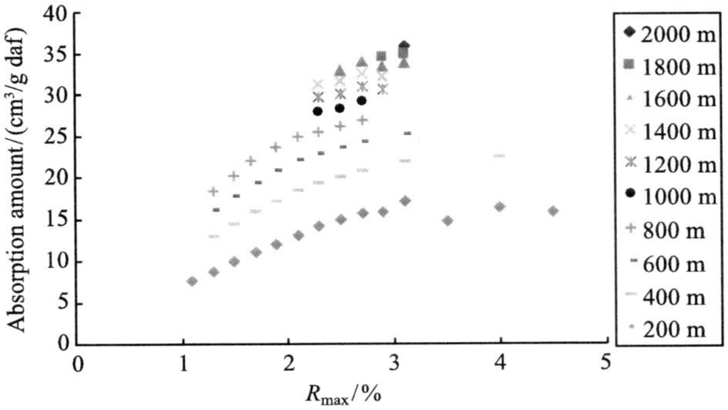

Fig. 4.25 Adsorption of different metamorphic coals at the same depth in Qinshui Basin

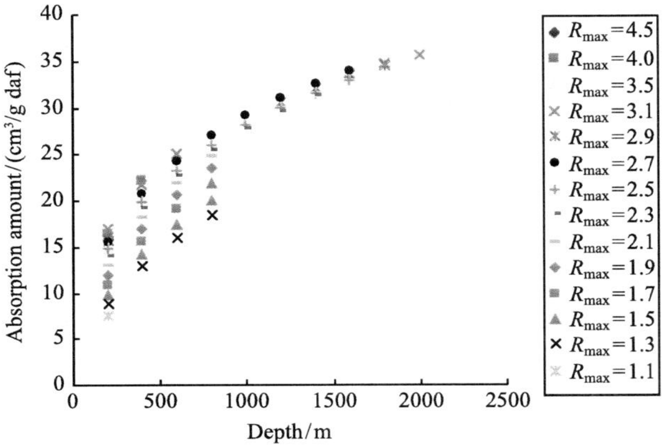

Fig. 4.26 Adsorption of the same metamorphic coal at different depths in Qinshui Basin

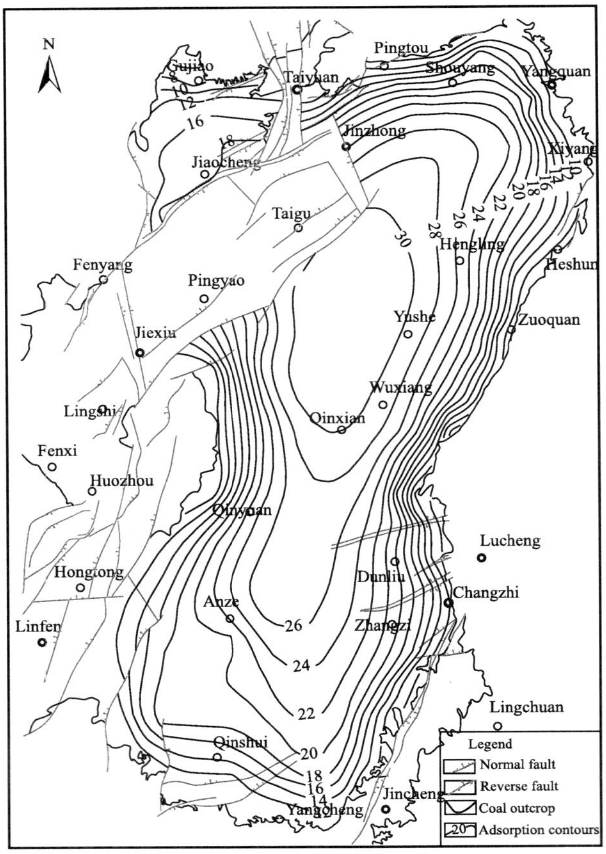

Fig. 4.27 Adsorption amount of no. 3 coal seam in Qinshui Basin under the influence of coal metamorphism, ash, equilibrium moisture, temperature, and pressure (unit: cm^3/g daf)

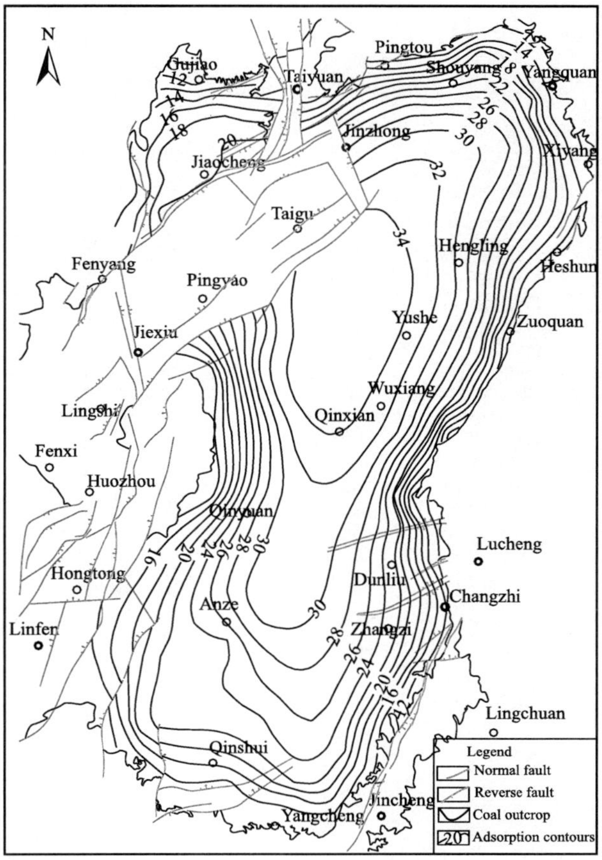

Fig. 4.28 Adsorption amount of no. 3 coal seam in Qinshui Basin under the combined influence of coal metamorphism, temperature and pressure (unit: cm^3/g daf)

References

Amankwah KAG, Schwarz JA (1995) A modified approach for estimating pseudo-vapor pressure in the application of the Dubinin-Astskhov equation. Carbon 33(9):1313–1319

Dubinin MM (1960) The potential theory of adsorption of gases and vapors for adsorbents with energetically nonuniform surface. Chem Rev 60(2):235–241

Joubert JI, Grein CT (1973) Sorption of methane in moist coal. Fuel 52:181–185

Zhang Q, Feng SL, Yang XL (1999) Residual gas content in coal and its influencing factors. Coalfield Geol Explorat 5:26–28 (in Chinese)

Zhang QL, Cao LG (2003) Research on data processing in isothermal adsorption experiment on coal. J China Coal Soc 28(2):131–135 (in Chinese)

5 Dynamic Conditions and Accumulation/Diffusion Mechanism of CBM

CBM accumulation is controlled by various geological conditions. In terms of the results, dynamic conditions are root causes, and accumulation effect is the reflection of CBM enrichment and high yield. However, in terms of process, dynamic conditions are only the reflection of CBM accumulation, and accumulation effect is the result of the dynamic balance of formation energy. How to profoundly understand the results of the accumulation process (the accumulation effect) from the reflection of dynamic conditions? In other words, what is the meaning of the connection between the reflection and the result? How to deeply understand the features and process of the connection (intermediate process), and then predict of the law of CBM enrichment and high-permeability zone distribution? Questions such as these can be attributed into "Dynamic conditions and accumulation/diffusion mechanism of CBM". This chapter describes the results of the discussion on this issue.

5.1 Structural Dynamics for CBM Accumulation

Structural dynamics plays a controlling role in CBM accumulation at different levels. At the basin level, regional structural setting and evolution are fundamental factors controlling the accumulation and distribution of CBM zones. At the structural level, structural styles and traps are dominant factors controlling the occurrence and enrichment of CBM. At the reservoir level, structures control the permeability and heterogeneity of coal reservoirs through the influence on reservoir pores and fracture systems.

5.1.1 Structural Evolution Laid the Foundation for CBM Accumulation

Studies have shown that the key period in the history of basin evolution has formed the basic conditions conducive to the formation of CBM. For the Qinshui Basin and the eastern margin of the Ordos Basin, the most important structural factor controlling CBM accumulation at the basin level is tectonic differentiation since the Yanshanian.

5.1.1.1 Control of Yanshanian Structural Inversion on CBM Accumulation

Since the Mesozoic and Cenozoic, the remote effects of the collision among the Indian plate, the Pacific plate and the Eurasian plate have caused the intracontinental orogeny in the North China plate and formed a series of basin-ridge structural units that are against and compatible. It is because of this transfer effect of tectonic dynamics that the tectonic deformation of the basin-ridge tectonic units in the central part of the North China plate is gradually weak from east to west.

In the coal-bearing area in the eastern Taihang Mountains, the Carboniferous and Permian systems were strongly reconstructed by late tectonic movements, and evolved into a modern tectonic pattern dominated by fault blocks, and the burial history of the coal seams is more complicated. The Qinshui Basin between the Taihang orogenic belt and the Lüliang orogenic belt is generally a large complex syncline basin, but its western margin shows a certain degree of fault depression. Multiple periods of uplifting movements since the Indosinian period made the burial history of the Carboniferous and Permian coal seams complex, and strong tectonic inversion occurred in the early to middle Yanshanian (Fig. 5.1). The Ordos Basin to the west of the Lüliang orogenic belt is a large complex syncline basin which eastern margin underwent large-scale settlement and sedimentation in the early to middle Yanshanian. Until the middle and late Yanshanian, tectonic inversion began, so the burial history of the coal seams is relatively simple (Fig. 5.2).

Obviously, although the Qinshui Basin and the Ordos Basin are common in relatively weak tectonic activities and primary tectonic inversion occurring in the Yanshanian, they differ in terms of tectonic evolution history and dynamic

Y. Song et al., *Coalbed Methane in China*,
https://doi.org/10.1007/978-981-33-4725-0_5

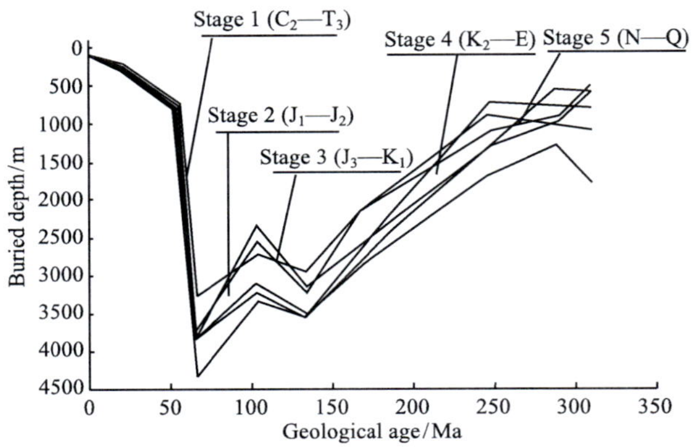

Fig. 5.1 The burial history of the Carboniferous–Permian coal seams in the southern part of Qinshui Basin

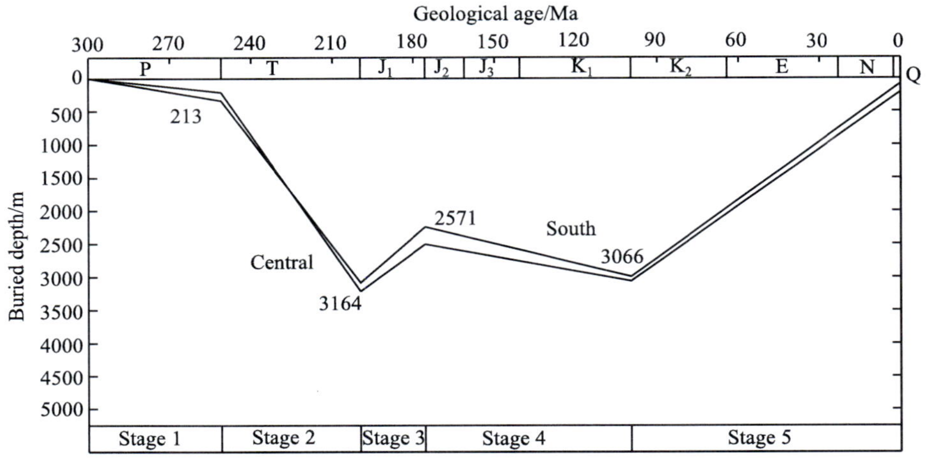

Fig. 5.2 The burial history of the Carboniferous–Permian coal seams in the southern segment of the eastern margin of Ordos Basin

conditions. Such tectonic background controls the accumulation characteristics of CBM from two aspects. On the one hand, the two basins may be conducive to the enrichment of CBM due to the relatively weak transformation, and may be beneficial to the permeability development of coal reservoirs due to structural uplift since the Yanshanian; on the other hand, the internal structural differences caused by different dynamic conditions may make the accumulation and distribution laws of CBM different in the two basins.

5.1.1.2 Profound Influence on CBM Accumulation from Intense Middle Yanshanian Magmatic Events

The regional survey data show that the Qinshui Basin, the eastern margin of the Ordos Basin and the surrounding orogenic belts have experienced tectonic magmatic activities in the Hercynian, Indosinian and Yanshanian, of which the Yanshanian activity was the strongest. The absolute age of the magmatic intrusion in the southern part of the Qinshui Basin and the Yanshanian Taihang Mountains is between 110 and 141 Ma, and the peak is 130–140 Ma. The absolute age of the Zijinshan complex and the kimberlite in Jianjiagou, Liujia, eastern Ordos Basin, is 132–158 Ma, and concentrated around 142 Ma. During that critical geological history, the control of magmatic activities made a profound impact on the characteristics of CBM reservoirs in the two basins by controlling the thermodynamic conditions.

5.1.1.3 Control of Yanshanian Tectonic Stress Field on CBM Accumulation

Analysis of secondary fold, fault and joint and finite element numerical simulation reveal that the tectonic stress fields in the Qinshui Basin and the eastern margin of the Ordos Basin have experienced three evolutionary stages and two transitional periods since the Late Paleozoic.

In the Qinshui Basin, the Indosinian tectonic movement produced reverse faults and a series of WE wide and gentle sub-folds in the southern part of the basin, showing a near NS tectonic compressive stress; the Yanshanian tectonic movement produced a series of NNE large-dip normal faults, and parallel, en echelon, asymmetric secondary folds are widely developed at the dominant orientation of 290°–330°, and the northwest wing is generally relatively steep, showing a strong NW–SE compressive and left-slip tectonic stress field (Fig. 5.3); the products of the Himalayan tectonic movement are mainly NW secondary folds and tectonic fractures, showing a horizontal NE-SW compressive stress field (Fig. 5.4).

In the Ordos Basin, the near NS horizontal compressive tectonic stress field in the Indosinian, the NW–SE one in the Himalayan and the NE-SW one in the Yanshanian have important influences on the CBM accumulation conditions. According to Zhang and Tao (2000), in the Indosinian, the

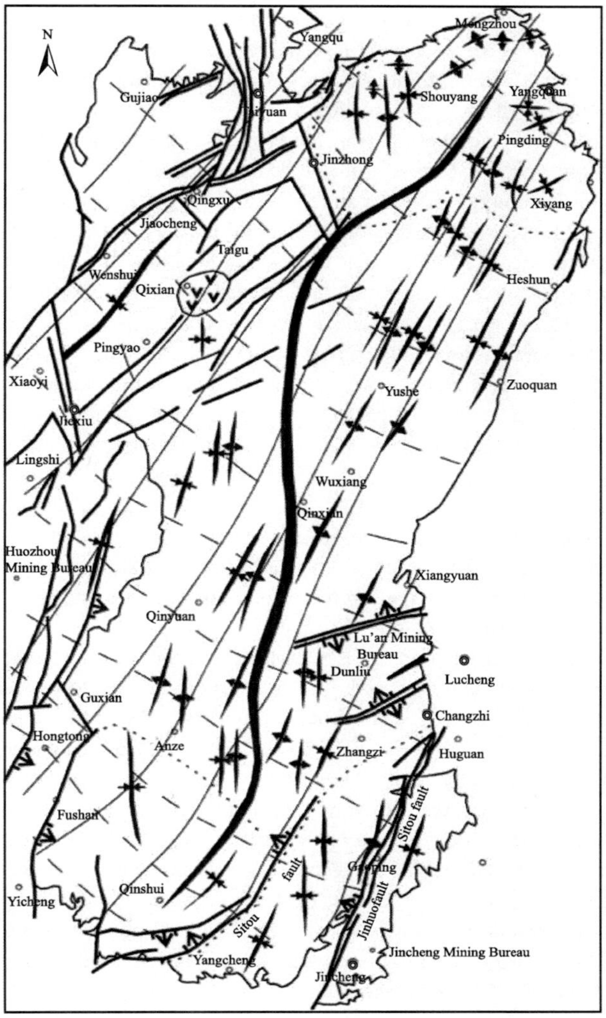

Fig. 5.3 Yanshanian principal stress trace in the Qinshui Basin, by finite element simulation

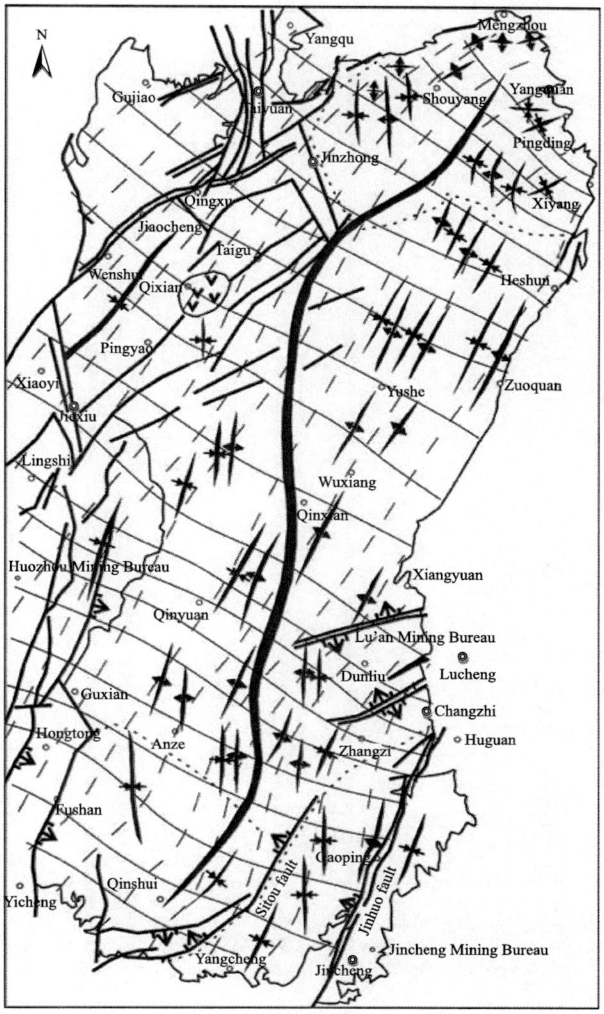

Fig. 5.4 Himalayan principal stress trace in the Qinshui Basin, by finite element simulation

tectonic deformation was not significant, and there were hardly faults and folds; the tectonic stress field was compressive in the north–south direction and extensive in the east–west direction; the maximum principal stress was 179°–359° and the inclination was 2°–3°; the minimum principal stress was 88°–268°, and the inclination was near horizontal; the intermediate principal stress was slightly skewed, and the average inclination was 83°. In the Yanshanian, the maximum principal stress was 130°–310° and the inclination was 2°–4°, the minimum principal stress was 220°–40° (nearly horizontal inclination), and the intermediate principal stress was slightly inclined, above 80° on average. The modern tectonic framework in the southern part of the eastern margin (Fig. 5.5) is the result of the tectonic stress field in the Yanshanian. In the Himalayan, the tectonic stress field changed fundamentally: the principal stress was 30°–210° (inclination 1°–2°), the minimum principal stress was 121°–301° (nearly horizontal inclination), and the intermediate principal stress was nearly upright or slightly inclined.

Obviously, the near NS horizontal compressive stress field in the Indosinian period had relatively weak influence on the CBM accumulation in the Ordos Basin. The Yanshanian was critical to controlling the CBM accumulation in the basin. The Himalayan tectonic stress field also affected the CBM accumulation. The NW–SE horizontal tectonic stress field in the Yanshanian was the primary tectonic dynamics of the formation of the gentle syncline basin. The NNE-NE secondary folds and the NNE high-angle normal faults controlled the occurrence of the CBM accumulation. The favorable configuration of tectonic evolution and

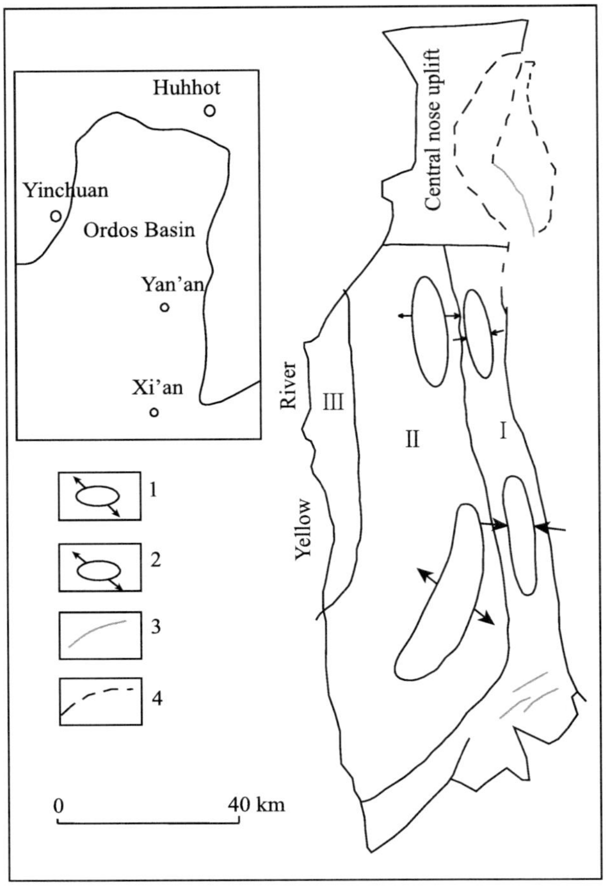

Fig. 5.5 Structure outline map of the southern segment of the eastern margin of Ordos Basin (according to Gui 1993) 1. anticline; 2. syncline; 3. fault; 4. coal outcrop; I. depression belt; II. slope sag belt; III. buffer zone

gasification made the NNE secondary folds become the primary gas-controlling structure. The Himalayan tectonic stress and framework superimposed on the Yanshanian tectonic products, and resulted in the modern pattern of CBM accumulation in the basin. In addition, the change in the orientation of the Himalayan principal stress field made the Yanshanian reverse faults to normal, and the closed to open, which might lead to strong escape of CBM.

5.1.2 Structural Differentiation Causes Complicated Dynamic Conditions

Field investigation and regional survey data reveal that the structural deformation strength of the Qinshui Basin and the eastern margin of the Ordos Basin is obviously weakened from basin edge to center, and the edge faults thrust outward, showing the characteristics of horizontal extrusion.

Although the structure of the Qinshui Basin, as a large complex syncline basin, is relatively simple, the internal structural differentiation is very significant, and results in complicated dynamic conditions of gas-controlling structures. In the southern part of the basin, taking the Sitou fault as the boundary, the Lishui-Yicheng area in the west of the fault is dominated by EW high-dip normal fault, and the east is characterized by superimposed NNE and EW folds. In the central part of the basin, the Lu'an area on the eastern complex syncline wing and the Qinyuan area on the northern part of the western wing have developed structures, and the Anze area in the southern part of the western wing has relatively weak structures. In the northern part of the basin, the normal faults in the form of collapse columns are relatively developed. In the northern Yangqu-Yangcheng area, NNE-NE folds superimpose on WE structures. In the southeast Zuoquan-Heshun area, the structure is relatively simple. Therefore, there may be structural conditions favorable for the enrichment of CBM in the Anze area in the southeastern part, the northern margin and the complex syncline turning point.

The southern part of the eastern margin of the Ordos Basin is transitional and has the structural development characteristics of North–South Division (Fig. 5.5). The southern structure is weakly deformed, with sparsely developed faults, and the eastern margin is a fault zone dominated by faults and subordinate folds; the central part is a sloped depression zone characterized by gentle folds; the west part is a relatively gentle monocline zone; and the northern part is the EW Lishi-Liulin tectonic belt which main structure is a westward nose (Liulin-Wubu nose-like structure) and secondary structure is also WE; the west wing of the NS Lüliang complex anticline and the WE complex uplift forms a series of short-axis folds, and the secondary, longitudinally extensive normal faults on the nose axis constitute graben structures (such as the Jucaita graben). Obviously, there are structural conditions favorable for CBM accumulation in the concave zone on the southern slope and the short-axis fold zone in the northern part.

5.1.3 Transformation from Tectonic Dynamics on Coal Seams Controls Permeable CBM Zones

5.1.3.1 Dynamic Conditions Reflected by Photosynthetic Structures of Vitrinite Reflectance

Under the action of tectonic stress, the vitrinite reflectance in coal is anisotropic, and its photonic structure is shaped by the end of coalification or gasification. In other words, the photostructural structure is the result of geotherm and stress. Therefore, by studying the photostructural structure based on vitrinite, the useful information of the relationship between tectonic dynamic conditions and the characteristics of coal reservoir transformation can be obtained.

Based on the measured 360° oil-immersed vitrinite reflectance of 3 orthogonal surfaces of the samples taken in 16 points on the primary coal seams of the Shanxi Formation in the Qinshui Basin, and using special software to calculate the photometric parameters and restore the tectonic occurrence of the indicative surface (ellipse) of the vitrinite reflectance on the surface of coal bricks, the tectonic stress field related to the finite strain in the Qinshui Basin was obtained (Fig. 5.6). The long axis of the photometric indicative surface (the maximum apparent reflectance) represents the direction of the laterally extensive strain or the relative tensile stress, and the short axis (the minimum apparent reflectance) represents the direction of the laterally shortened strain or the relative compressive stress. The relationship between the maximum and the minimum apparent reflectances may also provide useful data for investigating the strain.

The geological information contained in the results is of great significance for studying the dynamic conditions of CBM accumulation in the Qinshui Basin:

① The majority of coal samples (9 samples) have a long axis extending in the northeast direction, showing the overall characteristics of the southeastward compressive strain. It is not only consistent with the trend of the tectonic line near the sampling point, but also with the Yanshanian tectonic stress field, revealing that the gas-forming behavior of the Carboniferous-Permian coal ended in the Yanshanian, and meaning that the strain distribution related to the vitrinite reflectance is an important component of the regional strain field. In other words, the structural deformation and gasification mainly occurred in the Yanshanian.

② In the different parts of the basin, the photostructural characteristics of the NW vitrinite reflectance along the long axis are also different. The long axis of the indicating surface in the northern part is NNW, and gradually deviates westward in the southern part, and becomes NWW in the southern to central part. This trend is consistent with the distribution law of the Yanshanian tectonic stress trace (Fig. 5.3). On the one hand, it indicates that the local tectonic stress plays a more important role in the transformation of coal reservoirs; on the other hand, it further confirmed that the Carboniferous-Permian gasification was terminated in the Yanshanian.

③ The photometric indicative surface of the vitrinite reflectance of 7 coal samples taken in the southern margin and the middle section of the eastern margin of the basin shows a NE-NNE compressive strain, and the direction of the NS long axis slightly shifts northward. The "abnormal" distribution may be caused by two aspects, local change of the tectonic stress field in the Yanshanian, or the Carboniferous–Permian coalification (gasification continued in the Himalayan period. Comparative analysis of the Yanshanian and the Himalayan tectonic stress fields shows that the latter hypothesis is more likely to be established. Of the 7 "abnormal coal samples", 4 are in the southern margin, where thermodynamics is the strongest in the geological history, and the stress field indicated by the vitrinite photosynthetic structure is highly consistent with the Himalayan tectonic stress field (Fig. 5.4). Therefore, the southern margin might have a strong thermal source in the Cenozoic, and the tectonic-thermal conditions of CBM accumulation were better than other areas within the basin.

5.1.3.2 Structural Curvature of Coal Seams and Its Indicative Significance for CBM Accumulation

Curvature is a quantitative parameter that reflects the degree of bending of a line or surface. Structural curvature is the result of tectonic stress field. Curvature value reflects the relative development of tensile fractures in curved rock due to derived tensile stress. The structural curvature is negative at a syncline, and. positive at an anticline. High curvature is at the area where folds are relatively strong. Therefore, analyzing the distribution characteristics of the structural curvature of coal seams is important to study the dynamic conditions of CBM accumulation. Based on data such as formation thickness and elevation, the curvature values of 12,500 points in the primary coal seams in the Qinshui Basin were calculated by extreme principal curvatures.

The results show that the structural curvature of the primary coal seam in the Qinshui Basin is generally around 0.1×10^{-4}/m, up to 5×10^{-4}/m. Taking the general value of 0.1×10^{-4}/m as the standard, the coal seam in the structural belt with the curvature larger than this value has been reconstructed strongly, and may have structural fractures with high permeability. There are 8 structural belts that meet this standard: the eastern Zhanshang-northern Hengling nose-shaped flexure belt and the Xiyang deflected belt in the northern part have principal curvature of 0.1×10^{-4}/m to 0.5×10^{-4}/m; the central Hengling-Nanmahui anticline belt in the central-northern part has principal curvature of greater

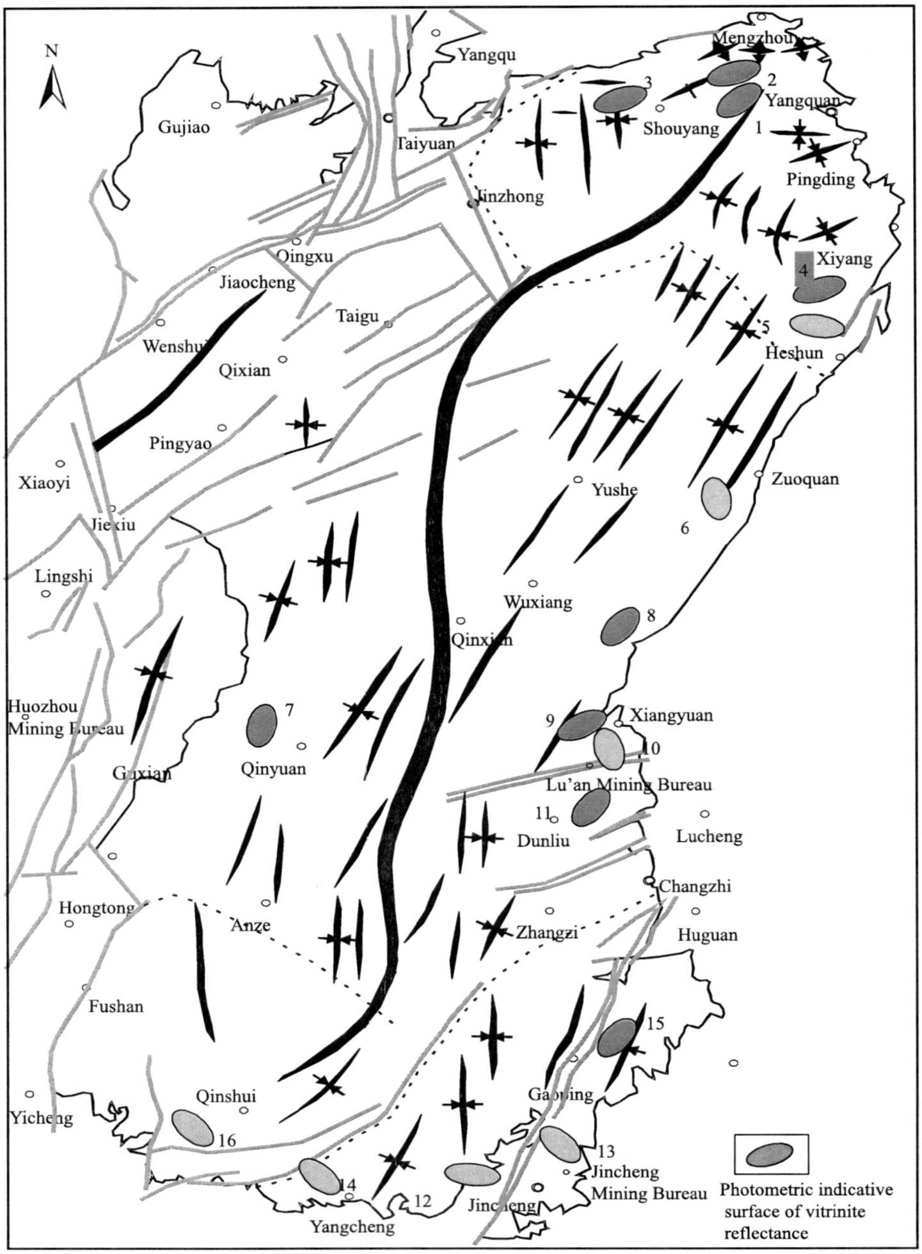

Fig. 5.6 The photostructure of the vitrinite reflectance of primary coal seams in Shanxi Formation and the strain distribution in Qianshui Basin

than 0.1×10^{-4}/m; the eastern Qiushuling-Langwogou anticline belt in the central-northern part has principal curvature of greater than 0.5×10^{-4}/m; the Jianzhang-Mopannao anticline structure on the east side has principal curvature greater than 0.1×10^{-4}/m, and the Watershed-Liuwan and Zhangyuan-Wangjiazhuang anticline belts on the west side have principal curvature of -0.3×10^{-4}/m to 0.5×10^{-4}/m in the central part; the Shuangmiaogou anticline belt on the west side of the central-southern part has principal curvature of greater than 0.1×10^{-4}/m; the Fengyi-Yuejiazhuang anticline belt has principal curvature greater than 0.1×10^{-4}/m on the east side of the southern part; and Yangcheng-Jincheng uplift has principal curvature of 0.1×10^{-4}/m–0.3×10^{-4}/m.

Obviously, the less the structural curvature, the weaker the coal seam reconstruction, the less developed fracture system, and the lower the coal seam permeability. But it doesn't mean that the larger the structural curvature, the higher the coal seam permeability. Comparison of the principal curvature with the well testing permeability of the primary coal seam shows that the coal seam moderately reconstructed or with middle structural curvature may provide favorable structural conditions for high CBM production. Statistical results show that the permeability of the coal seam is greater than 0.5 mD corresponds to structural curvature between 0.05×10^{-4}/m and 0.2×10^{-4}/m, and when the structural curvature is lower than 0.05×10^{-4}/m or higher than 0.2×10^{-4}/m, the permeability of the coal seam is low, and unfavorable for seepage of CBM (Fig. 5.7).

5.1.3.3 Control of Modern Principal Stress Difference on Coal Seam Permeability

The simulation results of finite element method show that the areas with high modern principal stress difference in the Qinshui Basin are distributed in three parts, namely the southern Yangcheng area, the southern-central Lucheng and Qinyuan areas, and the central Wuxiang and Zuoquan areas (Fig. 5.8).

Comparing the modern principal stress difference ($\Delta\sigma$) with the well test permeability (K) of the primary coal seam, it is found that, as the principal stress difference increases, the permeability increases exponentially. As for the the Shanxi Formation coal seam, they are related to each other as $K = 0.0147e^{0.0416\Delta\sigma}$ ($r = 0.75$), and the correlation is high. As for the Taiyuan Formation coal seam, $K = 0.0668e^{0.0178\Delta\sigma}$ ($r = 0.40$), and the correlation is relatively low. The relationship between the permeability and the modern principal stress difference in the Qinshui Basin can be divided into three levels: (i) both the permeability and the modern principal stress difference are high, $\Delta\sigma > 100$ MPa or > 150 MPa, $K > 1$ mD; (ii) both the permeability and the modern principal stress difference are moderate, 85 MPa $< \Delta\sigma <$ 100 MPa or 110 MPa $< \Delta\sigma <$ 150 MPa, 0.5 mD $< K <$ 1 mD; (iii) both the permeability and the modern principal stress difference are low, $\Delta\sigma <$ 85 MPa or < 110 MPa, $K <$ 0.5 mD.

The above law is mainly controlled by the relationship between the direction of the modern principal stress and the dominant direction of the natural fractures in the coal seam roof and the coal seam itself. The two directions are nearly parallel to the coal seam, and nearly orthogonal to the coal seam roof, which causes the natural fractures in the primary coal seam to be relatively tensile, while the natural fractures in the roof are compressive. The larger the primary stress difference, the more significant the relatively tensile (compressive) effect is. Based on this law and control mechanism, the Shanxi Formation coal seam with predicted permeability greater than 1 mD may be distributed among Yangcheng, Lu'an and Qinyuan and near Wuxiang.

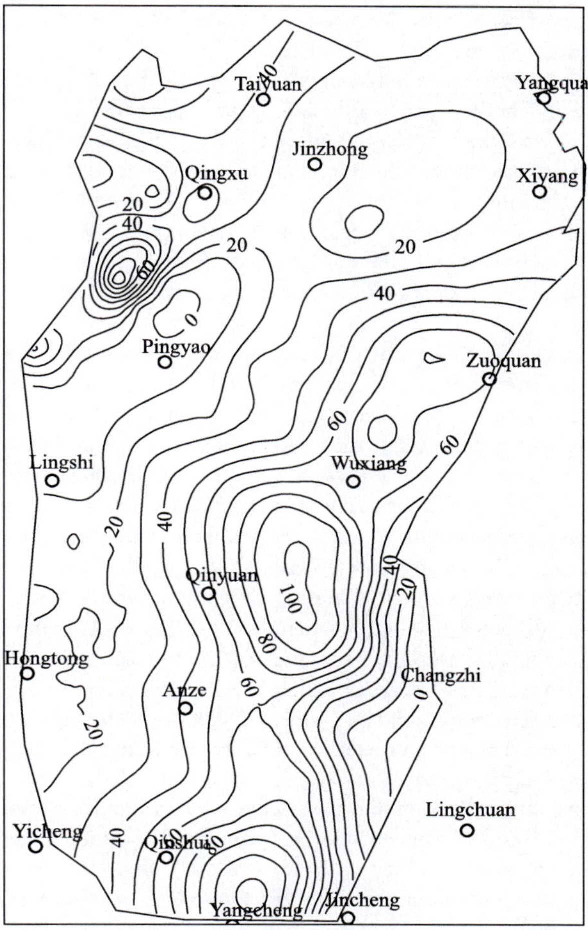

Fig. 5.8 Contours of the modern principal stress difference in Qinshui Basin

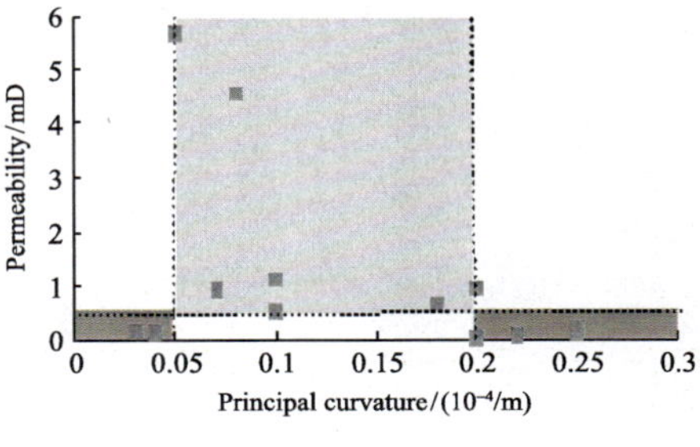

Fig. 5.7 Principal curvature vs. permeability in Qinshui Basin

5.1.4 Combined Structural Dynamic Conditions Control the Basic Pattern of CBM Accumulation and Distribution

Coupling analysis shows that the permeability of coal seams is determined by reasonable configuration among structural trace, structural curvature and modern principal stress difference. The combination of high tectonic curvature and tectonic setting indicates the area with relatively developed natural fractures, while high modern principal stress difference shows the area with relatively extensive fractures. If only structural curvature is high, but principal stress difference is relatively small, although fractures may be relatively developed, they are closed and the permeability is still relatively low. In conclusion, high structural curvature and tectonic setting are necessary conditions for the development of natural fractures in coal seams. High principal stress difference is the condition of open fractures. Only the favorable configuration of the three, the permeability may be high.

Based on this point of view, we can superimpose the distribution map of principal stress difference, the distribution map of structural curvature and structural outline map, to predict the favorable zone with abundant CBM and high permeability at the macroscopic level. In the Qinshui Basin, the dynamic conditions conducive to CBM enrichment and high permeability may be distributed in the Yangcheng-Jincheng area in the southern uplifted part, the Shuangmiaogou anticline belt in the western Anze-Qinshui area in the central-southern part, and the Liuwan anticline belt around the western Qinyuan in the central part. All three sections are secondary anticlines with high structural curvature and high principal stress difference. This understanding is almost consistent with the permeability from well tests in recent years.

As far as the southern part of the eastern margin of the Ordos Basin is concerned, the secondary anticline in the southern bending slope and the secondary short-axis anticline in the northern nose-like structure may have structural conditions favorable for CBM enrichment and high permeability.

5.2 Thermal Dynamic Conditions and Accumulation/Diffusion History of CBM

The characteristics of paleo-geothermal field directly affect the thermal evolution history, physicochemical properties, maturity and spatial distribution of organic matter in coal. It plays a decisive role in controlling the evolution of organic matter and the adsorption and desorption characteristics of coal. The study shows that in the Qinshui Basin and the Ordos Basin, no matter how many stages of the evolution history of the paleo-geothermal field, the anomalous paleo-geothermal field in the middle Yanshanian provided vital thermodynamic conditions to the Carboniferous-Permian CBM accumulation.

5.2.1 Thermal History of Carboniferous–Permian Coal Seams

5.2.1.1 Paleo-Geothermal Field and Thermal History of Qinshui Basin

According to the burial history of primary coal seams (Fig. 5.1), the analysis of coal vein inclusions, apatite and zircon fission track and the numerical simulation of thermal history, the paleo-geothermal field since the Late Paleozoic and the thermal history of the Upper Paleozoic coal seams in the Qinshui Basin underwent four evolutionary stages (Fig. 5.9).

Stage 1 is from the Late Carboniferous to the Late Triassic (C_3–T_3). The Late Paleozoic North China region was a unified and stable giant basin, and similar to a typical cratonic basin. During the Triassic, the North China platform began to activate, but the performance in the basin was not obvious, without evidence of magmatic activity, for example; however the whole performance was still characterized by the geothermal field in a continental cratonic basin. During that stage, the geothermal gradient of the sedimentary caprock was about 2.8 °C/hm, a normal paleo-geothermal field; the maximum buried depth of the Carboniferous–Permian coal seams was 3300–4400 m, and the highest temperature was about 110–140 °C; the coal seam was rapidly buried, coalification was mainly due to deep metamorphism, and the coal rank slowly increased, till to gas coal and fat coal by the end of the stage.

Stage 2 is from the Early Jurassic to the Middle Jurassic (J_1–J_2). After entering the Yanshanian, the structural differentiation of the North China region was further enhanced, the thermal fluid system in the deep crust was adjusted, and the characteristics of the paleo-geothermal field began to change. However, the internal magmatism activity was still

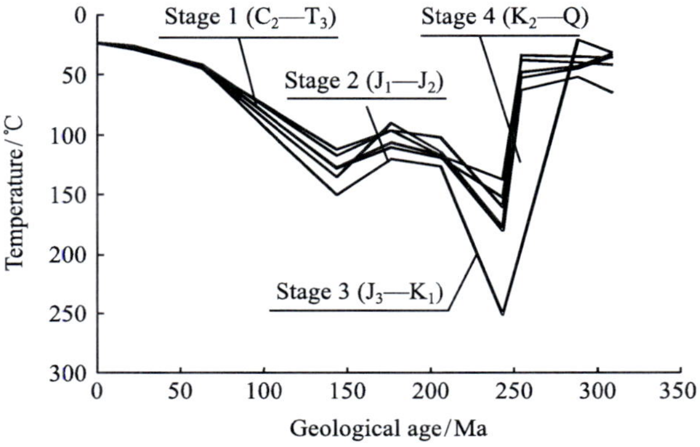

Fig. 5.9 Thermal history of the Carboniferous–Permian coal seams in Qinshui Basin

inactive, and the paleo-geothermal field was still in the category of normal paleo-geothermal field. As the Qinshui Basin was uplifted, the overlying layer on the coal seams was denuded. The maximum buried depth of the primary coal seam was 2300–3300 m, and the temperature reduced to 80–110 °C. The maturation of organic matter in the coal hardly progressed, and the gasification temporarily stopped.

Stage 3 is from the Late Jurassic to the Early Cretaceous (J_3–K_1). In the middle of the Yanshanian movement, the most important tectonic magmatic thermal event occurred in the Qinshui Basin since the Late Paleozoic. The isotopic age of the basin and the surrounding magmatic rocks is mainly distributed from 141 to 110 Ma, and thermal liquid veins (inclusions) were found widely in the coal-bearing strata in the basin. The strong tectonic magma thermal event formed an anomaly paleo-geothermal field, so that the paleo-geothermal gradient of the sedimentary caprock raised to 5.88–8.08 °C/hm. During that stage, although the coal seam became shallow, the high paleo-geothermal gradient made the temperature significantly increase, up to 182–263 °C. The rapid heating effect caused a sharp increase in coal rank, and large-scale secondary hydrocarbon generation (gas) appeared, which laid a decisive foundation for the modern distribution pattern of coal ranks in the basin.

Stage 4 is from the Late Cretaceous to the Cenozoic (K_2–Q). The basin gradually cooled, and the paleo-geothermal field gradually returned to normal. The overlying strata were denuded, the coal seams became shallow, and the coalification and gasification were basically terminated. According to the recent analysis of the fission track results of apatite and zircon, a rapid uplift occurred in the Qinshui Basin since the Miocene, and a rapid cooling event occurred in 26.2–11.5 Ma, which was early in the south and north and late in the central. According to the analysis of temperature in 20 boreholes, the modern geothermal gradient of the sedimentary caprock is 2.09–4.76 °C/hm, with an average of 2.82 ± 1.03 °C/hm. Combined with the measured thermal conductivity of the rock, the modern heat flow in the basin is between 44.75 and 101.81 mW/m^2, with an average of 62.69 ± 15.20 mW/m^2, which is slightly higher than the average heat flow in the mainland of China (61 ± 15.5 mW/m^2), and the distribution trend is high in the north and south and low in the central (Fig. 5.10).

5.2.1.2 Paleo-Geothermal Field Evolution and Coal Seam Heating History of Ordos Basin

The tectonic burial history of the Carboniferous–Permian coal seam in the southern part of the eastern margin of the Ordos Basin has also experienced four evolution stages. (Fig. 5.2), but the paleo-geothermal field history since the Late Paleozoic has shown three stages (Fig. 5.11). The coupling between the burial history of the coal seam and the thermal history of the basin determines the heated history of the coal seam.

At the end of the Indosinian period, the paleo-geothermal gradient was 2.2–2.4 °C/hm, the maximum buried depth of the coal seam was around 3100 m, the highest temperature of the coal seam was about 85 °C, and the coal rank reached gas coal. In the early Yanshanian, the paleothermal gradient was similar to that of the Indosinian, but the stratigraphic structure reverted, the stratum was denuded, the maximum buried depth of the coal seam reduced to about 2500 m, and the highest temperature of the coal seam was about 70 °C. The coalification did not continue. In the middle of the Yanshanian, the paleo-geothermal gradient rose to 3.3–4.5 °C/hm, and the basin accepted sediment again. The maximum buried depth of the coal seam increased to about 3000 m, and the highest temperature might reach 115–150 °C, and the coal rank was fat and lean, which has basically laid a modern distribution pattern of the coal rank. Since the Himalayan period, the basin has been uplifting, the geothermal field gradually returned to normal, the geothermal gradient reduced to 2.7–3.2 °C/hm (averaged 2.8 °C/hm), and coalification stopped.

5.2.2 Middle Yanshanian Tectonic Thermal Event and Its Thermodynamic Source

The research shows that the thermodynamic conditions generated by the abnormal geothermal field in the middle Yanshanian is characterized by high geothermal gradient, high heat flow, transient and heterogeneity.

The numerical simulation to the thermal history based on the burial history and the vitrinite reflectance of the coal seam (using the Thermodel simulator) shows that the northern temperature from the Late Jurassic to the Early Cretaceous was 252.68–263.34 °C, 182.32–188.26 °C in the central, and 246.68–252.03 °C in the south. Correspondingly, the paleo-geothermal gradient in the northern part was 7.30–7.38, 5.88–6.20 °C/hm in the central, and 8.03–8.08 °C/hm in the south, which are much higher than the average geothermal gradient of the normal geothermal field (about 3 °C/hm). Based on modern rock thermal conductivity, the geothermal flow during the period was 158.33 mW/m^2 in the northern part, 125.77 mW/m^2 in the central, and 169.38 mW/m^2 in the south.

The results of inclusions in the coal-bearing strata show that the forming temperature (uniform temperature) of calcite veins in the Jincheng-Yangcheng area in the south of the basin has undergone three phases, 129–140, 153–166 and 215–228 °C; the forming temperature of the quartz boundary in the middle of the basin has undergone two phases, 120–130 and 146–154 °C; the forming temperature of the quartz veins in the northern part of the basin has undergone two phases, 122.4 and 232.2 °C. The first phase of the inclusions

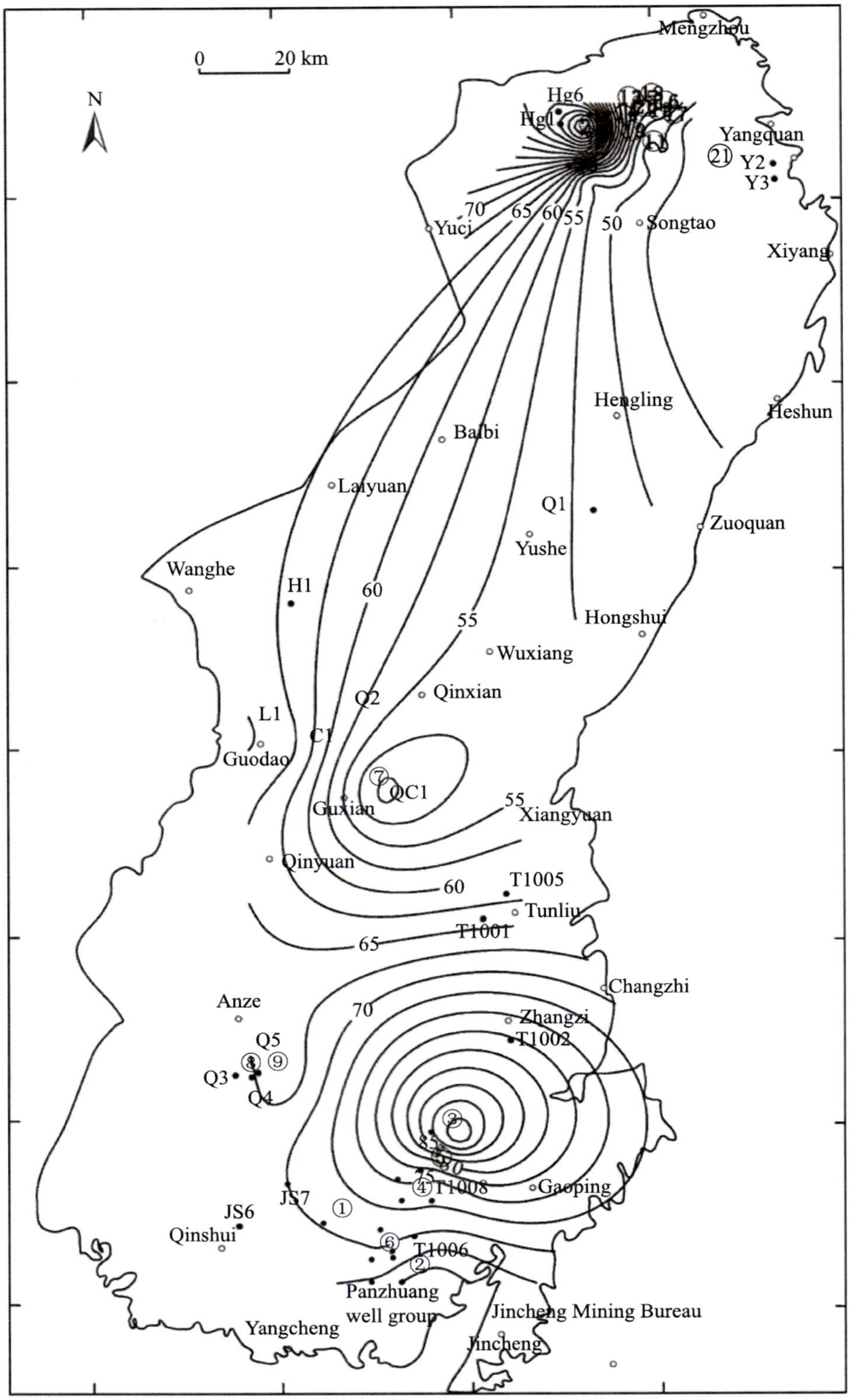

Fig. 5.10 Contours of modern heat flow in Qinshui Basin (unit: mW/m^2)

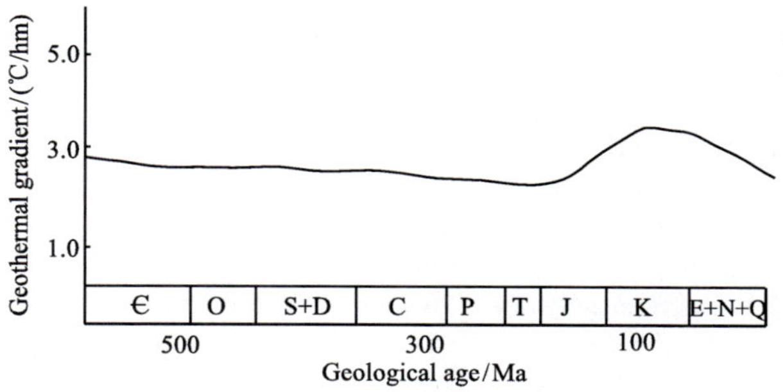

Fig. 5.11 Geothermal gradient evolution model in Ordos Basin (Ren 1996)

formed during the Indosinian and are generally distributed in the basin. The forming temperature is consistent with the numerical simulation results. They are the product of deep metamorphism. The second phase of the inclusions formed in the early Yanshanian, and are mainly distributed in the south and central parts of the basin. They are the product at the beginning of the Yanshanian tectonic thermal event. The third phase of the inclusion formed in the middle and late Yanshanian stages, and mainly developed in the southern and northern parts of the basin. They are the product of gas–water hydrothermal fluid after the magma period.

According to the results of apatite fission track, the average paleo-geothermal gradient at the Late Mesozoic was about 5.56 °C/hm in Well QC 1 in the central part of the basin (Fig. 5.12). Based on the paleo-geothermal gradient, and using the modern average thermal conductivity [1.73 W/(m K)] of the rock taken in the well, the heat flow value was above 96 mW/m^2. The fission track of the apatite in the Upper Paleozoic coal-bearing strata in the Jincheng-Yangcheng area at the southern part of the basin had been completely annealed and entered the cooling zone. The fission track age of the zircon is smaller than the stratigraphic age, indicating the paleotemperature, geothermal gradient and heat flow were higher.

The numerical simulation results further revealed that if the time corresponding to the highest temperature (±5 °C) was the continuous period of the peak of the tectonic thermal event, the duration of the peak period was about 10 Ma. It was obviously transient, and very consistent with the duration of the peak Yanshanian activity in the southern part of the basin.

The abnormal paleo-geothermal field originated from the magma activity under the control of the deep thermal fluid system in eastern China. There is a large area of diorite, more than 100 km^2, around Xiangfen, Fushan and Yicheng in the southwest corner of the Qinshui Basin. The rock mass invaded the Triassic, the potassium-argon (K–Ar) radioisotope age is distributed in 91–138 Ma, mainly in 130–140 Ma, that is, the magma activity peaked for about 10 Ma (Fig. 5.13). In the Jincheng and Pingshun areas on the southeastern edge of the basin and in the periphery, Yanshanian low-temperature hydrothermal lead and zinc deposits were found. In the second stage of the Yanshanian (Early Cretaceous), the medium-basic rock mass with mantle-derived magma properties is distributed in the Linxian and Huixian areas outside the southeastern margin of the basin. The middle Yanshanian magmatic rocks are also widely distributed outside the eastern and northwestern parts. There is no Mesozoic magmatic rock invading into the sedimentary strata since the Late Paleozoic, but there is EW distribution of intermittent, normal aeromagnetic anomalies, which may be a geophysical display of deep magmatic rock.

In nature, the carbon isotope composition (PDB) of methane from various genesis is –92‰ to –14‰. Test results show that CO_2 is the main contributor to the carbon isotope of the Carboniferous–Permian vein inclusions in the

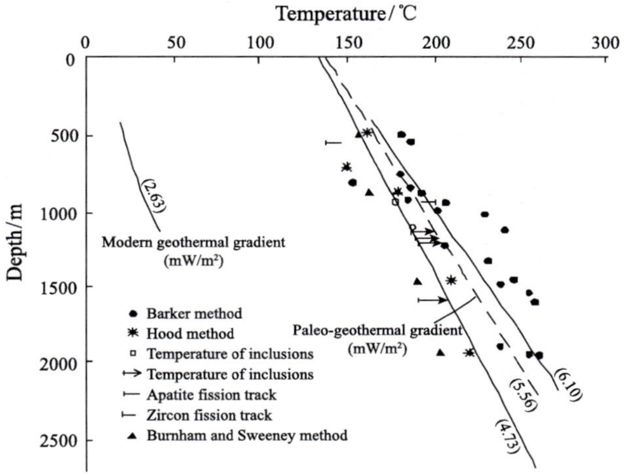

Fig. 5.12 Paleo-geothermal gradient during the Late Mesozoic in Well QC 1 in Qinshui Basin

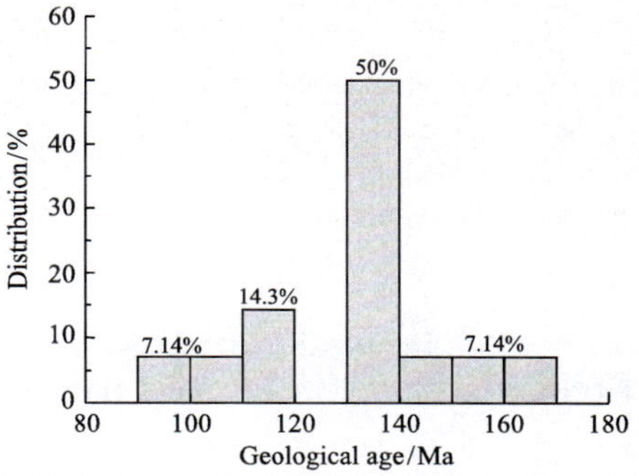

Fig. 5.13 Distribution of the absolute geological age of the igneous rock in Qinshui Basin and its periphery

Qinshui Basin. The CO_2 in the inclusions in the Yangcheng-Qinyuan area shows the characteristics of mixed (organic and inorganic) genesis. The CO_2 in the inclusions in the Yicheng area has the characteristics of inorganic genesis, which is consistent with the distribution law of the Yanshanian magmatic rock mass in the southern margin of the basin. It indicates that the underground fluids in the period were related to the later differentiation of the mantle magma.

The middle Yanshanian paleo-geothermal flow in the Qinshui Basin has a general distribution pattern of "higher in the north and the south, lower in the central, higher in the east, and the highest in the southeast". The span is very great. It is the inevitable result of heterogenous thermodynamic conditions and structural differentiation in the basin. The distribution of the strong thermodynamic condition in the Yanshanian is related to the regional deep structure, such as the 34° (hidden) deep fault in the south and the 38° (hidden) deep fault in the north. These faults led to the upwelling of the deep heat flow in the crust, which promoted the enhancement of large-area paleo-geothermal dynamic conditions, and resulted in strong secondary gasification of the Upper Paleozoic coal seam. This is an important geological dynamic condition for controlling the CBM accumulation in the Qinshui Basin.

The occurrence of the tectonically thermal event in the Qinshui Basin is consistent with the transition from lost to rich mantle and the whole lithospheric thinning in the North China Platform, and is controlled by the conversion process of the Mesozoic tectonic regime in eastern North China. The rich lithospheric mantle in the North China Platform took place from 140 to 120 Ma, which corresponds to the period of large-scale lithosphere demolition and thinning in North China. In the Late Mesozoic (110 Ma), the lithosphere in North China strongly extended in the NE-SW direction and slide left in the NNE direction. The large-scale magmatism in the Mesozoic was associated with regional lithospheric extension. During that period, large-area melting and material exchange between the crust and the mantle occurred in the North China continental block and adjacent blocks; the magma activities increased from east to west, the source area became deeper and the material of the mantle increased. It is generally believed that the thinning of the North China block occurred in the east of the Daxinganling-Taihang Mountains. The existence and time of the tectonic thermal event in the Qinshui Basin indicate that the western boundary of the thinning area of the North China block since the Mesozoic can extend to the west of the Qinshui Basin.

5.2.3 Numerical Simulation of CBM Accumulation/Diffusion History

In this study, the dynamic balance model and software for simulating CBM accumulation and diffusion were developed for the first time, and numerical simulation of CBM reservoir evolution in the Qinshui Basin was successfully realized, and some new understandings were obtained.

5.2.3.1 History of CBM Accumulation and Diffusion and Its Regional Differentiation in Qinshui Basin

Using the above software and considering the influence of structure (especially faults), the entire basin is divided into 80 grids (Fig. 5.14). Numerical simulation of the geological history of CBM accumulation and diffusion at every grid was carried out. Next we analyze the simulated results on G17.

Figure 5.15 shows that the geological evolution process on G17 has obvious stages:

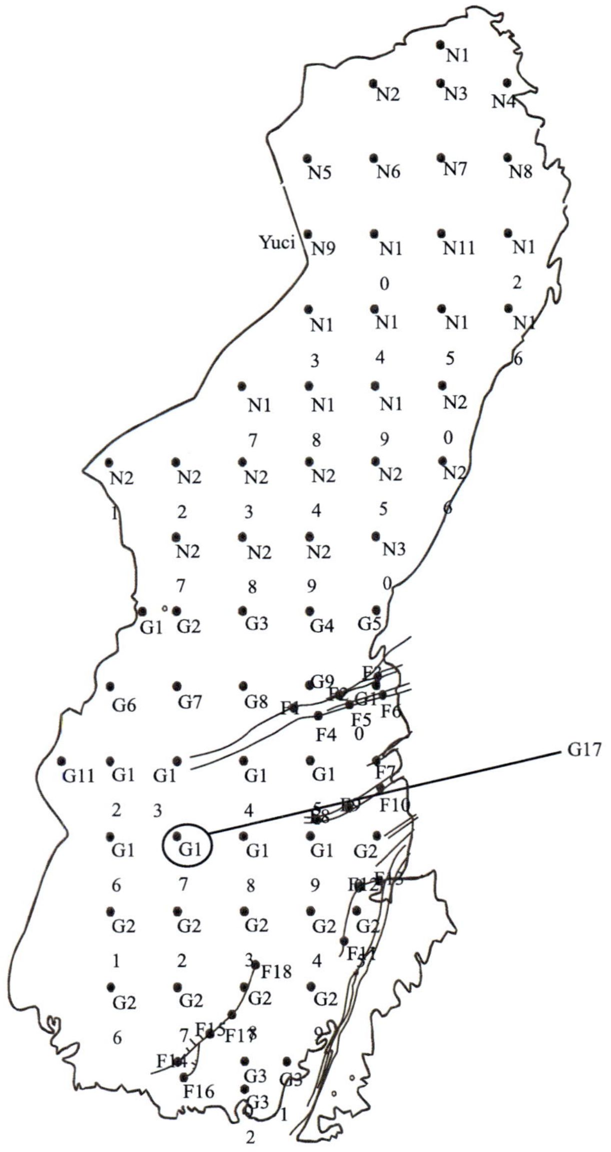

Fig. 5.14 Grids for geological history of CBM accumulation and diffusion in Qinshui Basin

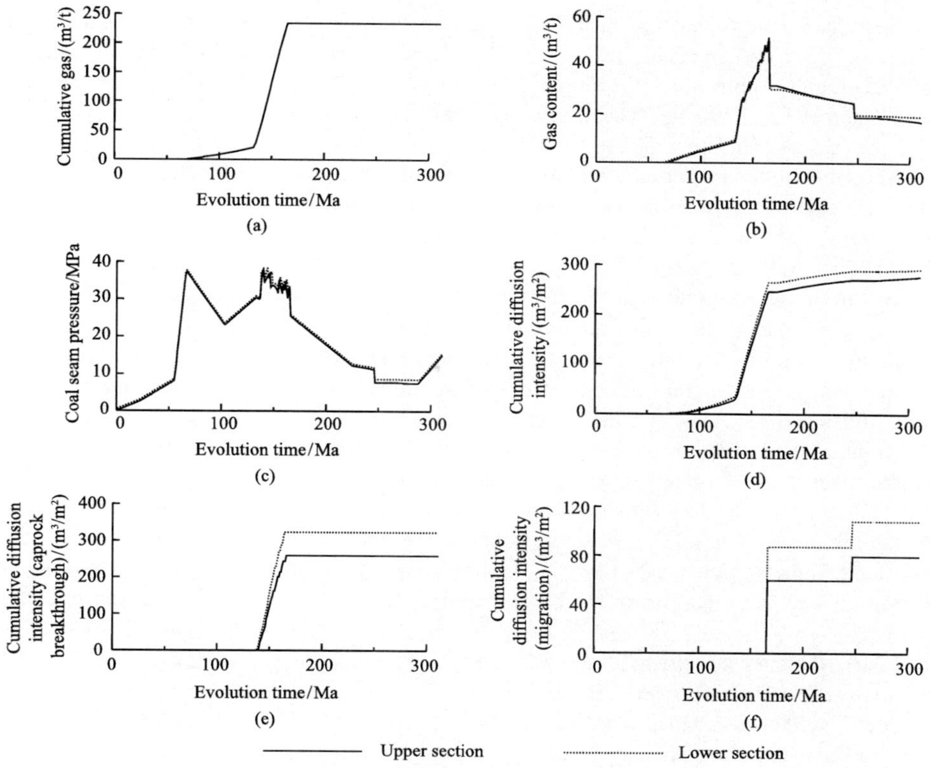

Fig. 5.15 Numerical simulation results of the CBM accumulation and diffusion history on G17 in Qinshui Basin

① Before the end of the Triassic (evolution from 0 to 66.3 Ma, 311.3–245 Ma B. P.), the cumulative gas generated was almost zero, and the gas content in the coal reservoir was extremely low, close to zero; the coal seam pressure increased with the burial depth; the cumulative diffusion intensity was also low, without breakthrough or migration to the caprock.

② From the Early to Middle Jurassic (evolution from 66.3 to 133.3 Ma, 245–178 Ma B. P.), coalbed organic matter basically stopped generating hydrocarbons, the coal seam pressure reached the maximum due to the burial depth at the end of the previous period, but then decreased. The cumulative diffusion intensity increased, but the overall magnitude was not large. Still no caprock breakthrough or gas migration occurred.

③ From the Late Jurassic to the Early Cretaceous (evolution from 133.3 to 165.7 Ma, 178–145.6 Ma B. P.), the geological evolution was the most active. During that stage, the hydrocarbon generation rate increased sharply, and the gas content increased sharply. The pressure of coal seams was higher than that of hydrostatic pressure due to the generation of hydrocarbons. Finally, the caprock was broken through and fractures opened, so that CBM began to migrate.

④ From the Late Jurassic to the present (evolution from 165.7 to 31.3 Ma, 145.6–0 Ma B. P.), the hydrocarbon generation ceased, and the gas content gradually decreased. To the early Late Tertiary (evolution at about 246.3 Ma, about 65 Ma B. P.), due to the occurrence of tensile tectonic stress field, fractures stretched and caused gas seepage and migration.

According to the above simulation results, the CBM accumulation and diffusion process in the Qinshui Basin has experienced three stages, namely the initial accumulation stage, the active accumulation and diffusion stage and the simple diffusion stage (Fig. 5.16).

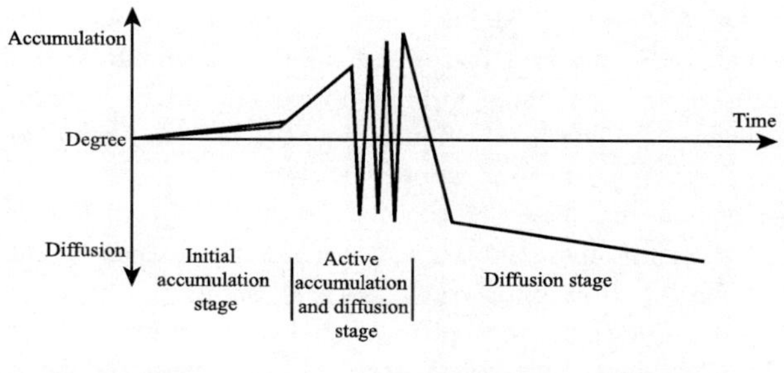

Fig. 5.16 Schematic evolution stages of CBM in Qinshui Basin

The study also found that in the sandstone in the coal seam roof in Well Fanzhuang (FZ) 1, the K–Ar isotope age of the authigenic illite which the ratio of I to S intervals is 20 is between 191 and 196 Ma. This may reflect CBM began to migrate into the sandstone in the Early Jurassic.

According to the numerical simulation results of 80 grids, the regional distribution of gas content and coal seam pressure was obtained in the above four stages in the Qinshui Basin (Figs. 5.17 and 5.18).

On this basis, according to the factors such as organic matter, gas content, the mode and strength of CBM accumulation and diffusion, and the depth of the coal seam, the entire basin is divided into different types of accumulation and diffusion zones, namely Class I, the most active zones in the southern and northern parts of the basin; Class II, more active zones in the southern and northern section of the basin axis; Class III, moderately active zones in the middle section of the basin axis; Class IV, less active zones in the eastern and western parts of the basin; Class V, inactive zones in the eastern and western margins of the basin (Fig. 5.19).

5.2.3.2 History of CBM Accumulation and Diffusion and Its Regional Differentiation in Ordos Basin

The faults in the study area are not very developed. Therefore, the whole area was divided into 109 uniform grids (Fig. 5.20), and the geological history of CBM accumulation and diffusion at each grid was numerically simulated. Next, we will analyze the simulation results of G69 in the western part of the study area.

As shown in Fig. 5.21, similar to the Qinshui Basin, the CBM geological evolution process at G69 also has obvious stages. During the first stage (Indosinian), as cumulative CBM increased slowly, the gas content and reservoir pressure increased rapidly, and diffusion loss increased accordingly; then later the caprock was weakly broken through, resulting in increasing diffusion loss, but no seepage loss. During the second stage (at early Yanshanian), gasification suspended temporarily, and diffusion continued, resulting in decreasing gas content and reservoir pressure, but no breakthrough loss and seepage loss. During the third stage (middle Yanshanian), the gas content and reservoir pressure increased sharply, and intermittent breakthrough was significant, resulting in a sharp fluctuation of gas content, and diffusion loss and breakthrough loss were strong, but no seepage loss. During the fourth stage (late Yanshanian to Himalayan), gasification stopped, the reservoir pressure and gas content first increased sharply and then declined slowly, and diffusion loss and strong seepage loss occurred. In addition, at some grids (such as G57), significant seepage loss occurred in the early Himalayan period (about 32.9 Ma), due to the tensile tectonic stress field and open fractures.

Based on the simulation results at 109 grids, a plane contour map was drawn to show the regional differentiation of CBM accumulation and diffusion history in the study area:

① During the Triassic, controlled by the Indosinian movement, the Late Paleozoic coal series was buried at about 3200 m at last, gradually a closed reservoir pressure system was formed. In most areas, the coal rank developed to B and C of moderately volatile bituminous coal, and locally developed to highly volatile coal. At the end, cumulative gas in the upper coal seam was 28.65–157.95 m^3/t, and staged gas was 28.65–157.95 m^3/t, and an area with annularly high values appeared near the Yellow River in the south. The gas content was 6.70–28.68 m^3/t (Fig. 5.22a), the diffusion loss was strong, at 20.96–139.99 m^3/m^2. The reservoir pressure exceeded hydrostatic pressure, and the caprock was broken through frequently in the southern part from the middle to late period. The pressure of the upper coal seam was 12.68–41.63 MPa, mainly caused by hydrostatic pressure.

② From the Early to Middle Jurassic, the coal seam was at 2600–3100 m, and the coal rank development was not obvious. By the end of the stage, the gas content of the upper coal seam was 3.90–17.53 m^3/t, higher in the south than in the north, and higher in the southwest than in the southeast. The cumulative diffusion intensity was larger than the previous stage, at 26.05–169.08 m^3/m^2. The gas content of the lower coal seam was 10.21–19.63 m^3/t, low in the south and the north, high in the central, and higher in the west than the east. The diffusion center was located to the south of Daning and north of Jixian and close to the Yellow River. The pressure in the upper coal seam was about 5.95–25.94 MPa, and about 18.49–30.62 MPa in the lower coal seam, caused by static pressure.

③ From the Late Jurassic to the Early Cretaceous, gas generated and gas content increased significantly, 7.89–41.49 m^3/t and up to the highest near the Yellow River in the central and southern parts, and the regional differentiation was significant. The coal reservoir pressure greatly exceeded the hydrostatic pressure, and the caprock was broken through frequently from the middle to late periods, covering almost the whole study area. The average gas content of the lower coal seam was 21.20 m^3/t, from 14.14–49.24 m^3/t. The area

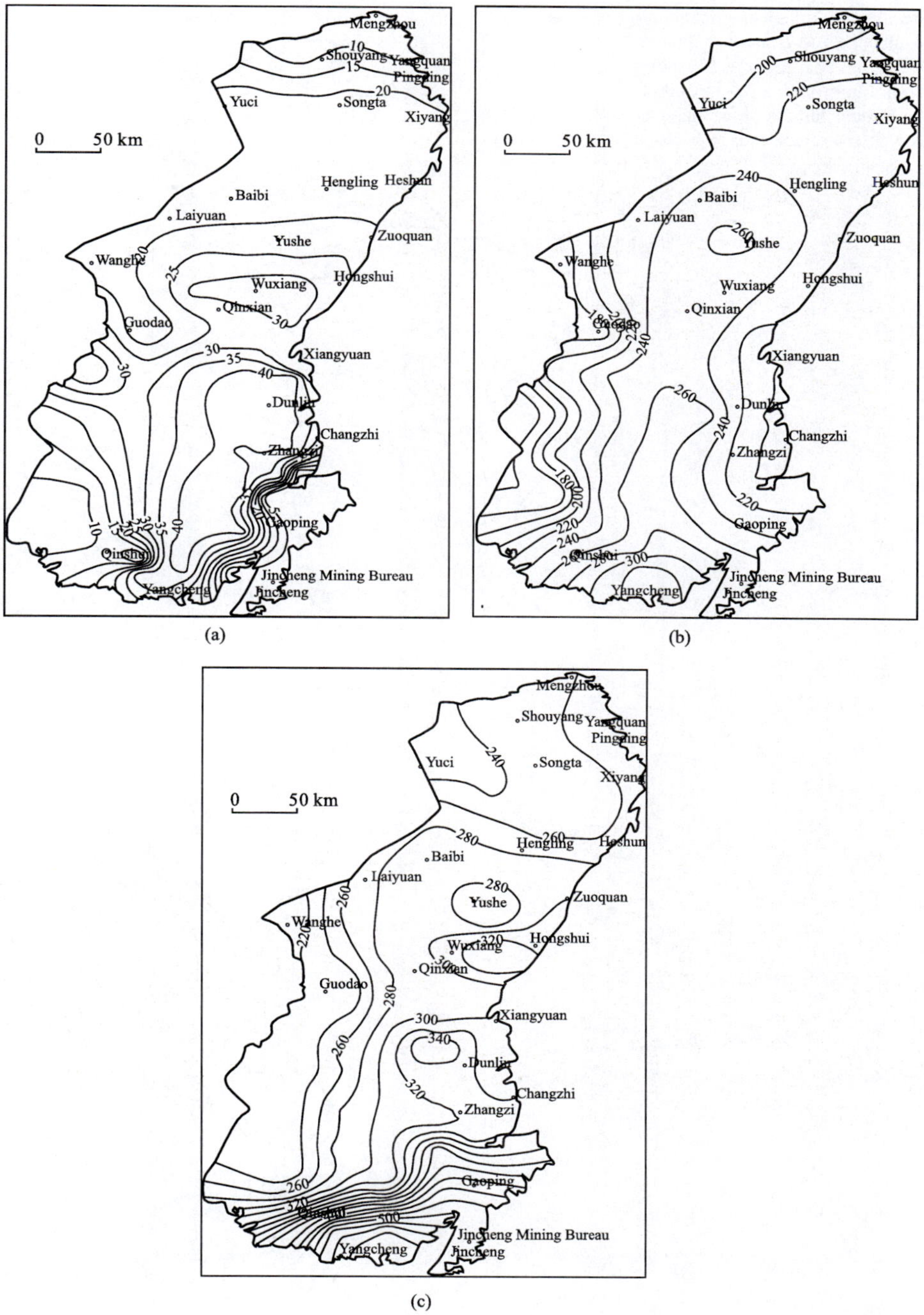

Fig. 5.17 Numerical simulation results of cumulative CBM diffusion intensity of Shanxi Formation coal seam in Qinshui Basin (unit: m^3/t) **a** by the end of the Middle Jurassic; **b** by the end of the Early Cretaceous; **c** modern

with high value was located to the west of Daning (Fig. 5.22b).

④ Since the end of the Late Cretaceous, the depth of the coal seam continued to decrease, the paleo-geothermal field returned to normal, gasification stopped, and CBM diffusion was dominant. From the simulation curve at a grid, the tensile structural stress suddenly appeared at the initial moment, which led to the fractures to open, the pressure to decrease, strong seepage occurred, and the gas content decreased sharply. In the early stage of the Himalayan movement, as the paleo-tectonic stress field temporarily changed from compressive to tensile, the fractures in the coal seam opened, and another CBM seepage occurred. Continuous diffusion resulted in the

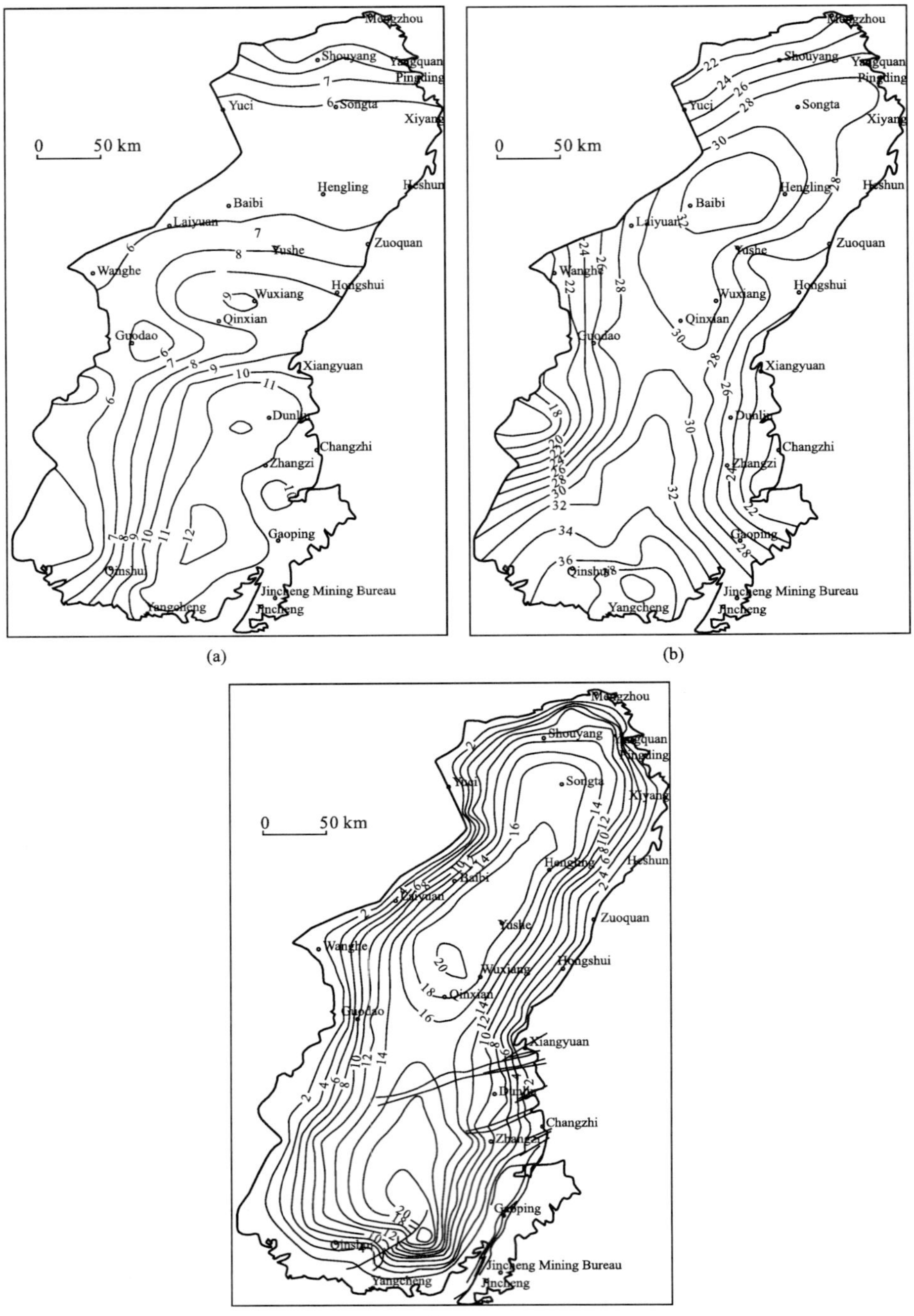

Fig. 5.18 Numerical simulation results of CBM content of Shanxi Formation coal seam in Qinshui Basin (unit: m^3/t) **a** by the end of the Middle Jurassic; **b** by the end of the Early Cretaceous; **c** modern

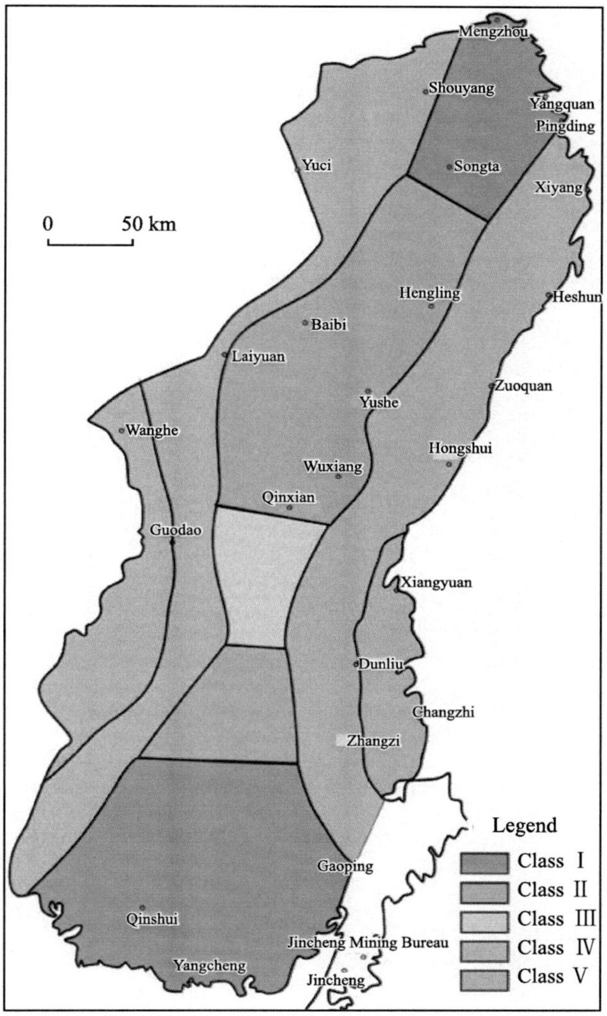

Fig. 5.19 Division of CBM accumulation and diffusion history in Qinshui Basin

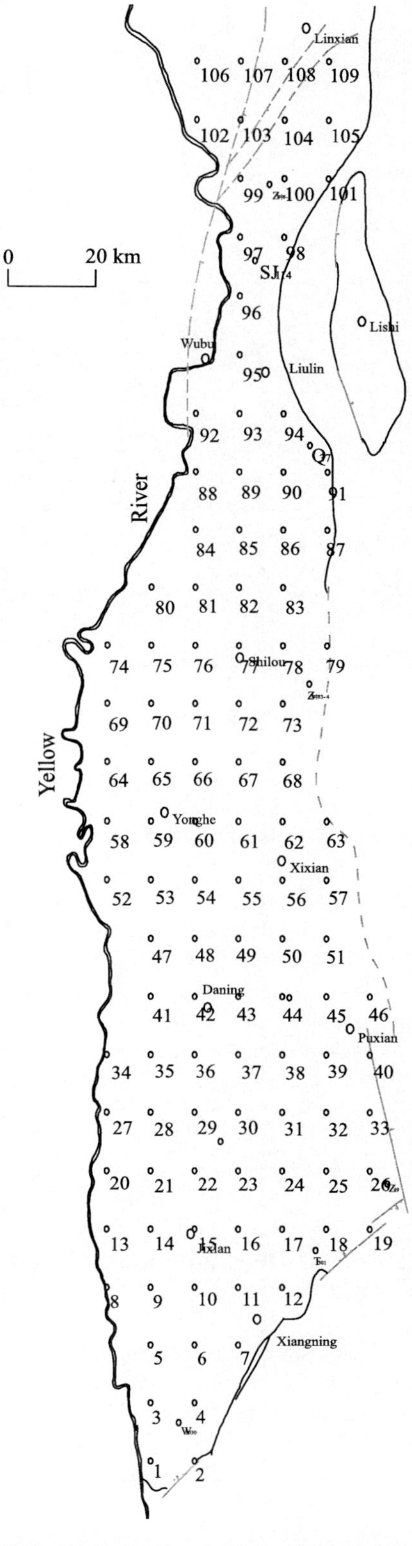

Fig. 5.20 Grids distribution for simulating the geological history of CMB accumulation and diffusion in the southern segment of the eastern margin of Ordos Basin

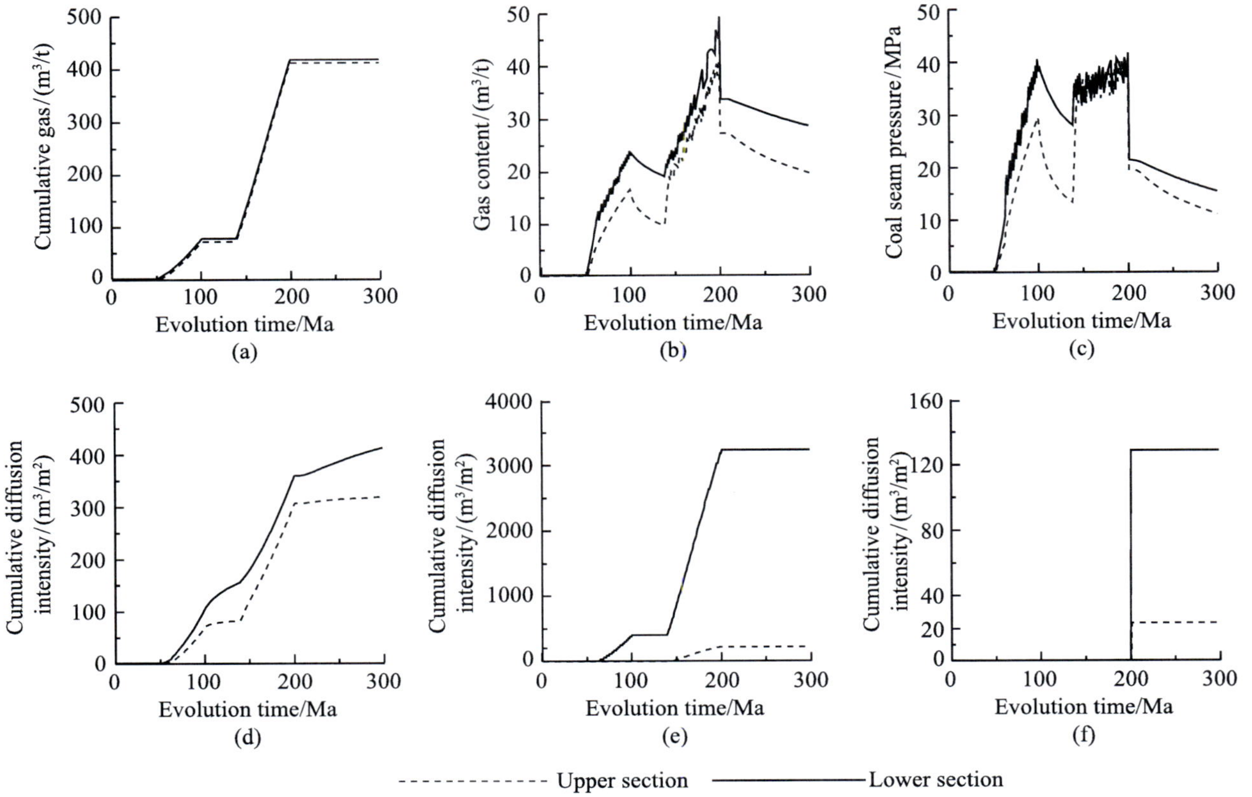

Fig. 5.21 Numerical stimulation results of the CBM accumulation and diffusion history on G69 in the southern segment of the eastern margin of Ordos Basin

pressure to decrease and the present regional distribution of the gas content developed. The gas content of the lower main coal seam is 0–28.77 m^3/t, which is higher in the central, the highest in the west of Yonghe county and the east of Yellow River, high in the north and lowest in the south (Fig. 5.22c).

5.2.4 Control of Thermodynamic Conditions on CBM Accumulation

The above results show that thermodynamic conditions have a significant control on CBM accumulation, and different zones have different gas generation and storage conditions.

In the Qinshui Basin, the effects of thermodynamic conditions on the generation and preservation of CBM can be roughly divided into three types. In the first type of area, the conditions for the generation and preservation of CBM are relatively good, including the Yangcheng-Yingcheng area in the southern part, the western Qinyuan-Qinxian area in the central part, and the Hanzhuang-Yangquan area in the north. Among them, the Yangcheng area is the most gas-bearing target currently known in the Qinshui Basin. In the second type of area, the conditions for the generation and preservation of CBM are moderate, mainly distributed in the western Anze area in the central to southern part. Further comprehensive research on the geological conditions of CBM is an important way to determine whether the area has exploration prospects. In the third type of area, the preservation conditions are relatively poor, and such type accounts for considerably large in the basin.

As far as the southern segment of the eastern margin of the Ordos Basin is concerned, the thermodynamic conditions conducive to the generation and preservation of CBM exist along Wubu–Shilou–Yonghe–Daning–Jixian east–Xiangning west (the specific location is at the west along G92–G77–G59–G25–G3), and can be roughly divided into two sections. The southern section is located around Jixian area, where the coal rank is higher, the amount of gas is larger, the coal seam is deeper, the thermodynamic conditions are strong, so gas diffusion has been weak since the Himalayan period. The northern section is located in the west of Shilou, where although the thermodynamic conditions are not very

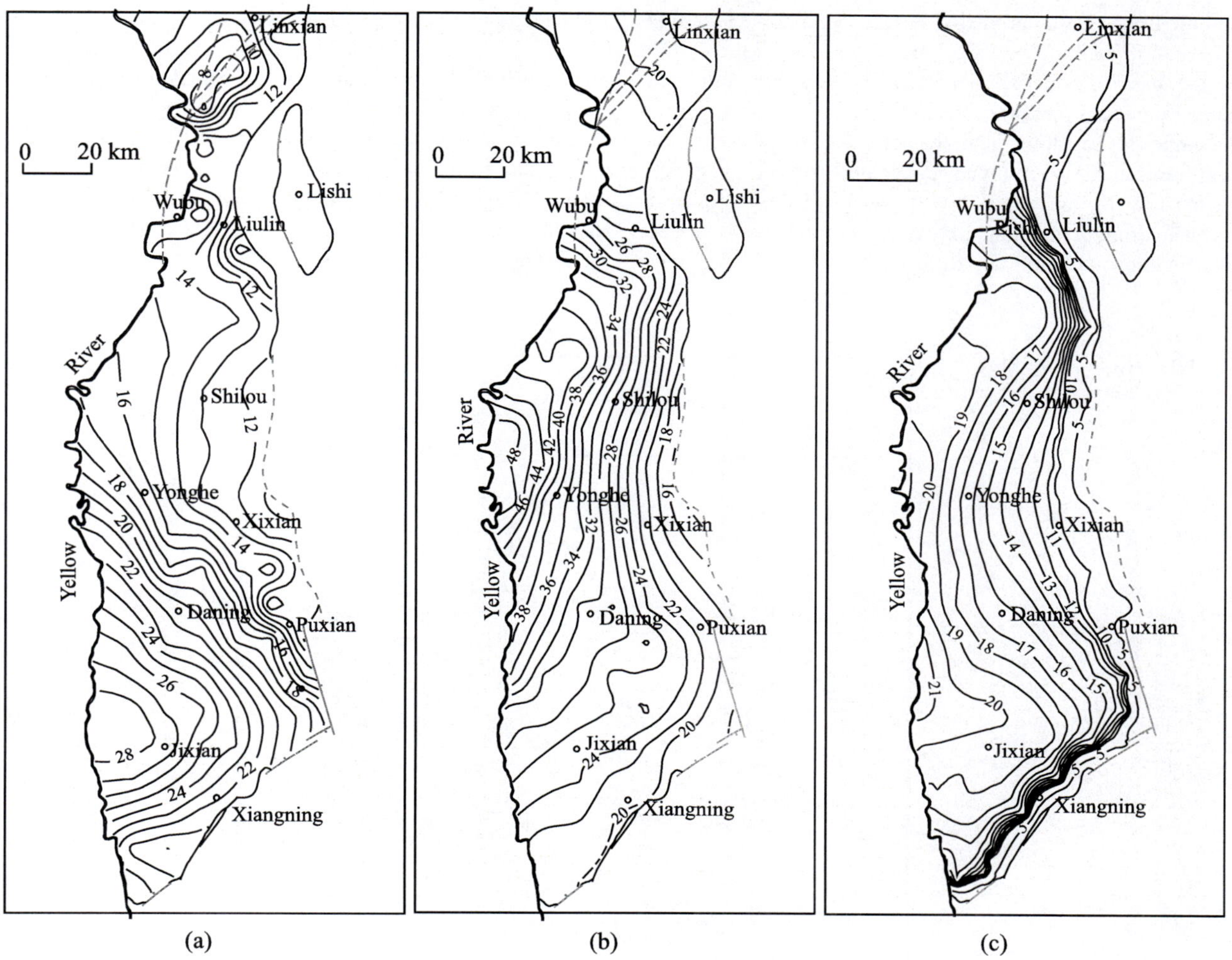

Fig. 5.22 Numerical simulation results of CBM content in the southern segment of the eastern margin of Ordos Basin (unit: m^3/t) **a** by the end of the Triassic; **b** by the end of the Early Cretaceous; **c** modern

strong, the coal seam is deep, and gas diffusion has been weaker since the Himalayan period.

5.3 Control and Mechanism of Underground Hydrodynamic System on CBM Accumulation and Diffusion

Studies show that hydrogeological unit boundary, modern groundwater zoning characteristics, groundwater geochemical field and other factors largely control CBM accumulation and diffusion. The incompatibility among these factors results in significant differences in the types of hydrodynamics-controlled gas.

5.3.1 Hydrogeological Unit Boundary and Its Internal Structural Differences and Characteristics of CBM Accumulation and Diffusion

5.3.1.1 Qinshui Basin

The surface water system in the Qinshui Basin consists of four hydrogeological units, namely the Niangziguan spring area in the north, the western Guangshengsi spring area in the central to southern part, the eastern Xin'an spring area in the central part, and the eastern Yanhe spring area in the southern part (Fig. 5.23). In terms of the basin scale, there is basically no hydraulic connection among the Middle Ordovician, the Upper Carboniferous Taiyuan Formation and the Lower Permian Shanxi Formation, and the latter two are weak to medium in water enrichment. Under the control of the basin's boundary structure, internal structure, and temperature and pressure, the groundwater system is further decomposed, resulting in more complicated hydrodynamic conditions such as groundwater dynamic field, groundwater chemical field, and coal seam water-bearing system. Different hydrodynamic units play different roles in the process of CBM accumulation and diffusion.

In terms of the hydrogeological characteristics around the Qinshui Basin, the eastern hydrogeological boundary is the Jincheng-Huojia fault belt; the southern boundary is a fault belt composed of a series of NE faults; the western boundary consists of a series of faults, complex anticlines and old strata; and the northeastern boundary is composed of the northeastern segment of the Jiaocheng large fault and the nearly NE fold belt dominated by the Xiakou fault (Fig. 5.23). The characteristics of conducting and blocking gas of each boundary are as follows:

Fig. 5.23 Hydrogeological zones in Qinshui Basin

The Jincheng-Huojia fault belt is 250 km long and 20–25 km wide. It is a NE complex anticline at 25°, where a series of parallel secondary folds and faults are developed, and most faults at 60° or higher. Different sections have different structural patterns, which make water conductivity and gas diffusion different. The northern section extends from Xiyang to Huguan, and the fault displacement reaches more than 300 m. It blocks the Middle Ordovician aquifers and is effective to block water laterally. The middle section extends from Hukou to Dayang, and the fault displacement is about 60 m. The groundwater level on both sides of the fault is basically the same, similar to HCO_3–Ca·Mg type water, and the calcite saturation index (β_c), the dolomite saturation index (β_d) and the gypsum saturation index (β_g) are all less than 1, indicating that the groundwater interaction between the two sides is strong and the water conductivity is very significant. The southern section extends from Dayang to Jincheng, and it is the eastern boundary of the Yanhe spring area from the Sangu spring area. The outcrop is Middle Ordovician limestone. The terrain is characterized by ridge-like mountain beams formed by structural extrusion and discontinuous structures such as faults and joints. The groundwater level on both sides is different by more than 50 m. Although the chemistry type is similar, there is a significant difference in the chemical reaction ratio constant. Therefore, the boundary has significant water blocking characteristics.

The southern hydrogeological boundary is divided into three sections according to its water-conducting characteristics, namely the eastern Qinhe–Xifan water-conducting section, the central Xifan–Xigou water-blocking section, and the western Xigou water-conducting section. The Xifan–Xigou water-blocking section consists of 8 faults which are very different in scale and fault throw, and make the coal-bearing sandy shale contact with the Ordovician limestone. The water conductivity coefficient of the fault zone is 0.0001 m^2/d in the direction of the vertical tectonic line. The water outflow and the aquifer permeable coefficient are extremely low in wellbores. The water level on both sides is different by more than 300 m, and the water blocking effect is quite remarkable. The faults in the eastern and western parts of the boundary are not developed, and the fault throw is relatively small. In the two walls, limestone contacts with limestone, the water chemistry types are similar, the salinity is not high, the $\delta^{18}O$ value changes a little, and the hydraulic connection is more obvious.

According to the water conductivity, the western hydrogeological boundary is divided into four sections from north to south: the northeast section is composed of the fault zone along Xingdao, Sijiaping and Dafangshan and the Guojiazhuang normal fault, and conductive; the northern-middle section is the Dongshan complex anticline, and relatively weak in water conductivity; the middle-northern section is the eastern fault in the Jinzhong fault depression, and the water conductivity is significant; the middle section is the Sinian-Cambrian strata of the Huoshan uplift, and the water blocking characteristics are very obvious; the southeastern section is the eastern margin fault of the Linfen-Yuncheng fault depression, and the water conductivity is significant.

The northern hydrogeological boundary is a nearly WE fault-fold structural belt, which shows the results of the Fuping uplift strongly compressing toward south and ancient block folding caprock. In the Mengxian area, it is generally a nearly WE short-axis, wavy and gentle fold formed in the Indosinian period, with late tensile normal faults. The terrain is relatively high. Generally, it is a boundary for groundwater recharge or flow.

In terms of boundary hydrogeological conditions, there are five important hydrogeological boundaries which form boundaries between spring area (Fig. 5.23). Four boundaries of them are surface or underground watershed formed by secondary uplifts, another one is the Sitou fault which has a significant influence on the CBM accumulation in the southern part of the basin.

The internal boundaries have the following characteristics:

The boundary between the northern-central Niangziguan spring area and the Jinzhong fault depression and the boundary between the central-southern Xin'an spring area and the Yanhe spring area are in the east–west direction. The internal boundary in the southern-central part consists of nearly WE secondary uplifts and distributed along the Zhaozhuang–Chushan–Changxing–Shuzhang line in the north of Gaoping. The groundwater level on the south and north sides shows low in the middle and high in the two sides or an equipotential surface, indicating there is an underground watershed which is basically consistent with the location of the surface watershed.

Two NNW-NS internal boundaries exist in the central and southern parts. One is the boundary between the Xin'an spring area and the Yangziguan spring area and the other is the boundary between the Guangshengsi spring area and the Xin'an spring area, as surface watersheds. The former is distributed along the Shengshan–Rentoushan–Zuoquan line, with elevation of 1000–1500 m, and gradually lowering from north to south. The latter is distributed along the Beilin–Baibi–Liaoyuan–Fangshan line, with elevation of 1300–1765 m, and gradually lowering from north to south.

The southern Sitou fault is the northern boundary between the Yanhe spring and the Guangshengsi spring area. It is exposed from the Sitou Village along the fault toward northeast. Although the Sitou fault appears as a normal fault in the tectonic pattern, the data of pumping test, geophysical logging, hydrogeochemistry, and coalbed gas content indicate that the fault is a closed fault which is poor in water and gas conductivity, blocks the flow of groundwater on the east and west sides, and significantly controls the enrichment of CBM in the Panzhuang-Fanzhuang area on the east side. However, the fault has a large displacement, extends far, and connects with other faults, so it may be conductive to water and gas locally.

5.3.1.2 he Southern Segment of the Eastern Margin of Ordos Basin

The area can be divided into four hydrogeological units, and the aquifers include the Middle Ordovician, the Upper Carboniferous Taiyuan Formation and the Lower Permian Shanxi Formation (Fig. 5.24). At the basin scale, there is essentially no hydraulic connection between the three aquifer systems, and two of which with coal are poor-medium in water. Under the control of structure, formation temperature and pressure conditions, the regional groundwater system is further decomposed, so that the controls of the groundwater dynamic conditions in different units are different on CBM preservation.

Boundary hydrogeological features are follows:

① The northern boundary is a EW tectonic belt composed of a series of EW or nearly WE normal faults and grabens and horst formed by the faults. The largest one is the Jucaita graben which consists of two nearly WE normal faults–generally compressive normal faults, but the water and gas conductivity changes greatly.

② The eastern boundary is a nearly NS fault zone. The northern section is the Lishi fault zone developed in the east wing of the Wangjiahui anticline, and composed of the Tanyaogou fault, the Zaolin fault, and the Zhujiadian fault. The induced structures on both sides of the fault are very developed, with local tensile faults. The middle and southern section is the Zijingshan fault zone which northern part consists of a series of compressive faults and folds that are parallel or oblique in the nearly NS direction; southern section is an unconformity contact of the Archaic basement complex, the Cambrian–Ordovician limestone and the Carboniferous–Permian system; and southern section is steep and has very developed tectonic breccia.

③ The southern boundary is a NE Xiangning faulted flexure zone. The eastern section is composed of a series of compressive fault arcs, and the main fault tends toward southeast, reflecting the southeast wall thrusts toward the northwest wall. The middle section is a large-scale knee-shaped flexure, and the stratigraphic occurrence changes from nearly horizontal to vertical, even reverse, and

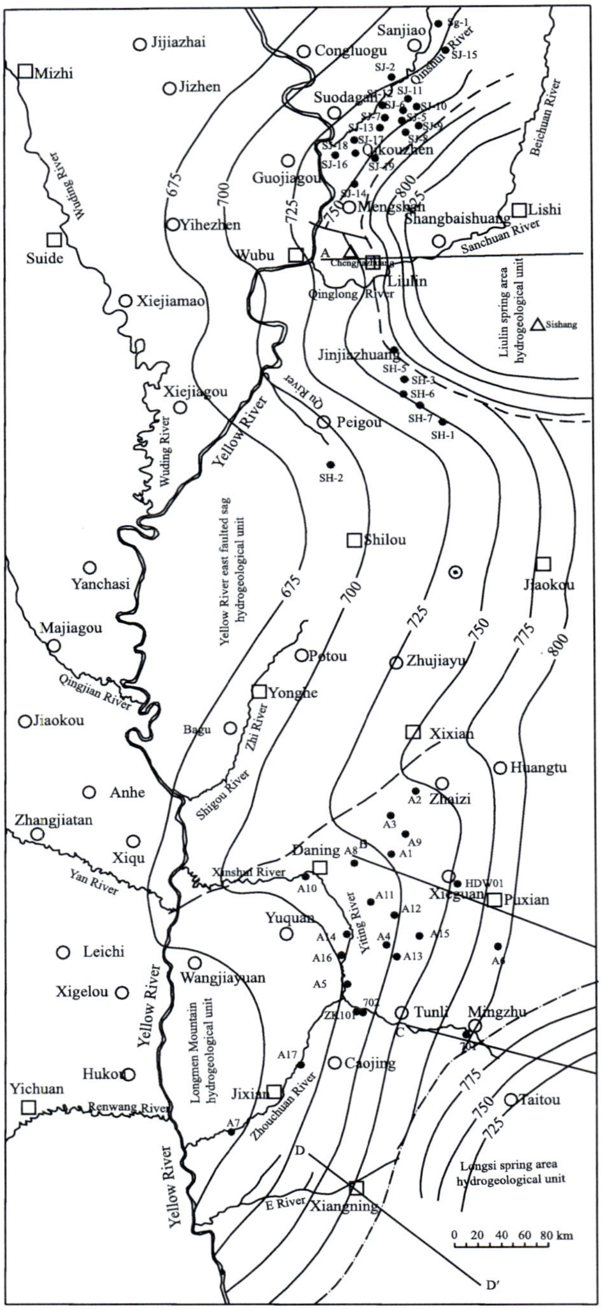

Fig. 5.24 Hydrogeological units and Ordovician water table contours in the southern segment of the eastern margin of Ordos Basin (unit: m)

then to nearly horizontal again within 1 km. The western section is characterized by thrust faults, and faults, folds and flexures make up of the southern gas barrier.

There are four hydrogeological units: the Liulin spring area, the Yellow River East faulted sag, the Longsi spring area and the Longmen Mountain.

① The Liulin spring area. Controlled by monoclinic structures, groundwater is replenished at the limestone outcrop and then converged to the Liulin area, where it flows out of the Liulin spring. There are two directions of strong runoff, one is from the exposed limestone area in the middle of the spring area to the Liulin spring, and the other is from the exposed limestone area in the southeast of the spring area to the Liulin spring after bypassing the Lishi syncline (a coal basin). From the recharge zone to the discharge zone, the groundwater basically has a uniform watertable, but different locally due to local structures and aquifuges, so that relatively independent groundwater systems are existent and control CBM preservation.

② The Yellow River east faulted sag. The northern boundary of the unit is adjacent to the hydrogeologic unit of the Liulin spring area. Paleozoic–Cenozoic strata are exposed. Most of the surface rivers flow into the Yellow River system. Groundwater recharge mainly comes from atmospheric precipitation and surface water, and discharges to river valleys in forms of springs and subsurface flows. The shallow Carboniferous and Permian aquifers discharge to rivers in the form of subsurface flows. The Early Paleozoic limestone water flows from east to west and from north to south in the form of confined water.

③ The Longsi spring area. The east and south boundaries extend beyond the unit. It is a syncline consisting of the nearly NS Qiaojiawan-Niuwangmiao complex syncline in the axis of the Shanxi faulted uplift. The Permian system is exposed in the syncline center, and only the west wing is in the unit. The groundwater is recharged by atmospheric precipitation, and converges from the two sides of the syncline to the axis and flows from north to south. The aquifers generally appear as confined water. The Longsi spring group are structurally uplifted springs and are discharge points of groundwater in the unit.

④ Longmen Mountain. The Yellow River flows from the western margin of the unit from north to south. The north and west are bordered by the Yellow River East faulted sag, the east connects with the Longsi spring area, and the south is bounded by the Matou Mountain fault. Strongly cut by the “V”-type gullies in the unit, favorable conditions are available to the discharge of surface water, but limits the recharge of primary aquifers. The groundwater is dominated by limestone water, and then the fissure water in the weathered mixtite zone. The limestone water flows from northeast to southwest, and one part of water flows into Yumenkou through the mixtite; and the other leaks into the Fenwei graben through the large Matoushan fault.

5.3.2 Underground Hydrodynamics Zoning and CBM-Bearing Characteristics

5.3.2.1 Qinshui Basin

The zoning of modern groundwater dynamic conditions in the Qinshui Basin is closely related to the CBM-bearing properties. From the wing to the axis of the complex syncline, the aquifer is from shallow to deep, the groundwater runs off from positively to slowly, and the salinity increases gradually, showing obvious zoning laterally. The relationship between the above zoning and the CBM-bearing properties is as follows:

① The strong runoff zone is 3–5 km from the basin edge, and the top of the Carboniferous system is 700–1000 m. In the zone, faults and secondary folds are relatively developed; fissures and karsts form a vein network; the water outflow per unit from a borehole is greater than 4 L/(s m); the water abundance is relatively high; the salinity is generally 300–600 mg/L; the water type is $SO_4{\cdot}HCO_3$–Ca·Mg; the karst water is in the state from no pressure to pressure; the hydraulic gradient is 0.4‰; the groundwater flow rate is about 1.1 km/a; the runoff conditions are strong; and the gas content is generally low.

② The medium runoff zone is located in the circular slope of the basin, about 3–8 km wide, and the top of the Carboniferous is 400–700 m. Influenced by faults and secondary folds, the runoff conditions are relatively strong; the karst water is pressured; the karsts and fissures are relatively developed; the water abundance is extremely uneven; the water outflow per unit from a borehole is generally 1–10 L/(s·m); the salinity is generally 600–1200 mg/L; and the water type is $SO_4{\cdot}HCO_3$–Ca·Mg; the hydraulic gradient is about 3.6‰, and the gas content varies greatly.

③ The weak runoff zone is located at the axis of the basin and is the retention boundary of groundwater. The water abundance is strong; the water inflow per unit from a borehole is less than 1 L/(s·m); the water quality is obviously deteriorated; the salinity is more than 1200 mg/L; and the water type is SO_4–Ca·Mg. The groundwater runoff is weak, but relatively strong in the axis of the secondary anticline with developed fractures and karsts. The gas content is generally high.

The zoning characteristics of modern groundwater dynamic conditions in the basin also obviously control the gas-bearing properties. The dynamic conditions of modern groundwater in different hydrogeological units or spring basins are different, and have profound impacts on the preservation conditions or gas-bearing characteristics of coal seams.

① The Niangziguan spring area is controlled by two structures: the Fuping uplift on the north wing and the eastern Taihang Mountains complex anticline on the east wing. It is an arc syncline dipping toward the center of the basin. Atmospheric precipitation is the main source of groundwater supply. The groundwater runs off consistently with the surface water. The groundwater level is higher in the Majiagou Formation than in the Taiyuan Formation, and higher in the Taiyuan Formation than in the Shanxi Formation, indicating that the hydrodynamic connections between the aquifers, and between the aquifers and the coal seams are weak. Water test by pumping from a borehole shows that the unit water outflow, permeability coefficient and discharge radius of the Taiyuan Formation aquifer is larger than that of the Shanxi Formation, which is one of the primary causes for the gas content of the Shanxi Formation is higher than that of the Taiyuan Formation.

② The groundwater dynamic field in the Xin'an spring area is controlled by southeast and northwest boundary faults, and controlled by northeast and southwest surface and underground watersheds. The main faults in the area are conductive to water and gas. They complicate the groundwater runoff conditions, and are an important geological cause for the obvious difference in gas content of coal seams in different sections of the area. In the layer domain, the hydrodynamic connections between the aquifers and between the aquifer and the coal seams are weak. The unit water outflow, permeability and discharge radius of the Taiyuan Formation aquifer are larger than those of the Shanxi Formation, resulting in the gas content of the main coal seam of the Shanxi Formation being higher than that of the Taiyuan Formation. In the area, the unit water outflow and the permeability coefficient increases, and the gas content decreases.

③ The groundwater in the Guangshengsi spring area flows from the exposed limestone zone of the Huoshan uplift in the northwest and the eastern underground watershed to the spring area. The former runs off from northwest to southeast, with enough supply from atmospheric precipitation. The latter runs off from northeast to southwest, with relatively weak supply. The groundwater runoff is blocked in the southern Sitou fault zone. The distribution of the groundwater level is the same as that in the Xin'an spring area in the stratigraphic domain, indicating that the hydrodynamic connection between the aquifers is weak. The unit water outflow and permeability coefficient of the Shanxi aquifer is larger than that of the Taiyuan Formation, resulting in the gas content of the main coal seam of the Taiyuan Formation being slightly larger than that of the Shanxi Formation. Regionally, the northwestern water equipotential is steeper than the southern, reflecting the weakening of the groundwater runoff from northwest to southeast, which is consistent with the law of the gas content.

④ The northwest and east of the Yanhe spring area are water-blocking fault boundaries, the north is an underground watershed, the western section of the south blocks water and the eastern part conducts water. The land elevation and water level are relatively low in the Qinshui Basin. The water equipotential is very gentle, reflecting weak groundwater runoff. The distribution of the groundwater level in the stratigraphic domain is the same as that in the Guangshengsi spring area, indicating that the hydrodynamic connection between the aquifers is weak. But the difference between the water level of the Taiyuan Formation and the water level of the Majiagou Formation is small, indicating that there is a certain hydrodynamic connection between the two. Regionally, the unit water outflow and permeability coefficient of the Taiyuan Formation and the Shanxi Formation are lower in the Panzhuang mine field and the Daning mine field, indicating that the groundwater is stagnant, which is consistent with the distribution of the CBM enrichment center in the southern part of the basin.

5.3.2.2 The Southern Segment of the Eastern Margin of Ordos Basin

The groundwater dynamic conditions are generally controlled by the monoclinic structure on the eastern margin of the Ordos Basin, and the influence of topography and geomorphology is also quite obvious. The direction of groundwater runoff is consistent with the surface water flow. However, the hydrodynamic fields in different sections are different. Not only the runoff intensity and direction are different, but also there are obvious differences in different aquifers. These differences are important causes for the heterogeneous gas content of coal seams in the area.

The Middle Ordovician aquifer is primarily in the Majiagou Formation. In the north-central part, the groundwater flows from east to west in general. In the southwest, the hydrogeological unit of the Longmen Mountains runs from northeast to southwest and runs from southeast to northwest. In the southeast, the Longsi springs run from northwest to southeast. The runoff intensity of the Liulin spring area, that is, the unit water outflow, permeability coefficient and discharge radius, are larger than the Longsi spring area, and that in the Longsi spring area larger than the Longmen Mountain hydrogeological unit (Table 5.1). Due to weak or virtually unrelated hydraulic linkages between the aquifers, the groundwater runoff from the aquifers has little effect on the preservation conditions of the Carboniferous–Permian CBM.

The Upper Carboniferous aquifer is dominated by Taiyuan Formation limestone. The roof of No. 8 coal seam is Miaogou limestone. The limestone and the Maoergou limestone of No. 8 coal seam are main aquifers in the Taiyuan Formation aquifer system and the main aquifers of CBM wells. According to borehole test data, the water level in the Taiyuan Formation limestone is 25–50 m larger than that in the Middle Ordovician, and the groundwater runoff intensity is weakened from east to west. Compared with the Middle Ordovician limestone aquifer, the hydrodynamics of the Taiyuan Formation aquifer is relatively weak in the Liulin spring area; the gradient of the water level in the Longsi spring area and the Longmen Mountain units is steep (Fig. 5.26); the unit water outflow, permeability coefficient and discharge radius are relatively large (Table 5.1); the runoff of groundwater is relatively strong; the gas content of No. 8 coal seam is generally lower than 5 m^3/t. The hydrogeological profile of the Taiyuan Formation in the Xiangning area shows that the water level in the watershed area is the highest, and the runoff capacity is relatively strong to east and west. Most of CBM diffused with runoff (Fig. 5.25). In the relatively stagnant groundwater area in the northwest of the southern Tunli Town, the coal seam has the highest gas content of 15 m^3/t.

The Lower Permian Shanxi Formation aquifer is the top and bottom sandstone of No. 5 coal seam. According to borehole water test data, the water level of the aquifer is about 25 m higher than that of the Taiyuan Formation, and the water level elevation gradually decreases from 850 to 775 m from east to west, forming a groundwater concentration zone in the Mingzhu Township, western Daning County. The runoff intensity becomes weak from east to west, which was basically consistent with the Taiyuan Formation. However, the hydrodynamic conditions of the Longmen Mountain hydrogeological unit have been significantly weakened, the gradient of the water level is generally slower, the runoff capacity is weak, and the unit water outflow, permeability coefficient, and discharge radius are relatively small (Table 5.1). As a result, the gas content of No. 5 coal seam is generally higher than that of No. 8 coal seam, especially in the relatively concentrated groundwater area in the west of Daning. The gradient of the water level is extremely gentle (Fig. 5.26), the gas content of No. 5 coal seam is as high as 20 m^3/t due to the extremely weak groundwater runoff, and gradually decreases to 10 m^3/t with the increase of runoff capacity toward east and two sides.

5.3.3 Groundwater Geochemical Field and CBM Preservation Conditions

The groundwater geochemical field is the result of the comprehensive action of water–rock-gas three-phase chemical equilibrium under the control of geological and hydrogeological conditions, reflecting the alternating conditions and runoff characteristics of groundwater, and has strong indicative significance for CBM accumulation and diffusion.

5.3.3.1 Qinshui Basin

In view of the karst fissure aquifer in the Majiagou Formation of the Central Ordovician, the runoff conditions in the western and southeastern parts are relatively strong, and the water type is generally HCO_3–Ca·Mg; in the medium runoff area, the water type is generally SO_4·HCO_3–Ca·Mg; in the deep weak runoff area or some local stagnant areas, the water type is mostly SO_4–Ca·Mg. Correspondingly, the salinity of groundwater shows an overall distribution trend: low in the basin circumference and high toward the center. However, the boundaries of the karst water in the four spring areas are relatively clear, the horizontal gradient of salinity is relatively gentle in the west and east, and is steep at the northern and southern margins, indicating regional groundwater recharge, runoff and drainage systems constraining the groundwater geochemistry.

Table 5.1 Groundwater dynamic characteristics in the southern segment of the eastern margin of Ordos Basin

Hydrogeological unit		Liulin spring area	Longmen Mountain	Longsi spring area
O_2	Unit water outflow/[L/(S·m)]	0.0085–1.232	0.0005–0.006	0.0016–0.053
	Permeability coefficient/(m/d)	0.0942	0.0017	0.0087
	Discharge radius/m	12.63–42.25	14–21	19–37
C_2	Unit water outflow/[L/(S·m)]	0.0008–0.0064	0.0004–0.208	0.00022–0.55
	Permeability coefficient/(m/d)	0.00378–0.02781	0.0029–0.15	0.0063–0.36
	Discharge radius/m	15.64–42.64	21–44	27–53
P_1s	Unit water outflow/[L/(S·m)]	0.00012–0.061	0.0005–0.108	0.0005–0.102
	Permeability coefficient/(m/d)	0.00037–0.00075	0.0023	0.0035
	Discharge radius/m	13–18	14–20	15–26

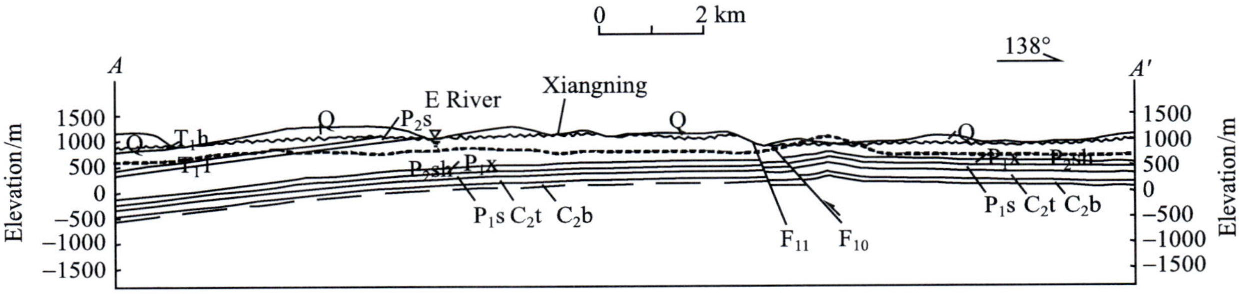

Fig. 5.25 Hydrogeological profile of the Taiyuan Formation in Xiangning, the southern segment of the eastern margin of Ordos Basin

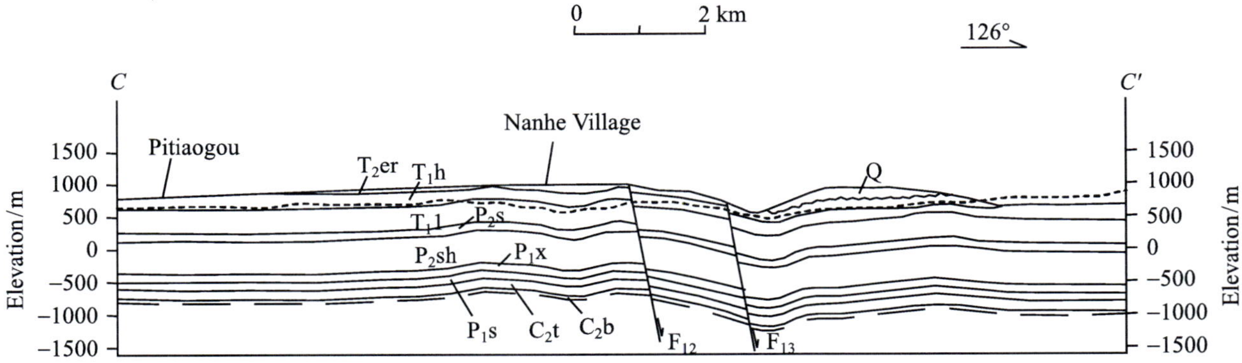

Fig. 5.26 Hydrogeological profile along Daning-Xueguan-Puxian in the southern segment of the eastern margin of Ordos Basin

The regional distribution of groundwater salinity in the karst fissure aquifer of the Upper Carboniferous Taiyuan Formation is similar to that of the Majiagou Formation (Fig. 5.27a), but the hydrochemical type is mainly $HCO_3 \cdot SO_4$–K·Na. From the slope to the axis of the basin, the water type changes from $HCO_3 \cdot SO_4$–Ca to $HCO_3 \cdot SO_4$–K·Na and $HCO_3 \cdot SO_4$–Ca·Mg, and the former is dominant. There are 1–2 groundwater centers with high salinity in the Niangziguan spring area and the Xin'an spring area. In the western Guangshengsi spring area, the runoff conditions are weakened and the water type changes from HCO_3–Ca to SO_4–Ca·Mg, and the salinity increases from 600 to 1200 mg/L. A stagnant zone is formed near the Sitou fault, which provides an important basis for explaining the gas content of the main coal seam of the Taiyuan Formation is higher than that of the Shanxi Formation coal seam in the Zhengzhuang area. In the southern Yanhe spring area, the Taiyuan Formation near the Daning mine field is dominated by karst fissure water, and the water abundance is weak. The water type is $HCO_3 \cdot SO_4$–Ca·Mg, and then converts to HCO_3–K·Na toward the deep. The high-salinity center is located in the Panzhuang mine field and its north, up to

2964 mg/L. The water type is $HCO_3{\cdot}SO_4$–Ca·Mg, and consistent with the dynamic conditions of stagnant groundwater and the southern gas enrichment centers.

The regional distribution of groundwater geochemical field in the sandstone fissure aquifer of the Lower Permian Shanxi Formation inherits the overall appearance of the underlying Taiyuan aquifer to some extent, for example, the general rule of salinity distribution is consistent with the Taiyuan Formation and the Majiagou Formation (Fig. 5.27 b). In the northern Niangziguan spring area, the water type in the Hanzhuang mine field in the south of Shouyang is HCO_3–K·Na, in the Shangzhuang mine field, No. 2 expansion area and No. 5 expansion area in the west of Yangquan, SO_4–K·Na and HCO_3–K·Na are dominant, and $CO_3{\cdot}Cl$–K·Na is dominant in the northern shallow part. In the eastern Xin'an spring area, the water type around Tunliu is mostly HCO_3–K·Na or $HCO_3{\cdot}SO_4$–K·Na; HCO_3–Ca·Na or Cl–Na around Zhaozhuang and Gaoping; and $HCO_3{\cdot}SO_4$–Ca·Mg or HCO_3–K·Na around Panzhuang and Daning. Along the change of the water type, the relationship between groundwater runoff and CBM accumulation and diffusion is consistent with the conclusions based on salinity distribution and water equipotential.

Based on the above results, the following general understanding can be obtained: (i) From the edge to the center of the basin, the groundwater runoff intensity in each aquifer is gradually weakened, which forms detention in the deep part of the basin or near the internal water-blocking boundary, forming a high-salinity center. (ii) Water in the weathered bedrock zone vertically infiltrates to the Shanxi aquifer, and the thick Taiyuan Formation limestone aquifer vertically infiltrates to the Middle Ordovician karst water. (iii) The runoff capacity of surface watershed is obviously stronger than underground watershed. (iv) The groundwater runoff characteristics reflected by changes in groundwater ion composition are highly consistent with the runoff characteristics reflected by groundwater equipotential surface.

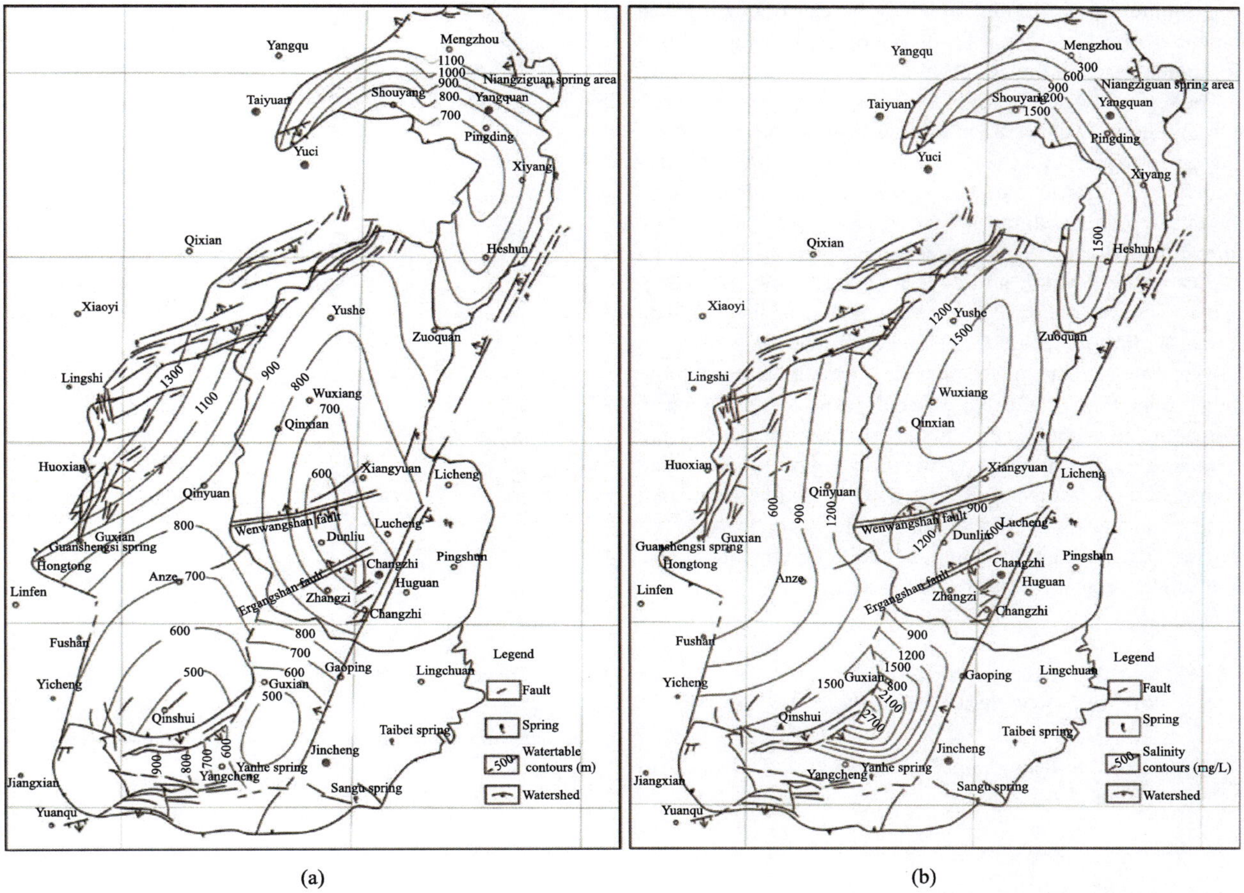

Fig. 5.27 Contours of groundwater salinity in coal-series aquifers in Qinshui Basin **a** upper Carboniferous Taiyuan Formation; **b** lower Permian Shanxi Formation

5.3.3.2 The Southern Segment of the Eastern Margin of Ordos Basin

Controlled by the structure and supply, runoff and discharge conditions, the geochemical field of the Carboniferous Taiyuan Formation aquifer shows regular changes in space. The water type is generally HCO_3–Ca·Na·Mg. With the increase of the aquifer depth, the runoff conditions of karst water become gradually weak, and the salinity gradually increases, so that the water type changes from SO_4·HCO_3–Ca·Na to Cl·SO_4–Ca·Na·Mg. In general, the runoff conditions in the southwestern part are relatively strong. The salinity in the Longsi spring area and the watershed with the Longmen Mountain hydrogeological unit is less than 1.0 g/L. In the Liulin spring area, the Yellow River fault and the Longmen Mountain hydrogeological unit, it is not zoned obviously, from 1.0 to 10.0 g/L from east to west. In the west of Daning and the Shaqu mine field, relatively stagnant zones appeared, where the salinity is greater than 10.0 and 5 g/L, respectively (Fig. 5.28). The former is consistent with the center of the Tunli Town gas-rich zone in the west of Daning. The latter is consistent with the gas-rich zone in the Shaqu mine field in Liulin, which is the cause why the gas content of the No. 8 coal seam is higher than that of the No. 5 coal seam in the Shaqu mine field.

The regional distribution of the groundwater geochemical field in the Lower Permian Shanxi Formation sandstone aquifer inherited the overall appearance of the Taiyuan Formation aquifer to a certain extent, and the plane distribution law of the salinity is similar to that of the Taiyuan Formation. However, the gradient of the groundwater salinity in the Shanxi aquifer is relatively slow, reflecting that its runoff capacity is weaker than that of the Taiyuan Formation limestone aquifer. This is why the gas content of the No. 5 coal seam in the Shanxi Formation is generally higher than that of the No. 8 coal seam in the Taiyuan Formation. In addition, the salinity contours of the Shanxi Formation aquifer moved southward its center in the Shaqu mine field, and formed another center which is highly consistent with the plane distribution of the gas-rich center of the No. 5 coal seam in the south, in addition to maintaining a high-salinity center in the west of Daning.

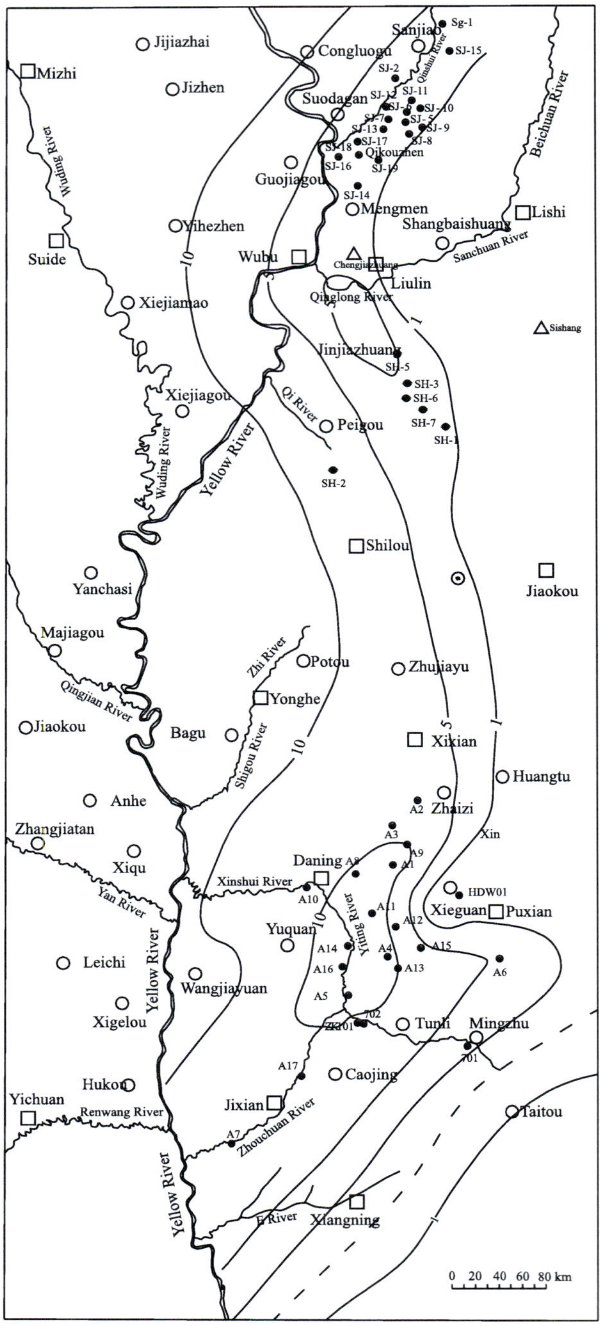

Fig. 5.28 Contours of groundwater salinity of the Taiyuan Formation in the southern segment of the eastern margin of Ordos Basin (unit: mg/L)

5.3.4 Groundwater Head Height and Gas-Bearing Properties

The groundwater head height in the Qinshui Basin has regular regional distribution:

① In the northern part of the basin, the watertables in different boreholes are not quite different, but the watertables in different aquifers in the same borehole are different, reflecting that the hydraulic linkages among boreholes and aquifers are weak. However, the watertable in the bedrock weathering zones in some boreholes is different from that in the Shanxi Formation by only 4.86 m, indicating that groundwater vertically infiltrates from the bedrock weathering zone to the Shanxi Formation.

② In the eastern part, the watertables in different boreholes are not quite different, reflecting that the groundwater flows from south to north in the south and north to south in the north. In the same borehole the watertable is the highest in the bedrock weathering zone, followed by the Shanxi Formation, the Taiyuan Formation and the Middle Ordovician formation. There is a tendency of vertical infiltration between the aquifers (especially the Taiyuan Formation and the Majiagou Formation).

③ In the western part, the watertable in the aquifer in the Qinyuan area is above 1100 m. Due to the shallow burial of the coal system, the reservoir pressure is low, and the watertable in different aquifers in the same borehole is different only 13.3–175.3 m, indicating weakly connected water.

④ In the southern part, the watertables in different aquifers are quite different, the highest in the Chengzhuang mine field, then the No. 154 hole in the Tingdian township, and relatively low in the Panzhuang-Daning mine field, indicating that the groundwater flows from east to west in the east and from southwest to northeast in the southwest. The watertables in different aquifers in the same borehole vary on the plane, and the lowest in the Panzhuang-Daning mine field, indicating that the hydraulic connections of the aquifers are stronger than other parts.

Many reservoir pressure data are available from CBM wells in the Qinshui Basin. In addition, according to the large amount of pumping experiments conducted during the exploration of coal resources, the equivalent pressure can be converted. Combining these two data, the contour map of the main coal seam pressure and its coefficient were plotted (Fig. 5.29).

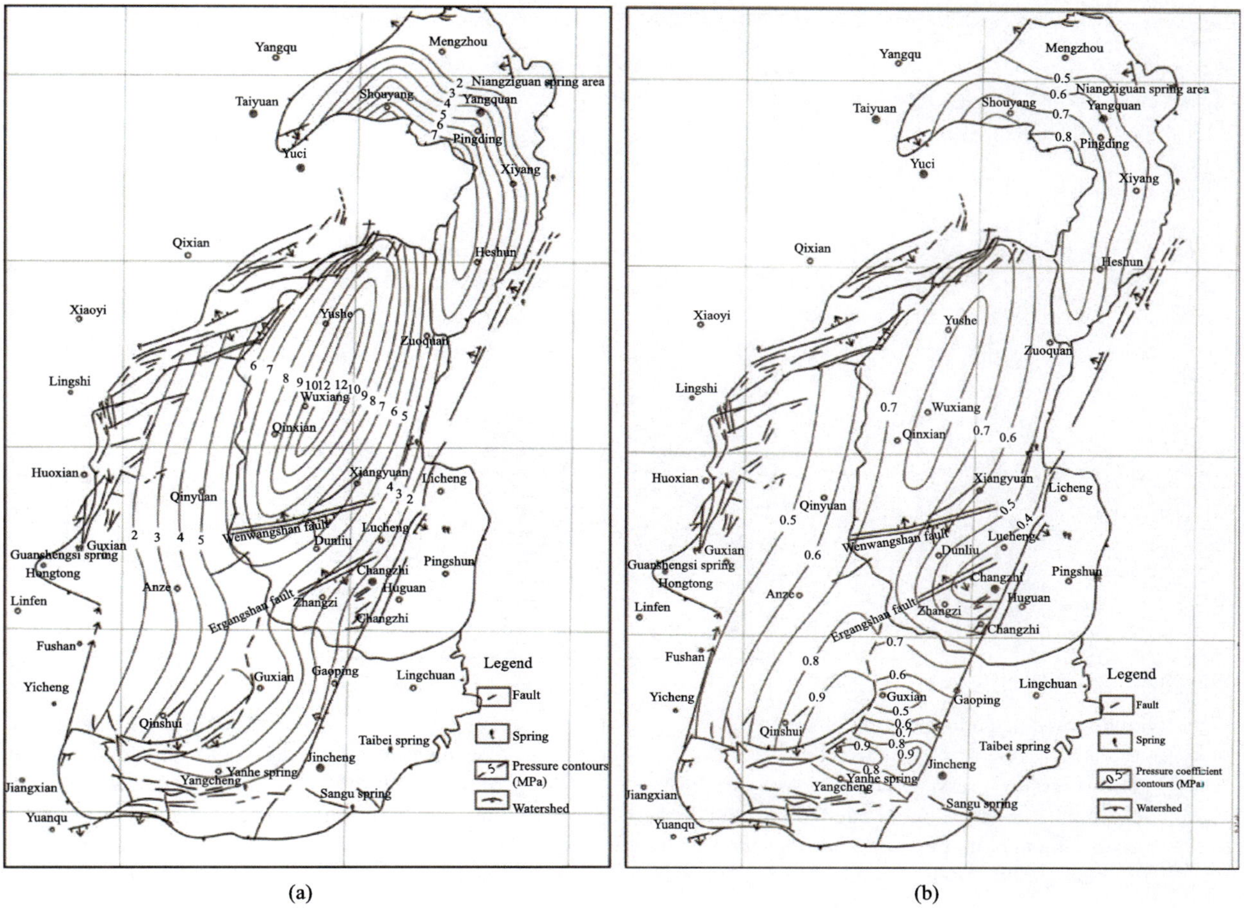

Fig. 5.29 Contours of Shanxi Formation pressure coefficient in Qinshui Basin **a** pressure; **b** pressure coefficient

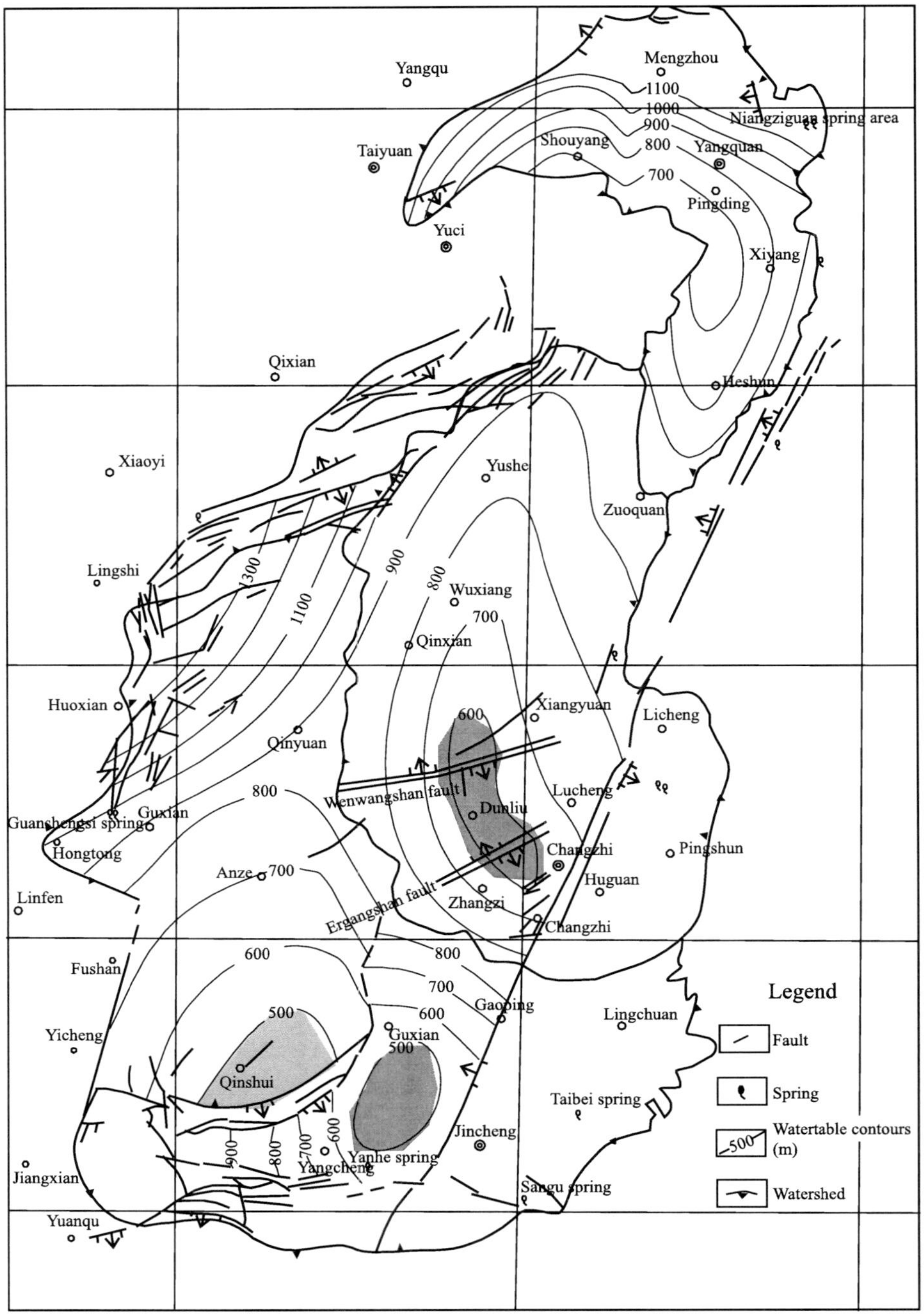

Fig. 5.30 Watertable contours of the Upper Carboniferous Taiyuan Formation aquifer in Qinshui Basin

The following understanding can be obtained:

① The reservoir pressure and pressure coefficient contours are distributed circularly around the main syncline in the basin, and generally distributed in the NNE direction, but in the EW direction in the north and south parts (Fig. 5.30). The measured pressure coefficient in the CBM wells is 0.39–1.02. Most of the coal seams are at underpressure (the pressure coefficient is 0.5–0.8). The reservoir at normal pressure (pressure coefficient > 0.8) is distributed in Qinxian,

Table 5.2 Measured pressure gradient of main coal seams in the southern segment of the eastern margin of Ordos Basin

Area	Pressure gradient/(MPa/hm)	
	No. 2–5 coal seams of Shanxi Formation	No. 8–10 coal seams of Taiyuan Formation
Linxing	–	0.99
Sanjiao	1.11	1.02
Shilou	0.61	0.74
Well Jin Test 1 in Jixian	1.1–1.2	
Wucheng, Daning	0.91	0.97

Wuxiang and Yushe in the center and the southern part. The reservoir at very low pressure (pressure coefficient < 0.5) is distributed in the Zhangzi and Guxian areas in the south-central part.

② The measured pressure of the Taiyuan Formation coal seam is greater than that of the Shanxi Formation coal seam. However, the pressure of the Taiyuan Formation after converting according to the water head height is mostly lower than that of the Shanxi Formation coal seam, indicating that the fluid connectivity is extremely poor, and hardly communicating between the two coal seams. However, the converted head pressure of two main coal seams in Well TL002 is the same, indicating that the two belong to the same fluid pressure system, and there is a possibility that the CBM in the coal seam of the Taiyuan Formation will flow to the coal seam of the Shanxi Formation.

③ The coal seam pressure is generally positively correlated with the burial depth, but the pressure coefficient is negatively correlated with the burial depth. That is, within the well testing depth, as the coal seam depth increases, the underpressure is serious. It is worth noting that the coal seam of the Shanxi Formation in Well QC 1 is 1008 m deep, but the pressure coefficient is as high as 0.96 MPa, which obviously deviates from the above fitting relationship. This phenomenon suggests that if the burial depth further increases, the pressure coefficient of coal reservoir may have a sudden increase. In other words, the controlling factors on the coal seam pressure in deep layers may be different from that in shallow layers. We should attach great importance to this problem in deep prediction

In the southern segment of the eastern margin of the Ordos Basin, the groundwater head height and the coal seam pressure also have significant impacts on coalbed gas content. According to test data, the pressure gradient in the western Sanjiao and the west of the central-southern Daning is greater than 1 MPa/hm, indicating overpressure reservoirs; 0.61–0.74 MPa/hm in the Shilou area, indicating underpressure reservoirs; close to hydrostatic pressure gradient in other areas, indicating normal pressure reservoirs (Table 5.2).

The groundwater head height converted from reservoir pressure is positively correlated with coalbed gas content, i.e. the higher the head height, the higher the gas content of the coal seam (Fig. 5.31). Specifically, when the water head height is less than 200 m, the gas content is less than 5 m^3/t; when the water head height is between 200 and 500 m, the gas content is 5–10 m^3/t; when the water head height is higher than 500 m, the gas content is generally greater than 10 m^3/t.

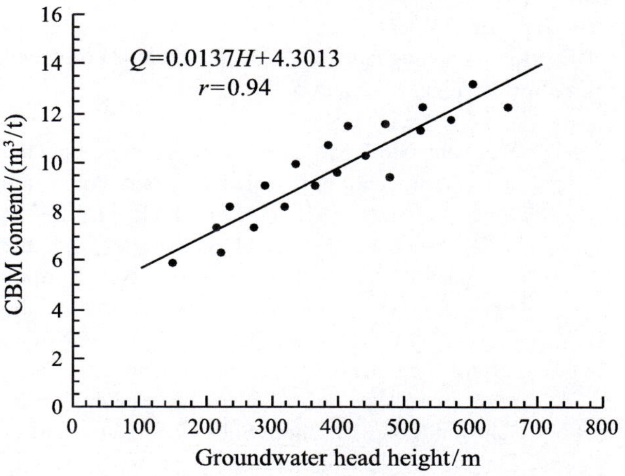

Fig. 5.31 Relationship between groundwater head height and CBM content in the southern segment of the eastern margin of Ordos Basin

5.3.5 Control Effect of Groundwater Dynamic Conditions and Its Manifestation

Groundwater dynamic conditions are concentrated on watertable potentials. Groundwater chemical conditions are important indicators for indicating groundwater dynamic conditions on the one hand and affected by the dynamic conditions on the other hand. That is to say, through detailed analysis of groundwater potentials and the characteristics of groundwater geochemical field, it is possible to understand the specific form of groundwater dynamic conditions controlling CBM accumulation and diffusion, so as to predict the possible areas of CBM enrichment.

Based on the comprehensive analysis of the above results, according to the dynamic and structural conditions of the aquifers, the relationship between the hydrogeological conditions of the Carboniferous and Permian and the CBM accumulation in the southern segment of the eastern margin of the Ordos Basin and the Qinshui Basin is reflected in two major categories: "strong runoff equipotential" plays the role of hydraulic dissipation of CBM, which is not conducive to the preservation of CBM. This situation is common. Secondly, "slow flow-stagnant flow" equipotential plays the role of hydraulic sealing, which is conducive to CBM accumulation, but it is often only distributed in local areas.

The "strong runoff equipotential" is reflected on the outer boundary of the spring system to different degrees, especially in the northern part of the Guangshengsi spring area in the Qinshui Basin (north of the Qinyuan area). At the same time, in the watershed on the inner boundary of each spring area or on both sides of the watershed, the groundwater runoff capacity is strong, which is unfavorable for the preservation of CBM.

The "slow flow-stagnant slow" equipotential is embodied in three forms (Figs. 5.30 and 5.32):

① Stagnant slow in "sag" equipotential. In the hydrogeological unit, local stagnant zones are formed by secondary folds and faults at different scales, such as the Hanzhuang mine field, the Yangquan No. 1 mine and the No. 3 mine expansion in the Niangziguan spring area, the Qixian, Wuxiang and Yushe areas in the Xin'an spring area, the Daning, Panzhuang and Fanzhuang areas in the Yanhe spring area in the Qinshui Basin; the south of Puxian and the east of Daning in the southern segment of the eastern margin of the Ordos Basin, i.e. around the Kemingzhu Town and the area surrounded by Yuquan, Zhaizi, Tunli and Caojing, where the groundwater is in a ring-shaped distribution, and less than 800 m. This type of equipotential is

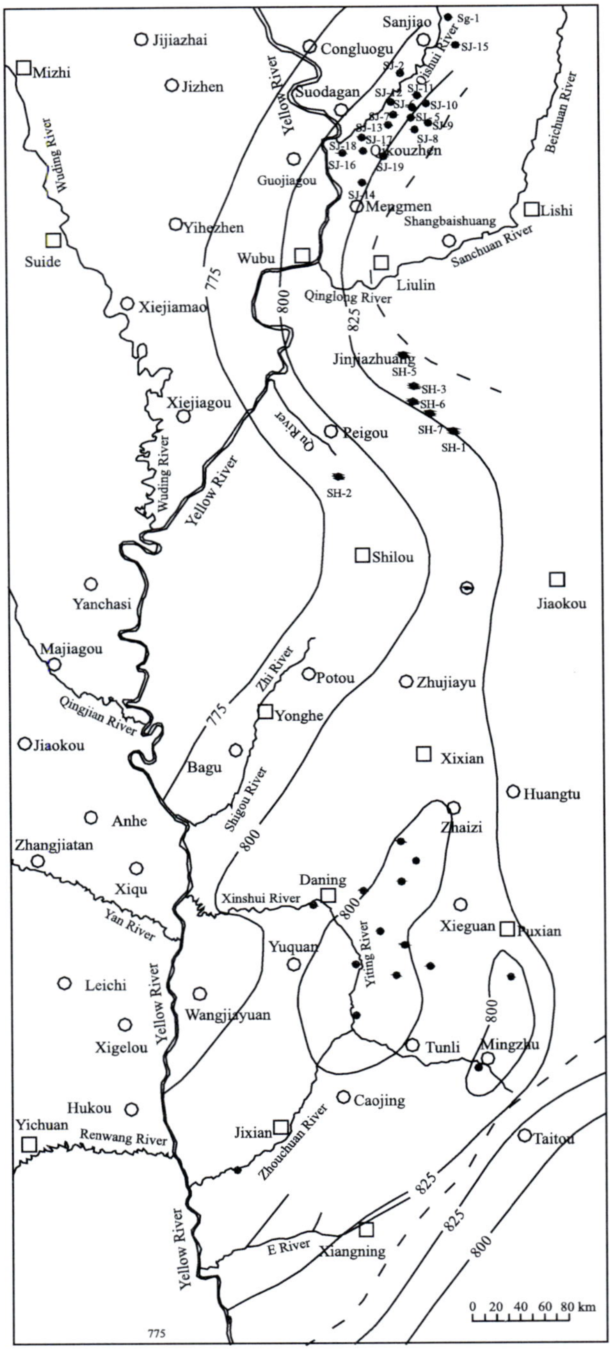

Fig. 5.32 Contours of the upper Carboniferous Taiyuan Formation aquifers in the southern segment of the eastern margin of Ordos Basin (unit: m)

clearly in a "sag" form, i.e. an almost closed form. The total solids and hardness are very large, the salinity is extremely high, and the strontium isotope value is low, indicating that groundwater flow is not smooth, the infiltration of surface water is weak, so CBM is enriched by hydraulic closure.

② Slow flow in dustpan equipotential. This type is only developed in local areas in the Qinshui Basin, such as the Tunliu, Qinyuan and Panzhuang north areas in the Niangzhiguan, Xin'an and Guangshengsi spring areas. It is characterized by high potential on three sides and low on one side with water supply, and the groundwater runoff is blocked to a certain extent. Since the terrain likes a dustpan, the equipotential is relatively flat, the hydrodynamic potential is weak and the hydrostatic pressure is dominant, so the groundwater flow is very slow driven by gravity, which is beneficial to the preservation of CBM.

③ Slow flow in fan equipotential. This type only appears in the Qinshui Basin, such as the Lishui area in the southwest, and is more typical in the Zhengzhuang area. The northern and western watertalbes are higher, and the eastern and southern are relatively low. The low potential may be blocked by the Sitou fault, or replenished by surface water and blocked in the outcrop area, so CBM diffusion with groundwater migration may be relatively weak.

Among the above three types, the first type is actually the manifestation of "hydraulic sealing effect", and the other two types show the characteristics of "hydraulic sealing effect" in different forms. Their common characteristics are that groundwater runoff is slow or even almost stagnant, hydrodynamic conditions are weak, and groundwater salinity is high. The common essence is that slow or stagnant groundwater greatly delays the escape of CBM and makes CBM relatively rich.

In addition, the salinity contours of the upper Carboniferous Taiyuan Formation in the eastern margin of the Ordos Basin is closedly distributed between the Chengjiazhuang and the Jinjiazhuang on the south side of the Liulin–Wubu line, and the salinity inside the circle is high in a large area. The salinity contours of the lower Permian Shanxi Formation is closed in the south of the Jinjiazhuang in a small area, and the salinity inside the circle is also relatively high. This phenomenon indicates that under the control of local structures, groundwater dynamic conditions near the Jinjiazhuang may be relatively favorable for CBM enrichment.

5.3.6 Relationship Between Hydrodynamic Conditions and CBM Enrichment

Based on the study of hydrogeology and CBM enrichment in the eastern margin of the Ordos Basin and the Qinshui Basin, and the hydrogeological conditions, CBM distribution and geochemical characteristics of other basins in China, it is found that in areas with strong dynamic conditions, not only the CBM content is relatively low, but also the carbon isotope of methane is lighter; in the area with weak hydrodynamics or stagnant water, the CBM content is relatively high, and the carbon isotope of methane is relatively heavy. Therefore, hydrodynamics has an important controlling effect on CBM accumulation, showing both preservation and destruction.

5.3.6.1 Control of Groundwater on CBM Content

On the plane, the gas content is low in areas with strong groundwater dynamic conditions, but high in areas with weak dynamic conditions. For example, in the southern part of the Qinshui Basin, the lower coal group of the Carboniferous Taiyuan Formation and the upper coal group of the Permian Shanxi Formation are developed widely. The No. 15 coal seam is primary in the lower group, and the No. 3 coal seam is primary in the upper group. As a monocline extends toward the basin center, faults are not developed, and groundwater flows confluently, and is supplied from atmospheric precipitation in the east and south, and the watershed in the north and west. Water concentrates from east, south, west and north to low equipotential. Whether in the No. 3 coal seam in the upper primary coal seam (Fig. 5.33) or in the No. 15 coal seam in the lower primary coal seam, CBM increases significantly along water flow. In the area where hydrogeological conditions are favorable for stagnant flow, the gas content is high, and CBM is enriched in southern Qinshui Basin.

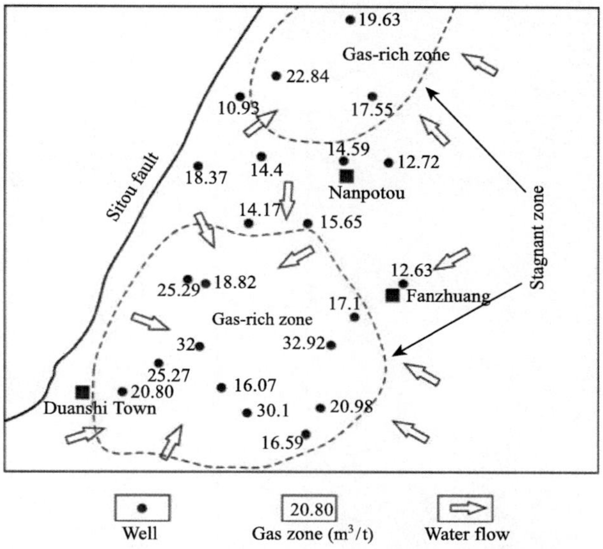

Fig. 5.33 Relationship between gas content and hydrogeology of No. 3 coal seam in southern Qinshui Basin

Another example is the Kaiping syncline, an asymmetrical syncline, which is steep in the northwest wing, and slow in the southeast wing, and the outcrop is high in the northwest wing and low in the southeast wing, so groundwater is replenished from the northwest wing and discharges to the southeast wing. Near the axis of the northwest wing, there is a stagnant zone with weak hydrodynamic conditions, where the groundwater runoff is relatively weaker than the southeast wing. A series of coalmines are distributed on the Kaiping syncline. The CBM content in the coalmine on the northeast wing is significantly higher than the CBM content at the similar depth on the southeast wing. For example, Zhaogezhuang Coalmine, Majiagou Coalmine and Tangshan Coalmine are distributed on the northeast wing. The gas content of these coalmines is significantly larger than that of the Linxi Coalmine, Lüjiatuo Coalmine and Qianjiaying Coalmine on the southwestern wing (Fig. 5.34).

Vertically, the gas content of coal seams with strong groundwater dynamic conditions is relatively low. It is agreed that in the entire North China, the groundwater dynamic conditions of the upper coal seam are generally weaker than those of the lower coal seam.

Theoretically, the gas content of the lower coal seam of the Taiyuan Formation should be higher than that of the upper coal seam of the Shanxi Formation. However, the actual situation is reversed. It is common that the gas content of the lower coal seam is lower than that of the upper coal seam. For example, in the Huozhou mining area, the gas content of the lower coal seam (the Carboniferous Taiyuan Formation) is significantly lower than that of the upper coal seam (the Permian Shanxi Formation) (Liu 1998); in the Yangquan mining area, the depths of No. 3 coal seam (Shanxi Formation), No. 12 coal seam (Taiyuan Formation) and No. 15 coal seam (Taiyuan Formation) increase in turn, but the gas contents gradually decrease, and the coal seam pressures gradually decrease (Zhang et al. 2002); in the Jincheng mining area, it is very common that the gas content of the lower coal seam is lower than that of the upper coal seam in most CBM wells, but the gas content of the No. 15 coal seam in the Taiyuan Formation is lower than that of the No. 3 coal seam in the Shanxi Formation.

5.3.6.2 Control of Groundwater on Carbon Isotope of CBM

In North China, the hydrodynamics of the lower coal group is greater than that of the upper coal group, and the carbon

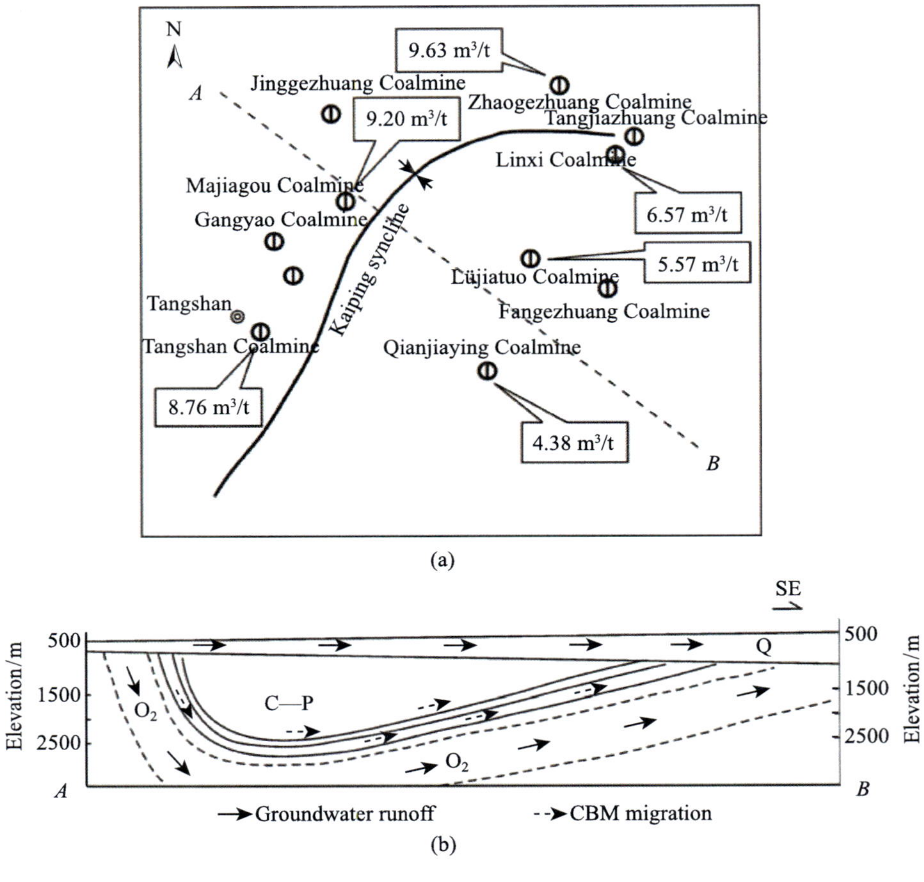

Fig. 5.34 Cross-sectional view of Kaiping syncline and groundwater runoff

isotope of the lower coal group is lighter than that of the upper coal group. For example, in the Wubu area in the Ordos Basin, the carbon isotope in the No. 3 coal seam of the upper coal group is −38.25‰, and that of the No. 10 coal seam of the lower Coal Group is −46.26‰, which is significantly lighter than the upper coal group. The carbon isotope in Well Test A-2 is −31.8‰, −32.1‰; and −33.56‰, −35.39‰ in Well Test A-3, respectively, in the No. 3 coal seam and No. 15 coal seam in the Qinshui Basin. It is obvious that the carbon isotope in the lower coal group is also significantly lighter than that in the upper coal group.

In the desorption gas of the No. 3 and No. 15 coal seams in Well Jin Test 2, and that of the No. 3, No. 10 and No. 15 coal seams in Well Jin Test 3 in the Qinshui Basin, desorbed during the same time, the carbon isotope in the lower coal group is significantly lighter than the upper coal seam (Zhang and Tao 2000).

5.3.6.3 Close Relation Between Carbon Isotope Values and CBM Content

As the maturity of conventional coal-formed gas source rocks increases, the methane carbon isotope becomes heavier, but the magnitude is not large, generally less than 10‰. For CBM, the carbon isotope varies greatly due to geological factors. Therefore, the correlation between CBM content and carbon isotope can be established without considering the influence of maturity (Fig. 5.35). It can be clearly seen that CBM content has a good correlation with carbon isotope value. If the CBM content is low, the carbon isotope value is light, which fully indicates that hydrodynamic conditions affect both CBM content and carbon isotope value.

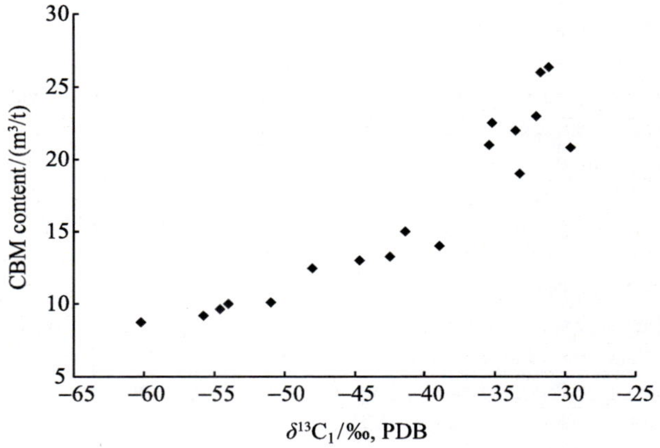

Fig. 5.35 Relationship between CBM content and methane carbon isotope (data provided by Wang Hongyan and Tang Xiuyi)

5.3.6.4 Control of Groundwater on CBM Reservoirs

According to the above phenomenon, hydrodynamic conditions in a coal system have a significant control on CBM reservoirs, and this control is manifested in two aspects. In a stagnant water area, water plays a role in preserving CBM reservoirs, that is, stagnant water is rich in gas; hydrodynamics in a runoff zone destroys CBM reservoirs.

(1) Gas enrichment mechanism in a stagnant zone

In coal seams, not only approximately orthogonal fracture systems consisting of end cleats and face cleats are developed, but also abundant matrix micropores are developed. Coalbed water mainly includes bound water in matrix pores and free water in fractures. CBM mainly exists in three phases (Su and Liu 1999; Su et al. 2001), namely adsorption phase (adsorption gas), dissolved phase (water soluble gas) and free phase (free gas). The three phases are in dynamic equilibrium and dominated by adsorption phase. Bound water is difficult to flow, and free water is always continuously alternately circulating, which causes the head and pressure in the fracture system to change, accordingly which changes the balance of the three phases, and methane in coal seams to dissolve and migrate.

The free water in coal systems dissolves the methane on the cleat surface of coal matrix, resulting in a decrease in the concentration of methane on the cleat surface, and the methane in the matrix to be converted from adsorbed gas to free gas by diffusion. Then the free gas will be dissolved and diffused when circulating with groundwater. In the long geological history, due to the long-term hydrodynamic effect, such dissolved amount plays an important role in the change of CBM content.

In addition, the movement of groundwater will cause changes in the head and water pressure fields in the coal seam fracture system. The pressure reduction will cause the CBM to desorb. The desorbed methane is first dissolved in water, then the dissolved methane and free methane flow with water from high head to low head areas under the action of hydrodynamic field.

In a stagnant water area, the desorption of CBM is not easy to occur due to the pressure of water, which plays an important role in the preservation of CBM reservoirs. In addition, due to the weaker circulation of groundwater, on the one hand, the amount of methane dissolved in water is very small, which is very beneficial to the preservation of CBM; on the other hand, the change of coal seam pressure caused by water flow is small, and the desorption of CBM is small, which is conducive to the preservation of CBM. Therefore, a stagnant water area is very beneficial to the preservation of CBM and is the best hydrodynamic condition for the preservation of CBM.

(2) Damage to CBM reservoirs in a runoff area

The damage mechanism of flowing groundwater to CBM reservoirs is to remove methane from coal by dissolving CBM, so that the CBM content is reduced, and the gas-bearing property is destroyed. In a runoff area, the hydrodynamic conditions are relatively strong, which promote the amount of methane taken away by water dissolution. The decrease of coal seam pressure caused by hydrodynamics also promotes the desorption of CBM, and the desorbed methane is dissolved and removed by water. The gas content of coal seams is greatly reduced.

Water-soluble effect not only reduces CBM content, but also makes methane carbon isotopes significantly fractionate through the exchange of free gas with adsorbed gas and the cumulative effect of methane carbon isotope.

Because water-soluble effect takes away $^{13}CH_4$ in free gas, more $^{12}CH_4$ is left. Then the $^{12}CH_4$ in free gas exchanges with the adsorbed gas in coal, and a part of $^{12}CH_4$ becomes adsorbed gas, and replaces the $^{13}CH_4$ in the adsorbed gas. Finally the replaced $^{13}CH_4$ is taken away by water. This process is constantly occurring. Through the cumulative effect, $^{12}CH_4$ is enriched in a large amount, so that the methane carbon isotope becomes lighter and the gas content is also reduced.

Water is a weakly polar solvent. The polarity of $^{13}CH_4$ is greater than that of $^{12}CH_4$. According to the principle of that the similar substance is more likely to be dissolved by each other, $^{13}CH_4$ is more likely to dissolve in water than $^{12}CH_4$. Water-soluble effect tends to take $^{13}CH_4$ away first, leaving more $^{12}CH_4$, and making the methane carbon isotope in free gas lighter. Then the $^{12}CH_4$ in free gas exchanges with the adsorbed gas in coal, and a part of $^{12}CH_4$ becomes adsorbed gas, and replaces the $^{13}CH_4$ in the adsorbed gas. Finally the replaced $^{13}CH_4$ is taken away by water. This process is constantly occurring. Through the cumulative effect, $^{12}CH_4$ is enriched in a large amount, so that the methane carbon isotope becomes lighter.

The stronger the hydrodynamic conditions, the greater the cumulative effect, the more obvious the control effect on CBM, the lower the gas content and the lighter the methane carbon isotope.

5.4 Coupling Controls of Dynamic Conditions on CBM Reservoirs

The above results provide understandings of macroscopic single dynamic factor controlling CBM reservoirs from structural, hydrological and thermal field conditions. However, CBM accumulation is the result of a favorable match in all aspects of dynamic conditions and needs to be analysed at a comprehensive level. More importantly, the macroscopic dynamic factors such as structure, groundwater and thermal field can only truly control the CBM system by converting to microscopic dynamic conditions. Based on this understanding, this project puts forward the research ideas of coupling control of dynamic conditions, discusses the superposed gas control of macroscopic and microscopic dynamic conditions, and then analyzes the geological selection process and essence of dynamic conditions controlling gas from the perspective of energy dynamic balance, and finally builds the CBM accumulation model based on dynamic conditions.

5.4.1 Geological Dynamic Conditions

The process of CBM accumulation is essentially a dynamic equilibrium process of geological kinetic energy to static energy conversion of gas-bearing systems. The realization of this process relies on a tangible carrier of the CBM system. The core of the carrier includes basic coal base block, formation water and CBM. Further, the result of CBM accumulation effect is a concrete manifestation of the coupling relationship among solid, liquid and gas phases in the coal seam under the action of kinetic energy.

The coupling relationship or interaction process among the solid elastic energy of coal base block, the liquid elastic energy of formation water and the gas elastic energy of CBM controls the effects of various macroscopic geological dynamic energies, and the effect of energy dynamic balance is expressed in various types of CBM accumulation. In other words, to explain the coupling effect and essence of dynamic conditions, we must first clarify the apparent characteristics of the effective transfer of CBM accumulation energy, the combination relationship among dynamic factors and their geological selection processes, and the focus is on scientifically describing how energy is transferred, concentrated and distributed in the effective pressure system.

From the results, the dynamic conditions are the root cause of CBM accumulation, and the accumulation effect is the representation of CBM enrichment and high yield. As far as the process is concerned, the dynamic conditions are only the representation of CBM accumulation, and the accumulation effect is the result of the dynamic balance of formation energy. The traditional way of connecting representations and effects is to mechanically superimpose single factors. Although this method has its advantages of being simple and easy to understand, there are essentially three inevitable drawbacks:

First, "representation + representation" is still a representation. We only know it, but don't know why, because various factors cannot be linked organically. It is by no means a true coupling method, or it is superficial.

Second, in the process of mechanical superposition of individual factors, favorable or unfavorable factors often cannot completely coincide. The results indicated by different factors often contradict each other. In this case, it can only rely on empirical judgment, resulting in artificially strong or scientifically weak results.

Third, mechanical superposition of individual factors is similar to "black box" operation. Only the input and output are known, but the process is not known, and the link between the cause and the result is absent, resulting in the so-called "dynamic" conditions illusory.

The true coupling method requires a causal relationship through a deep understanding of the action process or scientific connotation. As far as CBM accumulation is concerned, only through the energy dynamic equilibrium process can the geological dynamic conditions be coupled with the accumulation effect, and by analyzing the energy dynamic balance system and its geological selection process can solve the scientific issue of "dynamic conditions and CBM accumulation and diffusion mechanism".

Furthermore, the scientific meaning of analyzing how dynamic conditions control CBM accumulation through the energy dynamic balance system lies in three aspects: (i) clarify the representation, that is, clarify the control law of various individual dynamic factors on CBM accumulation by discussing representations; (ii) reveals the essence, that is, discuss how dynamic conditions control CBM accumulation by studying energy conversion and equilibrium process; (iii) establish a method, that is, build a scientific prediction method for CBM enrichment and high-yield zones based on accumulation process by understanding the scientific connotation.

5.4.2 Representational Dynamic Conditions

5.4.2.1 Superposition of Reservoir Forming Control Representational Dynamic Conditions

CBM is present in coal seams, so the dynamic conditions outside the coal seams can be regarded as representational dynamic conditions. These representational dynamic conditions are equivalent to the macroscopic dynamic conditions or macroscopic dynamic energy in the traditional sense, including structural energy, hydrodynamic energy and thermal power. Among them, the structural energy can be further decomposed into tectonic stress energy and formation gravity energy.

The analysis results reveal that the representational dynamic conditions for controlling the enrichment and high permeability of CBM in the eastern margin of the Ordos Basin are mainly reflected in four aspects: (i) tectonic stress energy of modern tectonic stress field; (ii) buried gravity caused by structural differentiation; (iii) hydrodynamic energy controlled by structure; (iv) Yanshanian thermodynamic energy and its differentiation characteristics controlled by deep thermal structures. Modern tectonic stress energy controls the permeability of coal seam through the coupling relationship with natural fractures. Hydrodynamic energy plays a key role in the enrichment characteristics of CBM through the change of groundwater runoff state. Thermal kinetic energy controls CBM enrichment pattern through coal adsorption and mechanical properties. In the deep part of the basin, gravity energy restricted by buried depth largely determines the permeability of the coal seam.

Obviously, different configurations of the above four representational dynamic factors determine the difference of CBM enrichment and high permeability in the different sections of the Qinshui Basin and the eastern margin of the Ordos Basin. Specifically, in the zone where the four factors reach a favorable configuration, the CBM is not only significantly enriched, but also has the dynamic conditions of high permeability and high yield; in the zone where three of the four factors reach a favorable configuration, the CBM has possible geological conditions of enrichment and high yield. If only the favorable configuration of the last two factors exists, there are possible conditions for significant enrichment of CBM, but lack of kinetic conditions for high permeability and high yield. If only the favorable configuration of the first two factors exists, there lack of dynamic conditions of CBM enrichment, but it is impossible for high yield (Fig. 5.36).

5.4.2.2 In Qinshui Basin

Based on the above analysis, according to the superimposed characteristics among structural dynamic conditions, thermodynamic conditions and groundwater dynamic conditions, the representational dynamic conditions and their control on CBM accumulation in the Qinshui Basin can be concluded as follows (Fig. 5.37):

① Structural dynamic energy and its evolution have gone through four stages, including the Indosinian, Yanshanian, Himalayan and modern tectonic stress fields. The thermodynamic energy conditions and their evolution and the coalification controlled by them have also experienced four stages. Among them, the middle Yanshanian violent tectonic magmatism is a critical period for the evolution of thermodynamic energy conditions. The groundwater dynamic energy conditions and their evolution include three stages,

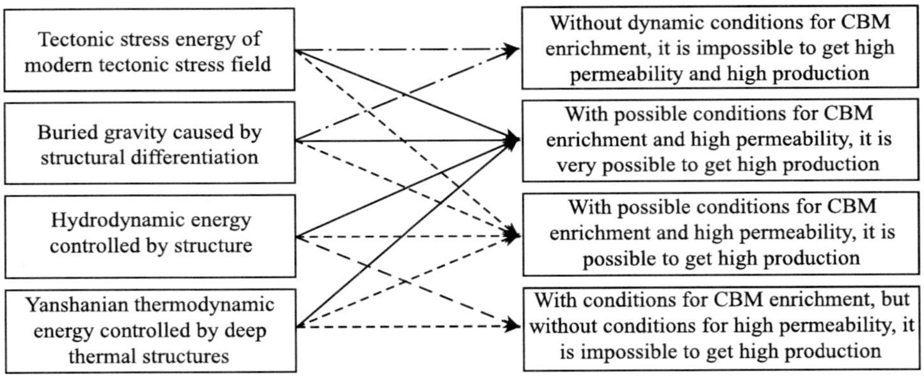

Fig. 5.36 Control of representational dynamic conditions on CBM enrichment and high permeability characteristics

namely the early-middle Yanshanian paleo-hydrogeological period and the middle-late Yanshanian paleo-hydrogeological period and modern groundwater dynamic conditions.

② Taking the tectonic stress energy as the main line of the superimposed relationship among the dynamic conditions of CBM accumulation, the tectonic energy field connects other energy fields together. The superimposed relationship is that the Yangquan-Shouyang area in the northern part of the basin is one of the favorable areas for CBM enrichment, but not conducive to high permeability and high yield of CBM; the southern Jincheng-Yangcheng area and the northern part are not only conducive to the enrichment of CBM, but also may be beneficial to high permeability and high yield of CBM; the central Qinyuan area is favorable for enrichment and high permeability of CBM; and the eastern Tunliu-Xiangyuan area is one of the favorable

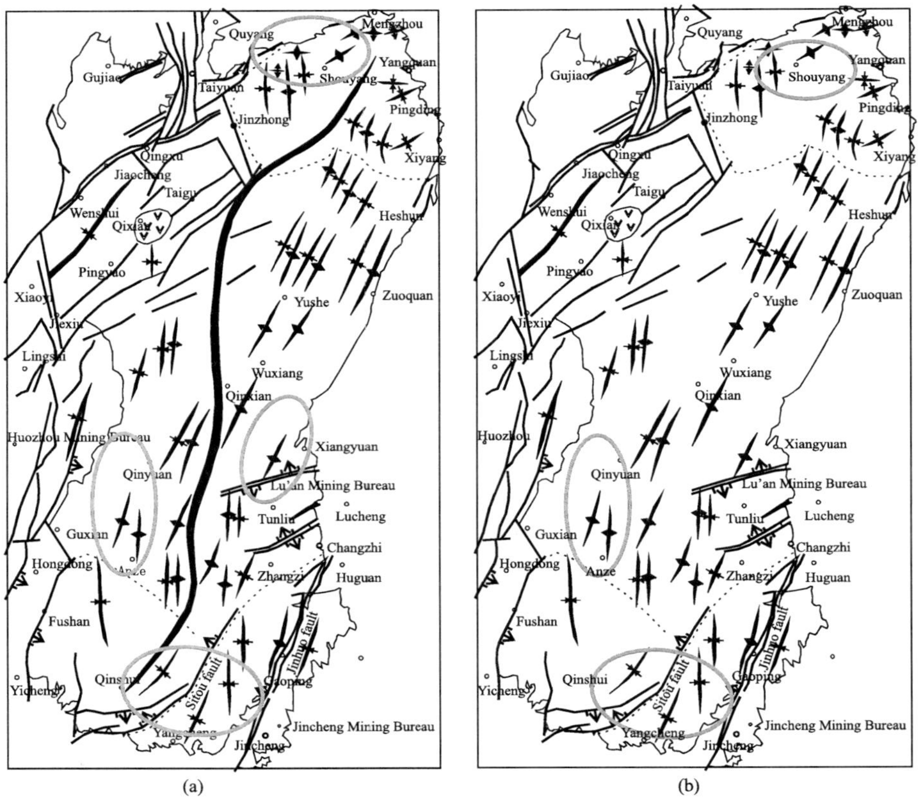

Fig. 5.37 Superimposition of favorable representational dynamic conditions of CBM enrichment and high permeability in Qinshui Basin **a** tectonic dynamic conditions; **b** thermodynamic and hydrodynamic conditions

areas for CBM enrichment, but not conducive to high permeability and high yield of CBM (Fig. 5.37a).

③ The high structural curvature indicates that the coal seam is strongly deformed and the structural fracture is relatively developed. The maximum principal stress difference of the high modern tectonic stress field shows that the coal seam fissure is relatively stretched, and the favorable sections are distributed in the middle and south of the basin, including the Jincheng-Yangcheng area, the Anze-Qinshui area and the Qinyuan area which are high-permeability coal seam development zones generally distributed in the north-northwest direction (Fig. 5.37a).

④ The gas generating and gas bearing capacities and the thermal kinetic energy, as well as the coalification and coal rank distribution controlled by them are closely related, which is one of the necessary conditions for the enrichment and high permeability of CBM. The abnormal paleo-geothermal field controlled by deep thermal structures in the middle Yanshanian caused intensified differentiation of coal-generated hydrocarbons in the Late Paleozoic; on the other hand, the strong secondary hydrocarbon generation enhanced pressure, which was the important dynamics to break through the cap rock, and controlled the overall pattern of gas-bearing distribution in the basin. As a result, the coal seams in the Jincheng-Yangcheng area in the southern part of the basin have the best gas-bearing properties, the central Anze-Tunliu-Qinxian-Qinyuan area has better gas-bearing properties (the high-value area is located in the Anze-Qinxian area), and the northern Shouyang-Yangquan-Xiyang area also has a high gas content (the south of Shouyang and the southwest of Yangquan have the highest gas content) (Fig. 5.37b).

⑤ The favorable thermal dynamics, hydrodynamic conditions and CBM enrichment in the Qinshui Basin have the following overlapping relationship. The most favorable zone in the southern part of the basin is Yangcheng and the northern part of Jincheng, distributed in the Panzhuang-Fanzhuang-Zhengzhuang area. The area is about 1300 km^2, and the CBM resource is about 2600 $\times$ 10^8 m^3. The favorable area in the southern part is the Anze-Qinyuan area located in the south-central part, with an area of about 1000 km^2 and CBM resource of about 200 $\times$ 10^8 m^3. The northern favorable area is the southeast of Shouyang, between the northeast of Yuci and the southwest of Yangquan. The area is about 600 km^2, and the CBM resource is about 700 $\times$ 10^8 m^3 (Fig. 5.37b).

According to the relationship between the above-mentioned superimposition of dynamic conditions and the conditions of CBM enrichment and high permeability, and from the perspective of the representational geological selection process, the most favorable zone of CBM enrichment and high permeability in the Qinshui Basin is in the southern Yangcheng-Qinyuan area, distributed as strips in the NNW direction and expanding tens of kilometers wide.

5.4.2.3 In the Southern Segment of the Eastern Margin of Ordos Basin

The superposition analysis shows that the dominant dynamic factors controlling the CBM enrichment and high permeability are the tectonic stress energy of the tectonic stress field, the buried gravity energy caused by structural differentiation, the thermal energy and groundwater dynamics controlled by the structure, and local areas are affected by the thermal energy of the Yanshanian paleo-geothermal field. In the area where the three factors are well configured, CBM is enriched and high permeability conditions are existent. In the zone where one of the first two factors and the last are well configured, dynamic conditions are possible for enrichment of CBM and high permeability. If only the last two factors are favorable, CBM may be enriched, but high permeability and high yield are impossible.

Based on the above analysis, the relationship between the geological dynamic conditions and the CBM accumulation in the southern segment of the eastern margin of the Ordos Basin is as follows (Fig. 5.38):

① During the tectonic movements since the Mesozoic, the Yanshanian movement laid the structural pattern, and its anomalous paleo-geothermal field had a significant impact on the gas-bearing differentiation; the Himalayan movement had a certain transformation effect on the early structure, and two short-term tensile stress fields caused a large loss of CBM. The controlled depth of the coal seam is the key factor determining the modern gas content, and is strongly influenced by the groundwater dynamic conditions, thus forming the modern enrichment distribution of CBM.

② The favorable superimposition area of 3 representational dynamic conditions is located in the area around Daning, Jixian, Puxian and Xixian, where a groundwater eguipotential surface, depression, and a high-salinity center was found,

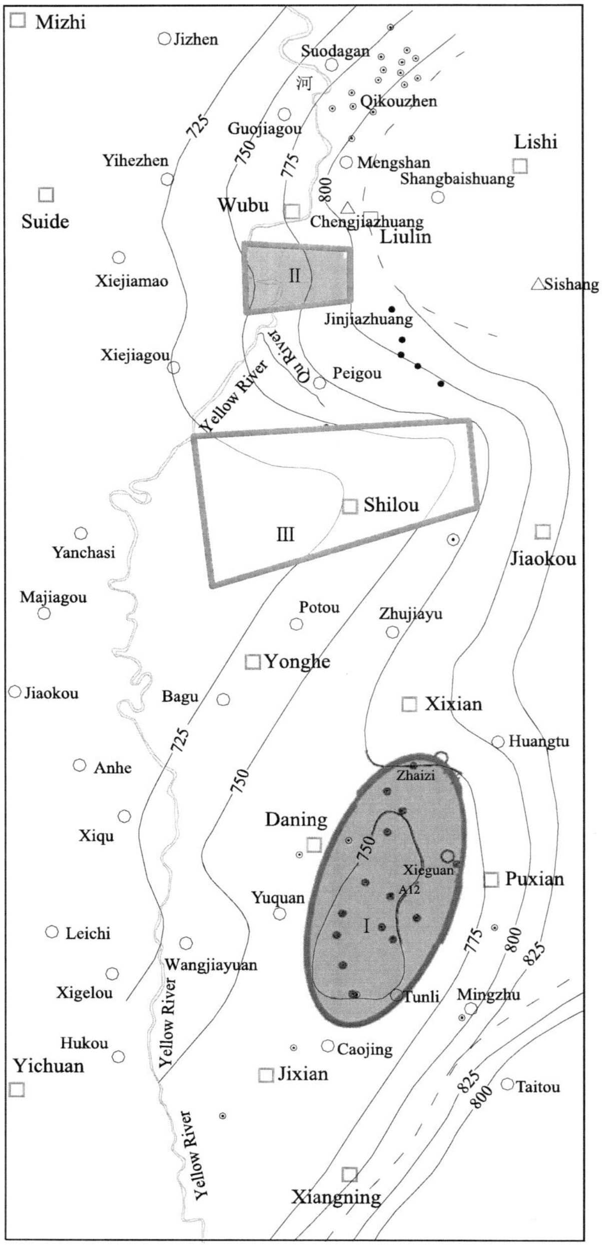

Fig. 5.38 Superposition of favorable representational dynamic conditions of CBM enrichment and high permeability in the southern segment of the eastern margin of Ordos Basin (unit: m) I. Most favorable; II. Favorable; III. More favorable

that is, dynamic conditions for enrichment and high permeability of CBM are existent. The favorable superimposition of structural and groundwater dynamic conditions are in the local area south of Wubu-Liulin, where is more conducive to enrichment and high permeability of CBM. In addition, in the vast area near the Shilou area, the favourablely coupling of structural and groundwater dynamic conditions makes possible enrichment and high permeability of CBM, which is worthy of attention.

5.4.3 Energy Dynamic Balance System and Its Geological Evolution Process

The energy dynamic balance system is the carrier of CBM accumulation effect. The elastic energy of coal seam is the link between dynamic conditions and accumulation effect, and controls CBM accumulation.

5.4.3.1 Elastic Energy and Its Relative Contribution to CBM Accumulation

The elastic energy of coal seam includes the elastic energy of coal base block, the elastic energy of water and the elastic energy of gas. The three are closely related to tectonic stress energy, thermal stress energy and groundwater dynamic energy. The elastic energy of coal base block is affected by the earth stress, the elastic modulus and Poisson's ratio of coal rock, and is closely related to the thickness of overlying formation (buried depth) and the degree of coalification (coal rank). The elastic energy of water is mainly affected by fluid pressure, water compression coefficient and thermal expansion coefficient, and the fluid potential is also a part of the elastic energy of water, which is also affected by buried depth, fluid pressure and fluid density. The elastic energy of gas is affected by fluid pressure, compressibility, thermal expansion coefficient, coal seam temperature, and CBM content.

The representational dynamic factors act on coal seams and force the coupling relationship among solid, liquid and gas in coal seams to change constantly. The dynamic balance variation of the energy system is embodied by the formation elastic properties caused by solid, liquid and gas, and controls CBM accumulation. Therefore, the elastic energy of coal seam intrinsically links dynamic conditions to CBM accumulation. It is also the key to interpreting the coupling characteristics of the conditions of CBM accumulation. It has not been deeply understood.

In this project, starting with the classical law of physics, and using the relevant theories and methods of solid mechanics, fluid mechanics, and physical chemistry, a mathematical model of elastic energy of coal seams is established, and the properties of any elastic energy and its relationship with CBM accumulation are explored.

Elastic Energy of Coal Base Block

This type of elastic energy is closely related to the mechanical properties and confining pressure of coal rock, i.e.

$$E_{coal} = \frac{\beta}{2}\left[\sigma_1^2 + \sigma_2^2 + \sigma_3^2 - 2\mu(\sigma_1\sigma_2 + \sigma_2\sigma_3 + \sigma_1\sigma_3)\right]$$

where, E_{coal} is the elastic energy of coal base block, β is the volume compression coefficient of coal base block; μ is the Poisson's ratio; σ_1, σ_2 and σ_3 are three axis stress, MPa. The calculation results show that at the same pressure, the elastic energy of coal base block (m^3 coal) from high to low is anthracite, lean-meager coal to coking coal (Fig. 5.39). In other words, at the same pressure, the higher the coalification (coal rank), the greater the elastic energy is.

Elastic Energy of Water

The main factors affecting the elastic energy of water include fluid pressure and temperature, the thermal expansion coefficient of water, and the compression coefficient of water, i.e.

$$W_{water} = RT_0\left[\frac{P_1}{P_0}(1 + \alpha\Delta T)(1 - \beta\Delta P)\right]$$

where P_1 is new fluid pressure, MPa; P_0 is original fluid pressure, MPa; T_0 is original fluid temperature; α is the initial thermal expansion coefficient of water; β is the initial compression coefficient of water; ΔT is temperature change; ΔP is pressure change; R is molar gas constant, and its value is 8.314 J/(mol k).

From the relationship between the elastic energy of water and coal ranks, at the same temperature and pressure, the elastic energy of anthracite (coal base block per cubic meter) is the largest, that of lean-meager coal is the second, and that of coking coal is the lowest. From the relation curve between the elastic energy of water and pressure, as pressure increases, the elastic energy of water increases, but the increase amplitude is relatively small (Fig. 5.39). It can also be seen from Fig. 5.40 that the elastic energy of water differs from the elastic energy of coal base block by more than two orders of magnitude. This shows that water is incompressible and not sensitive to pressure.

Elastic Energy of Gas

Gas elastic energy includes the elastic energy of free methane and the elastic energy of adsorbed methane. Its influencing factors include fluid pressure and temperature, thermal expansion coefficient and compression coefficient of methane, and are expressed as

$$W_{gas} = W_{free} + W_{absorbed} = W_{free}\left[1 + \frac{a}{v}\left(\sqrt{P_0} - \sqrt{P}\right)\right]$$

where W_{free} is the elastic energy of free methane, kJ/m^3; $W_{absorbed}$ is the elastic energy of absorbed methane, kJ/m^3; a is methane content coefficient, which is 3.16 × 10^{-3} m^3/(t Pa$^{1/2}$); v is the standard molar volume of methane, which is 22.4 × 10^{-3} m^3/mol; P_0 is original fluid pressure; P is new fluid pressure after gas state changes.

The elastic energy of methane gas increases with the increase of temperature and pressure (Fig. 5.41). Under the same temperature and pressure conditions, the overall trend of gas elastic energy in coal with different coal ranks is anthracite > lean-meager coal > coking coal. At 0.5 MPa, the increase in gas elastic energy is particularly obvious. That is to say, in this range, small changes in temperature and pressure have a great influence on gas elasticity.

Total Elastic Energy of Coal Seam

The total elastic energy of a coal seam is the sum of the above three types of elastic energy, namely $E = E_{coal} + E_{water} + E_{gas}$. The mathematical expressions of the elastic energy of coal base block, water and gas are taken into the above formula to obtain a mathematical model of the total elastic energy of the coal seam.

The numerical results show that the elastic energy of gas, water and coal accounts for different proportions in different CBM accumulation stages (Fig. 5.42). For example, at about 500 m deep, the relative contribution of the elastic energy of 1 m^3 coking coal, water and gas to CBM accumulation is 53.51%, 0.82%, and 45.67%, respectively; the relative contribution of the elastic energy of 1 m^3 lean-meager coal, water and gas is 3.72%, 0.47% and 95.81%, respectively; the relative contribution of the elastic energy of 1 m^3 anthracite, water and gas is 12.48%, 0.53% and 86.99%, respectively.

In the coking coal stage where CBM is accumulated and evolved, only the elastic energy of the coking coal base block accounts for more than 54%. That is, the energy storage in that period was mainly affected by the elastic energy of the coking coal. In the lean coal to anthracite stage, the total contribution of coal rock and water is only about 4%, and gas elastic energy becomes the main factor controlling energy storage.

The relative contributions of various elastic energies are concluded as follows:

① The elastic energy of coal rock accounts for a large proportion in the middle stage of CBM accumulation, which is closely related to the diagenesis in early accumulation and the compressive coefficient of coal rock decreasing

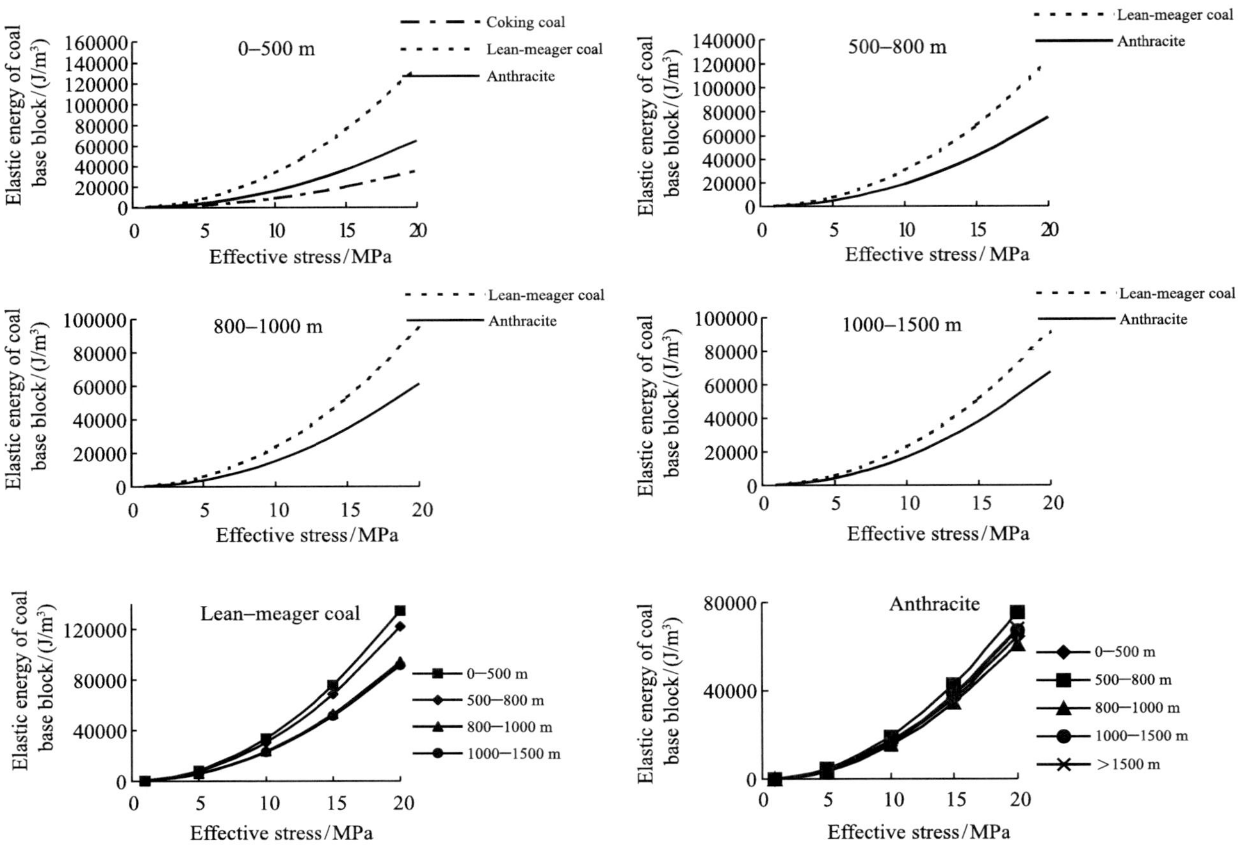

Fig. 5.39 Relationship between elastic energy of coal base block and average confining pressure and coal rank

Fig. 5.40 Relationship between water elastic energy and pressure and coal rank

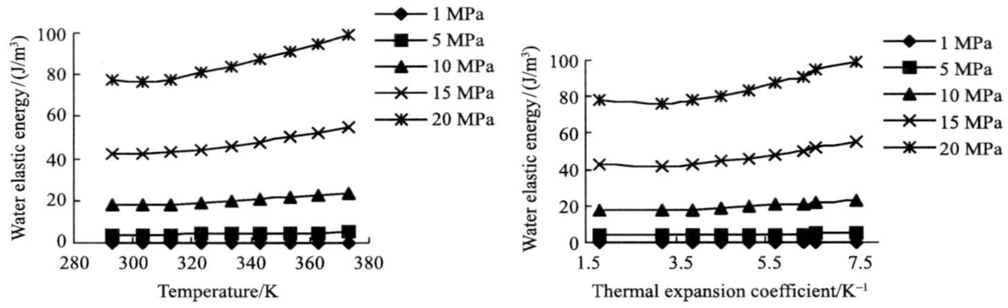

Fig. 5.41 Relationship between gas elastic energy and pressure and coal ranks

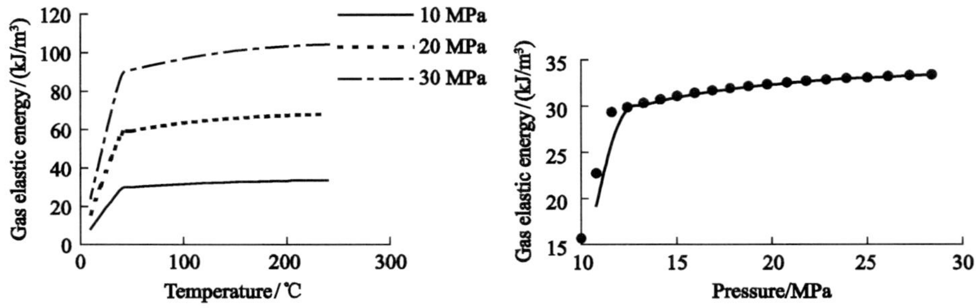

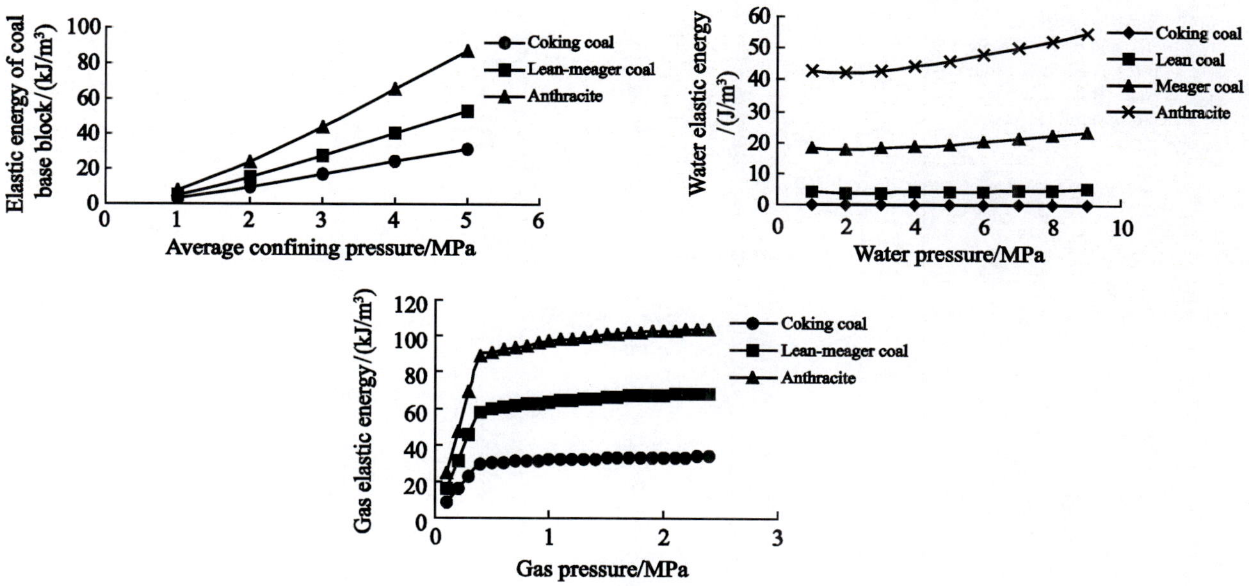

Fig. 5.42 Relative contribution of coal seam elastic energy to CBM accumulation

exponentially with increasing temperature and pressure, and then becoming gentle in late period.

② In the whole process of CBM accumulation, the contribution of water elastic energy is very small, because the compression coefficient of water changes little, generally 3×10^{-4}–6×10^{-4} MPa^{-1}, which leads to the contribution of water elastic energy is generally no more than 1%.

③ Methane gas elastic energy plays a major role in the middle stage of CBM accumulation, but in the late stage it directly determines the energy storage of coal seam.

5.4.3.2 Control of the Evolution of Coalbed Elastic Energy on CBM Accumulation in Qinshui Basin

At the end of the Middle Jurassic period, the elastic energy of the coal base block in the Qinshui, Jincheng and Lu'an mining areas was the largest, while the elastic energy of the coal base block in the Qinyuan and Qinxian areas at the syncline axis was relatively small, but the gas elastic energy was the maximum (Fig. 5.43). On the whole, the elastic properties during that period were mainly controlled by the elastic energy of the coal base block, and the gas elastic energy was generally low.

At the Late Jurassic to the Early Cretaceous, the elastic energy of coal base block decreased significantly compared with that at the end of the Middle Jurassic, and became a secondary factor on formation energy (Fig. 5.44). At that stage, the role of the anomalous paleo-geothermal field led to the generation of a large amount of organic matter, the gas content increased significantly, and the fluid pressure increased significantly, resulting in a increase of gas elastic energy by thousand times, which becomes a controlling factor on formation energy. At the syncline axis, the gas elastic energy was even more than 3200 kJ/m^3, making it possible to completely break through the caprock. The energy dynamic balance of gas accumulation and dispersion during that period determined the current scale and regional distribution of CBM accumulation.

From the middle Yanshanian to the present, the elastic energy of coal base block reduced significantly. The highest value was located in the Anze-Qinxian area, higher in the Yangcheng area and the Qinyuan area, and high in the Changzhi area. The gas elastic energy was the highest in the syncline axis, and gradually lower on two wings (Fig. 5.45). The formation energy was controlled by the gas elastic energy, and the distribution law of the two was basically the same. At that stage, the favorable dynamic conditions of CBM enrichment and high permeability were around in Duanshi, Qinshui, Anze, Qinyuan, Zhongcun, Tunliu and Fanzhuang.

5.4.3.3 Control of the Evolution of Coalbed Elastic Energy on CBM Accumulation Conditions in the Southern Segment of the Eastern Margin of Ordos Basin

From the end of the Late Triassic to the Early Jurassic, the elastic energy of coal base block gradually increased from west to east and from north to south, and was strictly

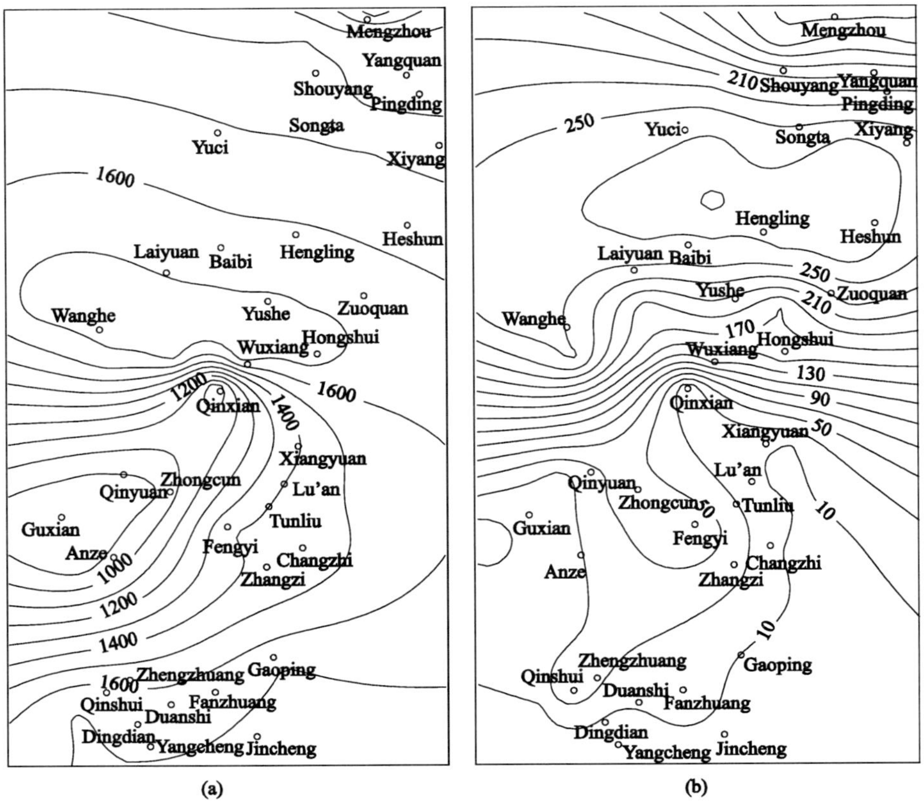

Fig. 5.43 Elastic energy contours by the end of the Middle Jurassic coal reservoir in Qinshui Basin (unit: kJ/m^3) **a** elastic energy of coal base block; **b** gas elastic energy

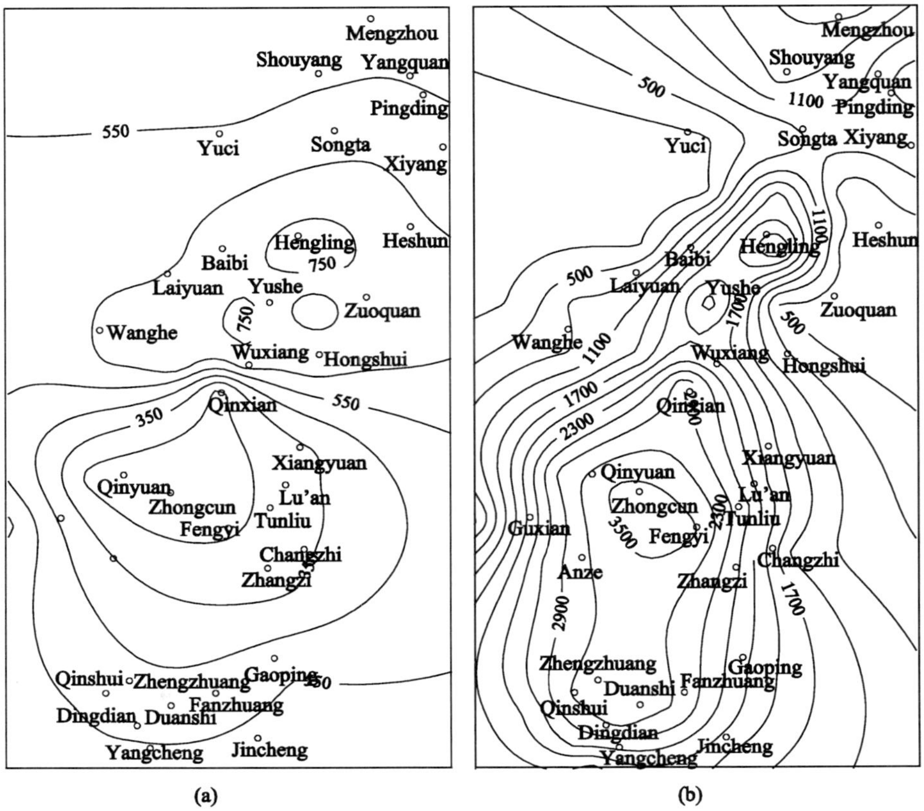

Fig. 5.44 Elastic energy contours at the end of the Early Cretaceous in Qinshui Basin (unit: kJ/m^3) **a** elastic energy of coal base block; **b** gas elastic energy

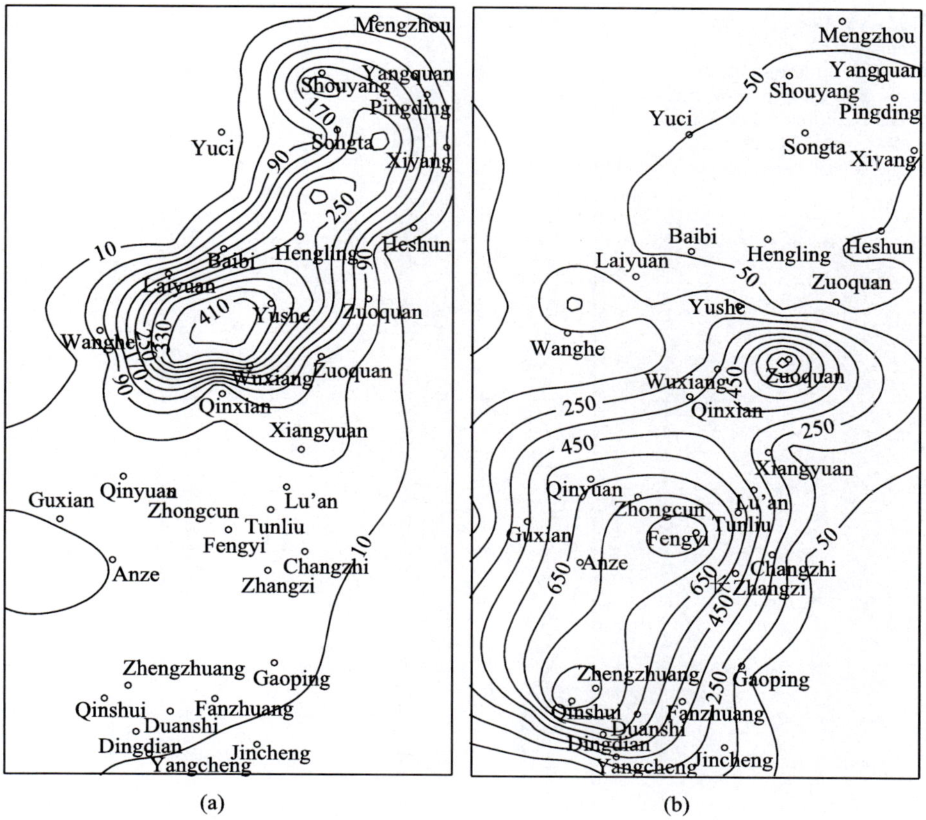

Fig. 5.45 Modern elastic energy contours in Qinshui Basin (unit: kJ/m^3) **a** elastic energy of coal base block; **b** gas elastic energy

controlled by the depth of the coal seam at that stage. High elastic energy was distributed near Daning, Puxian and Liulin. Gas elastic energy was relatively low, and directly related to the gas volume and content in that stage. The total elastic energy was mainly controlled by the elastic energy of coal base block, so its distribution law was basically consistent with that of coal base block.

From the Middle Jurassic to the end of the Early Cretaceous, the elastic energy of coal base block was slightly lower than before, but the overall distribution remained basically unchanged. The importance of gas elastic energy was gradually increasing. On plane, it was gradually increasing around the center of the Shilou area. The high value was found in two areas, the southern Puxian-Daning-Xiangning area, and the northern Liulin area (Fig. 5.46). At that stage, the total elastic energy was mainly controlled by gas elastic energy, so the regional distribution was consistent with the distribution of gas elastic energy.

At present, the elastic energy of coal base block is much lower than that of the Middle Jurassic to the Early Cretaceous, which is closely related to the shallower depth of the coal seam. There are only two high-value areas in the western and northern sides of the study area. Gas elastic energy is still dominant and gradually increases from west to east regionally, and the highest values found in the southern Daning-Puxian and the northern Liulin (Fig. 5.47). The distribution law of total elastic energy is still consistent with gas elastic energy.

5.4.4 CBM Accumulation Effect and Model

The CBM system is an energy dynamic balance system. Studies show that the elastic self-sealing effect of coal base block plays an important role in controlling the accumulation of CBM in high-rank coal. The distribution and evolution of energy in the process of CBM accumulation can be quantitatively expressed by two "systems" and three parameters. Such expression is a quantitative representation of CBM accumulation effect. Therefore, CBM accumulation effect can be divided into different types.

5.4.4.1 Elastic Self-Regulating Effect

CBM enrichment is controlled by many geodynamic factors, but depends on the relatively closed geological environment built by favorable configuration of various dynamic factors. Among them, tectonic stress field, groundwater dynamic field, thermal field and stratigraphic conditions all contribute to the

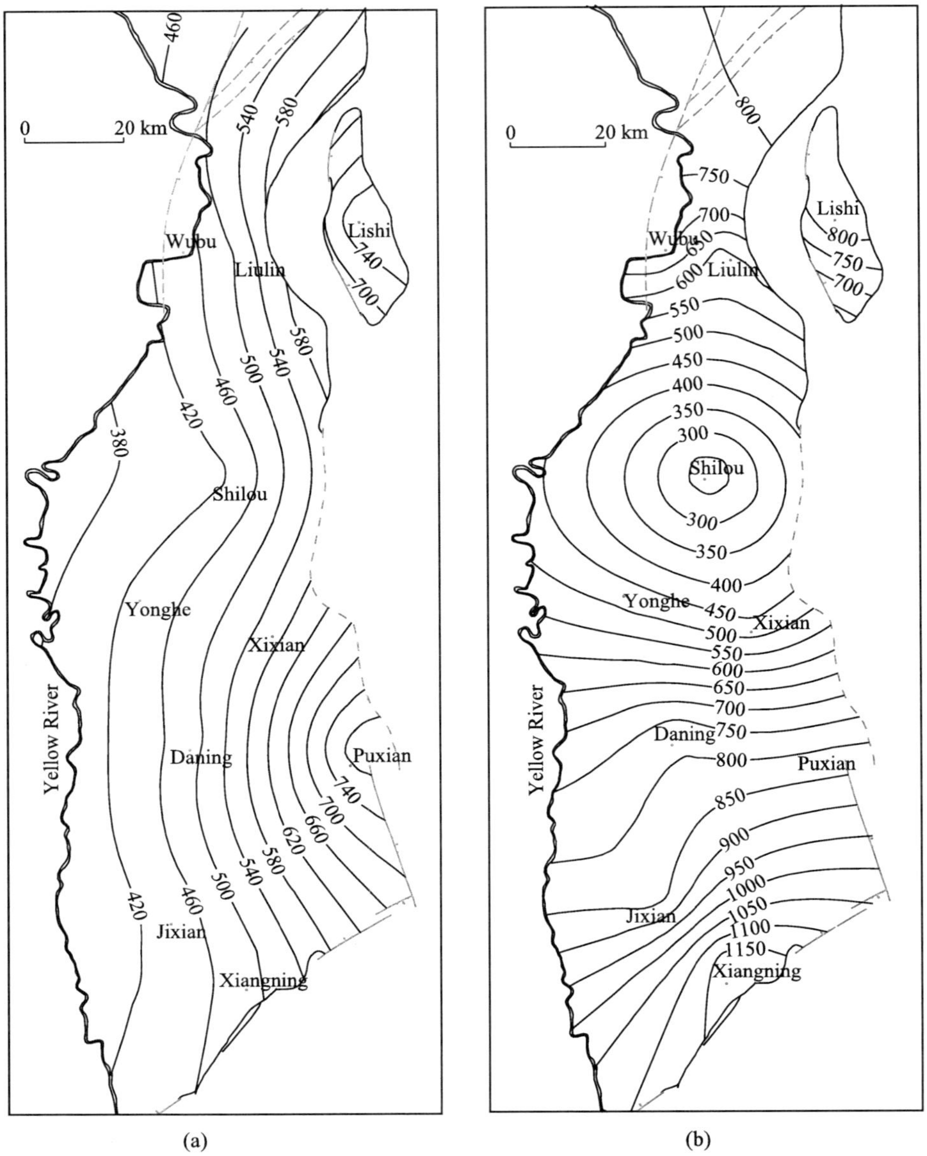

Fig. 5.46 Elastic energy contours at the end of the Late Cretaceous in the southern segment of the eastern margin of Ordos Basin (unit: kJ/m^3) **a** elastic energy of coal base block; **b** gas elastic energy

relatively "closed" environment. However, these factors first act on coal seam itself, which is reflected in the ability of coal seam to allow CBM to migrate or permeate. Coal rock itself has a very low ability to diffuse gas. In contrast, the developmental characteristics of natural fissures in coal seams, especially the open-close degree of natural fissures in geological history, are key factors in determining CBM enrichment.

Coal rock is an elastomer with considerable shrinkage or expansion properties before being deformed by force. As organic reservoirs, coal seams are more sensitive to stress than conventional reservoirs such as sandstone and carbonate rock. Therefore, the direct dynamic factors determining the opening degree of natural fractures in coal seams include two aspects. First, effective stress, that is the difference between the total stress perpendicular to the fracture direction and the fluid pressure in pores and fractures. As the effective stress increase, the width is reduced or even closed when fractures are compressed. Second, the expansion or contraction effect of the coal base block surrounded by natural fractures. The coal base block may expand after adsorbing gas, and force fractures to be narrow, while gas desorbing causes fractures wide. However, changes in effective stress and the occurrence of adsorption and desorption are affected by dynamic conditions such as structure, groundwater, and thermal field. These conditions are applied to the coal seam itself and act by elastic deformation of the coal base block.

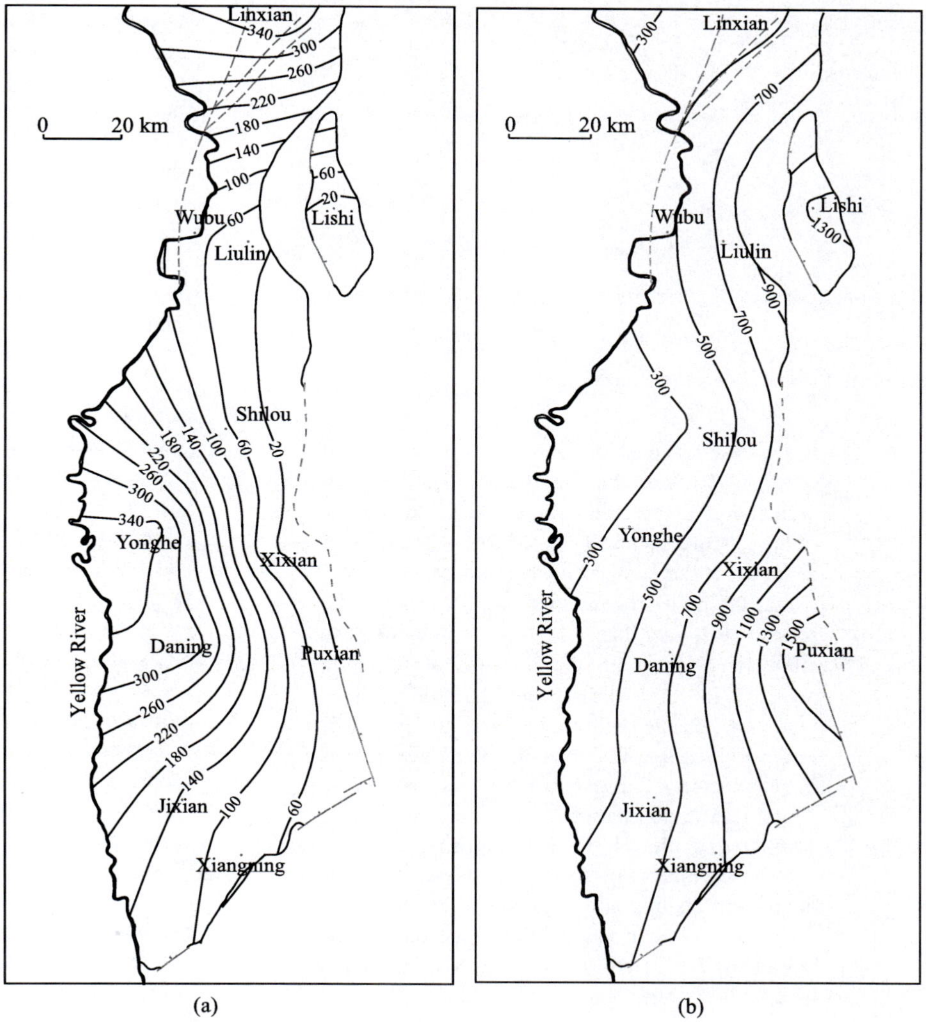

Fig. 5.47 Modern elastic energy contours in the southern segment of the eastern margin of Ordos Basin (unit: kJ/m^3) **a** elastic energy of coal base block; **b** gas elastic energy

The petrochemical and chemical structures of different coal ranks are different, resulting in different shrinkage or expansion characteristics under stress. Previous studies have done some researches on the relationship between effective stress and coal seam permeability or on adsorption, desorption and strain of coal base block. However, due to the difficulty in experiments, the relationship between the comprehensive effect of effective stress and adsorption and desorption, and the elastic deformation characteristics of coal has been less researched. Obviously, under the same stratum conditions, the open-close degree (even the direction) of natural fractures caused by effective stress and coal adsorption expansion may not be consistent, and only the comprehensive effect of the two can determine the relative "closing" or "opening" of the fractures.

During the process of CBM accumulation, coal base block adsorbs CBM and expands, and then compresses and makes fractures relatively close. The direct cause of adsorbing CBM is the increase in fluid pressure, so the effective stress on coal seams in the same tectonic stress field reduces, and the natural fracture surface is relatively stretched, resulting in the relative opening of the fracture. These two effects are the result of the response of coal's own elastic energy to different geological incentives, which is manifested by elastic self-regulating effect. The former is negative, and the latter is positive. Under formation conditions, the opening or closing of natural fractures is the result of the direct effect of these two factors, called the comprehensive self-regulating effect of coal seams.

Based on physical simulation experiments, the elastic, self-regulating, comprehensive effect mode of coal base block was established by referring to fracture plate models (Fig. 5.48), and the following basic understanding was obtained:

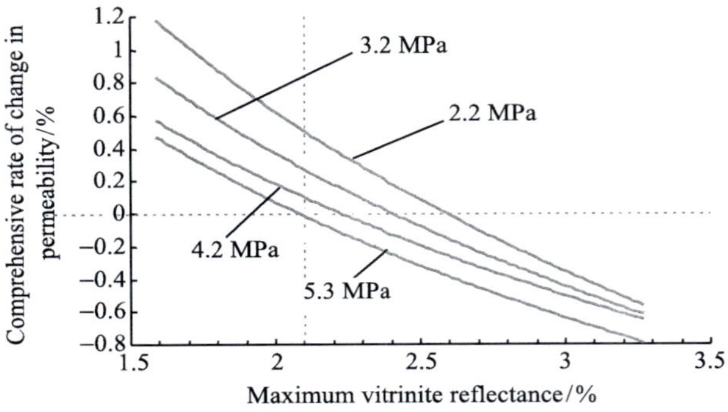

Fig. 5.48 Elastic self-regulating modes

① In the case of constant coal rank, the pressure of pore and fracture fluid increases, and the comprehensive effect gradually weakens. When the fluid pressure is lower than 3 MPa, the rate of decrease of the comprehensive effect is relatively fast as the fluid pressure increases. After the fluid pressure is higher than 3 MPa, the comprehensive effect decreases with the increase of the fluid pressure. In other words, when the fluid pressure is low, the elastic deformation of coal reservoir is relatively large, and when the fluid pressure is increased, the relative deformation is reduced.

② At constant fluid pressure, coal rank is negatively correlated to the elastic self-adjusting effect of coal reservoirs (Fig. 5.40). It shows that the degree of coalification controlled by structural dynamic and thermodynamic conditions is the key factor that determines the elastic deformation of coal reservoirs. Further, the higher the coal rank, the weaker the elastic deformation of the coal reservoir when adsorbing or desorbing CBM, especially when desorbing CBM, the natural fractures in the coal reservoir are less likely to open, so the internal conditions for accumulating CBM tend to get better.

③ The comprehensive elastic self-regulating effect of coal reservoirs is strictly controlled by the evolution of coal structure with the change of coal rank. When the experimental fluid pressure is about 5 MPa, once the maximum vitrinite reflectance exceeds 2.1%, the comprehensive effect changes from a positive value to a negative value. The position of coalification corresponding to this transformation is called the "balance point" of the elastic self-regulating effect of coal reservoirs. This point is not only the boundary between medium coal rank and high coal rank, but also corresponds to the jump point of the third coalification. Near the balance point, the physical and chemical structures of the coal change stepwise. For example, the maximum cutinite reflectance exceeds the maximum vitrinite reflectance; the relationship between the moisture content and the carbon content or the oxygen content reaches a minimum value; the extension of the macromolecular basic structural unit suddenly increases, indicating that coalification has entered from high coal rank to medium coal rank. Further, the balance point "drifts" in the direction of increasing coal rank as the fluid pressure decreases, but the final point does not exceed the maximum vitrinite reflectivity of 2.6%.

④ At constant coal reservoir pressure, as gas content increases, the comprehensive effect gradually decreases (Fig. 5.49). The greater the fluid pressure, the more significant the negative effect increases with gas content. When the fluid pressure exceeds 5.9 MPa, the comprehensive effect is always negative.
It is shown that under the dual effects of fluid pressure and adsorption expansion of coal base block on the degree of fracture open-close, when reservoir pressure is high (>5.9 MPa), the elastic self-sealing effect of the coal reservoir will be obvious, including the higher the gas content the more pronounced the elastic self-sealing effect.

⑤ At constant reservoir pressure, as reservoir permeability increases, the comprehensive effect gradually decreases; as fluid pressure increases, the negative effect of the elastic self-regulating of coal base block increases obviously with permeability. When permeability is much lower than 1 mD, the decrease of the elastic self-regulating effect is relatively fast. It is shown that coal reservoir permeability is controlled by the fracture system in the coal reservoir, and there is a positive correlation between the two.

The regional geological data shows that regardless of the differences in structure, hydrology and sedimentary conditions, most of the high-rank coal areas in China generally have the basic characteristics of high gas content and relatively shallow weathering zone (Ye et al. 1999), such as four high-rank coal distribution areas in southeastern Shanxi, northwestern Henan, central Hunan and northwestern Guizhou. This phenomenon implies that the coal reservoirs that evolved to higher ranks under the favorable matching of various dynamic conditions may have geological conditions conducive to the preservation of CBM.

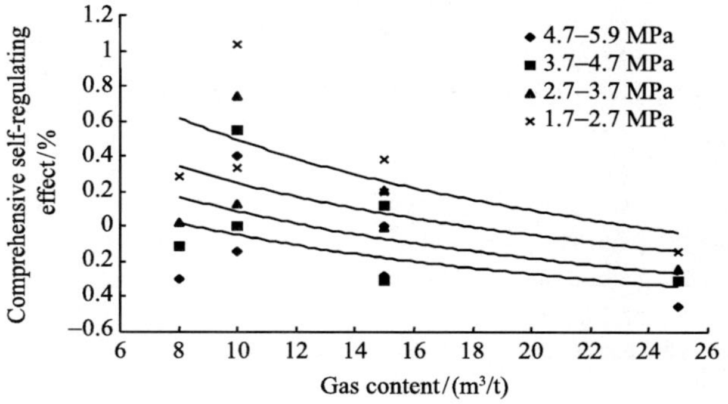

Fig. 5.49 Elastic self-regulating effect

5.4.4.2 Self-Sealing Effect

According to the traditional theory, the reason for the high gas content of high-rank coal reservoirs in China is mainly due to the developed micro-pores and high adsorption potential caused by abnormal paleo-geothermal field. However, the basic condition for CBM to be preserved in coal is to have sufficient reservoir pressure. If the reservoir pressure is greatly reduced due to structural damage, groundwater flushing, etc., no matter how high the adsorption potential is and how developed the micropores are, CBM would be unable to be enriched due to desorption. However, in the four high-rank coal areas where CBM is enriched in China, the structural, hydrological and sedimentary conditions vary greatly, or the coal reservoirs are very shallow (such as southeast Shanxi), or open faults are extremely developed and groundwater activities are very strong (such as northwestern Henan), or a number of independent small synclines are developed by structural damage (such as central Hunan and northwest Yunnan). This shows that external geological conditions are not the only main controlling factors for CBM enrichment.

Coal seam is a kind of geological body with relatively large elastoplasticity. The elastic self-regulating effect of coal seams reveals that under the premise that the petrochemical composition and pressure are the same, the mechanical properties of coal (especially the elastic self-regulating effect) are strongly controlled by coal rank. The result of elastic self-regulation of medium-rank coal seams shows a comprehensive positive effect, while high-rank coal seams show a comprehensive negative effect, resulting in a decrease in the conductivity of fractures in high-rank coal seams. The greater the reduction in reservoir pressure, the more significant the negative self-regulating effect of high-rank coal seams.

Further, the elastic self-regulating effect of coal base block is inversely proportional to its elastic energy. As coal rank increases, the elastic energy of coal base block increases, and the law of the elastic self-regulation reduces. Especially after entering the high-rank coal stage (the vitrinite reflectance is greater than 2.0%), the increase of the elastic energy of coal base block leads to the elastic self-regulating effect becoming negative, indicating that the elastic energy greatly weakens CBM to diffuse and seep, so that the CBM storage conditions become better due to the dynamic characteristics of high-rank coal seams, and under the same external geological conditions, it is more difficult for CBM diffusion in high-rank coal seams than in medium-rank coal seams. The higher the fluid pressure and the gas content of the coal seam, the more significant the negative effect of the elastic self-regulation, and the more difficult it is to desorb and diffuse CBM, thus strengthening CBM accumulation. In other words, the elastic self-sealing effect may be the key internal dynamic condition for the high gas content of high-rank coal seams.

Therefore, this project puts forward a new point of "elastic self-sealing effect" of high-rank CBM accumulation, and considers that this effect is an important geological reason for the widespread enrichment of CBM in high-rank coal areas in China, and to a considerable extent, offsets the damage to CBM reservoirs by other unfavorable geological conditions, and saves CBM in the unfavorable conditions of structure, hydrology and sedimentation. The main dynamic factor is the abnormal geothermal field, and the internal cause is the special mechanical properties controlled by the geochemical structure of coal. This understanding provides an entry point and scientific basis for further exploration of the dynamic factors and mechanisms of the "elastic self-sealing effect" and its role in the CBM energy balance system.

5.4.4.3 Effective Pressure System and Effective Migration System

The CBM accumulation process is a geological process in which the pressure system is gradually adjusted. The accumulation effect is the result of the dynamic balance of the pressure system. Aiming at this substantial problem, based on the comprehensive analysis of the coupling relationship of various dynamic conditions and the elastic energy of the formation, the concept of effective pressure system and effective migration system is proposed, and the development of pressure system ξ_2, the development of fracture system ξ_1 and the opening of fractures Δ are established, and their simulation calculation methods are defined, which quantitatively describe the energy distribution characteristics and evolution law in the process of CBM accumulation.

The study also found that the elastic energy of the formation is the link and interaction between the effective pressure system and the effective migration system. The migration, dispersion and redistribution of effective energy mainly act through the elastic energy of the formation. The constraint of energy system evolution on CBM accumulation is actually achieved by controlling the evolution of ξ_2, ξ_1

and Δ, and its acting mechanism can be summarized as a stress–strain coupling mechanism.

Effective Pressure System and Its Energy Evolution

The effective pressure system for CBM accumulation refers to the coal reservoir pressure system composed of tectonic stress, overburden gravity, thermal stress and groundwater dynamics under external dynamic conditions, which is reflected as the solid–gas-water coupling relationship facilitating CBM migration and accumulation in the coal reservoir.

The effective pressure system can be expressed by the ratio of gas elastic energy to comprehensive elastic energy of the coal seam (pressure system coefficient ξ_2). The higher the ξ_2, the greater the proportion of gas elastic energy in comprehensive elastic energy, the greater the coal seam pressure, the better the effective pressure system, and the better the CBM accumulation regardless of temperature change. On the contrary, the effective pressure system poorly developed is not conducive to CBM accumulation.

In the Qinshui Basin, the CBM effective pressure system has undergone four evolutionary stages:

The first stage is from the Late Carboniferous to the Triassic. The early energy dynamic balance factors mainly included hydrocarbon generation, methane adsorption and storage, paleo-hydrodynamics, etc. The late-stage gravity made a closed pressure system gradually form in the coal reservoir, and there were no conditions for coalbed gas breaking through the caprock and seeping out. Therefore, it was the gravity energy that controlled the pressure system (Fig. 5.50a).

The second stage is from the Early Jurassic to the Middle Jurassic. The gravity energy of overlying strata was in a process of wavy decreasing, and the differentiation effect was enhanced, resulting in increasing cumulative diffusion and differentiation of CBM. The reservoir pressure system was generally a semi-closed system, and the energy dynamic factors were the same as the first stage (Fig. 5.50b).

The third stage is from the Late Jurassic to the Early Cretaceous. The effective pressure system was extremely enhanced, but its differentiation increased, and organic hydrocarbon generation and hydrocarbon diffusion, caprock breakthrough and seepage, paleo-groundwater dynamics and tectonic stress became controlling factors on energy dynamic balance. Strong hydrocarbon generation pressure increased diffusion (Fig. 5.50c), gas elasticity energy was greatly improved, and hydrocarbon generation pressure became the primary driving force for the seepage and migration of CBM, so that the system experienced from a semi-open system to a open system, finally retuned to a semi-open system, and caprock breakthrough took place extensively on the basin slope. During that process, the coal reservoir pressure dropped rapidly, causing the pore-fracture system to close again. The high ξ_2 value was found at the south and north sides of the main syncline axis. The overall trend was high at the axis and gradually decreased toward the wings (Fig. 5.51a).

The fourth stage is from the Late Cretaceous to the Quaternary. The crustal movement was still dominated by uplifting, the paleo-geothermal field returned to normal, the gravity and thermal energy of the overlying strata continued to weaken, and the factors on energy dynamic balance changed to the elastic energy of coal base block and the elastic energy of groundwater. The overall result is that CBM was severely lost and the reservoir pressure system was weakened (Fig. 5.50d). According to the results of apatite fission track analysis (aforementioned), the Qinshui Basin experienced another significant uplifting in the Early Tertiary, and the paleo-tectonic stress field was temporarily transformed from compressive to tensile, which may lead the seepage and diffusion of CBM in a large scope. During that stage, the ξ_2 was higher than that in the Late Jurassic to the Early Cretaceous. The cause is that the effective stress of the coal reservoir was greatly reduced due to the crustal uplift, the elastic energy of the coal base block was significantly reduced, and the proportion of the gas elastic energy of the total elastic energy was significantly increased, indicating that modern pressure system is conducive to CBM accumulation (Fig. 5.51b).

Compared with the dynamic conditions of modern groundwater, the ξ_2 value corresponding to the sag-shaped stagnant hydrodynamic field in the Panzhuang-Fanzhuang area is greater than 0.96; the ξ_2 value corresponding to the half-graben-shaped stagnant hydrodynamic field in the Qinyuan-Anze area is greater than 0.90; the ξ_2 corresponding to the fan-shaped stagnant hydrodynamic field in the Qinshui area is between 0.8 and 0.9 (Fig. 5.51b). Therefore, the modern groundwater dynamic field is closely related to the development of the effective pressure system.

From the southern segment of the eastern margin of the Ordos Basin, the effective pressure system also shows four stages of evolution:

The first stage is from the Late Carboniferous to the end of the Triassic. The coal seam was mainly in the evolution process from peat to lignite, the reservoir pressure was extremely low, and the diffusion loss was extremely slow. After entering the Triassic, the buried depth of coal-bearing strata in the Late Paleozoic continued to increase, and the organic matter in the coal evolved rapidly, and the amount of gas generated increased greatly, gradually forming a closed reservoir pressure system. However, the reservoir pressure increased rapidly in the middle to late stages, and exceeded the breakthrough pressure of the coal seam roof, so that many breakthroughs occurred. The effective pressure system

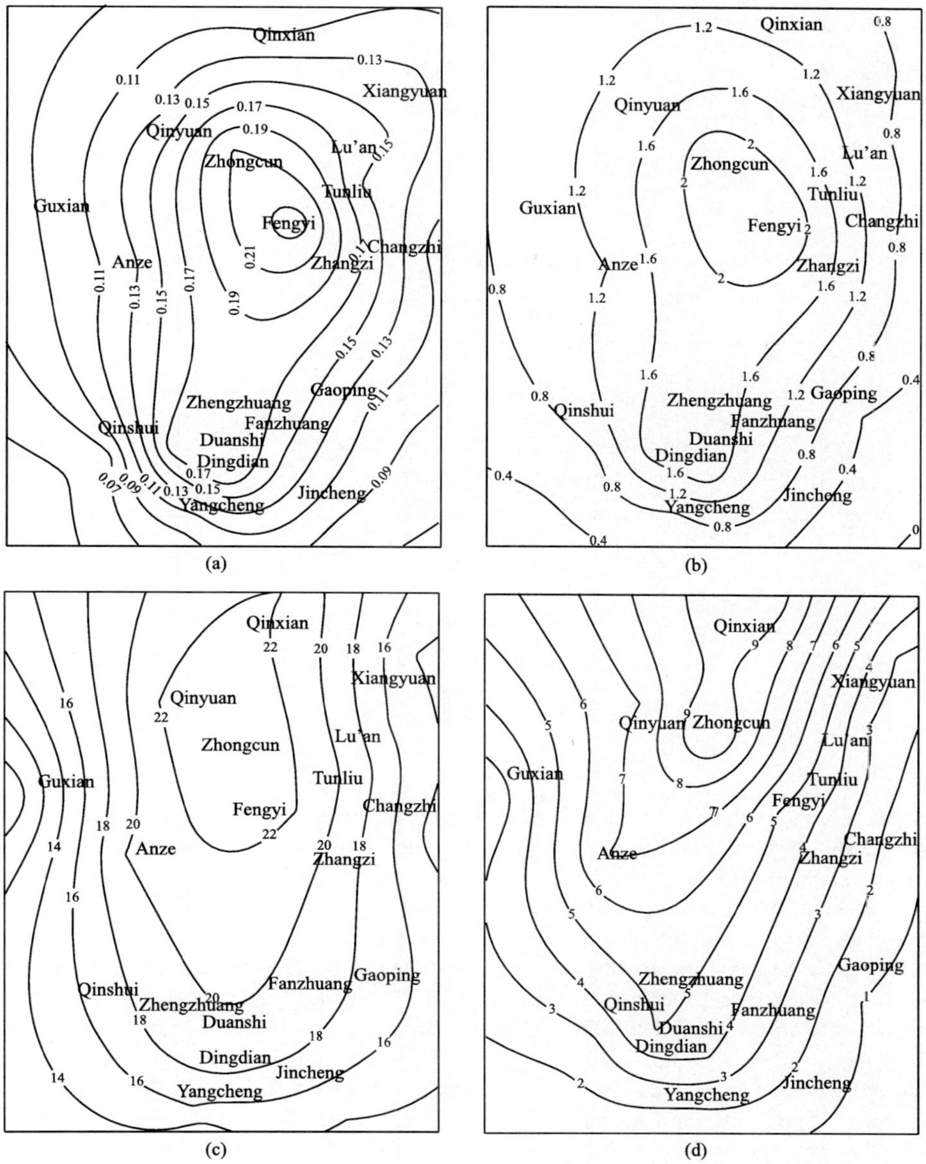

Fig. 5.50 Contours of the simulated pressure in the main coal seam of the Shanxi Formation in the Qinshui Basin (unit: MPa) **a** at the end of the Triassic; **b** at the Middle Jurassic; **c** at the end of the Early Cretaceous (after the seepage and migration of CBM); **d** modern

developed well, and formed a semi-closed pressure system which radiated radially around the Daning area.

The second stage is from the Early Jurassic to the Middle Jurassic. The basin was slightly uplifted during the Early Jurassic, and then declined during the Middle Jurassic. In the process, hydrocarbon generation stopped, the coal reservoir pressure decreased significantly, no caprock breakthrough and seepage loss occurred, and the CBM system was generally closed. The upper main coal seam pressure was 5.95–25.94 MPa, and the lower main coal seam pressure was 18.49–30.62 MPa. The coal seam pressure was generated by hydrostatic pressure, and the coal seam pressure was reduced due to diffusion loss.

The third stage is from the Late Jurassic to the end of the Early Cretaceous. The development of the anomalous paleo-geothermal field in the middle Yanshanian declared the most active CBM evolution. With the large increase in the gas content, the reservoir pressure increased rapidly, and multiple caprock breakthroughs occurred in the middle-late stages. Despite that, the coal reservoir pressure did not change much, indicating that the CBM system was generally closed. Controlled by the law of gas elastic energy distribution, the effective pressure system was the lowest in the Shilou area and increased radially around (Fig. 5.52a).

The fourth stage is from the Late Cretaceous to the present. The crustal movement was dominated by uplifting, the

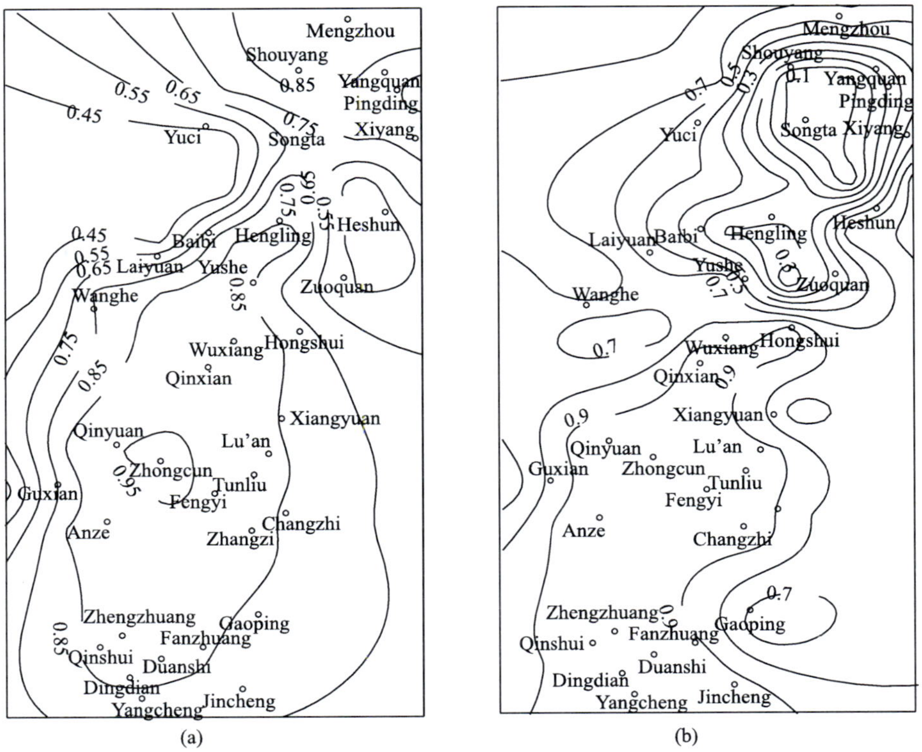

Fig. 5.51 Contours of the development of pressure system (ξ_2) in the main coal seam of the Shanxi Formation in Qinshui Basin **a** Early Cretaceous; **b** modern

depth of the coal seams continued to decrease, the geothermal field returned to normal, gas generation completely stopped, and the evolution of CBM was dominated by diffusion loss. From the numerical simulation results, the coal reservoir pressure suddenly dropped in the early stage, indicating that the tensile stress caused fractures to open, and strong seepage loss occurred, so that the CBM system was in an open state. Since then, the drop of the reservoir pressure has slowed down, the fracture system closed again to some extent, and the CBM system has been in a semi-open state until to the present. As a result, the effective pressure system gradually increases from east to west in the region, and the south is better than the north (Fig. 5.52b).

Efficient Migration System and Its Energy Evolution

In the process of CBM accumulation, the increase of effective stress leads to the decrease of reservoir permeability, while the shrinkage of coal base block leads to the increase of reservoir permeability. These two effects are the result of the elastic energy of the coal base block itself responding to different geological incentives, which is manifested by the elastic self-regulating effect of the coal base block, and reflects the opening of the fracture system in coal seams. Therefore, the elastic self-regulating effect can be defined as the fracture opening coefficient Δ.

The degree of fracture development can also be expressed as the ratio of the elastic energy per unit volume of coal seam to the fracturing energy of coal rock, which is called the fracture development coefficient ξ_1. The fracturing energy of coal rock is calculated based on the results of coal mechanics experiments (such as compressive strength, elastic modulus, Poisson's ratio, etc.), and using the applicable formula. The elastic energy of coal seam includes the elastic energy of coal base block, water elastic energy and gas elastic energy, which have been described above.

The effective migration system of CBM is the inevitable result of the micro-dynamic energy of the gas-bearing system. Macroscopically, it is controlled by the structural differentiation in the process of basin evolution and is closely related to the thermal and hydrocarbon generation histories of coal seams. Microscopically, it is controlled by the self-regulating effect of the coal base block and the elastic self-sealing effect of the coal seam, and is closely related to the evolution of reservoir physical properties and elastic energy. This geological selection process depends on the development and opening of the coal seam fractures, and determines CBM diffusion and seepage. Its external manifestation is mainly characterized by the pore-fracture system in the coal seam. Therefore, the effective migration system can be quantitatively described by the fracture development coefficient ξ_1 and the fracture opening coefficient Δ.

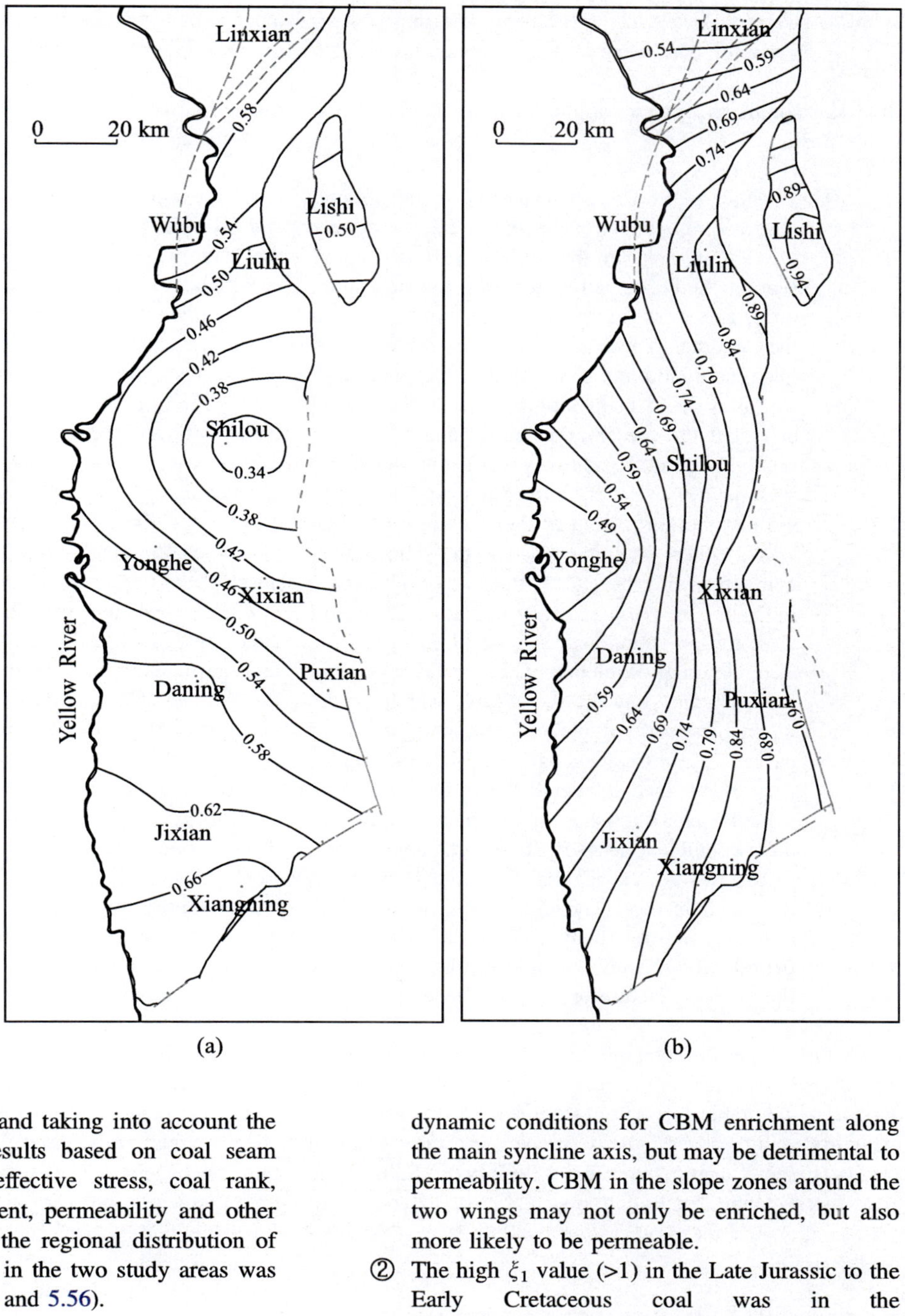

Fig. 5.52 Contoure of the development of pressure system (ξ_2) in the main coal seam of the Shanxi Formation in the southern segment of the eastern margin of Ordos Basin **a** Early Cretaceous; **b** modern

Based on the above model, and taking into account the above numerical simulation results based on coal seam burial depth, fluid pressure, effective stress, coal rank, reservoir temperature, gas content, permeability and other external geological conditions, the regional distribution of the elastic self-regulating effect in the two study areas was obtained (Figs. 5.53, 5.54, 5.55 and 5.56).

In the Qinshui Basin:

① The fracture opening coefficient Δ is negative in the main syncline axis, and the negative effect is the most significant in the Zhongcun-Fengyi and the Yangcheng areas. Toward the wings, the positive effect is getting stronger and stronger (Fig. 5.53). This feature indicates that there are dynamic conditions for CBM enrichment along the main syncline axis, but may be detrimental to permeability. CBM in the slope zones around the two wings may not only be enriched, but also more likely to be permeable.

② The high ξ_1 value (>1) in the Late Jurassic to the Early Cretaceous coal was in the Zhengzhuang-Duanshi-Fanzhuang area and the Qinyuan-Anze area, indicating that caprock breakthrough might occur in the middle Yanshanian, and the effective migration system of CBM was relatively developed (Fig. 5.54). Comparing the two contour maps in Fig. 5.54, it is found that the modern ξ_1 is significantly lower than that of the Late Jurassic to the Early

Cretaceous, which is basically less than 1, and generally conducive to the preservation of CBM.

In the southern segment of the eastern margin of the Ordos Basin:

① Whether at the Early Cretaceous or the present, the fracture opening coefficient is the lowest near the Shilou area in the eastern part of central section, and gradually increases toward south, north and west (Fig. 5.55). That is to say, from the middle Yanshanian to the present, the dynamic conditions favorable for coal seam permeability have been the worst near the Shilou area, but the preservation conditions have been relatively good; but those in the south, north and west have been better. Further analysis shows that the distribution pattern of the fracture coefficient was further strengthened in the Himalayan, compared with the end of the Early Cretaceous, indicating that the relevant dynamic conditions were further strengthened in the Himalayan.

② The fracturing development coefficient is "high in the east and low in the west", which is the inevitable result of the different pressure relief caused by different coal seam burial depths (Fig. 5.56). In that overall background, the distribution characteristics of the fracture development coefficient near the Shilou area are different in the middle Yanshanian from the Himalayan, that is, it is relatively high in the middle Yanshanian, and relatively low in the Himalayan period, which is consistent with that reflected by the modern fracture opening coefficient.

On the whole, the dynamic conditions for the preservation of CBM near the Shilou area are better, but the dynamic conditions favorable for the development of coal seam permeability may be relatively poor. In the southern Daning-Jixian-Xiangning area and the northern Liulin-Wubu area, the dynamic conditions are conducive to coal seam permeability, but the preservation conditions may be relatively poor.

5.4.4.4 Control Mechanism of Energy System Evolution on CBM Accumulation

The evolution of energy system can be attributed to the evolution of effective pressure system and effective migration system. The controlling factors on CBM accumulation include suitable pressure system and migration system and corresponding energy mechanism. The process of CBM reservoir formation or destruction is related to the dynamic balance of formation pressure field and elastic energy field (system internal energy). When the elastic energy field reaches an equilibrium and steady state, the elastic energy field in CBM reservoirs is high (Fig. 5.57). Therefore, the relationship between formation energy system and ξ_1, Δ and ξ_2 is a basis for revealing CBM accumulation effect.

The relationship between coal reservoir energy system and ξ_1, ξ_2 and Δ, as well as CBM accumulation reveals that the energy system evolution strongly constrains CBM accumulation.

In the whole geological evolution history of the Qinshui Basin, the elastic energy from the Late Jurassic to the Early Cretaceous was the highest, generally exceeding 2500 kJ/m^3, and the highest value was in the Zhengzhuang, Fanzhuang and Fengyi-Zhongcun areas (Fig. 5.44b). The higher ξ_1 was in the Anze-Qinyuan area, and the high in the Zhengzhuang-Fanzhuang area (Fig. 5.54a). At present, $\xi_1 > 1$ is mainly distributed in these two areas, low Δ is located in the south of the Yangcheng-Jincheng area and north of the Fengyi-Zhongcun area, while relatively high Δ in the Zhengzhuang-Fanzhuang area and the Anze-Qinyuan area (Figs. 5.54b and 5.45). High ξ_1, ξ_2 and Δ usually mean high gas content, high permeability and high fluid pressure. In such area, CBM is rich and of high production. This has been proved by early exploration and development of CBM.

In the southern segment of the eastern margin of the Ordos Basin, the area with total elastic energy of more than 1400 kJ/m^3 includes the southeastern Puxian-Jixian-Xiangning area and the northern Liulin area from the Late Jurassic to the Early Cretaceous. The total elastic energy might reach up to 1700 kJ/m^3 in the southeastern Puxian-Jixian-Xiangning area. The lowest is near the Shilou area, only 700–1000 kJ/m^3 (Figs. 5.46 and 5.47). ξ_2 and Δ also have the same distribution law on the plane (Figs. 5.52 and 5.55), but high ξ_1 is only distributed in the southeast (Fig. 5.56). It is worth noting that modern ξ_1, Δ and ξ_2 are consistent with those in the middle Yanshanian.

Furthermore, the effective pressure system and the effective migration system are actually a stress–strain coupling mechanism. The dynamic process of CBM accumulation is a fully coupling process involving structure, fluid (methane and water), thermodynamics and physical and chemical reactions. Among them, the stress comes from tectonic stress field, groundwater dynamic field and thermal stress field; the strain includes structural strain and volume strain zones. The stress–strain coupling system (state, structure and evolution) interacts with the energy system to constrain the formation and evolution of the accumulation system. On the one hand, the stress mechanism drives the fluid activity, resulting in the dynamic balance system for CBM accumulation. On the other hand, the tectonic strain mechanism provides external geological conditions for

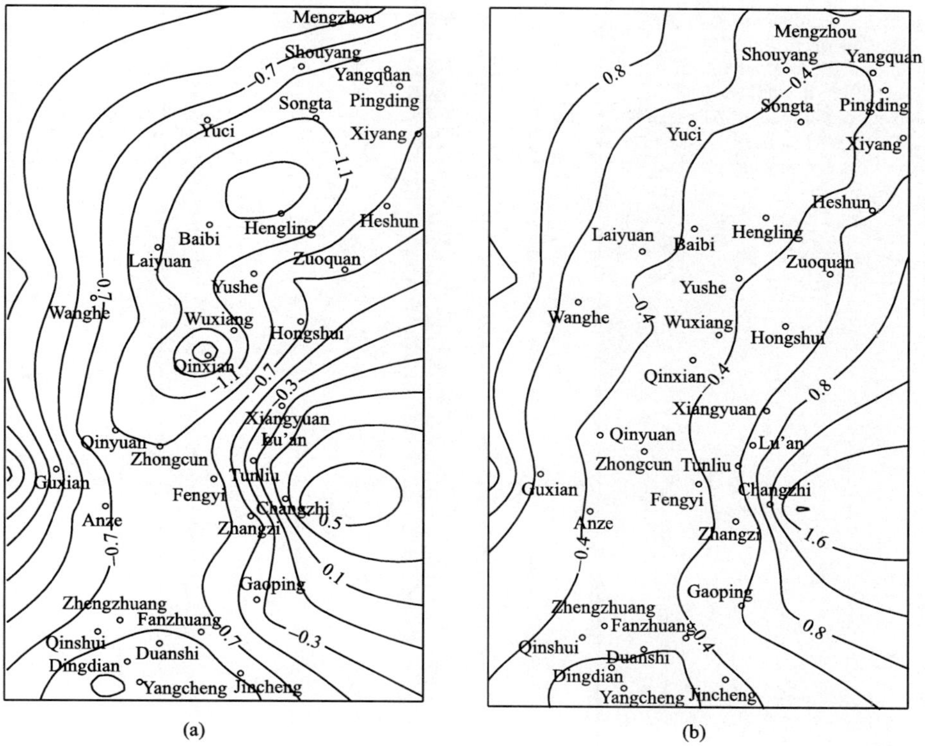

Fig. 5.53 Contours of the opening of fractures (Δ) in the main coal seam of the Shanxi Formation in Qinshui Basin **a** Late Jurassic to Early Cretaceous; **b** modern

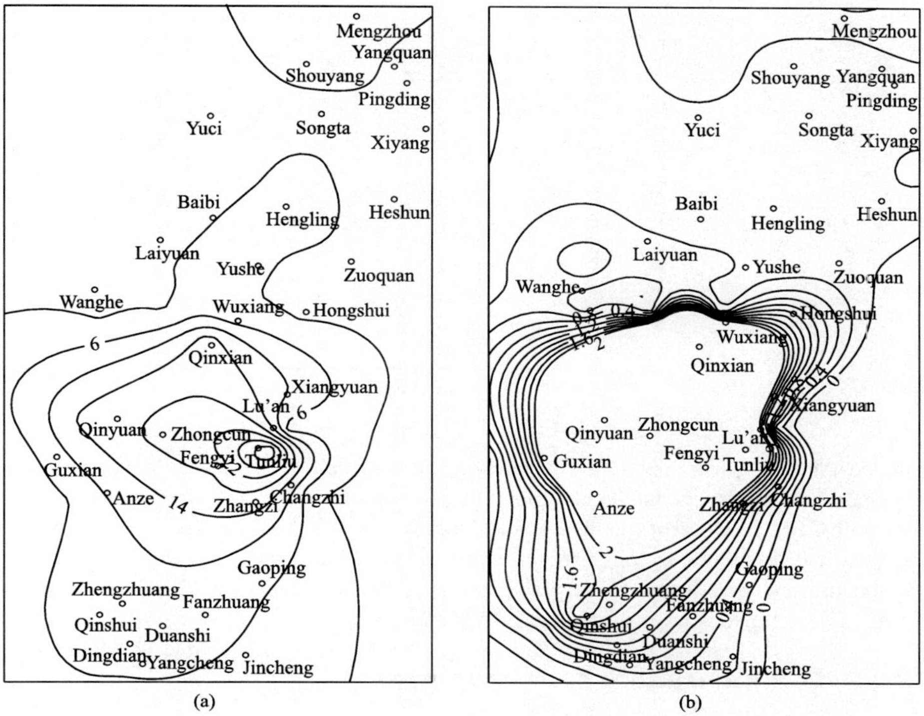

Fig. 5.54 Contours of the development of fracture system (ξ1) in the main coal seam of the Shanxi Formation in Qinshui Basin **a** Late Jurassic to Early Cretaceous; **b** modern

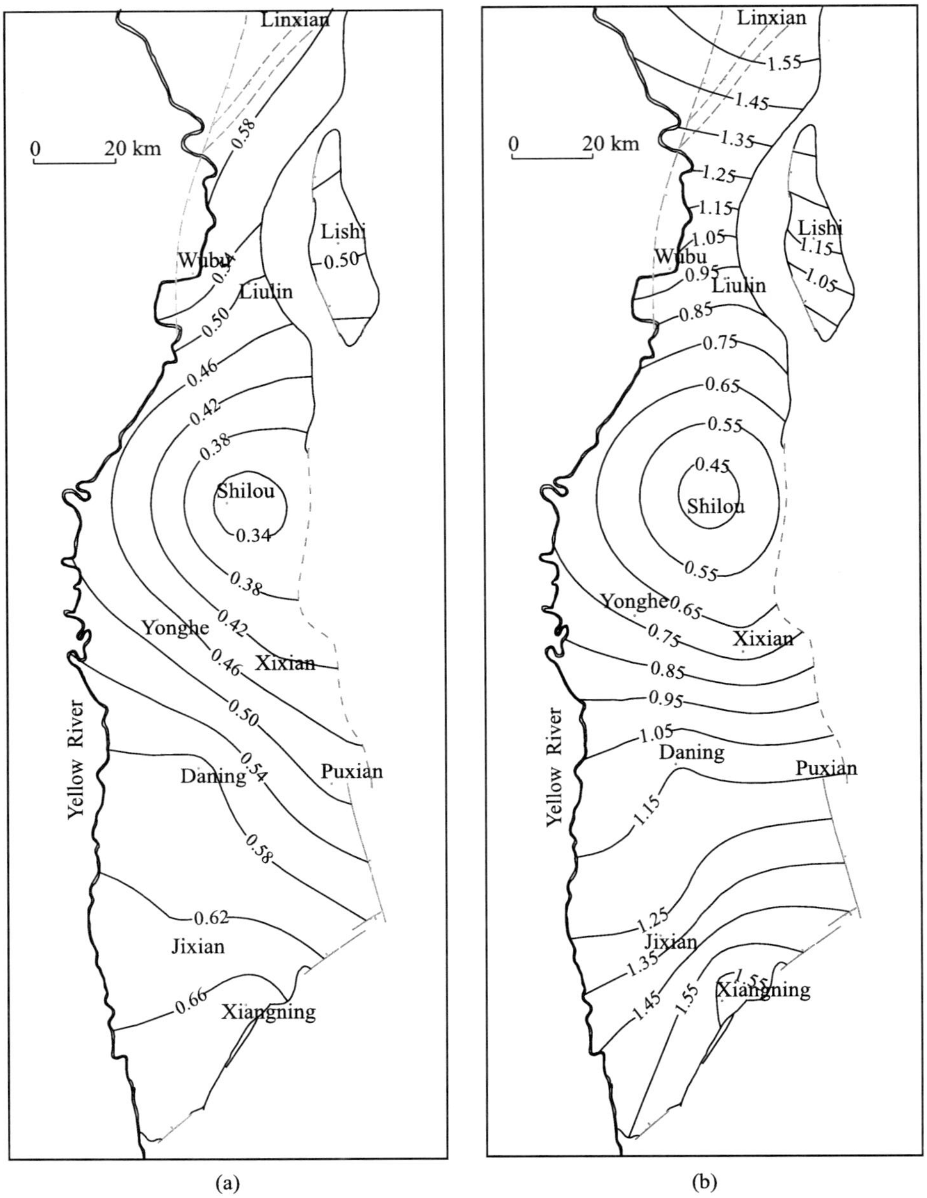

Fig. 5.55 Contours of the opening of fractures (Δ) in the main coal seam of the Shanxi Formation in the southern segment of the eastern margin of the Ordos Basin **a** Late Jurassic to Early Cretaceous; **b** modern

tectonic thermal events and CBM generation, occurrence, migration, accumulation and localization. In addition, the zone with CBM enrichment and high permeability is often in the area with extreme tectonic strain, volumetric strain, temperature and pressure, and is greatly restricted by the distribution characteristics of groundwater dynamic field.

5.4.4.5 CBM Accumulation Types and Energy Accumulation/diffusion Models

The above research results show that the controlling dynamic factors on the energy balance system for CBM accumulation are reflected in three aspects:

(i) ξ_1-the development of coal seam fractures controlled by the paleo-formation energy field constrained by paleo-tectonic stress field and thermal stress field;

(ii) Δ-the fracture opening controlled by tectonic stress field, thermal stress field and buried depth of coal seam;

(iii) ξ_2-the development of the effective pressure system controlled by the elastic energy constrained by groundwater dynamic conditions. Through further analysis of the coupling among the three factors, it is possible to establish effective marks to identify energy systems, and reveal how the energy balance system

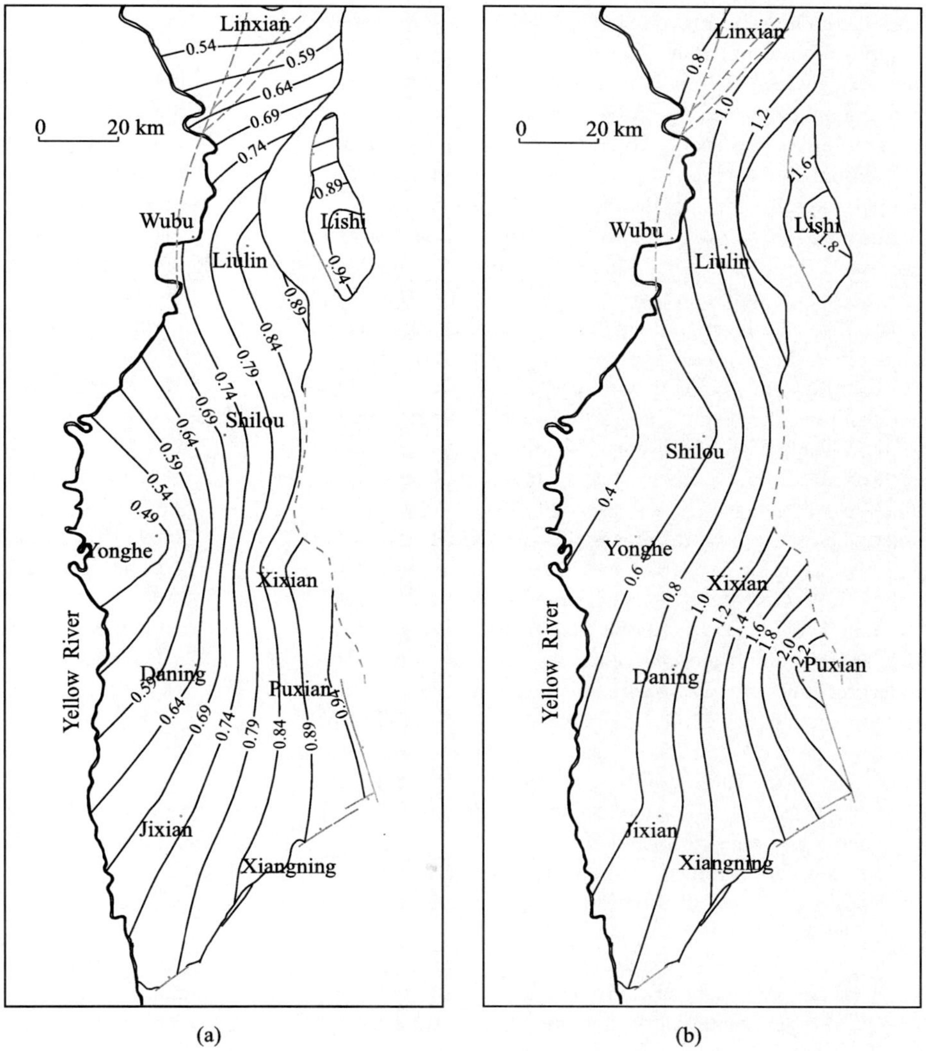

Fig. 5.56 Contours of the development of fracture system (ξ_1) in the main coal seam of the Shanxi Formation in the southern margin of the eastern margin of the Ordos Basin **a** Late Jurassic to Early Cretaceous; **b** modern

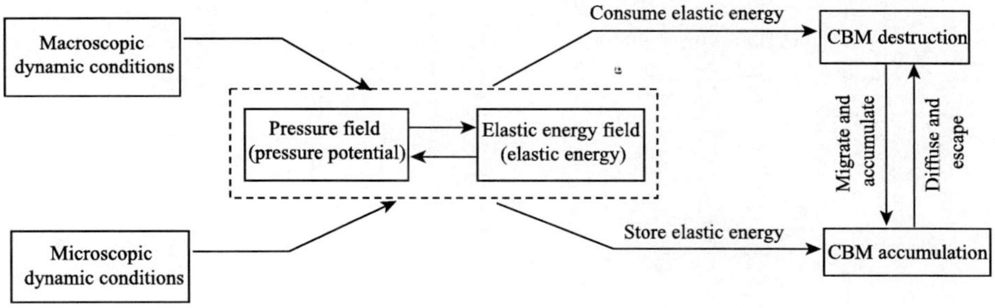

Fig. 5.57 Control of energy system on CBM accumulation

controls CBM accumulation, and predict the regional distribution law of CBM enrichment and high permeability zones from the perspective of dynamic conditions.

Types of CBM Accumulation

According to the above discussion on the ternary identification marks of the energy balance system, CBM accumulation effects can be divided into three levels: favorable, more favorable and unfavorable. The three levels constitute 27 types of accumulation. After eliminating low ξ_1, low Δ and low ξ_2 which are not conducive to CBM accumulation, there are only 8 types (Fig. 5.58). Combined with the above research results, the energy characteristics of the eight types of accumulation can be reflected by the effectiveness of migration system and pressure system (Table 5.3).

According to further analysis, the eight favorable and more favorable types can be grouped into four large categories. There is only one type in the first category (No. 1 in Table 5.3), where the ternary parameters are all more favorable, indicating effective pressure system and effective migration system favorable for CBM enrichment and high permeability. The second category includes 3 types (No. 2–No. 4 in Table 5.3), where only 2 of the ternary parameters are favorable, and the other is less favorable, indicating less effective pressure system and effective migration system, or effective pressure system and less effective migration system possibly favorable for CBM enrichment and high permeability. The third category includes three types (No. 5–No. 7 in Table 5.3), where only one of the ternary parameters is favorable, and the other two are less favorable, indicating favorable CBM enrichment or favorable fracture development, but not fully favorable for CBM enrichment and high permeability. The fourth category has only one type (No. 8 in Table 5.3), 3 parameters are only less favorable, indicating not very favorable conditions for CBM enrichment and high permeability.

The coupling analysis of modern ternary parameters reveals the regional distribution characteristics of CBM accumulation in the two basins:

① In the Qinshui Basin, there are actually only two types, namely, effective pressure system-effective migration system, and less effective pressure system-less effective migration system, distributed in the southern part of the Basin. The former is located in the Daning-Panzhuang-Fanzhuang area, and the latter is developed in the Zhengzhuang north area and the Qinyuan east-Anze east area. The results are highly consistent with the results of CBM exploration and development in the Basin over the years.

② Accumulation types in the southern segment of the eastern margin of the Ordos Basin differ from those in the Qinshui Basin. Although only two of the eight types in Table 5.3 exist in the southern segment of the eastern margin of the Ordos Basin, they are effective pressure system-less effective migration system, and less effective pressure system-less effective migration system. There is no effective pressure system-effective transport system. Therefore the dynamic conditions for CBM enrichment and high permeability are generally inferior to those in the Qinshui Basin. The former type is distributed in the Puxian-Xiangning area in the south; the latter type is developed in the south and north of the area. In other words, the conditions for CBM enrichment and high permeability in the south is relatively superior to those in the north, and those in the north is relatively better than those in the central part.

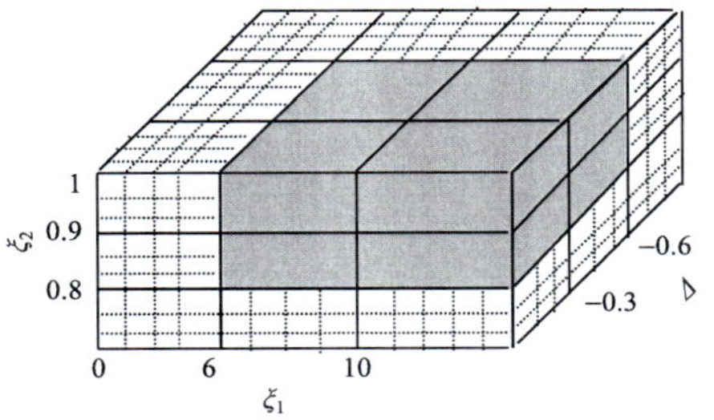

Fig. 5.58 Ternary modes of CBM accumulation conditions (the dark area means favorable-more favorable combination)

Energy Model for CBM Accumulation

CBM accumulation depends on the energy dynamic balance system. The configuration of macroscopic dynamic energy and microscopic dynamic energy determines the energy model for CBM accumulation. Each effective energy domain includes an effective pressure system, an effective migration system, and a formation energy accumulation mechanism. It has specific functions and relatively stable boundaries, and obtains substances, energy and information that evolve in an orderly direction by interacting with the outside world. The openness of an effective energy domain not only reflects the current state of the system, but also reflects the evolution history and evolution direction of the system. It is an effective, comprehensive indicator for characterizing the system. Due to the different degree of openness of different energy domain systems, the material flow and energy flow inside the system also have differences in strength and relative importance.

Therefore, the conceptual energy model can be expressed as:

$$S = (f, s, R) = (w, g, s, R)$$

where S means energy model; f means fluid dynamic system, representing the effective pressure system for CBM

Table 5.3 Favorable and more favorable types and dynamic conditions for CBM accumulation

No	Accumulation type	ξ_1		Δ		ξ_2	
		More favorable $\xi_1 > 10$	Favorable $6 < \xi_1 < 10$	More favorable $\Delta > -0.3$	Favorable $-0.6 < \Delta < -0.3$	More favorable $\xi_2 > 0.9$	Favorable $0.8 < \xi_2 < 0.9$
1	Effective pressure and migration systems	√		√		√	
2	Effective pressure system and less effective migration system	√			√	√	
3	Less effective pressure system and effective migration system	√		√			√
4	Effective pressure system and less effective migration system		√	√		√	
5	Less effective pressure and migration systems	√			√		√
6	Less effective pressure and migration systems		√	√			√
7	Effective pressure system and less effective migration system		√		7	√	
8	Less effective pressure and migration systems		√		√		√

accumulation; s is coal seam, representing the effective migration system for CBM accumulation; w is hydrodynamic system; g is gas chemical system; R means the relationship among the systems, representing energy mechanism.

According to the above results and discussion, the above conceptual mode can be transformed to the energy model for CBM accumulation, i.e.

$$S = (f, s, R) = (\xi_1, \Delta, \xi_2, R)$$

where ξ_1 and Δ are the identification marks for the effective migration system; ξ_2 is the identification mark for the effective pressure system.

Based on the above model, use fuzzy mathematics to establish the standard represented by "1" and "0", for identifying energy accumulation (see Chap. 10 for details). "1" represents favorable and "0" represents unfavorable. Thus, four sets of energy accumulation are obtained: set (1, 1) represents favorable migration system-favorable pressure system; set (1, 0) represents favorable migration system-unfavorable pressure system type; set (0, 1) represents unfavorable migration system-favorable pressure system; set (0, 0) represents unfavorable migration system-unfavorable pressure system.

According to the calculation results and the law of plane distribution, two typical energy accumulation models are summarized, and their dynamic conditions are shown in Table 5.4. The Daning-Panzhuang-Fanzhuang model is only distributed in the Qinshui Basin, and there are dynamic conditions conducive to the enrichment and high permeability of CBM; the Qinyuan-Anze model is distributed in the southern Qinshui-Zhengzhuang-Fanzhuang north area and the northwestern Qinyuan-Anze area in the Qinshui Basin, and the similar Puxian-Daning-Jixian-Xiangning area

Table 5.4 Typical energy models and dynamic conditions for CBM accumulation

Typical energy models	Evolution	Energy system	Energy state	Display		
				Accumulation		Diffusion
				Macroscopic dynamic energy	Microscopic dynamic energy	
Daning-Panzhuang-Fanzhuang model	Early	Close	Accumulating	Strong and dominant structural dynamic energy; weak thermal dynamic energy; and the weakest groundwater dynamic energy	Slow coalification and weak gas generation	Undeveloped fracture system; weak gas diffusion and seepage; and weak energy dissipation
	Critical	Open	Balanced-accumulating–unstable-diffusing	Dominant and strong thermal dynamic energy; weak structural dynamic energy and groundwater dynamic energy; and good pressure and migration systems	Fast coalification; significantly enhanced gasification; secondary hydrocarbon generation; and significant pressure increase by hydrocarbon generation and hydrotherm	Enhanced diffusion and seepage; significant caprock breakthrough; developed fracture system; and strong energy dissipation
	Modern	Almost close	Kept	Dominant groundwater dynamic energy controlled by structure; and weak thermal dynamic energy	No coalification; no gas generation; stagnant and closed groundwater	Almost terminating diffusion and seepage; weak energy dissipation
Qinyuan-Anze model	Early	Close	Accumulating	Strong and dominant structural dynamic energy; weak thermal dynamic energy; and the weakest groundwater dynamic energy	Slow coalification and weak gas generation	Undeveloped fracture system; weak gas diffusion and seepage; and weak energy dissipation
	Critical	Open	Balanced-accumulating–unstable-diffusing	Dominant and strong thermal dynamic energy; weak structural dynamic energy and	Fast coalification; significantly enhanced gasification; secondary hydrocarbon	Enhanced diffusion and seepage; significant caprock breakthrough; developed

(continued)

Table 5.4 (continued)

Typical energy models	Evolution	Energy system	Energy state	Display		
				Accumulation		Diffusion
				Macroscopic dynamic energy	Microscopic dynamic energy	
				groundwater dynamic energy; and better pressure and migration systems	generation; and significant pressure increase by hydrocarbon generation and hydrotherm	fracture system; and strong energy dissipation
	Modern	Almost open	Kept	Dominant groundwater dynamic energy controlled by structure; and weak thermal dynamic energy	No coalification; no gas generation; stagnant and closed groundwater	Weak and slow diffusion and seepage; weak energy dissipation

in the southern segment of the eastern margin of the Ordos Basin, where the dynamic conditions for CBM enrichment and high permeability are favorable.

References

Liu HJ (1998) Coalbed methane geology in southwestern Shanxi Province. China University of Mining and Technology Press, Xuzhou (in Chinese)

Ren ZL (1996) Study on the relationship between geothermal evolution history and oil and gas in Ordos Basin. Acta Petrolei Sinica 17 (1):17–24 (in Chinese)

Su XB, Liu BM (1999) Occurrence and influencing factors of coalbed methane. J Jiaozuo Inst Technol 18(3):157–160 (in Chinese)

Su XB, Chen JF, Sun JM (2001) Coalbed methane geology and exploration and development. Science Press, Beijing (in Chinese)

Ye JP, Shi BS, Zhang CC (1999) Coal reservoir permeability and influencing factors. J China Coal Soc 24(2):118–122 (in Chinese)

Zhang JB, Tao MX (2000) Geological significance of methane carbon isotope in coalbed methane exploration–a case study on Qinshui Basin. J. Sedimentol. 18(4):611–614 (in Chinese)

Zhang XM, Zhuang J, Zhang SA et al (2002) China coalbed methane geology and resources evaluation. Science Press, Beijing (in Chinese)

6 Formation and Distribution of CBM Reservoirs

CBM reservoir is an unconventional natural gas reservoir with different characteristics from conventional natural gas reservoirs, especially the accumulation conditions and processes are unique. Studying the formation and distribution of CBM reservoirs is helpful to prediction of CBM reservoirs, and favorable for exploration and development of CBM.

6.1 Meaning and Types of CBM Reservoir

Many scholars have given certain meanings and definitions to CBM reservoirs in researching CBM geology. Li et al. (1996) considered that CBM reservoirs are enrichment of methane in coal seam in a confined space when external conditions are appropriate. The gas enrichment degree and pressure in the confined space are generally higher than that outside. In other words, in a CBM field, there may be one or more CBM reservoirs. Qian et al. (1997) pointed out that the reservoir of methane in coal seam refer to the coal rock mass that "traps" a certain amount of gas under pressure (mainly water pressure). In addition, by limiting the concept of methane reservoirs in a broad sense, the concept of effective methane reservoir or economic methane reservoir is proposed, i.e., the CBM reservoir with commercial exploitation value. Zhang et al. (2002) defined CBM reservoirs as coal rock masses bearing coal and a certain amount of gas under formation pressure (water pressure and gas pressure) and having independent structures. Although some scholars use various terms to describe CBM reservoirs, it is generally believed that CBM reservoirs are not only difficult to give a specific definition, but also difficult to define in geological space.

6.1.1 Meaning of CBM Reservoir

According to the characteristics of CBM and the difference from conventional natural gas reservoirs, this book defines CBM reservoirs as the coal rocks controlled by similar geological factors, containing CBM at a large resource scale and in an adsorbing state, and having relatively independent fluid systems.

CBM reservoirs have the following three conditions. The CBM reservoir is the basic unit of CBM accumulation, with obvious boundary and isolated from surrounding geological structure; the independent fluid system refers to the basic fluid unit with similar evolution process and geological action; the coal rock mass is continuous and controlled by top and bottom plates. The CBM reservoirs that can be commercially developed under existing developmental conditions are called industrial CBM reservoirs, or called non-industrial CBM reservoirs. The industrial or non-industrial is a relative concept, depending on national requirements for resource abundance, economic policies, and technological advances.

6.1.2 Differences Between CBM Reservoirs and Conventional Natural Gas Reservoirs

CBM is an unconventional natural gas, which is significantly different from conventional natural gas.

Different reservoir mechanism. Conventional natural gas is stored in the reservoir pores in a free state. In the case of sufficient gas source, the amount of gas accumulation is mainly related to the size of the pore space, while CBM adsorbs to the surface of the reservoir pores (Ruppel et al. 1972; Yang and Saunder 1985; Liu et al. 2000), and the CBM amount is closely related to the adsorbing capability of coal seams (Guan et al. 1996; Qian et al. 1997; Su et al. 2001).

Different accumulation process. Conventional natural gas is generated by source rock and accumulated in reservoir after primary migration and secondary migration along a certain distance (Li 1987), and the migration direction is controlled by fluid dynamic field, that is, natural gas

Y. Song et al., *Coalbed Methane in China*,
https://doi.org/10.1007/978-981-33-4725-0_6

migrates mainly driven by buoyancy and fluid pressure (Hao et al. 1994; Song et al. 2002). CBM is directly adsorbed by coal reservoirs after generated in coal source rock, and CBM accumulation is not controlled by fluid dynamic field, but affected by temperature and pressure fields (Jolly et al. 1968; Ruppel et al. 1972). The CBM content in the coal with the same quality and rank increases with the increase of gas pressure, and decreases with the increase of temperature.

Different gas reservoir boundaries. Conventional natural gas reservoirs have obvious boundaries. The range and boundary of gas reservoirs are determined by trap conditions, and There is a qualitative difference between the inside and outside of the gas reservoir, that is, there is gas inside and no gas outside. CBM reservoirs have no boundaries. As long as there are coal seams, there is CBM. Under certain geological conditions, CBM is relatively enriched to form CBM reservoirs. It is the difference in gas abundance between inside and outside of a CBM reservoir, rather than "yes" or "no".

Different fluid state. Both conventional gas reservoirs and CBM reservoirs have gas and water phases, but the two states are different: conventional natural gas reservoirs generally contain gas, that is, the reservoir pore space is occupied by free gas, and less bound water mainly as edge and bottom water at the bottom or edge of the gas reservoir, and the gas-water contact is uniform. In the coal reservoir, the macropores are occupied by water, and the water contains a certain amount of dissolved gas; free gas exists in some pores; and most gas is adsorbed, accounting for more than 80% (Su and Liu 1999), that is, gas in coal reservoirs is adsorbed, free and dissolved.

6.1.3 Boundary and Types of CBM Reservoir

So far, the research on the boundary of CBM reservoirs is very weak, but the boundary is the premise of CBM reservoir division and the basis for the selection of CBM exploration and development process. Study on it has important theoretical and practical significance.

According to the definition of CBM reservoir, the boundary of CBM reservoir is divided into six categories: economic boundary, hydrodynamic boundary, weathered and oxidized boundary, physical property boundary, fault boundary and lithological boundary. CBM reservoirs in different regions have different dominant boundaries.

6.1.3.1 Economic Boundary

The economic boundary is only applicable to industrial CBM reservoirs, expressed in terms of the minimum gas content of the CBM reservoir with commercial development value, depending on the content of CBM, resource abundance, reservoir physical properties, hydrodynamic conditions, technical conditions, economy and policy, etc. The economic boundaries of different CBM reservoirs are different and the difference is significant. The shallow economic boundary is mainly expressed by the minimum gas content with commercial development value. For example, the CBM content in the southern part of the Qinshui Basin is generally high, the coal seam is not thick, and the reservoir permeability is poor, so its economic boundary is 8 m^3/t gas content; the coal seams in the Liujia and Wangying areas in the Fuxin region have generally low gas content, but there are many coal seams and they are thick, the resource abundance is high, so the boundary is 5 m^3/t; the economic boundary of the Powder River Basin in the United States is 1 m^3/t, or even lower, mainly because there are many and thick coal seams, the resource abundance is high, the permeability is good. The deep economic boundary depends mainly on technical development conditions. Under the current technical conditions, the CBM can be developed rarely exceeds 1000 m deep, that is, with the increase of resource demand and the advancement of technology, the CBM below 1500 m is difficult to develop, so 1500 m is taken as the deep economic boundary.

6.1.3.2 Hydrodynamic Boundary

The hydrodynamic boundary can be divided into two categories: watershed boundary and hydrodynamic sealing boundary.

A watershed boundary places the CBM reservoirs on both sides in different fluid units. Such boundaries were found in the central and eastern parts of Utah, USA, and in the southern part of the Qinshui Basin in China (Southern Qinshui) (Fig. 6.1). A hydrodynamic sealing boundary is the most common CBM boundary, and almost in all CBM reservoirs. The mechanism of hydrodynamic sealing is to maintain a certain amount of CBM in the reservoir, and it must have a certain reservoir pressure, that is, the static groundwater level (corresponding to the reservoir pressure) has a certain elevation. This elevation can be calculated from the gas volume corresponding to the economic boundary and the reservoir pressure calculated by the Langmuir equation. The eastern and southern boundaries of the southern Qinshui CBM reservoir are hydrodynamic. For the No. 15 coal seam, the maximum adsorption capacity is 39.91–46.84 m^3/t, with an average of 43.37 m^3/t; the Langmuir pressure is 3.034–3.184 MPa, with an average of 3.109 MPa. If the maximum adsorption capacity is 46.84 m^3/t, the Langmuir pressure 3.184 MPa, taking 8 m^3/t as the economic boundary, the corresponding reservoir pressure limit is about 1.3 MPa, that is, the height difference of groundwater level should be about 130 m.

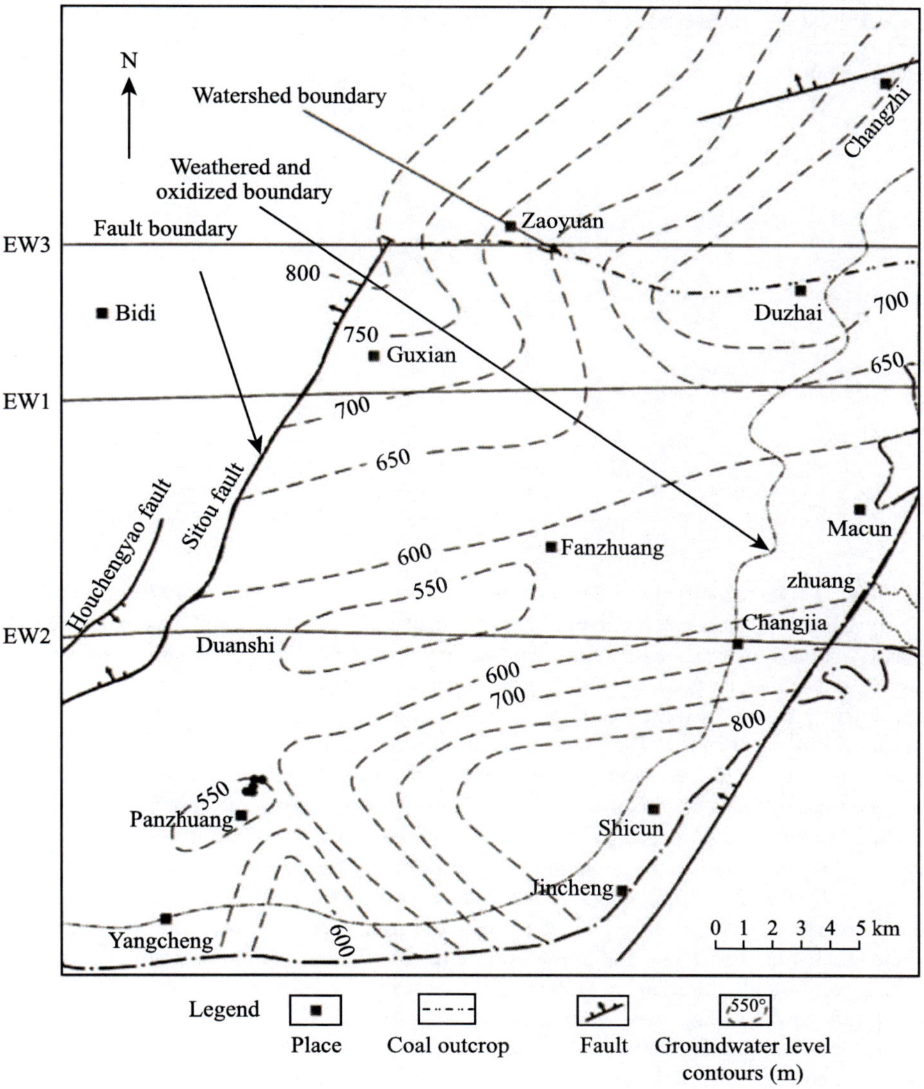

Fig. 6.1 Groundwater level contours of the Taiyuan Formation in southern Qinshui Basin

6.1.3.3 Weathered and Oxidized Boundary

As CBM is lost with outcrops and air is mixed into, the methane content in CBM reduces, and CO_2, N_2, etc. increase. Generally, 80% methane content as the bottom limit of an weathered and oxidized belt. Gas composition is analyzed using the samples from the boreholes drilled into effective overlying formation, CBM wells or coalmines at different depths, and then the analysis results are mapped to show the depth of the weathered and oxidized belt. For example, the map of CBM reservoirs in the southern Qinshui Basin (Fig. 6.2) shows that the methane content increases with the burial depth of the coal seam; the methane content greater than 80% corresponds to the coal seam at 180 m. Therefore, it can be determined that the depth of the weathered and oxidized belt is about 180 m.

The combination action of hydrodynamic sealing boundary, weathered and oxidized boundary and economic boundary is divided into three cases: (i) The bottom of the hydrodynamic sealing boundary is under the weathered and oxidized boundary, so the boundary of CBM reservoir is the hydrodynamic sealing boundary. (ii) The bottom of the hydrodynamic sealing boundary is above the weathered and oxidized boundary, so the boundary of the CBM reservoir is the bottom of the weathered and oxidized belt. The boundary of the southern Qinshui CBM reservoir belongs to the case, so the weathered and oxidized belt is the boundary of the CBM reservoir, that is, 180 m. (iii) The economic boundary is below the weathered and oxidized belt, and the boundary of the CBM reservoir is the economic boundary. This case has been found in local southern part of the Qinshui Basin.

Fig. 6.2 Coal seam depth versus methane content in southern Qinshui Basin

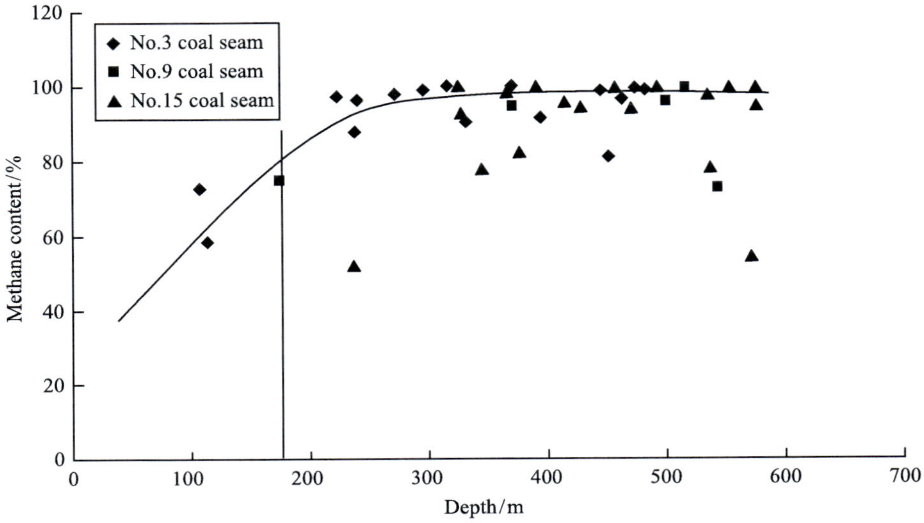

6.1.3.4 Physical Property Boundary

The principle of physical property sealing is that the coal mass becomes mylonite coal under the action of tectonic stress, the physical properties are deteriorated, the displacement pressure increases, and the diffusion and migration of CBM is prevented. The mylonite coal itself has high gas content and high reservoir pressure, and coal and gas outburst often occur in this type of coal. With high gas content, mylonite coal is sealed by gas concentration to a certain degree. In the Limin Coalmine in the north of the Shaolian Coalfield in Hunan, China, the coal seam experienced plastic rheology caused by the tectonic stress and became thick coal package; and at the same time, under the influence of small folds and wrinkles, the coal seam was subjected to kneading and sliding, which reduced the mechanical strength, and destroyed the structure of the coal, and formed lenticular structural coal with extremely poor permeability. Such structure prevents the diffusion of CBM and increases the adsorption capacity of the coal, causing gas accumulation which makes gas outburst easily take place. In the thin coal seam, the damage from gas outburst is more serious and the physical property of the coal is worse. That is the main cause for abnormal pressure compartment.

In the central and northern parts of the San Juan Basin in the United States, due to the deterioration of coal physical properties, low-permeability zones were formed, so that CBM that migrates with shallow groundwater accumulated here, and formed the abnormally high pressure of the CBM reservoir, which is a typical physical boundary (Fig. 6.3).

Gas geology research shows that on both sides of almost all faults, regardless of reversal or normal, damaged coal zones may appear, but the degree of damage is different. If it is damaged to mylonite coal, it may be a physical property boundary.

6.1.3.5 Fault Boundary

Fault boundaries can be divided into closed fault boundaries and open fault boundaries.

(1) Closed fault

The sealing of faults has been investigated relatively deep in oil and gas field, but less in CBM field. Closed faults are important boundaries of CBM reservoirs. Closed faults in the Sand Wash Basin, the Qinshui Basin and the Fuxin Basin are important boundaries of CBM reservoirs.

There are four mechanisms for the sealing of faults: lithologic configuration on both sides of the fault, mudstone smearing, granule fragmentation and diagenetic cementation.

The western boundary of the CBM reservoir in the southern Qinshui Basin is the Sitou fault which changes largely in the orientation. In the north of Duanshi, the strike is 10°–25°, and about 60° in the south of Duanshi. The fault relief is the largest around Sitou, reaching more than 360 m (Figs. 6.4 and 6.5). The Sitou fault has obvious controlling effects on the hydrogeological conditions, tectonic framework and CBM occurrence of the two CBM reservoirs on the east and the west.

The evidence for the sealing of the Sitou fault is as follows: (i) When drilling in the fault zone, the water level didn't change much, and water consumption was only 0.106 m^3/h. (ii) Water test of the Middle Ordovician aquifers on both sides of the fault shows the water quality is completely different. (iii) The salinity of the water on both sides of the fault is quite different. (iv) The calcite in the fractures in the breccia is not dissolved. (v) The gas content on both sides of the fault is significant. The gas content in the primary coal seams in the Panzhuang Minefield and Fanzhuang Minefield on the east side of the fault are high, up to 30 m^3/t, while the gas content

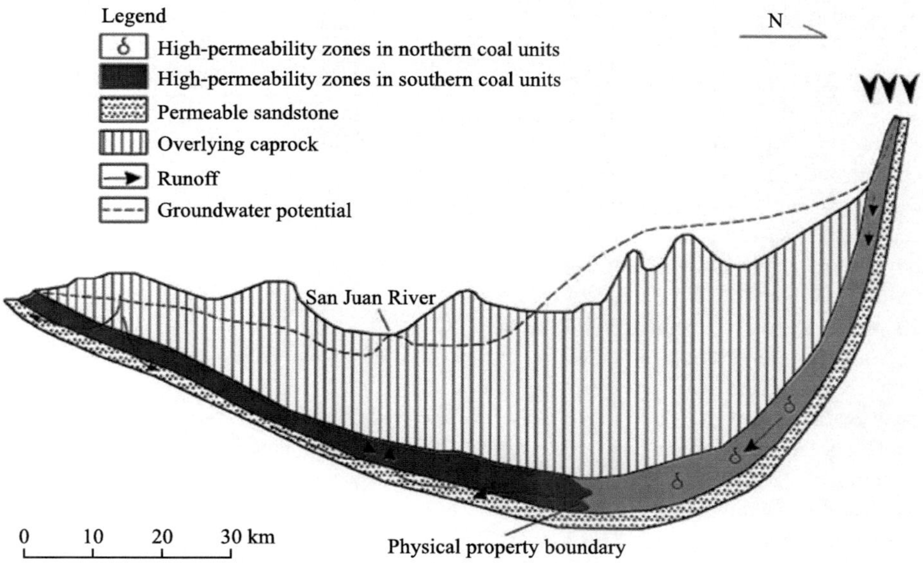

Fig. 6.3 Physical property boundary of CBM reservoirs developed in the San Juan Basin, North America

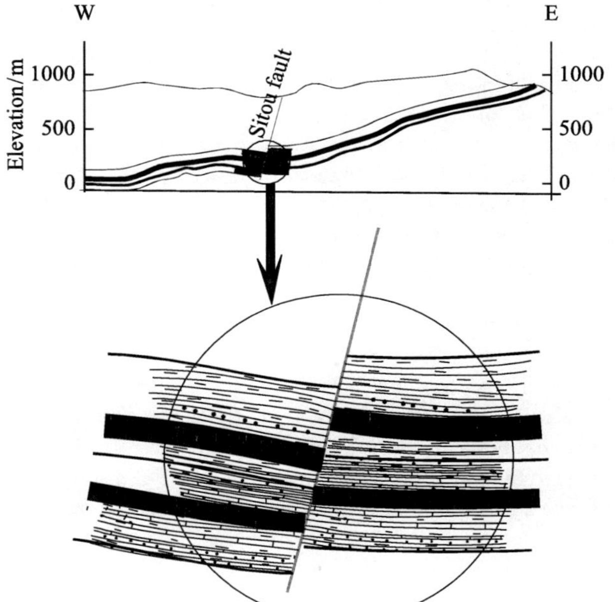

Fig. 6.4 The lithology between the two sides of the Sitou fault on the EW1 section

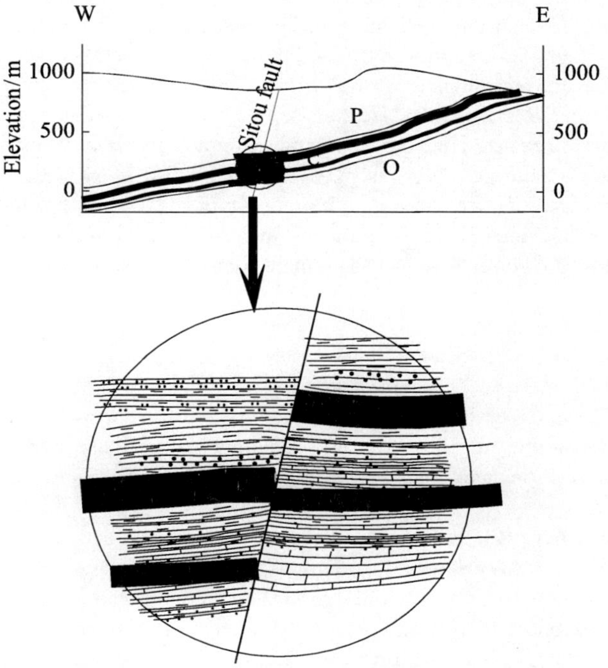

Fig. 6.5 The lithology between the two sides of the Sitou fault on the EW2 section

on the west side is relatively low, not more than 15 m^3/t at the same depth as the east side. (vi) The gas content is almost not affected by the distance from the Sitou fault.

In summary, it can be seen from the above analysis that the sealing of the Sitou fault is not only the result of mudstone smearing, but also closely related to the cementation of the fault zone and the existence of fault mud and the lithology on both sides of the fault. On the section with a smaller fault drop, mudstone smearing effect is dominant. On the section with a larger fault drop, fragmentation is the main factor, and the degree of cementation of the fault zone is very important.

(2) Open fault boundary

The open fault is divided into two cases as the boundary of CBM reservoirs: the open fault located in the groundwater recharge zone and the open fault located in the groundwater discharge zone. Groundwater enters along the open fault zone in the recharge zone and transports toward the coal seams and aquifers on both sides of the fault. The open fault prevents CBM from escaping. The aquifers and coal seams on both sides of the fault are different fluid flow units. Therefore, the fault plays the role of dividing CBM reservoirs and is conducive to the preservation of CBM.

6.1.3.6 Lithologic Boundaries

The lithologic boundary refers to the boundary of the pinchout zone. Such boundary can also be divided into two cases: (i) The lithology at the pinchout zone of the coal seam has large permeability, the displacement pressure is low, and CBM is difficult to accumulate, easy to escape, which is not conducive to the preservation of CBM. (ii) The lithology at the quenching zone of the coal seam has lower permeability, and the lithologic boundary is at higher displacement pressure, which is beneficial to the preservation of CBM. Such boundaries are more common. For example, in China's Tiefa Basin, groundwater flows from shallow runoff to deep stagnation, and there are non-permeability boundaries such as bifurcated pinchout zone and lithologic facies changing zone near the deep marginal fault, resulting in gas accumulating in deep layers, and forming typical CBM reservoirs in the Tiefa Basin. In addition, such boundaries within the Powder River Basin in the United States are also typical.

6.1.4 Types of CBM Reservoir

CBM reservoirs can be classified into various types according to groundwater dynamic conditions, CBM boundaries and coal ranks.

6.1.4.1 Classification Based on Groundwater Dynamic Conditions and Boundaries

Adsorbed CBM is mainly controlled by temperature, pressure and coal properties, and temperature changes regularly in space. Coal properties are the result of various factors in the geological history, and basically stable at present. The only factor is pressure that is constantly changing with the recharge, migration and discharge of groundwater. Therefore, in China, a CBM reservoir classification scheme was established mainly based on pressure, and considering boundary types and structural characteristics of CBM reservoirs (Table 6.1, Fig. 6.6).

First, according to the pressure forming mechanism of CBM reservoirs, it is divided into two classes: hydrodynamically sealed and self-sealing CBM reservoirs. The hydrodynamically sealed type can be further divided into two sub-classes: hydrodynamically blocked and hydrodynamically driven reservoirs. The hydrodynamic blocked reservoirs are subdivided into five types according to boundaries. The hydrodynamically driven reservoirs are subdivided into three types. Self-sealing CBM reservoirs can be divided into three sub-classes.

(1) Hydrodynamically sealed CBM reservoir

The hydrodynamically sealed CBM reservoir refers to the CBM reservoir that migration and enrichment of CBM is blocked or driven by groundwater, besides controlled by other geological boundaries; that is, such CBM reservoirs is closely related to the recharge, migration, retention and discharge of groundwater.

Groundwater or atmospheric precipitation migrates from the recharge area along the well permeable coal reservoir to the deep section of the basin, resulting in increased reservoir pressure and CBM enrichment after reaching the coal reservoir in the retention area, and finally forming hydrodynamically plugged CBM reservoirs. This phenomenon of "inversion" of gas and water is similar to deep basin gas. The boundary of the hydrodynamic closure is the groundwater level close to the outcrop area. The hydrostatic pressure caused by the water level should ensure the gas enclosed in the coal seam to the lowest content, which can be calculated according to the gas content and the Langmuir equation. There are many boundary types in the retention zone, which is the basis for the classification of CBM reservoirs.

For example, the CBM reservoirs of the Fruitland Formation in the San Juan Basin in the United States are bounded by physical properties; the CBM reservoirs of the Williams Fork Formation in the Sand Wash Basin are bounded by faults; the CBM reservoirs in the Qinshui Basin are blocked by groundwater in the shallow part of the syncline; The Hedong Coalfield in the eastern margin of the Ordos Basin is a monocline, where groundwater migrates to the deep retention zone, and has no obvious boundary; deep coal seam pinchout is present in many coal-bearing basins.

CBM reservoirs are classified into 8 types according to structural and hydrodynamic characteristics (Table 6.1).

(2) Self-sealing CBM reservoir

The characteristics and genesis of self-sealing CBM reservoirs are similar to those of conventional hydrocarbon anomalous pressure compartment. The formation of self-sealing CBM

Table 6.1 Classification of CBM reservoirs

Class	Sub-class	Type	Example
Hydrodynamically sealed CBM reservoirs	Plugging	Physically-hydrodynamically blocked	San Juan Basin
		Fault-hydrodynamically blocked	Sand Wash Basin
		Monocline-hydrodynamically blocked	Eastern margin of Ordos Basin
		Syncline-hydrodynamically blocked	Qinshui Basin
		Pinchout-hydrodynamically blocked	Powder River Basin
	Driving	Anticline-hydrodynamically driven	Powder River Basin
		Anticline-chopped-hydrodynamically driven	San Juan Basin
		Fault-anticline-hydrodynamically driven	Qinshui Basin
Self-sealing CBM reservoirs		Abnormal pressure compartment	Washakie Basin
		Low-permeability	Limin mine and other cropped coal
		Lenticular	Late Paleozoic coalfields in South China

reservoirs is related to temperature, hydrocarbon formation and tectonic stress (including uplifting, eroding, etc.), and is also related to coal mass deformation under tectonic stress. It can be divided into three types: (i) Abnormal pressure compartment. There is a tight layer related to hydrocarbon formation and diagenesis at around 3000 m, which is the top boundary of the compartment. Under it, the CBM reservoir is in a closed, independent pressure system. Such CBM reservoir isn't commercial due to its deep burial and low gas content. (ii) Low-permeability self-sealing CBM reservoir. This type of CBM reservoir is severely deformed by strong tectonic stress, called scale-like or powdery mylonite coal, which is compressed into a lenticular shape. This coal has extremely low permeability and often contains a large amount of CBM. Gas outburst disasters occur mostly in such coal reservoir, so that it is not suitable for CBM development due to poor permeability. (iii) Lenticular CBM reservoirs. This type of CBM reservoir is extremely poorly continuous, spatially lenticular and confined by surrounding rock. Such CBM reservoirs exist in many coal-bearing basins. If the surrounding rock is well sealed and the lens is large, it may be a development target for CBM.

6.1.4.2 Classification by Coal Rank

There are six types of coal metamorphism: deep metamorphism (regionally), regional magma thermal metamorphism, magma contact metamorphism, dynamic metamorphism, gas-water hydrothermal metamorphism, and combustion metamorphism. Deep metamorphism and regional magma thermal metamorphism contribute more to CBM generation. These six types of metamorphism create high-, medium- and low-ranked CBM reservoirs.

(1) High-ranked CBM reservoir

The coal rock with $R^o > 2\%$ is generally high-ranked CBM reservoir experiencing secondary metamorphism. After early deep metamorphism at normal paleotemperature and primary hydrocarbon generation, second metamorphism may occur under the action of magma heat, which causes the coal rank to increase and second hydrocarbon generation. Such CBM reservoirs are characterized by strong coal adsorption capacity, high gas content and poor permeability, such as the CBM reservoir in the southeastern part of the Qinshui Basin in Shanxi and the Erdaoling CBM reservoir in Ningxia.

(2) Medium-ranked CBM reservoir

The coal rock with $0.7\% < R^o < 2\%$ is almost the result of deep metamorphism, but secondary metamorphism may not occur, such as the Hedong Coalfield in Shanxi, the Piceance Basin in the United States. The CBM reservoirs are characterized by high gas content and good permeability.

(3) Low-ranked CBM reservoir

Low-ranked CBM reservoirs can be divided into two categories: (i) Low-maturity and low-ranked CBM reservoir. The coal rock has $0.5\% < R^o < 0.7\%$, mainly due to deep

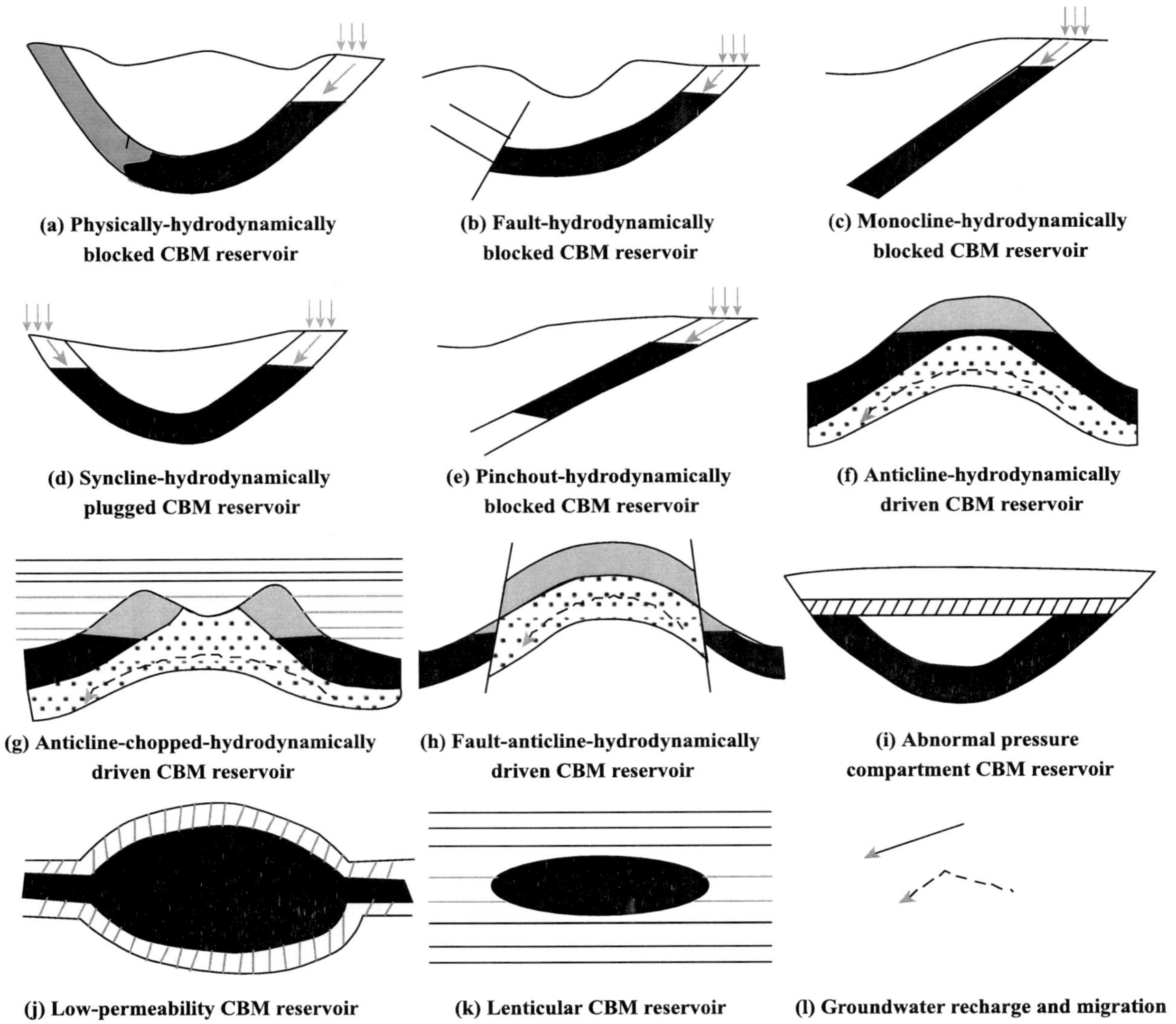

Fig. 6.6 Schematic CBM reservoir types

metamorphism, or contact deterioration, but the effect is limited. The CBM reservoir is characterized by moderate gas content and moderate permeability, and is represented by the Fuxin Basin. (ii) Immature and low-ranked CBM reservoir. The coal rock has $R^o < 0.5\%$, the reservoir did not undergo coal metamorphism, only diagenesis, so the maturity reaches lignite. The reservoir is characterized by low gas content and low permeability, and is represented by the Powder River Basin in the United States.

6.2 Accumulation Process and Accumulation Mechanism of Medium- to High-Ranked CBM Reservoir

6.2.1 Theoretical Basis of CBM Accumulation

The traditional view is that CBM reservoirs are unconventional natural gas reservoirs that are self-generating and self-storing, but recent studies have shown that the migration of CBM is widespread, especially for medium- and high-ranked CBM reservoirs, if there is no migration, There will be no accumulation of CBM. This study introduces the theory of adsorption potential for the first time, systematically discusses the fractionation mechanism of CBM diffusion and migration, and provides theoretical and methodological support for the recovery of CBM content history, and also provides basis for the prediction of CBM enrichment areas.

6.2.1.1 Application of Adsorption Potential Theory in the Change History of CBM Content

Taking the CBM reservoir in the southeast of the Qinshui Basin as an example, the No. 3 coal of the Shanxi Formation in the Sihe Coalmine was subjected to isothermal adsorption experiments at 30, 45 and 60 °C, using the equilibrium moisture and 6 points adsorption method. The maximum pressure is about 10 MPa. Figure 6.7 shows the adsorption isotherm. Figure 6.8 shows the characteristic adsorption curve established by these three groups of data. The mathematical expression of the characteristic curve is

$$\begin{aligned} \varepsilon = 1.43 \times 10^4 - 3.28 \times 10^5 \omega \\ + 3.49 \times 10^6 \omega^2 - 1.70 \times 10^7 \omega^3 \end{aligned} \tag{6.1}$$

The adsorption capacity at 30, 45 and 60 °C were calculated according to Eq. (6.1), and compared with the measured values. It was found that the two were in good agreement (Fig. 6.7), indicating the above theoretical analysis and calculation method is effective.

Using the adsorption potential theory to calculate the adsorption capacity of coal to methane under reservoir conditions means to quantitatively evaluate the theoretical gas content at each critical moment in the geological history. To calculate the gas content, the reservoir temperature and pressure must be known, which can be obtained in a variety

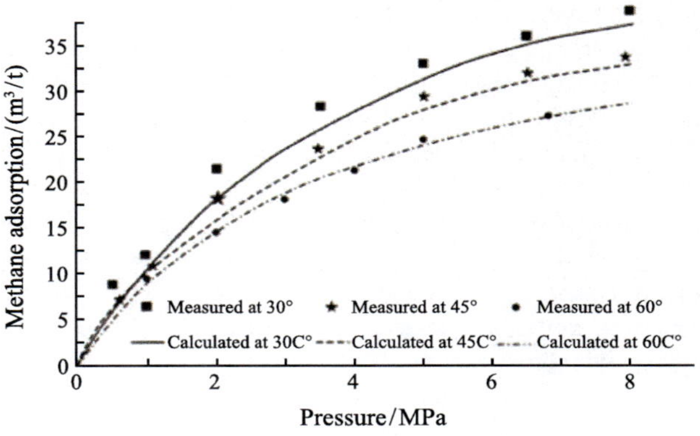

Fig. 6.7 Measured and calculated methane adsorption isotherm of southeastern Qinshui Basin

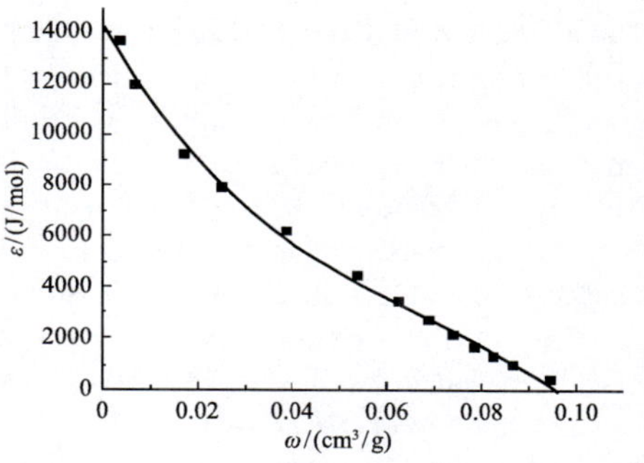

Fig. 6.8 Characteristic adsorption curve of southeastern Qinshui Basin

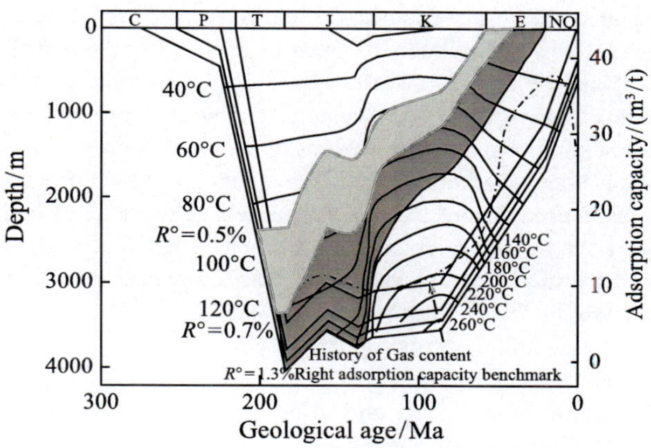

Fig. 6.9 The burial history, thermal history and gas content evolution of the Carboniferous–Permian coal reservoir in southern Qinshui Basin

of ways. In this study, the reservoir pressure is normal, and the reservoir temperature was calculated based on the geothermal gradient at different periods. Using them, the gas content in the Qinshui Basin during the geological history was calculated (Fig. 6.9). This Figure clearly reflects the burial history, thermal history, maturity history and evolution of gas content. Their specific geological implications will be discussed later.

6.2.1.2 Application of Adsorption Potential Theory in the Discussion of CBM Desorption and Fractionation Mechanism

Coalbed methane is self-generating and self-storing unconventional natural gas in the broad sense, but this is not the case in the narrow sense. Coalbed methane has been migrating since generated, and gas composition and isotope fractionation have kept changing throughout the process. The fractionation mechanism of coalbed methane was first proposed by Teichmuller, an internationally famous coal rock scientist. Teichmuller et al. studied the desorption zone and the original CBM zone in the Ruhr Coalfield and Saar

Coalfield in Germany, and found that the methane carbon isotope in the desorption zone is light, indicating that this phenomenon is the result of the desorption and diffusion of coalbed methane after the coal seam was uplifted. Teichmuller's theory was quickly accepted by many scholars. However, in recent years, research in the field of coalbed methane has shown that not all CBM carbon isotope distribution can be reasonably explained by this theory, so many researchers proposed mechanisms such as secondary biofractionation and pyrolysis (Hakan et al. 2002). These mechanisms can explain the distribution of carbon isotope in their specific coalfields, but can not reasonably explain the fact that the CBM carbon isotope distribution is uneven under strong hydrodynamic conditions. Recently, it has been reported that the mechanism of dissolved and fractionated natural gas is introduced (Hakan et al. 2002), which well explains the uneven distribution of carbon isotope in coal seams under strong hydrodynamic conditions. Similarly, the migration fractionation mechanism cannot explain the carbon isotope distribution under the conditions without or with weak hydrodynamics. Therefore, we believe that there is one or more migration fractionation mechanisms for coalbed methane migration in a coal basin. The following focuses on the fractionation mechanism in the coalbed methane adsorption and desorption process.

Fractionation of $^{12}CH_4$ and $^{13}CH_4$ in CBM Desorption-Diffusion Process

Previous studies have shown that $^{12}CH_4$ is first desorbed and then $^{13}CH_4$ desorbed during CBM adsorption and desorption. That is to say, as the desorption time goes on, the methane carbon isotope desorbed has a tendency to become heavier. At the same time, it is also found that the diffusion and migration rate of $^{12}CH_4$ is higher than that of $^{13}CH_4$, which results in light carbon isotope in the shallow coal seam and heavy in the middle and deep ones (Su et al. 2005). Here, the adsorption characteristics of $^{13}CH_4$ and $^{12}CH_4$ adsorbed on the pore surface of coal can be established to reasonably explain the above phenomenon. The adsorption curves of $^{13}CH_4$ and $^{12}CH_4$ from four different coal ranks (Fig. 6.10) show that the adsorption potential and adsorption space of $^{13}CH_4$ on the pore surface of coal are generally larger than $^{12}CH_4$, and there is a tendency to increase with increasing pressure, indicating at high pressure, first coal tends to adsorb $^{13}CH_4$. When the adsorption potential is the same, to adsorb the same volume of CH_4, the pressure required by $^{13}CH_4$ is lower than that of $^{12}CH_4$, indicating that the dispersion force of $^{13}CH_4$ acting on the coal surface is higher than $^{12}CH_4$. That is, $^{12}CH_4$ is preferentially desorbed under isostatic conditions. From the perspective of adsorption potential, it fully demonstrates that coal tends to preferentially adsorb and reluctantly desorb $^{13}CH_4$. This is the main cause of preferential desorption, diffusion and migration of $^{12}CH_4$ and the main mechanism of carbon isotope fractionation of methane. It can be seen that, both in the laboratory (Table 6.2) and at the natural state, $^{12}CH_4$ is preferentially desorbed, diffuses, migrates and causes isotope fractionation (Su et al. 2005).

This phenomenon generally occurs when the groundwater level drops and the reservoir pressure decreases. The $^{12}CH_4$ first desorbed diffuses into the shallow layer, causing the shallow methane carbon isotope to become lighter and the deep become heavier. This is one of the causes why the methane carbon isotopes are heavy in the deep CBM reservoirs in the southern Qinshui Basin.

Fractionation of Multi-component gas in CBM Desorption Process

In the past, like carbon isotope fractionation of methane, the fractionation of multi-component gas was qualitatively evaluated from experimental data. The introduction of the adsorption potential theory provides a method for the quantitative evaluation of these phenomena. There are three ways of fractionation of multi-component gases.

Method I:

It is represented by the adsorption-desorption of Argonne Premium coal in Germany. The desorption rates of CO_2 and CH_4 are both lower than their adsorption rates, and CH_4 behaves more significant at 1–3 MPa (Fig. 6.11). The desorption characteristics of CH_4 and CO_2 in this pressure interval are almost coincident, but the adsorption potential and adsorption space of CO_2 are larger than those of CH_4 at lower pressure and higher pressure, and there is a tendency to increase gradually (Fig. 6.12). It indicates that the fractional effect of CH_4 and CO_2 at 1–3 MPa is weak, but strong at low pressure and high pressure.

Method II:

It is represented by Lu'an coal. The desorption rate of CO_2 is greater than the adsorption rate. The adsorption and desorption isotherms of CH_4 intersect at point *A*, and the corresponding pressure is 2.5 MPa (Fig. 6.13). This point corresponds to point *A*′ on the characteristic curve (Fig. 6.14). From this point to lower pressure, the adsorption potential of CH_4 is generally greater than that of CO_2, but it is opposite from this point to higher pressure. This means that at below 2.5 MPa reservoir pressure, the output rate of CH_4 gradually decreases, that of CO_2 gradually increases, which is not conducive to fractionation; at 2.5 MPa or higher, the output rate of CH_4 gradually increases, and that of CO_2 decreases, which is favorable for fractionation.

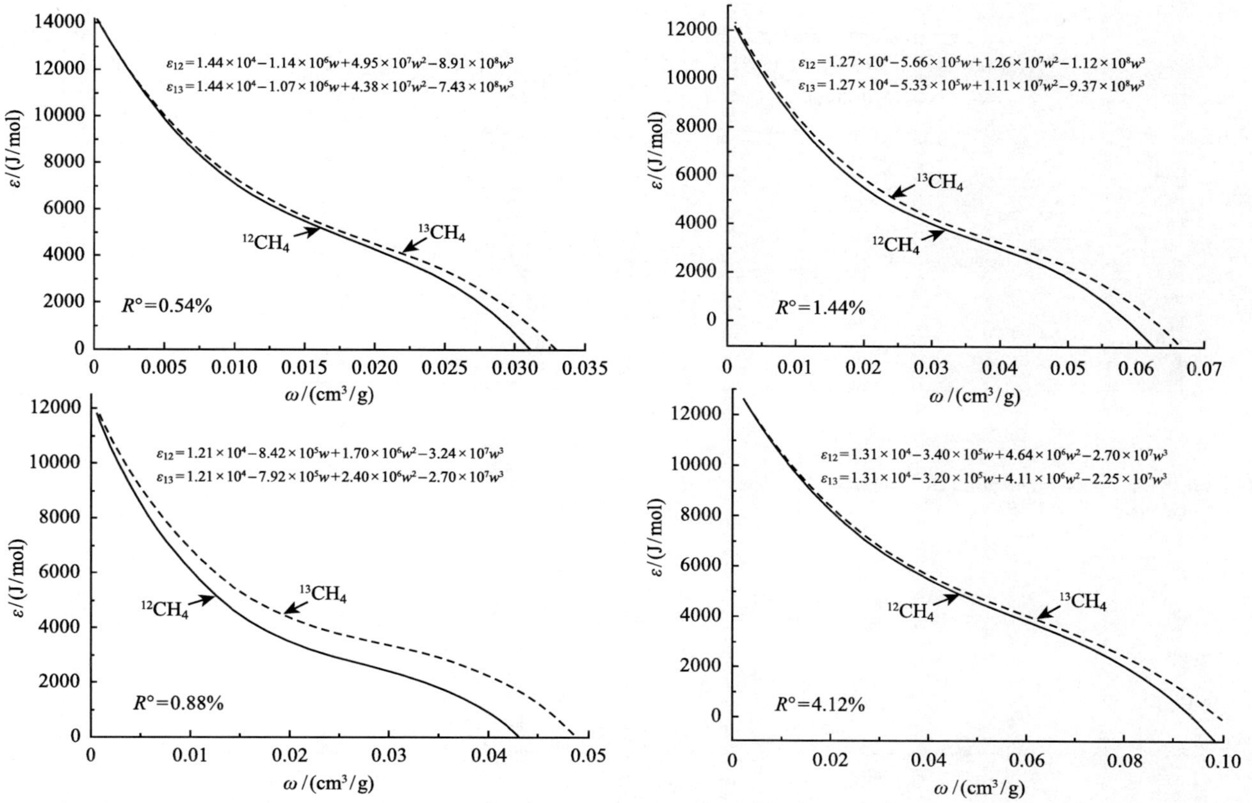

Fig. 6.10 Characteristic adsorption curves of $^{13}CH_4$ and $^{12}CH_4$ of different coal ranks

Table 6.2 CBM desorption results in Qinshui Basin

Coal sample	Desorption period	$\delta^{13}C$/‰	Coal sample	Desorption period	$\delta^{13}C$/‰	Coal sample	$\delta^{13}C$/‰	
							First desorption	Second desorption
HG2-4-1	4 h	−33.08	V-3/1 (off-line)	1d	−57.42	H-5	−64.35	−62.07
	24 h	−33.11		2d	−57.60	H-6	−58.09	−54.89
	96 h	−29.88		3d	−57.05	H-2	−72.27	−55.10
HG2-13-1	4 h	−37.76		5d	−57.03	H-17	−54.38	−48.95
	24 h	−33.98		7d	−56.70	H-19	−49.28	−46.66
	96 h	−29.56		8d	−56.23	H-10	−64.14	−62.02
HG3-3-1	4 h	−35.23		15d	−56.56	H-12	−67.97	−50.70
	24 h	−34.91		36d	−56.64	H-22	−58.60	−54.57
	96 h	−34.10		50d	−56.06	Y-2	−59.85	−59.09
HG3-9-2	4 h	−35.80		64d	−55.68	KZ-20	−68.58	−64.72
	24 h	−35.55	V-3/1 (on-line)	Id	−57.60	C-9	−45.98	−45.48
	96 h	−34.58		5d	−57.38	C-7	−47.70	−45.89
II-3/2 (off-line)	5d	−56.86		15d	−56.94			
	57d	−56.02		36d	−56.55			
	95d	−55.55		50d	−56.35			

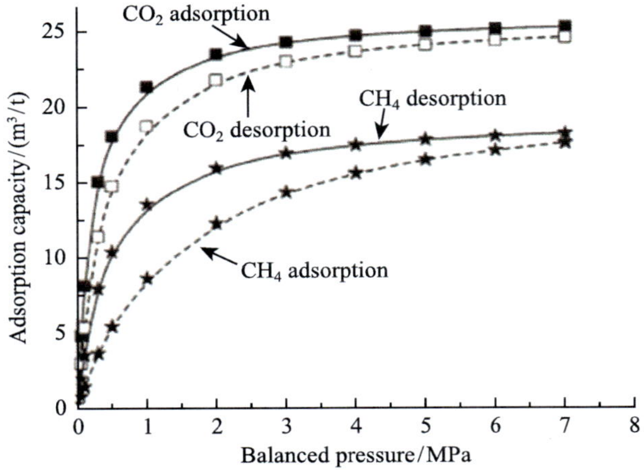

Fig. 6.11 Desorption isotherms of CH_4 and CO_2 of Argonne Premium coal in Germany (R^o = 1.16%)

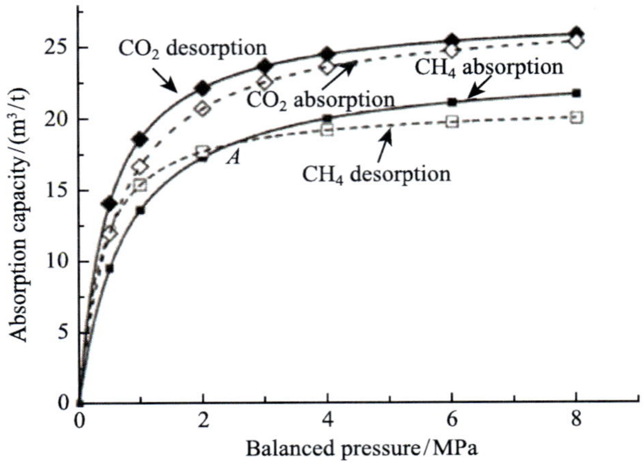

Fig. 6.13 Desorption isotherms of CH_4 and CO_2 of Lu'an coal (R^o = 1.68%)

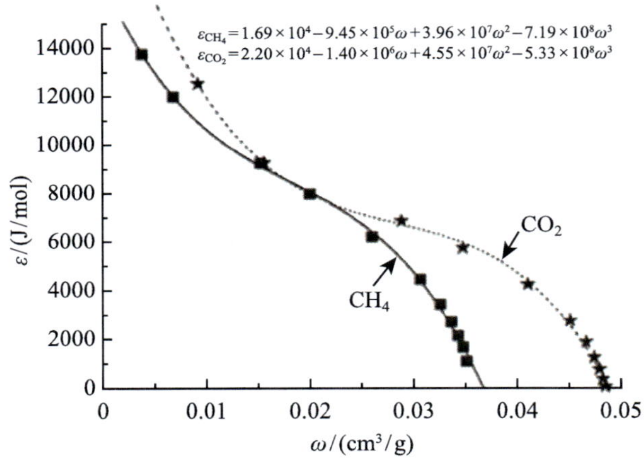

Fig. 6.12 Adsorption characteristics of CH_4 and CO_2 of Argonne Premium coal in Germany (R^o = 1.16%)

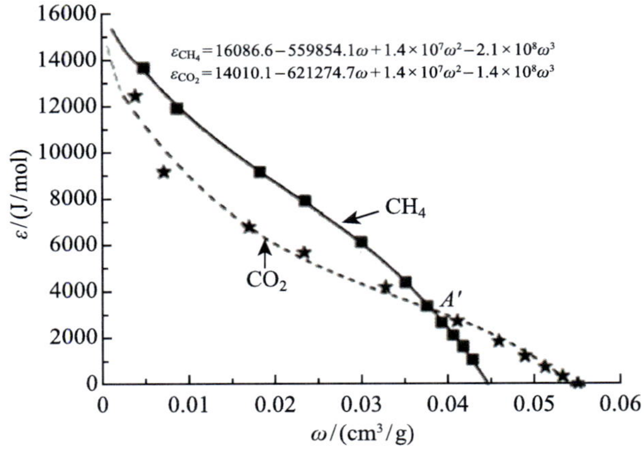

Fig. 6.14 Adsorption characteristics of CH_4 and CO_2 of Lu'an coal (R^o = 1.68%)

Method III:

It is represented by Jincheng coal. The desorption rate of CH_4 is consistent with its adsorption rate. The adsorption and desorption isotherms of CO_2 intersect at point *B*, and the corresponding pressure is 2.5 MPa (Fig. 6.15). This point corresponds to point *B*′ on the characteristic curve (Fig. 6.16). From this point to lower pressure, the adsorption potential of CO_2 increases, and that of CH_4 decreases, but it is opposite from this point to higher pressure.

In summary, in all three cases, fractionation is strong at high pressure, but weak at medium-low pressure. Therefore, fractionation of gas components in a CBM basin generally occurs at high reservoir pressure in deep layers.

6.2.2 Analysis of Reservoir Forming Mechanism of Typical CBM Reservoirs

As the smallest geological unit of CBM accumulation, it is of great significance for studying the hydrocarbon accumulation mechanism of CBM reservoir. Perfect theory and method for researching CBM accumulation mechanism are conducive to in-depth analysis and understanding of CBM accumulation rules, enrich CBM geology theory, promote the development of the world CBM industry, help learn from the experiences of selecting exploration and evaluation targets of conventional natural gas, and selecting favorable CBM zones and guide the exploration and development of CBM.

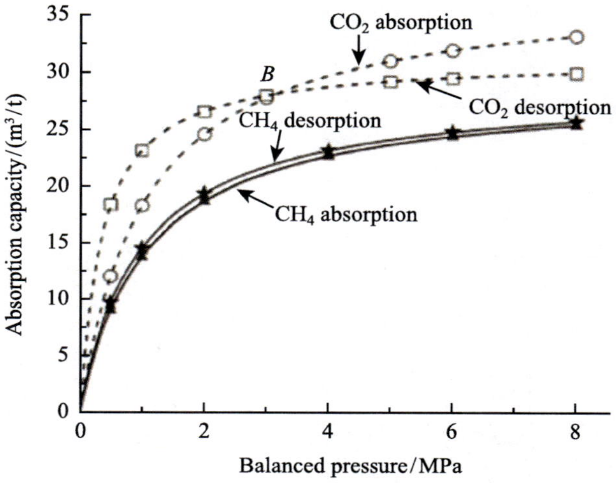

Fig. 6.15 Desorption isotherms of CH_4 and CO_2 of Jincheng coal (R^o = 4.21%)

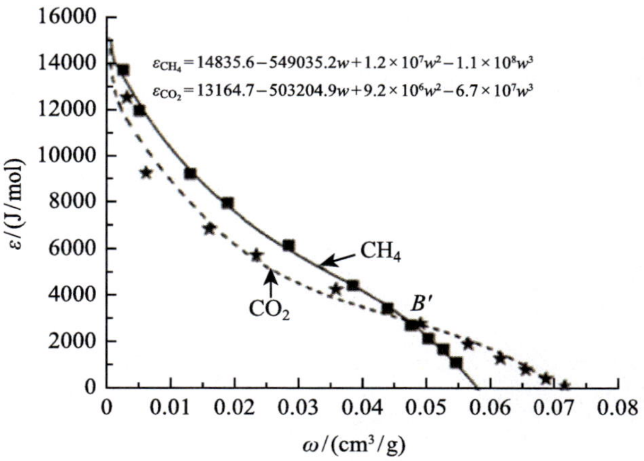

Fig. 6.16 Desorption characteristics of CH_4 and CO_2 of Jincheng coal (R^o = 4.21%)

Next we discuss the forming mechanism of the CBM reservoirs in the southern Qinshui Basin, the Liulin area in the eastern margin of the Ordos Basin, and the Fuxin Basin from accumulation conditions and accumulation process.

6.2.2.1 CBM Reservoir Forming Process and Accumulation Mechanism in Southern Qinshui Basin

Scope of CBM Reservoir

According to the classification scheme by boundary type, the CBM reservoirs in the southern part of the Qinshui Basin are bounded to the southern groundwater watershed, the eastern sealing Sitou fault, and the western and northern coal seam outcrops (Fig. 6.1). On the eastern and southern boundaries of the basin, since the weathering zone of the coal seam is at 180 m, the boundaries of the weathered and oxidized belts are the eastern and southern boundaries of the CBM reservoirs. In addition, because the No. 15 coal seam of the Taiyuan Formation and the No. 3 coal seam of the Shanxi Formation are comparted by tight rock and there is no hydraulic communication, the CBM reservoirs in the southern Qinshui Basin actually includes No. 15 and No. 3 coal seams.

Forming Conditions of CBM Reservoir

(1) Spatial distribution of coal seams

After experiencing long-term weathering and denudation, the Ordovician in the North China Basin began to settle and accept sediments in the Late Carboniferous, forming Carboniferous-Permian strata, such as Benxi Formation, Taiyuan Formation, Shanxi Formation, lower Shihezi Formation, Upper Shihezi Formation and Shiqianfeng Formation. The primary coal-bearing strata in the Qinshui Basin are the Taiyuan Formation and the Shanxi Formation, with an average thickness of about 150 m.

The Taiyuan Formation contains 5–10 coal seams, of which the No. 15 coal seam is primary. The Shanxi Formation contains 3 coal seams, of which the No. 3 coal seam is primary. The total thickness of the coal seams is about 15 m. The No. 15 coal seam is 1–6 m thick, and averaged 3 m, and the overall trend is thick on the east and thin on the west, thick on the north and thin on the south, which is relatively stable (Fig. 6.17). The No. 3 coal seam is 4–7 m thick, and averaged 6 m, and the overall trend is thick on the east and thin on the west, and stable in the distribution (Fig. 6.18). The buried depth of the No. 15 coal seam is about 0–900 m, and in the most of the region less than 700 m. The No. 3 coal seam is several ten meters shallower than the No. 15 coal seam. Its burial depth is very conducive to the development of CBM. The influencing factor on coal seam thickness is mainly the evolution of sedimentary environment.

(2) Gas content

The composition of CBM in the southern part of the Qinshui Basin is mainly CH_4, and its content is generally greater than 98%. In addition, there are a small amount of ethane, N_2 and CO_2, not more than 1%. To the margin of the basin, with the decrease of the buried depth of the coal seam, the content of CH_4 gradually decreases, and the content of N_2 and CO_2 increases.

The coal seam has high gas content. The No. 15 coal seam of the Taiyuan Formation is generally 10–20 m^3/t, and

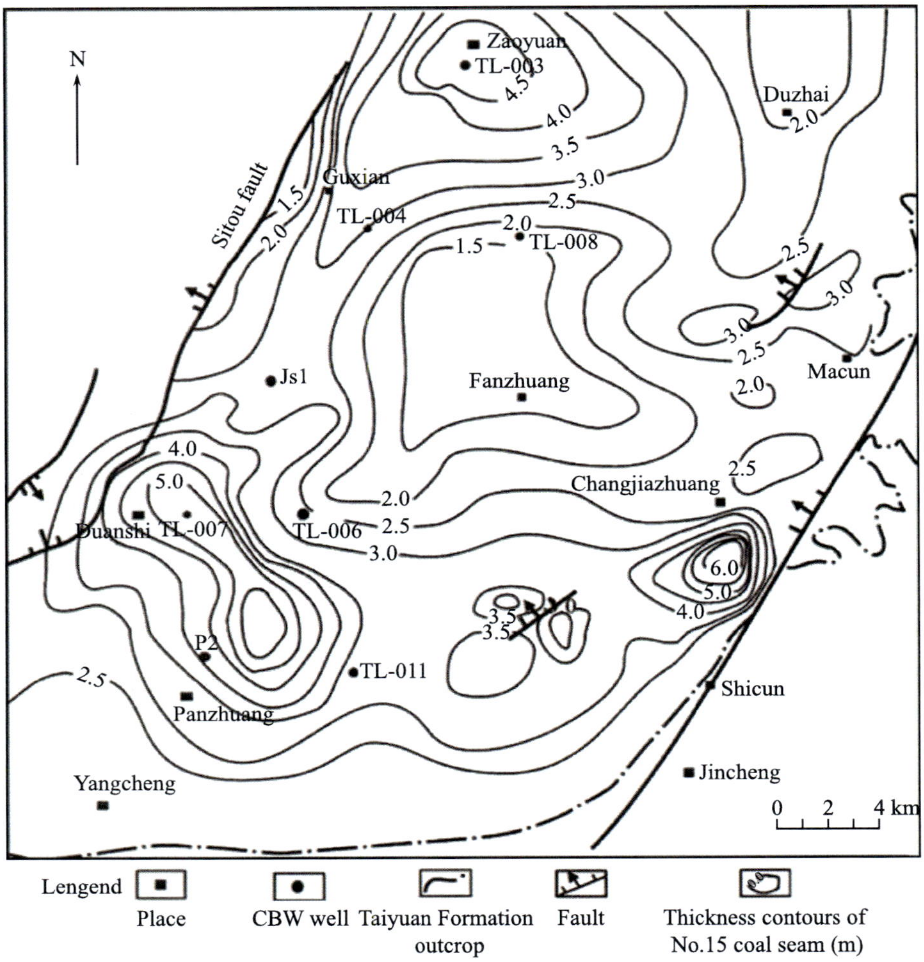

Fig. 6.17 The thickness contours of the No. 15 coal seam of Taiyuan Formation

up to 26 m^3/t (Fig. 6.19). The No. 3 coal seam of the Shanxi Formation is generally 8–30 m^3/t, up to 37 m^3/t (Fig. 6.20). The CBM content in the Duanshi-Panzhuang-Fanzhuang area is high, which is the main area of CBM enrichment. In addition, in the northern Zaoyuan area, the CBM content is relatively high, forming a small CBM enrichment center.

The gas saturation of the No. 3 coal seam is 87–98%, with an average of 93%. The gas saturation of the No. 15 coal seam is 71–76%, with an average of 74%. The gas saturation is low, mainly under-saturated, and locally saturated. The CBM resource abundance is high, up to 2×10^8 m^3/km^2 or higher. Taking CH_4 content of 80% as the lower limit, the shallow weathered and oxidized belt is generally about 180 m deep.

(3) Reservoir physical properties

The coal seams in the study area are the result of regional metamorphism. Under the high temperature and relatively low pressure environment, influenced by tectonic movement, cleats and fractures were developed. The reservoir properties are good, and the permeability is high. The pores of the coal seam are mainly micropores and transitional pores, and a small amount of mesopores and macropores. The pores have certain connectivity. The effective porosity is 1.15–7.69%, generally <5% (Su et al. 2001).

The permeability is 0.1–6.7 mD, generally not more than 2 mD, and has obvious directionality. The largest permeability is in the primary fracture direction. The permeability gradually decreases from the shallow layer to the deep layer, and decreases with the increase of the geostress. The reservoir heterogeneity is strong.

According to core and downhole observations, it is known that the permeability of the reservoir in the study area is controlled by structural fractures rather than cleats. It has been observed that the cleats are mostly "S" type, isolated tensile cracks, no more than 5 cm long, and mostly less than 2 cm, and completely filled with calcite (Su et al. 2001,

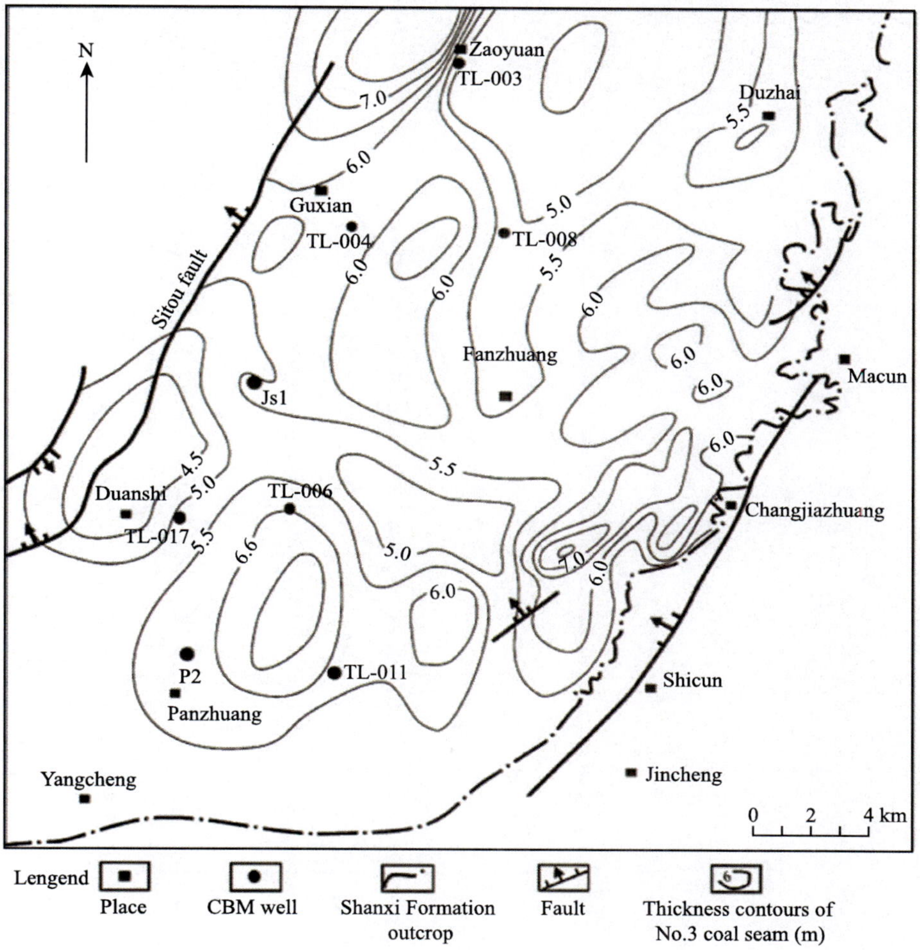

Fig. 6.18 The thickness contour of the No. 3 coal seam of the Shanxi Formation in the southeast of Qinshui Basin

2005). A group of cleats in the Shanxi Formation in the Jincheng mining area are less than 2 cm long, arranged in an en echelon shape, and completely filling with calcite. Four groups exogenous fissures were formed in two stages, three of which were relatively developed, which cut the coal rock into diamond or irregular blocks, indicating that cleats have little contribution to the permeability of the CBM reservoirs. Two groups of structural fractures are generally identified, with the strikes being northeast-southwest and northwest-southeast, and the former being the most developed. It is the two groups of fractures that build the network providing channels for CBM and water migration.

The degree of deformation of coal rock is another factor that affects the permeability of the coal reservoir. When the coal rock is not or slightly deformed, the original structure and structure preservation are relatively complete, the cleat or fracture extends to a large extent, the connectivity is good, and reservoir strengthening is beneficial to increase the permeability. Such coal is the development target available to present technology and process (Su et al. 2001). Under the action of strong tectonic stress (perhaps other forces, such as gravity), coal rock will be severely deformed and change to mylonite coal, whose original structure disappears, and is broken into scales or powders. In this type of reservoir, the cleats no longer exist, and the fractures only extend a short distance (several centimeters to tens of centimeters) or are undeveloped, which are mutually intercepted, resulting in very poor connectivity. The permeability of such reservoirs is very poor, and almost no more than 0.01 mD. More unfortunately, the broken coal cannot be strengthened effectively, and present CBM development process is ineffective (Su and Liu 1999). Although the degree of deformation of the coal is low in most parts of the study area, local strong deformation might occur. The degree of deformation may be a major factor controlling the heterogeneity of the reservoir permeability. According to the observation to the cores from CBM wells and exploration wells and to the coal structure under the coalmine, it is found that the coal structure in the southern part of the Qinshui Basin is relatively complete. Although cut by two major fractures, the

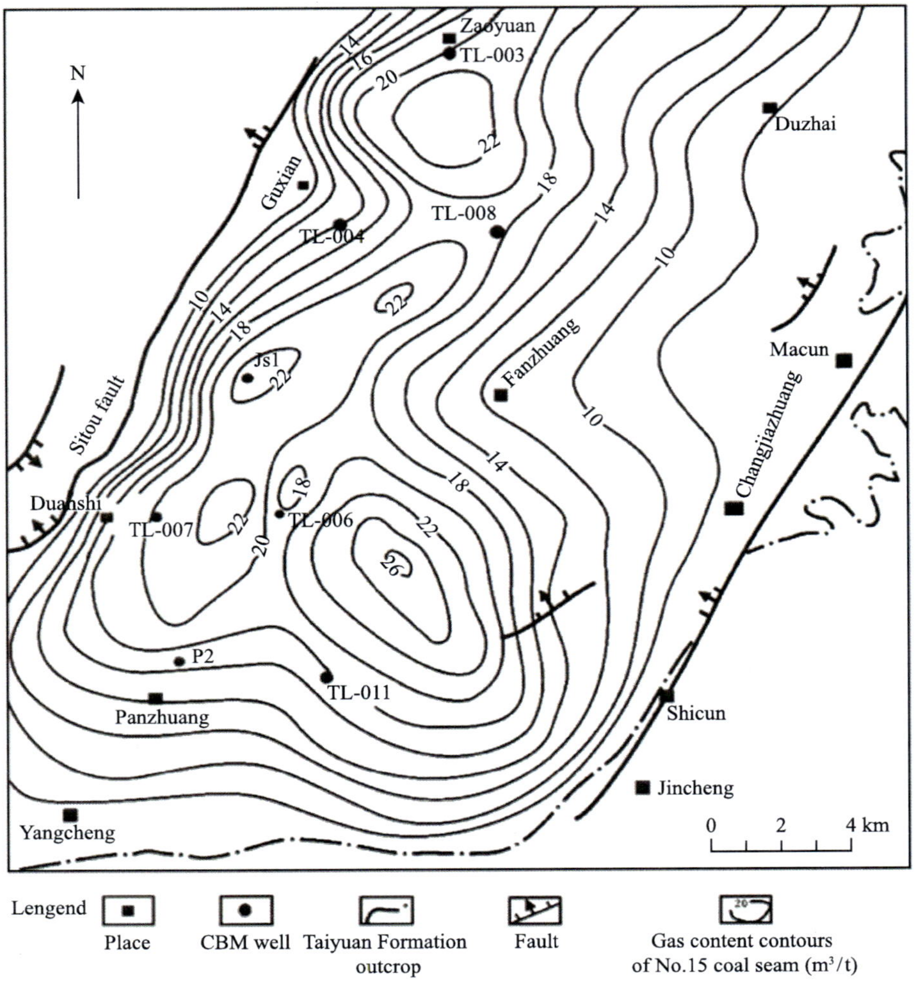

Fig. 6.19 Gas content contours of the No. 15 coal in southern Qinshui Basin

original structure is clearly visible, especially from the southern part of Fanzhuang which has the highest permeability in the study area.

(4) Adsorption characteristics

The sorption capacity of the coal in the southern part of the Qinshui Basin is large, the Langmuir volume is generally 28.08–57.87 m^3/t, and the Langmuir pressure is 1.91–3.99 MPa. The main causes for the large adsorption capacity are as follows: (i) With the increase of coal metamorphism degree, the adsorption capacity increases. When the vitrinite reflectance reaches 3.5–3.8%, the adsorption capacity reaches the peak during the whole metamorphic process. The metamorphism of the coal seam in the southern Qinshui Basin is high, generally lean coal-anthracite, mainly anthracite, and the vitrinite reflectance is 2.2–4.0% or so, so the adsorption capacity reaches the maximum. (ii) The micropores are developed, so the specific pore surface area of the coal seam is large, and the adsorption capacity is enhanced. (iii) The No. 3 coal seam and the No. 15 coal seam are mainly bright coal and semi-bright coal. The macroscopic coal type is good; the vitrinite reflectance is high, accounting for 84–96% of the organic matter microscopic composition; and the ash content is low, generally 3–23%, with an average of 11% (Zhao et al. 1999; Li and Druzhinin 2000). The maximum adsorption capacity of bituminous coal increases with the increase of vitrinite content, and increases with the decrease of ash content. (iv) The impact of reservoir pressure. Under isothermal conditions, the adsorption capacity is proportional to the reservoir pressure. As the reservoir pressure increases, the adsorption capacity increases. When the reservoir pressure is 0–1 MPa, the adsorption capacity and the reservoir pressure increase almost linearly, and then the increasing rate becomes smaller until the adsorption capacity reaches the maximum, and then the coal seam adsorption reached saturated.

Fig. 6.20 Gas content and $^{13}CH_4$ contours of No. 3 coal seam in southern Qinshui Basin

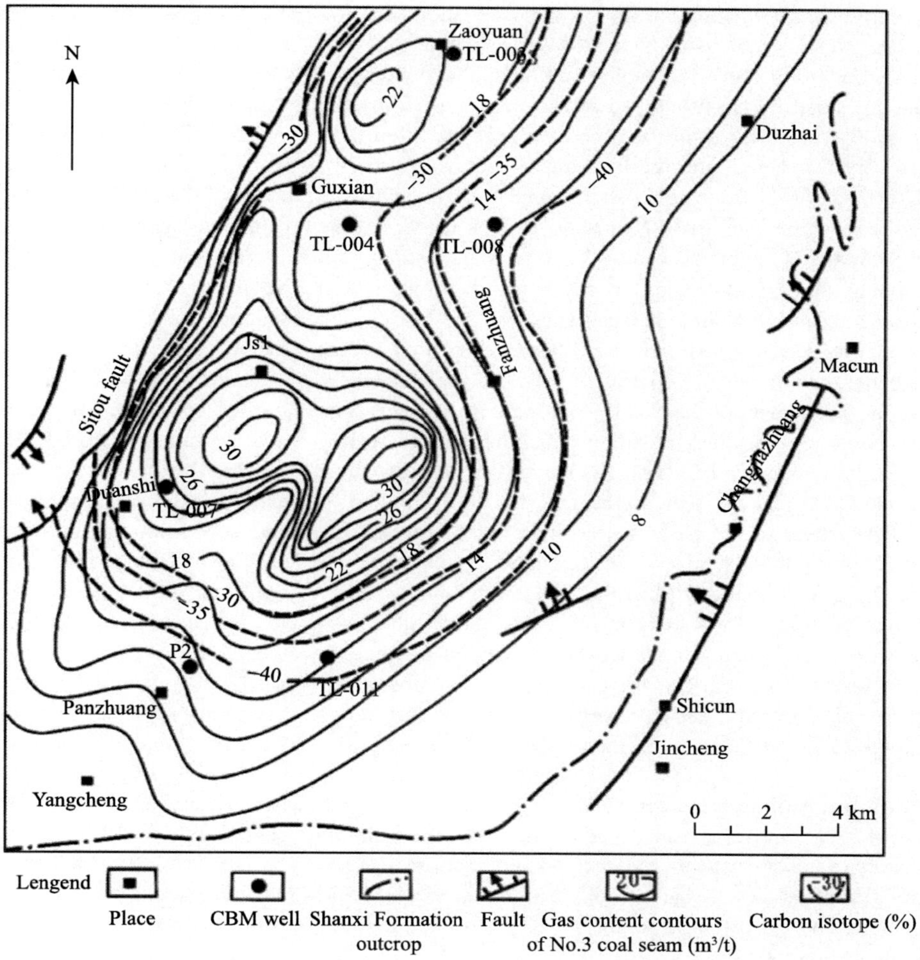

(5) Reservoir pressure

The pressure of the CBM reservoir is low in the southern part of the Qinshui Basin. In general, the pressure of the No. 3 coal seam is 0.08–3.36 MPa, that of the No. 15 coal seam is 2.24–6.09 MPa, and the pressure coefficient is less than 0.8, indicating under-pressure reservoirs. There is normal pressure in some areas, but abnormal high pressure is rare. The reservoir pressure has a tendency to increase with the increase of the burial depth of the coal seam. Normal pressure exists in the Fanzhuang-Panzhuang area in the middle of the basin. This depends mainly on groundwater dynamic conditions. The forming mechanism of reservoir pressure is the result of the pressure recovery event caused by groundwater recharge and migration along outcrops, similar to the dynamic abnormal pressure compartment (Ayers Jr and Kaiser 1994; Su and Zhang 2002).

(6) Sealing conditions

Sealing conditions are critical for the formation of a CBM reservoir. They involve the sealing abilities of the lateral, top and bottom of the coal seam. Tight and less permeable top and bottom can reduce the loss of CBM, maintain high reservoir pressure and maintain maximum adsorption capacity. The top of the No. 3 coal seam is mostly mudstone and silty mudstone, followed by siltstone and fine sandstone. The thickness of the immediate top of the No. 3 coal seam is more than 10 m, 24–55 m in the Fanzhuang-Panzhuang area, and 30 m in Well Jin Test 1. The thickness of the mudstone in the Shanxi Formation is 55.4 m, and the thickness of the regional caprock is 159 m. The bottom of the No. 3 coal seam is mainly silty mudstone. The fractures in the mudstone are not developed. The mudstone is continuously and stably distributed, and the sealing ability is strong, which are favorable for CBM preservation. The top of the No. 15 coal seam is shallow marine limestone (K_2 limestone) with stable regional distribution. The fractures in the limestone are not developed and the sealing performance is good. The fractures in the limestone near the Sitou fault are very developed, and the limestone is permeable to gas. The breakthrough pressure of the mudstone capping the No. 3 coal seam is 3–10 MPa, and that of the limestone capping the No. 15 coal seam is 2–15 MPa, indicating good capping ability. According to the type, distribution and structural and fracture characteristics of the No. 3 and No. 15 coal seams, it is considered that the No. 3 coal seam is better than the No. 15 coal seam, which is the main cause for the gas content of the No. 15 coal seam being lower than the No. 3 coal seam. The caprock distribution and sealing capacity around Well Jin Test 1 is the best in the study area, where CBM content is the highest.

The lateral of the CBM reservoir is mainly blocked by boundary fault and hydrodynamics. The western boundary is the Sitou fault, and the eastern and southern boundaries are hydrodynamics. The sealing mechanisms of the Sitou fault and the hydrodynamic boundary are described in Chap. 1. The northern boundary of the CBM reservoir is the watershed which is distributed east-west, to the coal seam outcrop on the east, and to the Sitou fault on the west (Fig. 6.18).

CBM Accumulation Process

CBM reservoir formation process involves the evolution history of the basin, the burial, thermal and maturity histories of CBM reservoirs, the genesis of CBM, the history of CBM content, and the dynamic conditions of groundwater and its influence on CBM reservoirs.

(1) The burial, thermal and maturity histories of CBM reservoirs in southern Qinshui Basin

The formation of CBM is controlled by its burial history and thermal history. The burial history and thermal history of the Carboniferous–Permian coal-bearing rock series in the Qinshui Basin have obvious stages (Fig. 6.9). The burial history can be divided into six stages:

The first stage is from the Late Carboniferous to the end of the Early Permian when the basement of the basin slowly settled. The average settling rate at that stage generally did not exceed 25 m/Ma. It was the slow subsidence that allowed the peat swamp to continue to develop, and form coal seam with stable thickness, continuity and wide distribution.

The second stage is the Late Permian to the end of the Late Triassic when the earth's crust rapidly settled. The buried depth of the coal seams in the Taiyuan Formation and the Shanxi Formation increased rapidly, and the maximum settlement range was above 4500 m. Along the Jincheng–Yangcheng–Houma line was the settlement center in the southern part of the basin, where and the average settling rate was 80–100 m/Ma.

The third stage is from the Early Jurassic when uplifting and denuding began. The Yanshanian movement caused the whole area to rise and be extensively denuded. The buried depth of the coal seam reduced, and the maximum elevation of the basement was uplifted more than 1000 m.

The fourth stage is from the Middle Jurassic when the basin slowly settled. The settlement is small and the average sedimentation rate is about 16 m/Ma, which is lower than the first stage. The Jurassic sedimentary range covers the real (modern) Qinshui Basin, that is, the Yanshanian movement formed the Carboniferous–Permian residual Qinshui Basin and the Jurassic Qinshui Basin.

The fifth stage is from the Late Jurassic to the Paleogene and the end of the Neogene when the basin continued to rise. The coal seam and its overburden were severely denuded.

The sixth stage is from the Neogene to the present when uplift and settlement are coexistent. The Himalayan movement formed a secondary fault depression in the basin, and locally the Neogene and Quaternary deposited, such as the Jinzhong graben in the northwestern part of the basin, which maximum settlement is more than 1500 m, but most areas were in a denuded state.

The thermal and maturity histories of coal seams in the southern Qinshui Basin can be divided into three stages (Fig. 6.9):

The first stage is from the Late Carboniferous to the Middle Jurassic when the paleotemperature was normal. The paleotemperature gradient was 2–3 °C/hm (Su et al. 2005). The coal seams of the Taiyuan Formation and the Shanxi Formation began to mature at the end of the Permian (the vitrinite reflectance reached 0.5%); at the end of the Triassic, the reflectance reached 1.2%, thus the first hydrocarbon generation peak started. The Early Jurassic uplifting caused the coalification to stop. The Middle Jurassic sedimentation did not reach the maximum depth in the burial history, and it was at the normal paleotemperature, so the thermal evolution of the coal did not deepen and the coal rank remained at the end of the Triassic.

The second stage is from the beginning of the Late Jurassic to the end of the Cretaceous when the Qinshui Basin was at anomalous paleotemperature due to the influence of the Yanshanian tectonic thermal event. The paleotemperature gradient was 4–6 °C/hm, or even higher. The evidence for the Yanshanian thermal event is the Yanshanian magmatic rock intrusion in the Pingyao area in the basin and the Linfen area in the southwestern part outside the basin, and the positive magnetic anomalies found in the southern and northern parts of the basin, indicating the existence of deep magmatic intrusions (Chen 1997). Although the basement of the basin was slowly rising at that stage due to the thermal events, the temperature of the Carboniferous–Permian coal seam was far above the temperature at the maximum buried depth at the end of the Triassic, so the second coalification began and reached the secondary peak of hydrocarbon generation. At the same time, the coalification caused by that thermal event finally determined the temporal and spatial pattern of the current coal rank, and then the coal rank no longer changed. High heat flow with magma intrusion events, or thermal convection with groundwater migration, caused unusually high thermal maturation at the north and south sides of the basin.

The third stage covers from the Paleogene and the Neogene when the Qinshui Basin returned to normal paleotemperature, with a geothermal gradient of 2–3 °C/hm. Most of the areas were in the stage of uplift and denudation. Although some grabens formed during the Himalayan settled greatly, they failed to make the coal seam exceed the historically highest temperature and the coalification was stagnant. Therefore, this stage is crucial for the preservation of CBM.

(2) CBM genesis

The genesis of CBM in the southern part of the Qinshui Basin is controlled by thermal history and burial history. The generation of CBM has experienced two stages. The first stage occurred at the end of the Triassic during which the coal seam was buried deepest, so that the regional metamorphism at normal paleotemperature caused the first generation of CBM, and the vitrinite reflectance was about 1.2%. Until the early Yanshanian, coalification was stagnant due to uplifting, and the generation was interrupted. The second stage occurred in the late Yanshanian during which the thermal events caused the temperature to far exceed the temperature at the maximum buried depth at the end of the Triassic, causing coalification to intensify and second generation of CBM. The second generation of CBM played a leading role. Liu (1998) pointed out that the CBM generated at the first stage accounted for 32%, and the CBM generated at the second stage accounted for 68% in southwestern Shanxi.

The test results of 10 CBM wells in the southern Qinshui Basin indicate that the methane carbon isotope of CBM is generally light, and the value of $\delta^{13}C_1$ is −36.7 to −26.6‰, and that in some gas samples exceeds −40‰. It becomes heavier with the increase of buried depth. For this phenomenon, the following three explanations can be given:

① According to the template established by Whiticar et al. (1986) (Fig. 6.21), the phenomenon is caused by isotope fractionation in the desorption-diffusion-migration process of coalbed methane. Since $\delta^{12}C$ is more preferentially desorbed than $\delta^{13}C$ (Zhang and Tao 2000), ^{12}C methane in the desorbed gas is relatively enriched. In the subsequent diffusion and migration process, the desorbed gas diffused to shallow layers and further increased the degree of isotope fractionation, resulting in the CBM in shallow layers becoming lighter. Therefore, in the groundwater runoff zone where desorption-diffusion-migration fractionation is strong, methane $\delta^{13}C$ becomes lighter. In the stagnant zone, the effect of desorption-diffusion-migration fractionation is less, so the original state remains.

② The phenomenon is caused by selective dissolution, migration and enrichment of groundwater. Since the polarity of $\delta^{12}C$ methane is weaker than that of $\delta^{13}C$ methane, $\delta^{13}C_{CH4}$ is preferentially dissolves in water, which causes heavier $\delta^{13}C$ to be dissolved in the recharge zone, and the $\delta^{13}C$ in CBM to become lighter. The heavier $\delta^{13}C$ dissolved is carried by groundwater to the retention area, and causes $\delta^{13}C$ to become heavier. According to the above two explanations, the southern part of the Qinshui Basin has secondary thermal CBM reservoirs, which is very common at home and abroad (Kaiser et al. 1991).

③ There may be secondary biogenetic gas in the shallow area. The main factors controlling the distribution of methane $\delta^{13}C_1$ are coal rank and secondary biogas. The methane $\delta^{13}C$ of thermal CBM is heavier than conventional natural gas and is mainly controlled by coal rank. Secondary

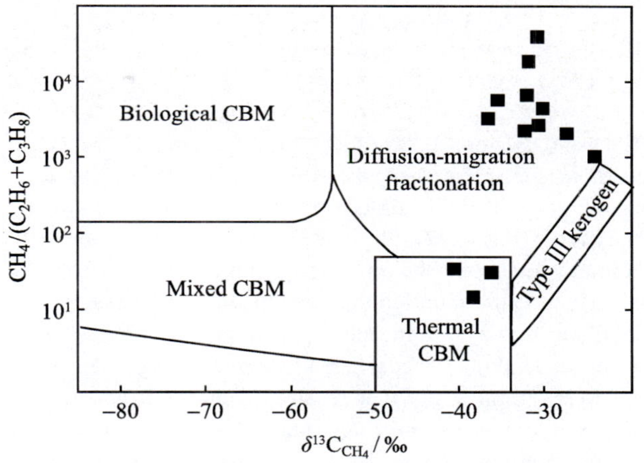

Fig. 6.21 Identification template of the genesis of CBM (template from Whiticar et al. 1986)

biogas is lighter than thermal CBM (Kaiser et al. 1991; Pashin 1998; Pashin et al. 2002; Pitman et al. 2003). In the southern and eastern parts of the basin, due to the recharge of atmospheric precipitation, bacteria that chemically degrade the moisture in the coal and the organic components in the coal seam were brought to the belt, producing secondary biogenetic gas, which is the main reason why the methane isotope of the coal seam on the edge of the basin gets lighter. The isotope analysis of the CBM in the Qinshui Basin shows that the CO_2 isotope in Well Fanzhuang 2 and Well Panzhuang 3 reaches +28.4‰, +23.2‰, respectively. The $\delta^{13}C$ of thermogenic CO_2 is from −25 to −15‰; the biogenetic CO_2 has rich $\delta^{13}C$ ranging from −20 to +30‰; and the magma-derived CO_2 has $\delta^{13}C$ greater than −8‰, mainly −8 to +3‰. New research shows that as long as the groundwater mineralization is appropriate (<2000 mg/L), even anthracite may be biogas production (Pashin et al. 2002). This also indicates that secondary biological CBM may exist in the shallow part of the southern Qinshui Basin.

(3) The evolution history of CBM content

The change of gas content in the Qinshui Basin during the geological history was calculated by the adsorption potential theory (Fig. 6.9). According to the history of gas evolution, it is known that during uplifting or after uplifting, the migration, loss and reaccumulation of CBM determine the spatial distribution pattern of current gas content. The adjustment, modification and favorable sealing conditions of CBM reservoirs make such CBM reservoirs the most favorable.

(4) Groundwater dynamic conditions and their effects on CBM reservoirs

The groundwater salinity of No. 15 coal seam and No. 3 coal seam in the southern Qinshui Basin is similar, the total salinity is 800–2400 mg/L, and the water type is mainly $NaHCO_3$ (Fig. 6.22). In the south of the watershed, groundwater from the northern boundary runs from the watershed to the south; in the outcrop area on the eastern and southern margins, atmospheric precipitation and surface water move along the coal seam to the deep stagnant area; in the west, groundwater that receives recharge from the outcrop is blocked by the Sitou fault and cann't enter. That forms a trend of convergence to the deep layer. Furthermore blocked in the vertical direction by the upper and lower less permeable rocks, water converges to a low-lying zone in the Panzhuang and Fanzhuang areas where the groundwater runoff conditions are the weakest (Fig. 6.23). The groundwater is replenished along the coal seam and aquifer, and migrates to the deep layer. The runoff intensity becomes weaker, and a runoff zone and a stagnation zone were formed. The shallow recharge zone is a CBM escape zone with low gas content; the deep stagnant zone has slow groundwater runoff and is a favorable zone for CBM accumulation.

Forming Characteristics of CBM Reservoir

① The favorable coal-forming environment created thick, continuous and stable coal seams, which provide the material basis for the accumulation of CBM. The No. 15 coal seam of the Taiyuan Formation and the No. 3 coal seam of the Shanxi Formation have large thickness, good continuity, shallow burial depth, medium permeability and good adsorption capacity. They are good reservoirs.

② The top and bottom plates are good at sealing, and laterally there are sealing faults as boundaries, such as the Sitou fault, groundwater watershed and hydrodynamic sealing along the outcrop.

③ The anthracite was formed by two metamorphisms in the Indosinian and Yanshanian. The vitrinite reflectance ranges from 2.2 to 4.0%. The two hydrocarbon generation processes generated a large amount of CBM, and the late hydrocarbon generation played a major role. The occurence of the CBM has undergone serious adjustment and transformation during the continuous uplifting in the Himalayan period, and finally formed the present CBM reservoirs in the southern Qinshui Basin. According to the evolution history of gas content, this adjustment and transformation is conducive to the formation of CBM reservoirs.

④ The CBM content is high, mainly methane. The CBM is mainly of thermal origin, and may be secondary biogenetic gas. According to the composition and isotopes, the CBM zones in the southern part of the Qinshui Basin are divided into two zones from top to bottom in the vertical direction: the thermogenic gas, and the secondary biogenetic-thermogenic CBM zone.

⑤ During the process of replenishment, migration and convergence, the recharge zone, runoff zone and stagnant zone were formed. The shallow

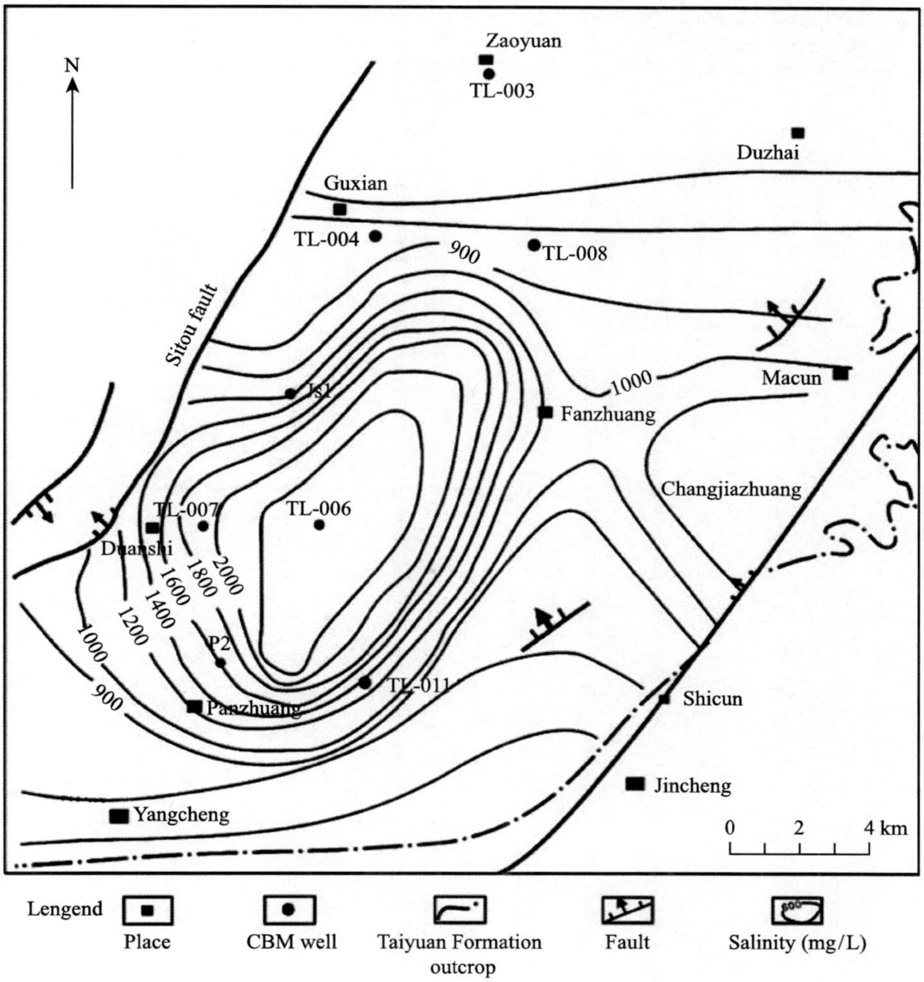

Fig. 6.22 Groundwater salinity contours of the Taiyuan Formation

recharge zone is a CBM escape zone with low gas content; the middle runoff zone has medium CBM content corresponding to the groundwater runoff zone; the deep stagnant zone is a favorable CBM accumulation zone with high gas content. The reservoir pressure is primarily dependent on groundwater dynamic conditions, which is the result of the pressure recovery event caused by groundwater recharge, migration, and stagnation along the outcrop, and similar to the dynamic abnormal pressure compartment. Therefore, the deep groundwater stagnant zone, especially the stagnant zone at abnormally high pressure or normal pressure, is a favorable zone for CBM accumulation. The current high and stable-production CBM wells are mostly located in the groundwater stagnant zone.

6.2.2.2 Process and Mechanism of CBM Accumulation in the Liulin Area, the Eastern Margin of Ordos Basin

The Liulin area is located in the eastern part of the Ordos Basin and the central part of the Hedong region. The regional structure belongs to the Lishi-Liulin nose-like structure bounded by the Lishi fault on the east. It is shallow on the east and deep on the west, and generally is a monocline dipping westward and striking north-south. The structure is simple, and the distribution direction of the structural line is basically consistent with the stratum. On the whole it strikes nearly north-south, inclined toward west at 3°–10°. The fault is rarely developed, and only three fracture zones are developed in the axis and southwest of the nose-like structure, each of which is a graben composed of two normal faults.

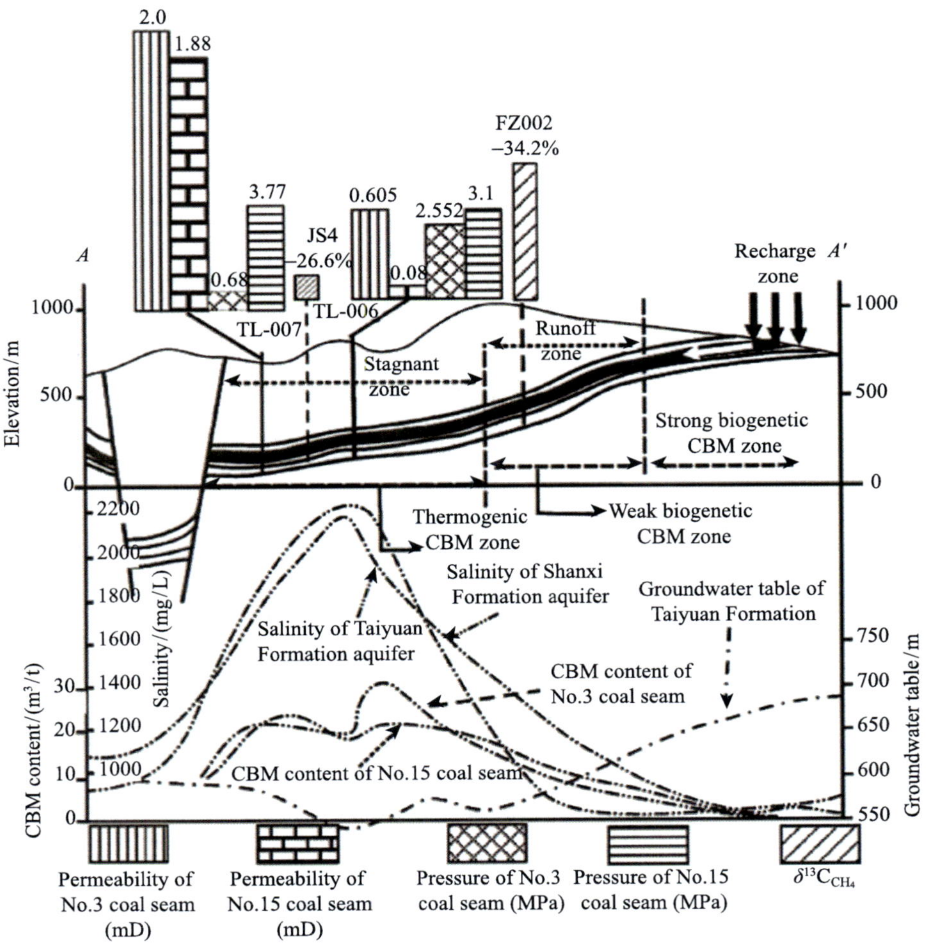

Fig. 6.23 CBM reservoir profile and CBM reservoir parameters in the southern Qinshui Basin

Division of CBM Reservoirs

The eastern boundary of the CBM reservoir in the eastern margin of the Ordos Basin is the weathered and oxidized belt, the western boundary is the groundwater dynamic boundary, and the southwestern boundary is the open fault (Figs. 6.24 and 6.25).

CBM Accumulation Conditions

(1) Spatial distribution

The primary coal-bearing strata in the Liulin area are the Taiyuan and Shanxi Formations of the Carboniferous–Permian (C–P). The upper Carboniferous Taiyuan Formation (C_2t) is 46–120 m thick and has a total of 7 coal seams. Most of the coal seams are stably distributed in the whole area. The No. 8 and No. 9 coal seams are primary ones. The lower Permian Shanxi Formation (P_1Sh) contains 4–6 coal seams. The recoverable coal seams are located in the lower part. The coal seams with stable distribution are No. 3, No. 4 and No. 5. The overlying layers are the Permian and Triassic Jurassic strata. The No. 8 coal seam in the middle of the Taiyuan Formation is stable and 3–10 m thick.

(2) CBM composition and content

The CBM in the Liulin test area has simple composition, mainly methane above 95%, and a small amount of N_2 (<5%). The No. 4 coal seam has a high gas content of 12–15 m^3/t; the No. 5 coal seam is lower, generally less than 10 m^3/t; the No. 8 coal seam has variable gas content, generally 5–20 m^3/t (Fig. 6.24).

(3) Reservoir physical properties

Due to the overall uplift or settlement during the geological history, there was no violent tectonic movement, so the original coal structure is relatively intact, that is, the cleat and fracture system are well developed. Exogenous fractures and

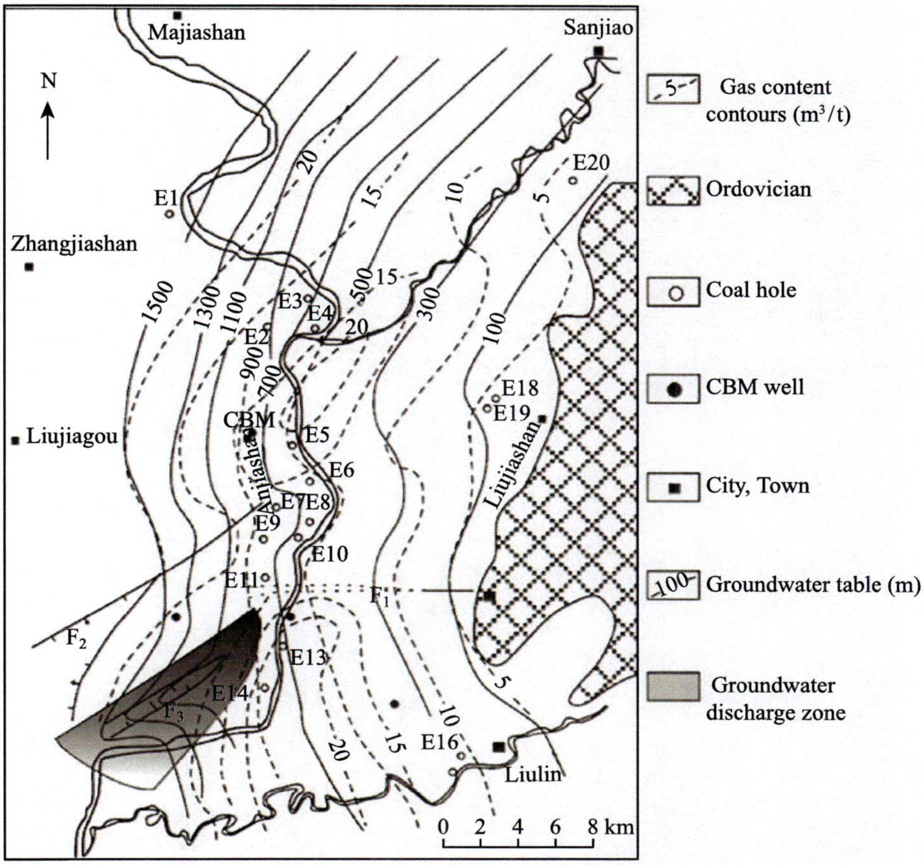

Fig. 6.24 The contours of the gas content of No. 8 coal seam and the watertable of Taiyuan Formation in Liulin area

cleats are mostly unfilled or semi-filled, or a little filled with calcite, indicating good reservoir permeability. Well test results indicate that the permeability of the No. 8 coal reservoir is 0.29–8.86 mD (Table 6.3). This high-permeability reservoir is rare in CBM reservoirs in China.

(4) Adsorption characteristics

The No. 8 coal seam has moderate metamorphism, the vitrinite reflectance is 1.1–1.8%, and the ash is generally around 12%. The No. 4 coal seam is at saturated or nearly saturated adsorption. The No. 5 coal seam and the No. 8 coal seam are at under-saturated adsorption, and the corresponding critical desorption pressure is 1.5 MPa and 2 MPa respectively. When the dry coal was restored to the Langmuir volume under the formation conditions, the No. 4 coal seam is average 19.1 m^3/t, the No. 5 is 19.77 m^3/t, and the No. 8 is 23.69 m^3/t.

(5) Reservoir pressure

According to the drilling data, CBM well data and shallow coal mining data collected in the exploration stage of the Liulin area, the static groundwater level of the top limestone aquifer with the close hydraulic connection to the Taiyuan Formation No. 8 coal seam was converted into hydrostatic pressure, and then the pressure of No. 8 coal seam was determined about 1–13 MPa, and increases westward with the increase of the coal seam depth. According to the buried depth and the distribution of the pressure gradient (Fig. 6.25), there are three kinds of pressure states in the No. 8 coal seam: underpressure (<9.30 kPa/m), normal pressure (9.30–10.30 kPa/m) and high pressure (10.30–14.70 kPa/m).

(6) Sealing conditions

Most of the roof of the No. 8 coal seam in the study area is Miaogou limestone (K_2 limestone), and locally mudstone. The limestone is relatively thick, generally several meters to ten meters, and stable. However, due to the uneven development of karst and fractures, it is unfavorable for he preservation of CBM. The overlying strata on the limestone are composed of relatively tight mudstone, sandy mudstone and carbonaceous mudstone at the bottom of the Taiyuan Formation and the Shanxi Formation. Their permeability is

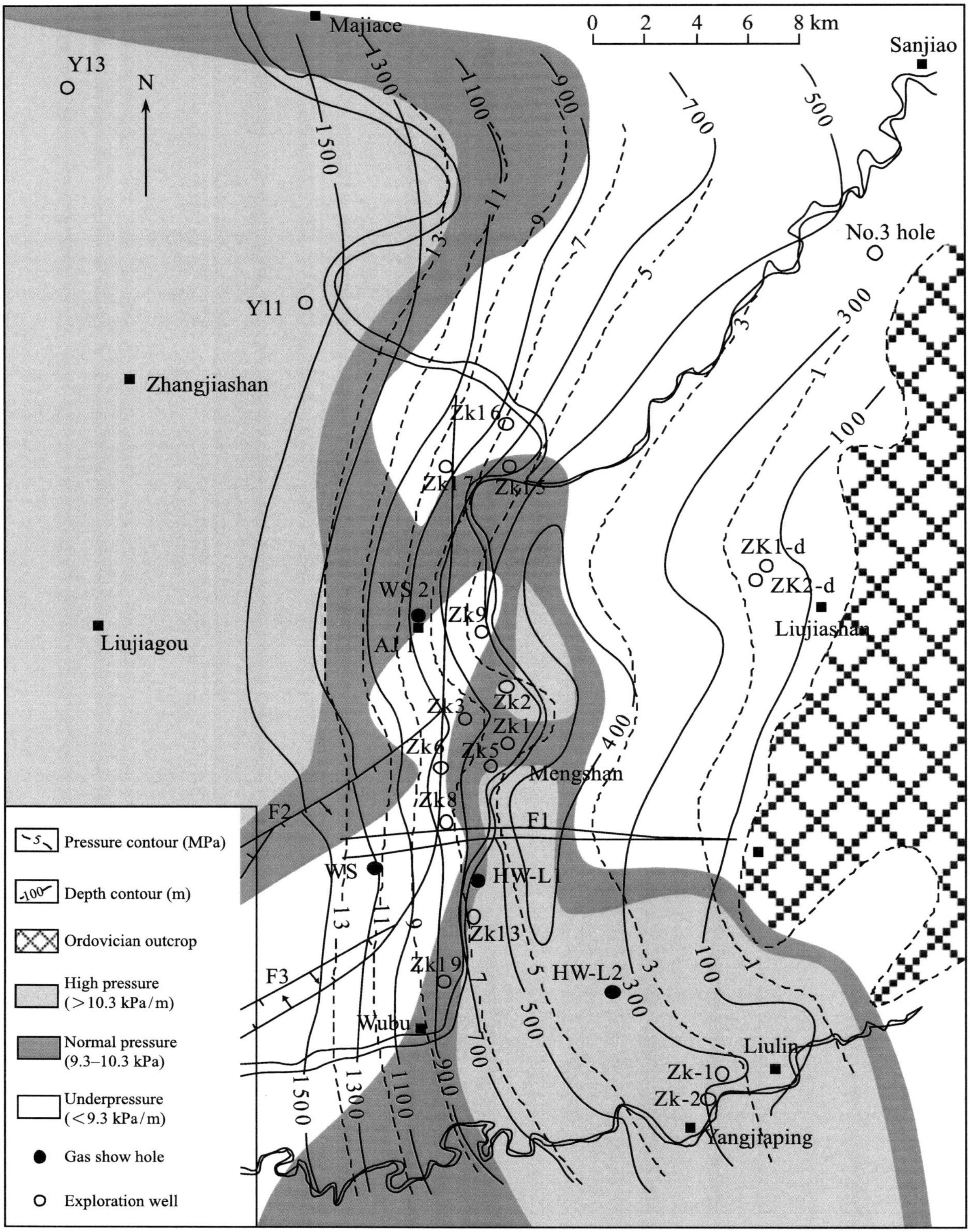

Fig. 6.25 Pressure and depth of No. 8 coal seam in Taiyuan Formation, Liulin area

Table 6.3 Basic data of No. 8 coal seam of Taiyuan Formation in Liulin-Wubu area

Well	L1	L2	L3	L4	L5	L6
Depth/m	408.1–412.2	408.1–411.7	399.0–403.1	394.1–398.4 398.6–399.3	402.5–407.0	404.2–408.4
Thickness/m	4.1	3.6	4.1	4.0 0.7	4.5	3.8
Permeability/mD	–	8.86	0.29	0.93	1.135	–
Pressure/MPa	–	4.45	3.297	4.03	3.917	–

poor, lateral continuity is good, and distribution is stable. They can effectively prevent the CBM in the No. 8 coal seam from escaping vertically, and act well as the regional caprock of the No. 8 coal seam. The floor of the No. 8 coal seam is dominated by mudstone.

Laterally, there are open faults as the southwest boundary of the coal seams, which leads to a decrease in the CBM content, is not conducive to the preservation of CBM.

Evolution History of CBM Reservoirs

(1) Evolution history of Ordos Basin

During the Indosinian period, the burial depth of the Late Paleozoic coal series continued to increase, the tectonic development was controlled by the differential settlement in the north-south direction, and the Triassic sedimentary and subsidence center was biased to the southern part of the basin. From the end of the Indosinian to the beginning of the Yanshanian, the overall uplift led to the erosion of the coal caprock. In the Yanshanian, the main part of the Ordos Basin accepted sediments and formed thick Jurassic–Cretaceous strata. Under the influence of the marginal uplift, the sedimentary process of the Yanshanian has limited impact on the eastern margin of the basin, failing to cause the depth of the coal series to significantly change. At the end of the Yanshanian, the development history of the basin was terminated by structural uplift. Near the marginal uplift, the strata including the coal series strongly were denuded, and the coal seams became shallow or exposed. At the late Himalayan, the coal system was lifted to the surface, and the Cenozoic strata was not enough to compensate the original caprock eroded. The CBM reservoir in the eastern margin of the Ordos Basin began to form when the coal system was buried deepest, and then by late uplifting and influenced by groundwater, developed to the present pattern.

(2) Burial, thermal and maturity histories of Taiyuan Formation

The restoration of the burial history shows that the No. 8 coal seam experienced four evolutionary stages during the geological history (Fig. 6.26):

The first stage is from the Carboniferous to the end of the Late Triassic when rapid subsidence happened. During the

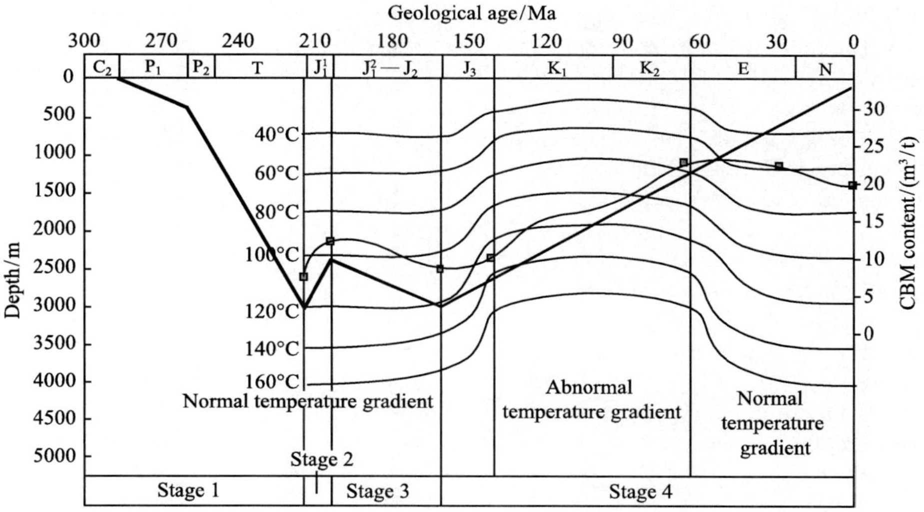

Fig. 6.26 The burial, thermal and gas content histories of the Taiyuan Formation in Liulin area

Carboniferous–Permian period, large-scale coalification took place and coal seams were buried rapidly, and entered the mature stage, so a large amount of hydrocarbon generated. Deep metamorphism caused Taiyuan Formation coal mature (R^o = 0.57–1.10%).

The second stage is from the early of Early Jurassic when gradual uplifting began. Affected by the Yanshanian movement, the strata were uplifted and eroded, the coal seams became shallow, and the maturity was terminated. Regionally, the east was uplifted strongly and the west relatively weakly.

The third stage is from the late of Early Jurassic to the Late Jurassic when slow settling was going. Influenced by the Yanshanian movement, the earth's crust fell again and sedimentation began again. The No. 8 coal seam became deeper, but the sedimentation rate was significantly lower than the first stage. The coal seam did not reach the maximum depth at the end of the Triassic, so the hydrocarbon generation was limited or no hydrocarbon generation.

The fourth stage started from the Late Jurassic. Sedimentation has basically stopped, and the earth's crust has been slowly rising, causing the overlying strata including coal seams to be denuded to different degrees.

In these four stages, the evolution of the paleo-temperature field is not uniform. During the Late Carboniferous to the Jurassic, the paleo-temperature gradient was 2.2–3.0 °C/hm. Then at the end of the Mesozoic (Cretaceous), due to strong tectonic movement and magma activities, the paleotemperature field became abnormal, reaching 3.6–6.2 °C/hm, mainly concentrated in the range of 4.0–4.5 °C/hm. Since the Neogene, the basin has been continuously uplifted, the crust became thick, and the temperature gradient reduced to 2.2–3.2 °C/hm.

(3) Formation of CBM

The No. 8 coal seam has experienced two hydrocarbon generation processes: the first from the burial of No. 8 coal seam to the end of the Triassic, which is the deep metamorphism process at normal geothermal gradient; the second is the magmatic thermal metamorphism process at the anomalous geothermal gradient caused by the thermal event during the Cretaceous period. Without isotope data, the CBM in this area was estimated to be thermal genesis and secondary biogas genesis according to the data of the San Juan Basin and the Piceance Basin.

(4) The history of CBM content

The history of the CBM gas content was restored using the same method as the southern part of the Qinshui Basin (Fig. 6.26):

① The uplifting from the Late Triassic to the Early Jurassic (T_3–J_1) caused an increase in the adsorption capacity to CBM, while the sedimentation of J_1–J_2 caused a decrease in the adsorption capacity.
② The thermal event from the Late Jurassic to the Early Cretaceous (J_3–K_1) increased coal maturity. After secondary hydrocarbon generation, coal reservoirs began to uplift continuously. The increase in maturity and uplift increased the adsorption capacity of coal.
③ The uplift after secondary hydrocarbon generation did not cause the theoretical gas content to decrease, and the CBM adsorption capacity increased, which determines the present high CBM saturation.
④ During uplifting or after uplifting, the migration, loss and re-accumulation of CBM determine the spatial distribution of present CBM gas content. The adjustment, modification and favorable sealing conditions make such CBM reservoirs the most favorable.
⑤ The key moment for CBM accumulation is at the end of the Triassic.

(5) Impact of groundwater dynamic conditions

The Paleozoic strata in the eastern margin of the Ordos Basin are generally exposed in the Lüliang Mountains to the east of the Yellow River. Due to the recharge of atmospheric precipitation, surface water flows along the strata to the deep of the basin, thus forming a condition of favorable confined water potential. The calculated Permian equivalent static watertable reduces obviously from east to west. The salinity of the formation water in the coal-bearing strata in the Liulin area is very regular. The total salinity and Cl^- content are both high in the west and low in the east. The total salinity is between 2800 and 5000 mg/L, with $NaHCO_3$ type, and locally $CaCl_2$ or $MgCl_2$ in local areas. The Cl^- content is between 700 and 2500 mg/L. It can be seen that the confined water in the coal seam in the Liulin area is high in the shallow part and low in the deep part. In the eastern outcrop recharge area, the atmospheric precipitation passes through the permeable coal-bearing strata toward west—the deepest coal seam, and the water salinity gradually increases from shallow to deep. As the depth increases, the water flow rate gradually slows down and forms a stagnation zone appears in the west. Regional abnormal pressure is present (Figs. 6.25 and 6.27).

(6) Characteristics of CBM accumulation

① The No. 8 coal seam of the Taiyuan Formation is the main recoverable coal seam. The coal seam is thick, continuous, shallow, permeable and strongly adsorbing. It is a good reservoir.

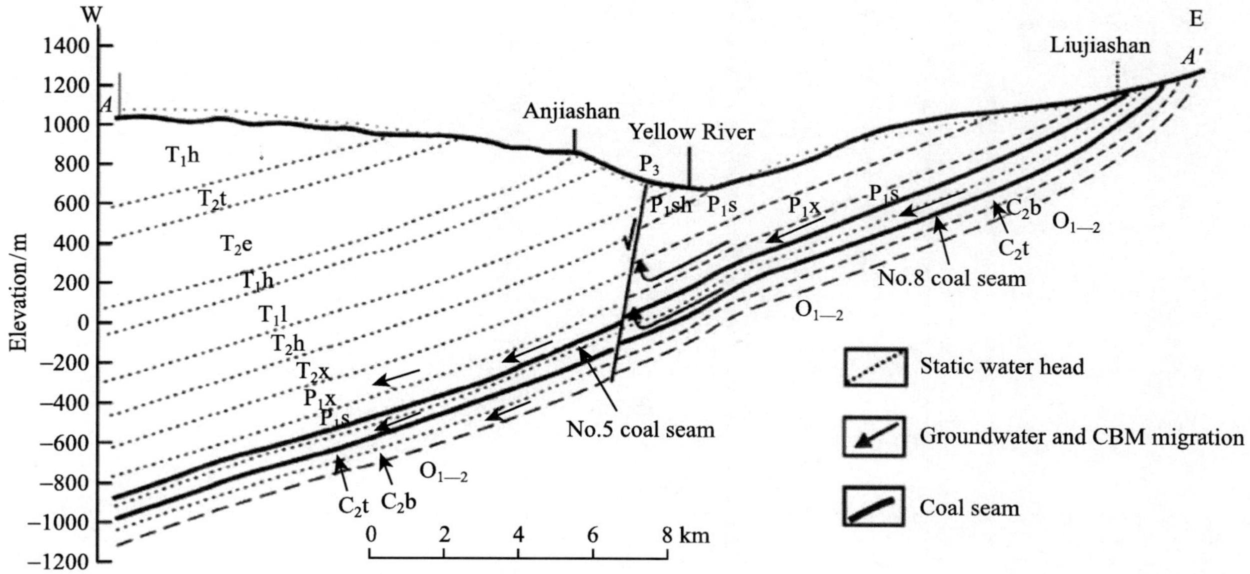

Fig. 6.27 Hydrogeological section in Liulin area

② The direct roof of the coal seam is unfavorable for the preservation of CBM, but the mudstone and sandy mudstone on the roof are continuous and tight, which are conducive to the preservation of CBM. In the lateral direction, there is an open fault boundary in the southwest, which leads to a decrease in the CBM content and is not conducive to the formation of gas reservoirs.

③ The CBM content is high, mainly methane. CBM is mainly of thermal origin. Secondary biogenetic gas is an important supplement and is continuous.

④ The deep metamorphism after reaching the maximum buried depth made the coal seam experience the first hydrocarbon generation, and the tectonic thermal event in the Indosinian period caused the second hydrocarbon generation, but the hydrocarbon generated is less and the impact is limited. The hydrocarbon generated and stored during the Yanshanian thermal event is lost, but the loss amount is not large. Today's CBM saturation is 80–100%. During and after uplifting, the migration, loss and reaccumulation of CBM determine the spatial distribution pattern of present gas content. The adjustment, modification and favorable sealing conditions make such CBM reservoirs the most favorable.

⑤ Due to the influence of groundwater dynamic conditions, formed an overpressure zones, normal pressure zones and underpressure zones were formed. Studies have shown that overpressure and normal pressure zones are not only conducive to the preservation of CBM but also favorable for CBM development.

6.2.2.3 Process and Mechanism of CBM Accumulation in the Wangying-Liujia Low-Ranked Coal in Fuxin Basin

The Fuxin Basin is a NNE-NE narrow strip basin located in the central and western part of Liaoning Province. It is under the jurisdictions of Fuxin City and Jinzhou City. The basin is surrounded by mountains on three sides and faces a river on the fourth side. It reaches the Erlang Mountain on the north and has an elevation of 300–600 m. It is adjacent to the Yiwulu Mountain on the east, with an elevation of 500–850 m. The west is to the Xiaosongling Mountains, with an elevation of 500–800 m. The south is bounded by the Daling River. Due to the serious disturbance of mining activities, there are only five blocks in the Fuxin Basin. The Wangying-Liujia CBM reservoir is a sub-sag with the least influence of mining activities and the most complete preservation in the basin, which is located in the northwestern part of Haizhou depression, bounded by the Pingxi F2 fault on the west, and the northwest and southeast are coal seam outcrops or anticline axis (Fig. 6.28). Faults are developed in the CBM reservoir. There are four major faults: Pingxi F2 fault, Ping'an F2 fault, Wangying F8 fault and Liujia F2 fault. At present, the commercial development of CBM has been realized in the Wangying-Liujia CBM reservoir.

Fig. 6.28 Sunjiawan bottom contours of Wangying-Liujia CBM reservoir

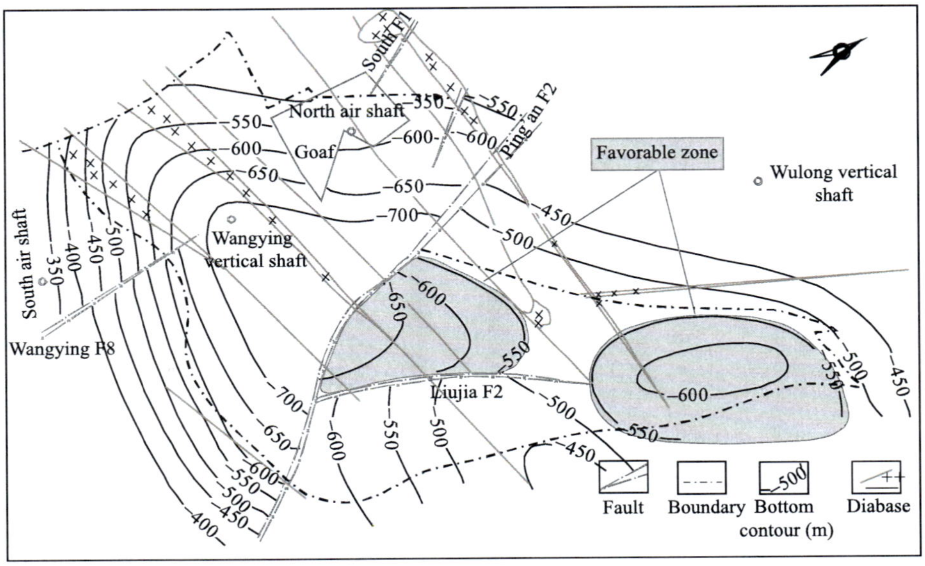

Division of CBM Reservoirs

The Wangying-Liujia CBM reservoirs are in a sub-sag in the Fuxin Basin, which is divided into two by the Ping'an F2 fault. Sealing by the Ping'an F2 fault, two reservoirs are distinguished: Wangying from Liujia CBM reservoirs. On the west, the Wangying CBM reservoir is bounded by the sealing Pingxi F2 fault, and on the north and the south are coal seam outcrops. The Liujia CBM reservoir shares the Ping'an F2 fault with the Wangying CBM reservoir (Fig. 6.28) on one side and the other sides are exposed. That is to say, in addition to the fault boundary, the other are weathered and oxidized belts or hydrodynamic boundary. The hydrodynamic boundary can be calculated from the Langmuir equation and the economic gas content. If the economical gas content is 5 m^3/t, the ultimate pressure for the Sunben group (the shallowest coal seam group) is 3.1 MPa, the corresponding hydrostatic height difference is about 310 m; the middle layer is at 1.44 MPa, corresponding to the hydrostatic height difference is about 244 m; the Taiping layer is at 3.32 MPa, and the difference is about 332 m. In general, to seal 5 m^3/t gas content, the water column elevation should be about 300 m. Since the depth of the weathered and oxidized belt is only 160 m, it is the hydrodynamic boundary that seal the gas. The Ping'an F2 fault and the Pingxi F2 fault are compression-torque faults, and the mylonite coal within the faults are strongly cemented, so they are good at sealing.

CBM Accumulation Conditions

(1) Spatial distribution

The Fuxin Formation developed in the middle of the Early Cretaceous is the primary coal-bearing strata (Fig. 6.29). From the bottom up, the Fuxin Formation can be divided into three coal seam groups: Taiping group, middle group and Sunben group. The coal seam groups tend to gradually thicken from the edge to the middle, and are characterized by multiple and thick coal seams. The maximum cumulative thickness of recoverable coal seams in the Liujia Coalfield is 96.07 m, and the average is 42.69 m. The maximum cumulative thickness of recoverable coal seams in the Wangying Coalfield is 35.21 m, and the average is 26.14 m.

(2) CBM composition and gas content

According to the composition of CBM in the Wangying-Liujia coal seam, there are three zones from the outcrop down: (i) an weathered and oxidized belt with methane content of lower than 80%, and higher N_2 and CO_2 contents; (ii) a transition zone with methane content of more than 80%, and less N_2 and CO_2 from air; (iii) an original methane zone with methane content of more than 98%, and no air. The CBM saturation is low. Most areas are under-saturated. The gas saturation of the Liujia CBM

Fig. 6.29 Regional strata in Fuxin Basin

Series	Formation	Member	Thickness / m	Lithology	Facies
Quaternary			0–20	Sandstone, gravel, clay, sub-clay, etc.	Alluvial, slope
Lower Cretaceous	K_1s		500–100	Red and variegated conglomerate, glutenite, sandstone, mudstone; lower green, graygreen sandstone	Piedmont
	K_1f	Upper	800–1200	Gray, grayyellow sandstone, siltstone, conglomerate, black mudstone, thick coal seams, rich plant fossils	River, lacustrine, delta
		Middle			
		Lower			
	K_1sh	4	800–1000	Black-gray mudstone, siltstone with thin glutenite interlayers	Lacustrine
		3		Dark grey sandstone, siltstone, mudstone, coal seam, carbonaceous mudstone, local conglomerate	Lacustrine
		2		Yellowish brown sandstone and coarse siltstone with thin siltstone interlayer, and thin coal seam	Lacustrine
		1		Aubergine and variegated conglomerate, sandstone and mudstone	Lacustrine
Upper Jurassic	J_3jf		1400–2000	Gray-yellow and white glutenite, sandstone, mudstone and thin layers, and local recoverable coal seams	Shallow lacustrine
	J_3y		2400	Variegated volcanic rock series with several layers of volcanic ash sandstone and mudstone, "Jehol Biota" fossils	—
Sinian				Dark gray siliceous rock, mudstone, siliceous limestone, quartzite	—
Pre-Sinian				Variegated granitic gneiss and schist, phyllite	—

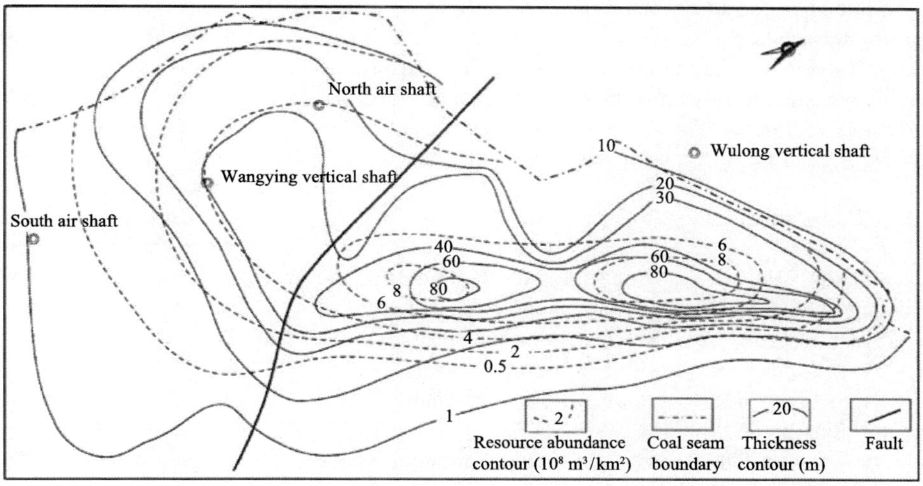

Fig. 6.30 Contours of the thickness and resource abundance of Wangying-Liujia coal seam

reservoir is 85–96%, generally 8–10 m^3/t. However, the coal seam is extremely thick, so the resource abundance is huge, up to 8 × 10^8 m^3/km^2 (Fig. 6.30).

(3) Reservoir physical properties

The porosity and permeability are mainly from four CBM wells in the Liujia exploration area.

The porosity of the Liujia coal seam has the following characteristics: (i) The porosity is generally low and contributes little to the permeability. The statistics of the true density and apparent density of 96 coal samples shows that the porosity is 4.7%, lower than the porosity of long-flame coal, 5.9–4.3%. According to the data of 31 samples, the porosity of natural coke is 7.4%. (ii) Close to the rock wall and metamorphic natural coke, the porosity of the coal seam increases, which is the contribution of the special pores in the natural coke. (iii) The coal seam porosity decreases with the increase of depth, and decreases with the increase of coal seam metamorphism.

The permeability of the Sunben coal seam group in Well LJ-1 is 0.428 mD, the tested section is 730–757 m; the permeability of the middle coal seam group is 0.469 mD, and the tested section is 818.82–836.80 m; The permeability of the Taiping coal seam group is 0.323 mD, and the tested section is 841.61–901.20 m.

The direct influencing factors on the permeability are cleat and fracture development characteristics. The cleats in the coal seams in the Fuxin Basin are rectangular networks and parallel, and the face cleats are larger in scale. The length of the face cleat is estimated to be tens of meters. End cleats are developed among face cleats, and perpendicular to the face cleats, but the density is low. The cleats in Well LJ-1 and the Wangying coal seam are relatively developed.

According to the statistics of Well LJ-1, the face cleats in the Sunjiawan coal seam is 8–10 pieces/5 cm, and 9–12 pieces/5 cm in the middle and Taiping coal seams; the corresponding end cleats are 2–3 pieces/5 cm, 2–4 pieces/5 cm. In addition, there is a cleat development zone near the rock wall with cleat density of 30 pieces/5 cm or more. There is no obvious change in the plane distribution of coal seam cleats. Vertically, the cleat density increases with depth because as the depth increases, the degree of coal metamorphism increases, resulting in the increase of vitrain and clarain, and endogenous fissures.

The fractures in the rock wall are developed with the density of 6 pieces/m, but the connectivity is poor, and they are closed, resulting in low permeability.

(4) Adsorption characteristics

According to the equilibrium water pressure isothermal adsorption test, the adsorption of the coal in the study area is strong, the isothermal adsorption line almost increases linearly (Table 6.4), and the dry ash-free saturated adsorption capacity is significantly larger than that under air drying, and increases with depth. The Langmuir volume is generally 20–35 cm^3/g, and the Langmuir pressure is 5–20 MPa.

(5) Sealing conditions

The Fuxin Basin is a continental fault basin where the lithology varies greatly in the lateral direction, and regional mudstone caprock is lost. The mudstone and siltstone in the middle and upper parts of the Fuxin Formation are 5–70 m thick and have good continuity, which is favorable for the preservation of CBM.

The upper sealing layer of the Fuxin Formation is the mudstone and siltstone in the upper Member of the Shahai Formation, formed during the lacustrine expansion period. They are compact and uniform, thick (250–450 m) and stable in the whole area, and can act as the regional caprock. The mudstone is normally compact, so its sealing mechanism is dominated by capillary. The sealing ability of the caprock varies in plane due to a variety of factors. The displacement pressure in the upper Member of the Shahai Formation is high in the east and low in the west, and the higher displacement pressure is distributed north-south along the eastern side of the basin. Besides the F8 fault sealing laterally, the other faults are open. The mylonite coal on both sides of the F8 fault plays a sealing effect. The faults in the area are generally not the channels through which CBM escapes.

Evolution History of CBM Reservoirs

(1) History of basin evolution

The evolution history of the Fuxin Basin can be roughly divided into four stages: Late Jurassic, Early Cretaceous, Late Cretaceous and Cenozoic.

① **Late Jurassic**. The Pacific plate subduct towards the Eurasian continent. On the one hand, under the strong NWW-SEE compressive stress, the Daliuhegou-Lüshan geanticline rose more strongly and created a large number of NNE longitudinal faults around its core; on the other hand, some pre-existing basement fault networks, such as NNE, NE and near SN faults, were transformed into the core of the geoanticline, slant and like fans, shown as composite “secondary longitudinal” faults. Under the action of the longitudinal NNE faults, in the core of the Daliuhegou-Lüshan geanticline, NNE grabens were formed, forming the prototype of the Fuxin Basin.

② **Early Cretaceous**. The Pacific plate continued to subduct toward the Eurasian continent, causing the temperature of the deep upper mantle on the continental margin (including the Fuxin area) to rise, prompting the Daliuhegou-Lüshan uplift to further stretch, and the basin to stretch and subduct (entering a period of extension), when the main coal-bearing strata were formed in the Fuxin Basin.

Shahai period. NWW-NW expansion continued and NNE faults (especially the basin marginal faults) activities were more active and the basin became deep. The NWW-NW compressive faults, such as the

Table 6.4 Comprehensive results of isothermal adsorption experiments on Liujia coal seams

Coal seam group	Langmuir volume V_L/(cm^3/g)	Langmuir pressure, P_L/MPa	Theoretical CBM content, V/(cm^3/g)	Reservoir pressure, P/MPa	Measure CBM content, V/(cm^3/g)	Critical desorption pressure, P_c/MPa	Saturation/%
Sunben	23.79	11.64	8.72	6.74	7.14	4.992	81.88
Middle	18.81	6.47	9.60	6.747	8.86	5.761	82.29
Taiping	33.71	19.07	1017	8.238	8.66	6.592	85.15

Fosi-Housanjiazi fault, its south plate thrusted upward and caused differential settlement of the Fuxin Basin who took the fault as a boundary. The south was relatively uplifted, and the north was relatively subducted, causing the difference in the sedimentary environment and coal accumulation on both sides of the fault. The tectonic movement during that period was WEE-NEE expansion and right-handed shear. Under the action of the stress field, NW-NNW synsedimentary folds were generated, and NNE, NE and NEE synsedimentary faults might be formed in the basin.

Fuxin period. The NNE-SSW squeezing activities were still going on, and the NW Fosi-Housanjiazi fault continued to thrust, causing the Fuxin Basin to uplift in the south and subduct in the north, thus triggering the NWW-NW synsedimentary folds. Bounded by the Fosi-Housanjiazi fault, the sedimentary environment and coal accumulation on both sides of the fault are different.

At the end of the Fuxin period, the regional stress field in the basin was gradually transformed from NWW-SEE extension to NWW-SEE compression and left-handed shear. The basin began to shrink.

③ **Late Cretaceous**. In the early stage, affected by the Yanshanian movement, the two sides of the basin were in a left-handed NS torsional and WE compressive stress field, and the NW-NE compressive maximum principal stress was dominant, causing axial, NNE-NE, right-type, oblique, long-axis folds, and the whole basin was uplifted. In the middle stage, the uplift of the basin caused the NW-NE extension of the upper crust, which caused the basin-marginal faults generally NE to break again. At that time, the Wangying-Liujia CBM reservoir was formed, but the boundary faults may be open. At the end stage, the NNE-NE folds were successively developed and engulfed in the Upper Cretaceous strata. At the same time, the original NNE, near NS and NE faults were eventually transformed into compressively thrusted faults. The Fuxin basin eventually shrink.

④ **Cenozoic**. In addition to the influence of the Pacific plate subduction toward the continental margin, the NE-SW compression from the Indian Ocean plate was active. These two forces caused the principal compressive stress to be transformed from NNW to NNE-NE. The boundary faults of the CBM reservoirs were transformed from tensile to compressive and torsional, and became sealing faults. Adjustment and transformation of CBM reservoirs went on during the period.

The Pacific plate dead beneath the continental margin, inducing the secondary thermal dynamics in the deep mantle of North China, causing it to rise and expand, and leading to the eastern uplift of the North China block. The deep magmatic material surged, leading to volcanic eruption, and its scale gradually increased from west to east. Although there was no volcanic eruption in this area, the magma intrusion activity was strong, especially the diabase wall and rock bed invaded during the Paleogene and Neogene. From the appearance of the rock wall, its orientation is generally NEE and near WE, which is the invasion of the basement fault along the NEE–near WE direction during the Late Cretaceous. The deep basement fault changed from compressive torsion before the Late Mesozoic to tensile torsion.

(2) Burial, thermal and maturity histories

The burial history of the Fuxin source rock can be divided into four stages (Fig. 6.31). **The first stage (Fuxin period)**. The marginal fault activities were weakened again, the lake water became shallow, and the plants were flourishing. Consequently, five large coal seam groups of the Fuxin Formation deposited. **The second stage (early Late Cretaceous)**. The whole basin was uplifted overall. **The third stage (middle and late Late Cretaceous)**. Local Sunjiawan Formation deposited. **The fourth stage (since the Tertiary)**. The Himalayan movement caused the whole area to rise and a large amount of magma invaded and extensively eroded coal-bearing strata.

According to the burial history of the Fuxin Formation, its thermal evolution history and maturity history (Fig. 6.31) can be divided into two stages.

The first stage is from the Fuxin period to the end of the Late Cretaceous, when normal thermal evolution was going and the coal seam underwent deep metamorphism. The paleotemperature gradient was 2.51–3 °C/hm. At the end of the Fuxin period, the depth of the coal seam of the Fuxin Formation reached the maximum, and the vitrinite reflectance reached about 0.7%, starting the peak of hydrocarbon generation. At the beginning of the Late Cretaceous, the Fuxin Basin was uplifted overall, causing the coalification to terminate or delay. The Sunjiawan Formation locally deposited during the middle of the Late Cretaceous, up to 900 m thick. However, the depth of the coal-bearing strata did not exceed the maximum depth at the end of the Fuxin period, and the maturity of the Fuxin Formation and its lower coal seam did not increase.

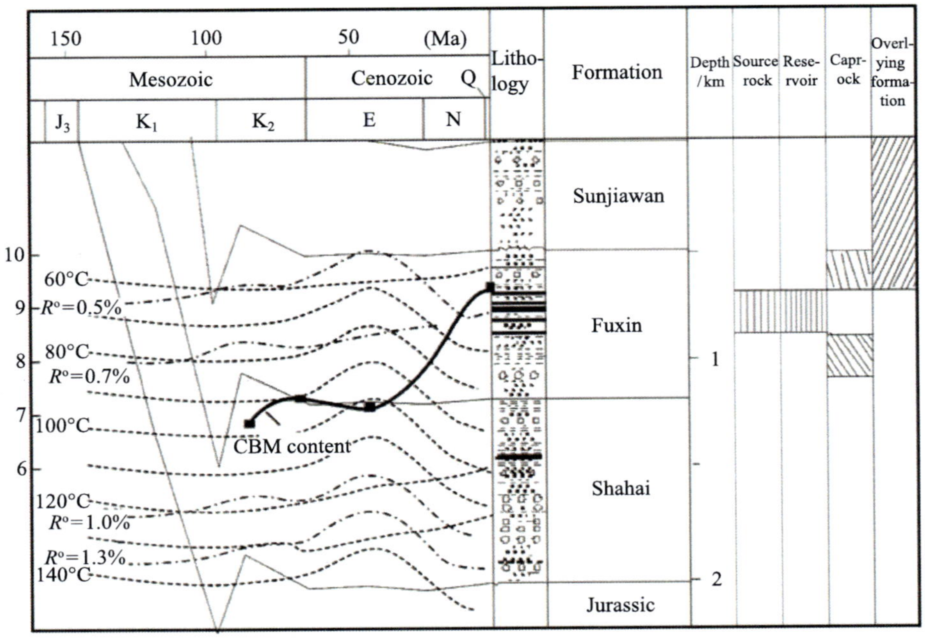

Fig. 6.31 The burial, thermal and maturity histories of the Fuxin Formation in Wangying-Liujia area

The second stage is from the Paleogene to the Neogene to the Quaternary when the paleotemperature was anomalous, the paleotemperature gradient exceeded 3.93 °C/hm, and the coal seam was transformed mainly by contact metamorphism. The Himalayan movement since the Paleogene and the Neogene led to the overall uplift of the Fuxin basin, and the magma at early Himalayan movement invaded in a large amount. The thermal maturity of local coal seams increased, the vitrinite reflectance was more than 4%, and the coal rank reached lean coal. Consequently the secondary peak of hydrocarbon generation started.

(3) Formation of CBM

From the above burial history and thermal history, the coal seam of the Fuxin Formation had entered the mature stage at the end of the Early Cretaceous and the first hydrocarbon generation happened (R^o = 0.5–0.8%); then the early Late Cretaceous uplift and late subduction didn't increase the degree of coalification; since the Cenozoic the coal seam was in the uplifting stage; the magmatism in the Himalayan period caused contact metamorphism, resulting in local coal rank to increase and trigger the secondary hydrocarbon generation, but the scope of impact was limited. The uplifting stage from the end of the Cretaceous to the present is the main stage of generating secondary biogenetic gas. In conclusion, the CBM is mainly thermogenic CBM formed by the first hydrocarbon generation in the Yanshanian, and the secondary hydrocarbon generated in the Himalayan period and the continuous biogenetic gas are the main supplements.

According to the CBM composition and carbon isotope of methane (−65 to −36‰) (Fig. 6.32), the genesis of the Wangying-Liujia CBM is very complicated, secondary biogenetic and thermal, and the thermal CBM is secondary (after diffusion, migration and re-accumulation). Presently available data does not show the existence of original thermal CBM. The closer to the rock wall, the heavier the carbon isotope is. In addition, according to the latest data statistics, the least inert gases and CO_2 in the Wangying-Liujia CBM reservoir are inorganic, so the methane may also be inorganic, which may be too least to be reflected in carbon isotope.

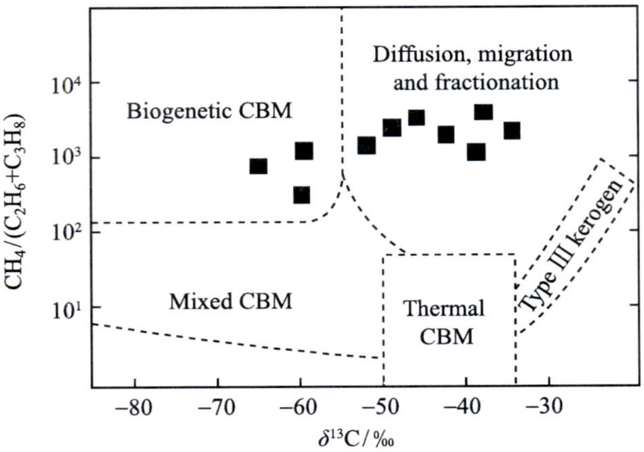

Fig. 6.32 Schematic identification model of CBM genesis

(4) Evolution history of gas content in CBM reservoirs

According to the methane adsorption characteristic curve, the evolution history of CBM content was calculated (Fig. 6.31):

① At the end of the Early Cretaceous (K_1), the maximum depth of the coal seam was not reached, but the main coal seam entered the mature stage, so the hydrocarbon generation reached its peak. The saturation at that time was 100%.
② The Late Cretaceous rise and settlement caused the gas content to fluctuate, but the change was not significant.
③ The burial depth has not changed much since the Cenozoic. The Paleogene magmatism caused the coal to undergo contact metamorphism, but the impact was limited and the gas content didn't change much.
④ Since the Neogene, due to the change of the thermal field, the geothermal gradient has decreased, which made the theoretical gas content increase significantly, and the saturation was only 80%.
⑤ The favorable preservation conditions and important supplement of secondary biogenetic CBM made the deep saturation of CBM reservoirs between 80 and 90%. Now a well group with 8 wells is producing commercial CBM. Therefore, CBM accumulation is not only dependent on the evolution history of gas content, but more importantly, later adjustment and transformation. If there are favorable storage conditions to maintain historical saturation, and supplement from secondary biogenetic CBM after migration and re-accumulation, the unfavorable evolution of gas content is also means commercial CBM reservoirs.
⑥ The key moment for the formation of CBM reservoirs is the end of the Early Cretaceous.

(5) Impact of groundwater dynamic conditions

The primary aquifers in the Fuxin Formation are glutenite and coal seam, and the water bearing zones are diabase walls. Although there lacks of a regional isolator between the aquifers, without vertical channels, the hydraulic linkage between these aquifers is very weak. However, the diabase walls are very developed. They cut through coal-bearing rocks and expose (or below the Quaternary). Fractures in the diabase wall are very developed. They not only make an aquifer permeable, but also communicate all aquifers, so that all coal seams are in the same flow unit, forming the CBM reservoirs with multiple coal seams (Fig. 6.33).

(6) Characteristics of CBM accumulation

① The coal reservoir is characterized by multiple and thick coal seams. The maximum cumulative thickness of recoverable coal seams exceeds 95 m. The permeability is moderately low, and mainly controlled by cleat fissures and rock walls and faults. The coal adsorption capacity is strong. The reservoir pressure is low and locally normal.
② The caprock is mainly mudstone and siltstone, and no regional mudstone caprock was found. Due to various factors, the sealing property of the caprock varies in the plane; the lateral sealing is more complicated, including weathered and oxidized belt, physical boundary, sealing fault and hydrodynamic boundary.
③ The Fuxin Basin began to form in the Late Jurassic, and in the Early Cretaceous the coal-bearing Fuxin Formation was formed under the action of the NWW-SEE tensile stress field. The tectonic stress field from the Late Cretaceous to the Paleogene should be NNE-SEE compressive, which produced the NNE boundary fault. At that time, the Wangying-Liujia CBM reservoir has been formed, but the boundary fault may be open. The tectonic stress field since the Neogene is horizontally NEE-SWW compressive. The boundary fault of the CBM reservoir was transformed from tensile to compressive or torsional, and it is a sealing fault. That is a stage when the CBM reservoirs was adjusted and transformed, which is conducive to the enrichment of CBM.
After the Early Cretaceous Fuxin Formation coal seam experienced hydrocarbon generation at the end of the Early Cretaceous, the Himalayan contact metamorphism made the local coal rank the highest and secondary hydrocarbon generation happen, but the impact range was limited. From the Cenozoic, the thermal generation was suspended, and the secondary biogenetic gas generation started and continued.
④ The intrusion of the rock wall destroyed the original state of groundwater. Therefore, the two CBM reservoirs entered the adjustment and transformation stage under the action of groundwater after intrusion of the rock wall. If the groundwater level in the rock wall is greater than or equal to the water level in the reservoir, the CBM is preserved, or lost along the rock wall. In the stagnant groundwater area, the block bounded by the rock wall is favorable for the enrichment of CBM, where producing CBM wells are located.

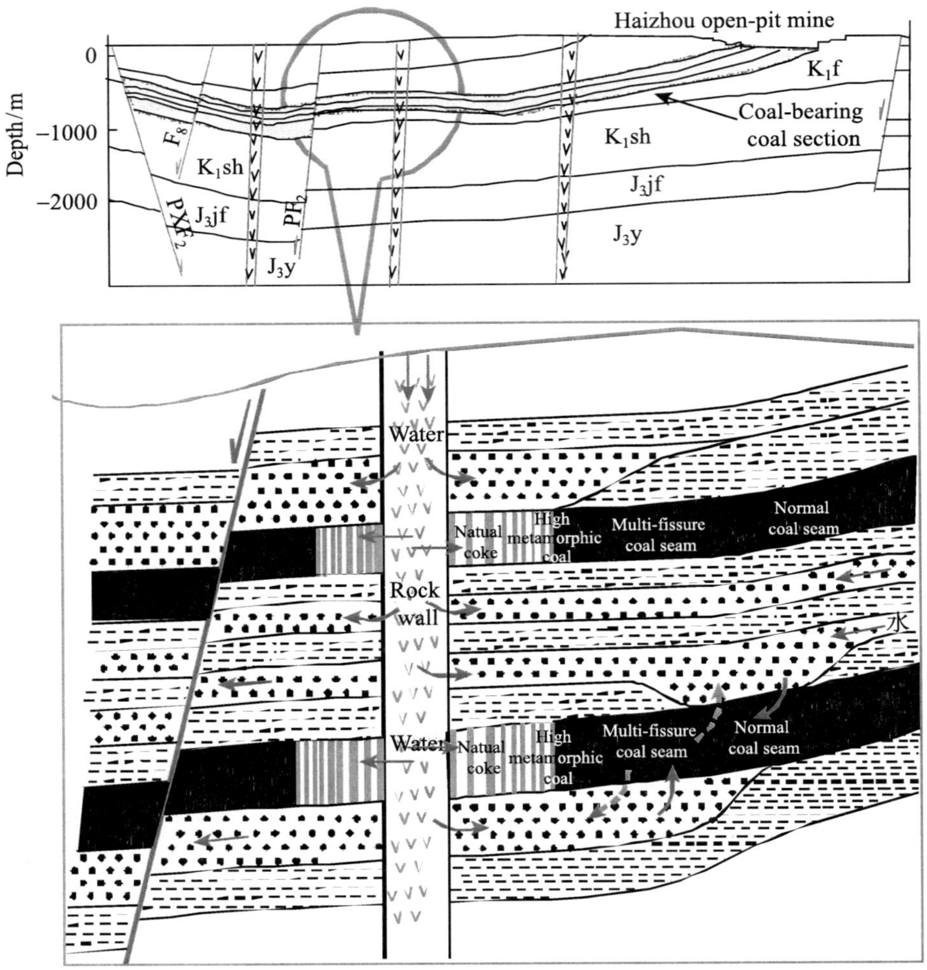

Fig. 6.33 Wangying-Liujia CBM reservoir section

⑤ The CBM content is relatively high, about 6–10 m^3/t, and the resource abundance is huge. The CBM may be secondary biogenetic and thermal, and maybe inorganic.

⑥ From the evolution history of gas content, it is known that the adjustment and transformation of the CBM reservoir are beneficial to the enrichment of CBM. Today's gas content generally does not exceed theoretical calculation.

6.2.3 Geological Models of CBM Reservoirs

According to the above theoretical analysis, combined with the other typical CBM reservoirs at home and abroad, three geological models of medium-and high-ranked CBM accumulation and a favorable model of low-ranked CBM accumulation were summarized.

6.2.3.1 Medium- and High-Ranked Coal

The current enrichment level of the gas reservoirs is the result of the superposition of the preservation and damage induced by reverse uplift and later evolution of the coal-bearing basin. The temperature and pressure changing with the thickness of the overlying strata during the process of uplifting controlled the change of the gas content in the coal seam. Therefore, the minimum thickness of the overlying strata in the geological history determines the gas content of the coal seam.

According to the recovery history of CBM accumulation, combined with the physical simulation experiment of CBM accumulation, three geological models of medium- and high-ranked CBM accumulation were summarized.

More Favorable Enrichment Model

The coal seam was uplifted below the weathering zone, and then settled, but did not exceed the depth before the uplift, or didn't settle. The settlement range is limited. The gas content

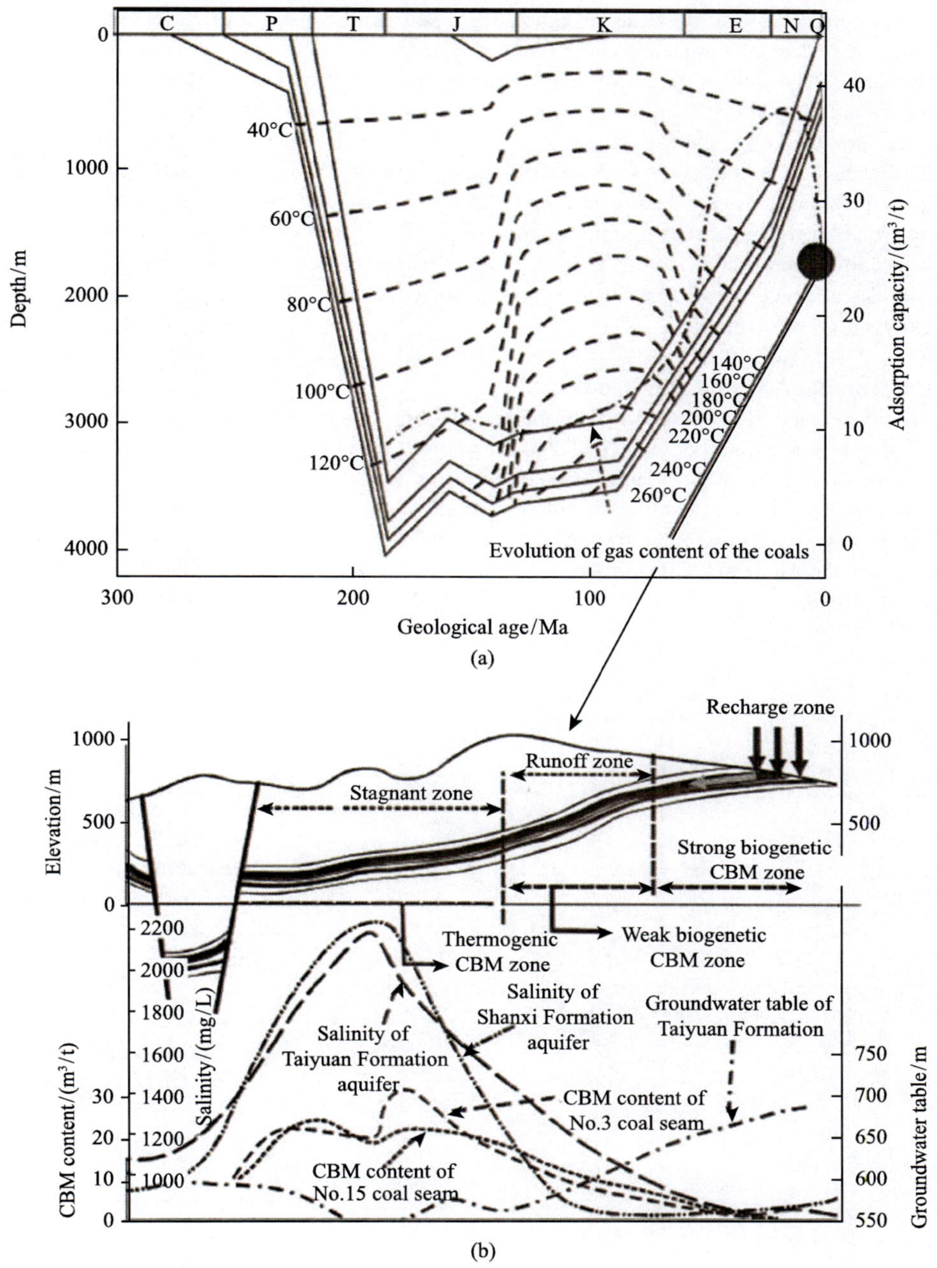

Fig. 6.34 Favorable CBM accumulation model. **a** Accumulation process; **b** modern gas reservoir section

of the coal seam depends on the minimum thickness of the overlying stratum during the geological history. The greater the thickness, the higher the gas content, and the higher the gas saturation. Most commercial CBM reservoirs belong to this category, such as those in the San Juan Basin, the Black Warrior Basin, the Piceance Basin in the United States, the Qinshui Basin and the Ordos Basin (some CBM reservoirs) in China. During the adjustment and transformation of such CBM reservoirs, favorable groundwater dynamic conditions caused CBM enrichment in the stagnant zone. At the same time, there was supplement from secondary biogas to the medium-ranked coal. It is a more favorable CBM accumulation model (Fig. 6.34).

Favorable Enrichment Mode

The coal seam was uplifted first and then settled, and exceeded the depth before the uplift. The CBM content depends on the minimum thickness of the overlying stratum in the geological history. Without external gas source, the saturation depends on the formation thickness after re-settling. The thicker of the settled formation, the higher the reservoir pressure and the lower the gas saturation. Therefore, the key to the enrichment of CBM of such model is the favorable gas source controlled by groundwater dynamic conditions. It depends on how much CBM carried by groundwater and enriched in the stagnant zone. This type of accumulation model is similar to the enrichment model and is a conditional enrichment model, such as the Qinyuan CBM reservoir.

Unfavorable Accumulation Model

The coal seam was uplifted into the weathering zone, causing a large amount of lost CBM, and the gas content and saturation significantly reduced. Generally, it is not the economic gas content, and there is not enough gas supply in the later stage. It is an unfavorable CBM enrichment model, such as the Huozhou CBM reservoir in western Qinshui Basin (Fig. 6.35).

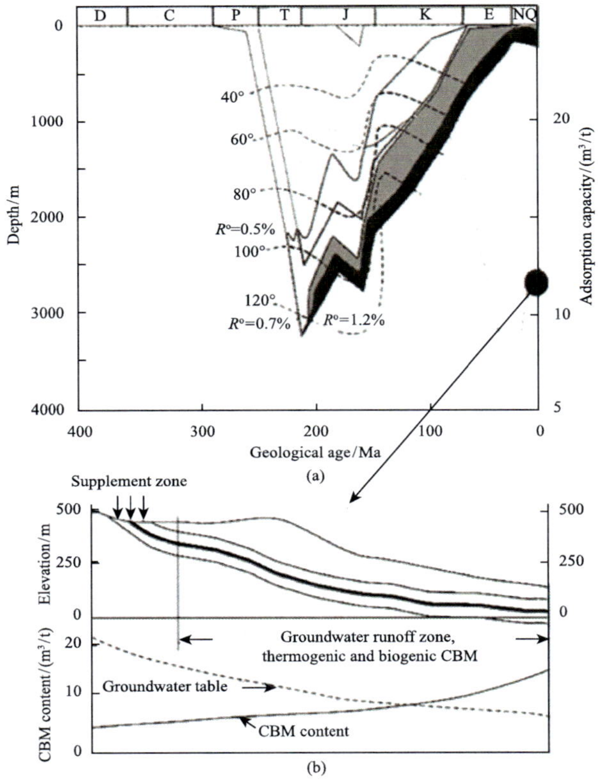

Fig. 6.35 CBM accumulation model. **a** Accumulation process; **b** modern CBM reservoir profile

6.2.3.2 Low-Ranked Coal

Compared with high-ranked coal, the formation process of low-ranked CBM reservoirs is simple, and the rise after hydrocarbon generation is small; there is generally no secondary hydrocarbon generation, even if there is magma intrusion causing contact deterioration, the scope of influence is also partial. The groundwater runoff belt is a favorable place for the generation of secondary biogas, providing continuous gas source to low-ranked CBM reservoirs. The CBM produced here can be preserved in place (i.e. the Powder River Basin) and can also be transported to the stagnant zone and enriched under the action of groundwater (i.e. the Fuxin Basin). It often appears as a thick coal seam or coal group, forming a high resource abundance, which in turn offsets the disadvantages of low adsorption capacity and low CBM content.

For example, the Wangying-Liujia low-ranked CBM reservoir in the Fuxin Basin reached its maximum depth at the end of the Early Cretaceous, and most coal seams entered a low mature stage, thermal CBM began to appear. The low-amplitude uplift and settlement during the Late Cretaceous and the relative stability of the Cenozoic only caused low fluctuations in coal adsorption capacity under reservoir conditions, which were basically maintained at 12 m^3/t (Fig. 6.36). Such CBM reservoir is very thick and the storage conditions are favorable, especially the important recharge of secondary biogas, which makes the saturation of modern CBM reservoirs between 80% and 90%, and the resource abundance exceed 2×10^8 m^3/km^2. In addition, the contribution of the matrix pores to the reservoir permeability means potential high production of such CBM reservoirs.

6.3 Mechanism and Favorable Conditions of CBM Accumulation

6.3.1 Comparison of CBM Source Conditions

CBM can be divided into primary CBM and secondary CBM according to the origin, which can be further divided into primary (early) biogenetic CBM and secondary (late) biogenetic CBM, and primary (early) thermogenic CBM, secondary (late) thermogenic CBM.

6.3.1.1 Genesis of High-Ranked CBM

Thermogenic gas means that when temperature exceeds 50 °C, coalification is enhanced, the carbon content in coal is rich, and a large amount of hydrogen-rich and oxygen-rich volatiles, mainly CH_4, CO_2 and water, are released (devolatilization).

The coal seams of high rank in China have experienced two stages of coalification. After passing through the typical

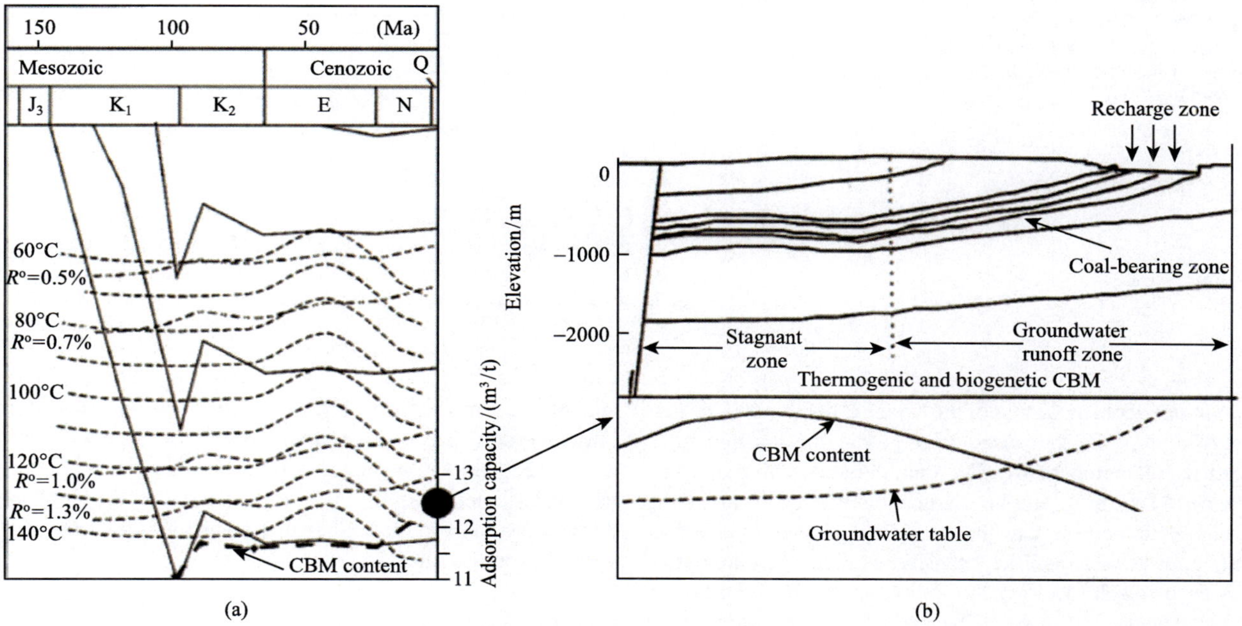

Fig. 6.36 Favorable model of low-ranked CBM accumulation. **a** Accumulation process; **b** modern CBM reservoir section

Fig. 6.37 Typical CBM accumulation model of high-ranked coal seams in China

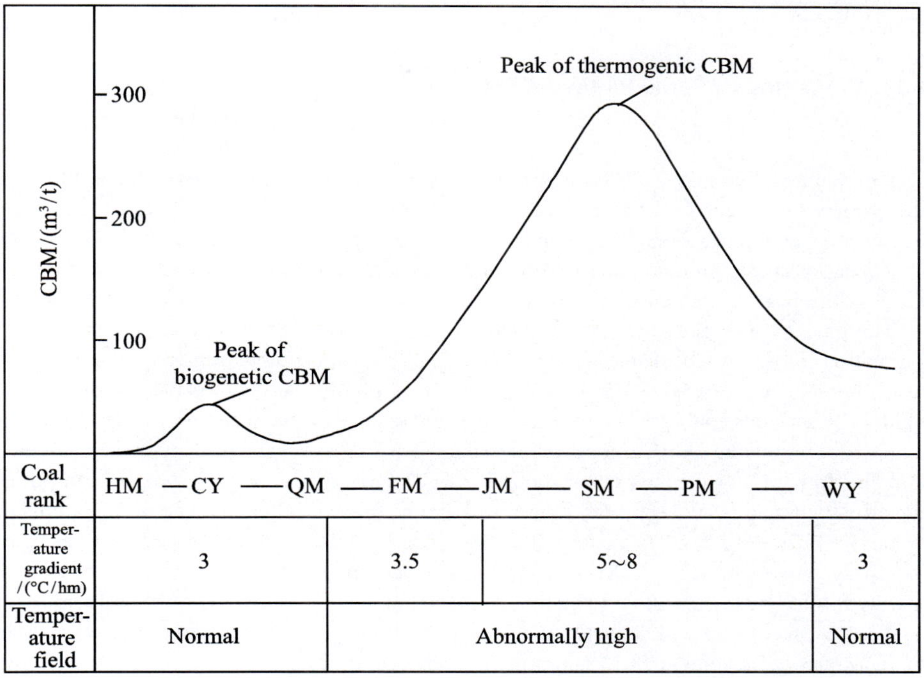

high-ranked CBM generation (Fig. 6.37), generally 1–2 peaks, secondary CBM generation happened at abnormally high paleotemperature. It is a powerful source for CBM accumulation. The high-ranked CBM is mainly thermogenic, because the coal becomes carbon-rich and hydrogen-rich volatile substance with the increase of buried depth, temperature, pressure and coalification, and CH_4, CO_2 and water are main products in the process of devolatilization.

The Upper Paleozoic coal seam in the Jincheng-Yicheng mining area in the southern Qinshui Basin reached its

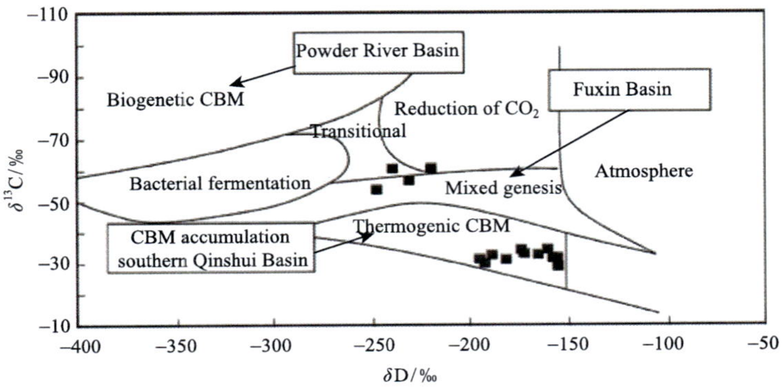

Fig. 6.38 Correlation between the genesis of CBM and carbon isotope of methane (according to Song et al. 2005)

maximum depth at the end of the Triassic, resulting in the first generation of CBM caused by regional metamorphism at normal paleotemperature. The cumulative CBM generation reached 81.45 m^3/t. The Yanshanian thermal event caused the coal rank to increase, causing the second generation of CBM. The range is wide and the intensity is large. The cumulative CBM may reach 359.10 m^3/t. Obviously, the late thermogenic CBM played a leading role. The methane carbon isotope in the southern part of the Qinshui Basin is between −15.39 and −29.63‰ (Fig. 6.38), indicating that high-ranked CBM reservoirs have mainly primary and secondary thermogenic CBM.

6.3.1.2 Genesis of Low-Ranked CBM

Genesis

The immature low-ranked CBM reservoir is dominated by the primary biogenetic CBM which is the CBM decomposed by bacteria and other organic matter in low metamorphic coal (peat to sub-bituminous coal) in the early coalification stage in the peat swamp environment.

The coal of the Paleogene and the Neogene Fort Union Formation in the United States is lignite (R^o = 0.3–0.4%) in most areas in the Powder River Basin, and there is highly volatile bituminous coal in the deep, which does not reach the maturity of producing a large amount of thermogenic methane, and thus formed biogas under peatization. The δ^{13}C value of the primary biogas is between −60.0 and −56.7‰, and the δD value is −315 to −307‰. It is indicated that the cause of CBM is mainly biogenetic (Fig. 6.38), and the CBM reservoir is mainly formed by microbial fermentation and metabolism.

The genesis of low-ranked mature CBM is very complex, both secondary biogenetic CBM and primary and secondary thermogenic CBM. It can be seen from the thermal history of the Fuxin Basin that the Cretaceous Fuxin Formation coal reached its maximum depth at the end of the Early Cretaceous. The regional metamorphism at normal paleotemperature caused the coal to reach a low maturity stage, causing the first hydrocarbon generation, and providing a sufficient CBM source for CBM reservoirs. The Cenozoic magmatic eruption and intrusion formed the Yixian Formation volcanic rock and intrusive rock, causing coal thermal metamorphism and producing second hydrocarbon generation, but the range of contact thermal metamorphism is limited. Therefore, the CBM in this area is mainly the secondary thermogenic, followed by the secondary biogenetic.

Physical Simulation Experiment on Low-Ranked Biogenetic CBM

(1) Methane bacteria CBM

The coal rock samples were mainly taken on the coal face in the Ili Basin and the central part of the Yanqi Basin. They were taken to laboratory in a sealed desorption tank (Table 6.5). In order to avoid the influence of the original methane adsorbed on the coal seam, the sample was naturally desorbed until no gas was released. Then methanogenic bacteria were injected into the coal core for methane production. Methane strains were cultivated, acclimated and inoculated in the Chengdu Biogas Research Institute, the Ministry of Agriculture. By preparing suspension inoculum, the majority of the inactive organic matter in the first enrichment was discarded, and the concentration and activity of the microorganism were increased by second enrichment. The results show that various coal rock samples can produce methane. Figure 6.39 shows the methane production curve of each rock sample. The amount of methane has been increasing continuously before 80 days, and the amount of methane produced after 80 days shows a downward trend. In short, the coal rock samples from both places produced a certain amount of biomethane. This is only the conclusion of simulation. Under the natural geological conditions, the living environment of the bacterial community is far less than that of the laboratory. The methanogenesis process cannot be completed within a few dozen days, but slowly in a very long geological history. However, the long-term cumulative effect at low yield can still produce a large amount of methane.

Table 6.5 Experimental samples used in the simulation to low-ranked biogenetic CBM accumulation

Sample No.	Original place	Formation	Temperature/°C	Sulfate-reducing bacteria/(pieces/g)	Methanogen/(pieces/g)	Zymophyte/(pieces/g)
1	Nantaizi, Yining	Badaowan	30	200	1400	5400
2	Qapqal, Yining	Xishanyao	30	320	1000	6100
3	Nilka	Xishanyao	30	4500	3600	20,000
4	Hamangou, Yanqi	Badaowan	30	2400	120	35,000

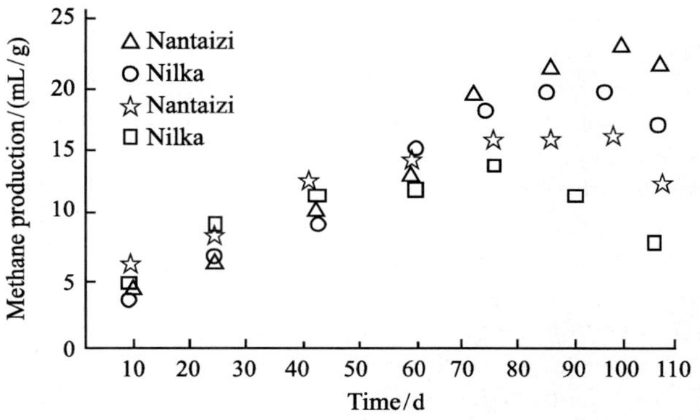

Fig. 6.39 Methane production of coal samples from Yining and Yanqi basins

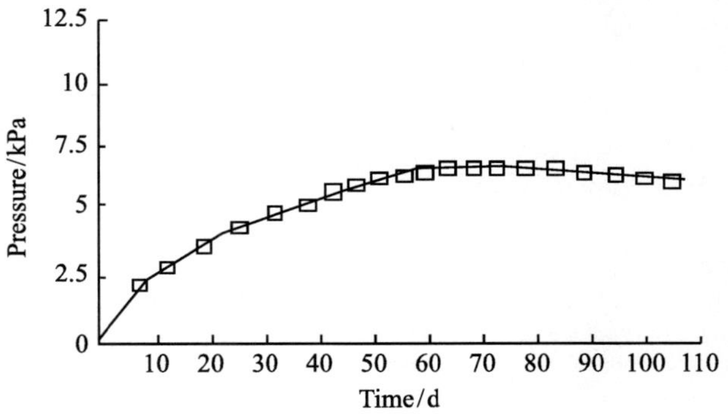

Fig. 6.40 Pressure versus time during adsorption after biogas generation

(2) Simulation of biomethane accumulation

The coal samples inoculated with methanogens were placed in a reservoir simulator, and cultivated at constant 35 °C. The gasification process was observed. There are two distinct stages on the curve. Fast gas production happened in the first stage and then in the second phase, gas production and adsorption balance reached (Fig. 6.40). By analyzing the components, they are mainly CH_4, N_2 and CO_2, and the content of heavy hydrocarbons (C_{2+}) in the gas produced by most samples is very low, and the carbon isotope values of methane are quite different, −67 to −56‰, indicating biogenetic gas.

According to the simulation experiment on low-ranked biogenetic gas accumulation, on the one hand, it shows the presence of methanogens in the low-ranked coal seams in Northwest China; on the other hand, it proves that the coal seams at low rank can be used as the source of secondary biogas. According to available data, the $\delta^{13}C_1$ in the CBM produced in the shallow Jurassic coal seam in the Ili Basin is −66.10 to −60.12‰, which is obviously biomethane.

6.3.2 Comparison of CBM Reservoir Conditions

CBM reservoir capacity includes adsorbed gas and free gas. The basic characteristics of coal reservoirs are pore structure, permeability and adsorption capacity. The differences in reservoir capacity of high and low coal ranks are described in three reservoir characteristics.

6.3.2.1 Pore Structure Comparison

On the basis of a large number of analytical experiments, and combined with the description of the pore characteristics in coal rock, the differences in pore characteristics at high and low coal ranks were studied. The results show that the pore size distribution is closely related to the degree of coal metamorphism. The distribution of pore size distribution in lignite is relatively uniform. The macropores of 9 × 10^3 nm to 9 × 10^4 nm and the micropores of 2–10 nm are obviously dominant. Microposes in high metamorphic coals such as lean coal and anthracite are dominant, while mesopores and macropores larger than 1000 nm account for only about 10%. Generally, micropores and transitional pores constitute the adsorption space, which correspond to the micropores inside the matrix, and have a large surface area. Small pores constitute capillary condensation and diffusion channels. Mesopores constitute slow seepage channels. Macropores constitute strong laminar flow channels, corresponding to cleat fractures and structural fractures.

6.3.2.2 Permeability Comparison

Permeability can be divided into cleat-fracture permeability and matrix permeability according to pore type. The pore type of high-ranked coal is dominated by cleat-fracture. The main factor affecting the adsorption performance of coal seam is cleat-fracture permeability. The pore in low-ranked coal is mainly matrix type, and the main factor affecting the coal seam adsorption performance is matrix permeability and porosity.

(1) Pressure difference

The migration-accumulation pressure difference is the pressure when gas migrates vertically. It is the difference between outlet pressure and inlet pressure. It is related to flow rate, fluid viscosity, sample length, sample cross-section, and inversely proportional to permeability.

$$P_1 - P_2 = \frac{Q_{\text{flow}} \times \mu_1 \times L}{10 \times A \times K} \tag{6.2}$$

where P_1, P_2 is pressure difference, P_1 is inlet pressure, P_2 is outlet pressure; Q_{flow} is flow rate; K is permeability; μ_1 is fluid viscosity; L is coal sample length; A is the cross-sectional area of coal sample.

(2) Simulation experiment

The experimental samples were obtained from the Qinshui Basin and the Turpan-Hami Basin (Tu-Ha Basin). They were taken by a wireline coring tool during drilling process and transported in a sealed can after natural desorption. The coal samples are anthracite and lignite. The purity of the methane sample is 99. 6%.

The coal samples from the Qinshui Basin are Jsh1-6-1 at 608.1–608.4 m, which field measured gas content is 22.39 m^3/t; Jsh2-12-1 at 713.15–713.45 m, which field gas content is 21.55 m^3/t; Jsh3-5-2 at 603.42–605.6 m, which field gas content is 12.70 m^3/t.

The core samples from the Tu-Ha Basin were taken in Well Ha 1 in the Shaer Lake area: Hsh1-1-1 at 605.22–605.52 m, which field gas content is 0.54 m^3/t; Hsh1-5-3 at 618.34–618.64 m, which field gas content is 1.57 m^3/t. The coal metamorphism is low, and the gas content is low, generally not exceeding 3 m^3/t.

Considering the depth of the samples and the temperature of the reservoir, the experiments were carried out at 8 MPa and 4 MPa, respectively, and constant confining pressure. The system pressure gradually increased from 0 MPa. The pressure difference on the sample was from 0 MPa. And by continuously changing the pressure at both ends of the sample, the pressure difference continuously increased. At the same time, the computer collected data and automatically plotted the relation curve of the pressure difference vs. the system pressure until the system pressure is balanced, that is, no methane was released from the sample.

(3) Analysis of experimental results

The pressure difference versus system pressure curves (Fig. 6.41) of high-ranked anthracite and low-ranked lignite under the same simulated conditions show the migration and accumulation of CBM occurred at 0.14–0.5 MPa in the anthracite; at high pressure, CBM breakthrough occurred at the pressure difference of 0.14 MPa; at low pressure (0–5 MPa), the breakthrough occurred at 0.5 MPa; with the increase of the system pressure, the pressure difference decreases, while the permeability is inversely proportional to the pressure difference [Eq. (6.2)]; in the lignite, at high pressure, CBM breakthrough occurred at 0.08 MPa; at low pressure, the breakthrough occurred at 0.3 MPa; and as the system pressure increases, the pressure difference increases, so the matrix permeability decreases.

(4) Law of permeability

Cleat-fracture type. As the system pressure reduces (i.e., the reservoir pressure increases), the matrix expands, the overburden pressure increases, the migration-accumulation pressure difference increases, $P_1^2 - P_2^2$ increases, cleat-fracture permeability decreases, which controls the adsorption capacity, so the flow rate, $Q_{\text{flow 0}} = Q_{\text{total 0}} - Q_{\text{adsorbed 0}}$ decreases [Eq. (6.2) adapted from Su et al. (2001)], and the adsorption capacity and volume increase

$$K = \frac{2 \times 10^5 \times (Q_{\text{total}\,0} - Q_{adsorbed0}) \times \mu \times L}{A \times (P_1^2 - P_2^2)} \tag{6.3}$$

where $Q_{\text{total0}} - Q_{\text{adsorbed0}} = Q_{\text{flow0}}$ is the flow rate at standard atmospheric pressure; K is fracture permeability; $Q_{\text{total 0}}$ is the total gas content at standard atmospheric pressure;

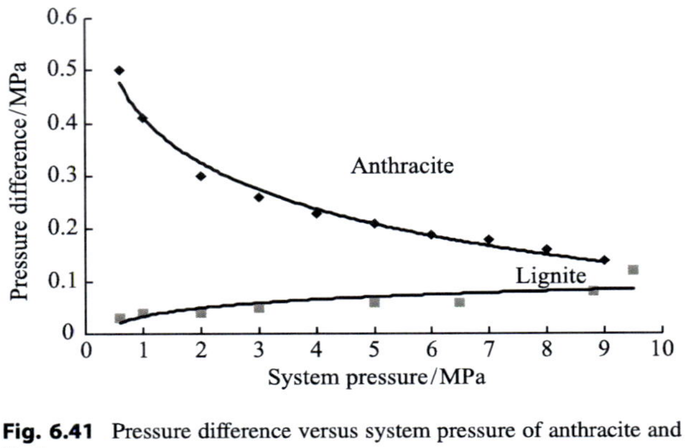

Fig. 6.41 Pressure difference versus system pressure of anthracite and lignite

$Q_{\text{adsorbed 0}}$ is the adsorbed gas at standard atmospheric pressure; μ is gas viscosity coefficient; L is the length of coal sample; P_1 is inlet pressure; P_2 is outlet pressure; A is the cross-sectional area of coal sample.

Matrix type. As the system pressure reduces (i.e., the reservoir pressure increases), the pressure difference decreases, and the permeability increases. The reduction of the system pressure makes the matrix expand, the porosity increases, the pore volume increases [Eq. (6.4)], and the matrix permeability increases. Therefore, in the case that the system pressure reduces, the permeability of low-ranked coal reservoirs increases.

$$\Phi = \frac{V_b - V_g}{V_b} = \frac{V_p}{V_b} \tag{6.4}$$

where V_b is the total volume of coal sample; Φ is matrix porosity; V_g is matrix particle volume; V_p is pore volume.

(5) Influence of permeability on CBM accumulation

Cleat-fracture type. Being uplifted or severely denuded to open surface means that formation pressure reduces, and overburden pressure reduces, causing fractures to open and reservoir permeability to increase. When formation pressure reduces to critical desorption pressure, CBM begins to desorb. If there is no good sealing condition, the increase of permeability will increase the lost gas rate until a new adsorption-desorption balance reaches. With a good capping layer, CBM is desorbed into free gas and stored in CBM reservoirs or migrated to reaccumulate. The increase of permeability increases the desorption rate and finally reaches an adsorption-desorption balance. Obviously, good capping conditions are an important factor in the accumulation of CBM, and reservoir permeability is the driving force for the preservation and destruction of CBM reservoirs. However, in the geological history of continuous uplift of coal-bearing strata, the increase of permeability promotes the increase of free gas content. When the gas pressure is greater than the breakthrough pressure of the capping layer, the seepage flow will increase, the adsorption amount will decrease correspondingly, and gas will be saturated. The large loss of CBM have an adverse effect on the accumulation of CBM.

Matrix type. Being uplifted or severely denuded to open surface means that formation pressure reduces, migration-accumulation pressure difference increases, and permeability reduces. When the reservoir pressure reduces to the critical desorption pressure, CBM begins to desorb, the matrix shrinks, the porosity reduces, and the matrix permeability reduces. If there is no good capping conditions, the reduction of permeability will reduce the lost gas rate until a new adsorption-desorption balance reaches. If there is a good capping layer, desorbed CBM will be free and preserved in CBM reservoirs. The decrease in permeability reduces the desorption rate and finally reaches a desorption-adsorption balance.

(6) The essence of affecting permeability

Harpalani and Shraufnagel (1990) further studied the effects of effective stress, gas slip, volumetric strain and other factors on the permeability of coal samples. The results show that the coal sample permeability has increased by 17 times in the process of reducing the pressure from 6.12 to 0.17 MPa, of which the volume strain effect is 12 times and the gas slip effect is 5 times.

The pressure drop causes the shrinkage of the coal reservoir matrix, and due to the difference in pore structure, low-ranked coal rock (matrix type) is mainly controlled by volumetric strain, followed by gas slip; high-ranked coal (cleat-fracture type) is controlled by gas slip, followed by volumetric strain. Therefore, this also confirms that it is harder to improve the permeability of high-ranked coal than that of low-ranked coal.

6.3.2.3 Adsorption Characteristics

Coal adsorbing CBM belongs to the category of solid-gas physical adsorption, and its adsorption capacity is affected by various factors such as temperature, pressure, ash, moisture and coal rock composition. Therefore, the adsorption capacity of coal seams with different coal ranks is different.

Influencing Factors

Coal adsorbing methane is closely related to physical and chemical properties. In the natural state, such adsorption is a dynamic process.

(1) Temperature and pressure (buried depth)

Kim's study showed that under experimental conditions, the adsorption of methane decreased by 26.32–29.88% when the temperature was raised from 0 to 50 °C. The isothermal adsorption experiments carried out by domestic scholars at 8 MPa and by increasing from 25 to 50 °C showed that the adsorption amount decreased by 10.59–13.53%. Levy et al. found that at 5 MPa, and increasing temperature from 20 to 65 °C, the adsorption capacity of coal to methane decreased by 0.12 m^3/t when increasing 1 °C, and showed a linear decreasing trend. Accordingly, Yee et al. believe that the increase in temperature helps the methane in coal seams become free rather than adsorbed.

In fact, the combined effects of temperature and pressure on the adsorption capacity of coal are interdependent. On the high-pressure methane isotherm, when the pressure is less than 10 MPa, it is characterized by normal Langmuir type; as the pressure increases, the temperature isotherms of dry and equilibrium water samples cross, indicating that at high pressure, the absorption capacity at high temperature is greater than that at low temperature.

(2) Ash and moisture

The ash yield and ash content in coal are closely related to the coal-forming environment and the original type of the peat swamp. In addition to the intrinsic inorganic matter in the coal-forming plant itself, the ash has another primary source which is soluble and insoluble inorganic substances carried by water into the original peat swamp. The adsorption capacity to methane has a significant negative correlation with the ash yield in coal. In terms of adsorption properties, methane is adsorbed on coal surface rather than the ash or the dispersed inorganic substance in coal layers. Even the gas stored in the carbonaceous mudstone associated with coal seams is adsorbed on organic particles, rather than inorganic particles or minerals.

(3) Coal composition

In the case of the same coal rank, the adsorption capacity of the coal with rich vitrinite is higher than that with rich inertinite. Although some scholars believe that the adsorption capacity of coal increases with the increase of vitrinite, the true relationship between microscopic composition and adsorption capacity has not been fully established. In some cases, the coal with rich inertinite has a strong ability to adsorb methane, and the adsorption capacity of fusinite is even twice that of vitrinite.

Comparison of Adsorption Characteristics

The adsorption capacity of coal has experienced three stages, from low to high, then to low, with the increase of coal metamorphism (Fig. 6.42). In the first stage, R^o = 1.3%, the adsorption capacity increased rapidly with the increase of coal rank, the rising rate is the fastest of the three stages. In the second stage, R^o = 1.3–2.5%, the adsorption capacity continued to increase, but the rising rate is significantly lower than that of the first stage. In the third stage, R^o > 2.5%, the adsorption capacity began to decline, and from 2.5 to 4.0%, the adsorption capacity is strong, and up to the strongest at R^o = 5%, while the rate of change is the lowest.

Isothermal adsorption experiments on the No. 3 coal seams in four wells (Wells CQ 9, Pan 3 and Jin Test 1 from souther to north, and the northern TL-001, in the southern Qinshui Basin) show that the Langmuir volume of the No. 3 coal seam from the Panzhuang Coalfield to the Tunliu Coalfield is 33.99–43.11 m^3/t, with an average of 38.76 m^3/t; the average Langmuir volume of the No. 3 coal seam in Well Jin Test 1 is 42.71 m^3/t, which indicates the high-ranked Coal reservoir has a very high adsorption capacity.

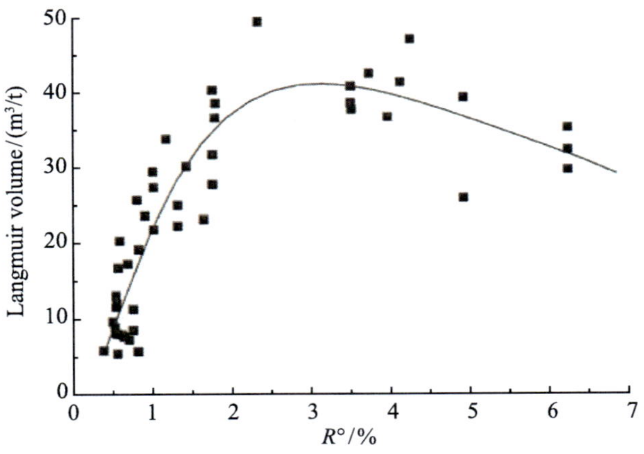

Fig. 6.42 Relationship between adsorption capacity and coal rank (according to Su et al. 2005)

The coal seam of the Sunjiawan Formation in the Wangying-Liujia mining area in the Fuxin Basin is long-flame coal. From the results of isothermal adsorption experiments, the Langmuir volume of the Sunjiawan coal seam is 21.59 m^3/t (Fig. 6.43). The results show that the adsorption capacity of high-ranked anthracite is much larger than that of low-ranked long-flame coal. It explains why the gas content of high-ranked CBM reservoirs is much higher than that of low-ranked CBM reservoirs.

6.3.3 Features of CBM Occurrence Comparison

CBM exists coal seams in three states: adsorbed, free, and dissolved.

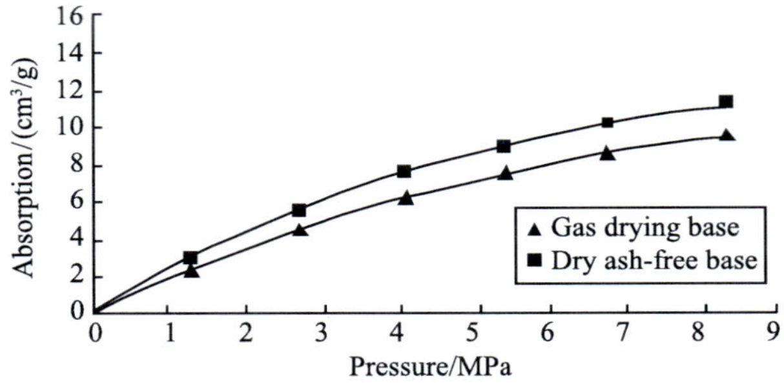

Fig. 6.43 Isothermal adsorption curve of the coal seam of the Sunjiawan Formation, Fuxin Basin

6.3.3.1 Occurrence State Comparison

Adsorbed CBM

The adsorption potential of coal reservoirs can be simulated by isothermal adsorption experiment. Under present experimental and theoretical conditions, the Langmuir monolayer adsorption dynamics model based on a homogeneous porous medium is still the theoretical basis for experimental research. CBM is a mixed gas dominated by methane, and can be described by the Langmuir isotherm adsorption equation.

(1) Single gas component

According to the Langmuir adsorption dynamic model, when gas mixture with multiple components are adsorbed at constant temperature, the adsorption volume of any single component is still controlled by the partial pressure of the single component, and is affected by the interaction between the gas components. It can be described by the extended Langmuir isotherm adsorption equation.

$$V_i = \frac{V_{Li}P_i}{P_{Li}\left(1+\sum_{j=1}^{m}\frac{P_j}{P_{Lj}}\right)} \quad i = 1,2,3,\cdots,m \tag{6.5}$$

where V_i is the adsorption amount of gas component i, m^3/t; V_{Li} is the Langmuir volume of gas component i, m^3/t; P_{Li} is the Langmuir pressure of gas component i, MPa; P_i is the partial pressure of gas component i, which is related to the molar ratio or volume concentration of the gas component, MPa; i, j, m are gas components, and the number of the gas components.

When CBM is considered to be composed entirely of methane (m = 1), the extended Langmuir isotherm adsorption equation is the general form.

$$V = \frac{V_{LCH_4}P}{P_{LCH_4}+P} \tag{6.6}$$

Real mixed CBM is mainly composed of three kinds of gases, CH_4, CO_2 and N_2 (m = 3), of which the amount of adsorbed CH_4 is

$$V_{CH_4} = \frac{V_{LCH_4}P_{CH_4}}{P_{LCH_4}\left(1+\frac{P_{CH_4}}{P_{LCH_4}}+\frac{P_{CO_2}}{P_{LCO_2}}+\frac{P_{N_2}}{P_{LN_2}}\right)} \tag{6.7}$$

The total adsorption potential of CBM ($V_{adsorbed}$) is equal to the sum of the adsorption volumes of the three kinds of gases, i.e.,

$$V_{adsorbed} = V_{CH_4} + V_{CO_2} + V_{N_2} \tag{6.8}$$

The total Langmuir constant of a single gas can be obtained by single gas isotherm adsorption experiment.

(2) Real mixed gas

According to the gas composition of CBM, the experimental gas was prepared, and the isothermal adsorption experiment was carried out by simulating reservoir temperature. The volume of the adsorbed gas at a balanced pressure point was obtained by component and compressive factor analysis. The relationship between the balanced pressure and the volume of adsorbed gas at a measured point is

$$\frac{P}{V_{adsorbed}} = \frac{P_{Lmix}}{V_{Lmix}} + \frac{P}{V_{Lmix}} \tag{6.9}$$

Linear fitting was performed to determine the Langmuir adsorption constants V_{Lmix} and P_{Lmix} of the mixed gas at reservoir temperature by linear slope and intercept. The isothermal adsorption equation of the mixed gas can be further established.

$$V_{adsorbed} = \frac{V_{Lmix}P}{P+P_{Lmix}} \tag{6.10}$$

According to reservoir pressure P, the adsorption potential $V_{absorbed}$ can be obtained.

Free CBM

For CBM reservoirs, the pore volume provides space for free gas and is also the space for liquid water. Theses pores for storing gas and water should be seepage pores, which are basically equivalent to the macropores (>1000 nm) and the mesopores (100–1000 nm) in coal reservoirs. According to the research results of Fu et al., the size threshold between adsorption pores and seepage pores is 75 nm. Seepage pore volume generally accounts for 50–80%, which is related to coal rank and coal rock composition. At initial temperature and pressure, the potential of free gas is mainly related to pore volume and gas (water) saturation (the percentage of gas and water occupying connected pores). The potential of free gas is the sum of the volumes of gas components stored, i.e.,

$$V_{free} = \sum_{i=1}^{m} V_{freei} \tag{6.11}$$

where V_{free} is the potential of free gas, cm^3/g; V_{freei} is the potential of free gas component i, cm^3/g.

At normal temperature and pressure conditions, CBM can be regarded as an ideal gas. However, at reservoir temperature and pressure, CBM can be regarded as a real gas, which has a certain deviation from the ideal gas. The larger the temperature and pressure, the larger the deviation, and the ideal gas state equation needs to be corrected by the gas compression factor (deviation factor). The formula for calculating n_i is

$$n_i = \frac{p_i V_p S_g}{Z_i RT} \tag{6.12}$$

where p_i is reservoir pressure, Pa; T is reservoir temperature, K; V_p is seepage pore volume, cm^3/g; S_g is gas (water) saturation,%; Z_i is the compression factor of component i, which is a parameter related to gas temperature and gram molecule density.

Dissolved Gas

The dissolved CBM, that is, the water-soluble gas in coal seams, depends on the amount of water in the coal seams and the amount of gas dissolved in water. The water content in coal seams is generally less than 5% except for peat and lignite, that is to say the CBM dissolved in water is less. When water is discharged and conditions change, such as a decrease in pressure, the CBM dissolved in the water may become free. In addition, in the process of hydrocarbon generation (R^o = 0.5–1.3%), there is oil-soluble gas, although the ability of oil to solve gas is greater than water, ultimately the amount of oil-soluble gas is not large due to the oil production and storage capacity of coal is limited.

The dissolved gas is a function of gas solubility, pore volume and water saturation, which can be expressed as

$$V_{\text{dissolved}} = \sum_{i=1}^{m} V_p S_w R_{si} \tag{6.13}$$

where $V_{\text{dissolved}}$ is the potential of dissolved gas, cm^3/g; V_p is pore volume, cm^3/g; S_w is water saturation, cm^3/cm^3; R_{si} is the solubility of component i (volume ratio), cm^3/cm^3; i and m are components and the number of gas components, respectively. When methane concentration is high or approximate, it is considered that the CBM is all methane.

Based on the data measured in laboratory, and using the above formula, the CBM amount and occurrence in the low-ranked Jurassic coal reservoirs in the Junggar Basin and the Tu-Ha Basin and those of the high-ranked coal reservoirs in the Yangquan mining area in the Qinshui Basin were analyzed and calculated. The results show that CBM in the low-ranked coal reservoir is mainly adsorbed gas and free gas, and a little dissolved gas. Adsorbed gas may be up to 100%, free gas up to 60%, and dissolved gas up to 2.1%. The occurrence changes with temperature and pressure and gas (water) saturation with the depth of the coal reservoir. For the deep coal reservoir, in the case of similar gas (water) saturation, with the increase of the buried depth of the coal reservoir, the importance of the occurrence states of free gas and dissolved gas are enhanced accordingly, especially the importance of free gas. The CBM in the high-ranked coal reservoirs is mainly composed of adsorbed gas, free gas is less, almost no dissolved gas. Adsorbed gas may be up to 100%, free gas up to 29%, and dissolved gas up to 0.35%. The occurrence changes with the depth of the coal reservoir similarly to that of low-ranked coal reservoirs, but the importance of free gas is relatively large.

6.3.3.2 Comparison of CBM Content

Gas content is the test result of the above three kinds of occurrence state, and it has a very important influence on the prediction of CBM recoverable resources, the deployment and management of development wells, and the development of CBM industrialization.

The gas content of high-ranked CBM reservoirs is generally far higher than that of low-ranked CBM reservoirs. In the Jincheng mining area dominated by high-ranked anthracite, the gas content can reach 25.13 m^3/t. In Shenbei, Tiefa, Junggar, Uintah and Powder River Basins where low-ranked lignite and long-flame coal are dominant, the minimum gas content is 4.0 m^3/t and the maximum is 11.2 m^3/t. The gas content increases with the increase of coal rank, especially in low-ranked coal, it is very obvious.

The high-ranked CBM is represented by the Jincheng mining area in the Qinshui Basin, and the gas content gradually increases from east to west or from the basin edge to the center of the basin. Except for the No. 3 coal seam, the gas content of the outcrop at the edge of the mine field is relatively low, the remaining gas content is generally higher, and the gas content is greater than 8 m^3/t. The gas content of the main coal seam is mainly concentrated at 4–20 m^3/t. In this interval, the No. 3 coal seam accounts for more than 80%, especially 8–16 m^3/t. In the Pan 1 and Pan 2 mine fields, the coal seam has the highest gas content, and that of the No. 3 coal seam can reach 25.74 m^3/t, with an average of 13.47 m^3/t. It can be seen that the CBM distribution in this area is generally balanced.

For low-ranked immature gas reservoirs, the CBM content in the Powder River Basin in the United States is generally low, 0.78–1.6 m^3/t, the highest is not more than 4 m^3/t. The CBM reservoirs in the Junggar Basin are very similar to those in the Powder River Basin. The gas-bearing gradient method, pressure-adsorption curve and coal-ash-gas content analogy were used to predict the deep CBM in the northwestern margin of the Junggar Basin. The result shows that the gas content gradually increases from the margin to the center of the basin with the increase of depth. The gas content of the Xishanyao Formation is 2 m^3/t at the margin, and gradually increases toward the basin center, up to the highest 14 m^3/t. The gas content of the basin is significantly higher than that of the Powder River Basin in the United States, and the CBM accumulation conditions in the Junggar

Basin are similar to those in the Powder River Basin. The commercial CBM development in the Powder River Basin provides a good idea for the exploration and development of CBM in the Junggar Basin.

The low-ranked mature CBM reservoir is represented by the Upper Cretaceous Ferron sandstone in central Utah, USA. The gas content of the gas reservoir varies greatly from north to south, with the minimum of 0.37 m^3/t and the maximum of 14.3 m^3/t, generally 5–10 m^3/t. The Wangying-Liujia CBM reservoir in the Fuxin Basin is a typical low-ranked mature CBM reservoir. The gas content is between 8.42 and 12.6 m^3/t. It reflects the low gas content of low-ranked CBM reservoirs, and also indicates that the Wangying-Liujia CBM reservoir in the Fuxin Basin is a relatively low-ranked CBM reservoir with relatively high gas content. The CBM reservoir belongs to low-ranked coal seam, but more favorable for exploration and development and having better prospects.

6.3.4 Reservoir Forming Process

The formation process of high-ranked CBM reservoirs is complicated. Whether there is secondary thermogenic gas or secondary hydrocarbon generation, the thermal metamorphism caused by structural anomalous thermal events is a necessary condition for the formation of high-ranked CBM reservoirs. The formation of high-ranked CBM reservoirs has obvious stages. After reaching the highest level of evolution, there is no longer the generation of CBM, and it enters the stage of adjustment and transformation of CBM reservoirs.

Taking the Qinshui Basin, a typical high-ranked coal-bearing basin, as an example, the burial history of the main coal seam is divided into five stages according to the nature and characteristics of the tectonic movement stages and the recovery results of the ancient burial depth. The Upper Paleozoic coal seams have undergone four major stages since the Late Paleozoic evolution of thermal-maturation-hydrocarbon generation. The burial history of the coal seam and the history of hydrocarbon generation also reflect the accumulation process of CBM reservoirs in the southern Qinshui Basin.

The first stage is from the formation of the main coal seam to the Late Triassic during which the area was at a normal paleo-geothermal field, the paleo-temperature gradient was about 3 °C/hm, the organic matter underwent deep metamorphism (maturation), and the thermal evolution of the coal was mainly controlled by the depth of the coal seam or the thickness of the Permian-Triassic formation. At the end of the phase, the maximum vitrinite reflectance (R^{o}_{max}) was 0.6–0.7%, generally reached low coalified bituminous coal, mainly corresponding to a biogas stage.

The second stage is the Early-Middle Jurassic when the earth's crust reversed, the coal seam became shallower, the temperature decreased, and the maturity of organic matter didn't enhance or a little, but basically maintained the coal rank distribution in the previous stage. During the stage, the hydrocarbon generation from organic matter in the coal mainly corresponded to the initial thermal degradation, and a certain amount of hydrocarbons were produced.

The third stage is from the Late Jurassic to the Early Cretaceous. The middle Yanshanian magma thermal event superimposed a strong additional thermal flow on the normal geothermal heat flow, forming an anomalous paleo-geothermal field. The paleo-temperature gradient at generally above 5.5 °C/hm caused thermal degradation of the sedimentary organic matter, and triggered the Late Paleozoic coal to generate gas. Due to the heterogeneity of the paleo-geothermal field and the difference in the depth of the coal seam caused by structural differences, the degree of coalification in different regions was significantly different, and finally formed modern coal-rank distribution pattern. In the middle and south of the Qinshui Basin, the organic matter in the coal has entered the stage of thermal cracking gas, and secondary hydrocarbon generation has occurred, while the east and west are still in the stage of thermal degradation gas.

The fourth stage is from the Late Cretaceous to the Quaternary when the earth's crust rose, the maturation was terminated as the temperature reduced, and the hydrocarbon generation ceased.

According to the development and evolution of the burial history, thermal history and gas-bearing history of the main coal seam (Fig. 6.9), it can be seen that the formation of CBM reservoirs in the southern Qinshui Basin has experienced two hydrocarbon generations. The first period occurred at the end of the Triassic during which the burial depth of the coal seam reached the deepest, the normal regional metamorphism led to the first generation of CBM, and the vitrinite reflectance of the coal rock reached about 1.2%. At the early Yanshanian, the coalification was stagnant due to the uplift of the coal-bearing strata, and the hydrocarbon generation was relatively interrupted. The second stage occurred in the late Yanshanian during which the thermal event caused the temperature of the coal seam to far exceed the temperature at the maximum buried depth in the Late Triassic, the coalification to intensify, and the second generation of CBM start. The secondary generation of CBM played a decisive role in the formation of CBM reservoirs. Then the CBM reservoirs underwent serious adjustment and transformation during the Himalayan period, and finally evolved to the modern pattern.

In addition, the modern groundwater pattern also has an important impact on the formation of CBM. Vertically blocked by low-permeability surrounding rock at top and bottom, low-lying zones appeared in the Panzhuang and Fanzhuang areas in the southern Qinyuan Basin. The groundwater runoff is the weakest, so they are the most favorable zones for CBM accumulation.

The formation process of low-ranked mature CBM reservoirs is relatively simple. The generation of CBM is mainly affected by deep metamorphism. Even if there is thermal metamorphism of the rock shelf, it is only contact metamorphism, and has limited influence on the formation of CBM reservoirs. The modern tectonic pattern and the occurrence of groundwater are the main controlling factors for controlling the adjustment and reconstruction of CBM reservoirs. The formation of CBM reservoirs is staged and continuous. The formation of thermogenic CBM reservoirs is determined in the period of maximum buried depth and highest thermal evolution. Therefore, the formation of thermogenic gas reservoirs is staged. From the uplift of the coal seam to the depth at which microorganisms can be active, secondary biogas began to generate and has been continuing to the present, which indicates that the generation of secondary biogas is continuous.

The Wangying-Liujia CBM reservoir in the Fuxin Basin is an example of low-ranked mature gas reservoirs. According to the burial evolution and thermal history, the CBM accumulation process was discussed.

According to the burial history and thermal history of the Wangying-Liujia CBM reservoir (Fig. 6.31), the coal seam of the Fuxin Formation became mature after the end of the Early Cretaceous, and the first hydrocarbon generation occurred (R^{o}_{max} = 0.5–0.8%). The early uplift and late settlement at the Late Cretaceous failed to enhance coalification. Then entering the Cenozoic coal seam was uplifted, and the Himalayan magma intrusion caused contact metamorphism, resulting in local coal rank to increase and secondary hydrocarbon generation, but the scope of impact was extremely limited. The uplifting stage since the end of the Cretaceous is the main generating stage of secondary biogas. So the CBM is mainly thermogenic CBM from the first generation in the Yanshanian. The secondary hydrocarbon generated in the Himalayan period and the continuous secondary hydrocarbon become the main supplement of CBM.

From the hydrodynamic profile of the Wangying-Liujia CBM reservoir in the Fuxin Basin, the primary aquifers of the Fuxin Formation are glutenite and coal seam, and the water bearing zone is the diabase rock wall. Despite the lack of regional isolators between the aquifers, the rock walls are very developed, and these walls cut through the entire coal-bearing rock and expose (or below the Quaternary). The fissures in the diabase rock wall are very developed, which is not only an aquifer, but also communicates all aquifers, so that all coal seams in the CBM reservoir are in the same flow unit, forming multi-coal-seam CBM reservoirs, similar to the open fault in the groundwater recharge area. Groundwater flows along the fissures and moves to the coal seams and the aquifers on both sides of the rock wall, and prevents the loss of CBM. The aquifer and coal seam on both sides of the rock wall are different fluid flow units, so they isolate the CBM reservoirs.

The formation process of low-ranked immature CBM reservoirs is simple. After formed, they generally only undergo a sedimentation and then rise to form into reservoirs. Today's groundwater recharge, migration, drainage and stagnation play a decisive role in the adjustment and transformation of CBM reservoirs. The formation of CBM reservoirs is persistent, but today's tectonic framework and groundwater hydrogeological features are key factors affecting CBM generation, and also key factors controlling the accumulation of low-ranked immature CBM reservoirs.

Take the Powder River Basin in the United States as an example. The Powder River Basin was formed after the Late Cretaceous Laramie movement. During the Paleocene, the main coal-bearing Fort Union Formation was formed, and then another coal-bearing Wasatch Formation formed during the Eocene. The favorable environment created very thick coal seam sections, including many coal seams. In the early or middle Miocene, the buried depth of the coal seams reached the maximum, but the vitrinite reflectance was only 0.3–0.4%, which did not reach the threshold of thermogenic CBM, or there may be thermogenic CBM at deep layers with higher coal metamorphism, which is primary biogas generated by microbial degradation. Since then, the regional tectonic uplift that has been continuing to today has made the Powder River Basin tilt toward north in a wide range, and identified the modern tectonic pattern. The tectonic pattern affected the recharge, migration and discharge of groundwater, and played a decisive role in the adjustment and transformation of CBM reservoirs. It is the key to the generation of CBM, and also the key to controlling the accumulation of CBM. Groundwater flows westward or northward along the aquifer (Fig. 6.44), and blocks CBM with the support from the upper and lower shale or the impermeable network-like roof. Since formed, the coal

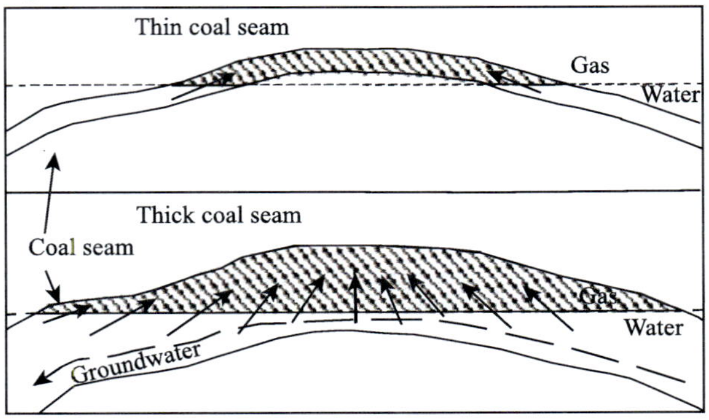

Fig. 6.44 Hydrogeological section of the anticline trap associated with biogas generation

seams have been producing gas which is influencing the gas composition and isotope of the CBM reservoirs.

In short, the accumulation process of high- and low-ranked CBM reservoirs is very different. The process of low-ranked immature CBM reservoirs is simple, while that of high-ranked CBM reservoirs is complicated, and that of low-ranked mature CBM reservoir is between the above two.

After formed, low-ranked CBM reservoirs generally experienced only a uplifting process. Today's groundwater recharge, migration, drainage and stagnant flow play a decisive role in the adjustment and transformation of CBM reservoirs. For high-ranked CBM reservoirs, whether secondary hydrocarbon generation happens or not, regional magma thermal metamorphism is the necessary condition. Today's groundwater recharge, migration, discharge and stagnation have certain influences on high-ranked CBM reservoirs. The formation of low-ranked CBM reservoirs is a continuous process, while the formation of high-ranked CBM reservoirs is staged.

6.3.5 Hydrogeological Conditions

Domestic and foreign scholars believe that the sedimentary, tectonic, and thermal evolutionary backgrounds determine the maturity and distribution of the source rocks in the San Juan Basin, the Sand Wash Basin, and the Great Green River Basin, and determine the relative orientation of the groundwater flow. Groundwater dynamic conditions directly reflect reservoir pressure, and directly affect the preservation and dissipation of CBM.

6.3.5.1 Hydrodynamic Conditions Comparison

Mainly by physical simulation experiments, the damage of hydrodynamics to CBM reservoirs was analyzed, and the dynamic forming mechanism of CBM reservoirs that the dynamic equilibrium is established, broken and restored continuously was established. This provides a theoretical basis for CBM exploration and development.

6.3.5.2 Hydrodynamic Simulation Experiment

(1) Sample collection

The high-ranked coal and CBM samples are from the No. 15 coal seam in Well Jin Test 7 in the Zhengzhuang block, the Qinshui Basin. The low-ranked coal and CBM samples are mainly from the No. 6 coal seam in Well ZG 3 well in the southeastern margin of the Ordos Basin, and the No. 8 coal seam of the Xishanyao Formation in Well Ha Test 1 in the Tu-Ha Basin. During the drilling process, cores were taken by a wireline coring tool and sealed in a can. The coal and CBM samples were collected after natural desorption at the reservoir temperature of 38 °C. The coal samples are anthracite, long-flame coal and lignite.

(2) Experimental device and principles

The experimental device is a CBM reservoir simulation device developed by the Langfang Branch of Research Institute of Petroleum Exploration & Development. The coal cores were flooded by water around the cores and into the pores and fractures in the cores at 0.5 m/s and 0.1572 m^3/min.

Considering the buried depth of the sample and the reservoir temperature, the groundwater migration process was simulated, and gas samples were taken on day 4, day 8, day 12 and day 16, respectively, and H_2S, gas composition and methane carbon isotope were analyzed.

(3) Experimental phenomena and mechanism analysis

① Hydrodynamic physical simulation experiments on high-ranked coal samples.

A. The methane carbon isotope became lighter and staged.

The initial value of the carbon isotope of the gas sample is −29.50‰. As waterflooding went on, the methane isotope of the Sample 4 became −36.60‰, indicating that the methane isotope became light (Fig. 6.45), and the gas reservoir was damaged. The damaged process experienced three stages: a slow damage stage during which free gas dissolved, diffused and migrated; a rapid damage stage during which adsorbed gas dissolved, gas concentration was very different, gas diffusion was significant, and the gas reservoir was heavily damaged; a slow damage stage during which the damage became less as the gas concentration difference decreased, and a final balance of adsorbed, dissolved and free gas was established in the reservoir.

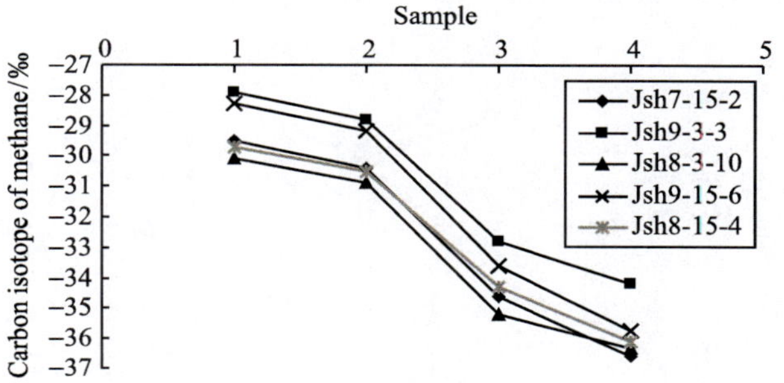

Fig. 6.45 Changes of carbon isotope of methane in waterflooding experiments

B. The methane content reduced.

Under the action of waterflooding, the methane in the gas sample was reduced (Fig. 6.46), from initial −96.35% in the Sample 1 to −12.42% in the Sample 3. The methane content decreased sharply, indicating that the gas reservoir was severely damaged. The high-ranked CBM is mainly adsorbed. Under the action of waterflooding, the effect of water on gas is to dissolve gas, and the dissolved gas will be carried away by water. The process of "dissolving, transporting and then dissolving happens again and again" until the CBM reservoir is destroyed.

C. The content of CO_2 decreased first and then increased; the content of N_2 increased.

The N_2 content increased from 2.9% of the gas Sample 1 to 86.45% of the Sample 3 (Fig. 6.47); the CO_2 content changed from 0.75% of the Sample 1 to 0.68% of the Sample 2, then increased to 1.13% of the Sample 3 (Fig. 6.48), showing a trend of decreasing first and then increasing. N_2 content is closely related to the preservation conditions of CBM. Therefore, the preservation conditions can be judged according to N_2 content. When N_2 content is >10%, the gas reservoir is completely destroyed; when the content is <5%, the preservation conditions are considered to be better. Combined with experimental data, it indicates that the gas reservoir has suffered severe damage.

② Hydrodynamic physical simulation experiments on low-ranked coal samples

The original content of methane isotope in the gas sample is −40.50‰. Under the action of waterflooding, the methane isotope in the Sample 4 became −48.26‰, indicating that the methane isotope became light (Fig. 6.49), the methane content reduced. Under the action of waterflooding, the methane component in the gas sample reduced from the original 79.45–7.93%, and the total trend is reducing; the N_2 component increased from the original 19.89–90.79%, indicating that the N_2 component increased significantly; the CO_2 content decreased first and then increased (Table 6.6), indicating that the components of the CBM reservoir changed under the action of waterflooding.

③ Theoretical analysis.

During the geological history, the changes in gas composition in CBM reservoirs under hydrodynamic washing are mainly due to the following causes. Since methane is partially oxidized in water and dissolved in water under hydraulically sufficient conditions, it diffuses and dissipates with hydraulic migration. The methane content greatly

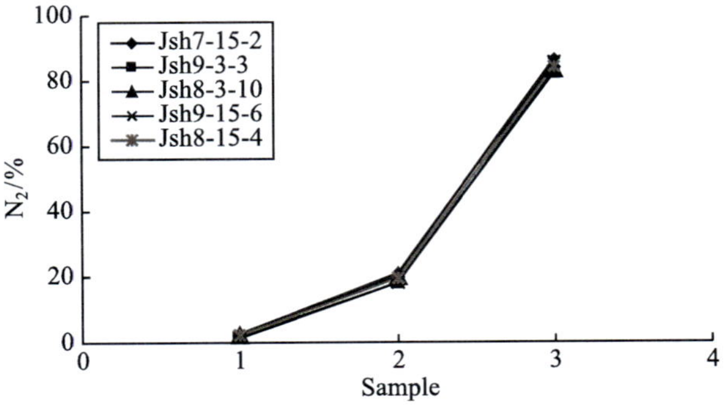

Fig. 6.47 Change of N_2 content in waterflooding experiments

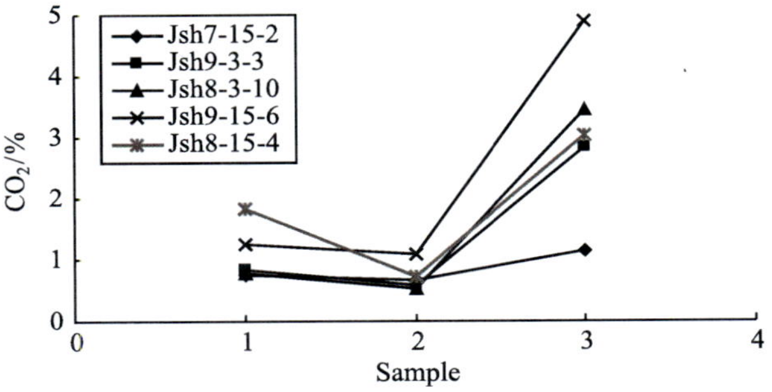

Fig. 6.48 Change of CO_2 content in waterflooding experiments

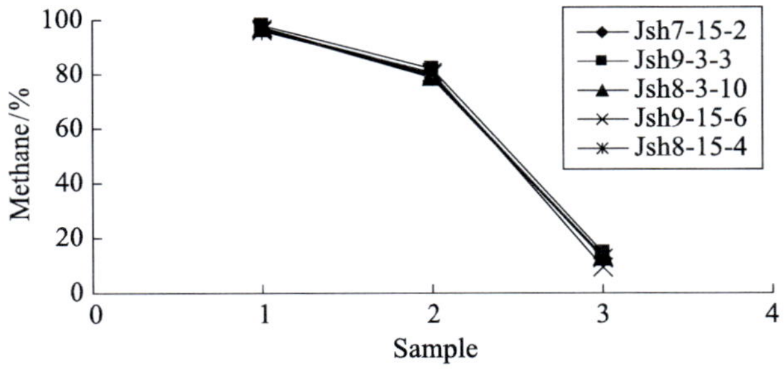

Fig. 6.46 Change of methane content in waterflooding experiments

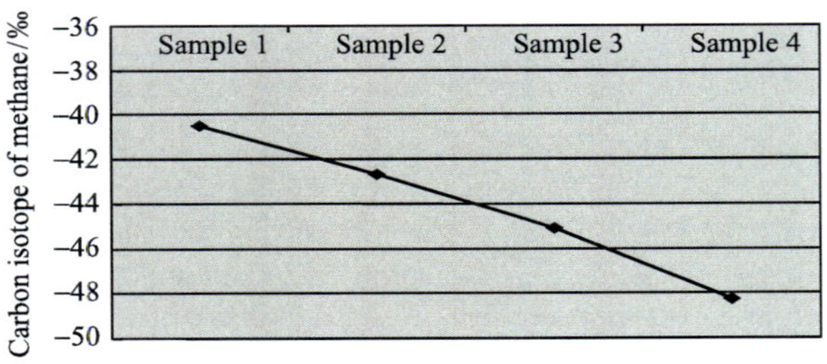

Fig. 6.49 Change of methane carbon isotope in waterflooding experiments

Table 6.6 CBM composition during waterflooding experiments

Sample composition	Sample 1 (original)	Sample 2 (after 4 days)	Sample 3 (after 8 days)
CH_4/%	79.45	69.23	7.93
N_2/%	19.89	30.26	90.79
CO_2/%	0.66	0.52	1.28

reduces, and the adsorbed gas is continuously desorbed, dissolved and transported. $^{13}CH_4$ is preferentially dissolved and transported, but $^{12}CH_4$ remains in place, making the $\delta^{13}C_1$ value lighter. Through the cumulative effect, $^{12}CH_4$ is enriched, and the carbon isotope of methane gradually becomes lighter. Under the same partial pressure conditions, the solubility is $CO_2 > CH_4 > N_2$. CO_2 preferentially is desorbed and decreases after transported with water. N_2 is almost insoluble in water, but due to the oxidation of some methane, the content of N_2 and CO_2 increases. Under the action of oxygen molecules and sulfate dissolved water in contact with the gas reservoir, some methane in the epigenetic zone is oxidized, resulting in a large amount of N_2, CO_2 and the like.

Different Influence of Hydrodynamics

(1) Generation of CBM

Low-ranked CBM is mainly biogas. Active hydrodynamics are primary geological dynamics in the generation of CBM, and the appropriate formation water salinity is important for the generation of low-ranked CBM. Only methanogens are transported with hydrodynamics to the medium where oxygen, nitrate and most of sulfates are reduced can grow and produce methane (Wang et al. 2007). A large number of simulation tests show that: (i) The optimum temperature for methanogens is 36–42 °C. (ii) The metabolism of methanogens is restricted by the pH of the aqueous medium. The most suitable pH is 7.0–7.2. When the pH is below 6.2, most of methanogens stop growing or produce no methane. When the pH is close to 8.8, the solubility of CO_2 is close to zero, and most of methanogens cannot survive. Tests show when the salinity of the aqueous medium is less than 0.4×10^4 mg/L, methanogens are the most active, gas production rate is the highest, and too high or too low salinity is unfavorable for the growth and development of methanogens. Therefore, active hydrodynamics carry a large number of methanogens to a suitable salinity environment, which is conducive to methane reproduction, and a large amount of biogenetic CBM may be produced.

(2) Forming of CBM reservoir

Low-ranked CBM is mainly composed of free and adsorbed gas. The hydrodynamic action is to dissolve gas. The organic matter in water is to adsorb methane and carry and transport free gas (Wang et al. 2007). High-ranked CBM is mainly adsorbed gas. The hydrodynamic action is to dissolve gas. The dissolved gas is carried away by hydrodynamics. The process of "dissolving, migrating, dissolving and re-migrating" is continuous until the CBM reservoir is destroyed.

For high-ranked CBM reservoirs, the reduction of gas content is one of the main features of damaged CBM reservoirs. For example, in the strong runoff zone on the southern margin of the Qinshui Basin, the water flow rate is large, and the hydrodynamic alternating action is strong, which adversely affects the preservation of the coal seams. The gas content of the coal seams is generally low. Whether the No. 3 coal seam of the Shanxi Formation or the No. 15 coal seam of the Taiyuan Formation has the gas content between 6 and 8 m^3/t. Due to the slower change of the hydraulic gradient, the gas content does not change much. Under the influence of water-conducting faults and secondary folds, the hydrodynamic alternating effect in the slope zone is strong, which is not conducive to CBM accumulation. The development of the normal fault network in the central and western parts greatly enhances the water conductivity, and the hydraulic interaction is particularly strong. The minimum gas content of the No. 3 coal seam is 4 m^3/t, and the minimum gas content of the No. 15 coal seam is 3 m^3/t. With strongly changing hydraulic gradient, the gas content varies greatly. The highest gas content of the No. 3 coal seam is 14 m^3/t, and that of the No. 15 coal seam is 16 m^3/t. The water-conducting faults are less in the eastern part of the Basin, so the hydrodynamic alternating action is weaker. The minimum gas content of the No. 3 coal seam is 6 m^3/t, and that of the No. 15 coal seam is 8 m^3/t. The small hydraulic gradient makes the gas content change little, and the highest gas content is 10 m^3/t in the two main coal seams. The hydrodynamic alternating action is weak in the deep weak runoff zone, the permeability is low, and it is easy to induce a sealing condition caused by water pressure. The gas content increases significantly with the weakening of the hydraulic interaction. The gas content of the No. 3 coal seam can reach 26 m^3/t, and that of the No. 15 coal seam up to 28 m^3/t. Obviously, the runoff intensity has a certain negative correlation with the gas content. Weak or stagnant water is the best hydrogeological condition for the preservation, enrichment and accumulation of CBM reservoirs.

For low-ranked CBM reservoirs, hydrodynamics controls the formation and containment of gas reservoirs. For example, the marginal slope of the Junggar Basin is cut by faults, and the faults are weakly water-conducting. The slope is dipped upward to the coal seam outcrop or connected with the Cenozoic aquifer (Sang et al. 1999), which is a groundwater recharge area. The water-conducting fault constitutes a drainage zone. The strong sandstone aquifer has a good hydraulic connection with the weak coal aquifer in the fault block, and becomes a groundwater runoff zone or a stagnant zone, building a relatively complete hydrodynamic system. In the southeastern margin of the basin, there is a strong drainage zone and a strong runoff zone. The gas content of the main coal seam in this area increases with the content of methane, indicating that groundwater continuously dissolves CBM and floods coal in the process of migration, so that the content of methane continuously reduces, and CBM escapes. It can be seen that strong hydraulic interaction is not conducive to the accumulation of low-ranked CBM. In the northwestern margin and the northeastern margin, the generation, development and sealing conditions are better, runoff is weak or stagnant, the hydrodynamic storage conditions are better, the hydraulic alternating action is weak, and the CBM composition changes little, so that CBM is accumulated and preserved.

6.3.5.3 Comparison of Geochemical Characteristics of Groundwater

Influence of groundwater geochemistry of high-ranked CBM on accumulation, the geochemical characteristics of groundwater are also important factors affecting the accumulation of CBM. The water salinity influenced by different groundwater ions in different regions have different effects on the accumulation of CBM in high-ranked coal, under normal geological conditions. The stronger the rock erosivity, the better the hydrodynamic alternating condition. The strength depends on the amount of erosive chemicals in water. In addition, the groundwater chemical field reflects the characteristics of groundwater alternation and runoff, and has a certain indication for the enrichment conditions of CBM. Here taking the Qinshui Basin as an example, we analyze how the distribution characteristics of groundwater chemical field influence the accumulation of CBM reservoir.

(1) Geochemical field of the Upper Carboniferous Taiyuan Formation

The groundwater in the Carboniferous aquifer is mainly $HCO_3{\cdot}SO_4$-K·Na. From the two wings to the axis of the basin, the Carboniferous is covered by the Permian and the Triassic, and is open, semi-closed to closed. The water type is from $HCO_3{\cdot}SO_4$-Ca to $HCO_3{\cdot}SO_4$-K·Na and $HCO_3{\cdot}SO_4$-Ca·Mg, and dominated by $HCO_3{\cdot}SO_4$-K·Na.

The regional distribution of the groundwater salinity in the Upper Carboniferous Taiyuan Formation aquifer is similar to that of the Middle Ordovician aquifer, showing an overall trend of increasing from northwest to southeast, indicating that groundwater recharge mainly comes from northwest. Groundwater in local areas is obviously stagnant or slow.

(2) Geochemical field of the Lower Permian Shanxi Formation aquifer

The most important transformation feature of the groundwater geochemical field in the Shanxi Formation is the change in the relative importance of the main recharge areas reflected by the regional distribution of groundwater salinity. In the middle, south and north of the Qinshui Basin, the orientation of the salinity contours are transformed from NNE-SN in the Taiyuan Formation to NNW in the Shanxi Formation. The contours are arc-shaped with the top convex toward north. The arc top line is biased to the west of the axis of the basin. The salinity in the northeast is obviously higher than that in the northwest.

In addition, the high salinity zone in the middle and south of the Qinshui Basin not only exists, but also the distribution extends further to the surrounding area. The area with a salinity greater than 1000 mg/L cover Zhengzhuang, Daning, Panzhuang, Fanzhuang and the south of Zhaozhuang. The high-salinity groundwater is the inevitable result of the extremely gentle equipotential or a "low-lying" zone, which together reflects very stagnant groundwater. It is very important to the preservation of CBM in the upper main coal seam.

(3) Influence of groundwater geochemical field on high-ranked CBM reservoirs

In the southern part of the Qinshui Basin, the salinity distribution shows an overall trend increasing from northwest to southeast. The gas content is 4–8 m^3/t in the northwest main coal seam, and 8–14 m^3/t in the southeast, indicating that the distribution of the gas content also shows an overall trend increasing from northwest to southeast.

In addition, the most significant salinity center in the southern part of the Qinshui Basin appears in the Daning-Panzhuang-Fanzhuang area where the salinity is 1800 mg/L and the highest is 2600 mg/L, the gas content is

16 m^3/t, and the highest 22.5 m^3/t. It shows that with the increase of salinity, the gas content shows an increasing trend, indicating a good positive correlation between the two.

In short, both the global and the local salinity centers are consistent with the gas content, indicating that high salinity is conducive to the accumulation of high-ranked CBM reservoirs.

6.3.5.4 Influence of Groundwater Geochemistry on CBM Accumulation

Influence of Groundwater Geochemistry on Low-Ranked CBM Accumulation

The influence of groundwater geochemical characteristics on low-ranked CBM accumulation were analyzed by simulation experiments on the Tu-Ha Basin.

(1) Physical simulation experiment on groundwater geochemical characteristics

The experimental samples were taken from the No. 8 coal seam of the Xishanyao Formation in Well Ha Test 1 in the Tu-Ha Basin. During the drilling process, the cores were taken by a wireline coring tool and stored in a sealed can, and then naturally desorbed at the reservoir temperature of 38 °C. The methane is pure up to 99.6%. The coal samples are long-flame coal and lignite with a low degree of deterioration.

Saturated brine with a salinity of 15,000 mg/L, steamed water and dry coal samples were used in the simulation experiment. The coal core was charged with methane and flooded with different types of aqueous solution, and then the absorption capacity was analyzed to explore how salinity affects the formation of low-ranked CBM.

(2) Experimental phenomena and analysis

The simulation with saturated saline water shows that when the pressure reaches 1.7 MPa, the gas content reaches 2 m^3/t. The simulation with steamed water shows that when the pressure reaches 2.5 MPa, the gas content reaches 2 m^3/t.

The pressure-time curve (Fig. 6.50) shows that as the experimental time goes on, the pressure curves at the three media tend to decrease. The higher the salinity, the smaller the pressure drop, the faster the formation pressure gradient decreases, the lower the reservoir pressure, the lower the adsorption capacity, the higher the gas saturation, and the greater the desorption of gas.

The adsorption capacity of lignite is low, so the pressure change is not obvious. The higher the salinity, the lower the adsorption capacity and the lower the gas content. During

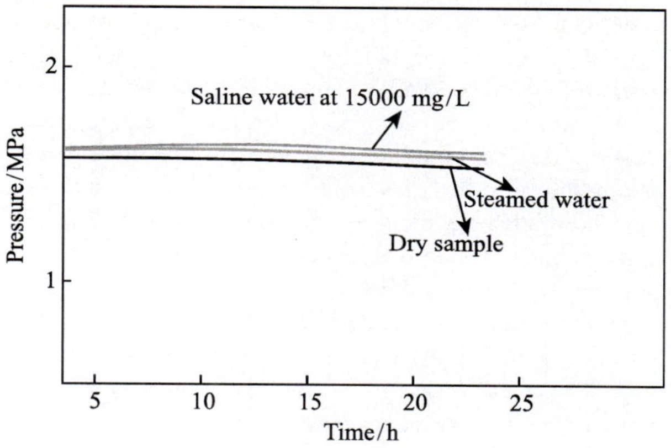

Fig. 6.50 Pressure-time curve of the simulation experiment on formation water salinity

the geological history, the salinity was increasing, and the high salinity resulted in decreases in adsorption capacity, formation pressure gradient and reservoir pressure, and increases in gas saturation and gas desorption. In addition, with a large amount of water supply from the outcrop, a large amount of methanogens were carried by water to suitable depth and temperature, where they degraded low-ranked lignite. In short, low-salinity formation water was beneficial to the growth of methanogens in geological history, and coal seams could be degraded and produce methane.

(3) Influence of groundwater chemistry on CBM accumulation

For low-ranked CBM reservoirs, the accumulation requires an environment conducive to methane production, growth and enrichment—suitable temperature and groundwater salinity. If the burial depth is generally at the stagnant groundwater zone, the salinity is very high, which is not conducive to the activity of methanogens. From the simulation experiment on groundwater chemical field, it can be seen that high salinity causes the adsorption capacity of low-ranked coal reservoirs to decrease. The free gas migrates and dissipates with the hydraulic action. At the same time, as the reservoir pressure drops to the critical desorption pressure, the adsorbed gas continuously desorbs, diffuses, seeps and migrates, eventually leading to a decrease in the CBM content and serious damage to the CBM reservoir (Wang et al. 2007).

Methanogens are strict anaerobic bacteria. In the nature, methanogens can only grow if the oxygen, nitrate and most of the sulfate contained in the medium are reduced. In addition, as the salinity of the formation water increases, the solubility of natural gas decreases significantly. Moreover, it can be seen from the simulation experiment on groundwater chemical field that high salinity also causes the adsorption capacity of the low-ranked coal reservoir to decrease, and the free gas migrates and dissipates with the hydraulic force. When the reservoir pressure is reduced to the critical desorption, the adsorbed gas continuously desorbs, diffuses, seeps and migrates, which eventually leads to a decrease in the gas content and serious damage to the CBM reservoir.

In the Tu-ha Basin, the low-ranked lignite is shallow at 800 m, the thickness is greater than 50 m, the water salinity is as high as 16,000 mg/L, and the gas content is less than 3 m^3/t. The main cause for the low gas content is the high salinity of the groundwater, which on the one hand destroys the growth of low-ranked coal methanogens and the production of biogas; on the other hand, reduces the CBM adsorption. It is extremely detrimental to the preservation of CBM reservoirs because CBM dissipates with hydraulic action.

Differences in Groundwater Geochemistry on CBM Accumulation

High-ranked coal means high salinity, indicating good storage conditions. High-salinity water mainly exists in the pressure zone where the CBM reservoir has high gas content and good storage conditions. High salinity is conducive to the accumulation of high-ranked CBM reservoir. Low-ranked lignite means low salinity. Low-salinity water is mainly located in the zone with a large amount of supply from outcrop or a water supply area, and the water can carry a large amount of methanogens. In addition, conditions such as relatively low salinity, active, not strong but favorable hydraulic interaction provide a good living environment for the growth of a large number of methanogens, which facilitates and makes these methanogens to degrade coal seams and produce and enrich secondary biogas. Finally low-ranked CBM reservoirs accumulate.

According to the above analysis, the high-ranked CBM reservoir has three significant advantages: (i) High coal metamorphism, large amount of gas generated, strong adsorption capacity and high gas content. (ii) Tectonic thermal events and tectonic stress field greatly influence on physical properties of the coal seam. Tectonic thermal events promote the massive generation of CBM and improve reservoir physical properties. Tectonic stress exerts influence on the original reservoir permeability by controlling the opening of natural fractures. (iii) The stagnant and high-salinity zones have good preservation conditions for CBM, which are conducive to CBM preservation and production by draining water and reducing pressure. The low-ranked CBM has two significant advantages: (i) The accumulation process is simple, and it is easy to form biological gas. With sufficient external gas source, multi-source CBM reservoirs can be formed. The cumulative production may be higher than the original resources. (ii) The matrix is loose, and the reservoir permeability is high, so it is easy to produce CBM by depressurizing and desorbing.

6.4 Controlling Factors and Distribution Rules of CBM Enrichment

Because of the occurence state and accumulation mechanism different from conventional gas reservoirs, CBM has its own characteristics in controlling factors and enrichment rules. According to the analysis of tectonic evolution characteristics of coal-bearing basins in China, except for coal-bearing basins with low metamorphism, most of the basins have undergone subsidence and reverse uplift, and the coal seams have undergone burying and uplifting processes, and some basins even experienced many cycles. The burying and uplifting processes determine CBM accumulation and evolution. With the burial and uplift of coal seams, CBM reservoirs undergoes the generation and adsorption of CBM, the increase of adsorption capacity and the desorption, diffusion and preservation of CBM. The generation and adsorption stages include the generation and adsorption stage caused by the burial of the coal seam and that caused by abnormal thermal events. The desorption, diffusion and preservation stage mainly involves caprock diffusion mechanism and groundwater dissolution mechanism. The CBM preserved under various mechanisms accumulates in the modern CBM reservoirs (Fig. 6.51). Therefore, tectonic evolution, hydrogeological conditions and sealable surrounding rock are main controlling factors on CBM accumulation. Syncline is comprehensive expression of these controlling factors. CBM tends to accumulate in synclines. Hydrogeological conditions are described in Chap. 5.

6.4.1 Structural Control and Key Period

CBM is unconventional natural gas generated and stored in place. It is adsorbed on the pore surface of coal seams. Its accumulation is controlled by various factors. The formation of CBM reservoirs is not only related to the environment and characteristics of modern coal seams, but also closely related to the tectonic evolution process. During the geological history, the shallowest period of the overburden strata after the coal seams stopped producing CBM was the critical period when forming CBM reservoirs. During the period, the amount of CBM stored in the coal seams is critical to the formation of today's CBM reservoirs.

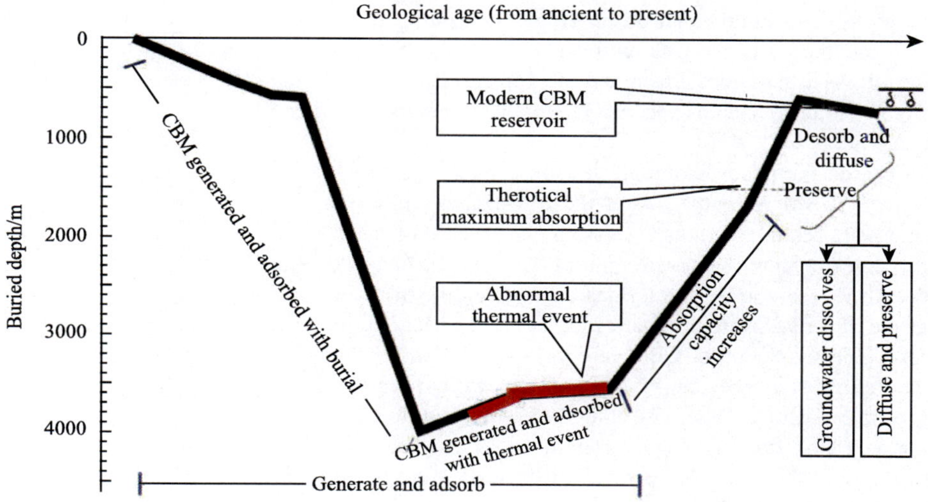

Fig. 6.51 Schematic evolution and accumulation mechanisms of CBM reservoirs

6.4.1.1 Controlling Mechanism and Key Controlling Period of Tectonic Evolution on CBM Accumulation

The controlling mechanism of tectonic evolution on CBM accumulation is based on the monolayer adsorption theory-Langmuir equation. The adsorption capacity of coal is affected by three factors: coal properties, temperature and pressure (Zhang and Tao 2000; Su et al. 2001). Once formed and reaching a certain maturity, coal properties are basically stable. Only temperature and pressure affect the adsorption capacity. The higher the temperature and the lower the pressure, the smaller the adsorption capacity. At normal geothermal gradient, the combined effect of pressure and temperature makes the adsorption capacity reach maximum when the coal seam is buried at about 1500 m, and then the gas content decreases with the depth (Zhang and Tao 2000). On the basis of a large amount of data, the change of the adsorption capacity of two key coal ranks in the No. 3 coal seam was established with the buried depth. The solid line less than 1000 m is supported by actual exploration data, and the dotted line larger than 1000 m is mainly prediction of CBM gas content controlled by temperature. The coal rank with $R^{\circ} = 1.0\%$ is roughly equivalent to the evolution of the coal seam in the southern part of the Qinshui Basin at the end of the Triassic. The coal rank with $R^{\circ} = 3.5\%$ is roughly equivalent to the evolution of the coal seam in the southern part of the Qinshui Basin at the end of the Early Cretaceous (Fig. 6.52).

Except for those with low coal metamorphism, most coal-bearing basins in China have experienced reverse uplift and late evolution. The current enrichment of CBM reservoirs is a comprehensive superposition of reverse uplift and evolution preserving and destructing CBM reservoirs. Therefore, the control on CBM reservoirs by structure is

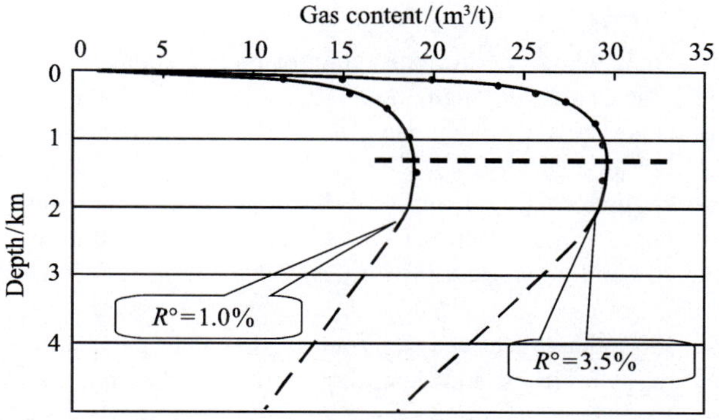

Fig. 6.52 Relationship between adsorption capacity and depth of No. 3 coal seam in Qinshui Basin

mainly manifested by the change of the overburden thickness happening during the reverse uplift. The temperature and pressure changes with the overburden thickness control the gas content. The key period of CBM accumulation is the period when the overburden was the thinnest, during which the pressure and temperature corresponding to the "effective thickness" determine the gas content of the modern coal seam.

The timing and term of the reverse uplift of the coal-bearing basin is important for today's enrichment of CBM. If the uplift is late and short, CBM dissipation is short, which is beneficial to the preservation of CBM. For example, the reverse uplifts in the eastern and western regions of North China are different, the degree of CBM enrichment is significantly different (Wang et al. 2003). From the Indosinian period, the North China block was disintegrated, and the western region continued to sink, and then began to rise until the Late Cretaceous Yanshanian movement. Most areas in the east of the Taihang Mountains began to rise from the end of the Triassic and suffered erosion, and then began to sink until the Quaternary. Therefore, the CBM in the Ordos Basin and the Qinshui Basin is relatively enriched and has become a favorable target for CBM exploration.

The degree of reverse uplift directly controls the enrichment of CBM reservoirs. After the Yanshanian uplift, the coal seam ceased to produce gas, so in the absence of special geological events such as tectonic thermal events, whether the CBM reservoir can be formed depends on the thickness of the effective overburden after the uplift. The key period for the accumulation of CBM reservoirs is the period when the "effective overburden" was the shallowest during the geological history. When selecting a CBM exploration area, not only the current geological conditions such as coal seam thickness, coal rank, pore permeability, cleats, caprock and hydrodynamics are to be considered, but more importantly, whether the shallowest burial is conducive to the preservation of CBM after stopping gas generation. For example, in two areas where the current geological conditions are basically the same, although today's burial depths of the coal seams are the same, they were different after uplifted during their geological histories. One is under the weathered zone and the preservation condition is better; another is above the weathered zone and most CBM are lost. If the current geological conditions are used for CBM evaluation, errors will occur. For example, in the Dacheng area of Hebei Province, due to frequent tectonic activities in the late Yanshanian, long-term uplifting and denuding made the effective overburden thinner, and the preservation conditions are not good, so that the CBM in most of the area is lost (Wang et al. 2004). CBM drilling results show that the CBM production is low, and the gas saturation and the desorption pressure are low, which is not conducive to the development of CBM.

Figure 6.53 shows three types of reverse uplifting models in the case of the same depth. Type a and Type b have the same critical period. Type a means the critical period showing severe CBM destruction. Type b means the critical period showing CBM accumulation. The biggest difference between Type b and Type c is that although their effective thickness is consistent today, the critical period of CBM accumulation is different. Type b was uplifted earlier, so the critical period is earlier, but the CBM loss is more than Type c. During the evolution of the three coal seams, Type c is the most favorable, Type b is more favorable, and Type a is the last.

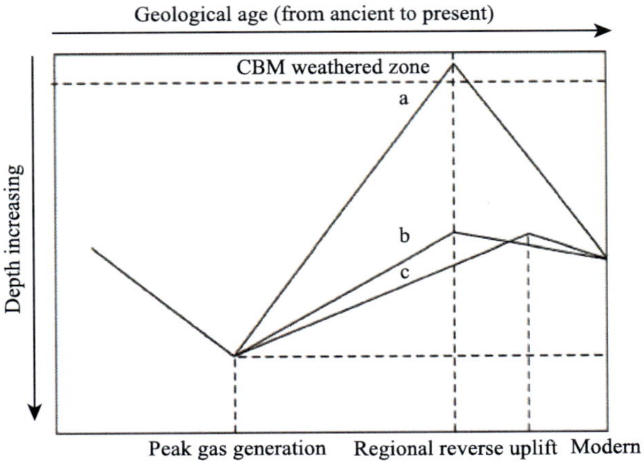

Fig. 6.53 Three types of reverse uplifting models in the case of the same depth

6.4.1.2 Three Geological Models of Tectonic Evolution Control

The tectonic evolution in the Himalayan period after the reverse uplift of the coal-bearing basin has an important influence on the enrichment of CBM. The CBM in the zone being uplifted and eroded will keep lost, while the CBM in the zone subsiding can be preserved. But it is easy to cause the decrease of CBM saturation.

Theoretically, in the absent supply of gas source, when coal seams rise, the overburden pressure decreases, and the fluid pressure in the coal seam decreases, causing the CBM to desorb and dissipate, and some remain as free gas and water-soluble gas (Sang et al. 1997). As a result, the CBM content is reduced, but the saturation is increased to 100%. When the basin settles, the thickness of the overburden increases, the fluid pressure in the coal seam increases accordingly, and the adsorption capacity of the coal seam increases with the increase of pressure (Zhao et al. 2005). In other words, theoretically, the adsorption capacity increases, but because there is no gas source, the gas content still keeps the level before the basin settles, thus causing a decrease in the gas saturation. Using the newly established FY-type CBM reservoir simulation device (Wang et al. 2004), the relationship between pressure, adsorption capacity and saturation of high-ranked coal was simulated. The simulation results are consistent with the results of the above theoretical analysis.

Through theoretical analysis and experimental simulation, three geological models of tectonic evolution control were summarized (Fig. 6.54):

Destructive type (Type I). The coal seam continues to rise into the weathered zone after the regional uplift, so that the gas content and saturation in the coal seam are very low, and generally can't meet the standard for CMB reservoir.

Enriching type (Type II). The coal seam continues to rise into the weathered zone after the regional uplift. The gas content depends on the thickness of the overburden in the critical period. The thicker the effective overburden, the higher the gas content, but the saturation is high.

Re-expanding type (Type III): The coal seam settles after the uplift. The gas content depends on the thickness of the

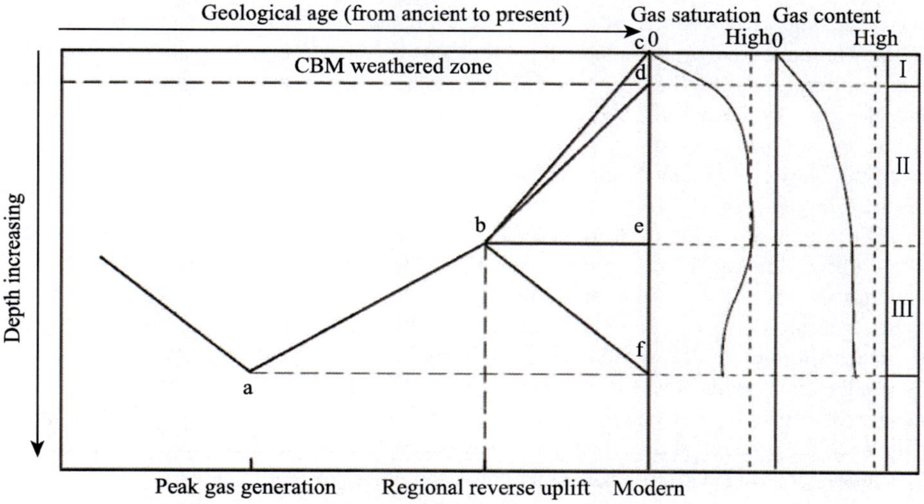

Fig. 6.54 Modern CBM enrichment of three tectonic evolution models after reverse uplift

overburden in the critical period. The saturation depends on the thickness of the re-settled formation. The thicker re-settled formation means lower saturation.

It is worth noting that the above description only analyzes the influence of single factor related to structural evolution. If thermal event, biogas and deep temperature are considered, the results may be more complicated.

6.4.1.3 Case Analysis and Application

Three geological models of tectonic evolution control are of great significance for CBM exploration. For example, CBM accumulation in the southern Qinshui Basin is of enriching type, the western Qinshui Basin is of destructive type.

The burial history of the Taiyuan Formation and the Shanxi Formation in the Qinshui Basin has undergone five stages. Stage 1 is from the Late Carboniferous to the end of the Early Permian when the crust settled slowly; at the end of the Early Permian, the No. 15 coal seam is the deepest 300 m, and the average settling rate did not exceed 25 m/Ma. Stage 2 is from the Late Permian to the end of the Late Triassic when the crust settled rapidly, the buried depth of the coal seam increased rapidly; and by the end of the period, the maximum settlement range reached 4500 m, with the Jincheng–Yangcheng–Houma line as the settlement center. The average settlement rate is 80–100 m/Ma. At the end of the Late Triassic, the maximum buried depth of the No. 15 coal seam in the Jincheng-Yangcheng area is 4800 m. Stage 3 is the Early Jurassic when uplifting and denuding were going, and the maximum uplift exceeds 1000 m. Stage 4 is the Middle Jurassic when slow subsidence was going, and the Yanshanian movement formed the Qinshui synclinoria and a basal shape of the Qinshui Basin. Centered on the syncline axis, the maximum subsidence is more than 400 m and the average rate is about 16 m/Ma. Stage 5 is since the Late Jurassic. Form then on, uplifting has been going. From the Late Jurassic to the end of the Paleogene, the entire basin was in a state of uplift for a long time, the coal systems and the overlying strata were severely denuded; and the Himalayan movement formed secondary fault basins. The Neogene and the Quaternary sediments appeared locally, and the maximum settlement in the northwestern part of the basin is more than 1000 m. The general feature of the sedimentary and burial history is slowly denuding during the middle Yanshanian (J_3–K_1), and quickly denuding from K_2 to N_1. In addition, the middle Yanshanian anomalous thermal event from the Late Jurassic to the Early Cretaceous (Chen 1997; Sang et al. 1997; Ren et al. 1999) made the geothermal gradient anomalously high. The highest paleotemperature gradient might reach 5.0–6.1 °C/hm at J_3–K_1, which is crucial for the determination of the temperature at critical moments.

In the Jincheng area in the southern part of the Qinshui Basin, the No. 3 coal seam (enriching type) was buried at about 4000 m at the end of the Triassic. Although the coal seam reached the threshold of natural gas generation at a large amount, its adsorption capacity is the lowest, only about 15 m^3/t. At the middle Yanshanian highest temperature, the absorption capacity is not the highest. When the depth of coal seam is about 1500 m, the absorption capacity is the highest. The absorption capacity is of great geological significance when the coal seam is buried the shallowest, because the coal seam has no ability to produce gas. Therefore, the amount of adsorbed gas in today's coal seam depends mainly on the geological history when the coal seam was the shallowest after experiencing the highest temperature. The coal seam in this area was the shallowest at the end of the Neogene, only 550 m, and then accepted 50 m Quaternary sediments, so the caprock was almost not destroyed. In addition, the late Neogene pressure and the present pressure is almost similar, the adsorption capacity was little impacted, even if there was a little impact, the CBM was preserved by the superior overburden. In short, the adsorption capacity had been determined at the end of the Neogene which is the key geological moment for the formation of CBM reservoirs.

In the western Qinshui Basin, the No. 3 coal seam (the destructive type) in the Huozhou area at the end of the Triassic was nearly 4000 m deep, and had a relatively low CBM absorption capacity, roughly 17 m^3/t. The Huozhou area has not experienced the middle Yanshanian anomalous thermal event, so the coal seam basically maintains the thermal evolution degree at the end of the Triassic period. When the coal seam was about 1500 m deep, the CBM adsorption capacity was the highest, reaching nearly 20 m^3/t. Similarly, the amount of adsorbed gas when the coal seam was the shallowest determines the amount of adsorbed gas in the present coal seam. The coal seam was the shallowest at the late Paleogene, only 150 m, and the adsorption capacity was 8.05 m^3/t. The gas greater than 8.05 m^3/t is mainly free. After accepting 350 m Neogene and Quaternary sediments, on the one hand, the CBM cannot be regenerated; on the other hand, the caprock is weak and unfavorable for the preservation of CBM. In other words, the Paleogene pressure is very inconsistent with the present pressure, so the absorption capacity changed a lot. In short, the end of the Paleogene is the key geological moment for determining the CBM content in the Huozhou uplift. Because the burial was the shallowest and the pressure was lowest at the end of the Paleogene, the adsorption capacity decreased greatly. In addition, due to late diffusion, the CBM content measured is lower than that when the coal seam was the shallowest.

6.4.2 Control of Effective Overburden and Coal Seam Top and Bottom on CBM Accumulation

Coal seam top and bottom and effective overburden are essential for CBM preservation and have important controlling effects on CBM enrichment. CBM enrichment depends largely on the sealability of coal seam roof (i.e. top caprock), bottom (i.e. bottom caprock) and the effective thickness of overburden. The difference between CBM reservoirs and conventional gas reservoirs in terms of sealing conditions depends on gas occurrence, how gas diffuses, and what drives gas to lose.

6.4.2.1 The Sealability of Coal Seam Top and Bottom and Their Control on CBM Enrichment

The understanding to the sealability of coal seam top and bottom begins with how CBM loses. CBM is stored in coal seams as adsorbed gas, free gas and water-soluble gas. The gas generated by coal seams first is adsorbed, and then dissolved in water and free, and the three states are in a dynamic equilibrium system. The amount of adsorbed gas is related to temperature and pressure. When temperature and pressure change, CBM may desorb. Therefore, lost CBM must first be desorbed, then dissipates as free gas and dissolved gas. There are three ways for CBM to dissipate and lose: the free gas in pores is lost through the cap layer (by overcoming capillary pressure); CBM diffuses driving by the different hydrocarbon concentration in the reservoir and the cap layer; and the gas dissolved in water is directly carried away by water (Fig. 6.55).

By investigating how CBM loses, the sealability of coal seam top and bottom depends on free CBM loss by overcoming capillary pressure and diffusing CBM driven by different concentration. In theory, the coal seam should have enough gas. The ideal storage condition is that the coal seam is in a closed system, which means that the top and bottom of the coal seam are also very important for the preservation of CBM. If the top cover is good and the bottom is permeable, gas generated by the coal seam will diffuse through the underlying permeable layer, causing CBM to be lost and affecting the adsorption capacity. Similarly, if the bottom is tight, and the top is permeable, CBM will diffuse too (Fig. 6.56).

After analyzing previous researches on CBM preservation conditions (Li 1987; Qin et al. 2005), and understanding CBM loss mechanism and control factors, we built three types of geological model for CBM loss in China's coal seams (Fig. 6.57): (i) Closed system, mainly located in the syncline axis (basin bottom), where the groundwater is stagnant (within the closed watertable contours); the water is not connected with the water outside; the formation water salinity is high (more than 3000 mg/L in the Qinshui Basin); and the solubility of methane in water is low. The loss of CBM is controlled by the top and the bottom whose sealability is very important for the preservation of CBM. Good top and bottom provide high sealability for enriched CBM. For example, CBM accumulates in the axis of the synclines in the Qinshui Basin and the Xishan Coalfield. (ii) Lateral water system located at syncline wings or monoclinics where the coal seams are exposed; the groundwater is weakly

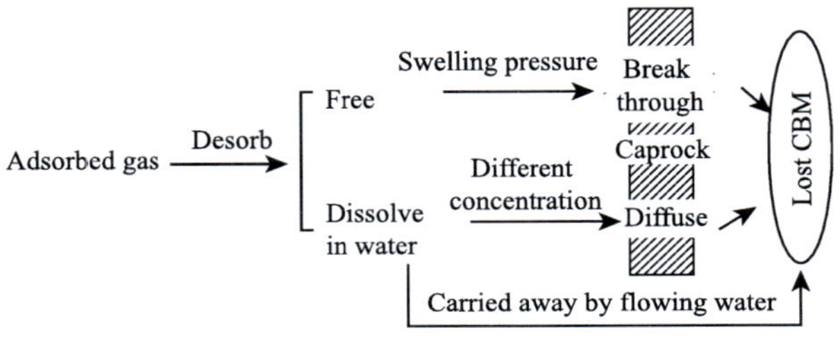

Fig. 6.55 Modes of CBM diffusion

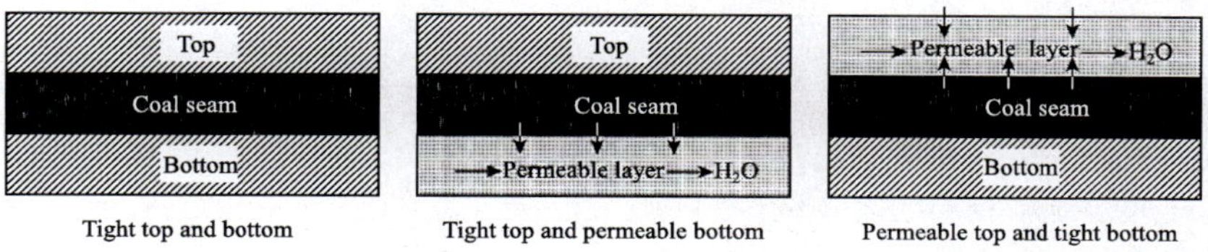

Fig. 6.56 How top and bottom affect the preservation of CBM

running off or stagnant; the water may connect with the water outside; the formation water salinity is high (1000–3000 mg/L in the Qinshui Basin); and the solubility of methane in water is low. The coal seam is locally open, and the boundary conditions such as faults on one side limit the circulation of water along the coal seam, and the preservation conditions include not only the top and the bottom, but also the water in the updip direction. For example in the Fanzhuang block in the southern part of the Qinshui Basin, the No. 3 coal seam at 300–1500 m is a highly saturated adsorption zone, and CBM is rich. (iii) Open system located at the edge of a syncline or a monocline (basin margin) where the coal seam is exposed or shallow (coal weathered zone); the hydrological unit belongs to water supply (discharge) or strong runoff; the water is active; the formation water salinity is low (less than 300 mg/L in the Qinshui Basin); and the solubility of methane in water is high. The coal seam is connected with external water and strong runoff occurs. The preservation conditions are closely related to the hydrological conditions. In this case, the preservation conditions are poor and not good for the formation of gas reservoirs. It is hard to form CBM reservoirs. For example, in the Fanzhuang block in the Qinshui Basin, water at 300 m and shallower is active. It is a weathered zone, so natural gas loss is large and the gas content is low.

Further analysis of the relationship between the sealing top and bottom and the gas content of coal-bearing reservoirs proves their importance to CBM accumulation. Generally, thicker top and bottom, high clay content and high breakthrough pressure are beneficial to the preservation of CBM. For example, in the southern part of the Qinshui Basin, the top cover of the No. 3 coal seam in Well Jin Test 1 is dark mudstone, 55 m thick and its breakthrough pressure is 8–15 MPa, so the gas content is 25.29 m^3/t (Zhang and Tao 2000); The top cover of the No. 3 coal seam in Well Jin Test 4 is sandstone and mudstone interbeds, 70 m thick, and the breakthrough pressure is 5–8 MPa, so the gas content is high, up to 25.3 m^3/t; the top cover of the No. 3 coal

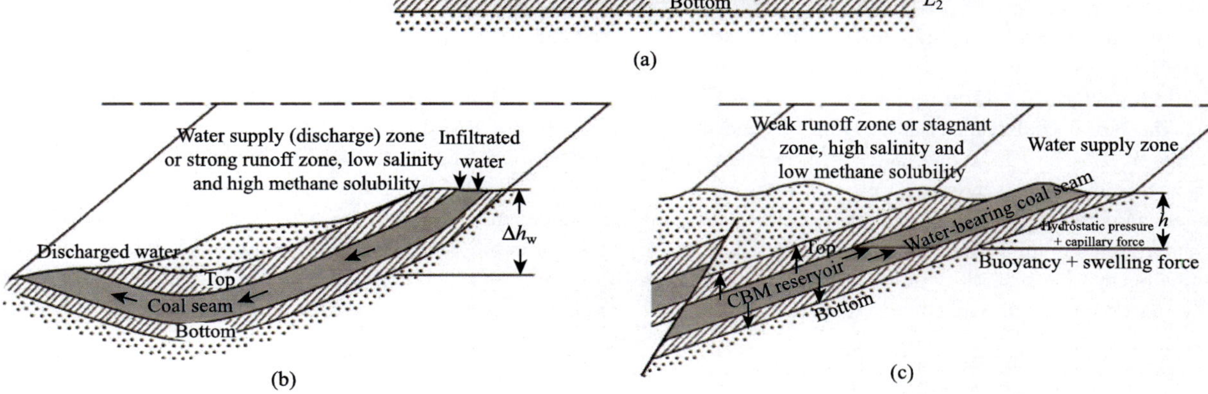

Fig. 6.57 Schematic geological models of CBM loss. **a** Closed system; **b** lateral water sealing system; **c** open system

seam in Well Jin Test 3 is mainly composed of sandstone and mudstone interbeds, only 20 m thick, so the gas content is 17.1 m^3/t, obviously lower than that in Well Jin Test 1 and that in Well Jin Test 4. The statistical analysis of the lithology of the top and bottom at the similar depth and the gas content of the No. 3 coal seam and the No. 15 coal seam also shows that the CBM content is generally higher in the case of mudstone top and bottom than the case of sandstone top and bottom. The average gas content is about 18 m^3/t with a mudstone top, and only 5 m^3/t with a siltstone top. The average gas content is about 19 m^3/t with a mudstone bottom, 9 m^3/t with a sandstone bottom and 1.7 m^3/t with a siltstone bottom (Figs. 6.58 and 6.59). The average gas content of 9 coal seams with mudstone tops, 7 coal seams with limestone tops and 4 coal seams with sandstone tops in the No. 15 coal seam group is 18.87 m^3/t, 15.33 m^3/t and 12.7 m^3/t, respectively. This also shows that the gas content with a mudstone top is higher than that with a limestone or sandstone top. In the Daning area of the Ordos Basin, the No. 5 coal seam at 635 m, 1194 m, 1027 m, 920 m in Wells Ji Test 1, 3, 2, 6 has gas content of 20 m^3/t, 18.5 m^3/t, 12 m^3/t, 14 m^3/t, respectively. In Wells Ji Test 1 and 3, there are mudstone top and bottom, so the gas content is high. In Wells Ji Test 2 and 6, there are sandstone top, so the gas content is obviously low (Sun and Wang 2003). By comparison, the gas content of the No. 3 coal seams at different depths in the Panzhuang, Zhengzhuang and Fanzhuang areas is outstanding due to different lithology of the top and bottom. Theoretically, the burial in the Zhengzhuang area and in Well 0801 is relatively deep (>600 m), the gas content should also be high, but in fact, the coal seams in the Zhengzhuang area is 4.6–5.7 m thick and 5.24 m on average, the gas content is 6.33–6.81 m^3/t and only 6.57 m^3/t on average. The gas content of the No. 3 coal seam in Well 0801 is only 10.87 m^3/t, lower than that in shallow Well 1 in the Panzhuang area and that in the Fanzhuang area. For example, the No. 3 coal seam in Well 1 in the Panzhuang area are at 300–350 m, 4.81–6.79 m thick, and 5.58 m on average; the gas content is 3.74–21.86 m^3/t, and 10.89 m^3/t on average. In the Fanzhuang area, the coal seams are 4.83–6.56 m thick and 5.56 m on average; the gas content is 7.59–22.96 m^3/t, and 13.43 m^3/t on average; and in Well Jin Test 1, the No. 3 coal seam is at 450–500 m, and the gas content is 21.97–27.17 m^3/t, and 25.29 m^3/t on average. The cause is that there are sandstone tops and bottoms in the Zhengzhuang area and Well 0801, while there are mudstone tops and bottoms in the Panzhuang area and Well Jin Test 1. It is obvious that the sealability of coal seam top and bottom is an important factor on gas content (Fig. 6.60).

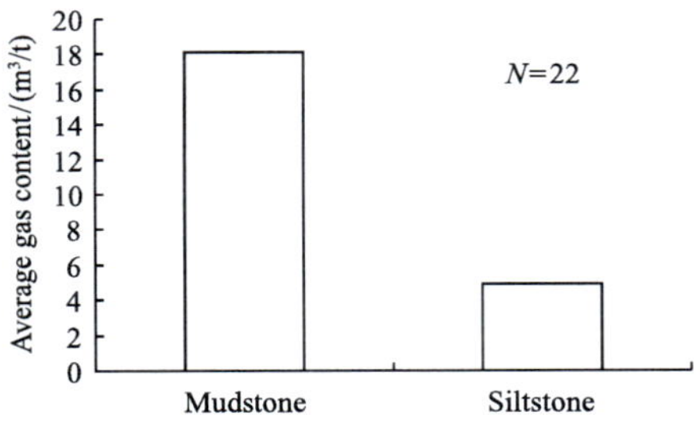

Fig. 6.58 Top lithology and gas content of No. 3 coal seam in Qinshui Basin

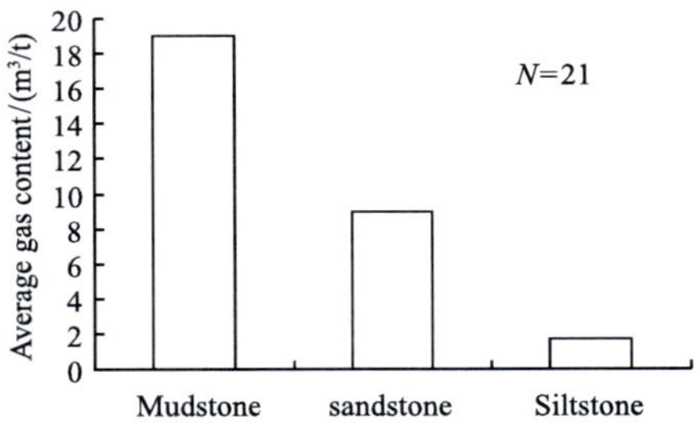

Fig. 6.59 Bottom lithology and gas content of No. 3 coal seam in Qinshui Basin

6.4.2.2 Control of Effective Overburden Thickness on CBM Enrichment

The effective thickness of overburden refers to the thickness of the first unconformity after coal seams generate a large amount of gas. Generally, thicker effective overburden provides good preservation conditions, while thinner effective overburden means the formation has been seriously uplifted and eroded, the formation pressure dropped largely and CBM is easy to desorb and lose.

The coal seams in North China began to settle and reached the peak of gas generation in the Triassic. Then they were first uplifted and denuded, and then settled in the Paleogene and the Neogene, but secondary gas generation didn't take place, the pattern of the coal seams has been keeping to the present. The tectonic movement lifted and denuded the coal seam that passed the peak of gas generation. The denudation did not reach the gas weathering zone. Therefore, CBM didn't lose more, or destroyed, so that the coal seam is rich in CBM. For example, in the Dacheng bulge of North China, after sedimentation, the coal-bearing strata began to sink and reached the peak of gas generation. Then after the Triassic post-structural uplift, the stratum was denuded, and the Permian system (effective overburden) continuously deposited on the top of the main coal seam

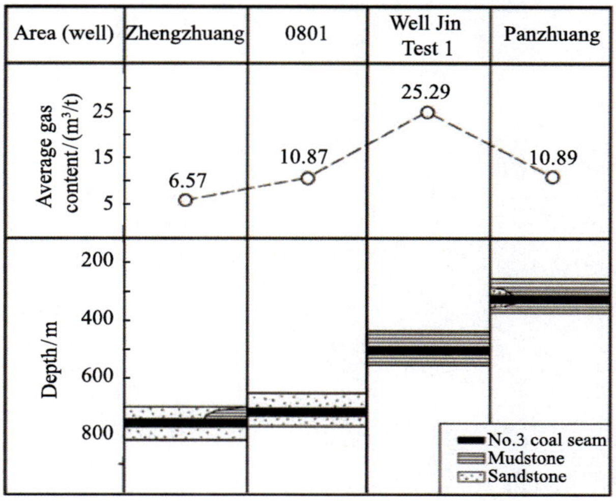

Fig. 6.60 Top and bottom lithology and gas content of No. 3 coal seam in southern Qinshui Basin

around Well Da Test 1 is only 100 m, and the gas content is less than 2 m^3/t, which is in a gas weathering zone. The continuous deposition on the main coal seam in Well Da 1 is 200 m, and its gas content is relatively high, generally greater than 100 m^3/t (Fig. 6.61). In the Jianhe area, Kaiping, although the coal seam is deep (>1000 m), the effective overburden, especially in the Xihe sag, is only about 100 m, so the well logging interpretation shows poor gas-bearing property in Well Xi 2. Poor; in the Kaiping syncline, the continuous deposition of the overburden is relatively thick, generally 200–2000 m, so the storage condition is good in this sense. In the Jiaozuo coal-bearing area and the Jincheng area of the Qinshui Basin, there are examples in which the effective overburden is thick and the gas content of the coal seam is generally higher.

6.4.3 Theory of CBM Enrichment in Synclines

The enrichment of CBM mainly depends on geological factors such as tectonic evolution, hydrodynamic action and sealing conditions in the coal-bearing basin. The comprehensive effect of the three factors determines that a syncline can keep pressure in nature, that is, it is easy to attract CBM to accumulate. A syncline generally has an thickest effective overburden, which is favorable for maintaining a relatively stable formation pressure. It generally attracts formation water to its center, so a high formation pressure system appears in its core. Generally fractures take place in the core, but are not developed, so CBM is dissolved and washed weakly by hydrodynamics.

The occurrence of CBM in coal-bearing basins or enriched zones has a phenomenon or characteristic that as the depth of the coal seam increases, the gas content generally increases; whether in new or old basins, the coal-bearing strata are mostly synclines or composite synclines which axes are generally deep and have high gas content. For example, the profile of the Qinshui Basin shows a complete complex syncline basin, where there is a geological mechanism to maintain a high potential in the syncline core, and the gas content in the core is significantly higher than that in the two wings. This phenomenon is common in the northeastern margin and in the Jincheng area on the southeastern margin of the Qinshui Basin (Figs. 6.62 and 6.63). In addition, there are similar rules in other coalfields or secondary structural belts. For example, in the San Juan Basin in the United States, regardless of the influence of coal ranks, the gas content of coal seams is higher in the syncline core (Fig. 6.64). In the Chengzhuang Coalmine in the Jincheng mining area on the southern margin of the Qinshui Basin, Wang Chunxin et al. conducted a detailed study on the relationship between CBM content and structural morphology in the area, and found that the tectonic form has a significant control on the enrichment of CBM. The gas content of the coal seam at the syncline is generally much higher than that at the anticline. It can be seen that CBM tends to accumulate in synclines, whether from the structural section or plane.

The enrichment of CBM is mainly shown by gas content. For the same coal rank, high gas content means CBM enrichment somewhere, and low gas content means no enrichment. Through careful analysis in the laboratory, it is found that there are six static factors affecting the gas content of coal seams, namely coal rank, temperature, pressure, moisture, coal rock composition and ash. In a geological sense, there are 5 fields, i.e. geothermal field, pressure field, stress field, hydrodynamic field and biological field (Fig. 6.65). CBM is adsorbed on coal matrix under the control of formation pressure. Therefore, under certain geological conditions, for a coal rank, as long as the formation pressure can maintain enough, CBM can be adsorbed in the coal seam, so CBM is enriched in the zone at high formation pressure. The enrichment zone in a syncline is at high pressure, and CBM is mainly adsorbed in the coal matrix driven by the pressure. The theory of CBM enrichment in synclines can be explained by fluid potential.

Hubbert defines the sum of the mechanical energy of underground unit mass fluid as fluid potential (Φ):

$$\Phi = gZ + \int_0^p dp/\rho + q^2/2$$

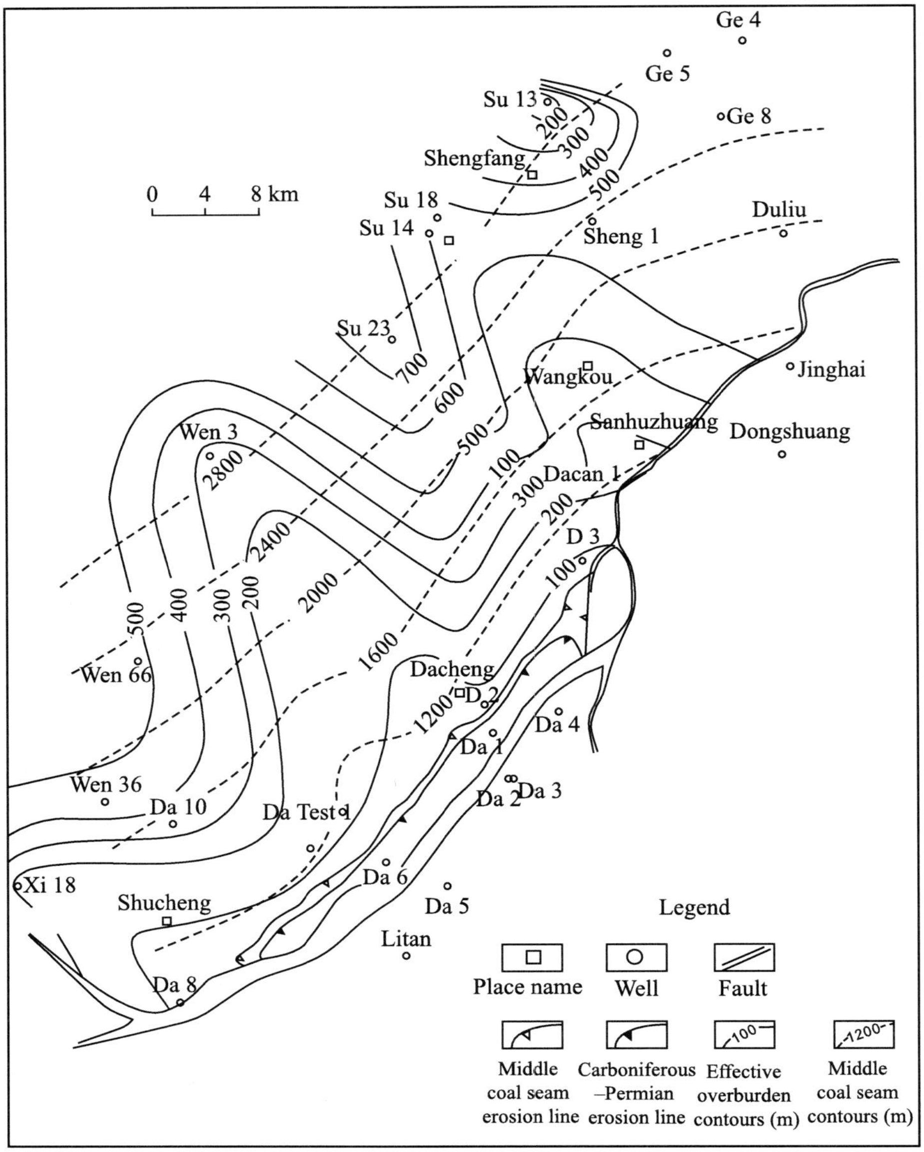

Fig. 6.61 Effective overburden contours in the Dacheng area

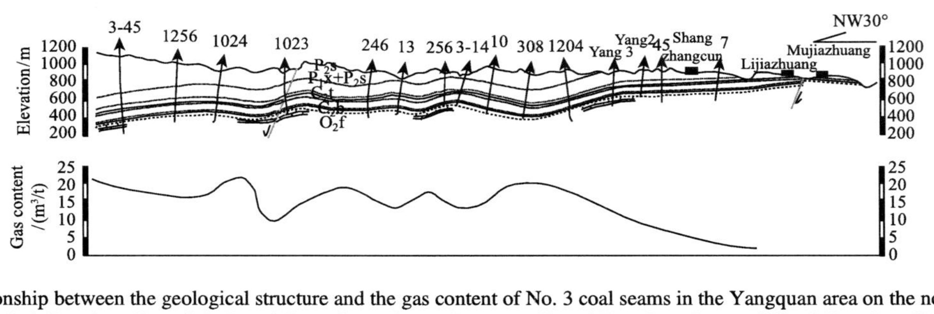

Fig. 6.62 Relationship between the geological structure and the gas content of No. 3 coal seams in the Yangquan area on the northeastern margin of Qinshui Basin. O_2f. Fengfeng Fm.; C_2b. Benxi Fm.; C_3t. Taiyuan Fm.; P_1s. Shanxi Fm.; P_1x. Lower Shihezi Fm.; P_2s. Upper Shihezi Fm

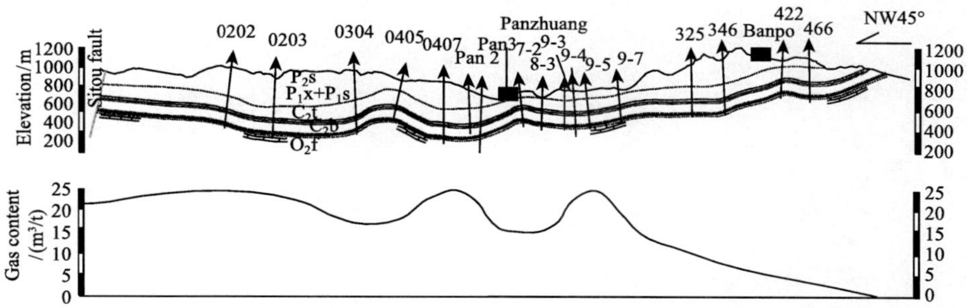

Fig. 6.63 Relationship between the geological structure and the gas content of No. 3 coal seams in the Jincheng area on the southeastern margin of Qinshui Basin. O_2f. Fengfeng Fm.; C_2b. Benxi Fm.; C_3t. Taiyuan Fm.; P_1s. Shanxi Fm.; P_1x. Lower Shihezi Fm.; P_2s. Upper Shihezi Fm

Fig. 6.64 CBM content contours in the Fruitland Formation in the San Juan Basin

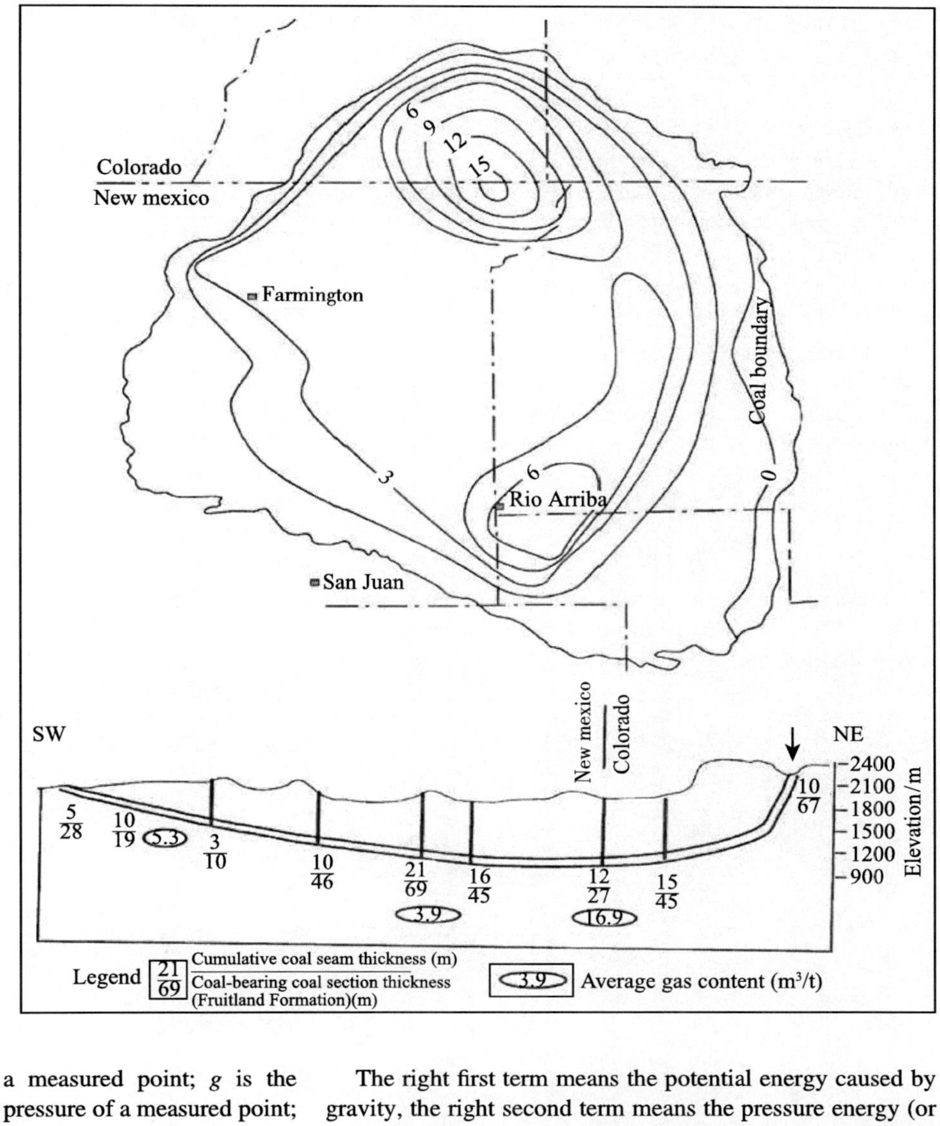

where Z is the elevation of a measured point; g is the acceleration of gravity; P is the pressure of a measured point; ρ is the density of fluid; q is flow rate.

The right first term means the potential energy caused by gravity, the right second term means the pressure energy (or elastic energy) of fluid, and the right third term indicates

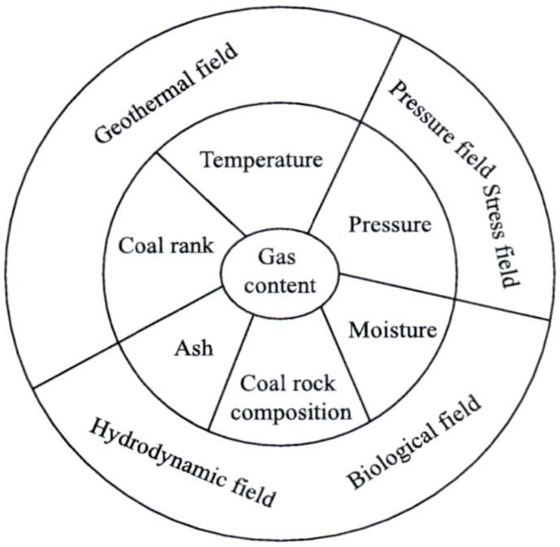

Fig. 6.65 Controlling factors on CBM content

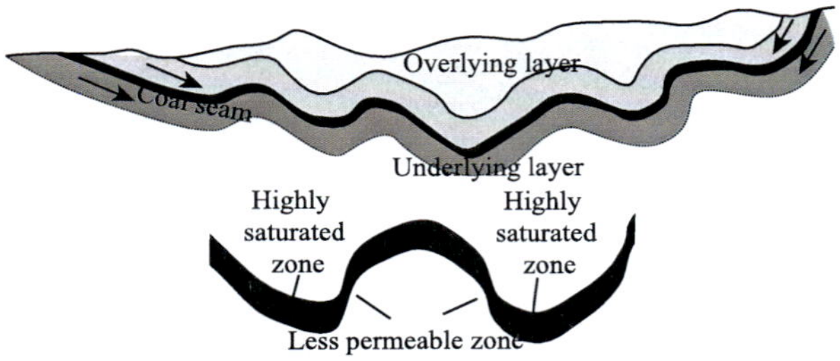

Fig. 6.66 CBM control in a syncline

kinetic energy. In a still water environment or when fluid flow is very slow, $q^2/2$ is negligible, so the sum can be simply the potential energy and the pressure energy of unit mass fluid under formation conditions.

$$\Phi = gZ + \int_0^p \mathrm{d}p/\rho$$

In general, oil and water are incompressible, that is, the density does not change with pressure. The density of gas can also be regarded as a constant when pressure does not change much. The water and gas potentials are written as

$$\Phi_w = gZ + p/\rho_w$$

$$\Phi_g = gZ + p/\rho_g$$

The water potential Φ_w can be expressed by piezometric head h_w. The piezometric head at a measured point is the sum of the elevation and the pressure head of the measured point

$$h_w = Z + p/(g\rho_w)$$

The water potential Φ_w can be rewritten as

$$\begin{aligned}\Phi_w &= gZ + p/\rho_w \\ &= g[h_w - p/(g\rho_w)] + p/\rho_w \\ &= gh_w\end{aligned}$$

In a syncline, from the wing to the core, the buried depth of the coal seam is increasing, and the elevation is reducing continuously, resulting in a decrease in the formation water potential caused by gravity. For the syncline CBM reservoirs in the same fluid system, if there is no external water supply, the fluid system is balanced, the water heads are at the same level, and the water potential is equal. The deep coal seam buried at the axis has a higher pressure energy (elastic energy), and it is easier to adsorb CBM and enrich it. This is so called the theory of CBM enrichment in synclines.

The syncline has natural conditions for maintaining formation pressure and enriching CBM. First, the overlying layer of a syncline is generally thick and the pressure is high, which are favorable for CBM adsorption, and can effectively prevent CBM losing vertically. Second, Synclines generally have the mechanism of centripetal flow of formation water, and maintain a high formation pressure system in the core of syncline, which is easy to form confined sealing of retained water. In addition, fractures in the core are not developed, so the good sealing conditions prevent CBM escaping (Figs. 6.66 and 6.67).

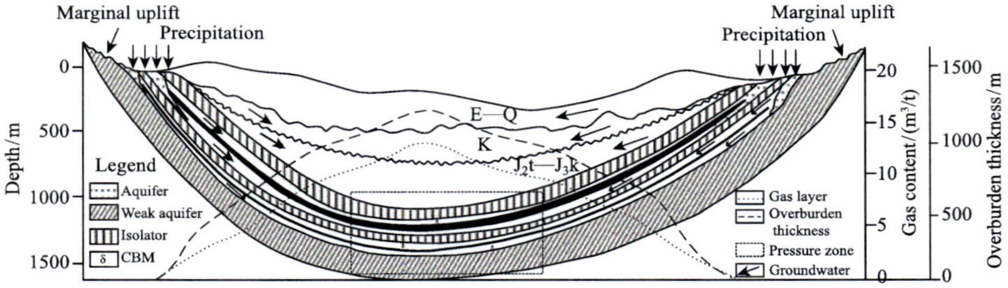

Fig. 6.67 CBM enrichment in a syncline

References

Ayers Jr WB, Kaiser WA (1994) Coalbed methane in the Upper Cretaceous Fruitland Formation, San Juan Basin, Colorado and New Mexico. Texas Bureau Econ Geol Rep Investig 218:216

Chen G (1997) The Yanshanian tectonic thermal event in Qinshui Basin and its petroleum geological significance. Northwest Geol Sci 18(2):63–67 (in Chinese)

Guan DS, Niu JY, Guo LN (1996) Unconventional oil and gas geology in China. Petroleum Industry Press, Beijing (in Chinese)

Hakan H, Namik MY, Cramer B et al (2002) Isotopic and molecular composition of coal-bed gas in the Amasra region (Zonguldak Basin–western Black Sea). Org Geochem 33:1429–1439

Hao SS, Huang ZL, Yang JT (1994) Dynamic balance of natural gas migration and accumulation and its application. Petroleum Industry Press, Beijing (in Chinese)

Harpalani S, Shraufnagel RA (1990) Shrinkage of coal matrix with release of gas and its impact on permeability of coal. Fuel 69:551–556

Jolly DC, Morris LH, Hinsley F (1968) Investigation into the relationship between the methane sorption capacity of coal and gas pressure. Trans Min Eng 127

Kaiser WR, Ayers WB, Ambrose WA et al (1991) Geologic and hydrologic characterization of coalbed methane production, Fruitland Formation, San Juan basin. In: Ayers WB et al (eds) Geologic and hydrologic controls on the occurrence and production of coalbed methane, Fruitland Formation, San Juan Basin. Gas Research Institute Topical Report, GRI 91/0072, pp 273–301

Li MC (1987) Oil and gas migration. Petroleum Industry Press, Beijing (in Chinese)

Li A, Druzhinin XY (2000) A practical approach to P-SV prestack migration and velocity analysis for transverse isotropy, 70th Ann. Internat Mtg Soc Expl Geophys, Expanded Abstracts

Li MC, Liang SZ, Zhao KJ (1996) Coalbed methane and exploration and development. Geological Publishing House, Beijing, pp 1–131 (in Chinese)

Liu HJ (1998) Coalbed methane geology in Southwestern Shanxi Province. China University of Mining and Technology Press, Xuzhou (in Chinese)

Liu HL, Wang HY, Zhang JB (2000) Calculation of coalbed methane adsorption time and influencing factors. Pet Exp Geol 22(4):365–367 (in Chinese)

Pashin JC (1998) Stratigraphy and structure of coalbed methane reservoirs in the United States: an overview. Int J Coal Geol 35:207–238

Pashin JC, Payton JW, Jin G (2002) Application of discrete fracture network models to coalbed methane reservoirs in the Black Warrior Basin. AAPG Annu Conv Off Program 11:137

Pitman JC, Pashin JC, Hatch JR et al (2003) Origin of minerals in joint and cleat systems of the Pottsville Formation, Black Warrior Basin, Alabama: implications for coalbed methane generation and production. Am Assoc Petrol Geol Bull 87:713–731

Qian K, Zhao QB, Wang ZC (1997) Theory and experimental technology for coalbed methane exploration and development. Petroleum Industry Press, Beijing (in Chinese)

Qin SF, Song Y, Tang XY et al (2005) The destructive mechanism of movable groundwater on gas-bearing coal seams. Sci Bull 50 (Supplement):99–104 (in Chinese)

Ren ZL, Zhao CY, Chen G (1999) Late Mesozoic tectonic thermal events in the Qinshui Basin. Oil Gas Geol 20(1):46–48 (in Chinese)

Ruppel TC, Grein CT, Bienstock D (1972) Adsorption of methane/ethane mixtures on dry coal at elevated pressure. Fuel 51 (4):297–303

Sang SX, Liu HJ, Li GZ (1997) Coalbed methane generation and enrichment I: effective volume and enrichment. Coalf Geol Explor 25(6):14–17 (in Chinese)

Sang SX, Fan BH, Qin Y (1999) Sequestration and enrichment conditions of coalbed methane. Pet Nat Gas Geol 20(2):104–107 (in Chinese)

Song Y, Wang Y, Wang ZL (2002) Natural gas migration and accumulation dynamics and gas reservoir formation. Petroleum Industry Press, Beijing (in Chinese)

Song Y, Zhang XM et al (2005) CBM reservoir accumulation mechanism and theoretical basis of economic exploitation. Science Press, Beijing (in Chinese)

Su XB, Liu BM (1999) Occurrence and influencing factors of coalbed methane. J Jiaozuo Inst Technol 18(3):157–160 (in Chinese)

Su XB, Zhang LP (2002) Forming mechanism of abnormal pressure in coalbed methane reservoirs. Nat Gas Ind 22(4):15–18 (in Chinese)

Su XB, Chen JF, Sun JM (2001) Coalbed methane geology and exploration and development. Science Press, Beijing (in Chinese)

Su XB, Lin XY, Zhao MJ, Song Y, Liu SB (2005) The upper paleozoic coalbed methane system in the Qinshui Basin, China. AAPG Bull 89(1):81–100 (in Chinese)

Sun B, Wang YB (2003) Distribution characteristics of coalbed methane in Daning-Ji County, Ordos Basin. In: Li WY et al (eds) China coalbed methane exploration and development. China University of Mining and Technology Press, Beijing, pp 65–73 (in Chinese)

Wang FG, Li LJ, Xu DH (2003) Influencing factors of coalbed gas content in North China. J Jiaozuo Inst Technol (Nat Sci Ed) 22 (2):88–90 (in Chinese)

Wang HY, Liu HL, Liu HQ, Li GZ (2004) Simulation technology and application of coalbed methane accumulation. 15(4):349–351 (in Chinese)

Wang B, Jiang B, Wang HY et al (2007) Physical simulation of hydrogeological conditions of low-ranked coalbed methane reservoirs. J China Coal Soc 32(3):258–260 (in Chinese)

Whiticar MJ, Faber E, Schoell M (1986) Biogene methane formation in marine and freshwater environments, CO_2 reduction vs. acetate fermentation-isotopic evidence. Geochim Cosmochim Acta 50:693–709

Yang RT, Saunders JT (1985) Adsorption of gases on coals and heat-treated coals at elevated temperature and pressure L adsorption from hydrogen and methane as single gases. Fuel 64(5):616–620

Zhang JB, Tao MX (2000) Geological significance of methane carbon isotope in coalbed methane exploration—a case study on Qinshui Basin. J Sedimentol 18(4):611–614 (in Chinese)

Zhang XM, Zheng YZ et al (2002) Distribution of coalbed methane resources in China. In: Academic collection of the 80th anniversary of the Chinese Geological Society. Geological Publishing House, Beijing, pp 458–463 (in Chinese)

Zhao QB et al (1999) Coalbed methane geology and exploration and development technology. Petroleum Industry Press, Beijing (in Chinese)

Zhao MJ, Song Y, Su XB et al (2005) The key geological period for determining the geochemical characteristics of coalbed methane. Nat Gas Ind 25(1):51–54 (in Chinese)

7 Evaluation and Prediction of Technically Recoverable CBM Resources

Abstract

Recoverability evaluation of coalbed methane (CBM) resources belongs to the category of mineral resources evaluation. The so-called mineral resources evaluation is the systematic cognition and comprehensive analysis of the geological, technical, economic attributes and quantitative relations of mineral resources. The quantitative recoverable evaluation of CBM resources should and must follow the normative and systematic theories and comprehensive analysis methods. Evaluation methods of recoverable CBM resources are systematic and theoretical methods about CBM resources sequence, properties and categories of CBM resources evaluation objects, resources calculation parameters and assignments, recoverable CBM resources calculation methods, calculation results evaluation and disclosure, etc. It is used to guide scientific and reasonable quantitative evaluation on the recoverability of CBM resources.

7.1 Classification System

The different understandings of the meaning of CBM resource reflect the different understandings and solutions of the issues like the concept of CBM reservoirs, the type and purpose of resource prediction, the classification of CBM resource and the estimation methods of CBM resource. The establishment of scientific and standardized CBM resource sequence can meet various requirements and is the premise of the prediction and evaluation of CBM resource.

On the basis of sufficient investigation and comparison, extensive consultation and consideration of the characteristics of CBM, we put forward a new CBM resource classification scheme and established a "CBM resource classification system" that is brought in line with conventional gas and international practice (Fig. 7.1).

Total resources CBM: the estimated volume of methane (including heavy hydrocarbons) in underground coal reservoirs and the surrounding rocks. Depending on its occurrence, quantity and quality, the methane can be expected to be technically and economically feasible to eventually produce. According to the degrees of geological understandings, it can be divided into three types: reserves, potential resources and inferred resources.

① Reserves are the CBM volume in the discovered CBM reservoirs with clear calculation boundaries, which is calculated based on the exploration results of the CBM exploration projects that have been implemented.

According to economic feasibility, CBM reserves can be divided into geological reserves, technically recoverable reserves and economically recoverable reserves.

(i) Geological reserves are the in-place CMB volume in the discovered CBM reservoirs with clear calculation boundaries, which is calculated based on the exploration results.
(ii) Technically recoverable reserves are the estimated CBM volume within geological reserves that can be produced using applicable production technologies, whose economic implications are uncertain.
(iii) Economically recoverable reserves are the estimated CBM volume that can be commercially produced from the discovered CBM reserves (The economically recoverable production in China is equivalent to the term of Reserves internationally used).

According to degree of certainty and project maturity, reserves can be further divided into proven reserves, controlled reserves and predicted reserves, and proven reserves can be sub-divided into developed and undeveloped reserves. For this point is out of the scope of the subject research, it will not be discussed in details here.

After the detailed divisions according to economic feasibility, degree of certainty, project maturity and other

Y. Song et al., *Coalbed Methane in China*,
https://doi.org/10.1007/978-981-33-4725-0_7

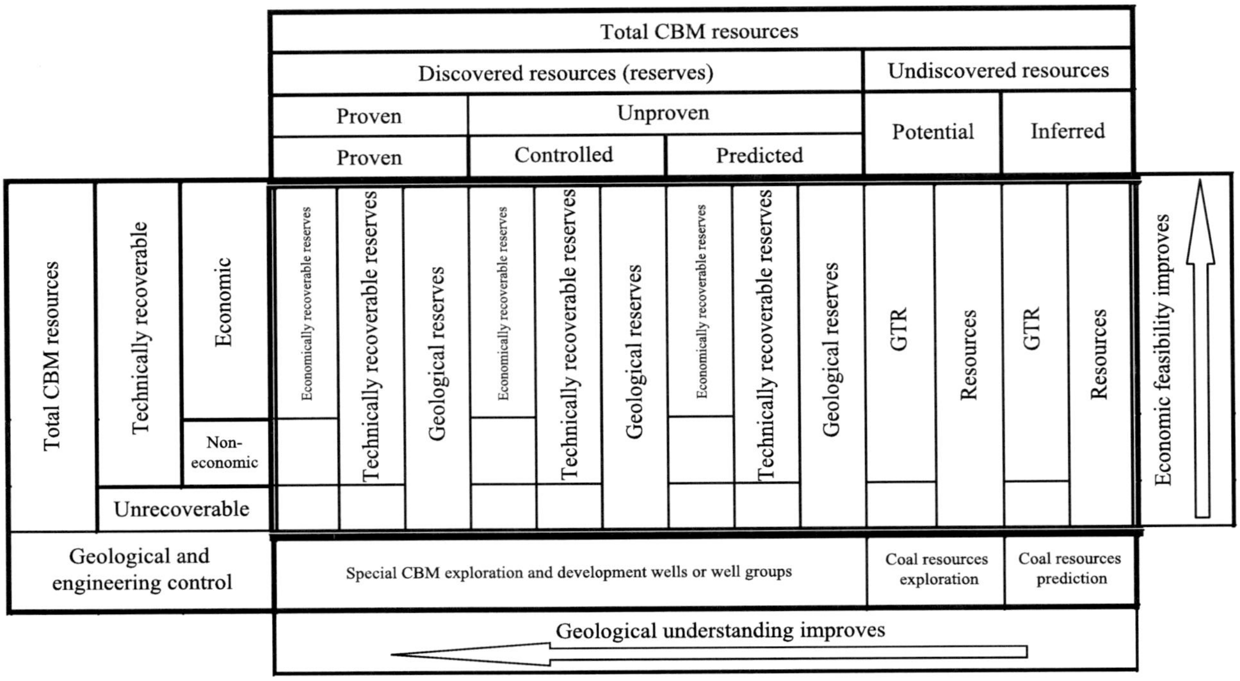

Fig. 7.1 CBM resource classification system

indicators, reserves can be named in a compound way, such as proven geological reserves, controlled and technically recoverable reserves (Fig. 7.1).

② Potential resources are in the level of general survey and other more detailed survey, but no special CBM exploration. They are the in-place CBM volume estimated only based on comprehensive analysis of the coalfield and avaiable exploration results.

③ Inferred resources are in the level of coal resource prediction, but no special CBM exploration. They are the estimated in-place CBM volume based on comprehensive analysis of coal resource prediction data, oil and gas exploration data, theoretical research results, etc.

Gas technically recoverable (GTR) are the estimated CBM volume within undiscovered resources that can be produced using applicable production technologies, whose economic implications are uncertain. It is generally believed that, the GTR of CBM exist in coal reservoirs at a considerable scale (thickness, area), which is suitable for production using applicable technologies.

The main features of this classification system are as follows: First, the CBM resources classification is established based on geological understanding and economic feasibility, which are the basic principles of international and conventional resources classification in oil and gas industry. It makes the classification of CBM resources consistent with that of conventional oil and gas. Secondly, categories as GTR, technically recoverable reserves and economically recoverable reserves (equivalent to the Reserves internationally used) are divided, and the classification sequence of CBM resources is refined. It has realized the basic conformity with the international conventions. Thirdly, according to the practice and habit of China, categories as geological reserves and (geological) resources have been reserved to ensure the convergence of new and old classification schemes. Fourthly, during categorizing CBM resource types according to geological understandings, it not only emphasizes the necessity of special CBM exploration projects, but also takes full account of the geological exploration stages and scientific research results of coalfields, which reflects the natural connections between CBM and coalfields, and shows the principle of efficiency and economy.

In this classification system, the concept of technically recoverable resource (reserves) of CBM was put forward for the first time, and the specific meaning was defined, which is conducive to deepening CBM resource evaluation in China. The issues related to CBM GTR resources are further explained as follows.

As mentioned above, there are many factors influencing the recoverability of CBM. For resources evaluations of regional and large-scale CBM, it is impossible to quantitatively estimate the recoverable volume of CBM while taking geological conditions, price, laws and regulations, technology and other factors all into consideration, especially in the case of low degree of CBM exploration and development. In addition, most of CBM resources in China are undiscovered. Therefore, in quantitative evaluation of the recoverability of country-wide and regional CBM resources, CBM GTR resources are a good choice. Correspondingly, for discovered CBM resources, GTR reserves can be used.

Since the evaluation of CBM GTR resources (reserves) requires to investigate the adaptability of CBM geological conditions to applicable development technologies, which means even for same CBM geological conditions, the recoverability may be different due to different development technologies, therefore during evaluation, the selection and application of development technologies are key factors. In other words, using different development technologies, the predicted CBM GTR resources (reserves) are different.

At present, pressure depletion has been widely used in the global CBM industry as a mature development method. It tends to reduce the gas pressure in coalbed by using effective methods. With the decrease of the gas pressure, CBM changes from adsorbed to free, and then the free CBM flows into CBM wells through cracks (fractures) until the gas pressure in the coalbed is very low. Pressure depletion realizes in surface vertical wells, GOB wells and underground horizontal wells. All these three methods have been applied in China to different degrees.

As a result of technical progress, there are many advanced CBM development methods emerged, such as gas injection (N_2, CO_2) and pinnate horizontal well, which can greatly improve CBM recovery. These new technologies have only been tested locally in a few countries, such as the United States and Canada, and have not been adopted on a large scale. Based on these conditions, CBM resources evaluation at this time used pressure depletion in surface vertical wells. Meanwhile, the important roles of gas injection (N_2, CO_2) and pinnate horizontal wells were considered too, as they can significantly improve the recovery of CBM resources and overcome the interference of terrain factors.

7.2 Prediction Methods of CBM GTR

7.2.1 Controlling Factors on CBM Recoverability

The quantitative evaluation of CBM recoverability is to estimate CBM recoverable resources. There are many factors influencing the recoverability of CBM resources, including factors in geology, engineering and regulations. Through the sensitivity analysis of these factors on CBM production, we can acquire some useful inspiration for quantitative evaluation of CBM resource recoverability. The estimation of CBM recoverable resources is to examine technical and economic effectiveness of the resources from the perspective of recoverability and to estimate resource volume. Therefore, various factors that may influence CBM well production should be analyzed. These factors can be divided into natural factors and human factors. Natural factors refer to geological conditions. Human factors refer to well pattern, drilling operations, stimulations, market, price, etc. In this part, the influences of relevant factors are discussed by COMET2, a numerical simulation software for CBM reservoirs.

7.2.1.1 Geological Factors

(1) Resource volume and abundance

CBM resource volume and abundance are important factors to determine gas well productivity. Resource volume and abundance are controlled by coal seam thickness and gas content which are key factors on stable and sustainable gas production. Resource abundance is one of the controlling factors to ensure sustainable gas supply and high production. The sensitivity simulation experiment of the relationship between coal seam thickness and gas production suggests that gas wells penetrate thick coalseam reservoirs can obtain high production.

(2) Reservoir pressure

Reservoir pressure is the energy source for fluid flow in coal seam. CBM is produced by draining water from coal seams to reduce the pressure of coal reservoirs and then desorb CBM. The process of CBM production is the process of decreasing reservoir pressure and depleting formation energy. Therefore, the higher the original reservoir pressure is, the greater the room for pressure decline will be, the more effective the drainage and pressure reduction operation will be, and the greater the gas well production will be.

(3) Coal seam permeability

Both domestic and foreign CBM exploration and development show that permeability is the most important controlling factor on CBM gas well productivity. Regardless of the thickness of coal seam and the CBM gas content, the higher the permeability is, the higher the CBM productivity is, and the higher the potential productivity of the gas wells is (Fig. 7.2).

In addition, the adsorption performance, desorption rate and free gas content of CBM directly control the CBM concentration in fractures, and influence gas well productivity. But the desorption rate and free gas content of CBM only affect the early productivity of a CBM well, and for a long-term, the influence is small.

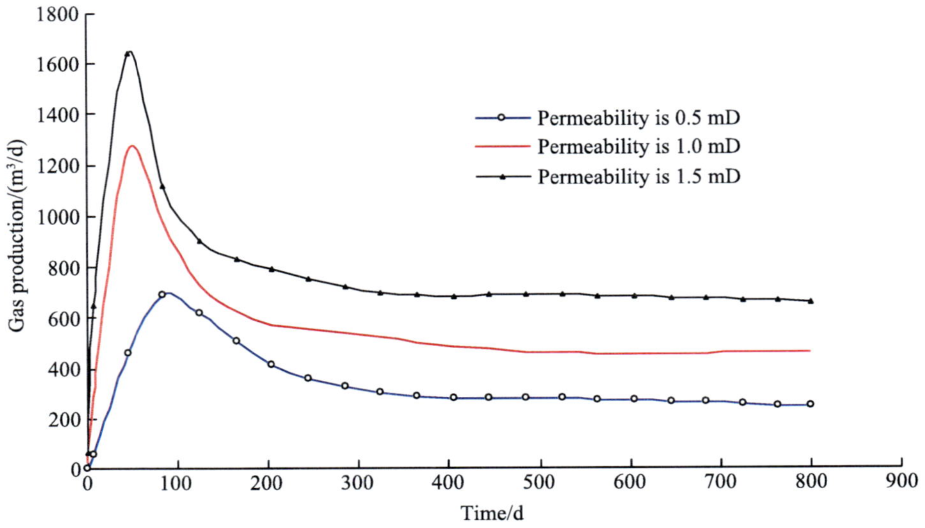

Fig. 7.2 Simulation on how permeability influences gas production

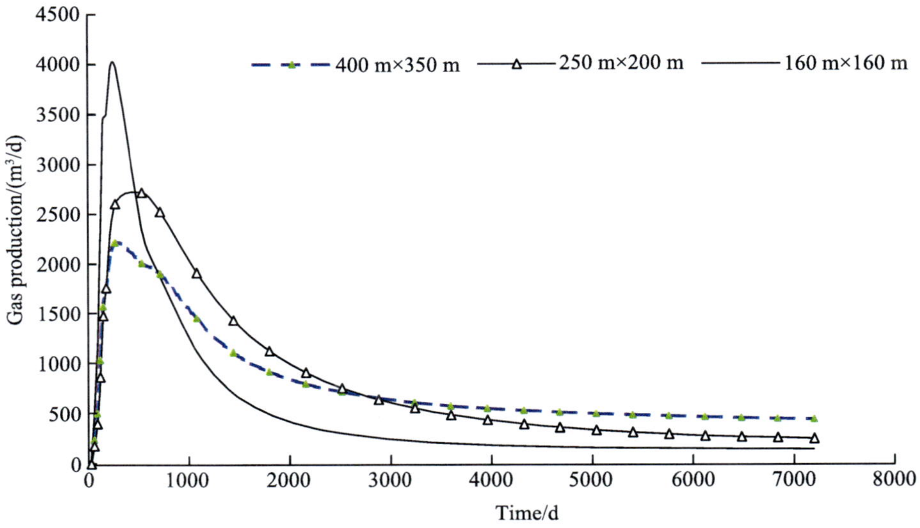

Fig. 7.3 Prediction of gas well production on different well spacing

7.2.1.2 Well Pattern

(1) Well location

In order to maximize the development effect, well spacing should be appropriately increased along the direction perpendicular to the minimum principal stress and decreased along the direction parallel to the minimum principal stress.

(2) Well spacing

Well spacing is an important factor affecting gas well productivity. If well spacing is large, initial gas production would be low, but later gas production would be high, and vice versa (Fig. 7.3).

7.2.1.3 Development Technology

(1) Drilling operation

Long drilling cycle or improper drilling fluid will cause pollution to reservoirs. Reservoir pollution has a significant influence on gas well productivity. Well XS-02 in the Xinji Coalmine was drilled from March 15, 1998, and completed on November 4, 1998. The drilling time is nearly 8 months and the total depth is 840.03 m. The mud may cause serious

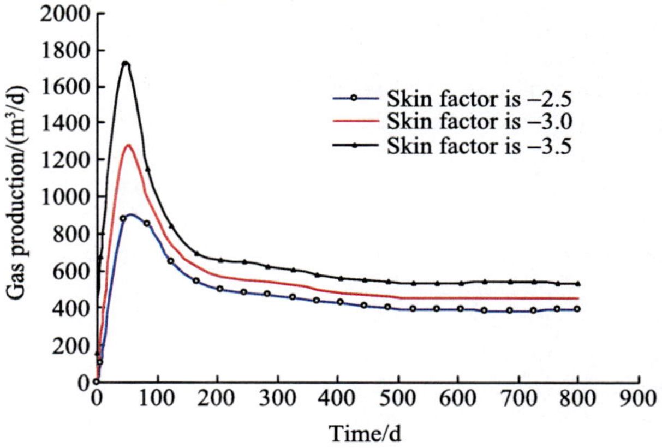

Fig. 7.4 Simulation on how fracturing stimulation on production

pollution to the reservoirs so that the gas production was low. Well XS-03 was drilled from November 14, 1999 and completed on January 6, 2000. The drilling time is less than 2 months, the total depth is 1050 m. The gas production is relatively high.

(2) Stimulation measures (fracturing)

Fracturing scale as well as fracture shape and extension of induced fractures all influence gas well productivity. In fracturing operation, it is required that the fracturing fractures should be extended horizontally along the coal seam as far as possible to expand effective scope. The longer the induced fractures extend along the coal seam, the better the CBM production will be, and the better the gas well productivity will be naturally (Fig. 7.4).

(3) Drainage operation

Control of flow-back of fracturing fluid, control of fluid level and well flushing during draining water period all influence gas well productivity. Fracturing fluid flow-back directly influence propping of fractures and further influence gas well productivity. The degree of liquid level reduction is an important factor to determine gas well production. The more the fluid level decreases, the lower the bottom hole pressure is, the better the gas well productivity is, and the higher the gas production is. Frequently flushing well not only influences propping of fracture, but also causes reservoir pollution. Figure 7.5 shows the daily and cumulative CBM productions with the fluid level in a CBM gas well in the Xinji Coalmine in the Huainan Coalfield.

7.2.1.4 Market, Gas Price, Policies and Regulations, Etc.

The market, macroscopic government decisions and gas policies regulations have potential influences on CBM development. Influenced by CBM market and government decisions, the development plan of CBM must be adjusted in order to effectively control gas production (usually by controlling development scale and drainage operations) and achieve stable and high production. At present, in the Qinshui Basin, the existing CBM production wells are unable to give a full play to their normal deliver abilities due to market problems. And the price is not standardized. There are 15 production wells in the Shizhuang CBM project, and the processing capacity of the compression station is 2×10^4 m^3/d. As the market sales volume is only 5000 m^3/d, the CBM production is limited by the market demand. The wellhead pressure is high, usually within 0.4–0.5 MPa. Part of the CBM

Fig. 7.5 Prediction of gas production of a well

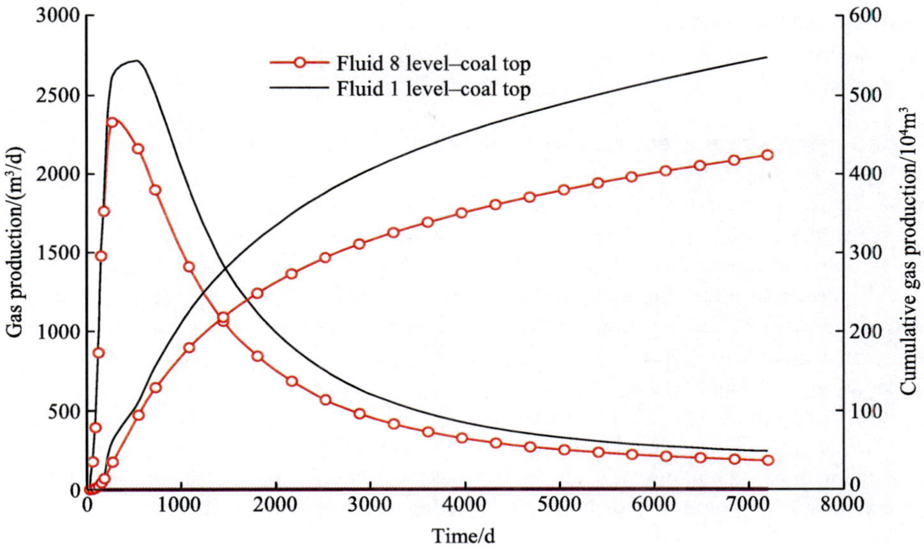

gas is made into compressed natural gas (CNG). At present, 5000 m^3 of CNG is sold out every day for residential fuel gas, at the sales price of 0.7 yuan/m^3. The other part supplies a 400 kW natural gas generator for generating electricity. There are more than 200 production wells in the Panzhuang CBM project. The productivity is about 40×10^4–50×10^4 m^3/d. The processing capacity of the CBM compression station is 50×10^4 m^3/d. Limited by market demand, the actual production of compressed natural gas (CNG) is about 20×10^4 m^3/d, which is mainly used as domestic gas for residents in the Jincheng mining area and fuel gas for urban CNG vehicles at the sales price of 2.2 yuan/m^3.

Of course, strictly speaking, market, gas price and other economic factors cannot directly affect the daily production of CBM wells. However, due to economic considerations, operators are very concerned about the economic limit production, which will affect the ultimate cumulative production of CBM wells.

7.2.2 Determination Methods for Important CBM Parameters

In the calculation process of CBM GTR, regardless of gas reservoir numerical simulation or loss analysis method, the parameters substituted into the models and formulas must be known, but not logical symbols or interval values. Some parameters, such as block area, coal seam thickness, gas content, can be accurately measured; some parameters, such as coal seam permeability, CBM recovery, are difficult to be accurately measured. Based on in-depth mechanism researches and investigations and analyses of domestic and foreign CBM exploration and development practices, for the important CBM parameters which are difficult to be accurately measured, including coal seam permeability, CBM recovery and gas content of lignite, we established a standard determination method to meet the requirements of the prediction of CBM GTR.

7.2.2.1 Statistical Prediction of Coal Seam Permeability

Coal seam permeability is a reservoir parameter that describes the seepage performance of gas, water and other fluids in coal reservoir. The permeability of coal seam directly determines the migration and production of CBM, and is the main reservoir parameter that influences production and recovery of CBM wells. Coal seam permeability depends on the number, the degree of opening and connectivity of fractures (cleats) in the coal seam.

The determination methods of coal seam permeability mainly include laboratory test of coal (rock) samples, instantaneous pressure analysis (injecting-pressure drawdown test, drill-stem test, slug test and water tank test, etc.), and fitting CBM production history.

In the formulas and models of CBM GTR resources calculation and CBM production prediction, the coal seam permeability is an indispensable parameter, and it must be a definite value instead of qualitative evaluation. Due to obvious heterogeneity of coal reservoirs and other cause, such as man-made fractures in the process of sampling and sample preparation, small size and poor representativeness of coal (rock) samples, laboratory test on coal (rock) samples for coal permeability is rarely used. In China, most of coal reservoir permeability data were obtained by injecting-pressure drawdown test. Although the permeability obtained by injecting-pressure drawdown test can overall reflect the change of coal reservoir permeability, there is a large error of the obtained permeability due to various causes, so it cannot represent the real coal seam permeability. The permeability determined by fitting production history is the most accurate, and it can represent the real permeability of coal reservoirs. However, due to the limitation of the degree of CBM exploration and development, it cannot be widely used. In order to meet engineering requirements, this Project proposes statistical prediction for coal seam permeability. According to the statistical variation of coal seam permeability obtained by well test, the range of coal seam permeability determined by fitting production history is used to correct well test permeability. And then, the coal seam permeability in the unexplored area can be determined.

Controlling Factors on Coal Seam Permeability

The characteristics and controlling factors of coal seam permeability have been extensively studied and reported in China (Zhang and Li 1996; Li et al. 1998; Ye et al. 1999; He et al. 2000; Fu et al. 2001; Zhang et al. 2002a; Li et al. 2002). In general, there are many internal and external factors controlling coal seam permeability. Internal controlling factors include fracture, coal rank, maceral and coal structure. External controlling factors include in-situ stress, effective stress, burial depth, Klinkenberg effect and geological tectonism. In different geological environments, dominant factors are different. Sometimes a factor plays a major role, and sometimes it is the result of a combination of factors. It is worth pointing out that, most of available researches are qualitative descriptions and analyses, and there are less researches on quantitative relationships between coal seam permeability applicable in engineering and primary (single) controlling factors.

Both scientific researches and production data show with constant internal factors, the influence of in-situ stress on coal seam permeability is the most significant. Fracture is the main channel for CBM migrating in coal seams and the prerequisite for the permeability of coal reservoirs. Without

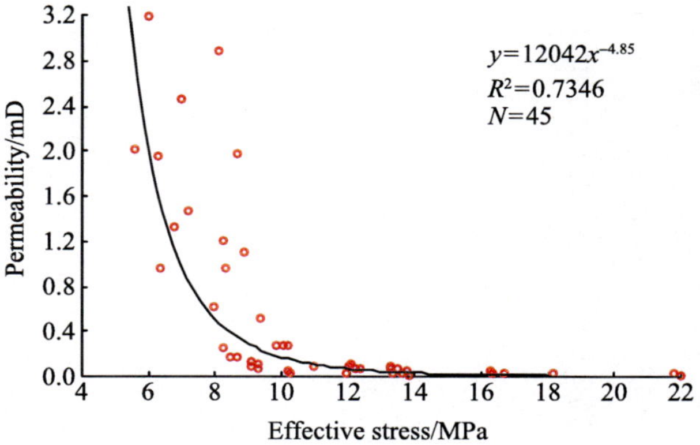

Fig. 7.6 Correlation between permeability and effective stress

fractures, there is no good permeability. But even there are fractures, coal reservoirs do not necessarily have good permeability. It also depends on the degree of fracture opening. The degree of fracture opening depends on effective stress. Effective stress is the difference between vertical in-situ stress and formation pressure. The relationship between the closure pressure (effective stress) measured by injecting-pressure drawdown test and coal seam permeability (Fig. 7.6) shows that the permeability of coal reservoir is negatively correlated with the effective stress exponentially. In general, the permeability of coal reservoir decreases with the increase of effective stress.

In an unexplored area, the in-situ stress is unknown. Although both vertical in-situ stress and formation pressure increase linearly with the increase of burial depth, the effective stress increases with the increase of burial depth as the density of formation is much higher than that of fluids (mainly water) in the formation. Mckee et al. (1988) studied the relationship between coal seam permeability and burial depth in the Piceance Basin, San Juan Basin and Black Warrior Basin in the United States, and Zhang et al. (2002a, b) studied the eastern margin of the Ordos basin and the Lianghuai Coalfield in China. Both found that coal seam permeability decreases exponentially with the increase of burial depth.

The above analyses provides a theoretical and practical basis for the prediction of coal seam permeability in an unexplored area by using the statistical law of coal seam permeability changing with burial depth.

Statistical Prediction of Coal Seam Permeability Changing with Burial Depth

(1) Regression analysis of well test permeability and buried depth of coal seam

Measured data of coal seam permeability is the basis of the statistical prediction of coal reservoir permeability. The reliability of measured data and the rationality of the data selected play an important role in predicted results. Due to the limitations of well test methods and the limitations of field operating conditions, the accuracy of the coal seam permeability measured by well test is greatly different, which leads to a large dispersion on the scatter plot of coal seam permeability and burial depth. It is difficult to establish a qualified mathematical relation. Therefore, the collected permeability must be filtered to "discard the false and retain the true", and to eliminate those obviously abnormal data. Meanwhile, we should pay attention to the "uniformity" of the measured coal seam permeability in depth distribution. Regression analysis would be conducted on the measured permeability selected and the buried depth of coal seam, and then the regression equation would be established, and the standard deviation and correlation coefficient R would be calculated. Using relevant data, the regression equation should be tested to verify the significance of the regression equation. If the regression equation is significant, it indicates that there is a certain correlation between the buried depth of coal seam and the permeability, then the permeability of coal seam can be predicted by the regression equation.

(2) Prediction of well test permeability

The regression equation established can be used to quantitatively predict the permeability at some depth. Due to heterogeneity and variability, the permeability at the same depth varies to some extent. The permeability calculated by the regression equation represents the general case (i.e. the average value) in some depth range, namely the medium permeability. According to the distribution and standard deviation of data points on the scatter diagram of the coal seam permeability and the buried depth, the better and worse permeability values can be determined, so that the average value and variation range of the well test permeability can be determined at a certain buried depth.

(3) Comparison of well test permeability with real permeability

Due to the influence of test equipment, test technology, wellbore damage, injection and shut-in time and other factors, the permeability obtained by injecting-pressuring drawdown test only shows the situation around borehole, and cannot represent the actual situation of whole coal reservoirs, and the measured values are often lower[1] (Zhang et al. 2002a, b). Therefore, the predicted permeability based on the above-mentioned regression equation is also smaller

[1]ARI. 1993. Reservoir characterization of Mary Lee and Black Creek coals at the Rock Creek Field Laboratory, Black Warrior Basin.

Table 7.1 Permeability from well test and production history (Zhang et al. 2002a, b, supplement data)

Well	Coal seam no	Tested permeability/mD	Fitted permeability/mD	Fitted/tested	Remarks
Wu Test 1	10	0.100	1.789	17.890	–
Meiliu 1	4	1.310	4.200	3.206	–
Meiliu 2	8	8.860	12.200	1.377	–
Meiliu 3	4	1.200	4.200	3.500	–
Meiliu 4	8	0.960	12.200	12.708	–
Dacan 1	6	0.061	1.000	16.393	–
TL-003	3	0.946	3.546	3.748	–
	15	0.257	0.707	2.751	–
FZ-003	3	2.87	2.870	1.000	–
	15	0.11	1.260	11.450	–
Pz-1	15	1.525	1.950	1.280	–
PIB	Mary Lee	3–11	15.000	1.4–5.0	Black Warrior Basin, USA
PIC	Black Creek	2.300	3.000	1.3	Black Warrior Basin, USA
P2	Mary Lee	2.000	25.000	12.5	Black Warrior Basin, USA
P3	Mary Lee	1–8	25.000	3.1–25	Black Warrior Basin, USA
P3	Black Creek	0.2–2.3	2. 500	1.1–12.5	Black Warrior Basin, USA
P6	Black Creek	0.1–0.9	1. 500	1.7–15.0	Black Warrior Basin, USA
Hamilton3	Fruitland	6.700	10.000*	–	San Juan Basin, USA

*Data from multi well simulation results

than the actual permeability, which must be corrected before application.

As CBM production wells are generally enhanced by stimulation, reservoir damage around borehole during drilling and completion operations would be eliminated. And the production time is long, usually more than 3 months, some wells may produce for years, expanding pressure reduction and causing inter-well interference. Therefore, actual CBM production can be used to fit production history, and the permeability obtained can represent actual permeability.

Table 7.1 compares the permeability values from well test and fitting production history to the same well. The permeability from production history is generally higher than that from well test, generally by 1–25 times, from 2 to 10 times.

Therefore, by comparing fitted permeability and tested permeability, we can determine a reasonable proportion for the specific CBM field (reservoir), which then is multiplied by the tested permeability calculated from the regression equation. The permeability close to the real permeability is determined and used for calculating CBM recoverable resources.

Case Analysis

(1) Predicted the coal seam permeability in the Qinshui Basin

Permeability data from 36 CBM well test were collected in the Qinshui Basin, and abnormal values (≤ 0.01 mD or > 100 mD and data that are too dense at the same depth) were excluded. Then linear regression analysis was conducted on 27 groups of burial depth and well test permeability, and the regression equation was obtained (Fig. 7.7).

$$y = 8 \times 10^7 x^{-2.9631} \tag{7.1}$$

$$R^2 = 0.3596 \tag{7.2}$$

$$R = 0.5997 \tag{7.3}$$

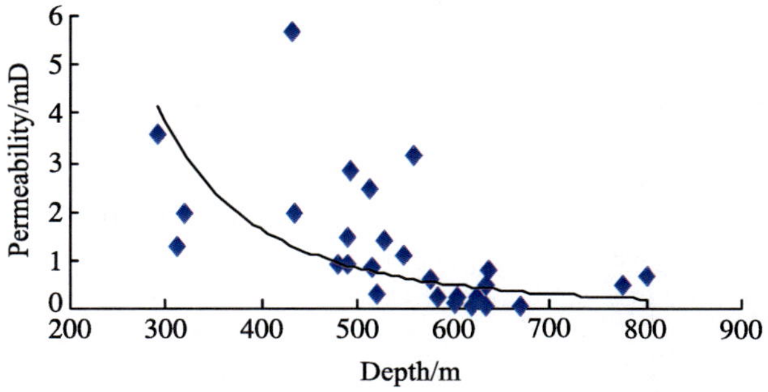

Fig. 7.7 Burial depth vs. permeability in the Qinshui Basin

After significance test, $R = 0.5997 > \alpha_{0.01}(27) = 0.470$ (when the significance level is 0.01), indicating that the above regression equation has a certain significance.

From the relationship of depth vs. permeability in the Qinshui Basin (Fig. 7.7), there is a negative correlation, which reflects that with the increase of coal seam depth, fractures are gradually closed and the permeability gradually decreased. The distribution of permeability data above the regression line is scattered, and the measured permeability is generally 2 times the regressed value. The permeability under the regression line is relatively concentrated, and the measured permeability is generally half the regressed value. That is to say, the good predicted permeability is twice the medium (regressed value), and the poor predicted permeability is half the regressed value.

Table 7.2 shows the coal seam permeability from fitting actual CBM production data and the permeability measured by well test in the Qinshui Basin. The fitted permeability is greater than the tested permeability, and the ratio of the two varies from 1 to 24, with an average of 6.16. In other words, the ratio of the fitted permeability to the tested permeability is 6.16 times for the CBM fields in the Qinshui Basin.

(2) Predicted coal seam permeability in the Ordos Basin

Similarly, by linear regression of buried depth and permeability of 47 groups of coal reservoirs in the Ordos Basin, the regression equation between coal seam depth and permeability was obtained (Fig. 7.8):

$$y = 7 \times 10^7 x^{-2.8045} \tag{7.4}$$

$$R^2 = 0.4438 \tag{7.5}$$

$$R = 0.6662 \tag{7.6}$$

According to the significance test, since $R = 0.6662 > \alpha_{0.01}(27) = 0.470$ (when the significance level is 0.01), the above regression equation shows a certain significance, which means, there is a power function between the buried depth and the permeability.

Figure 7.8 shows that there is a negative correlation between the depth and the permeability, which reflects that the permeability gradually decreases with the increase of the depth. The permeability above the regression line is relatively scattered, while below the regression line is relatively concentrated. At the same depth, the measured permeability above the regression line is generally two times the regressed permeability, and the measured permeability below the regression line is generally half the regressed permeability. According to this, we can get good, medium and bad permeability values.

Table 7.2 Fitted coal seam permeability and tested coal seam permeability in Qinshui Basin

Well	Coal seam no	Tested permeability/mD	Fitted permeability/mD	Tested/fitted
TL-003	3	0.95	1.81	1.91
	15	0.26	0.73	2.81
FZ-003	3	2.87	2. 87	1.00
	15	0.11	1.26	11.45
FZ-004	3	1.46	3.72	2.55
	15	0.07	1.68	24.00
FZ-007	3	3.18	3.72	1.17
FZ-008	3	0.91	3.72	4.09
	15	0.26	1.68	6.46
Pz-1	3	3.61*	3.95	1.09
	15	1.525**	1.95	1.28
Pz-4	3	3.61*	3.71	1.03
	15	1.525**	1.79	1.17

*Data from Well CQ 9 is the result of the injecting-pressuring drawdown test
**Data from Well Pz-2 is the result of flooding test

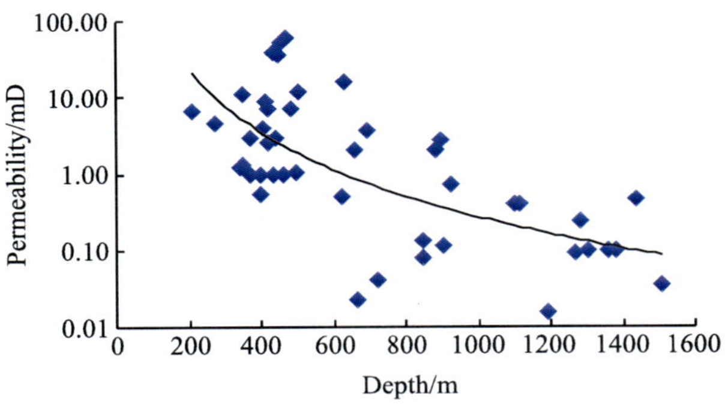

Fig. 7.8 Coal seam depth versus permeability in the Ordos Basin

As for the comparison between the fitted permeability and the tested permeability in the Ordos Basin, the data of the top 5 wells in Table 7.1 (all of which are located in the eastern margin of the Ordos Basin) are referable, with an average of 7.74.

7.2.2.2 Determination Methods for CBM GRF

GRF (Gas Recovery Factor) is an important parameter for calculating reserves and recoverable resources. According to the industry standard of *CBM Resources (Reserves) Specification*, CBM GRF refers to "under current economic and technical conditions and government regulations, and from a specific time, the ratio of the CBM volume that can be or expected to be commercially produced from a gas reservoir to the total CBM volume discovered in the gas reservoir" (Yang et al. 2003). In available references and data of CBM researches in the United States, CBM GRF is the percentage of the CBM that can be economically recovered by using applicable technologies over in-place CBM (Mavor and Nelson 1997).

There are many ways to determine CBM GRF, such as desorption experiment, isothermal adsorption curve, gas reservoir numerical simulation, CBM gas content attenuation rate, material balance and analogy methods, etc.

Desorption Experiment

The gas content determined by direct measurement is composed of three parts, lost gas, measured gas and residual gas. Among them, lost gas and measured gas are naturally desorbed, which can be produced from gas reservoirs. Residual gas cannot be produced and may be left underground. Therefore, CBM GRF can be determined according to the gas content measured by the direct method (Fu et al. 2002). The specific calculation formula is as follows:

$$R_f = \frac{Q_S + Q_J}{Q_S + Q_J + Q_C} \tag{7.7}$$

where R_f is theoretical CBM GRF,%; Q_J is the gas content measured in CBM desorption experiment, m^3/t; Q_S is the lost gas content in CBM desorption experiment, m^3/t; Q_C is the residual gas content in CBM desorption experiment, m^3/t.

According to the statistics of 325 measurements in more than 40 CBM wells in China (Table 7.3), the CBM GRF is mostly above 80%, the GRF of gas coal is the lowest, followed by lignite, long-flame coal, coking coal and others of higher coal rank. This trend is generally consistent with the trend of CBM content distribution in the United States (Table 7.4).

It is worth noting that the GRF is not only affected by gas content and coal adsorption characteristics, but also by various human factors during the development process. The GRF obtained by desorption experiment is theoretically the highest possible, which is referred to as the theoretical GRF. It is an ideal state and cannot represent actual GRF.

Desorption experiment is at a relatively low cost, but the result is relatively rough.This method can be used in areas explored to some extent. It is generally applicable to the initial stage of CBM exploration and development or to provide conceptual data, and the degree of quantification is low.

Isothermal Adsorption Curve

With an isothermal adsorption curve, GRF of a gas reservoir can be estimated based on the isothermal adsorption curve, the original gas content and the assumed abandonment pressure. The formula is:

$$R_f = (C_i - C_a)/C_i \tag{7.8}$$

where R_f is CBM GRF,%; C_i is original CBM gas content, m^3/t; C_a is the CBM gas content at abandonment pressure, m^3/t.

The advantage of this method is that it is easy to operate, while the disadvantage is that the abandonment pressure is difficult to determine accurately. In the initial stage of CBM exploration, the abandonment pressure is usually estimated based on experience in combination with local geological conditions[2] (Zhang 1996). The isothermal adsorption curve is used to calculate GRF, so the GRF calculated by estimated abandonment pressure is highly uncertain.

[2]Schraufnagel et al., 1993, *Well completion and production experience in Rock Creek coal reservoirs.* Ma Dongying (trans.): Collection of CBM Surface Development (II), the Department of Comprehensive Planning, Ministry of Energy and Xi'an Branch of China Coal Research Institute.

Table 7.3 Statistics of gas content measured by the direct method in China

Coal rank R_{max}/%	Number of samples	Gas content/(m^3/t)				GRF/%
		Measured gas	Lost gas	Residual gas	Total volume	
Lignite (<0.5)	10	1.32	0.22	0.27	1.81	85.05
Long-flame coal (0.5–0.65)	11	5. 95	5.95	0.48	6.44	92.40
Gas coal (0.65–0.9)	56	5.01	1.30	2.40	8.71	74.73
Fat coal (0.90–1.2)	27	4.19	1.59	0.96	6.75	80.96
Coking coal (1.2–1.7)	59	9.99	1.20	1.13	12.31	91.67
Lean coal (1.7–1.9)	14	13.05	1.70	1.88	16.63	89.89
Meager coal (1.9–2.5)	54	11.39	0.94	1.29	13.63	87.08
Anthracite III (2.5–4.0)	64	16.18	1 53	2.55	20.88	87.56
Anthracite II (4.0–6.0)	44	15.81	0.91	1.82	18.54	90.21

Table 7.4 Statistics of gas content measured by the direct method in the United States

Coal rank	Number of samples	Gas content/(m^3/g)				GRF/%
		Measured gas	Lost gas	Residual gas	Total volume	
Anthracite	9	8.10	0.98	0.61	9.69	93.69
Low-volatile bituminite	21	11.97	1.21	0.25	13.43	98.14
Medium-volatile bituminite	22	6.31	1.33	0.32	7.96	95.98
High-volatile bituminite A	217	2.77	0.21	1.38	4.36	68.35
High-volatile bituminite B	86	2.01	0.31	0.47	2.79	83.15
High-volatile bituminite C	42	1.09	0.12	0.07	1.28	94.53

Gas Reservoir Numerical Simulation

Gas reservoir numerical simulation uses the measured initial reservoir properties (or estimates) to predict the future production curve (ARI 2002) of an "average well", and then to to determine the CBM GRF.

The numerical simulation of coal reservoir requires a large number of reservoir parameters for support, which is more suitable for the areas where CBM exploration test has been conducted and there are drainage and production data. Due to the complexity of geological conditions, the measured reservoir parameters are often different from those of actual reservoirs. Therefore, during the simulation prediction, some reservoir parameters can only be used after the correction by history data. Using the fitted results as input, the production is predicted according to the proposed production system, and the cumulative gas production at the end of simulation divided by the estimated in-place resource volume can be used to obtain the CBM GRF of a given block.

In order to simulate a CBM reservoir, the mathematical description of the reservoir must be established. Then according to the sensitivity, the production system with appropriate accuracy should be determined to simulate the CBM reservoir. In the simulation calculation of GRF, it is also necessary to consider the limit production, that is, in the actual CBM development process, under what conditions, the production should stop.

Gas reservoir numerical simulation is an ideal method, and has high reliability, but the prediction process is relatively complex, and it requires special simulation software and a large number of effective production data. It is the most suitable in areas highly CBM explored and having a certain scale of production well pattern. This method is often used in the middle stage of CBM exploration and development. It provides theoretical support for the formation of CBM industry.

CBM Gas Content Attenuation Rate

The gas content attenuation rate method determines CBM GRF by calculating the percent of the decrease of CBM content based on measured CBM content before and after CBM production. If measured gas content during the development is used, a dynamic GRF can be obtained. If the measured data obtained after the development is used, an ultimate actual GRF will be obtained. The calculation formula is

$$R = (Q_{\rm i} - Q_{\rm t})/Q_{\rm i} \tag{7.9}$$

In the formula, R is dynamic or ultimate GRF; $Q_{\rm i}$ is original in-place gas content, m^3/t; $Q_{\rm t}$ is the CBM content at the time of calculation, m^3/t.

According to the data of the Oak Grove CBM project in the Black Warrior Basin in Alabama, USA, the initial CBM gas content before producing (1976) was 455 ft^3/t. And the measured CBM content decreased to 215 ft^3/t after producing for 4 years (1981), and 122 ft^3/t after producing for 10 years (1988) (Nelson 2004). Therefore, by calculating the reduction of the CBM content, the GRFs in different periods of the project are 52.7% and 73.2%, respectively. More information about the Oak Grove CBM project will be introduced in the next part.

Material Balance Method

Material balance method calculates the proportion of total cumulative gas production of all CBM wells at the time shut in, to the original in-place CBM resources (within the scope of all gas supply areas) after the completion of the CBM development project. This method is directly derived from the concept of CBM GRF, and is the most accurate and reliable calculation method. The calculation formula is

$$R_{\rm f} = Q/Q_{\rm i} \tag{7.10}$$

In the formula, Q is the total cumulative gas production at the end of the development project, m^3; $Q_{\rm i}$ is the initial in-place resources, m^3.

The premise of application of the material balance method for GRF measurement is that the CBM development project is mature, which means the development of CBM reservoir has been completed, and the gas supply ranges (vertical and horizontal) of CBM wells have achieved an accurate understanding. At present, there isn't mature CBM project in China. The long-term follow-up study on the CBM demonstration project of the Oak Grove field in the Black Warrior Basin conducted by the CBM experts in the United States (Nelson 2004) provides a good example for the determination of CBM GRF using the material balance method.

The Oak Grove CBM demonstration project in the Black Warrior Basin was started in 1976 to assess the CBM production effectiveness of vertical wells and the impact of gas production process on underground coal mining activities. The target coal reservoir for CBM production is the Mary Lee coal seam group of the Pennsylvanian. The coal metamorphism is medium-volatile bituminite (equivalent to the metamorphic stage of fat coal and coking coal in China). In 1976, 23 vertical CBM production wells were drilled on a well spacing of 1000 ft (a 5 wells × 5 wells square pattern; 2 wells missing near the southern boundary). In 1981, 3 coring boreholes (M-1, M-2 and M-3) were drilled, and in 1988, 2 coring boreholes (M-862 and M-863) were drilled. These 5 coring boreholes were arranged inside and outside the square production well pattern. The purpose was to evaluate the reduction of CBM content caused by CBM production through coal core desorption experiments. Figure 7.9 shows the surrounding conditions of the Oak Grove CBM demonstration project block, location of production wells, location of coring boreholes and vertical distribution of coal reservoirs on the stratigraphic column.

The CBM production wells of the demonstration project were completed in the Blue Creek coal seam (average 1090 ft) with open holes, and then stimulated by hydraulic fracturing operation. 23 production wells began producing gas in 1977, and most of them were shut down in 1994–1996, which means the CBM demonstration project was completed in 1996 (Fig. 7.10). During 11 years from 1977 to 1987, 3.20 × 10^9 ft^3 of gas was produced from 23 wells; during 20 years from 1977 to 1996, totally 4.33 × 10^9 ft^3 of gas was produced. A total of 4.33 × 10^9 ft^3 of gas was produced over 20 years, significantly exceeding the 3.12 × 10^9 ft^3 of in-place resources estimated within the 365-acre area delineated by 23 wells.

Based on the comprehensive analyses of CBM development and coalmine production status, geological conditions, hydraulic fracturing effect, CBM content experiment and monitored formation pressure in and around the demonstration block, the horizontal scope of gas supply to the 23 CBM production wells is as follows: In addition to the 365-acre (4000 × 4000 ft) area within the well pattern delineated by the boundary wells, it also includes a 1000ft-wide surrounding belt outside the well pattern boundary, with an area of 459 acres. The vertical scope is as follows: Apart from the Blue Creek coal seam and the Mary Lee coal seam developed in the whole block, the Jaggar coal seam and the New Castle coal seam which are locally developed are gas reservoirs. These two local coal reservoirs are located 5–10 ft below and 45 ft above the Blue Creek coal seam, respectively.

Based on the above-mentioned production data and comprehensive analyses results, according to the calculation results based on the principle of material balance, during the 20-year CBM exploitation period, the CBM GRF within the scope of the well pattern is 90%, and the CBM GRF of the

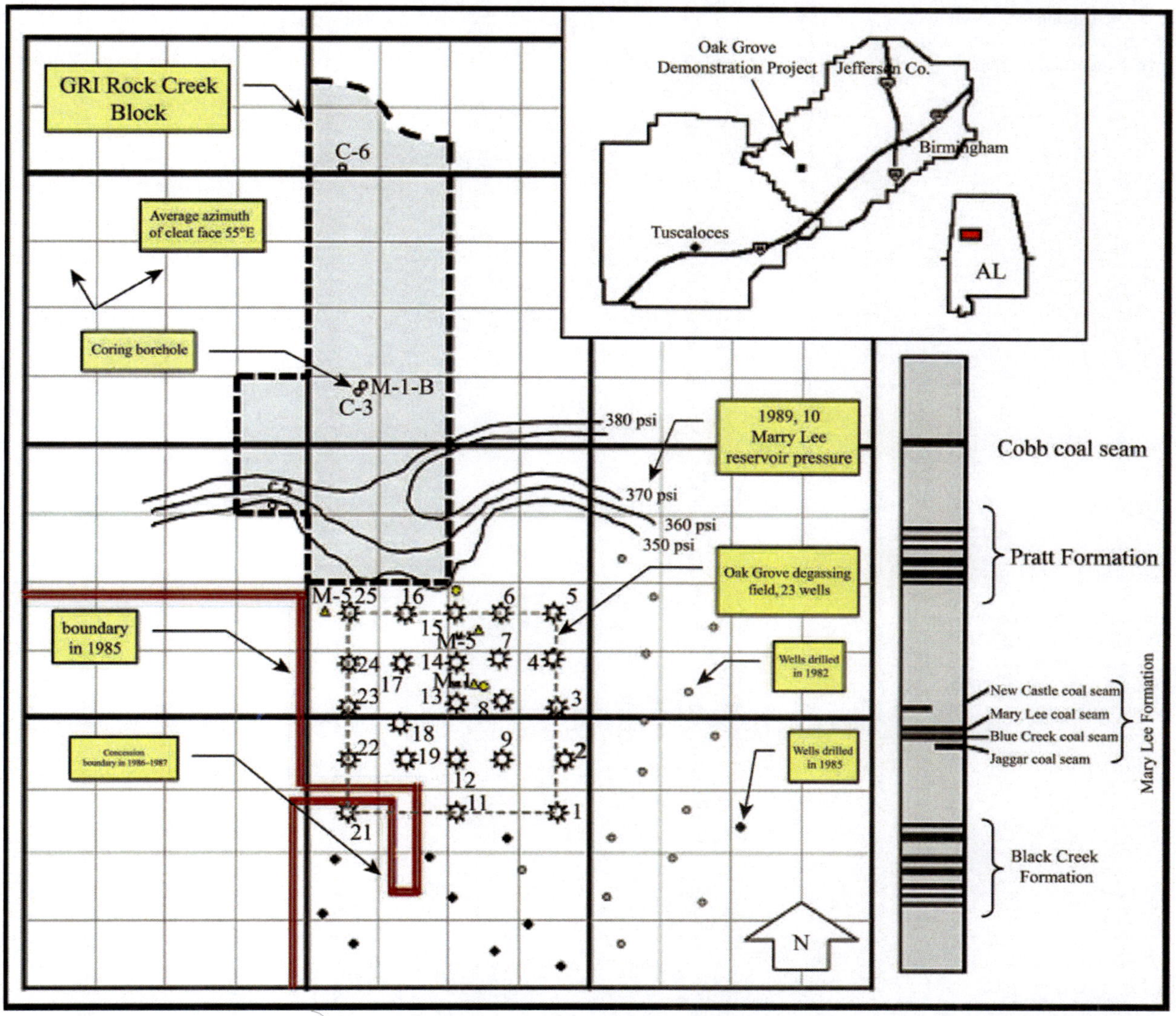

Fig. 7.9 Locations of CBM wells and the stratigraphic column in the Oak Grove demonstration project block (according to Nelson 2004)

external zone of the well pattern is 38%. The balance relationship is shown in Table 7.5.

Analogy Method

Analogy method determineS the GRF of the studied CBM reservoir conditionally by using the calculation and statistical analyses of some CBM reservoirs which have been maturely developed or are close to the end of production. The application condition is that, the to-be-evaluated gas reservoirs and the analogue reservoirs are similar or basically similar in geological parameters, reservoir performance, development and stimulation mode, well spacing and operation, etc.

Due to the different geological conditions of different CBM reservoirs, and various development technologies, methods and economic conditions, the application of analogy method to determine CBM GRF should be cautious.

In the conventional oil and gas industry, generally accepted calibrated GRF are used for analogies, such as the parameters and GRFs of 312 reservoirs published by the American Petroleum Institute (API) and the empirical formula proposed by CNPC for GRFs of reservoirs with different driving types (Zha 1999). But at present, no CBM reservoirs have been produced for a long time in China, and the experience values are insufficient. There are less CBM projects completed in the United States. The Oak Grove

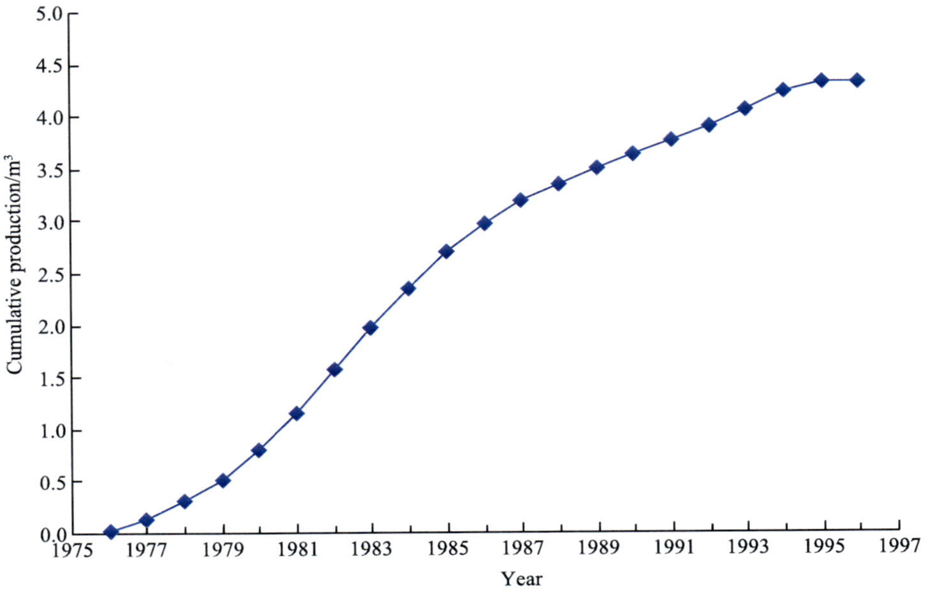

Fig. 7.10 Cumulative gas production of 23 wells in the Oak Grove demonstration block

Table 7.5 Estimation of CBM GRF in the Oak Grove project block from 1977 to 1996 (Adapted from Nelson 2004)

	Coal seam	Area /acre	Average coal seam thickness/ft	Coal density /[t/(acre·ft)]	Initial gas content / (ft^3t)	Original GIP /MMcf①	GRF /%	Gas production /Bcf②
Inside the demonstration block	New Castle	365	1.7	1900	455	536.4	0.90	0.48
	Mary Lee	365	2.3	1900	466	743.3	0.90	0.67
	Blue Creek	365	5.5	1900	466	1777.4	0.90	1.60
	Jagger	50	1.5	1900	466	66.4	0.90	0.06
	Subtotal					3123.5	–	2.81
Surrounding the demonstration block	New Castle	459	1.7	1900	455	674.6	0.38	0.26
	Mary Lee	459	2.3	1900	466	934.7	0.38	0.36
	Blue Creek	459	5.5	1900	466	2235.2	0.38	0.85
	Jagger	122	1.8	1900	466	194.7	0.38	0.07
	Subtotal					4038.9	–	1.54
Total						–	–	4.35

Note: ① 1 MMcf = 1×10^6 ft^3 = 2.83168×10^4 m^3; ② 1 Bcf = $10^9 ft^3$ = $2.83168 \times 10^7 m^3$

demonstration project is a rare example. According to the practical experiences and understandings from the United States, the range of the CBM GRF is wide, from 10 to 75% and generally from 30 to 50% (Sun et al. 1998).

According to the result from the Oak Grove demonstration project, the ultimate CBM GRF within the production well pattern can be as high as 90%. It is quite encouraging that such a high GRF has been achieved in the practical case of CBM development and production.

7.2.2.3 Determination Methods for Lignite Gas Content

Traditional CBM gas content measurement methods, such as the direct measurement method from U.S. Bureau of Mines

(USBM), the desorption method from China coal industry standard (MT/T 77–94), etc., are mainly based on the theoretical assumptions that most CBM gas exists in the adsorbed state. The measured gas content only represents the adsorbed gas. Free gas and water soluble gas are not included. This theoretical hypothesis and practical applications are suitable for coal seams of medium and high coal ranks. But for coal seams of low rank, especially for lignite, the error is obvious.

Occurrence of CBM in lignite, the adsorption performance of lignite and the control factors of gas content are the basises for CBM resources prediction and potential evaluation. Reasonable determination of lignite CBM content is one of the key factors. In the past, little attention has been paid to lignite CBM resources in China with little researches and understandings. Through the recent work, we have achieved some preliminary understandings on properties of lignite CBM reservoirs and determination methods of the gas content.

CBM Reservoir Properties

(1) Pore structure

Coal seam is a kind of reservoir with both fractures and pores. Fractures (also known as cleats) in coal reservoirs have small porosity and are usually saturated with water, thus having little influence on gas storage capacity. The pores in coal matrix between the cleat network are the main part of the pore volume. The porosity controls the CBM gas storage capacity. Therefore, only pore structures in coal matrix are discussed here.

Researches on reservoir performance of bituminite and anthracite show that pores in the reservoirs are mainly micropores (<10 nm) and small pores (10–100 nm), accounting for about 60%–85% of the total porosity. Bituminite and anthracite have large internal surface area and strong adsorption capacities. Adsorbed gas is the main part of CBM, and the proportion of free gas is generally no more than 10% (Zhang et al. 1991). This is the theoretical basis for only considering adsorbed gas and ignoring free gas in gas content test and resources calculation of CBM.

Lignite is a product during diagenetic coalification and is unmetamorphic coal. As the overburden pressure is small and the geothermal temperature is low (less than 40 to 60°C) (Yang 1987), the aromatic cluster in lignite is very small and randomly distributed, and the change of molecular structure is little, mainly including compaction, dehydration and the shedding of hydrophilic and oxygen-rich groups. The pores are mainly mesopores (100–1000 nm) and macropores (>1000 nm), and the surface areas of micropores and pores are relatively small. SEM observation found (Fig. 7.11), pores in lignite are mainly plant tissue pores (Fig. 7.11a), clastic intergranular pores Fig. 7.11b, c, gas pores Fig. 7.11b, c, various kinds of fractures (Fig. 7.11d), etc. The pore size (fracture width) is relatively large, generally greater than 100 nm, and is obviously different from bituminite and anthracite (Zhang et al. 1991, 2003; Bustin and Clarkson 1999).

(2) Isothermal adsorption curves

At 30°C and equilibrium water content, isothermal adsorption experiments were conducted using lignite samples from the Hailaer coal basin group (Figs. 7.12 and 7.13). The correlation coefficients (*R*) corresponding to the Langmuir equation are only 0.37 and 0.98, respectively. According to the general standard, only the correlation coefficient (*R*) greater than 0.99 is deemed conforming to the Langmuir equation. Therefore, the isothermal adsorption curves of both two coal samples cannot be described by the Langmuir equation.

In addition, the isothermal adsorption curves of lignite from Hami of Xinjiang and Xiaolongtan of Yunnan also have the same characteristics, which can be considered to be universal. At present, the adsorption characteristics of lignite are only descriptions of experimental phenomena. The mechanism explanation of the above-mentioned phenomena needs to be further studied. And researches should be carried out from the organic molecular structure, pore structure, adsorption kinetics and other aspects of lignite. On the basis of experiments and mechanism researches, the mathematical equation that describes the isothermal adsorption curve of lignite will be established.

Isothermal adsorption experiments on coal samples also show that the adsorption capacity of lignite is very low. At experiment pressure of 8 MPa, the methane adsorption capacity of lignite samples from different regions in China are generally only about 4 cm^3/g (Figs. 7.12 and 7.13), which is significantly lower than those of bituminite and anthracite. According to the comprehensive isothermal adsorption curves from 41 sub-bituminite (equivalent to lignite in China) samples in the Powder River Basin of the United States (ARI, 2002), the adsorption capacity is very low. When the pressure is 3.4 MPa (500 psia[3]), the adsorption capacity is only 1.70 m^3/t (60 scf/t), which is similar to the adsorption capacity of the lignite in the Hailaer coal basin group in China.

[3]Psia is absolute pressure. 1 psia = 11 b/in^2 = 6.5948 kPa.

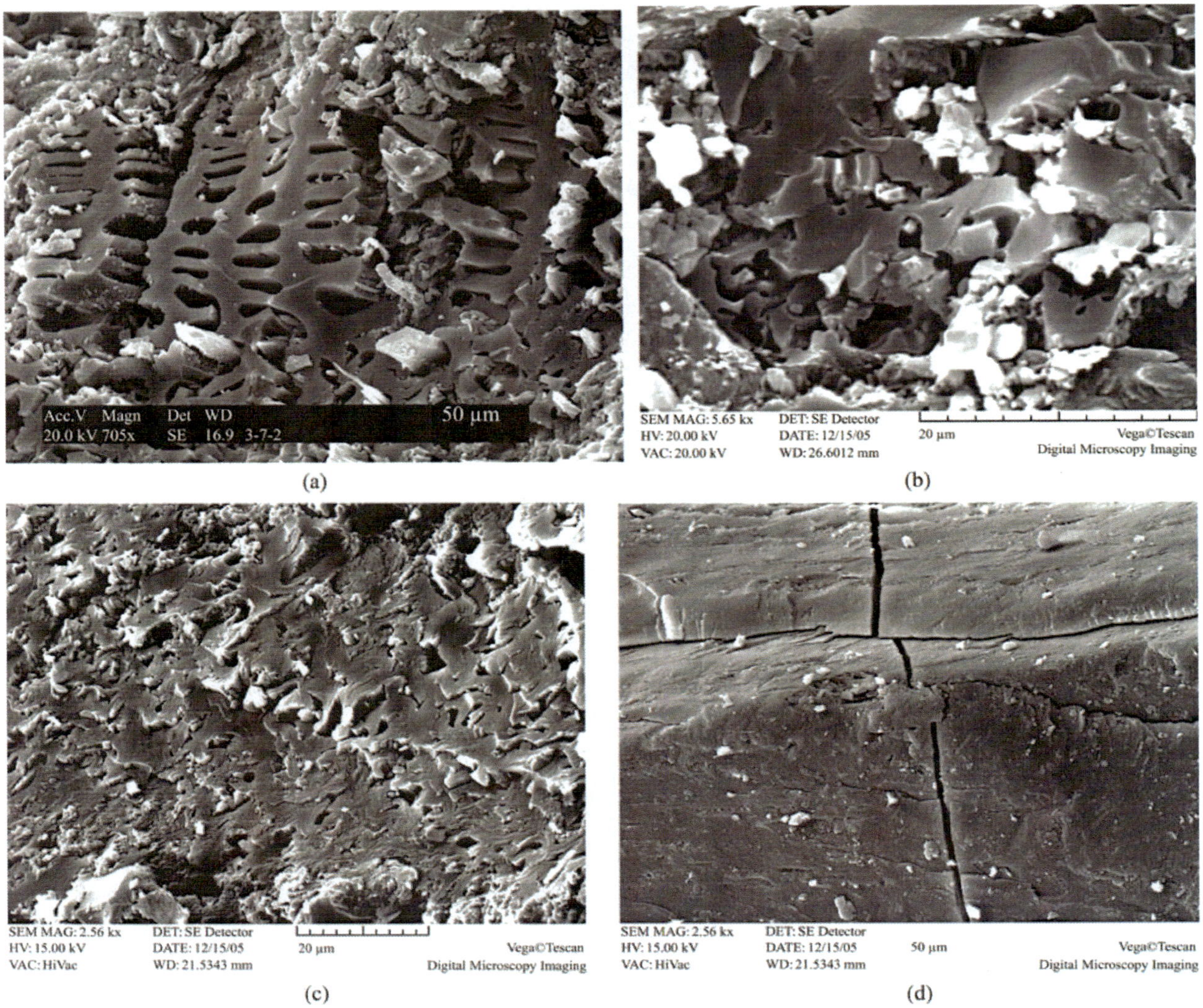

Fig. 7.11 Various pores in lignite under scanning electron microscope

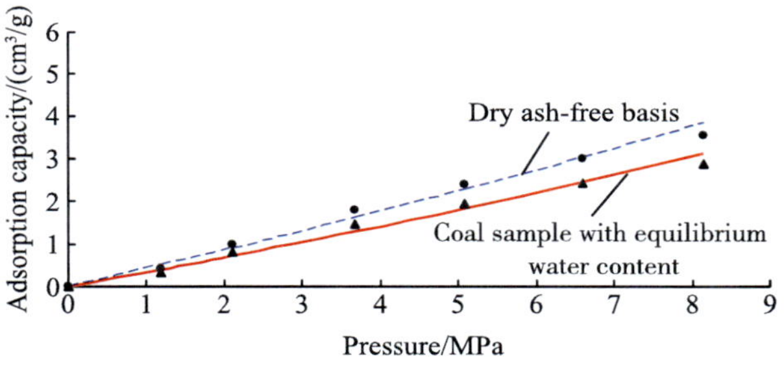

Fig. 7.12 Isothermal adsorption curve of lignite ($R^{o}_{max} = 0.38\%$) in the dayan No.2 coalmine

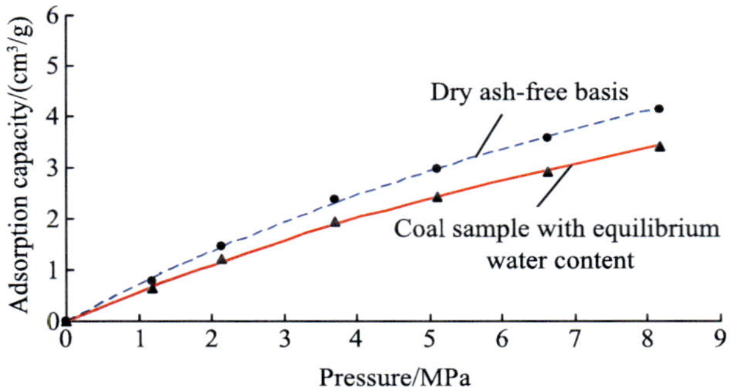

Fig. 7.13 Isothermal adsorption curve of lignite ($R^{o}_{max} = 0.28\%$) in the Jalainur Lingquan open pit coalmine

(3) High free gas content

According to the pore structure and isothermal adsorption experiment, it can be fully inferred that the content of free CBM in lignite is significantly higher than those in bituminite and anthracite. Bustin and Clarkson (1999 combined the analysis of high-pressure methane isothermal adsorption curves with measured porosity to study free gas capacity. They believed that, at typical reservoir pressure and temperature (1–3 MPa, 20–25 °C), free gas in the matrix of bituminite and highly volatile bituminite can be 70% of the total gas capacity, while in the medium volatile bituminite and semianthracite, the free gas is about 5%. Pratt et al. (1999) analyzed the test results of gas content from coal cores of Well Triton in the Powder River Basin, and believed that the gas content was underestimated by 22% as free gas and dissolved gas were not included.

According to the gas and water production data from the early stages of the CBM wells in the Powder River Basin, by gas reservoir numerical simulation and history fitting methods, US CBM geologists proved that many matrix pores and fracture systems in lignite are gas-saturated and contain free gas. They have considered the resource contributions of free gas (ARI 2002), when predicted the CBM Gas in-place volume (G_{IP}) and Gas technically recoverable volume (G_{TR}).

The above theories and facts show that the content of free gas in lignite is very high, and it cannot be ignored in the estimation of lignite CBM resources.

Determination Methods for CBM Gas Content

Pore structures and adsorption properties of lignite suggest that free gas in lignite cannot be ignored. Free gas in the pores can neither be determined by isothermal adsorption curve, nor predicted by estimation for the lost gas in desorption test on coal cores. Therefore, the determination method of CBM gas content of lignite should be different from those of bituminite and anthracite. According to this study, the CBM gas content of lignite should include the adsorbed gas in micro pores and the free gas in coal matrix pores (mainly medium and large pores). This is the basis for establishing the determination methods for CBM gas content of lignite.

(1) Adsorbed gas content

The adsorbed gas content in the micro-pores of lignite can be either measured by coal core desorption or estimated by typical isothermal adsorption curve. Since the isothermal adsorption curve is barely fitted to the Langmuir equation, the Langmuir equation cannot be directly used to calculate the adsorbed gas content at different pressures. Until a precise description equation is established, it can only be determined by graphical method.

(2) Free gas content

Usually, the fracture (cleat) system in coal reservoirs is saturated with water, so only the free gas in matrix pores is considered. The specific calculation is carried out according to the gas equation, that is

$$C_y = \phi \cdot P \cdot K \cdot S \quad (7.11)$$

where C_y is the content of free gas in coal matrix, cm^3/g; ϕ is effective pore volume for free gas in coal matrix, cm^3/g; P is gas pressure, MPa; K is compressibility coefficient of methane, MPa^{-1}; S is gas saturation,%.

The key is to determine the effective pore volume and the gas saturation of free gas in coal matrix. The effective pore volume of free gas refers to the remaining part after deducting the volume occupied by adsorbed gas from the total pore volume in the coal matrix. The total pore volume can be determined by nitrogen density method, mercury injection experiment and true or false specific gravity, etc. The volume occupied by adsorbed gas can be calculated using the density of methane adsorption phase (0.375 g/cm^3) (Cui et al. 2003) and based on methane isothermal adsorption curve of coal samples.

As the pore size of lignite matrix is micron to nanometer and CBM has characteristics as "self-source and self-reservoir", unlike the fracture (cleat) system, gas saturation in matrix pores is high.

(3) Total gas content

By adding adsorbed gas content to free gas content at reservoir pressure and temperature, the isothermal curve of the total gas content can be obtained. Using this isothermal curve, the gas content at the temperature and different pressures can be obtained (Fig. 7.14).

When the reservoir pressure is 5 MPa (equivalent to 500 m deep), the total gas content (lignite matrix porosity 7%) is 4.3 m^3/t, of which adsorbed gas is 2.1 m^3/t, accounting for 48.8%, and free gas is 2.2 m^3/t, accounting for 51.2%.

Based on the above discussion, for lignite CBM reservoir characteristics and gas content determination, the following conclusions can be drawn:

① The pores in lignite matrix are mainly medium and large pores. Experiments show that adsorption capacity of lignite to methane is very low, and the matching degree of isothermal adsorption curve to Langmuir equation is very low. Free gas in lignite usually accounts for more than 50%,

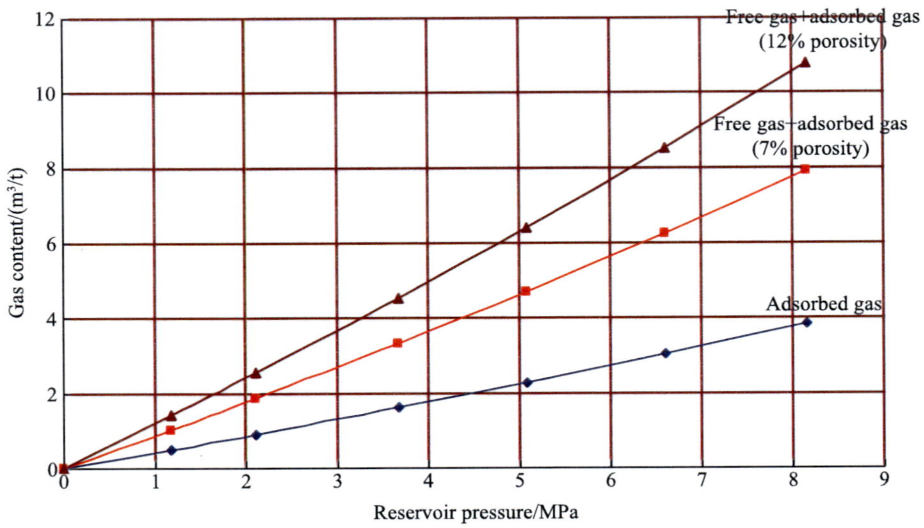

Fig. 7.14 Total CBM content (free gas + adsorbed gas) versus pressure

which cannot be ignored in the evaluation of CBM resources.This is obviously different from coals of medium and high ranks.

② Traditional determination methods for CBM content is not suitable for lignite. As a result, a new determination method suitable for lignite should be established.

③ By adding adsorbed gas to free gas at reservoir pressure and temperature, the isothermal curve of total gas content can be obtained. Using this isothermal curve, the gas content at this temperature and different pressures can be obtained. The CBM content of lignite determined by this method is greater than that determined by traditional coal core desorption experiment. And the results are more scientific, reasonable and close to the actual. It is of great significance for objective evaluation of lignite CBM resources.

7.2.3 Predicting Methods of CBM GTR

7.2.3.1 Calculating Methods

Before discussing the predicting methods for CBM GTR, it is better to know some information about the methods for calculating recoverable CBM resources. Theoretically, volume method, production decline analysis method, material balance method and reservoir numerical simulation method, etc., are widely used for calculating conventional oil and gas reserves, can be used to calculate recoverable CBM resources. However, due to the unconventional properties of CBM and the short development history of CBM industry, the applications of production decline analysis method and material balance method are greatly restricted, and even cannot be used.

Volume method is the simplest method and can be used in any situation where the volume of in-place resources (G_{IP}, the product of four parameters-coal seam thickness, bulk density, block area and initial CBM gas content) and the GRF (E_R, determined by analogy method, isothermal adsorption curve method or gas reservoir numerical simulation method, etc.) are known. Especially in the early stage of a CBM exploration and development project, volume method is the only available method in the absence of coal reservoir parameters and gas reservoir engineering data. It is a complicated process to calculate CBM recoverable resources by using production decline method. It requires CBM production wells to have a stable trend of "declining" gas production. Production of a CBM well usually takes several years to begin to decline. Therefore, production decline is only suitable in the middle and later stages of CBM reservoir (field) development. On the other hand, strong reservoir heterogeneity, well spacing, stimulation measures and other factors all have important influences on the CBM production curve, and make it very difficult to select a representative production decline curve and determine decline rate. Material balance method is based on the unification of three effects-CBM desorbed from coal matrix and dynamic changes of gas and water permeability in coal seam fractures. To use the method, the precondition is that assuming that free gas and adsorbed gas reach a balance (that is, the saturation state of isothermal adsorption curve) (King 1993). This precondition greatly limits its application. At present, this method has not been widely used, although it is theoretically correct under specific boundary conditions. CBM reservoir simulation is the best method to analyze the long-term dynamic data of CBM wells, as well as an important method to calculate CBM recoverable resources. It

Table 7.6 Calculating methods for CBM reserves

Method	Application condition	Application scope	Note
Volume	Known G_{IP} and E_R	Anytime	Necessary to determine GRF
Production decline analysis	Stable "declining" gas production curve	Middle and later project stages	Consistent decline rate mathematical model
Material balance	Assuming free gas and adsorbed gas is balanced—100% gas saturation	Middle and later project stages	Reservoir pressure, permeability and adsorption properties estimated accurately
Reservoir numerical simulation	Complete reservoir parameters, CBM production wells	Early and middle project stages	Special CBM reservoir simulation program required

can not only "correct" coal reservoir parameters to improve the description model of CBM reservoir, but also be used for sensitivity analysis (discussed in the next section). The characteristics of these four methods are summarized in Table 7.6.

Based on the above analysis, in this study, we used volume and gas reservoir numerical simulation methods to calculate CBM recoverable resources. When using these methods, it is necessary to establish specific criteria and procedures to ensure the normalization and accuracy of calculated results, which should be done.

To evaluate CBM resources in China, foreign experiences can be referred to (Cook 2003; Henry and Finn 2003), and suitable methods should be established in line with the principles of scientificity, applicability and operability.

We quantitatively evaluated the recoverability of CBM resources through CBM enrichment zone analysis, that is, taking CBM enrichment zone as a basic geological unit. The so-called CBM enrichment zone refers to the area of continuous distribution of coal-bearing strata within a CBM basin, where the conditions of CBM generation, enrichment and preservation are basically the same due to the influence of structure and sedimentary factors. According to the degree of CBM exploration and development and the richness of data, resources prediction should be discussed in two cases: CBM enrichment zone with and without CBM exploration and development.

7.2.3.2 Gas Reservoir Numerical Simulation

For gas enrichment zone with CBM exploration and development, gas reservoir numerical simulation should be used for calculating CBM GTR. As CBM mainly exists in adsorb state, and porosity and permeability have multiplicity and variability, fluid (gas and water) flow in coal seams is subject to different mechanisms such as desorption, diffusion and seepage. The black oil simulator widely used in conventional oil and gas cannot simulate CBM reservoir very well. Special simulators which can accurately simulate the characteristics of CBM reservoir are required.

COMET2 CBM reservoir simulator from ARI company is a numerical simulation system, 3D, two-phase, multi-component and full-implicit finite difference, for unconventional reservoirs. It is a commercial computer application specially developed for CBM well production and reserve assessments.

Many institutes and units in China (Xi'an Research Institute, Langfang Branch of Research Institute of Petroleum Exploration & Development, etc.) have introduced and used the software. It provides a practical tool for popularization and application of gas reservoir numerical simulation on prediction of CBM GTR in China.

The calculation method of recoverable resources based on numerical simulation on CBM reservoir is to calculate and evaluate Estimated Ultimate Recovery (EUR) of all CBM wells expected to be drilled in the target area by gas reservoir numerical simulation. Specifically, it includes 7 steps described below.

Analysis of Basic Characteristics

A CBM enrichment zone involves its location and distribution, stratigraphic sequence, development of coal-bearing strata and coal seams, tectonic setting and geological structure, geological data of coal and CBM, CBM exploration and development practice and achievements, etc.

Division and Description

According to coal seam depth, CBM and coal development, topographic conditions, mining right, resource (reserve) type, etc., the object (CBM enrichment zone) to be evaluated should be divided into several blocks. Coal seams below the lower limit of the weathering zone, and at 1000, 1500 and 2000 m are first considered.

Description involves block scope and area, geological and engineering data, and key reservoir parameters (coal seam depth, thickness, gas content, reservoir pressure and permeability) of each block. Reservoir parameters should be defined average and interval values.

Key contents include the changing rule and prediction method of gas content, reservoir pressure (gradient) model, trend and determined principle of permeability, etc.

Determination of Grids in a Block

Grids in a block means the number of CBM wells expected to be drilled in the block. It is equal to the block area divided by the drainage area of a well. The drainage area of a well can be determined according to the actual well spacing of the well pattern or empirical data. In the southern Qinshui Basin, the well spacing is usually 300–400 m.

Selection of Typical CBM Production Wells in a Block

A typical CBM production wells reflects the overall characteristics of coal reservoirs in the block, including reliable and complete primary reservoir parameters and gas well production data. The well is producing in a good state and has been producing for over 1–2 years. It's better the well is at peak production. Typical CBM production wells may be several as required.

Numerical Simulation (Establish a CBM Reservoir Model and Calculate EUR)

(1) Establish a CBM reservoir model

To the typical CBM wells selected, COMET2 CBM reservoir simulator is used for fitting production history. This can determine the reservoir parameters such as permeability, porosity and relative permeability which cannot be accurately measured in laboratory, and further confirm the test results such as reservoir pressure and CBM gas content. All of these are key parameters for CBM production prediction.

By adjusting reservoir parameters several times, until the simulated gas and water production performance is consistent with the actual gas well performance, a representative CBM reservoir model is obtained.

(2) Calculation of estimated ultimate recovery

Estimated ultimate recovery (EUR) is defined as "the estimated recoverable oil and gas plus the recovered from a reservoir during a given period". It is applicable to any reservoir in any state or maturity (discovered or not) (according to SPE/WPC/AAPG oil resource classification and definition). Here a reservoir means a CBM well.

Using the established CBM reservoir model and inputting relevant geological, engineering and work system data according to the requirements and format of the CBM reservoir simulator, the simulator can predict gas and water production in 20 years, and provide the history curves of predicted daily and cumulative productions. Then the curves are "truncated" by the law of "economic limit production", and the remaining cumulative CBM production is the EUR of the well. The meaning of "economic limit production" is that when the sales revenue of CBM production is exactly equal to the operating cost of the CBM well, the CBM well will be abandoned. The calculation method of "economic limit production" is: economic limit production = direct operating cost/net gas price.

The method to determine EUR by using the prediction curve of a CBM well production through gas reservoir numerical simulation is shown in Fig. 7.15.

EUR is estimated based on a well (single grid mode) or unit coal thickness (coal thickness per meter).

In order to reflect the entire EUR change range within a block, the above simulation typically needs to run at least 11 times. Coal depth and permeability can be grouped by the maximum, minimum and average in the block (Table 7.7), so as to fully reflect the CBM geological conditions and the reservoir heterogeneity. Then the data selected are loaded into computer to calculate EUR. The data types and formats can be listed in a table.

Establish EUR Probability Distribution

Numerical simulation provides the basis for establishing EUR probability distribution. But However, the EUR probability distribution model of a CBM well should be determined first.

The probability distribution model is established by analyzing the whole distribution and trend of EUR of a CBM production well. At present, CBM development in China is in the experimental stage, the number of CBM production wells is very limited, and the production history is very short. There is no CBM well that has been completed the production cycle. As a result, the probability distribution model cannot be established. Based on the CBM production cases in the San Juan Basin and the Black Warrior Basin of the United States, it is generally assumed that EUR probability distribution is lognormal.

After establishing EUR probability distribution, the logarithms of at least 11 EURs is calculated and then the average and the mean variance, and finally a lognormal normal distribution ln.

The probability distribution of EUR in the block is obtained by multiplying the ln by the number of grids (CBM wells) in the block.

Calibrate CBM GTR

EUR obtained by the above method is equal to CBM GTR of unproductive wells (grids) in meaning and volume. For

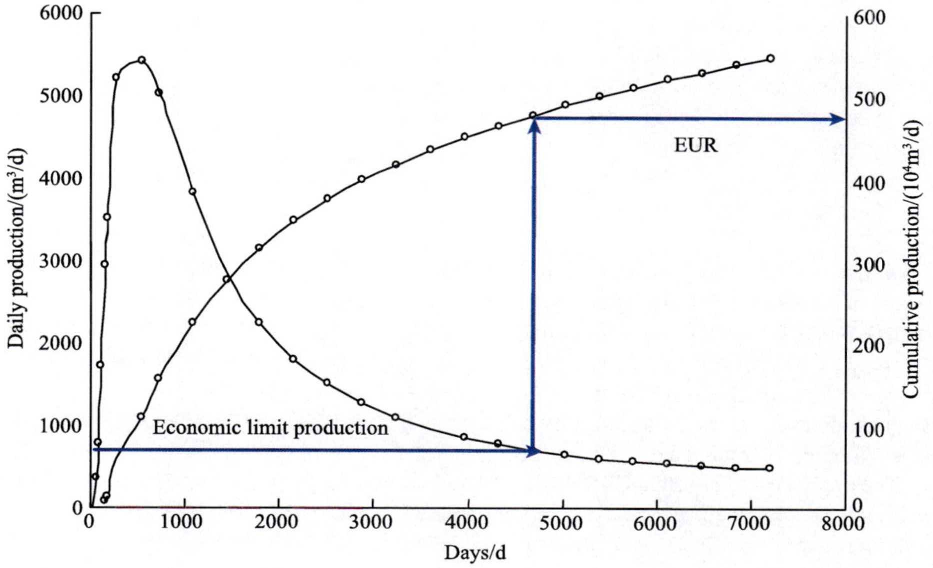

Fig. 7.15 Schematic diagram of CBM EUR

Table 7.7 Parameter groups for calculating EUR

Group no	Reservoir depth	Permeability
1	Shallow: average thickness and corresponding gas content and reservoir pressure	Good
2		Medium
3		Poor
4	Middle: average thickness and corresponding gas content and reservoir pressure	Good
5		Medium
6		Poor
7	Deep: average thickness and corresponding gas content and reservoir pressure	Good
8		Medium
9		Poor
10	Middle: the maximum thickness, gas content and reservoir pressure at the middle interval	Medium
11	Middle: the minimum thickness, gas content and reservoir pressure at the middle interval	Medium

produced wells, CBM GTR is equal to the EUR minus the cumulative gas production before the estimated time.

Add all CBM GTRs of all wells to get the GTR of a target (a gas enrichment zone). GTR is expressed to be the maximum (5% probability), median (50% probability), minimum (95% probability) and average (mathematical expectation of the distribution function).

7.2.3.3 Loss Analysis

For basins without CBM exploration, CBM enrichment zone analysis using reservoir numerical simulation is not applicable. In China, most of CBM gas enrichment zones have not been explored. For these zones, analogy method is useful by selecting the EUR probability distribution of the zone that has been explored and developed and with similar geological conditions. In China, there are only a few CBM wells with a long and continuous production history, so that it is impossible to establish a sufficient number of "standard CBM wells" with representativeness at the enrichment zone level. For these kinds of gas enrichment zones, gas reservoir numerical simulation cannot be used for quantitative evaluation of CBM recoverability.

In the past, when calculating CBM in-place resources in China, the CBM in-place volume in all minable coal seams (a coal seam is thicker than 0.7 m) was calculated by taking a gas bearing zone or a coal reservoir intervals as a basic unit (Ye et al. 1998; Zhang et al. 2002b). According to the calculating practice of CBM geological resources (gas in-place) in China, when changing from geological resources to GTR, three parts of lost volume should be deducted: lost reservoir, lost thickness and lost area. In other words, loss analysis is useful to estimate CBM GTR in zones (belts) without CBM exploration and development.

(1) Lost reservoir

Like other mineral deposits, natural gas stored in coal reservoirs cannot be fully extracted, and some remains underground forever. This lost volume is known as lost reservoir. Lost reservoir can be estimated by commonly used CBM GRF. GRF is mainly controlled by geological factors such as reservoir permeability, then by artificial factors such as stimulation measures and well spacing.

The methods to determine CBM GRF include: desorption rate statistics, analog, abandoned reservoir pressure, reservoir numerical simulation and physical simulation, which will be discussed in other parts in the paper.

(2) Lost thickness

Lost thickness refers to that, in the scope of a CBM well or a reservoir unit for CBM resources (reserves) calculation, and in all coal layers (any coal seam thicker than 0.5 m or 0.7 m) that participates in the calculation of CBM geological resources, the coal seams without gas production because wells were not completed due to geological and technical causes cannot participate in the calculation of CBM GTR, which are deemed as lost thickness. For example, in the Panzhuang Coalmine and Shizhuang Coalmine in the southern Qinshui Basin, there are 3 minable coal seams (No. 3, No. 9 and No. 15) of the Shanxi Formation and Taiyuan Formation, with a cumulative minable thickness of 10.31 m, but almost all CBM development wells were only completed in the No.3 and No.15 coal seams. The average completion thickness of a well is 9.3 m. The lost coal seams are nearly 10%. The Xinji CBM project area is another example in Huainan Coalfield. there are 3 CBM production wells penetrated more than 10 Carboniferous and Permian coal seams, of which the cumulative thickness of coal seams thicker than 0.6 m a layer is 32.36 m on average. Three wells were completed only in 3 or 4 primary coal seams, and the average completion thickness of a well is 19.2 m. The lost thickness is nearly 41%.

Lost thickness in any CBM well should be calculated, and lost thickness of different types of CBM wells should be statistically analyzed to study the distribution rule.

In the Powder River Basin of the United States, a CBM well was completed in only one coal seam. By the end of 2001, more than 12,000 CBM wells were drilled in the basin. In that way, lost thickness can be minimized. Similar designs have been used in the CBM development plan in the Enhong Coalmine in Yunnan Province in China.

(3) Lost area

In an evaluation unit, due to the CBM volume (gas content and composition), resource scale, terrain conditions and other factors, there are some areas not suitable for industrial CBM development, so CBM gas volume in these areas should not be calculated as CBM GTR, which is known as lost area.

Lost area has been taken into account in predicting national CBM resources in China. For example, the gas weathering zones and the coal areas in Fujian, Guangdong and Tibet were deducted. This work is mainly aimed at the gas-bearing zones where CBM resources has been calculated. The areas that are not suitable for industrial CBM development (isolated blocks with limited resources volume) are removed.

The isolated blocks with limited resources refer to the CBM gas enrichment blocks with CBM resources volume less than 30×10^8 m^3 (equivalent to small CBM reservoirs) and blocks far away from adjacent CBM reservoirs (fields).

(4) GTR calculation

In the calculation, lost area is deducted first to obtain the geological resources (G_{IP}) suitable for industrial CBM development, then lost thickness (C_h) is deducted according to the development and distribution characteristics of coal seams and the possibility of completion technology, and finally CBM GTR (G_{TR}) are calculated based on GRF (E_R). The calculation formula is

$$G_{TR} = G_{IP} \times (1 - C_h) \times E_R. \quad (7.12)$$

7.3 Potential Analysis of CBM GTR in China

CBM GTR is the quantitative evaluation of the recoverability of CBM resources. The purpose for establishing an evaluation system of CBM GTR is to evaluate the potential of CBM GTR in China.

Based on the established system, and according to unified and standard procedures and detailed data, we predicted CBM GTR in China from the perspectives of basins, gas enrichment zones and burial depth, and analyzed the distribution characteristics of CBM GTR in China.

7.3.1 Division of CBM Enrichment Units

7.3.1.1 Sequence of CBM Enrichment Units

Definition of CBM Enrichment Unit

Domestic and foreign coal and CBM exploration practices have proved that the distribution of coal in the Earth crust and CBM gas in coal seams are uneven. The CBM distribution in coal seams is restricted by coal seam conditions, structure and preservation. And there are natural CBM units with different scales or ranks in CBM distribution. In conventional oil and gas exploration, various natural units of hydrocarbon accumulation are called accumulation units, such as hydrocarbon systems, zones and so on. Compared with conventional oil and gas, CBM is self-sourc and self-reservoir, without obvious migration and re-accumulation process. Therefore, the natural units of CBM are called CBM enrichment units.

CBM is unconventional gas self-sourc and self-reservoir. Although it is different from conventional natural gas, there are enrichment units of different scales or ranks which are similar to conventional oil and gas accumulation units, which have been proved. But currently, researches on CBM enrichment unit and division are still very weak and have not even attracted enough attentions. Current divisions are quite confused. Some divisions even directly apply the ideas of coal accumulation unit or conventional oil and gas accumulation unit. It is not completely consistent with the actual conditions of CBM enrichment. Especially for the divisions of CBM enrichment units in gas-bearing areas or coal basins, the opinions are still inconsistent at present, and the researches are very weak, which to some extent restricts deepening CBM exploration and evaluation. Therefore, it is important and necessary to study characteristics of CBM enrichment units thoroughly and systematically and to divide CBM enrichment units scientifically.

The geological environment of CBM formation in China is complex. As the continent of China is a composite continent formed by some small cratons, intermediate massifs and the fold zones between them, tectonic stabilities in most coal-accumulation basins are poor, so the CBM accumulation conditions are very complex (Zhang and Lin 1998). Studies on formation and characteristics of CBM enrichment units in China, correct divisions of CBM enrichment units, and in-depth researches on CBM reservoir types, accumulation patterns and rules are important for the exploration and development of CBM in China. This is not only of great practical significance to select favorite CBM areas and improve successful CBM exploration, but also of great scientific significance to establish scientific CBM exploration procedures and develop and improve CBM geological theories.

Sequence Divisions of CBM Enrichment Units

Under the controls of various geological factors such as structure, sedimentation, burial depth, pressure and hydrodynamic forces (Zhang et al. 2002a, b), CBM distribution in coal seams is uneven, and in different scales, controlling factors are not always the same. Therefore, there are CBM accumulation units or gas units of different scales in the earth's crust.

As unconventional natural gas, the "unconventional" properties of CBM is mainly reflected in its occurrence state. It is well known that, conventional natural gas is mainly in free state and exists in the pores or fractures in sandstones as primary reservoirs, while CBM is mainly (more than 90%) adsorbed to the inner surface of coals in adsorbed state. Only a small volume of CBM gas is stored in the cleats, fractures and pores of coals in free state, or dissolved in coal seam water. The special occurrence determines the particularity of CBM in reservoir formation and the exploration and development characteristics. Therefore, the occurrence state of CBM is the most essential characteristic that distinguishes it from conventional natural gas.

The characteristic of CBM occurrence state is essentially different from those of conventional natural gas, which determines that the rules and control factors of CBM accumulation are fundamentally different from conventional oil and gas, and the CBM enrichment units are different from conventional gas accumulation units.

The close relationship between CBM and coal is self-evident. The gas adsorption characteristics of coal rocks, especially the adsorption of methane generated by coal rock itself, is one of the basic properties of coal rocks. Generally, as long as there is a coal seam, there will be a certain volume of CBM in the coal seam, which has been confirmed by coal mining. Coal-bearing strata and coal seams are the material basis for CBM formation and provide reservoir spaces for CBM accumulation. Due to the universality of CBM existence in coal seams, in the early stage of CBM exploration, it was once considered that a large coalfield was a large CBM field (Zhang et al. 2002b). However, later exploration proved that the distributions of economically recoverable CBM fields (reservoirs) and coal seams were not identical. Although in a large scope, the distribution of CBM is inevitably controlled by the distribution of coal seams, within the range of coal seam distribution, only the coal seams with good gas content and high permeability are recoverable for CBM. Researches have proved that controlling factors on CBM recoverability include coal seam thickness, coal volume, coal metamorphism degree, coal seam depth, overlying thickness, the lithology of roof and floor and hydrogeology, etc. Only when good coal seam conditions, pressure sealing conditions and preservation conditions match with each other in both time and space, can form a recoverable CBM gas reservoir (field) or CBM enrichment zone.

A large number of researches have proved that during the coal forming process, the gas generated is enough to meet the CBM adsorption capacity. But what factors can cause the inconsistency of CBM gas and coal seam? The author thinks that, besides the difference in the adsorption performance of coal seam itself, another cause is the basic characteristics of CBM itself. Although CBM is adsorbed to coal seam, essentially it is a fluid mineral with strong mobility, and its enrichment and accumulation must have good "trap" conditions. The difference between solid mineral as coal and fluid mineral as CBM makes the existence of coal seam a necessary but not a sufficient condition for CBM enrichment. In other words, the formation of industrial CBM resources is not only controlled by coal seam, but also by other factors related to gas mobility, such as "trap" and preservation conditions. Different controlling factors on coalmine and CBM accumulation determine that their accumulation units are by no means the same, different exploration procedures should be selected for CBM and coal exploration.

Due to the adsorption characteristics of CBM and complex geological conditions of CBM in China, correct understandings and divisions of CBM enrichment units that objectively exist are not only the basis for studying the CBM geological rules, but also the key to formulating correct exploration procedures and improving exploration success.

CBM enrichment units can be divided into 5 levels: gas-bearing region, gas-bearing basin, gas-enriched zone, gas-enriched belt and CBM reservoir (field).

(1) Gas-bearing region

Gas-bearing region is the first-order unit in the sequence of CBM enrichment units, which refers to a regional geological tectonic unit with clear boundary and CBM geological conditions. CBM gas-bearing region is a regional geological tectonic unit formed in a geological period. It has a history of negative crustal activities and sedimentary filling. Similar coal accumulation occurred in each negative unit in the gas-bearing region.

(2) Gas-bearing basin

Gas-bearing basin is the second-order unit in the sequence of CBM enrichment units. It refers to a secondary negative sedimentary unit within a certain gas-bearing area, with clear boundaries, coal accumulation during the geological history, and formation conditions and environment of CBM reservoirs. The classification of gas-bearing basins is mainly based on the present coal-bearing basins and their CBM geological conditions. A gas-bearing basin must be a coal-bearing basin, but a coal-bearing basin is not necessarily a gas-bearing basin. Gas-bearing basin is an important grade of CBM enrichment unit, which plays an important role in controlling the distribution of CBM in the Earth crust. The material basis of CBM formation is coal and coal-bearing strata, which must be formed in basins. The type, structure and configuration of basins control the properties and quantity of coal-forming materials, the development and distribution of coal-bearing strata. Later sedimentary and tectonic evolutions control the coal-forming process (such as the thermal evolution degree of coal) and the CBM generation process, as well as the distribution of coal seams and the geological conditions of CBM occurrence at present.

(3) Gas-enriched zone

Gas-enriched zone is the third-order unit in the sequence of CBM enrichment units. It refers to the area with relatively good CBM-bearing properties within a coal seam area in a basin under the influences of structure and sedimentation.

(4) Gas-enriched belt

Gas-enriched belt is the forth-order unit in the sequence of CBM enrichment units. It refers to a combination of several CBM reservoirs (fields) which are of same coal property and similar evolution process, and are adjacent to each other horizontally and controlled by similar geological factors.

CBM reservoir (field)

CBM reservoir (field) is the basic unit of CBM enrichment units. It refers to the coal rock or coal seams that can absorb CBM to a considerable volume sealed by pressure.

Although the above-mentioned division scheme includes 5 orders of CBM enrichment units—gas-bearing region, gas-bearing basin, gas-enriched zone, gas-enriched belt and CBM reservoir (field), these 5 orders may not exist at the same time in a study area, especially in a basin. There may be more than one gas-enriched belt in a middle to large basin with complex tectonic deformation. But, in a small basin with simple structure, there may be only one gas-enriched belt, which means the gas-bearing basin and the gas-enriched belt is an order.

Corresponding to the five orders, exploration procedures should follow five steps, from region to basin to zone to belt to CBM reservoir, and favorable targets are selected according to the research content of each step (Table 7.8).

Table 7.8 Characteristics study and exploration stages division of sequence of CBM enrichment units

CBM enrichment unit	Research contents	Exploration evaluation stage
Gas-bearing region	Regional paleostructure, paleogeography and paleoclimate; regional coal accumulation and coal accumulation rules; evaluation and prediction of CBM resources in a gas-bearing region	Evaluation of gas-bearing region
Gas-bearing basin	Basin type, tectonic paleogeography and sedimentary background; coal accumulation rules and coal seam distribution; potential evaluation and prediction of CBM resources in a gas-bearing basin	Evaluation of gas-bearing basin
Gas-enrich zone	Evaluation of CBM basic geological conditions; analysis of main gas-controlling factors in a gas-enriched zone; resources prediction and economic factor analysis in a gas-enriched zone	Evaluation of gas-enriched zone
Gas-enriched belt	Evaluation of CBM basic geological conditions; analysis of main gas-controlling factors a in gas-enriched belt; prediction of CBM accumulation model; resources estimation and economic factor analysis in a gas-enriched belt	Evaluation of gas-enriched belt
CBM reservoir	Accumulation condition evaluation; summary of accumulation pattern and rules; reserve calculation and economic evaluation	Evaluation of CBM reservoir

7.3.1.2 Division Scheme of CBM Enrichment Units in China

The generation, reservoir and production of CBM are closely related to coal seams, so CBM enrichment units are inevitably controlled by coal seam distribution and coal seam properties. The formation and distribution of coal seams in the Earth crust are controlled by geological structures.

The formation and distribution patterns of coal seams in China are obviously controlled by tectonic systems. The first-order depression zone in a giant or large-scale tectonic system controls the overall distribution of large-scale coal accumulation zone or large coal accumulation basin (group) in a certain geological period. While the lower-order tectonics often control the distribution of a coal-bearing basin and the variation of thickness, lithology, lithofacies and coal-bearing properties in the basin. The interlacing of giant latitudinal tectonic systems, longitudinal tectonic systems and various shear tectonic systems constitutes the basic tectonic pattern and determines the morphology, occurrence, development and distribution of coal accumulation depressions in China.

Compared with the regional tectonic characteristics of coalfields in the United States, the tectonic characteristics of coalfields in China are much more complicated. The United States is part of the North American platform. The north American platform is a giant platform developed outwards in concentric circles with the Canadian and the Greenland shield as the center. The part where the United States is located is called the "central platform", which is relatively stable in geological history. China is a composite continent formed by some small platforms, intermediate massifs of different sizes and fold zones between them, which determines the poor tectonic stabilities in most coal-bearing basins in China, the complication and variety of structure types, and the complexity of geological conditions of CBM occurrence. Due to the geological background of multi-landmasses and the multi-stage tectonic movements in China, the sedimentation, coal accumulation rules and coal deformation characteristics in late stage in coal basins are obviously regional. Different types of gas-bearing basins and gas-enriched belts are developed in different regions, which have different probability of CBM distribution. Therefore, CBM enrichment units in China are only divided into gas-bearing region, gas-bearing basin and gas-bearing belt, and gas-bearing basins are classified.

(1) Gas-bearing region

According to the CBM geological background, regional CBM resources, economic and geographical conditions and other factors, and with reference to the scheme of Zhang et al. (2002a, b), except sea regions, the gas-bearing regions on the continent of China are divided into eastern regions including Heilongjiang, Jilin, Liaoning (I), Hebei, Shandong, Henan, Anhui (II) and South China (III); central regions including eastern Inner Mongolia (IV), Shanxi, Shaanxi and Inner Mongolia (V) and Yunnan, Guizhou, Sichuan and Chongqing (VI); and western regions including northern Xinjiang (VII), southern Xinjiang, Gansu, Qinghai (VIII), Yunnan-Tibet (IX).

① Gas bearing regions: Heilongjiang, Jilin and Liaoning (I). The north and east boundaries are the Border Line of China, the south boundary is the eastern segment of the Yinshan-Yanshan fold belt, and the west is bounded by the Greater Khingan Range tectonic belt. The coal-bearing strata in these regions are mainly Lower Cretaceous, Neogene and Paleogene, and Carboniferous and Permian.

② Gas bearing regions: Shandong, Henan and Anhui (II) in the east of the Taihang Mountain, roughly equivalent to the east of North China landmass. From the west to the east is from the Taihang Mountain tectonic belt to the Tanlu fault zone; from the north to the south covers the southern boundary of Heilongjiang, Jilin and Liaoning gas-bearing regions to the eastern segment of the Qinling-Dabie Mountain fold belt. The coal-bearing strata are mainly Carboniferous–Permian, and less Lower and Middle Jurassic.

③ South China gas bearing region (III), tectonically equivalent to the eastern part of the Yangtze landmass and the South China mobile belt. It is located in the south of the Qinling-Dabie Mountain fold belt and the east of the Wuling Mountain tectonic belt, including vast Southeast and South China. Late Permian coal-bearing strata are mainly developed.

④ Gas bearing region—eastern Inner Mongolia (IV)—is from the northern Border Line to the Yinshan-Yanshan fold belt in the south, from the Greater Khingan Range tectonic belt in the east to the north Xinjiang in the west.

⑤ Gas bearing regions—Shanxi, Shaanxi and Inner Mongolia (V)—cover the western Taihang Mountain, roughly equivalent to the western part of North China landmass. The area is from the Helanshan-Liupanshan fault zone in the west to the western boundary of Hebei, Shandong, Henan and Anhui gas bearing regions in the east, from the western segment of the Yinshan-Yanshan fold belt in the north to the western segment of the Qinling-Dabie Mountain fold belt in the south. They have the most abundant CBM resources in China.

⑥ Gas bearing regions—Yunnan, Guizhou, Sichuan and Chongqing (VI)—the west of South China coal-bearing area. They are from the Longmen-Ailaoshan fault zone in the west to the western boundary of South China gas-bearing area in the east, and from the southern boundary of Shanxi, Shaanxi and Inner Mongolia gas-bearing regions in the north, to the southern Border Line. Permian strata are dominant.

⑦ Gas bearing area of northern Xinjiang (VII): The geographical distribution is in and to the north of the Tianshan fold belt in Xinjiang. There are many Early–Middle Jurassic coal-bearing basins in the area.

⑧ Gas bearing region—southern Xinjiang-Gansu-Qinghai (VIII)—the south of the Tianshan Mountain. The area is from the western segment of the Tianshan-Yinshan fold belt in the north, to the western segment of the Kunlun-Qinling fold belt in the south, and from the Border Line in the west, to the western boundary of Shanxi, Shaanxi and Inner Mongolia gas-bearing regions in the east. There are Early–Middle Jurassic coal-bearing basins and Carboniferous–Permian coal-bearing basins.

⑨ Gas-bearing regions—Yunnan and Tibet (IX)—the south to the southern Xinjiang-Gansu-Qinghai gas bearing region. Coal seams of all ages in this area are well developed, including more Late Triassic, Neogene and Paleogene, less Carboniferous and Permian, and sporadic Jurassic and Cretaceous basins. Strong crustal activities made the coal seams underdeveloped, thin and variable. Since the Yanshanian, the basin structures have become more complex after many tectonic movements, so the CBM geological conditions are poor.

(2) CBM basins

Geological environments of coal-bearing basins in China are complex and the reformation of Mesozoic–Cenozoic and Paleozoic basins is different.

Mesozoic–Cenozoic coal-accumulation basins are mainly distributed in northeastern China, and Ordos, Xinjiang and Inner Mongolia. The coal accumulating basins are abundant and well preserved (Wang 2004). Paleozoic coal basins in South and North China have been through many tectonic movements such as Indosinian movement, Yanshanian movement and Himalayan movement. Structure patterns are complicated in most areas and the original basin features no longer exist. This makes the basic geological structure units of CBM formation and occurrence—CBM basins—have different degrees of reconstruction. It includes not only the original sedimentary basins (coal accumulating basins) with obvious boundaries, which are less affected by later reconstruction, but also the residual coal-bearing basins (groups) or tectonic basins (groups), which have been cut by late tectonic changes and denudation, and whose boundaries are not easy to be determined. Therefore, basin boundary demarcation principles and schemes are divided into two categories, as follows:

① Coal-bearing basins dominated by Mesozoic and Cenozoic coal seams. It mainly refers to coal accumulating basins formed in Mesozoic and Cenozoic. Of course, some basins may contain Paleozoic coal seams. The reconstructions of coal accumulation basins are weak with simple structure patterns. The original characteristics of coal accumulation basins are basically maintained. The determination of basin boundaries make full use of petroleum basin division results to delineate basin scope. For Mesozoic and Cenozoic coal-bearing basins with concealed or partially concealed boundaries, according to the types of the basins, the boundary faults and their properties should be determined by comprehensively using the geophysical data, and the basin scopes should be roughly delineated, such as the Erenhot basin group. The Paleozoic coal-bearing strata within the basins are also included in CBM basins.

② Coal-bearing basins dominated by Paleozoic coal seams. It mainly refers to Carboniferous and Permian coal-bearing basins in South and North China, and some basins may contain Mesozoic and Cenozoic coal seams. Paleozoic coal basins were reconstructed by multiple stages of tectonic movements such as Indosinian movement, Yanshanian movement and Himalayan movement. The original basin structures have been destroyed or broken into several modern coal-bearing basins, with orogenic belts or mountains as basin boundaries. Therefore, the boundary determination of Paleozoic CBM-bearing basins is based on the basin tectonic reconstruction and give consideration to the depression-uplift tectonic patterns during late Hercynian. Referring to the results of petroleum basin division, in practice, large-scale orogenic belts or mountain ranges with partition function would be used as the rough boundaries of basins. If there are coal seam distribution outside the basin margin, use the maximum geological boundary of coalseam distribution as the extensive basin boundary. The concealed boundaries are similar as those of Mesozoic and Cenozoic coal-bearing basins, determined by comprehensive analysis of geological and geophysical data.

According to the above principles, we determined 65 CBM basins (areas) in China.

Gas-enriched characteristics of CBM basins are controlled by the tectonic characteristics of both coal accumulation period and later reconstruction period. Therefore, the following 4 principles should be specifically considered when classifying CBM basins:

① Since coal seams are both CBM source rocks and reservoirs, the accumulation and evolution characteristics of coal seams are directly related to the accumulation characteristics of CBM resources. Therefore, the classification of coal accumulating basins aiming at CBM studies must not only reflect the characteristics of coal accumulation rules, but also to some extent represent the characteristics of coal accumulation and preservation.

② The types of coal-bearing formations depend on geotectonic conditions and sedimentary environment. Globally, marine, continental and transitional sedimentary formations are developed. Mesozoic and Cenozoic formations in China are dominated by continental facies, only some are marine facies. It is necessary to illustrate the sedimentary formation developed in the basin and summarize the deposit and material composition.

③ Four basic criteria are followed for sedimentary basin classification by modern Chinese and foreign geologists: the genetic mechanism of coal accumulating basin, that is, the basement crust type of the basin; the geotectonic position of the basin in plate tectonics; the geodynamic environment of coal accumulating basin and the relationship between plate interactions during the formation and development of coal accumulating basin; and the age of basin development.

④ For the classification of coal-accumulation basins aiming at CBM studies, as CBM gas content is controlled by later reconstruction, characteristics and mechanisms of later reconstruction must be considered.

In order to reveal and understand the rules of CBM generation and enrichment at basin scale, based on basin division, the classification of CBM basin should start from studies of basin dynamic evolution and reconstruction, follow the simple, practical and conventional classification principles, and adopt the double classification and naming scheme considering both coal accumulation environment and later reconstruction mechanisms and characteristics.

(3) Gas-enriched belt

Gas-enriched belt is the forth-order unit in the sequence of CBM enrichment units. It refers to a combination of several CBM reservoirs (fields) which are of same coal properties and similar evolution process, and are adjacent to each other horizontally and controlled by similar geological factors. Gas-enriched belt usually contains coal-bearing strata of a geological age, but it may also contain coal-bearing strata of two successive geological ages. Refering to the division scheme of coal-bearing areas proposed by Mao Bijie et al. (1999), only some CBM basins with large areas are divided into CBM gas-enriched belts. As for some small basins, they can be considered as gas-enriched belts because the coal seam, coal quality and evolution are similar. Generally, gas-rich belts are delineated according to regional tectonic lines or sedimentary (denudation) boundaries, and jurisdiction is consider in special regions. 108 gas-rich units (Table 7.9) have been determined.

7.3.2 Predicted CBM GTR in China

7.3.2.1 Scope and Units Calculated

Scope

The calculation of national CBM technically recoverable resources is based on the evaluation of CBM GIP (gas in-place resources). Therefore, the scope of CBM GTR is basically the same as that of CBM GIP (Zhang et al. 2002a, b), but the CBM resources in lignite is included. The following are not included in the calculation:

Table 7.9 CBM enrichment units and their characteristics in China

CBM Gas-bearing region			Gas-bearing basin		Gas-enriched belt			CBM basin type		Tectonic position	Basement	Cap rock
Region	Zone		No.	Name	Name	No.	Age	Original	Reconstructed			
	No.	Description										
Eastern region	I	Heilongjiang, Jilin and Liaoning	1	Songliao Basin	Southwest	I-1	K_1	Rift basin	Extensional fault block, superimposed and buried (broad folds or monocline)	Songliao massif	Neoproterozoic metamorphic basement	Pz^2-Kz
					East	I-2						
			2	Hegang Basin	Hegang	I-3	K_1	Intracontinental depression basin	Extensional fault block, thermal (monocline)	Jiamusi massif		Mz-Kz
			3	Sanjiang-Mulenghe Basin	Sanjiang-Mulenghe	I-4	K_1	Intracontinental depression basin	Extensional fault block, thermal (broad folds)			Mz-Kz
			4	Yanji Basin	Yanji	I-5	K_1	Intracontinental depression basin	Extensional fault block, thermal (broad folds)	Tianshan–Chifeng active tectonic zone	Hercynian fold basement	Mz-Kz
			5	Yilan-Yitong Basin	Yilan-Yitong	I-6	E_{2-3}	Rift basin	Extensional fault block, thermal (broad folds)	Songliao, Jiamusi massifs and Tianshan–Chifeng active tectonic zone	Proterozoic and Hercynian mixed basement	Kz
			6	Dunhua-Meihe Basin	Dunhua-Meihe	I-7	E_1	Rift basin	Medium compressional, fold thrust nappe, superimposed and buried (Dunhua: monocline; Meihe: syncline)			Kz
			7	Tieling-Changtu Basin	Tiefa	I-8	K_1	Rift basin	Extensional fault block, thermal (asymmetric syncline)	North China landmass	Archean–Proterozoic metamorphic basement	Pz^2-Kz
			8	Fuxin-Yixian Basin	Fuxin	I-9	K_1	Rift basin	Extensional fault block, thermal (fault blocks)			Pz^2-Kz
			9	Lower Liaohe Basin	Shenbei-Dawa	I-10	C_2–P_1	Epicontinental marine basin	Extensional fault block, superimposed and buried, thermal (broad folds)			Pz^1-Kz
							E_{1-2}	Rift basin	Extensional fault block, thermal (broad folds)			
			10	Fushun Basin	Fushun	I-11	E_1	Rift basin	Medium compressional, fold thrust nappe, superimposed and buried (asymmetric syncline)			Mz-Kz
			11	Hongyang region	Hongyang	I-12	C_2–P_1	Epicontinental marine basin	Strongly compressional thrust nappe (synclinorium)			Pz^1-Kz

(continued)

Table 7.9 (continued)

CBM Gas-bearing region			Gas-bearing basin		Gas-enriched belt			CBM basin type		Tectonic position	Basement	Cap rock
Region	Zone		No.	Name	Name	No.	Age	Original	Reconstructed			
	No.	Description										
Central region	II	Hebei, Shandong, Henan and Anhui	12	East of northern Hebei Basin group	Pingquan-Shouwangfen	II-1	C_2–P_1	Epicontinental marine basin	Medium compressional, fold thrust nappe, superimposed and buried (synclinorium)	North China landmass	Archean–Proterozoic metamorphic basement	Pz^1-Kz
					Luanping-Chengde	II-2	J_3	Rift basin				
			13	West Beijing Basin	West Beijing	II-3	J_{1-2}	Rift basin	Medium compressional, fold thrust nappe, superimposed and buried (tight folds)			Pz^1-Kz
			14	Tangshan Basin	Kaiping	II-4	C_2–P_1	Epicontinental marine basin	Medium compressional, fold thrust nappe, superimposed and buried (echelon folds)			Pz^1-Kz
					Chezhoushan	II-5	J_{1-2}	Rift basin				
			15	South Bohai Bay Basin	North Henan-Northwest Shandong	II-6	C_2–P_1	Epicontinental marine basin	Extensional fault block superimposed and buried (fault blocks)			Pz^1-Kz
					Central Huabei Plain	II-7						
			16	East of Taihang Mountain region	Xingtai-Jiaozuo	II-8	C_2–P_1	Epicontinental marine basin	Extensional fault block, thermal (monocline)			Pz^1-Kz
			17	Huangxian Basin	Huangxian	II-9	E_1	Rift basin	Extensional fault block (monocline)			Pz^1-Kz
			18	Central Shandong Basin group	Central Shandong	II-10	C_2–P_1	Epicontinental marine basin	Extensional fault block (monocline)			Pz^1-Kz
			19	Southwest Shandong Basin group	Southwest Shandong	II-11	C_2–P_1	Epicontinental marine basin	Extensional fault block, buried, thermal (fault blocks)			Pz^1-Kz
			20	South of North China Basin	East Henan	II-12	C_2–P_2	Epicontinental marine basin	Extensional fault block, buried (fault blocks)			Pz^1-Kz
					Gongyi-Pingdingshan	II-13	C_2–P_2	Epicontinental marine basin	Strongly compressional fold thrust nappe (tight folds)			
					Xuzhou-Huaibei	II-14	C_2–P_2	Epicontinental marine basin	Strongly compressional thrust nappe (syncline)			
					Huainan	II-15	C_2–P_2	Epicontinental marine basin	Medium compressional, fold thrust nappe, superimposed and buried (synclinorium)			
	III	South China	21	Southeast Hubei-north Jiangxi Basin	Southeast Hubei-north Jiangxi	III-1	P_2	Marginal marine basin	Medium compressional, fold thrust nappe, superimposed and buried (synclinorium)	Yangzi landmass	Late Archean–Palaeoproterozoic metamorphic basement	Pz^1-Kz
			22	Zhenjiang-Anqing Basin	Zhenjiang-Anqing	III-2	P_2	Marginal marine basin	Strongly compressional thrust nappe (synclinorium)			Pz^1-Kz
			23	Changzhou Basin	Changzhou	III-3	P_2	Marginal marine basin	Strongly compressional thrust nappe (synclinorium)			Pz^1-Kz
			24	Central Hunan Basin	Central Hunan Permian	III-4	P_2	Marginal marine basin	Medium compressional, fold thrust nappe, superimposed and buried (syncline)			Pz^1-Kz
					Central Hunan Carboniferous	III-5	C_1					
			25	Chenlei Basin	Chenlei	III-6	P_2	Marginal marine basin	Medium compressional, fold thrust nappe, superimposed and buried (syncline)	South China active tectonic zone	Caledonian metamorphic basement	Pz^2-Kz
			26	Pingle Basin	Pingxiang	III-7	T_3	Intracontinental depression basin	Medium compressional, fold thrust nappe, superimposed and buried (syncline)	Joint part of Yangzi landmass and South China active tectonic zone	Late Archean, Palaeoproterozoic and Caledonian mixed basement	Pz^2-Kz
					Leping-Yichun	III-8	P_1	Marginal marine basin				
			27	Shangrao Basin	Shangrao	III-9	P_1	Marginal marine basin	Medium compressional, fold thrust nappe, superimposed and buried (synclinorium)			Pz^2-Kz
			28	Central Guangxi Basin	Central Guangxi	III-10	C_1–P_2	Marginal marine basin	Medium compressional, fold thrust nappe, superimposed and buried (syncline)	Yangzi landmass	Late Archean – Palaeoproterozoic metamorphic basement	Pz^1-Kz
			29	Baise Basin	Baise	III-11	E_1	Rift basin	Extensional fault block (asymmetry syncline)	South China active tectonic zone	Caledonian metamorphic basement	Pz^2-Kz

(continued)

Table 7.9 (continued)

CBM Gas-bearing region			Gas-bearing basin		Gas-enriched belt			CBM basin type		Tectonic position	Basement	Cap rock
Region	Zone		No.	Name	Name	No.	Age	Original	Reconstructed			
	No.	Description										
Central region	IV	Eastern Inner Mongolia	30	Hailaer Basin group	Jalainur	IV-1	K_1	Rift basin	Extensional fault block, superimposed and buried (fault blocks)	Jalainur-Xing'an active tectonic zone	Hercynian fold basement	Mz-Kz
					Beier	IV-2			Extensional fault block, superimposed and buried (broad folds)			
					Bayanshan	IV-3						
					Huhehu	IV-4						
			31	Erenhot Basin group	North Bayan Baolige	IV-5	K_1	Rift basin	Extensional fault block, superimposed and buried (fault blocks)	Joint part of Jalainur-Xing'an active tectonic zone, Tianshan-Chifeng active tectonic zone and North China landmass	Archean–Proterozoic crystalline basement and Hercynian fold mixed basement	Mz-Kz
					Daxinganling	IV-6			Extensional fault block, superimposed and buried (broad folds or monocline)			
					Wunite	IV-7						
					Manite	IV-8						
					West Bayan Baolige	IV-9						
					Sunite	IV-10						
					Tenggeer	IV-11						
					Wenduermiao	IV-12						
					Ulanqab	IV-13						
					Chuanjing	IV-14						
	V	Shanxi, Shaanxi and Inner Mongolia	32	Yinchuan Basin	Shizuishan	V-1	C_2–P_1	Epicontinental marine basin	Medium compressional, fold thrust nappe, superimposed and buried (syncline)	North China landmass	Archean–Proterozoic crystalline basement	Pz^1-Kz
							$J_{1\text{-}2}$	Intracontinental depression basin				
					Weizhou	V-2	C_2–P_1	Epicontinental marine basin				
			33	Ordos Basin	Jungar	V-3	C_2–P_1	Intracontinental depression basin	Weakly compressional, fold (monocline)			Pz^1-Kz
					Eastern margin	V-4		Epicontinental marine basin	Medium compressional, fold thrust, uplifted and denuded (monocline)			
					Weibei	V-5		Epicontinental marine basin				
					North	V-6	J_2	Intracontinental depression basin	Weakly compressional, fold (broad folds)			
					West	V-7	$J_{1\text{-}2}$		Medium compressional, fold thrust, uplifted and denuded type (tight fold)			
					Huanglong	V-8	J_2		Weakly compressional, fold (monocline)			
			34	Zhangjiakou Basin	Zhangjiakou	V-9	$J_{1\text{-}2}$	Rift basin	Medium compressional, fold thrust nappe, superimposed and buried (broad syncline)			Pz^1-Kz
			35	Xiahuayuan Basin	Xiahuayuan	V-10	$J_{1\text{-}2}$	Rift basin	Medium compressional, fold thrust nappe, superimposed and buried (broad syncline)			Pz^1-Kz
			36	Weixian Basin	Weixian	V-11	$J_{1\text{-}2}$	Rift basin	Medium-compressional, fold thrust nappe, superimposed and buried (synclinorium)			Pz^1-Kz
			37	Datong Basin	Carboniferous- Permian	V-12	C_2–P_1	Epicontinental marine basin	Strongly compressional, thrust nappe (syncline)			Pz^1-Kz
			38	Ningwu Basin	Carboniferous- Permian	V-13	C_2–P_1	Epicontinental marine basin	Strongly compressional, thrust nappe (syncline)			Pz^1-Kz
			39	Huoxi Basin	Huoxi	V-14	C_2–P_1	Epicontinental marine basin	Weakly compressional, thermal (broad syncline)			Pz^1-Kz
			40	Qinshui Basin	Qinnan	V-15	C_2–P_1	Epicontinental marine basin	Medium-compressional, fold thrust, uplifted and denuded (synclinorium)			Pz^1-Kz
					Huodong	V-16						

(continued)

Table 7.9 (continued)

CBM Gas-bearing region			Gas-bearing basin		Gas-enriched belt			CBM basin type		Tectonic position	Basement	Cap rock
Region	Zone		No.	Name	Name	No.	Age	Original	Reconstructed			
	No.	Description										
					Yangquan-Xiangtan	V-17						
					Xishan	V-18						
					Gaoping-Jincheng	V-19						
			41	Sanmenxia-Luoyang Basin	Sanmenxia-Luoyang	V-20	J_2	Rift basin	Weakly compressional, fold (broad folds)			Pz^1-Kz
Central region	VI	Yunnan, Guizhou, Sichuan and Chongqing	42	Sichuan Basin	Huayingshan Triassic	VI-1	T_3	Foreland basin	Medium compressional, fold thrust, uplifted and denuded (trough-like folds, ejective folds)	Yangzi landmass	Late Archean–Palaeoproterozoic metamorphic- hypometamorphicrock	Pz^1-Kz
					Huayingshan Permian	VI-2	P_2	Epicontinental marine basin				
					Yongrong Triassic	VI-3	T_3	Foreland basin				
					Yongrong Permian	VI-4	P_2	Epicontinental marine basin				
					Yale Triassic	VI-5	T_3	Foreland basin				
					Yale Permian	VI-6	P_2	Epicontinental marine basin				
			43	South Sichuan and North Guizhou Basin	South Sichuan and North Guizhou	VI-7	C_1–P_2	Epicontinental marine basin	Medium compressional, fold thrust, uplifted and denuded (trough-like folds, ejective folds)			Pz^1-Kz
			44	South Guizhou Basin	Guiyang	VI-8	C_1–P_2	Epicontinental marine basin	Medium compressional, fold thrust, uplifted and denuded (superimposed folds)			Kz
			45	Zhaotong Basin	Zhaotong	VI-9	E	Rift basin	Extensional fault block (broad synclinorium)			Pz^1-Kz
			46	Nanpanjiang Basin	Liupanshui	VI-10	C_1–P_2	Epicontinental marine basin	Medium compressional, fold thrust, uplifted and denuded (superimposed folds)			Pz^1-Kz
			47	Chuxiong Basin	Dukou-chuxiong	VI-11	T_3	Foreland basin	Medium compressional, fold thrust, uplifted and denuded (synclinorium)		-	Pz^1-Kz
Western region	VII	Northern Xinjiang	48	Fuhai	Fuyun	VII-1	$J_{1\text{-}2}$	Rift basin	Medium-compressional, fold thrust, uplifted and denuded (monocline)	Jalainur-Xing'an active tectonic zone	Hercynian fold basement	Mz-Kz
			49	Junggar	Karamay gas	VII-2	$J_{1\text{-}2}$	Rift basin	Medium-compressional, fold thrust nappe, superimposed and buried (monocline)	Jalainur massif	Proterozoic crystalline basement	Mz-Kz
					East Junggar gas	VII-3						
					South Junggar	VII-4						
			50	Santanghu	Santanghu	VII-5	$J_{1\text{-}2}$	Rift basin	Medium-compressional, fold thrust nappe, superimposed and buried (synclinorium)	Jalainur- Xing'an active tectonic zone	Hercynian fold basement	Mz-Kz
			51	Tu-Ha	Keyayi-Bujiaer	VII-6	$J_{1\text{-}2}$	Rift basin	Medium-compressional, fold thrust nappe, superimposed and buried (monocline)	Tianshan-Chifeng active tectonic zone	Hercynian fold basement	Mz-Kz
					Takequan bulge	VII-7						
					Southeast margin of Taibei sag	VII-8						
					Tuokexun	VII-9			Medium-compressional, fold thrust nappe, superimposed and buried (syncline)			
					Shaerhu	VII-10						
					Dananhu	VII-11						
					Hami depression	VII-12						
			52	Yili	Yining	VII-13	$J_{1\text{-}2}$	Rift basin	Medium-compressional, fold thrust nappe, superimposed and buried (syncline)	Yili massif	Archean–Proterozoic metamorphic basement	Mz-Kz
			53	Youerdusi	Youerdusi	VII-14	$J_{1\text{-}2}$	Rift basin	Medium-compressional, fold thrust nappe, superimposed and buried (syncline)	Tianshan-Chifeng active tectonic zone	Hercynian fold basement	Mz-Kz
			54	Yanqi	Yanqi	VII-15	$J_{1\text{-}2}$	Rift basin	Medium-compressional, fold thrust nappe, superimposed and buried (synclinorium)	Tianshan-Chifeng active tectonic zone	Hercynian fold basement	Mz-Kz
	VIII		55	Tarim	North Tarim	VIII-1	$J_{1\text{-}2}$	Rift basin	Strong compressional, thrust nappe (tight fold)	Tarim landmass		Pz^1-Kz

(continued)

Table 7.9 (continued)

CBM Gas-bearing region			Gas-bearing basin		Gas-enriched belt			CBM basin type		Tectonic position	Basement	Cap rock
Region	Zone		No.	Name	Name	No.	Age	Original	Reconstructed			
	No.	Description										
Western region		Southern Xinjiang – Gansu – Qinghai			East Tarim	VIII-2			Extensional fault block (monocline)		Archean–Proterozoic metamorphic basement	
			56	Qaidam	North Qaidam Jurassic	VIII-3	J_1	Rift basin	Medium-compressional, fold thrust nappe, superimposed and buried (syncline, local inversion)	Qaidam micro landmass	Early Proterozoic, Caledonian and Hercynian mixed basement	Pz^2-Kz
			57	Chaoshui	Chaoshui	VIII-4	J_2	Rift basin	Medium-compressional, fold thrust nappe, superimposed and buried (syncline or monocline)	North China landmass	Archean–Proterozoic metamorphic basement	Mz-Kz
			58	Central Qilian	Muli-Menyuan	VIII-5	J_2	Rift basin	Medium-compressional, fold thrust nappe, superimposed and buried (synclinorium)	Kunlun-Qinling active tectonic zone	Caledonian metamorphic basement	Mz-Kz
					Tuole-Donggoukou	VIII-6	C_1–P_2	Epicontinental marine				
			59	Minle	Minle	VIII-7	J_2	Rift basin	Medium-compressional, fold thrust nappe, superimposed and buried (syncline)			Mz-Kz
							C_1–P_2	Epicontinental marine				
			60	Margin of Gasu-	Jingyuan	VIII-8	$J_{2\text{-}3}$	Rift basin	Medium-compressional, fold thrust nappe, superimposed and buried (synclinorium)			Mz-Kz
				Ningxia-Inner Mongolia	Margin of Gasu-Ningxia-Inner Mongolia	VIII-9	C_1–P_2	Epicontinental marine				
			61	Xining-Lanzhou	Xining-Lanzhou	VIII-10	$J_{2\text{-}3}$	Rift	Medium-compressional, fold thrust nappe, superimposed and buried (syncline)			Mz-Kz
	IX	Yunnan and Tibet	62	Qiangtang	No data, undivided		T_3	Rift	Strongly compressional, thrust nappe, thermal (tight fold)	Qiangbei-Simao (micro) landmass	Caledonian metamorphic basement	Mz-Kz
			63	Tuotuohe			P_2	Marginal marine	Medium-compressional, fold thrust, uplifted and denuded (anticlinorium)			Mz-Kz
			64	Zhakang-Changdu			C_1–P_2	Marginal marine	Medium-compressional, fold thrust, uplifted and denuded (synclinorium)			Mz-Kz
			65	Rikaze-Angren			E_2	Rift (back-arc rift)	Strong compressional, thrust nappe (synclinorium)	Gangdese-Tengchong active tectonic zone	Himalayan fault zone	Kz
Marine region	X	Taiwan	Undivided									

① No.1 Anthracite coal seam;
② Coal seams in CBM weathering zones;
③ Small, complex and sporadical coalfields or coal areas in Fujian, Guangdong, Hainan and Taiwan where coals are highly metamorphism and have reached anthracite No.1 rank;
④ Sporadical coalfields in Tibet less explored, no CBM;
⑤ Coal seams not recoverable and coal roof and floor;
⑥ Coal seams below 2000 m.

Unit

Taking gas-enriched belt as a basic unit to calculate CBM GTR, totally 108 gas-enriched belts were determined in this study (Table 7.9). 11 gas-enriched belts were not involved in the calculation as they were not included in the calculation of CBM GIP or the CBM GIP is less than 30×10^8 m^3. The 11 gas-enriched belts are Fuhai gas-enriched belt, Central Shandong gas-enriched belt, Southwest Shandong gas-enriched belt, Gaoping-Jincheng gas-enriched belt, Southwest Songliao Basin gas-enriched belt, Yanji gas-enriched belt, Shangrao gas-enriched belt, Pingquan-Shouwangfen gas-enriched belt, Pingxiang gas-enriched belt, West Bayan Baolige gas-enriched belt and Wenduermiao gas-enriched belt. At last 97 gas-enriched belts were calculated.

7.3.2.2 Calculating Methods

Gas-enriched belts were used for calculation of CBM GTR in this study. According to CBM exploration and development in China and CBM data acquisition in every gas-enriched belt, the 97 gas-enriched belts for calculating CBM GTR in China can be divided into two types: gas-enriched belts explored and developed and gas-enriched belts not explored or developed. Different methods were used to calculate CBM GTR.

Gas-Enriched Belts Explored and Developed

For gas-enriched belts explored and developed, CBM reservoir numerical simulation was conducted for the calculation. Up to now, China has only carried out small-scale CBM commercial productions in 4 gas-enriched belts: the south part of the Qinshui Basin, eastern margin of the Ordos Basin, the Weibei Coalfield and the Liujiajing Coalfield in the Fuxin Basin. For these gas-enriched belts, recoverable resources were calculated by CBM gas reservoir numerical simulation. Taking the gas-enriched belt in the southern Qinshui Basin as an example, the calculation of recoverable resources using CBM gas reservoir numerical simulation is briefly described below.

(1) Basic characteristics

The gas-enriched belt is located at the southern margin of the large Qinshui synclinorium. It is a northward monocline consisting of simple structures—NNE and NS broad and gentle folds. The strata are continuous, complete and flat. The formation dip is commonly 5°–10°. The C–P coal seams are at 300–1200 m. The CBM gas content is 8–20 m^3/t. The coal evolution was influenced by plutonic metamorphism and regional magmatic metamorphism. From south to north, the coal ranks are anthracite, lean coal and meagre coal. As a belt more explored, there have found three CBM well groups for comercial CBM production: Shizhuang, Panzhuang and Panhe well groups.

(2) Intervals

The gas-enriched belt is divided into 2 intervals in depth, 300–1000 and 1000–1500 m. According to the resources calculation method mentioned above, the interval 300–1000 m is subdivided into three sections: 300–500, 500–700 and 700–1000 m. The interval 1000–1500 m is further subdivided into two sections: 1000–1150 and 1150–1300 m (the coal seams in this belt are shallower than 1300 m). Totally there are 5 sections.

The CBM well spacing in southern Qinshui Basin is usually 300–400 m.

(3) Reservoir parameter combination

According to different reservoir parameter combinations, reservoir numerical simulation were carried out for the five depth segments. In the parameter combinations, coal seam permeability is combined with other parameters at good, medium and poor conditions. For the coal seam permeability at medium condition, in the Sect. 300–500 m, the permeability is determined directly by fitting gas well production history. For the other sections, the permeability is determined by permeability prediction. At good and poor conditions, the coal seam permeability is twice and 1/2, respectively, of that at medium condition. The coal seam permeability of the five depth sections is shown in Table 7.10. Geological parameters such as coal seam thickness, depth and top and gas content, etc., are average values of the unit. Statistical results are shown in Tables 7.11 and 7.12.

(4) Production prediction of CBM wells

According to prepared data and resources calculating methods mentioned above, COMET 2.1 (CBM reservoir

Table 7.10 Coal seam permeability

No	Section/m	Medium permeability/mD		Good permeability/mD		Poor permeability/mD	
		Coal No.3	Coal No.15	Coal No.3	Coal No.15	Coal No.3	Coal No.15
1	300–500	3.36	1.52	6.72	3.04	1.68	0.76
2	500–700	2.73	1.50	5.46	2.99	1.37	0.75
3	700–1000	1.12	0.82	2.24	1.63	0.56	0.41
4	1000–1150	0.47	0.35	0.94	0.69	0.23	0.17
5	1150–1300	0.34	0.26	0.68	0.53	0.17	0.13

Table 7.11 Statistics of geological parameters

No	Section/m	Top of Coal No.3/m	Top of Coal No.15/m	Depth of Coal No.3/m	Depth of Coal No.15/m	Thickness of Coal No. 3/m	Thickness of Coal No.15/m	Gas content of Coal No.3/(m^3/t)	Gas content of Coal No.15/(m^3/t)
1	300–500	464.93	372.61	435.91	525.91	4.93	2.61	12.85	14.92
2	500–700	371.03	231.08	611.50	749.22	4.53	2.30	15.97	17.48
3	700–1000	254.09	157.20	825.56	919.44	5.76	2.76	18.97	21.06
4	1000–1150	23.35	−99.85	1107.50	1227.50	5.85	2.65	21.98	24.28
5	1150–1300	33.00	−72.48	1232.50	1347.50	5.50	2.53	22.18	25.85

Table 7.12 Statistics of coal seam thicknesses

No	Section/m	Thickness of Coal No.3/m			Thickness of Coal No.15/m		
		Minimum	Maximum	Average	Minimum	Maximum	Average
1	300–500	2.7	7.0	4.93	0.8	4.6	2.61
2	500–700	0.8	6.3	4.53	0.8	3.7	2.30
3	700–1000	4.5	6.3	5.76	1.4	3.5	2.76
4	1000–1150	5.3	6.2	5.85	1.8	3.1	2.65
5	1150–1300	5.4	5.6	5.50	2.4	2.7	2.53

numerical simulation software) was used to predict production by taking the control area of 0.14 km^2/well. Using reservoir parameter combination, 19 simulation calculations were carried out, and 19 CBM production performance curves were predicted. Table 7.13 shows the 20-year predicted well production. Figure 7.16 shows the predicted production curve.

(5) EUR

The drainage and production period of a CBM well is usually long, but commercial CBM development aims at economic benefits, so it is impossible for a CBM well to drain and produce for a long time without limit. Production of a CBM well is the primary criterion to measure whether CBM development can be continued. In early CBM development, it is usually necessary to estimate gas well productiviey, and estimate the ultimate production with economic benefits, that is, the so-called "EUR".

EUR is not only related to gas well productiviey, but also closely related to market demand, gas price and production and operation costs. According to the latest survey, the price of CBM gas and natural gas is generally around 1.0 yuan/m^3 (Table 7.14).

In CBM production cost, operating cost not related to CBM production is called fixed cost, while that related to CBM production is called variable cost. Surveys show fixed costs and variable costs in CBM production process (Table 7.15).

By comprehensive consideration, the latest market price of CBM should be 1.0 yuan/m^3. Affected by variable cost,

Table 7.13 20-year predicted single well CBM production in the gas-enriched belt in southern Qinshui Basin.

Parameter combination No	Section/m	Gas production /(m^3/d)	Cumulative production /10^4m^3	Recovery /%	Section/m	Gas production /(m^3/d)	Cumulative production /10^4m^3	Recovery /%
1	300–500	656.49	1572.91	75.76	1000–1150	1007.81	2099.13	53.61
2	300–500	700.06	1246.04	60.02	1000–1150	970.43	1619.66	41.37
3	300–500	681.37	912.21	43.94	1000–1150	870.49	1172.01	29.93
4	500–700	659.42	1901.86	83.24	1150–1300	639.98	1279.85	33.67
5	500–700	705.82	1586.77	69.45	1150–1300	611.83	974.54	25.64
6	500–700	723.77	1193.48	52.24	1150–1300	530.22	693.59	18.25
7	700–1000	964.97	2327.84	68.59	–	–	–	–
8	700–1000	979.57	1834.97	54.07	–	–	–	–
9	700–1000	923.45	1385.17	40.81	–	–	–	–
10	500–700	1092.31	2317.67	69.07	1150–1300	623.44	991.94	35.76
11	500–700	161.34	375.28	69.08	1150–1300	600.26	957.12	36.78
Average			1514.02	62.39			1223.48	34.38

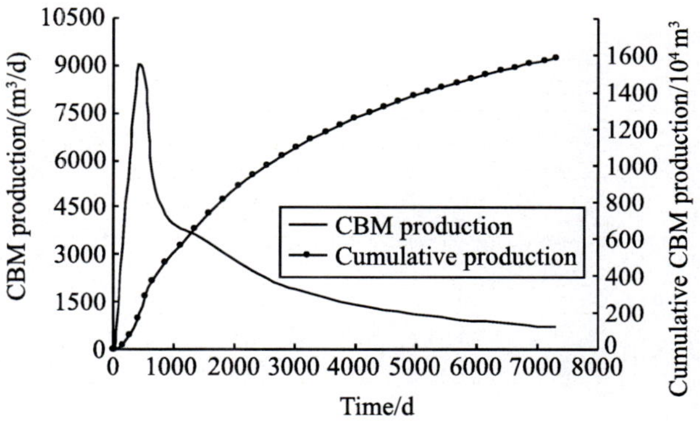

Fig. 7.16 Predicted CBM production for Combination No.5

net CBM gas price is the difference between the market price and the variable cost, i.e. 950.9 yuan/km^3.

When the sales revenue is just equal to the CBM operating cost, the CBM gas well should be abandoned, and the production of the CBM well is called "economic limit production". "Economic limit production" is the ratio of direct operating cost to net gas price. According to the above data, the economic limit production is 208 m^3/d per well.

Table 7.13 shows that, only at 300–1000 m, the daily production will be less than the economic limit production in the 20th year, therefore, the EUR of CBM can be determined by economic limit production. According to the experience of the United States, the production life of a CBM well is generally around 15 years, so the production life of a CBM well in the southern Qinshui Basin will be longer because of the high CBM content and low permeability, therefore, the economic limit production is determined by 20-year production life. Table 7.16 shows the EUR of CBM in the gas-enriched belt in the southern Qinshui Basin.

Since there are less CBM production wells and the production history is short, the calculated EUR is the CBM GTR per well.

GTR

CBM reservoir numerical simulation provides a basis for establishing the probability distribution of EUR. But at present, CBM development in China is in the experimental stage, there are less CBM production wells, the production history is very short, and no CBM well has finished its production life. So it is hard to establish an EUR probability distribution model. Based on the CBM production in the San Juan Basin in the United States, it is generally assumed that the EUR of a CBM well obeys lognormal distribution. Following the method, we built a EUR probability distribution model, then the logarithms of 11 EUR values from reservoir numerical simulation were calculated, followed by the average (μ) and mean square error (δ^2), and finally a log normal distribution function ln(μ, δ^2) was obtained.

By multiplying ln(μ, δ^2) by the number of grids (number of CBM wells) in a section, the EUR probability distribution in the section can be obtained. By adding up the probability of every section, a probability distribution curve of the evaluated target can be obtained (Fig. 7.17). CBM GTR is expressed by maximum (5% probability), median (50% probability), minimum (95% probability) and average (expected mathematical value) (Table 7.17).

Table 7.14 Statistics of CBM / natural gas prices

Area	Price/(yuan/m^3)	Data source
Fuxin	Wellhead CBM at 1.10	Field survey, 2004–03
Liaohe Oilfield	Wellhead natural gas at 1.16	Field survey, 2004–03
Panzhuang Coalmine	Wellhead CBM at 1.20	Lanyan CBM Company, 2004–09
Shizhuang Coalmine	Wellhead CBM at 0.7	Field survey, 2003–09
Jingbian in Changqing Oilfiled	Station natural gas at 0.56–0.58	Changqing Oilfield, 2004–09
Sichuan (marketing)	Wellhead natural gas at 0.90	1996

Table 7.15 Statistics of CBM operating costs

Variable cost		Fixed cost	
Item	Price /(yuan/km^3)	Item	Price/[10000 yuan/(well·year)]
Materials	12	Salary	1.5
Fuel and power	8.5	Welfare	0.21
Gas treatment	0.6	Workover	1.8
Repair/maintenance	10	Re-fracturing stimulation	2.5
Others related to gas production	8	Management, etc	1.2
Resources compensation	10	–	–
Total	49.1	Total	7.21

Table 7.16 EUR and GRF of CBM in the gas-enriched belt in southern Qinshui Basin

Parameter combination No	300–1000 m		1000–1500 m	
	EUR/$10^4 m^3$	GRF/%	EUR/$10^4 m^3$	GRF/%
1	1572.91	75.76	2099.13	53.61
2	1246.04	60.02	1619.66	41.37
3	912.21	43.94	1172.01	29.93
4	1901.86	83.24	1279.85	33.67
5	1586.77	69.45	974.54	25.64
6	1193.48	52.24	693.59	18.25
7	2327.84	68.59	–	–
8	1834.97	54.07	–	–
9	1385.17	40.81	–	–
10	2317.67	69.07	991.94	35.76
11	348.32	64.11	957.12	36.78
Average	1511.57	61.94	1223.48	34.38

The total CBM GTR are 6064.32×10^8 m^3 with a probability of 50%, the CBM GTR at 300–1000 m are 5436.92×10^8 m^3, and that at 1000–1500 m are 617.34×10^8 m^3 in the gas-enriched belt in the southern Qinshui Basin.

Gas-Enriched Belts not Explored or Developed

Of the 97 gas-enriched belts that have been calculated CBM GTR in China, 93 belts haven't been explored or developed for CBM. For these 93 belts, based on the evaluated results of CBM GIP, loss analysis method was used to calculate CBM GTR.

In addition to accurate and reliable CBM GIP, it is also very important to scientifically and rationally determine the CBM GRF, lost thickness and lost area, which are key to calculating CBM GTR by using loss analysis method. The determining principles and results of the parameters will be described later.

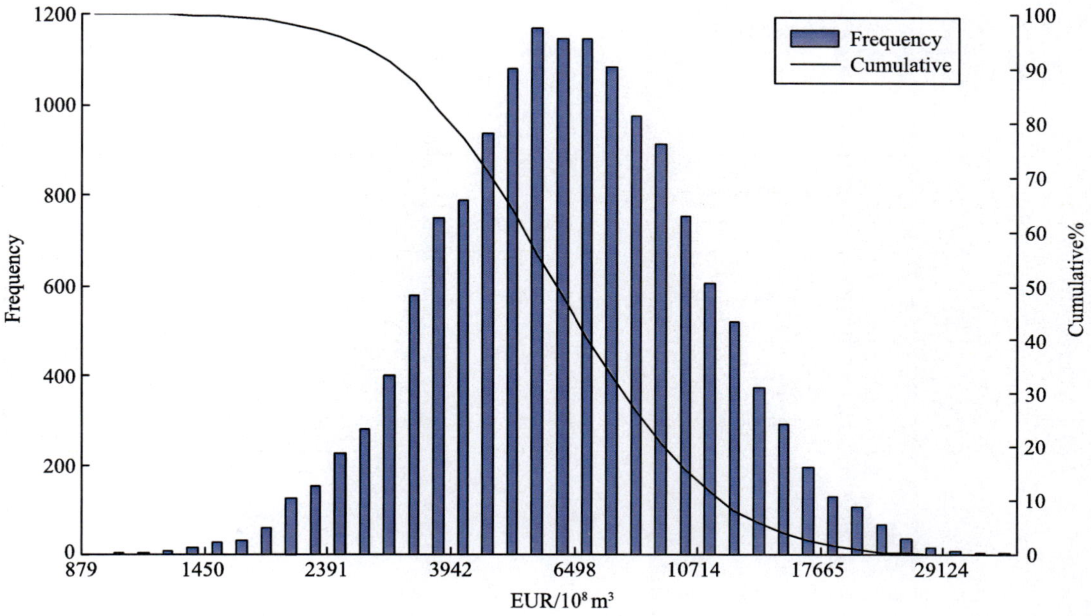

Fig. 7.17 Probability distribution of CBM GTR in the Qinnan gas-enriched belt

Table 7.17 Calculation results of CBM GTR in the gas-enriched belt in southern Qinshui Basin

Probability/%	CBM GTR/10^8 m^3		
	300–1000 m	1000–1500 m	Total
5	15,925.29	1232.87	17,074.12
50	5436.92	617.34	6064.32
95	1856.17	309.12	2153.90
Average	5436.92	617.34	6064.32

(1) Basic data of CBM GIP

The basic data of CBM GIP (G_{IP}) used in the calculation of CBM GTR (G_{TR}) by loss analysis are from two research results: (i) The project of "National CBM Resource Evaluation", the first-class geological exploration project of National Development and Reform Commission, completed in 1999–2000 by Xi'an Research Institute who was entrusted by China United Coalbed Methane Co., Ltd. The project passed the appraisal organized by CUCBM. (ii) The special research on "Prediction of CBM Resources in Major Lignite Areas in China" completed by research team 06 of the CBM Project of the 973 Program from 2005 to 2006. The research passed the appraisal organized by Research Institute of Petroleum Exploration & Development, CNPC.

On the basis of the latest and most authoritative systematic prediction of coal resources in China, the CBM resources for the project "National CBM Resource Evaluation" were collected as detailed as possible over the years from coalfield exploration and CBM exploration and development. The results from the project were calculated with unified methods and standards, which are widely recognized and adopted by authoritative state departments, industry and academia. Similarly, on the basis of the latest and most authoritative systematic prediction of coal resources in China, according to the latest research results on the occurrence mechanism of lignite CBM at home and abroad and the CBM exploration and development practice in the United States, the gas content of lignite CBM was determined by scientific methods in the special research "Prediction of CBM Resources in Major Lignite Areas in China". The CBM resources in major lignite areas in China were systematically calculated for the first time, to be $13{,}954.65 \times 10^8$ m^3.

(2) CBM GRF by analogy method

The adaptability, advantages and disadvantages of various methods for CBM GRF determination were discussed in details in Part 2 in this chapter. In the calculation process,

Table 7.18 CBM GRF by analogy

CBM reservoir type	Burial depth/m	Analogue basin (area)	Value/%
Unmetamorphic coal (lignite)	< 1000	Powder River Basin, USA	72.0
	1000–1500	Speculative	55.0
	1500–2000	Speculative	45.0
Low metamorphic coal	< 1000	Liujia CBM well group in Fuxin	80.0
	1000–1500	Liujia CBM well group in Fuxin	65.0
	1500–2000	Speculative	55.0
Medium metamorphic coal	< 1000	Oak Grove block in Black Warrior Basin, USA	90.0
	1000–1500	Speculative	57.8
	1500–2000	Speculative	47.4
High metamorphic coal	< 1000	South of Qinshui Basin	62.4
	1000–1500	South of Qinshui Basin	34.4
	1500–2000	Speculative	30.0

according to the actual conditions of China, analogy method was used to determine the CBM GRF of each gas enrichment unit. Firstly, in the coal metamorphism series, according to the degree of coal metamorphism, the CBM gas reservoirs were divided into four types: unmetamorphic coal (lignite), low metamorphic coal (including long-flame coal and gas coal), medium metamorphic coal (including fat coal, coking coal and lean coal) and high metamorphic coal (including meagre coal, anthracite No.3 and anthracite No.2). Then, according to the burial depth, each type of CBM reservoirs was divided into three sections: < 1000, 1000–1500 and 1500–2000 m. Thus, a total of 12 CBM GRFs under different geological conditions were used for analogy (Table 7.18).

(3) Lost thickness

On a coal-bearing column, which coal seam can be a target for CBM resource development depends on both geological (coal seam thickness and distribution stability, vertical distribution of coal seam on strata column, distance between coal seams, etc.) and engineering factors (gas production technology, well pattern, stimulation measures, etc.). In order to be uniform and comparable, it is necessary to establish corresponding principles and standardize determination methods.

Combined researches and experts' suggestions, considering the development technology "one pad with multi-wells (multiple CBM wells drilled into different target coal seams or coal seam groups on a pad to effectively reduce lost thickness)" and pinnate horizontal well, we sought advices from some domestic experts with rich practical experience in CBM hydraulic fracturing, and then established the principles of determining lost thickness of coal seams.

The principles for calculating lost thickness: (i) Only the coal seams with relatively stable development can be selected as targets for CBM development. (ii) An individual coal seam with interlayer spacing more than 15 m isn't deemed as a target when its thickness is less than 1 m, so it is regarded as lost thickness. (iii) The interval with thickness less than 15 m is a completion interval (when there are several coal seams are targets, their development thickness should be calculated together). The spacing between two completion intervals should be more than 20 m.

According to the above principles, taking the coalmines from the *Atlas of China's Primary Coal Resources* (1996) as units, based on the distribution of coal seams in each coalmine on the strata section, the lost thickness of coal seam was determined for 161 coal-bearing strata sections. The workload of this project is huge, so it was shared based on regions by China National Administration of Coal Geology, Shandong University of Science and Technology and Xi'an Research Institute, and finally summarized by Xi'an Research Institute. The average lost thickness in every coalmine was taken as the lost thickness of the gas-enriched belt. Next are examples.

Table 7.19 shows the basic characteristics of coal seams in the Jixi Coalmine of Heilongjiang province. The coal-bearing strata in the Jixi Coalmine are the Muleng Formation and Chengzihe Formation in Early Cretaceous, and 220–2060 m thicker. There are 17–57 coal seams, including 5–24 coal seams that can be mined or partially mined, with a cumulative mining thickness of 14.45 m. The

Table 7.19 Lost thickness of coal seam in the Jixi Coalmine

Coal seam No		Thickness/m $\left(\frac{\text{Min.–Max.}}{\text{Aver.}}\right)$	Interlayer spacing/m	Stability	Targets/m	Lost thickness/m	Lost thickness/%
Muleng Fm	1	0.50–0.80	–	More stable	–	0.70	–
	2	0.70–1.30 / 1.00	16	Stable	1.0	–	
	3_{up}	0.20–1.00	25	More stable	–	0.80	
	3_{low}	0.70–1.50 / 1.10	6	Stable	1.1	–	
	4	0.60–0.85 / 0.70	12	More stable	0.70	–	
	5	0.70	14	More stable	0.7	–	
	6	0.80	13	More stable	0.8	–	
Chengzihe Fm	13	0.75	–	More stable	–	0.75	
	14	0.80	12	More stable	0.8	–	
	15	1.00	13	More stable	1.0	–	
	22	0.90	173	More stable	0.9	–	
	23	1.40	1.4	More stable	1.4	–	
	25	0.80	8	More stable	0.8	–	
	29	0.70	20	More stable	–	0.70	
	32	1.20	62	More stable	1.2	–	
	33	1.10	33	More stable	1.1	–	
Total thickness/m		14.45	–	–	11.5	2.95	20.42

thickness per coal seam is less than 1 m. According to the development stability, coal seam thickness and interlayer spacing, there are 12 coal seams which can be selected as targets for CBM development, including coal seam No. 2, No. 3_{upper}, No. 4, No.5, No. 6, No. 14, No. 15, No. 22, No. 23, No. 25, No. 32 and No. 33. The cumulative thickness is 11.5 m. For coal seams No. 1, No. 3_{upper}, No.13 and No.29, the average thickness is less than 1 m and the interlayer spacing is relatively large. They cannot be selected targets for CBM development, and regarded as lost thickness (totally 2.95 m). The lost thickness of coal seam in the Jixi Coalmine is 20.4%.

Table 7.20 shows the basic characteristics of coal seams in the Huainan Coalmine and Panxie Coalmine of Anhui Province. The coal-bearing strata in the Huainan Coalmine and Panxie Coalmine are the upper Shihezi Formation, lower Shihezi Formation and Shanxi Formation in Permian, with the thickness of 925 m. There are a total of 40–56 coal seams, of which, the mining and partially mining coal seams are mainly distributed in the Shanxi Formation, lower Shihezi Formation and lower part of upper Shihezi Formation, i.e., the coal seams below No.13–1. The strata thickness of this section is about 450 m, including 12 mining coal seams with mining thickness of 27.53 m. Within this 450 m, according to the coal seam stability, single coal seam thickness and interlayer spacing, there are 10 coal seams which can be selected as targets for CBM development, including coal seams No.13–1, No.11–2, No.8, No.7–1, No.6–1, No.5–1, No.4–1, No.4–2, No.3 and No.1. Their cumulative thickness is 24.02 m. And coal seams No.7–2 and No.6–2 are unstable coal seams with cumulative thickness of 3.51 m. They cannot be selected as targets for CBM development, and are regarded as lost thickness. The lost thickness of coal seam in the Huainan Coalmine and Panxie Coalmine is 12.8%.

The average lost thickness of coal seam in all coalmines in a gas-enriched belt is taken as the lost thickness rate of coal seam in the gas-enriched belt.

(4) Lost area

Lost area is mainly considered in the following aspects:

① CBM resource scale. A area which is of small CBM resource scale, and have no development value or very low development value of CBM resources is regarded as a lost area for CBM GTR. The gas-enriched belt whose CBM resource abundance can not meet the development requirement is also regarded as a lost area. During the actual evaluation process, there are some gas-enriched belts with CBM GIP less than 30×10^8 m^3, which have no

Table 7.20 Lost thickness of coal seam in the Huainan Coalmine and Panxie Coalmine

Coal seam No	Thickness/m (Min.–Max. / Aver.)	Interlayer spacing/m	Stability	Targets/m	Lost thickness/m	Lost thickness/%
13–1	0.83–14.12 / 4.87	–	Stable	4.87	–	
11–2	0–7.58 / 2.62	65	Stable	2.62	–	
8	0–7.15 / 2.37	78	Stable	2.37	–	
7–2	0–8.30 / 1.50	9	Unstable	–	1.50	
7–1	0–6.80 / 1.73	7	More stable	1.73	–	
6–2	0–8.00 / 2.01	11	Unstable	–	2.01	–
6–1	0–8.19 / 1.35	3.5	More stable	1.35	–	
5–1	0–8.06 / 1.33	20	More stable	1.33	–	
4–2	0–4.80 / 1.58	8	More stable	1.58	–	
4–1	0–9.35 / 1.19	4	More stable	1.19	–	
3	0–8.35 / 3.44	69	More stable	3.44	–	
1	0–11.19 / 3.54	6	More stable	3. 54	–	
Total thickness/m	27.53	–	–	24.02	3.51	12.75

development value after analysis. GTR of these gas-enriched belts are not included.

② Terrain condition. Terrain complexity is considered. In a mountainous area, the lost area is large; in plain and desert areas, the lost area is small. For hilly areas, it is determined according to terrain elevation difference and incision. In the actual determination process, terrain data are obtained through field investigation and survey or by refering to topographic maps and relevant topographic data.

③ Structural complexity. Lost area is large in a gas-enriched belt with complex structures. In practice, it is determined according to the structure development of a gas-enriched belt.

7.3.2.3 Predicted Results

Results

For gas-enriched belts explored and developed for CBM, CBM GTR have been calculated by using CBM reservoir numerical simulation method. For gas-enriched belts not explored or developed for CBM, according to the above determining principles and methods of GRF, lost thickness and lost area of coal seam, lost thickness and lost area of coal seam were determined with a gas-enriched belt as the unit. The GRFs of all sections— < 1000, 1000–1500 and 1500–2000 m—were determined by analogy method. CBM GTR were calculated by using loss analysis method. Thus, the CBM GTR were calculate all over the country.

Based on the calculation and summary of 58 CBM basins (regions) in China, the CBM GIP in China is 32.86×10^{12} m^3 (GIP in lignite areas is 13954.65×10^8 m^3, in other areas is 314612.48×10^8 m^3). CBM GTR were calculated for 97 gas-enriched belts in which CBM GIP are more than 30×10^8 m^3 (The 97 gas-enriched belts are located in 56 CBM basins. As CBM GIP in the Yanji basin and the Shangrao basin are less than 30×10^8 m^3, the GTR were not calculated), CBM GTR in China are 13.90×10^{12} m^3. Therefore, from a macro perspective, CBM GTR in China are about 42% of CBM GIP.

Table 7.21 Statistics of CBM GTR in China

Gas-bearing region	CBM GTR/10^8 m^3	Percent/%
Heilongjiang, Jilin and Liaoning	1775.03	1.28
Hebei, Shandong, Henan and Anhui	11,209.27	8.07
South China	475.22	0.34
Eastern Inner Mongolia	6991.38	5.03
Shanxi, Shaanxi and Inner Mongolia	66,541.85	47.88
Yunnan, Guizhou, Sichuan and Chongqing	9189.96	6.61
Northern Xinjiang	37,501.34	26.98
Southern Xinjiang – Gansu, Qinghai	5292.69	3.81
Total	138,976.75	100.00

Result Evaluation

It is the first time to quantitatively evaluate the technical feasibility of CBM resources in China. 13.90 × 10^{12} m^3 is the first prediction of CBM GTR, which can meet the urgent needs of the industry. Based on comprehensive analysis, this work has the following characteristics:

① In the classification sequence of CBM resources in line with international rules and conventional oil and gas industry, CBM GTR has its definite position, clear definition and precise meaning. According to this definition, the calculation methods are scientific and standard, the parameter systems are matched, and the predicted results are universal and comparable.
② The lignite areas were included in the evaluation scope, and a complete evaluation result of the national CBM resources was obtained, which filled the blank of CBM resource evaluation in the lignite areas in China.
③ According to the exploration and development status and data acquisition of CBM, GTR were calculated by using CBM reservoir numerical simulation method and loss analysis method respectively. The latest achievements of CBM exploration and development in China are reflected by making full use of the various existing data.
④ Due to solid work, detailed data and scientific methods, the basic data (coal resources, CBM gas content, CBM GIP resources, CBM well production performance and engineering data, coal-bearing strata column, etc.) for calculation has been generally accepted with high authority.
⑤ Based on the above work, rich experiences in CBM evaluation and responsible attitude of the researchers, the prediction results are accurate and highly reliable.
⑥ Based on division sequence of CBM enrichment units, CBM resources were calculated by basin (region), belt and depth. The division and naming of CBM basins is consistent with that of conventional oil and gas. In this way, statistical data and information are easy to be shared, used and exchanged in various industries.

7.3.3 Distribution of CBM GTR in China

There are abundant CBM resources in China with obvious characteristics in time and space distribution. These characteristics have great influences on the exploration and development of CBM resources. The following is a brief analysis of the characteristics in terms of region, burial depth and types of CBM reservoirs.

7.3.3.1 CBM-Bearing Regions

The regional distribution of CBM GTR in China is extremely uneven. This is mainly reflected in the fact that CBM GTR in Shanxi, Shaanxi and Inner Mongolia are the largest, 66,541.85 × 10^8 m^3, accounting for 47.88% of the total CBM GTR in China; then in northern Xinjiang, 37,501.34 × 10^8 m^3, accounting for 26.98%; and in South China, 475.22 × 10^8 m^3. CBM GTR are mainly distributed in Shanxi, Shaanxi and Inner Mongolia in north China and northern Xinjiang in northwest China. The total volume of these two Gas-bearing regions is 104043.19 × 10^8 m^3, accounting for 75%; the total volume of other six Gas-bearing regions is only 34,933.56 × 10^8 m^3, accounting for 25%. CBM GTR in these regions are shown in Table 7.21.

7.3.3.2 CBM-Bearing Basins

In China, there are 4 basins with CBM GTR greater than 1 × 10^{12} m^3, including the Ordos Basin, the Qinshui Basin, the Tu-Ha Basin and the Junggar Basin, accounting for 62% of

Table 7.22 Statistics of CBM GTR in CBM basins (regions) in China

Basins (regions)	CBM GTR/10^8 m^3	Percent/%	Basins (regions)	CBM GTR/10^8 m^3	Percent/%
Songliao Basin	12.60	0.01	Ordos Basin	42,346.78	30.47
Hegang Basin	258.34	0.19	Zhangjiakou Basin	27.91	0.02
Sanjiang-Mulenghe Basin	945.54	0.68	Xiahuayuan Basin	16.37	0.01
Yilan-Yitong Basin	36.19	0.03	Weixian Basin	68.25	0.05
Dunhua-Meihe Basin	24.52	0.02	Datong Basin	125.10	0.09
Tieling-Changtu Basin	63.88	0.05	Ningwu Basin	1685.99	1.21
Fuxin-Yixian Basin	28.71	0.02	Huoxi Basin	3877.51	2.79
Lower Liaohe Basin	152.33	0.11	Qinshui Basin	15,939.60	11.47
Fushun Basin	28.23	0.02	Sanmenxia-Luoyang Basin	251.53	0.18
Hongyang region	224.68	0.16	Sichuan Basin	588.07	0.42
East of northern Hebei	31.28	0.02	South Sichuan and North Guizhou region	2323.50	1.67
West Beijing Basin	68.67	0.05	South Guizhou Basin	117.23	0.08
Tangshan region	459.79	0.33	Zhaotong Basin	346.84	0.25
South Bohai Bay Basin	2480.15	1.78	Nanpanjiang Basin	5724.65	4.12
East of Taihang Mountain region	2080.81	1.50	Chuxiong Basin	89.66	0.06
Huangxian Basin	46.06	0.03	Junggar Basin	13,263.96	9.54
South of North China Basin	6042.51	4.35	Santanghu Basin	1685.58	1.21
Southeast Hubei-north Jiangxi Basin	15.89	0.01	Tu-Ha Basin	14,275.56	10.27
Zhenjiang-Anqing Basin	22.01	0.02	Yili Basin	6346.60	4.57
Changzhou Basin	38.85	0.03	Youerdusi Basin	224.44	0.16
Central Hunan Basin	105.50	0.08	Yanqi Basin	1705.20	1.23
Chenlei Basin	96.30	0.07	Tarim Basin	3488.44	2.51
Pingle Basin	131.27	0.09	Qaidam Basin	253.46	0.18
Central Guangxi Basin	48.37	0.03	Chaoshui Basin	108.98	0.08
Baise Basin	17.04	0.01	Central Qilian Basin group	849.93	0.61
Hailaer Basin group	2745.77	1.98	Minle Basin group	134.77	0.10
Erenhot Basin group	4245.61	3.05	Margin area of Gasu-Ningxia-Inner Mongolia	408.35	0.29
Yinchuan Basin	2202.81	1.59	Xining-Lanzhou Basin	48.76	0.04

total CBM GTR in China, while the other basins (or regions) accounting for only 38%. CBM GTR in the Ordos Basin are the highest, 42,346.78 × 10^8 m^3, accounting for 30.47% of the total CBM GTR in China; then the Qinshui Basin, 15,939.60 × 10^8 m^3, accounting for 11.47%; the third the Tu-Ha Basin, 14,275.56 × 10^8 m^3, 10.27%; finally 13,263.96 × 10^8 m^3 in the Junggar Basin, and 12.6 × 10^8 m^3 in the Songliao Basin, the lowest (Table 7.22 and Fig. 7.18).

7.3.3.3 GTR Statistics by Types of CBM Reservoir

Figure 7.19 shows the GTRs of different types of CBM reservoirs. The CBM GTR of low metamorphic reservoirs is the largest, 81,699.14 × 10^8 m^3, and accounting for 58.79%; then the middle metamorphic, 30,682.13 × 10^8 m^3, and 22.08%, and the last lignite, 6381.96 × 10^8 m^3 and 4.59%.

7.3.3.4 GTR Statistics by Reservoir Depth

Figure 7.20 shows CBM GTR at different depths. CBM GTR above 1000 m is the largest, 53,206.88 × 10^8 m^3, and accounting for 38.28%; then between 1500 and 2000 m, 45,083.86 × 10^8 m^3, and 32.44%; and the smallest from 1000 to 1500 m, 40,686.01 × 10^8 m^3, and 29.28%.

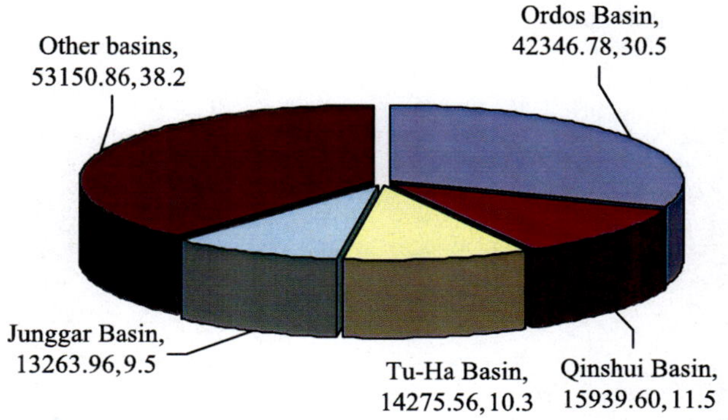

Fig. 7.18 Distribution of CBM GTR in Coal basins (regions) in China (unit: 10^8 m^3, %)

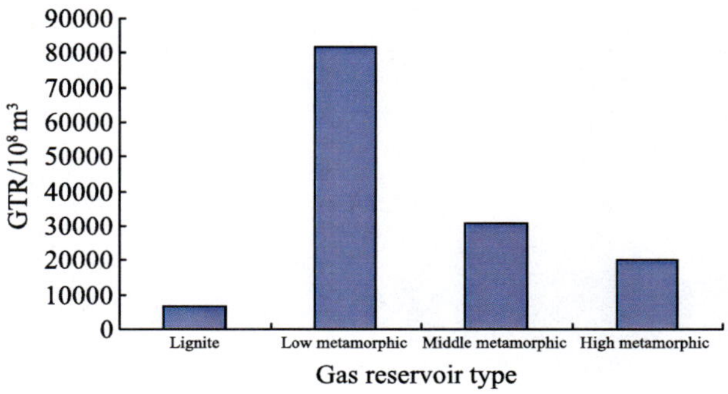

Fig. 7.19 Histogram of CBM GTR of different reservoir types

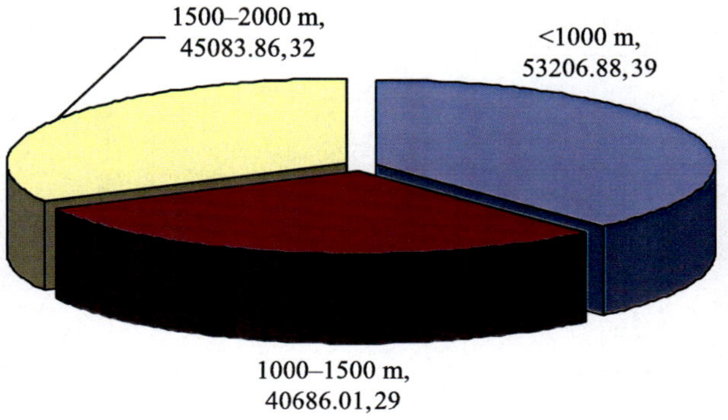

Fig. 7.20 Distribution of CBM GTR at different depths (unit: 10^8 m^3, %)

References

ARI (2002) Powder river basin coalbed methane development and produced water management study. U S department of energy

Bustin RM, Clarkson CR (1999) Free gas storage in matrix porosity: a potentially significant coalbed resource in low rank coals. In: Proceedings of the 1999 international coalbed methane symposium. Tuscaloosa, pp 197–214

Cook T (2003) Calculation of estimated ultimate recovery for wells in continuous-type oil and gas accumulation of the uinta-piceance province. Int J Coal Geol 56:39–44

Cui YJ, Zhang QL, Yang XL (2003) Adsorption properties and varying law of equal adsorption heat of different coals. Nat Gas Ind 23 (4):130–131 (in Chinese)

Fu XH, Qin Y, Li GZ (2001) Influencing factors on coal reservoir permeability in the middle and south of Qinshui Basin. Shanxi J Geomech 7(1):45–52 (in Chinese)

Fu XH, Qin Y, Li GZ (2002) Adsorption experiment at equilibrium water on extra high-ranked coal. Pet Exp Geol 24(2):177–180 (in Chinese)

He WG, Tang SH, Xie XD (2000) Influence of geostress on coal seam permeability. J Liaoning Tech Univ (Nat Sci Ed) 19(4):353–355 (in Chinese)

Henry ME, Finn TM (2003) Evaluation of undiscovered natural gas in the upper cretaceous ferron coal/wasatch plateau total petroleum system, wasatch plateau and castle valley, utah. Int J Coal Geol 56:3–37

King GR (1993) Material-balance techniques for coal-seams and devonian shale gas reservoirs with limited, water influx. SPE Reserv Eng 2:67–72

Li GY, Lu MG et al (2002) Atlas of China's petroleum-bearing basins. Petroleum Industry Press, Beijing (in Chinese)

Li XY, Li J, Yang LJ et al (1998) Prediction of coal reservoir permeability in tiefa coalfield. Coalf Geol Explor 26(1):34–36 (in Chinese)

Mavor M, Nelson CR (1997) Coal reservoir gas-in-place analysis. US Gas Research Institute, Chicago

McKee CR, Bumb AC, Koening RA (1988) Stress-dependent permeability and porosity of coal. Rocky Mountain Association of Geologist, pp 143–153

Nelson CR (2004) Effect of vertical degasification wells on coal seam gas content reduction at the oak grove field, black warrior basin, Alabama. Paper 0428. In: Proceedings of the 2004 international coalbed methane symposium. The University of Alabama, Tuscaloosa, pp 4–6

Pratt TJ, Mavor MJ, DeBruyn RP (1999) Coal gas resource and production potential of subbituminous coal in the powder river basin. In: Proceedings of the 1999 international coalbed methane symposium. Tuscaloosa, pp 23–34

Sun MY, Huang SC et al (1998) CBM development and utilization manual. Coal Industry Press, Beijing (in Chinese)

Wang SW (2004) Evaluation of coal reservoirs in coalbed methane exploration and development. Nat Gas Ind 24(5):82–84 (in Chinese)

Yang LW, Feng SL, Hu AM et al (2003) CBM resources/reserve specifications. Geological Publishing House, Beijing (in Chinese)

Yang Q (1987) Progress in coal geology. Science Press, Beijing (in Chinese)

Ye JP, Qin Y, Lin DY et al (1998) China's coalbed methane resources. China University of Mining and Technology Press, Xuzhou, pp 38–39 (in Chinese)

Ye JP, Shi BS, Zhang CC (1999) Coal reservoir permeability and influencing factors. J China Coal Soc 24(2):118–122 (in Chinese)

Zha QH (1999) Foundation of oil and gas resources management. Petroleum Industry Press, Beijing (in Chinese)

Zhang DM, Lin DY (1998) Tectonic characteristics and potential for coalbed methane development of coal basins in China. China Coalf Geol 10(Supplement):3–40 (in Chinese)

Zhang H, Li XY, Hao Q et al (2003) Scanning electron microscopy study for Chinese coal. Geological Publishing House, Beijing (in Chinese)

Zhang SL, Li BF (1996) Forming mechanism of cleats in coal seams and the significance in evaluating coalbed methane exploration and development. China Coalf Geol 8(1):72–77 (in Chinese)

Zhang XM (1996) Discussion on several issues in the calculation of coalbed methane resources. In: Proceedings of the second national coalbed methane symposium. China's Coalbed Methane, vol 2, pp 98–101 (in Chinese)

Zhang XM, Zhang SA, Zhong LW et al (1991) China's coalbed methane. Shaanxi Science and Technology Press, Xi'an, pp 1–110 (in Chinese)

Zhang XM, Zheng YZ et al (2002a) Distribution of coalbed methane resources in China. In: Academic collection of the 80th anniversary of the Chinese geological society. Geological Publishing House, Beijing, pp 458–463 (in Chinese)

Zhang XM, Zhuang J, Zhang SA et al (2002b) China coalbed methane geology and resources evaluation. Science Press, Beijing (in Chinese)

8 Seismic Prediction Technology of Favorable CBM Zones

In order to achieve the specific goal of using seismic data to predict favorable CBM (coalbed methane) zones, we used precise inversion to control key geological parameters of CBM occurrence as a breakthrough point, systematically researched the theory and technologies for seismic data acquisition, processing, interpretation and inversion of CBM reservoirs, and used these technologies to obtain high-precision inversion results of the distribution, thickness, internal structure, roof and floor, density, elastic modulus, shear modulus and fracture development of coal seams, and finally realized the prediction of favorable CBM zones by seismic data and comprehensive analysis of these inversion results.

8.1 Primary Geological Attributes of CBM Occurrence and Prediction

Geological conditions are important factors affecting CBM enrichment. The sedimentary environment and evolution of coal-bearing series, the geotectonic location and structural characteristics of coal basins, regional tectonic evolution and the degree of coalification play a leading role in the formation and preservation of CBM. Geological structure and combination, surrounding rocks and boundary condition of coal seams, coal thickness and change, coal structure and micro-composition, coal damage type and performance, and coal seam burial depth are primary geological factors affecting CBM enrichment.

Taking the Huainan Coalmine as a target, we studied and applied the new seismic-logging inversion technology to invert primary parameters controlling CBM enrichment (such as coal seam structure, burial depth, thickness, and roof and floor lithologies), and predicted fracture development zones and favorable CBM zones using waveform classification technology.

8.1.1 Seismic Survey to Coal Seam Depth

The burial depth of a coal seam is closely related to the enrichment degree of CBM. The increase of burial depth leads to the increase of formation pressure, and at the same time, the increase of burial depth not only makes the permeability of the coal seam and its surrounding rock worse with the increase of in-situ stress, but also enlarges the distance of CBM migration to the surface, all of which are beneficial to the preservation of CBM. Therefore, the determination of the burial depth of a coal seam is of great significance for CBM exploration and development. Using 3D seismic migration, some projects first conducted time-depth conversion and calibrated the reflection wave group of the target coal seam based on well logging data, then conducted fine structural interpretation, and finally obtained burial depth of the coal seam. The distribution of the bedrock thickness of the No. 13-1 coal seam (the vertical distance from the roof to loose Neogene, Paleogene or Quaternary sedimentary bottom) in the Panji No. 3 Coalmine (Fig. 8.1) describes the burial conditions and structural forms of the coal seams during a long geological history, and better characterizes the preservation conditions of CBM. The bedrock thickness of the No. 13-1 coal seam is estimated between 15 and 550 m, and thicker in the south and thinner in the north.

8.1.2 Coal Seam Thickness from Seismic Inversion

Coal seam thickness can be inverted by seismic multi-attribute quantitative analysis. Seismic attributes refer to the geometric, kinematic, dynamic and statistical characteristics of seismic waves, derived from pre-stack or post-stack seismic data through mathematical transformation. They are descriptive and quantifiable features contained

Y. Song et al., *Coalbed Methane in China*,
https://doi.org/10.1007/978-981-33-4725-0_8

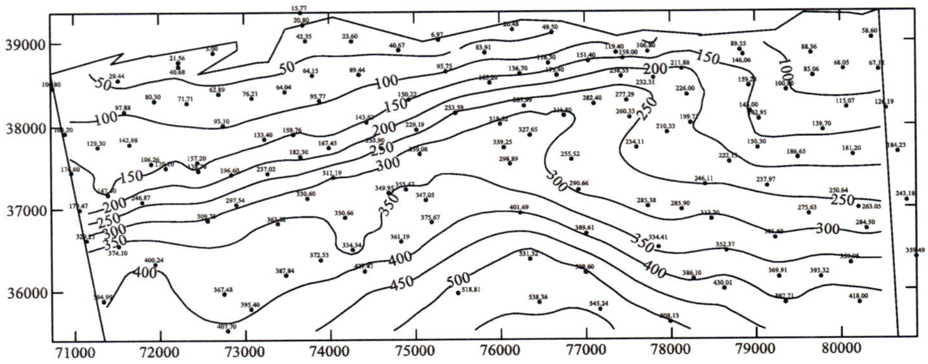

Fig. 8.1 Bedrock thickness distribution of the No. 13-1 coal seam in the Panji No. 3 Coalmine, Huainan Coalfield (unit: m)

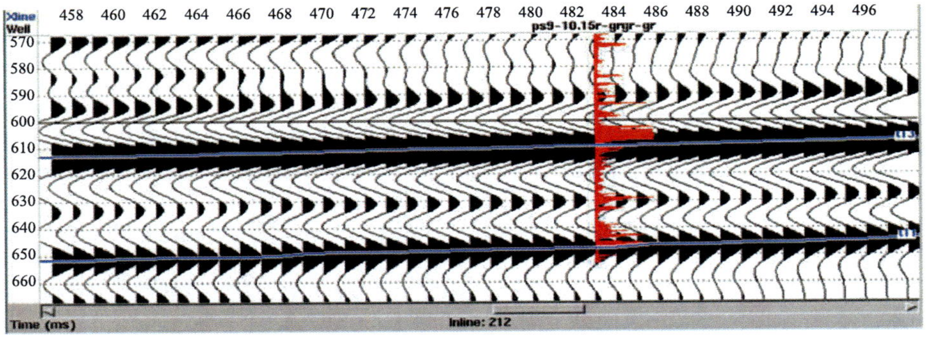

Fig. 8.2 Original seismic section across borehole IX-X-15 in Panji No. 3 Coalmine

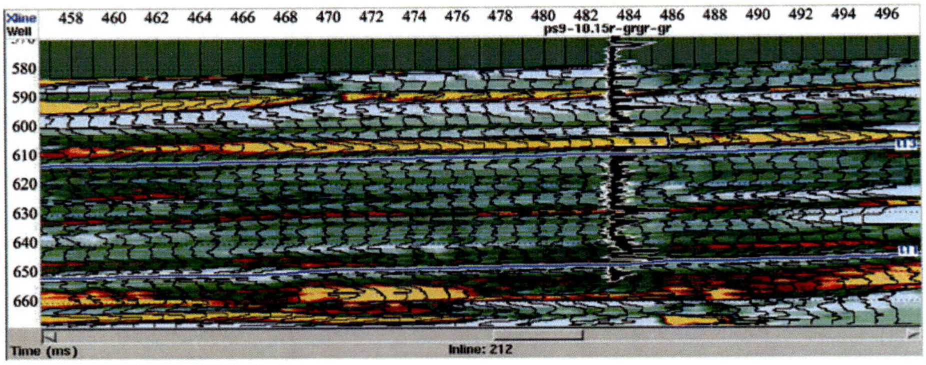

Fig. 8.3 Predicted apparent resistivity section across borehole IX-X-15 in Panji No. 3 Coalmine

in seismic data and can be displayed in the same scale as the original data. Seismic attribute analysis has been successfully applied in oil and gas exploration, but there are only a few successful examples in CBM exploration. This project broke through the traditional research methods for coalfields and applied seismic multi-attribute quantitative analysis to predict CBM enrichment. By combining well logging and seismic data, we obtained a 3D pseudo-logging response data volume with high vertical resolution and accurate vertical and lateral distribution, and then obtained the distributions of coal seam thickness, roof lithology, CBM content and structural coals, etc. They are primary geological factors affecting the enrichment and distribution of CBM, and should be solved for safe production in coalfields.

Figure 8.2 is the original seismic section across borehole IX-X-15, and Fig. 8.3 is the corresponding predicted apparent resistivity section. The two horizons are the wave peaks of the seismic events corresponding to No. 13-1 coal seam and No. 11-2 coal seam, respectively. The logging curve shown on the section is the apparent resistivity curve.

By comprehensively interpreting the 3D pseudo- logging data volume, we obtained the plane thickness prediction of the No. 13-1 coal seam (Fig. 8.4). It shows the distribution of this coal seam is stable in the whole area, generally thin in

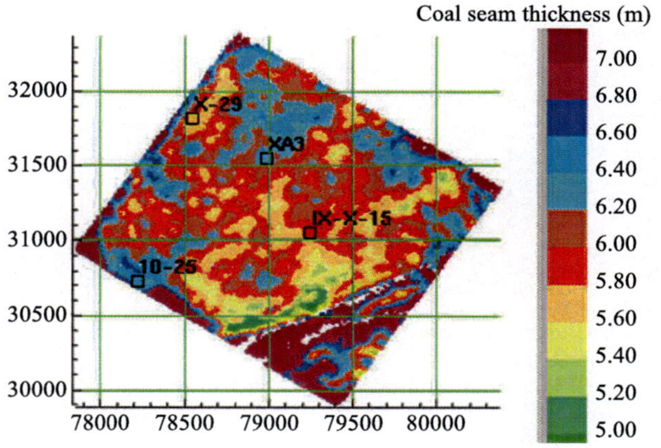

Fig. 8.4 Predicted thickness distribution of No. 13-1 coal seam in Panji No.3 Coalmine

the south and thick in the north. There is a narrow thin coal belt near the southern fault, which is getting thicker in the direction of borehole 10–25 to the west. The distribution of the coal seam in the north can be roughly divided into thin—thick—thin—thick irregular strips from the northwest to the southeast. Table 8.1 lists the errors between predicted coal seam thickness and the measured coal thickness in boreholes XA3, X-29 and IX-X-15. The absolute errors are less than ±0.07 m and the relative errors are less than 1.10%.

8.1.3 Roof Lithology from Seismic Inversion

Firstly, linear relationships between well logging curves and single seismic attributes were established to get a correlation list. Secondly, linear relationships between well logging curves and multiple attributes were established according to the attributes selected. Finally, the characteristics of well logging curves were synthetically predicted by the multiple attributes. The basic principles are as follows:

8.1.3.1 Linear Regression Method for Single Attribute

Assuming that there is a linear relationship between a well logging curve and a seismic attribute, the following linear equation is used for regression.

$$y = a + bx \tag{8.1}$$

where a and b can be derived from the prediction error in the least square sense.

$$E^2 = \frac{1}{N}\sum_{i=1}^{N}(y_i - a - bx_i)^2 \tag{8.2}$$

From this, the single attribute linear regression equations of various seismic attributes and well logging curves can be obtained. Then the seismic attribute values can be converted into logging curve values, and the converted curves can be correlated with the actual curves to obtain the correlation coefficient.

8.1.3.2 Linear Regression Method for Multiple Attributes

On each time sample point of seismic attributes, well logging curves are modeled as a multivariate linear equation:

$$L(t) = \omega_0 + \omega_1 A_1(t) + \omega_2 A_2(t) + \omega_3 A_3(t) \tag{8.3}$$

The weight of this equation can be derived from the prediction error in the least square sense.

$$E^2 = \frac{1}{N}\sum_{i=1}^{N}(L_i - \omega_0 - \omega_1 A_{1i} - \omega_2 A_{2i} - \omega_3 A_{3i})^2 \tag{8.4}$$

where A_i is the ith seismic attribute; ω_i is the weight of the ith seismic attribute; i is the number of attributes.

After applying the obtained linear regression equation of multiple attributes to convert seismic attributes to logging curves, the curves converted can be correlated with the actual curves to obtain the correlation coefficient.

Table 8.1 Errors of predicted thickness of No.13–1 coal seam

Borehole	Actual thickness/m	Predicted thickness/m	Absolute error/m	Relative error/%
10–25	6.54	6.6	0.06	0.92
XA3	6.37	6.3	−0.07	1.10
IX-X-15	5.95	5.9	−0.05	0.84
X-29	5.69	5.7	0.01	0.18

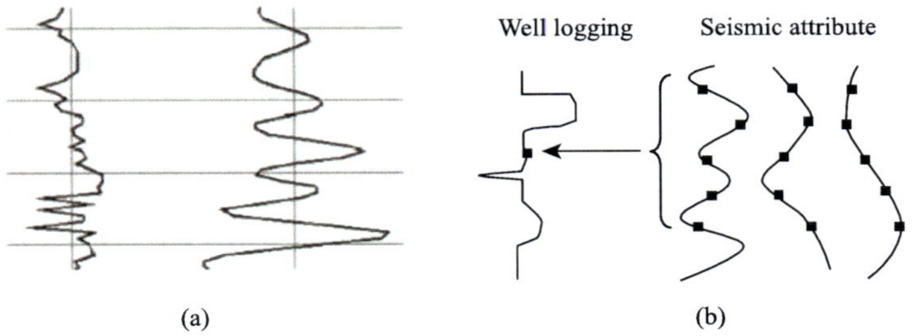

Fig. 8.5 **a** Difference in frequency between actual curve (left) and seismic attribute (right); **b** schematic diagram of correlation between seismic attribute and actual curve using a 5-point convolution operator

8.1.3.3 Regression Method of Multiple Attributes with a Convolution Operator

Figure 8.5a shows that the frequency of the well logging curve is obviously higher than that of seismic attribute. Therefore, we should not correlate the logging curve with the seismic attribute point by point, but assume that one sample of the well logging curve is correlated with a group of adjacent sample points of the seismic attribute. As shown in Fig. 8.5b, five adjacent sample points of the seismic attribute were correlated with the well logging curve.

The purpose of using a convolution operator is not to eliminate the influence of every sample on the logging curve, which is beyond the range of adjacent seismic sample points. The attribute regression formula of using a convolution operator is as follows:

$$L(t) = \omega_0 + \omega_1^* A_1(t) + \omega_2^* A_2(t) + \omega_3^* A_3(t) \quad (8.5)$$

where ω_i^* is the operator with a specified length, and its coefficient can be derived from prediction errors.

$$E^2 = \frac{1}{N}\sum_{i=1}^{N}(L_i - \omega_0 - \omega_1^* A_{1i} - \omega_2^* A_{2i} - \omega_3^* A_{3i})^2 \quad (8.6)$$

The convolution operator is the time shift of seismic attributes, used to obtain the correlation coefficient between the converted curve and the actual curve.

8.1.3.4 Nonlinear Statistical Methods

Neural network is used to simulate the structure and function of a biological neural network to obtain intelligent information processing functions. A neural network processes complex pattern information by means of complex connections of a large number of neurons, and bottom-to-top parallel distribution built by self-learning, self-organization and non-linear dynamics. In this study, the neural network was used to carry out the non-linear statistics of seismic attributes and well logging curves, and the results are better correlated with the actual curves.

(1) Multilayer feed forward neural network

The structural model of a multilayer feed forward neural network includes an input layer, an output layer and several hidden layers. Each layer contains a number of neurons. There is no connection between neurons in the same layer. Weighted connections are used to output to the next layer, and the weights are determined by the output layer. The neurons have computational functions. Each computational unit can have any number of inputs, but only one output. The output can be sent to multiple computational units in the next layer as input.

The training is completed by providing training samples to the neural network, and each sample contains data sampled at the corresponding time $\{A_1, A_2, A_3, L\}$, where, A_i is an attribute; L is the actual curve value. Weight estimation is a non-linear optimization problem. The solution is to use conjugate gradient and simulated annealing to converge quickly to avoid local minimum, so as to realize the least square error between well logging curve and predicted curve values.

(2) Probabilistic neural network

Probabilistic neural network (PNN) is a mathematical interpolation method. Given enough training samples, PNN can provide consistent estimates of probability density attributes and predictive variables that converge gradually to the internal joint. If there are n training samples and 3 attributes, L_i is the measured value of each sample point, then every sample point value in the analysis window for all wells is:

$$\begin{matrix} \{A_{11}, & A_{21}, & A_{31}, & L_1\} \\ \{A_{12}, & A_{22}, & A_{32}, & L_2\} \\ \{A_{13}, & A_{23}, & A_{33}, & L_3\} \\ \vdots & \vdots & \vdots & \vdots \\ \{A_{1n}, & A_{2n}, & A_{3n}, & L_n\} \end{matrix}$$

Assuming that each new output curve value is written as a linear combination of curve values in training data, a new data sample containing attribute information is presented as:

$$x = \{A_{1j}, A_{2j}, A_{3j}\}$$

The estimated value of the new curve is:

$$\hat{L}(x) = \frac{\sum_{i=1}^{n} L_i \exp(-D(x, x_i))}{\sum_{i=1}^{n} \exp(-D(x, x_i))} \quad (8.7)$$

$$D(x, x_i) = \sum_{j=1}^{3} \left(\frac{x_j - x_{ij}}{\sigma_j}\right)^2 \quad (8.8)$$

where $D(x, x_i)$ is the distance between the input point and training point x_i, whose value is determined in the multi-dimensional space of attributes.

Define the prediction error of the whole training data as follows:

$$E_V(\sigma_1, \sigma_2, \sigma_3) = \sum_{i=1}^{N} (L_i - \hat{L}_i)^2 \quad (8.9)$$

Taking Xi 1 producing zone in the Xieqiao mine area as an example to illustrate the process of quantitative seismic multi-attribute analysis. We selected pseudo-acoustic logging curves of seven boreholes (1703, 8-9-3, D3, L3, buV4, jian1, jia2). The selected seismic attributes include: phase-weighted amplitude, average frequency, apparent polarity, cosine instantaneous phase, first derivative of seismic trace, first derivative of instantaneous amplitude, principal frequency, instantaneous frequency, instantaneous phase, integral of seismic trace, integral of absolute amplitude, second derivative of seismic trace, second derivative of instantaneous amplitude and time value, etc. Wave impedance was used as an external attribute to participate in statistical analysis. Mathematical operations of external attributes include reciprocal, logarithmic, exponential, root mean square, etc. We totally finished analysis to 115 types of single attribute correlation. Table 8.2 shows the results of correlation analysis for some single attributes. Among them, the highest correlation coefficient is wave resistance (0.739), followed by seismic trace integral (0.645). Figures 8.6 and 8.7 were used to compare well logging curves predicted by these two attributes with actual acoustic logging curves. The black lines in the Figures represent the actual curve, and the red lines represent the predicted curves. The prediction errors are 250.52 and 284.19, respectively. Figures 8.8 and 8.9 are the results of multi-attribute regression and three-point convolution operator regression respectively, with correlation coefficients of 0.829 and 0.861. It can be concluded that from single attribute linear regression, multi-attribute linear regression to multi-attribute regression using a convolution

Table 8.2 Correlation coefficient and prediction error of well logging curves predicted with single seismic attributes

Attribute	Predicted error	Correlation coefficient
Wave impedance	250.528	0.739
Integral of seismic trace	284.190	0.645
Frequency weighted amplitude	285.139	0.642
First derivative of seismic trace	295.743	−0.606
Amplitude envelope	301.090	−0.587
Phase weighted amplitude	314.500	0.534
Integral of absolute amplitude	317.760	−0.520
Second derivative of seismic trace	357.511	0.277
Instantaneous phase	358.599	0.266
Average frequency	358.977	0.262
Dominant frequency	360.808	0.244
Instantaneous frequency	368.322	−0.140
Time	370.201	0.099
Apparent polarity	371.853	0.030
Cosine instantaneous phase	371.907	−0.026
First derivative of instantaneous amplitude	371.927	−0.023
Weighted amplitude of cosine phase	372.000	0.012
Seismic trace	372.001	0.012
Second derivative of seismic trace	372.018	−0.008

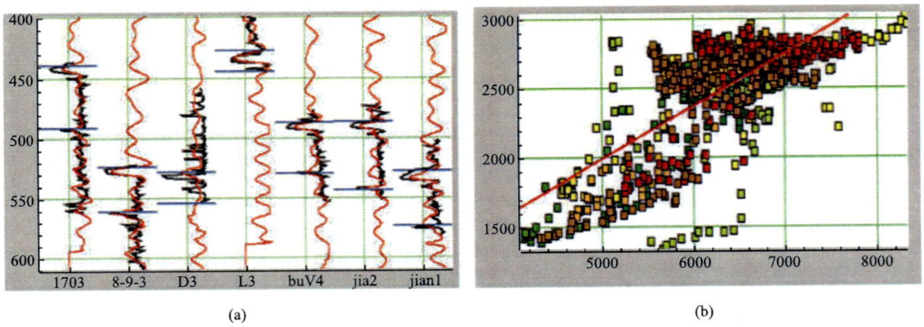

Fig. 8.6 **a** Comparison of logging curves predicted by wave impedance (red lines) with actual acoustic logging curves (black lines); **b** crossplot of wave impedance and actual sonic velocity

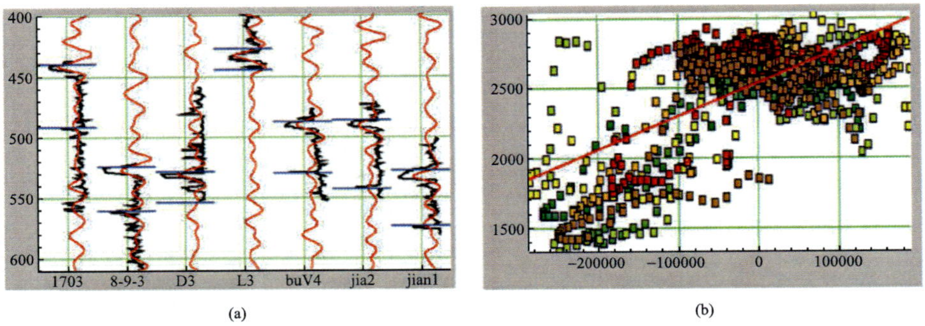

Fig. 8.7 **a** Comparison of logging curves predicted by seismic trace integration (red lines) with actual acoustic logging curves (black lines); **b** crossplot of trace integration and actual sonic velocity

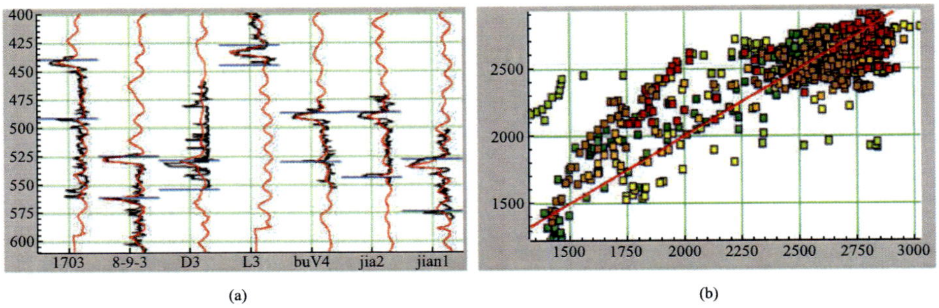

Fig. 8.8 **a** Comparison of logging curves predicted by multiple seismic attributes (red lines) with actual acoustic logging curves (black lines); **b** crossplot of predicted and actual sonic velocities

operator, the correlation coefficient between the converted curve and the actual curve increases gradually.

Figures 8.10 and 8.11 are the prediction results of two kinds of neural networks. The correlation coefficient of MLFN is 0.899, and that of PNN is 0.940. Obviously, PNN prediction is better. Finally, to compare the acoustic velocity curves predicted by the above five methods. The comparison range is between the top of No. 13-1 coal seam and the top of No. 11-2 coal seam (Fig. 8.12). The first three curves are linear statistical predictions, being rougher; the third one using a convolution operator is relatively close to the actual; the last two curves are neural network predictions. Compared with linear statistics, the neural network reflects higher frequency and is closer to the actual curve.

Actual curves (black); Predicted curves (red). The prediction range is between the top of No. 13-1 coal seam and the top of No. 11-2 coal seam.

Figure 8.13 shows the planar lithologic prediction of the roof of the No. 13-1 coal seam in Panji No.3 Coalmine by using the above methods. There is mudstone roof in the

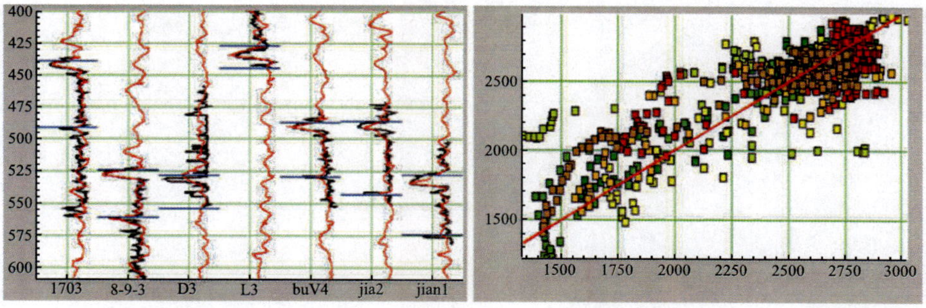

(a) (b)

Fig. 8.9 **a** Comparison of logging curves predicted by 3-point convolution operator (red lines) with actual acoustic logging curves (black lines); **b** crossplot of predicted and actual sonic velocities

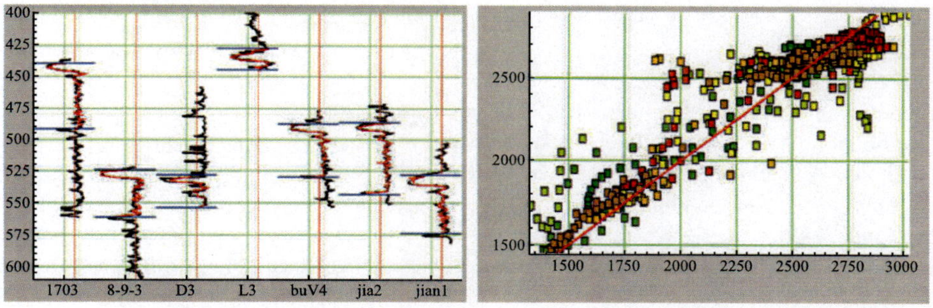

(a) (b)

Fig. 8.10 **a** Comparison of logging curves predicted by MLFN (red lines) with actual acoustic logging curves (black lines); **b** crossplot of predicted and actual sonic velocities

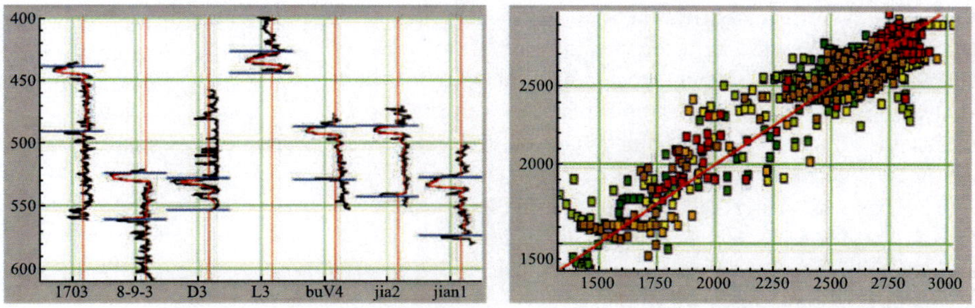

(a) (b)

Fig. 8.11 **a** Comparison of logging curves predicted by PNN (red lines) with actual acoustic logging curves (black lines); **b** crossplot of predicted and actual sonic velocities

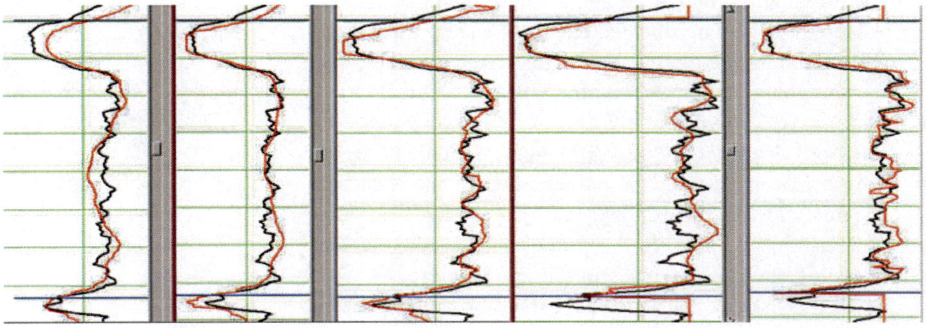

Fig. 8.12 Sonic velocity curves of borehole 8-9-3 predicted by five methods

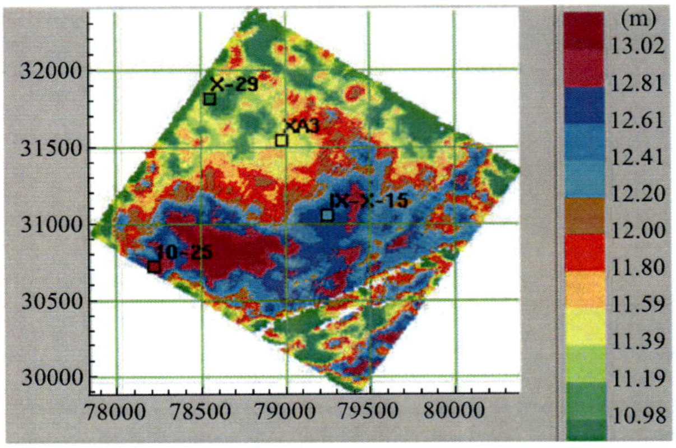

Fig. 8.13 Planar lithologic prediction of No. 13-1 coal seam roof in Panji No.3 Coalmine

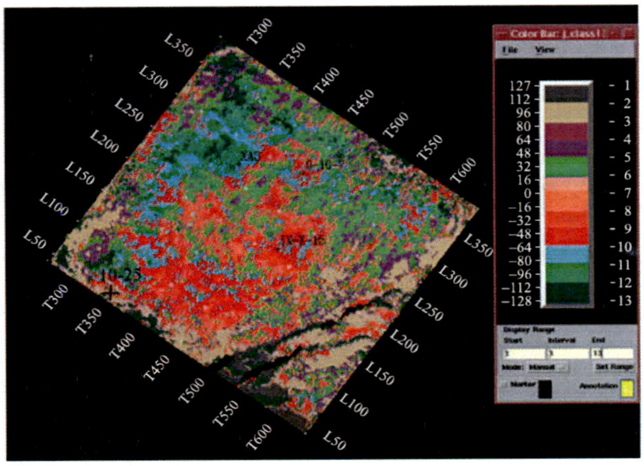

Fig. 8.14 CBM enrichment zones predicted by seismic waveform classification in the No. 13-1 coal seam

south of borehole IX-X-15, a narrow mudstone roof strip near borehole X-29 in near NE direction, sandstone roof near borehole XA3 and the most of the area in NEE direction, and siltstone roof from borehole XA3 toward the southwest borehole 10–25.

8.1.4 Seismic Waveform Classification for Predicting CBM Content

Seismic waveform classification is a neural network technology based on the characteristics seismic waveforms to process and classify. After processing, classification and analysis of the seismic facies represented by seismic waveforms, the differences in lithology, petrophysics and fluid can be studied in exploration target strata.

The basis of using seismic waveform classification to detect CBM enrichment zones is as follows: (i) CBM enrichment is related to coal seam structure. Thicker coal seams and development of fractures in coal seams are favorable conditions for CBM enrichment, which directly change the characteristics of seismic reflection waveform. (ii) The lithological changes of coal seam roof and floor may cause larger changes in reflected energy. Seismic waveform classification does not directly distinguish these changes, but allows researchers to extrapolate the CBM geological characteristics revealed by boreholes to the area not drilled. In the process of production, the CBM geological characteristics in the mined area can be extrapolated to the area not mined, but with the same waveform characteristics as the area mined, thus to achieve the purpose for predicting the CBM geological characteristics in the unmined areas.

Figure 8.14 shows the detected CBM enrichment based on 3D3C seismic data in the Dongsixiashan mining area of Panji No.3 Coalmine by using seismic waveform classification. Using various classification methods and grouping training methods of neural network, 13 seismic waveforms were classified, and classes 7, 8 and 9 were considered to be CBM enrichment zones.

According to the characteristics of seismic waveforms, when coal seams are divided into seven categories, faults and folds are more developed in the first and second categories. In the areas where faults and folds are developed, general fractures are also more developed, so they are considered to be the areas with developed coal seams. From Fig. 8.15, it can be seen that the fracture development areas are mainly distributed in the deep fault and fold areas, but not in the shallow continuous coal seam areas.

Figure 8.16 shows the detected CBM enrichment from 3D3C seismic area in the Xi 1 mining area in the Xieqiao

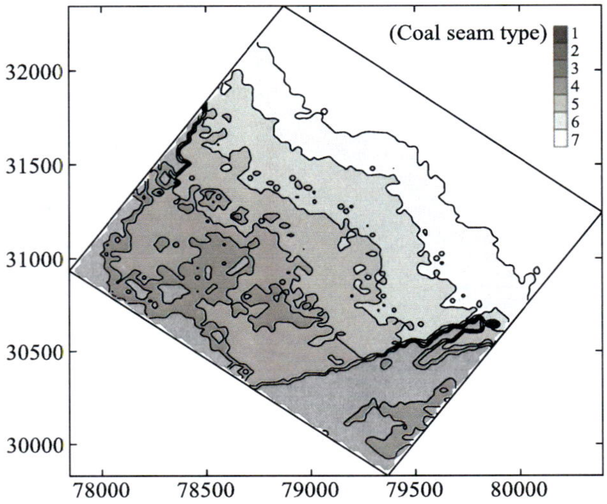

Fig. 8.15 Distribution of fracture development areas predicted by seismic waveform classification

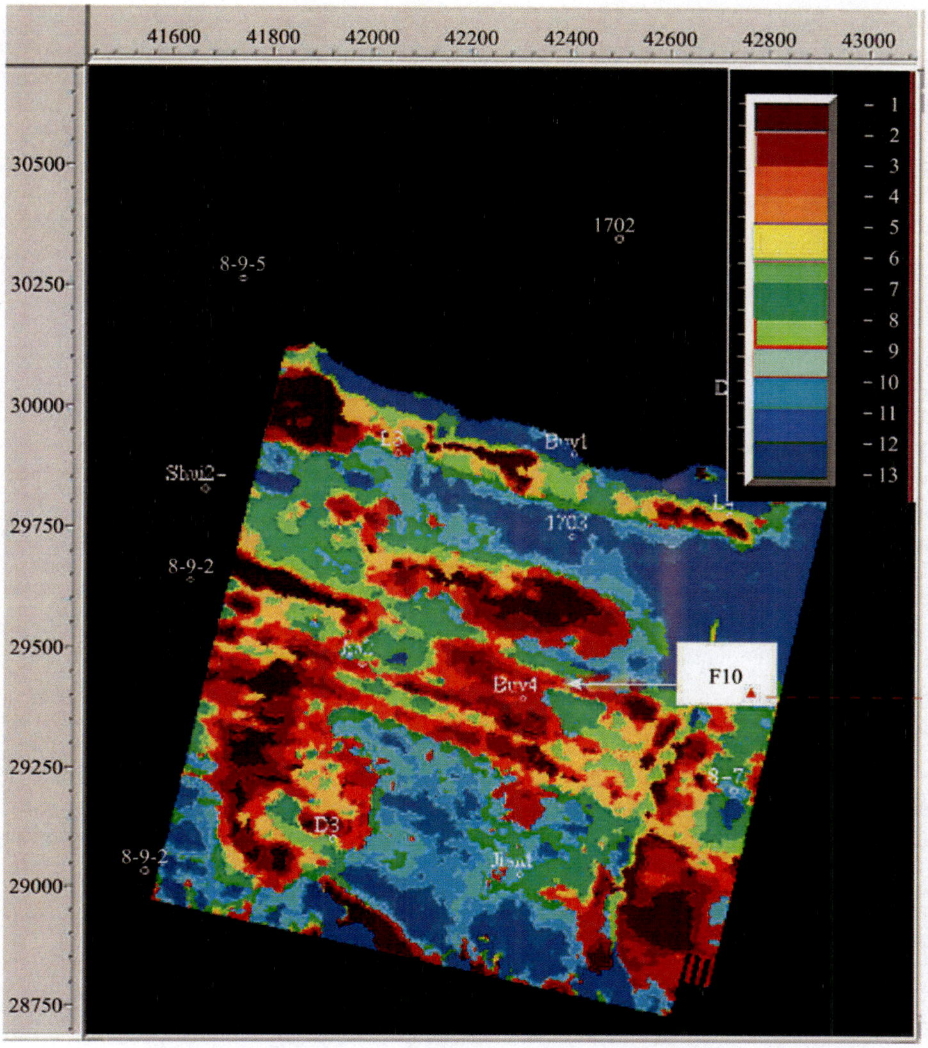

Fig. 8.16 CBM enrichment by seismic waveform classification in the Xi 1 mining area in Xieqiao Coalmine

Coalmine by using seismic waveform classification. 13 classes of waveforms were used. The seismic waveforms corresponding to the two through roadways and cut holes on the 1221(3) working face are red, yellow and brown respectively. The areas indicated by these colors have well-developed fractures and possible CBM enrichment. There are extensive fractures in the downthrown wall of the F10 fault, which may be the result of the positive traction by the F10 fault. There are extensive fractures in the western part of the study area, which are related to the neighboring F6 normal fault. A large number of fractures and tectonic coals on the 1221(3) working face were caused by excavating the roadways. Some fractures are developed on the 1211(3) working face and shallow coal seams. Deep fractures are enriched with CBM. Because of well-developed tectonic coals in shallow layers, most CBM has effused from there, and no CBM enrichment was found.

8.2 Multiwave Seismic Responses and Prediction of Fractures in Coal Seams

Theoretical researches and field observations show that there are many fractures with different development degrees in coal seams. Therefore, accurate determination of fracture development parameters is of great significance to the prediction of CBM enrichment zones. Regular development of vertical fractures leads to the anisotropy of coal seams, so to invert fracture development parameters is to detect rock physical properties and anisotropy.

8.2.1 Multiwave Seismic Responses of Vertical Fractures

Seismic waves have a unique propagating law in anisotropic media. The most obvious feature is S-wave splitting, i.e. when S-waves pass through azimuthally anisotropic media, they will split into two S-waves with nearly orthogonal polarization directions. Fast S-wave has the polarization direction parallel to fracture direction, and slow S-wave has the polarization direction perpendicular to fracture direction. When other conditions are the same, the more fractures develop in coal seams, the stronger anisotropy and the greater time difference between fast and slow S-waves. Therefore, the orientation and density of vertical fractures can be determined by estimating the polarization direction and time difference of fast and slow S-waves, and then favorable CBM enrichment zones can be determined based on other data.

The multi-wave seismic response of a medium with vertical fractures is studied by forward record. The second layer of the model (Fig. 8.17) is a HTI medium, and the surrounding rock above and below is homogeneous and isotropic.

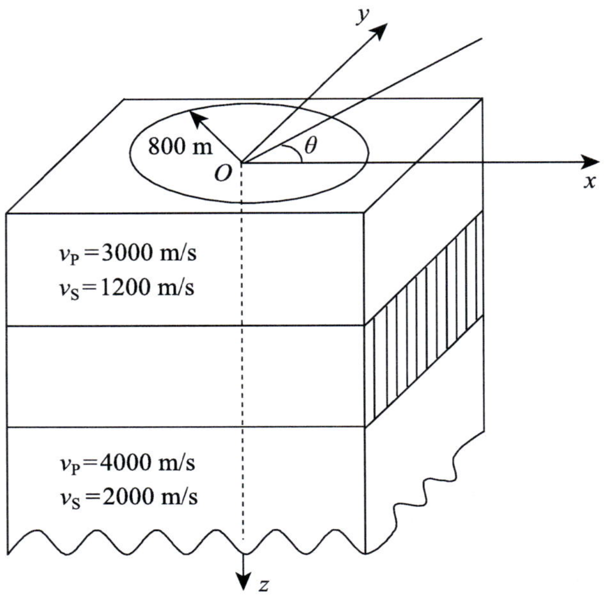

Fig. 8.17 Schematic graph of a theoretical model and recording system

Figure 8.18 is the 2D3C seismic record of the model in Fig. 8.17, at different azimuth. Although the model is very simple—only two reflectors and an anisotropic layer, the 3C record is much more complex than P-wave, and at least 12 strong waveforms were recorded. These waveforms overlap each other and build a very complex set of wave fields. It can be seen from Fig. 8.18 that when the survey line is parallel to the fracture orientation (Fig. 8.18a), the y component is zero, and the energy of reflected and converted waves mainly concentrates on the x and z components; the P-wave is stronger in the z component, and the converted S-wave is stronger in the x component. According to S-wave splitting mechanism an in anisotropic medium, converted wave is fast S-wave, and its polarization is parallel to the fracture orientation. When the survey line is perpendicular to the fracture orientation (Fig. 8.18d), the y component is still zero, the converted waves in the x and z components are slow S-waves, and the polarization is perpendicular to the fracture orientation. In other cases (Fig. 8.18b, c), there are converted waves from the top and bottom of the anisotropic layer in the y component, and the three components contain both fast and slow S-waves. According to the theory of elastic wave, the y component in a 2D3C record is always zero in isotropic horizontal layered media, no matter how the azimuth of the survey line changes. A preliminary conclusion can be drawn from Fig. 8.18b and c that an underground layer is anisotropic, but its specific anisotropy, i.e. fractures, is difficult to obtain directly from 2D records.

In order to study the multi-wave seismic response of vertical fractures more intuitively, the following recording system was designed: Point O was a shot point, 72 receiver points were placed every 5°; 3C seismic data were recorded with 800 m around Point O (r = 800). Figure 8.19 shows 3C synthetic records, on which the horizontal axis is the angle between the line from the shot point to the receiver point and the x axis; the x component of every receiver point is parallel to the line from the shot point to the receiver point, that is, the x component is radial and the y component is transverse; the arrow points to the converted wave (x component) or reflected wave (z component) from the bottom of the anisotropic layer.

Analysis of Fig. 8.19 leads to conclusions: (i) When the converted S-wave passes through a medium containing vertical fractures, different splitting conditions occur according to the relative relationship between the incident azimuth and the orientation of fractures: slow waves are only produced at vertical incidence, fast waves are only produced at parallel incidence, and fast and slow shear waves can be observed at other azimuths. The time difference between the two kinds of S-waves reaches its maximum when the angle between the incident azimuth and the orientation of fractures is 45°, and becomes zero in the vertical and horizontal directions. When the incident azimuth changes in the vertical

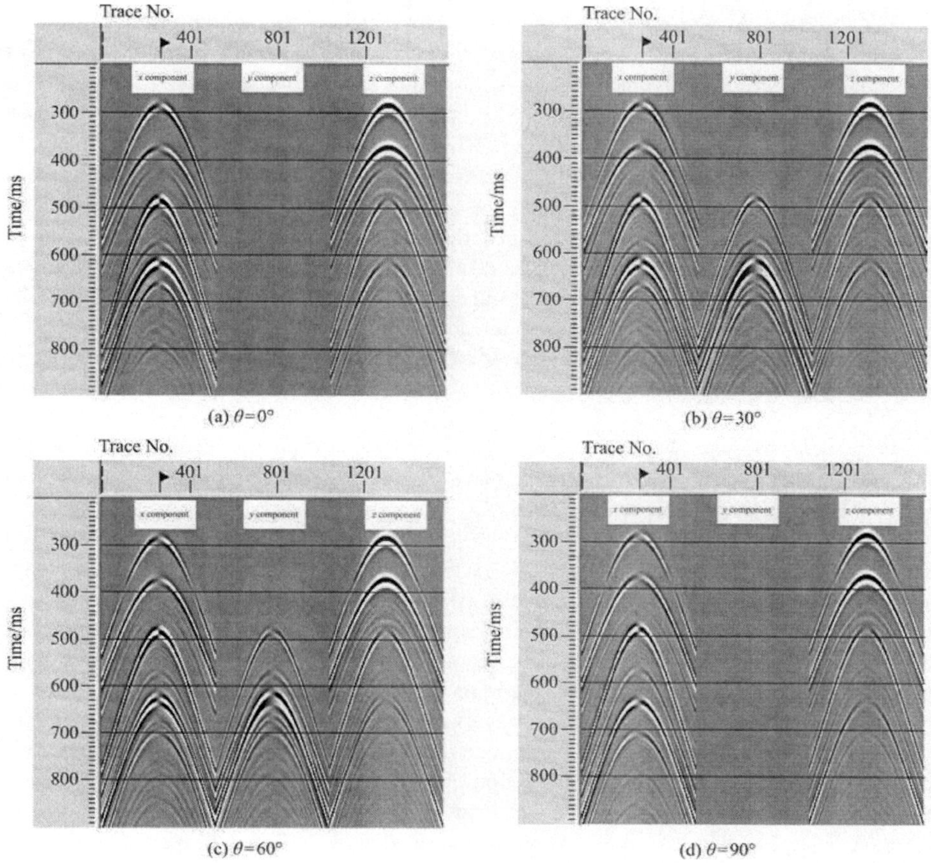

Fig. 8.18. 2D3C synthetic seismogram with different lateral azimuth

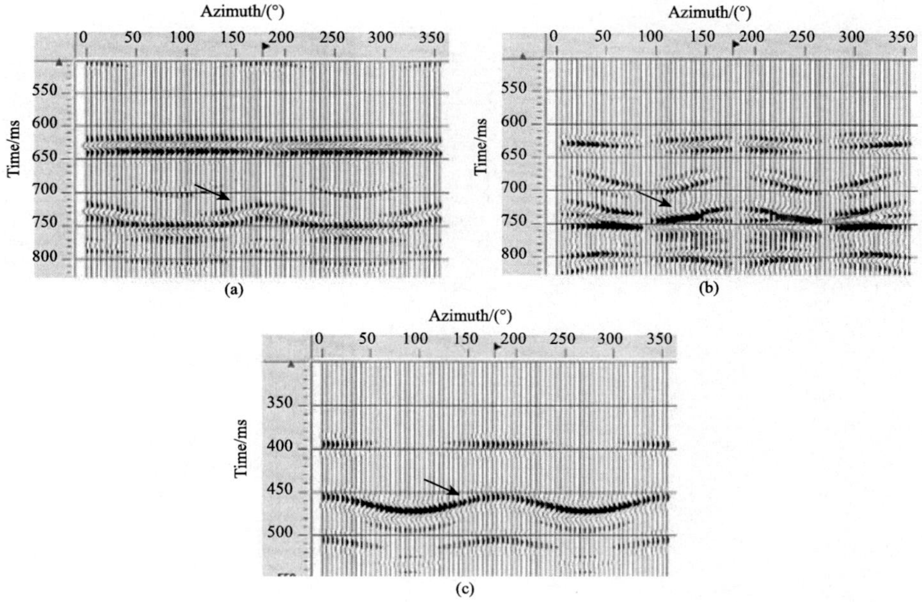

Fig. 8.19 Forward 3C synthetic records. **a** radial component; **b** radial component; **c** z component

or parallel directions, the polarity of the lateral component reverses. (ii) When the thickness of the target layer is fixed, the greater the time difference between fast and slow S-waves, the stronger the anisotropy of the layer and the more developed the fractures are. (iii) The azimuthal changes of velocity and amplitude will occur when P-wave passes through anisotropic media. When the incident azimuth is parallel to the orientation of fractures, the velocity is faster, and the amplitude is weaker; while it is vertical, the velocity is slower and the amplitude is stronger. (iv) When the anisotropic layer is thicker, the orientation and density of fractures can be determined by azimuthal stack of 3C data after NMO correction in CCP trace gathers.

8.2.2 Detection of Vertical Fractures by Multiwave Seismic Data

The development parameters of vertical fractures in coal seams are predicted by the following ideas: (i) on the premise of fidelity, improve the signal-to-noise ratio of original converted waves; (ii) conduct azimuth correction to receiver points to eliminate acquisition footprints; (iii) use model-based three-parameter converted wave velocity analysis and non-distortion correction to eliminate the influence of non-zero offset. Non-distortion correction can eliminate the time difference of non-zero offset and ensure the dynamic characteristics of converted wave not to distort; (iv) use angle scanning and angle spectrum to find the main orientation of underground fractures; (v) verify the principal azimuth obtained from azimuthally stacked trace gathers; (vi) separate fast from slow S-waves by coordinate rotation, and image the separated fast and slow S-waves; (vii) determine the development of fractures by the time difference between fast and slow S-waves.

The primary task of detecting coal seam fractures based on the above ideas is to carry out fine pre-processing of converted wave data. For this reason, the first task is to solve the main technical problems such as wavelet estimation, denoising, velocity analysis, NMO and NMO stretching, static correction and wave field separation, so as to provide basic fidelity data for converted wave fracture inversion. On the premise of acquiring high-quality multi-wave data and taking fidelity pretreatment, the project focused on some key technologies, such as inversion of fracture orientation and development by using converted wave. On the basis of solving many technical problems, a software system for multi-wave and multi-component processing and inversion was developed with independent intellectual property rights. Then the software system was used to separate fast from slow S-waves and imaging processing, and finally obtaining fracture parameters. Figure 8.20 shows the workflow for fast and slow S-wave separation and imaging.

In the above process, the key steps, such as denoising, model-based velocity analysis, non-distortion NMO correction, fast and slow wave separation and NMO correction, cannot be performed in available seismic data processing systems. The project team broke through traditional seismic

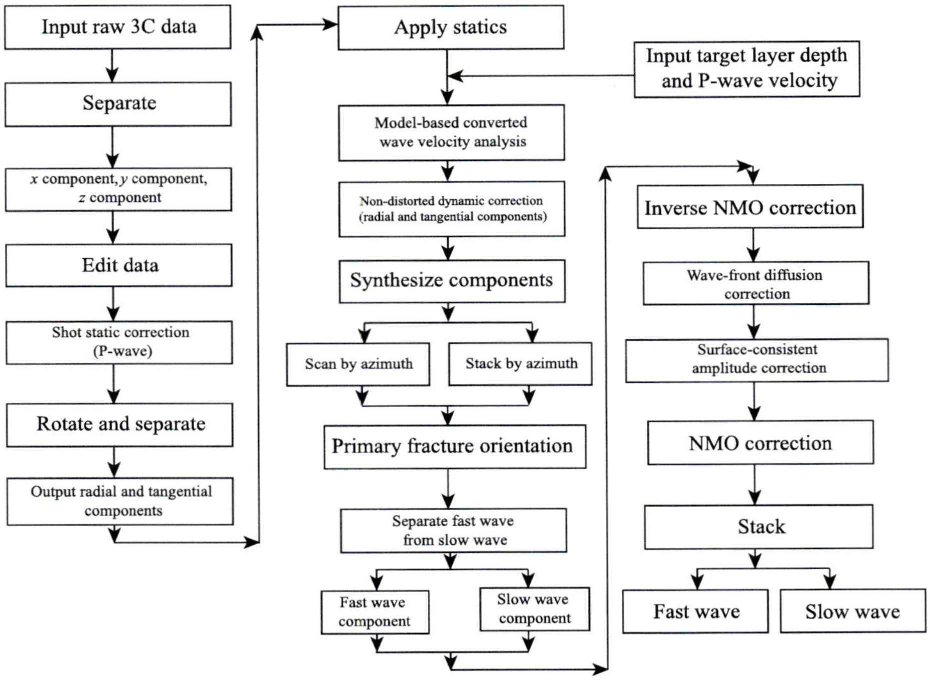

Fig. 8.20 Workflow for separation of fast from slow S-waves

data processing ideas, and carried out targeted research. At present, most of the technical problems have been almost solved, and separation and imaging of fast and slow S-waves can be done. The achievements of the project in fracture inversion based on converted wave are as follows.

8.2.2.1 Extract Mixed-Phase Wavelet

The relationship between seismic records, seismic wavelet and reflection coefficient sequence can be expressed by a convolution model:

$$x(t) = \omega(t)\xi(t) + n(t) \tag{8.10}$$

where $x(t)$, $\omega(t)$, $\xi(t)$ and $n(t)$ are seismic record, seismic wavelet, reflection coefficient and noise, respectively.

The purpose for extracting wavelet is to obtain seismic wavelet $\omega(t)$ from multiple raw data. Seismic wavelet had been collected by recording near-field direct wave in the past, but the effect was not good. With density logging and acoustic logging data, seismic wavelet can be inverted by using seismic traces near a borehole. However, the sparse spatial distribution of well logging data and the variation of wavelet in transverse space limit the application of this method. Based on the systematic analysis of the previous methods of extracting mixed-phase wavelet, we proposed to use wavelet amplitude spectrum of to determine the maximum and minimum components of mixed-phase wavelet in cepstrum domain.

(1) Decompose maximum and minimum phases

Ulrych introduced homomorphic deconvolution into seismic exploration. Ignoring the noise term, Eq. (8.10) can be expressed in frequency domain as follows:

$$x(\omega) = \omega(\omega)\xi(\omega) \tag{8.11}$$

After calculating logarithms to both sides, Eq. (8.11) is transformed into a linear system:

$$\ln x(\omega) = \ln \omega(\omega) + \ln \xi(\omega) \tag{8.12}$$

By applying the inverse Fourier transformation to the upper formula,

$$\bar{x}(t) = \bar{\varpi}(t) + \bar{\xi}(t) \tag{8.13}$$

where $\bar{x}(t), \bar{x}(t)$ and $\bar{\xi}(t)$ are the cepstrum of $x(t), \omega(t)$ *and* $\xi(t)$, respectively.

Because of the difference of "smoothness" between wavelet and reflection coefficient sequence, the cepstrum of wavelet is generally near the zero point, while the cepstrum of reflection sequence is far from the zero point. Based on this understanding, the design of a low-pass filter in the cepstrum can achieve the separation of wavelet and reflection coefficient in the cepstrum, and achieve the purpose of wavelet extraction.

While calculating the logarithmic spectrum, the phase spectrum becomes the imaginary part of the logarithmic spectrum, but the multi-valued property of the phase makes it necessary to expand the phase spectrum continuously. Many people have studied it and given the derivative method, recursive algorithm and other solutions, but these theoretical solutions have many difficulties in numerical implementation.

Although the phase spectrum of a wavelet is difficult to obtain, the amplitude spectrum of the wavelet can be obtained accurately and conveniently. Multichannel statistical autocorrelation is one of the most commonly used methods. This research is how to decompose the maximum and minimum phase components of a wavelet in the cepstrum when the amplitude spectrum is known.

Let the maximum and minimum components of the wavelet $\omega w(t)$ be $u(t)$ and $v(t)$:

$$\omega(t) = u(t) * v(t) \tag{8.14}$$

It can be expressed in the Fourier domain as:

$$|\omega(\omega)| \exp(i\phi_\omega(\omega)) = |u(\omega)| \exp(i\phi_u(\omega)) + |v(\omega)| \exp(i\phi_v(\omega)) \tag{8.15}$$

It can be expressed by the logarithmic spectrum as:

$$\ln|\omega(\omega)| + i\phi_\omega(\omega) = \ln|u(\omega)| + i\phi_u(\omega) + \ln|v(\omega)| + i\phi_v(\omega) \tag{8.16}$$

For eliminating the phase spectrum:

$$\begin{aligned} 2\ln|\omega(\omega)| &= \ln|u(\omega)| + i\phi_u(\omega) \\ &+ \ln|u(\omega)| - i\phi_u(\omega) + \ln|v(\omega)| \\ &+ i\phi_v(\omega) + \ln|v(\omega)| - i\phi_v(\omega) \end{aligned}$$

It can be expressed by the cepstrum as:

$$2\bar{\varpi}(t) = \bar{u}(t) + \bar{v}(t) + \bar{u}(-t) + \bar{v}(-t) \tag{8.17}$$

where $\bar{\varpi}(t)$ is the cepstrum of the amplitude spectrum, appearing symmetrically on the positive and negative axes of the cepstrum; $\bar{u}(-t)$ is the cepstrum of the maximum phase function corresponding to the minimum phase component $u(t)$ of the wavelet; $\bar{v}(-t)$ is the cepstrum of the minimum phase function corresponding to the maximum phase component $v(t)$ of the wavelet; $\bar{u}(t), \bar{v}(t)$ appear on the positive and negative axes of the cepstrum, respectively. Correspondingly, $\bar{u}(-t), \bar{v}(-t)$ appear on the negative and positive axes of the cepstrum respectively. Equation (8.17) can be used to determine the minimum phase component of the wavelet on the cepstrum, and accordingly the maximum phase component of the wavelet is determined. A set of wavelets with the same amplitude spectrum but different phase spectrum can be determined by scanning.

(2) Determine wavelet

Assuming that the output of the deconvolution of seismic records by the deconvolution filter determined by the k th wavelet to deconvolute seismic records is $y_k(t)$, select the wavelet that can make the following equation to be the maximum as the desired wavelet:

$$e[y_k(t)] = E[y_k^4(t)] - 3E^2[y_k^2(t)] \tag{8.18}$$

The above formula is the theoretical value under the condition of infinite samples. In practice, it is impossible to have infinite samples, and estimates are often used instead, then Eq. (8.18) becomes:

$$e[y_k(i)] = \frac{1}{M}\left\{\sum_{i=1}^{M} y_k^4(i) - 3\left[\sum_{i=1}^{M} y_k^2(i)\right]^2\right\} \tag{8.19}$$

After reorganizing, it is equivalent to maximizing the following formula:

$$\tilde{e}[y_k(i)] = \frac{\sum_{i=1}^{M} y_k^4(i)}{\left[\sum_{i=0}^{M} y_k^2(i)\right]^2} \tag{8.20}$$

Equation (8.20) is actually the variance modulus norm. Its essence is the peak value of the probability density distribution function of non-Gaussian sequence. The larger the value, the sharper the probability density distribution. For seismic reflection sequence, sparse reflection coefficient sequence is required. It is a commonly used criterion for wavelet extraction.

On the basis of applying the maximum variance modulus criterion, this project used interactive technology to display the results of wavelet and deconvolution, and the transformation of variance modulus in real time. Combining with geological knowledge and the characteristics of first break waveform, the relationship between the best variance modulus and seismic wavelet suitable for different regions is determined.

(3) Analysis of results

The correctness and validity of this method were verified by simulated data. Figure 8.21a is a local display of the simulated white noise reflection sequence. Figure 8.21f is the input mixed-phase seismic wavelet. Figure 8.21b is a synthetic seismogram formed by the convolution of the input wavelet in Fig. 8.21f and the white noise reflection sequence in Fig. 8.21a. Figure 8.21c is the minimum phase component of mixed phase wavelet extracted by the method proposed in this book, and Fig. 8.21d is the extracted maximum phase component. Figure 8.21e is the final output seismic wavelet synthesized by the minimum and the maximum phase components. Comparing the input wavelet in

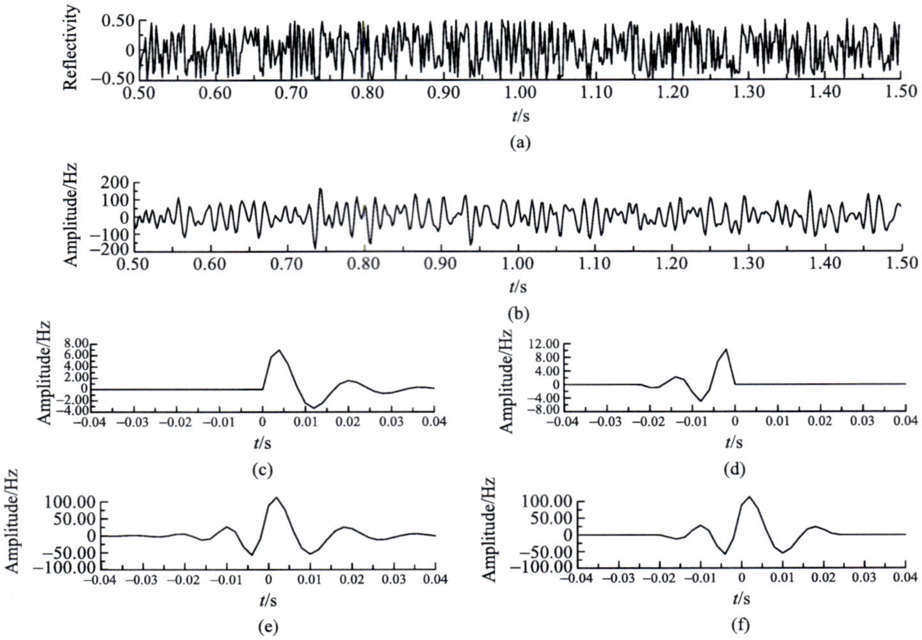

Fig. 8.21 Comparison of wavelet extraction effects from synthetic data. **a** white noise reflection coefficient sequence; **b** synthetic seismogram formed by the convolution of the white noise reflection sequence **a** and the input wavelet **f**; **c** extracted minimum phase component; **d** extracted maximum phase component; **e** extracted final wavelet; **f** input mixed-phase wavelet

Fig. 8.21f with the extracted wavelet of Fig. 8.21e, it can be seen that they are very close.

Using the interactive software compiled by this method, we carried out the experiments for extracting wavelets from shot records acquired using a physical model and actual raw data. Figure 8.22a is a shot record acquired by a physical model, and Fig. 8.22b is a wavelet extracting template. The upper part in Fig. 8.22b shows the cepstrum sequence of multi-trace average amplitude spectrum, and the wavelet amplitude spectrum in the middle (the wavelet amplitude spectrum can be simulated by multi-trace amplitude spectrum, or can be mutually adjusted by truncated cepstrum sequence). The lower part in Fig. 8.22b shows the extracted wavelet (Wavelet scanning can be carried out automatically or interactively. When adjusting interactively, the input records on the right side of the template, the output results of the deconvolution and the transformation of variance modulus can be referred to). By analyzing the final seismic source wavelet, we can see that a better extraction effect has been achieved.

8.2.2.2 Suppress Surface Wave

At present, the surface wave suppression methods commonly used for processing converted wave all assume that surface wave has spatial coherence, and the suppression effect depends on the linear correlation of the surface wave. If the linear interference is completely correlated, the methods work well. But in fact, because surface wave does not have complete mathematical coherence, it is often not ideal to suppress surface wave directly by using linear noise suppression methods such as *f-k* filtering. Based on the analysis of the influence of surface wave energy distribution and static time shift on *f-k* filtering effect, we proposed a set of better suppression method for surface wave in converted wave processing.

(1) Principles

In converted wave record, surface wave is characterized by seismic events with large angle and large time difference between traces. This feature causes the *f-k* sector digital filter to fold at Nyquist wave number, and pseudo-seismic events with the direction opposite to the noises to be suppressed appear in filtered seismic records, which makes the effect of *f-k* filtering worse. In this project, we first made use of surface wave velocity to make linear NMO correction to seismic records, increased the apparent velocity of surface wave events, and carried out *f-k* filtering, and then made linear inverse NMO correction to avoid alias.

In addition, due to the difference of surface wave energy at different offsets, the direct use of a *f-k* filter on the original record can't achieve the desired denoising effect. Therefore, the inter-trace energy should be normalized to centralize the energy in (*f, k*) spectra into a straight line with a certain slope. Figure 8.23a is the converted wave shot record after linear NMO correction. Figure 8.23b is the record after *f-k* filtering based on the record in Fig. 8.23a, the residual surface waves is still strong. Figure 8.23c is the record after energy balance and then *f-k* filtering based on the record in Fig. 8.23a. It can be seen that energy balance well improved the suppression by *f-k* filtering to surface wave.

(2) Application

We used the above method to process the converted wave data in the Huainan Coalfield in Anhui Province. Figure 8.24 shows the original and the denoised sections, and the surface wave suppressed. The target coal seam is at about 1100 ms. The converted wave record has been preserved after suppressing surface wave. Because small offsets are not

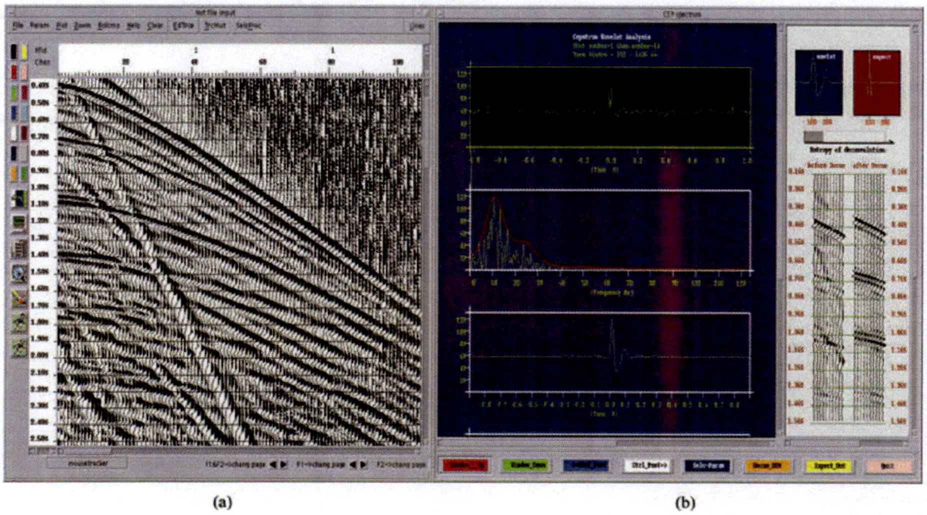

Fig. 8.22 Wavelet extraction of a shot record from a physical model. **a** shot record from a physical model; **b** interactive wavelet extracting template and extracted wavelet

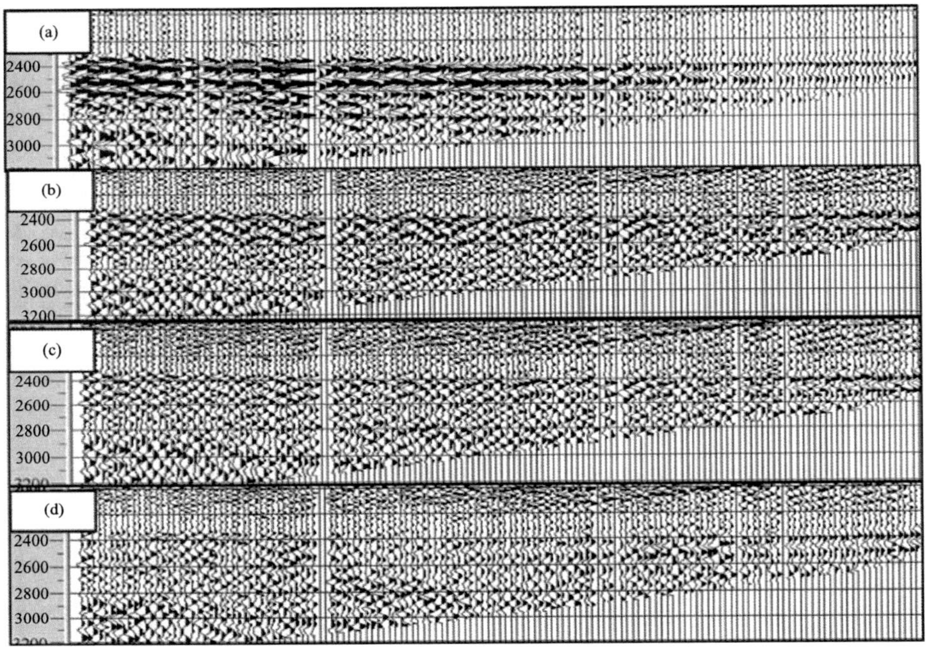

Fig. 8.23 Surface wave suppression. **a** shot record after linear NMO correction; **b** shot record after *f-k* filtering; **c** shot record after energy balance and *f-k* filtering; **d** shot record after balancing energy and adjusting time shift, and then *f-k* filtering

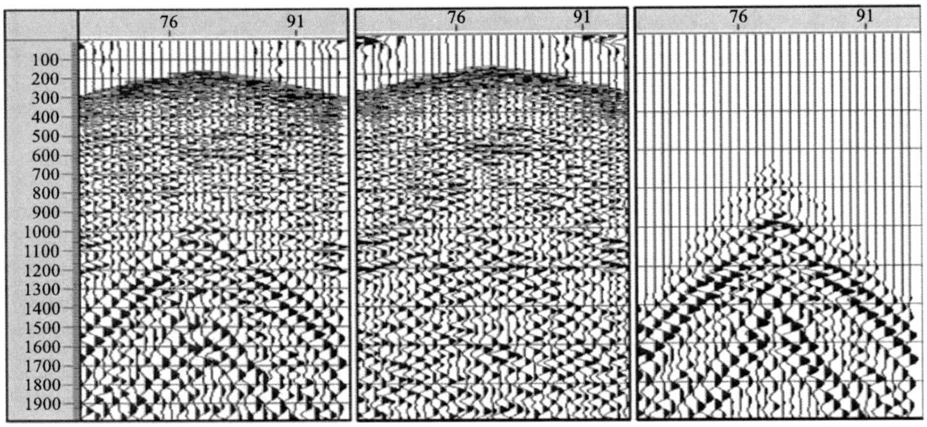

Fig. 8.24 Original record (left), denoised record (middle) and surface wave suppressed (right) in Huainan Coalfield

within the effective aperture, no converted wave appears on the denoised record in small offsets.

8.2.2.3 Model-Driven CCP Three-Parameter Velocity Analysis

Because converted wave shows asymmetric ray paths, the velocity analysis method commonly used for P-wave processing is no longer suitable. We adopted model-driven velocity analysis to pre-processing of converted wave for fracture inversion by following the procedures below:

① Analyze P-wave velocity by a conventional processing technology to obtain stacked velocity, and finally migrated P-wave.

② Conduct structural interpretation to the migrated P-wave to build a depth model of the target layer.

③ Based on available data, estimate the ratio range (γ_{max} and γ_{min}) of P-wave velocity to S-wave velocity in the target layer, and calculate the corresponding conversion points, (x_{c1}, y_{c1}) and (x_{c2}, y_{c2}), on the model trace. Assuming that the vertical and horizontal coordinates of the receiver point are larger than those of the shot point, no matter how the γ (γ is the ratio of P-wave velocity to S-wave velocity) at a conversion point changes, if $\gamma_{min} \leq \gamma \leq \gamma_{max}$, the location of the real conversion point (x_c, y_c) must meet the following equation:

$$\begin{cases} x_{c1} \leq x_c \leq x_{c2} \\ y_{c1} \leq y_c \leq y_{c2} \end{cases} \tag{8.21}$$

④ For a given conversion point, extract all seismic traces that any conversion point may fall at the point to form a large trace gather.
⑤ Dynamically extract traces in this large trace gather to realize model-based CCP (CCP) velocity scanning. The steps are:

i. Define γ_{min}, γ_{max} and $\Delta\gamma$ (scanning step);
ii. Scan from $\gamma = \gamma_{min}$, calculate the position of the conversion point on each trace in the target layer at this velocity ratio, and then calculate the ray path of the converted wave. Using this ray path to calculate the NMO in the target layer, and conduct NMO correction to all seismic traces in this large trace gather. At this time, the NMO correction is similar to the static correction, that is, the correction of each sampling point is the NMO correction of the converted wave of the target layer, which can flatten the seismic event of the converted wave of the target layer, and avoid the influence of the stretching distortion of the NMO correction on subsequent processing.
iii. In this large trace gather, extract all the seismic traces whose conversion points are located in the large bin for velocity analysis, and then obtain velocity traces.
iv. Rearrange the above seismic traces by offset, and calculate the similarity coefficient.
v. Change the scanning velocity ratio (γ) and repeat steps (ii) to (iv), until $\gamma = \gamma_{max}$.
vi. Draw velocity spectra from the similarity coefficient curves at different velocity ratios.
vii. The velocity ratios in the above steps correspond to different CCP gathers. It is a true model-based CCP velocity ratio analysis method.

The software developed according to the above principles can fully meet the needs of industrial production. Figure 8.25 is an example of velocity analysis for 3D converted wave data in a study area of Huainan Coalfield. The left part on this figure is the velocity ratio spectra, and the right part shows the NMO correction result at various velocity ratios. The better the NMO correction result is, the stronger the spectrum energy is.

The method mentioned above is based on the following assumptions: (i) the kinematic characteristics of P-wave strictly satisfy the hyperbolic law; (ii) uniform single-layer medium; (iii) the change of the S-wave velocity of the overlying layer does not affect the position of the conversion point on the underlying reflector. In fact, the above assumptions are inadaptable to varying degrees in practical work. Therefore, in order to improve the imaging quality of converted wave, it is necessary to study the method of converted wave velocity analysis and stacking which can overcome the above limitations.

Assuming that the ratio of P-wave velocity to S-wave velocity is known, we studied three-parameter velocity analysis, NMO and stacking methods, and developed software, which overcomes the shortcomings of conventional

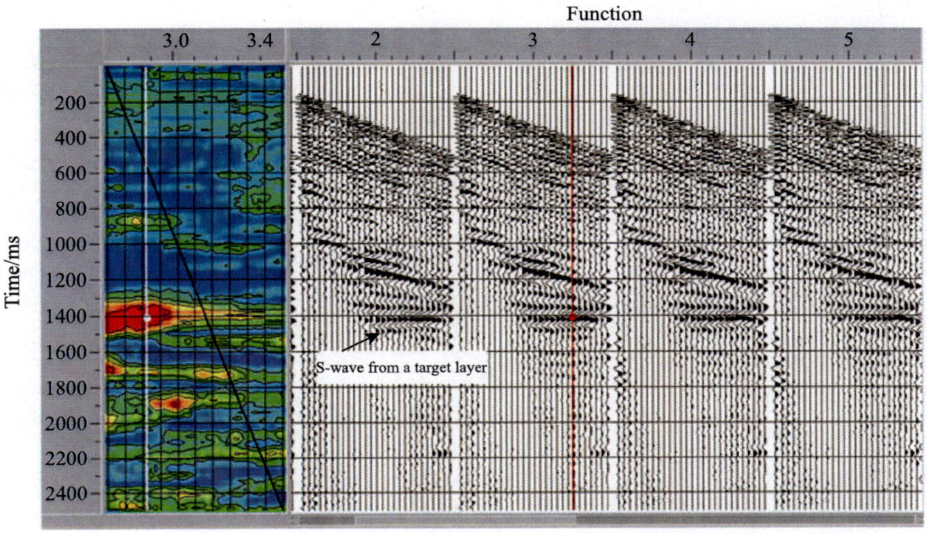

Fig. 8.25 An example of CCP velocity analysis to a trace gather

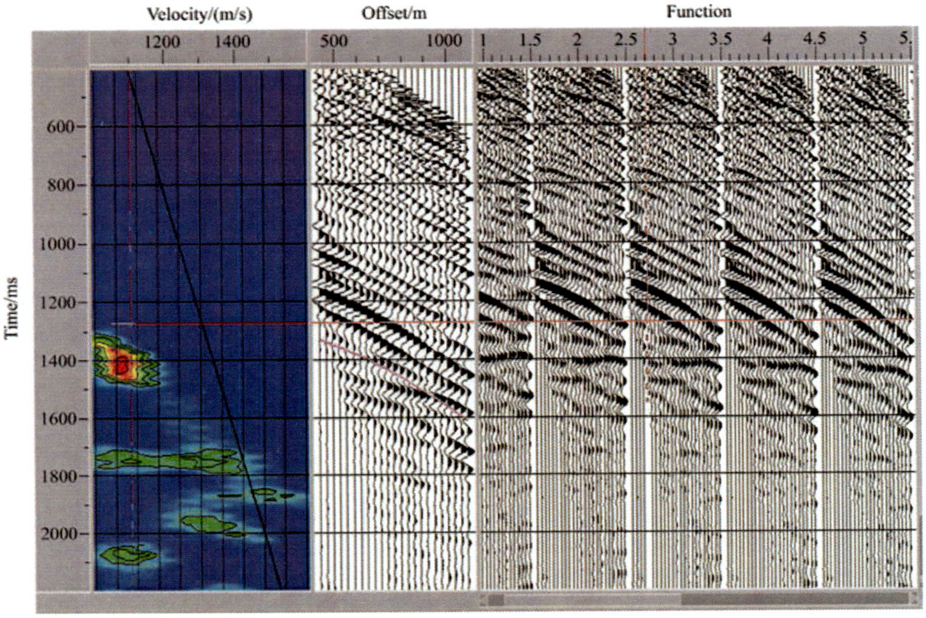

Fig. 8.26 An example of three-parameter velocity analysis to converted wave

two-parameter velocity analysis to a certain extent. Figure 8.26 is an example of three-parameter velocity analysis for 3D converted wave in a study area of Huainan Coalfield. The left part on this figure is the velocity ratio spectra, and the right part shows the NMO correction result at various velocity ratios. The better the NMO correction result is, the stronger the spectrum energy is.

8.2.2.4 Model-Based Distortion-Free NMO Correction to Converted Wave

After obtaining the velocity ratio, the converted wave can be corrected by NMO. As some applications of converted wave require very strict dynamic characteristics of seismic waves (such as to obtain the development orientation and degree of underground fractures from converted wave), it is necessary to ensure that the dynamic characteristics of seismic waves do not distort after NMO correction. Conventional point-by-point NMO correction always destroys the dynamic characteristics of original converted wave, while model-based distortion-free NMO correction doesn't. Instead of NMO correction point by point, distortion-free NMO correction means to first calculate the position of the conversion point of the target layer, and then the NMO value of a trace to the target layer according to the ray path of the converted wave, and finally correct all points on the trace by the NMO value. In the sense of RMS velocity, the NMO value to converted wave can be calculated by the following formula:

$$\Delta t = t_{\mathrm{ps}} - t_{\mathrm{ps0}} \tag{8.22}$$

where $t_{\mathrm{ps}} = \frac{A+\gamma B}{v_{\mathrm{P}}}$,$A = \sqrt{D^2 + x_{\mathrm{P}}^2}$, $B = \sqrt{D^2 + (x - x_{\mathrm{P}})^2}$, $t_{\mathrm{ps0}} = \frac{(1+\gamma)D}{v_{\mathrm{P}}}$, D is the target layer depth at the conversion point, m; x_{p} is the distance between the shot point and the conversion point, m; x is offset, m; v_{P} is P-wave velocity, m/s; γ is the velocity ratio of P-wave to S-wave.

Figure 8.27a is the original CCP gather of a target layer, and Fig. 8.27b is the NMO correction result obtained by this method. The converted wave of the target layer has been corrected better, which shows that this method can get good results in improving the image of a coal seam by converted wave.

8.2.2.5 Use Multi-wave Seismic Data to Obtain Primary Fracture Orientation

After obtaining the CCP gather of the target layer, the fracture orientation in the target can be obtained from the gather by an appropriate mathematical method. We developed an azimuth inversion method to estimate fracture development based on multi-trace scanning statistics. The specific principles are as follows.

Select a trace from the two horizontal components in Fig. 8.18b to analyze, given different β angles, and rotate the coordinates according to Eq. (8.23). The rotated result is shown in Fig. 8.28.

$$\begin{cases} Y = x\sin\beta - y\cos\beta \\ X = x\cos\beta + y\sin\beta \end{cases} \tag{8.23}$$

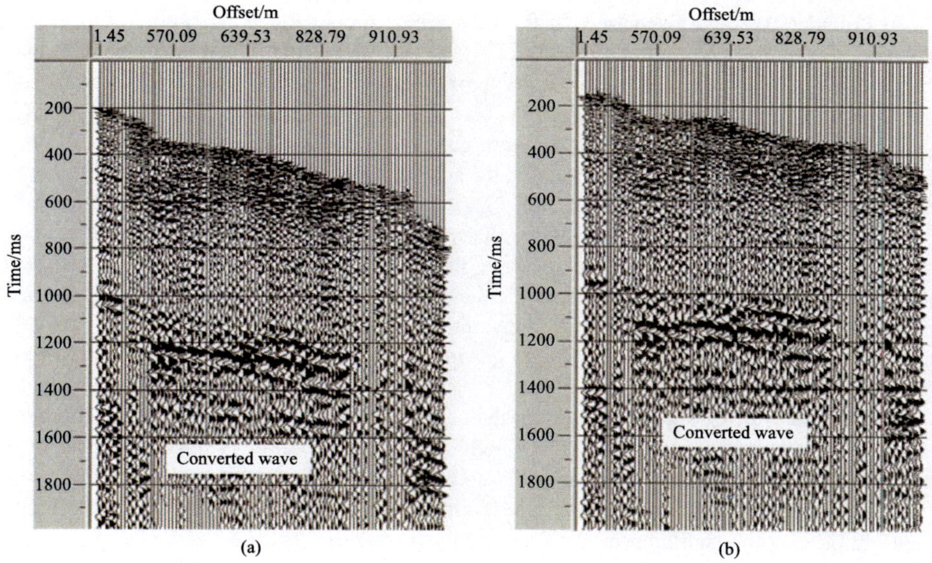

Fig. 8.27 Distortion-free NMO correction to converted wave of a target layer. **a** original record; **b** NMO corrected record

where, x and y is radial component and transverse component, respectively; X and Y is rotated fast wave and slow wave, respectively.

In the Fig. 8.28, the horizontal axis is the rotated angle β, and the effective wave is between 660 and 720 ms. In each trace gather, the first trace is fast S-wave, and the second trace is slow S-wave. The analysis of rotation results shows that when the rotated angle $\beta = 30°$, the time difference of fast and slow waves reaches the maximum, which is exactly fracture orientation. At the same time, the energy of them satisfies the following relationship:

$$\frac{Y}{X} = \tan(\beta) \tag{8.24}$$

The rotated results of other angles do not satisfy the above relationship. The bigger the difference between β and fracture azimuth, the bigger the difference of the ratio of slow to fast wave energy and $\tan(\beta)$. This shows that fracture orientation can be obtained by angle scanning to two horizontal components.

There are a lot of regular and random noises in raw data. It is impossible to obtain accurate results by inverting fracture parameters with only a trace. The following ideas can be adopted to improve the inversion accuracy.

① By comprehensive utilization of geological, well logging and P-wave data, determine the depth and structure of the target coal seam, and establish its depth model;

② After high-fidelity pretreatment to converted S-wave data, use model-driven velocity analysis to obtain the velocity ratio of P-wave and S-wave of the target layer, and extract the CCP gather;

③ Based on the CCP gathers of two horizontal components, conduct angle scanning to all data in the gather, and stack the scanning results;

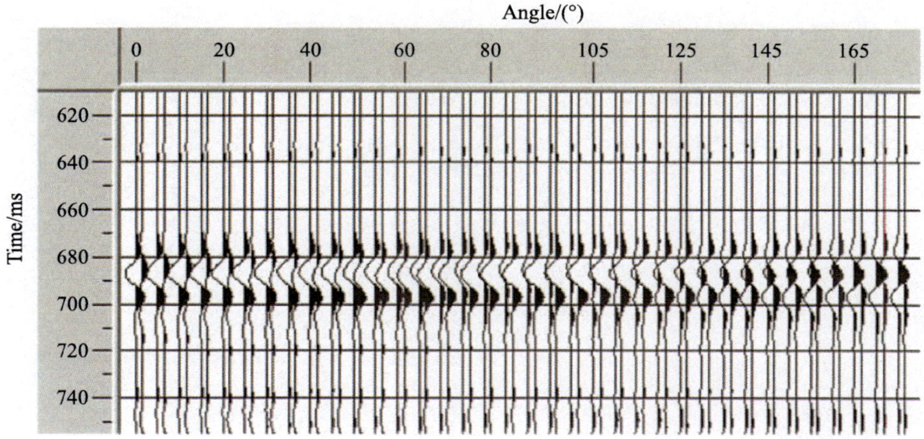

Fig. 8.28 Angle scanning result of horizontal component

④ Calculate the angle spectrum in the time window on the target layer using Eq. (8.24);

⑤ Interactively pick up the orientation of fracture at every point. 0° means when no fracture;

⑥ Extract and stack the common azimuth gathers on the CCP with higher folds and uniform azimuth, and draw the stacked profiles at different azimuths (Fig. 8.29). Modify the angle scanning results by the stacked profiles, and obtain final fracture azimuth. In order to increase the accuracy, structural changes, structural coal distribution and stress field of the target coal seam can be used as constraints to reduce the error of scanned fracture azimuth, to get more reliable fracture azimuth.

Figure 8.29 is an example provided by the angle analysis software programed according to this method. The left part on the Figure is angle spectrum, the middle part is CRP (common reflection point) gather after NMO correction, and the right part is the result after rotating and stacking seismic traces at different angles. The reflection event of the target layer is within 1300–1600 ms, and only the data in this time window were processed. The strong energy group in the angle spectrum appears near 110° and the stacked results after rotating at this angle also confirm the existence of fast and slow waves.

According to the basic propagating mechanism of fast and slow waves in underground media, the energy and time difference of seismic waves from different incident angles are inevitably different. Therefore, the fracture azimuth can be determined by the stacked azimuth of multi-wave data. Figure 8.30 is the stacked result of various azimuth gathers of a CCP gather. According to the theory of seismic wave, the polarization azimuth of the fast wave at this point is about 90°. In practical production, the calculation results of the above two methods can be used for comprehensive analysis, so as to obtain fracture azimuth with higher accuracy.

8.2.2.6 Separate Fast from Slow Waves

After obtaining the primary orientation of underground fractures, separation of fast from slow waves can be carried out by Eq. (8.23). Figure 8.31a and b are the shot records of original radial and transverse components, and Fig. 8.31c and d are fast and slow wave after separation.

After obtaining original fast and slow waves, conduct inverse NMO correction to the fast and slow waves, and apply similar processing flow and parameters (such as velocity, velocity ratio, etc.) to image fast and slow wave, respectively, to obtain migrated fast and slow waves. Figure 8.32 is the fast and slow wave sections processed by this project. By picking up the time difference between the two, underground fracture development can be obtained. Using well logging data as control, pick the time difference of fast and slow waves of the converted wave from the target coal seam. After removing the influence of other factors, this time difference represents underground fracture development. Then use the fast and slow wave data to build a time difference map in the study area. Inversion results show that the fractures are better developed in the central part of the study area. The inversion results are the same as those revealed by Well 5–24. Combining with other relevant data, the central part of the study area was finally determined to be the favorable area for CBM.

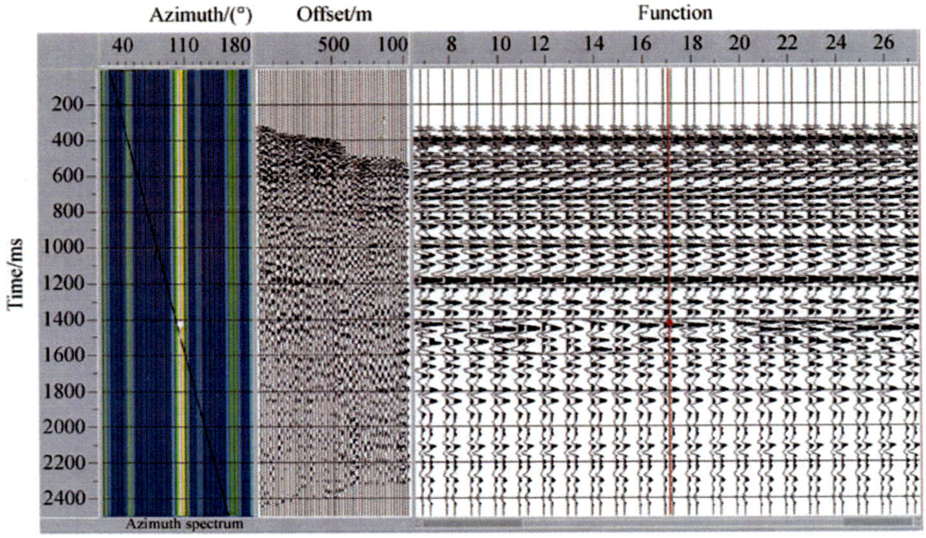

Fig. 8.29 An example of fracture azimuth analysis by angle scanning method

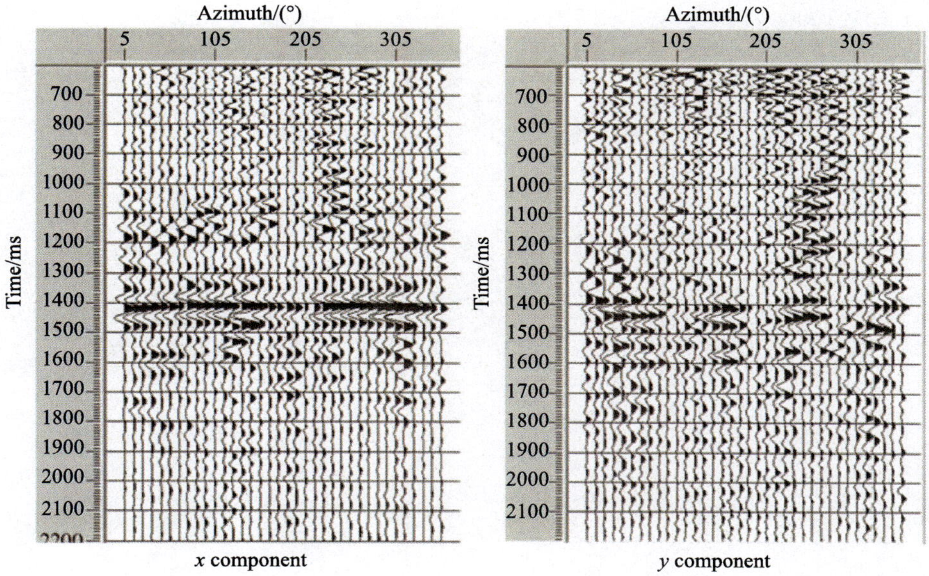

Fig. 8.30 Stacked azimuth result of a CCP gather

8.3 AVO Inversion and Prediction of CBM Enrichment Zones

According to the theory of reflection and transmission of seismic waves, the variation of amplitude with incident angle is related to the seismic parameters of the medium on both sides of the interface. This fact contains two meanings: (i) The characteristics of amplitude coefficient varying with incident angle are different for different lithologic parameter combinations. Using a AVO forward model to analyze AVO characteristics of known oil, gas, water and lithology is helpful to identify lithology and hydrocarbons from actual seismic records and to qualitatively describe reservoirs by seismic data. (ii) The variation of amplitude coefficient with incident angle itself implies the lithologic parameters. Using AVO relation, rock density ρ, P-wave velocity v_P and S-wave velocity v_S can be directly inverted to quantitatively describe reservoirs by seismic data. These two meanings reflect the basic idea of AVO analysis, and represent two basic AVO analysis methods. The former is called forward method, and the latter is called inversion method. Post-stack inversion can only obtain P-wave impedance, while pre-stack inversion can simultaneously obtain rock density, P-wave velocity and S-wave velocity, which is the reason why AVO inversion has been paid attention to. The key of AVO analysis is to fully tap and utilize the potential of non-zero offset seismic information in pre-stack seismic records.

8.3.1 Basic Dynamics of AVO Theory

The reflection and transmission amplitudes of elastic waves are functions of elastic parameters of media on both sides of the interface. Knott (1899) and Zoeppritz (1919) gave the reflection and transmission equations in potential and displacement style. Aki and Richards (1980) extended the work of Knott and Zoeppritz to incident P and SV waves from both sides of the interface. Assuming that the density of the media on both sides of the elastic interface is ρ_1 and ρ_2, P-wave velocity is v_{P1} and v_{P2}, S-wave velocity is v_{S1} and v_{S2}, incident angle of elastic P-wave is φ (Fig. 8.33), transmission angle of transmitted P-wave is φ, reflection angle of reflected S-wave (converted wave) is ϑ, transmission angle of transmitted S-wave (converted wave) is ϑ_2, then the Zoeppritz equation of incident P-wave is:

$$\left.\begin{aligned}
&A_{PR}\cos\varphi - A_{SR}\sin\vartheta + A_{PT}\cos\varphi_2 + A_{ST}\sin\vartheta_2 \\
&\qquad = A_P\cos\varphi \\
&A_{PR}\sin\varphi + A_{SR}\cos\vartheta - A_{PT}\sin\varphi_2 + A_{ST}\cos\vartheta_2 \\
&\qquad = -A_P\sin\varphi \\
&A_{PR}Z_1\cos 2\vartheta - A_{SR}\sin 2\vartheta - A_{PT}Z_2\cos 2\vartheta_2 \\
&\qquad -A_{ST}W_2\sin 2\vartheta_2 = -A_P Z_1\cos 2\vartheta \\
&A_{PR}(v_{S1}/v_{P1})W_1\sin 2\varphi + A_{SR}W_1\cos 2\vartheta \\
&\qquad + A_{PT}(v_{S2}/v_{P2})W_2\sin 2\vartheta_2 \\
&A_{ST}(v_{S2}/v_{P2})W_2\cos 2\vartheta_2 = A_P(v_{S1}/v_{P1})W_1\sin 2\varphi
\end{aligned}\right\} \tag{8.25}$$

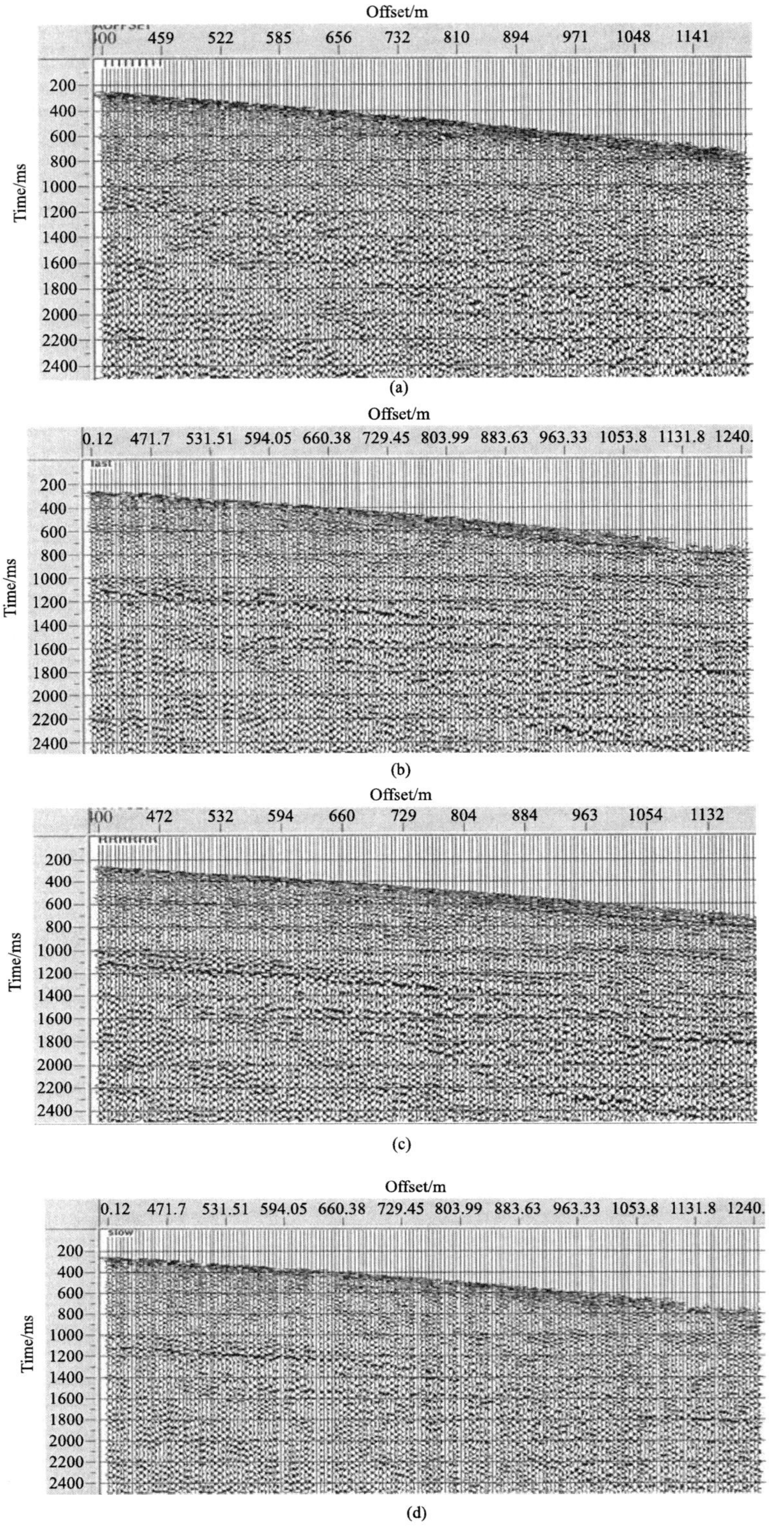

Fig. 8.31 Transverse and radial components and of fast and slow wave components of an original shot gather. **a** original transverse component; **b** original radial component; **c** fast wave component; **d** slow wave component

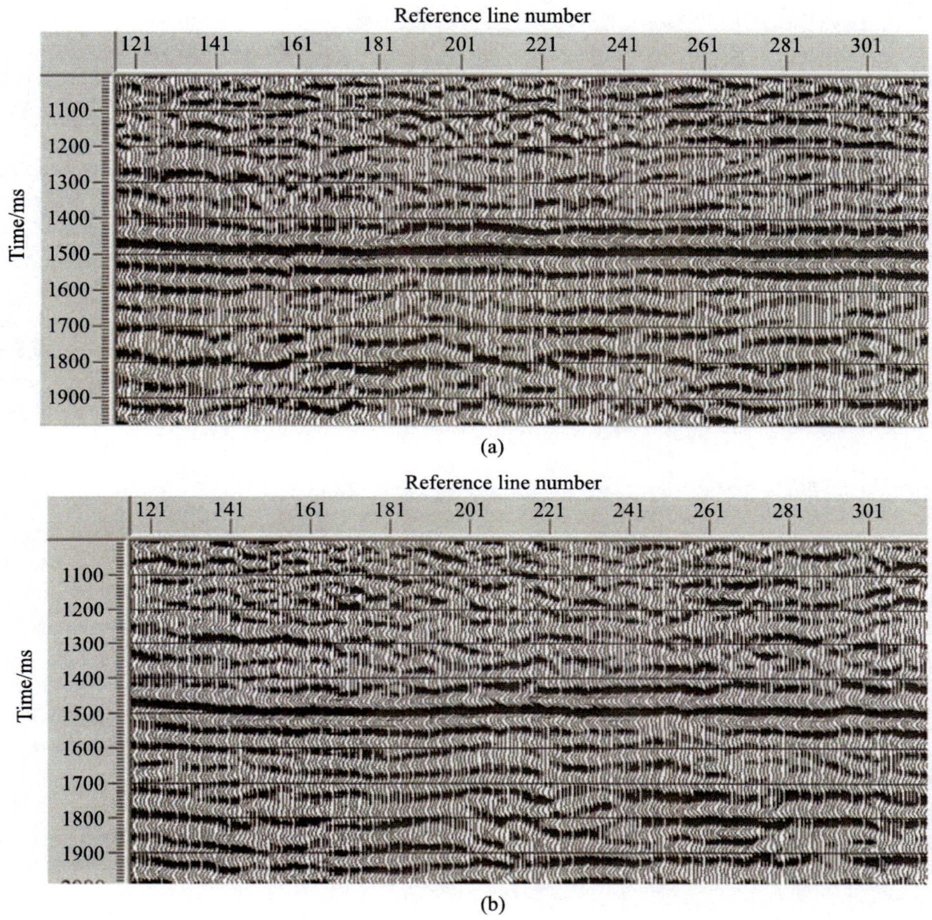

Fig. 8.32 Stacked sections of fast and slow waves. **a** stacked section of fast wave; **b** stacked section of slow wave

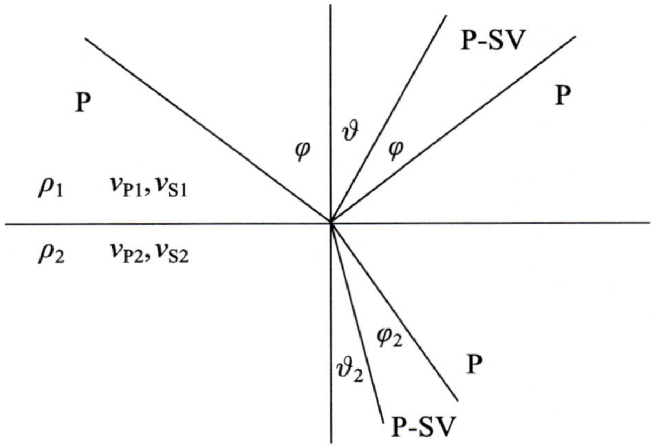

Fig. 8.33 Reflection and transmission of incident P-wave

where, A_P, A_{PR}, A_{PT}, A_{SR} and A_{ST} are the amplitude values of incident P-wave, reflected P-wave, transmitted P-wave, reflected S-wave and transmitted S-wave respectively. $Z_i = {}_iv_{Pi}$, $W_i = \rho_i v_{Si}$. Equation (8.25) can be used to derive the exact relationship between the amplitude coefficient and lithologic parameters with the change of incident angle. Obviously, this relationship is very complex, and it is not convenient for practical application. It needs to be simplified or approximated. At present, the popular approximation formulas have the following main forms.

$$R_P = A_0^R + A_2^R \sin^2\varphi + A_4^R \sum_{n=2}^{\infty} \sin^{2n}\varphi \quad (8.26)$$

$$R_P = A_0^R + A_2^R \sin^2\varphi + A_4^R(\tan^2\varphi - \sin^2\varphi) \quad (8.27)$$

$$R_P = A_0^R + A_2^R \sin^2\varphi + A_4^R \sin^2\varphi\left[\frac{1}{\cos^2\varphi} - 1\right] \quad (8.28)$$

$$R_P = (A_0^R - A_4^R) + (A_2^R - A_4^R)\sin^2\varphi + \frac{A_4^R}{\cos^2\varphi} \quad (8.29)$$

$$R_P = (A_0^R + A_4^R \tan^2\varphi) + (A_2^R - A_4^R)\sin^2\varphi \quad (8.30)$$

where, $A_0^R = \frac{1}{2}\left[\frac{\Delta v_P}{v_P} + \frac{\Delta\rho}{\rho}\right]$;

$$A_2^{\mathrm{R}} = A_0^{\mathrm{R}}\left[Q - \frac{2(1+Q)(1-2\sigma)}{1-\sigma}\right] + \frac{\Delta\sigma}{(1-\sigma)^2};$$

$$Q = \frac{\frac{\Delta v_{\mathrm{P}}}{v_{\mathrm{P}}}}{\frac{\Delta v_{\mathrm{P}}}{v_{\mathrm{P}}} + \frac{\Delta\rho}{\rho}};\sigma = \frac{\left(1-2\gamma^2\right)}{2\left(1-\gamma^2\right)};\gamma = \frac{v_{\mathrm{S}}}{v_{\mathrm{P}}};\ A_4^{\mathrm{R}} = \frac{\Delta v_{\mathrm{P}}}{2v_{\mathrm{P}}}.$$

It can be proved that the above formulas are equivalent. Equation (8.29) is Shuey's approximate formula. In this formula, the first term is regarded as a small angle, the second term as a medium angle, and the third term as a large angle. In practice, the large angle term is usually ignored and the following formula is adopted:

$$R_{\mathrm{P}} = A_0^{\mathrm{R}} + A_2^{\mathrm{R}} \sin^2 \varphi \tag{8.31}$$

Shuey's approximation formula is the most widely used in AVO processing.

8.3.2 The Basis of Using AVO to Detect CBM

According to the mechanism of coalbed methane enrichment, using AVO to detect coalbed methane is to use amplitude to search for intensive joint and fissure zones in coal seams (i.e. structural coal zones). The main basis includes the following four aspects:

① Under the same conditions, the larger the pore surface area of coal, the greater the possibility of adsorbed CBM enrichment. It is generally believed that the dense zone of joints and fractures is the location for adsorbed CBM enrichment.

② The decreasing rate of S-wave velocity in coal seams with well-developed joints and fractures is greater than that of P-wave velocity. Therefore, the Poisson's ratio of coal seams with well-developed joints and fractures increases. Laboratory measurements show that there is a positive correlation between fracture density and Poisson's ratio in coal samples, as shown in Fig. 8.34.

③ As the Poisson's ratio of coal seams with well-developed joints and fractures increase, the difference of the Poisson's ratio between coal seams and roof and floor changes. This change of the Poisson's ratio may cause observable amplitude variation with offset. In other words, according to the amplitude versus offset of the reflection events on CDP gather, the development of joints and fractures in coal seams can be predicted, and then the CBM enrichment in coal seams can be predicted according to the development of joints and fractures.

④ Generally speaking, fractures are directional, so it is possible that the velocity of the S-wave being transverse to fractures may be lower than that being parallel to fractures, i.e., the Poisson's ratio being vertical to fractures is higher than that being parallel to fractures. The difference of the Poisson's ratio at different azimuths may cause the variation of amplitude with offset at different azimuths.

8.3.3 AVO Responses of Controlling Geological Parameters on CBM Enrichment

8.3.3.1 AVO Responses of Coal Structure, and Roof and Floor Lithology

Using AVO forward modeling to study the AVO response law of coal structure and the lithology of roof and floor, and applying Zoeppritz equation to calculate the P-wave

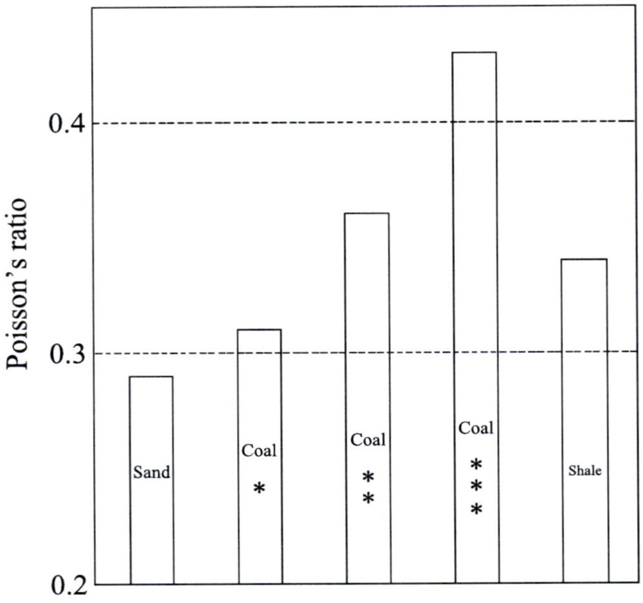

Fig. 8.34 Fracture development on the Poisson's ratio of coal samples. * rare fractures; ** developed fractures; *** more developed fractures

Table 8.3 Parameters in the AVO forward model

Strata		P-wave velocity /(m/s)	S-wave velocity /(m/s)	Density / (g/cm^3)	Poisson's ratio
Coal seam	Primary coal I	1960	1090.00	1.390	0.276
	Primary coal II	2400	1259.40	1.500	0.310
	Tectonic coal I	1500	681.39	1.350	0.370
	Tectonic coal II (soft layering)	650	195.98	1.250	0.450
Mudstone		3170	1585.00	2.360	0.333
Sandstone		3601	2172.00	2.562	0.214

reflection coefficients of a layer of a three-layer model (sandstone-primary coal-sandstone, or mudstone-primary coal-mudstone, or sandstone- tectonic coal-sandstone, mudstone-tectonic coal- mudstone) with the variation of incidence angle. The parameters in the AVO forward model are listed in Table 8.3.

Figure 8.35 is the AVO response curve of the coal seam roof, which is featured by:

(1) Roof lithology has a great influence on AVO characteristics. The curves in the Figure are obviously divided into two groups according to roof lithology (sandstone and shale).
(2) When the incident angle is small (<18°), the change of the amplitude with the incident angle is not obvious, and the most significant range of the incident angle is 15°–40°.
(3) The degree of coal fragmentation influences AVO characteristics. However, because the influencing factors are not single, the factors need to be separated in the inversion.
(4) The AVO characteristics of the soft layer (i.e. very fractured structural coal) are prominent in both sandstone roof and shale roof. This is beneficial to the prediction of gas enrichment using AVO technology, and is a positive result of the forward modeling.

Figure 8.36 shows the variation of reflection coefficient of coal seam floor with incident angle. Due to the interface from a low-velocity medium to a high-velocity medium, full reflection is very prominent. Especially, for soft layered coal, full reflection occurs when the incident angle is less than 15°; when the incident angle is between 0° and 15°, although the absolute value of the reflection coefficient is different, the

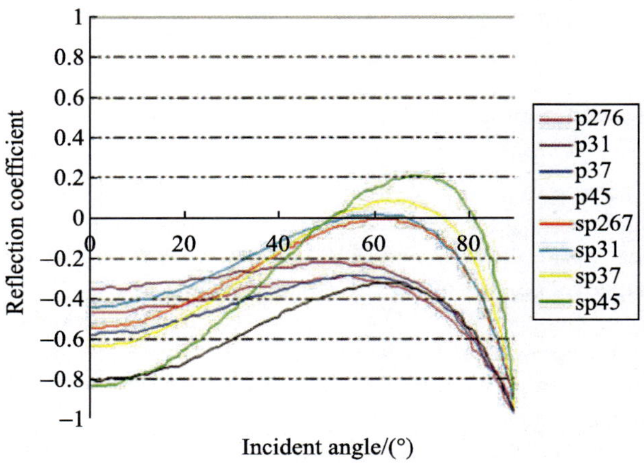

Fig. 8.35 Change of reflection coefficient of coal seam roof with incident angle, sp. sandstone roof; p. shale roof; Poisson's ratio expressed by number, Poisson's ratio changing from 0.267 to 0.45 indicates that coal structure changes from primary coal to soft layers (rank IV tectonic coal, i.e., mylonitic coal): sp45. sandstone-soft layer interface (roof), 45 indicates that Poisson's ratio of soft layered coal is equal to 0.45; p276. shale-primary coal interface (roof), 276 indicates that Poisson's ratio of primary coal is 0.276

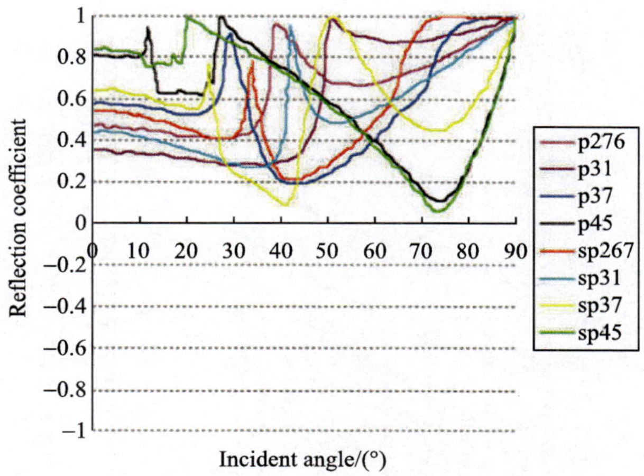

Fig. 8.36 Reflection coefficient of coal seam floor vs. incident angle, sp. sandstone roof; p. shale roof; number indicates Poisson's ratio, Poisson's ratio changing from 0.267 to 0.45 indicates that coal structure changes from primary coal to soft layers (Rank IV tectonic coal, i.e., mylonitic coal): sp45. sandstone-soft layer interface (roof), 45 indicates that Poisson's ratio of soft layered coal is equal to 0.45; p276. shale-primary coal interface (roof), 276 indicates that Poisson's ratio of primary coal is 0.276

gradient of the reflection coefficient varying with the incident angle is not significantly different. Therefore, for AVO inversion to coal seams, the reflection signal of coal seam floor should not be used. This is another important understanding of the forward modeling in this study. Although this phenomenon has been predicted beforehand according to the general law of elastic wave reflection, the result of forward modeling is so distinct that we must use the reflection information of roof for AVO inversion of coal seams, which is very instructive.

According to the crossplots of AVO gradient (*G*) and intercept (*P*) of coal seams with different structures (Figs. 8.37, 8.38 and 8.39), the absolute values of AVO gradient and intercept of structural coals are larger than that of primary coal when the roof lithology is the same, and the difference is obvious. The AVO response of coal seam top also changes greatly when the coal structure is the same and the roof lithology is different. Therefore, for coal seams with large thickness, according to AVO response, it can not only distinguish different structures of coal, but also reflect the change of roof lithology.

8.3.3.2 AVO Response of Coal Seam Thickness

In theory, the AVO response of a layer in the model can truly reflect the elastic parameters of a medium. However, a coal seam is usually thinner (<10 m). On an actual seismic section, the top and bottom of a coal seam are not corresponding peak and trough (or trough and peak), but complex waves (or from adjacent strata). The amplitude response of this complex wave not only reflects the coal structure and the lithology of roof and floor, but also is affected by the thickness of the coal seam. Therefore, the influence of target coal seam thickness on AVO characteristics is also an important content of AVO forward modeling.

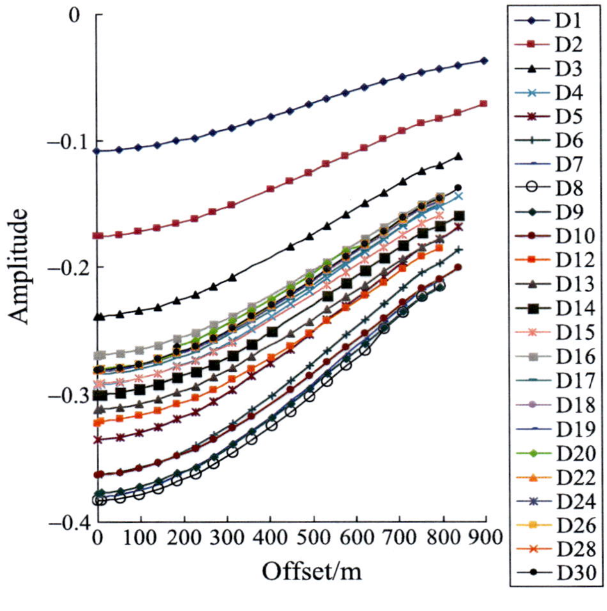

Fig. 8.37 Relationship between amplitude and offset of roof

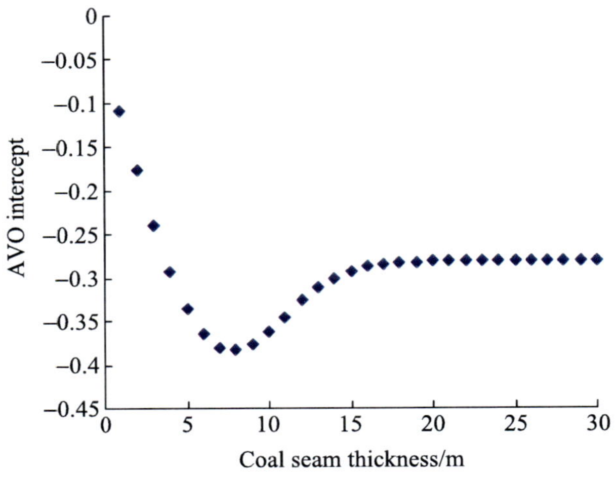

Fig. 8.38 Scatter of coal seam thickness and AVO intercept

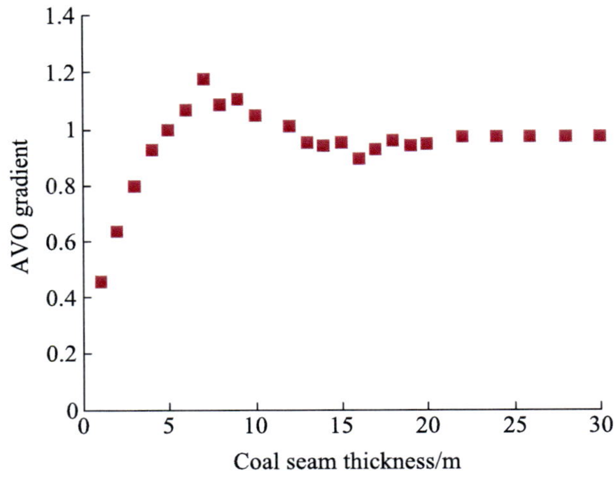

Fig. 8.39 Scatter of coal seam thickness and AVO gradient

In the study of AVO response of coal seam thickness, 58 Hz zero-phase Rick wavelet was selected and Zoeppritz equation algorithm was used to simulate the theoretical model of three-layer media (Fig. 8.40).

The relationship between amplitude and offset for coal seams with different thicknesses (Fig. 8.37) shows that the reflection amplitude (absolute value) of the roof decreases with the increase of offset, and the thickness of the coal seam has obvious influence on the reflection amplitude (AVO intercept) at zero offset: when the thickness of the coal seam is less than 9 m (1/4 wavelength), the AVO intercept increases with the increase of the thickness; when the

Mudstone roof (thickness>50 m) v_P=3200 m/s, v_S=1700 m/s, ρ=2.45 g/cm^3
Coal seam (thickness=1–50 m) v_P=2200 m/s, v_S=1050 m/s, ρ=1.45 g/cm^3
Mudstone floor (thickness>50 m) v_P=3200 m/s, v_S=1700 m/s, ρ=2.45 g/cm^3

Fig. 8.40 Theoretical model and parameters of three-layer media

thickness of the coal seam is 9–15 m, the AVO intercept decreases with the increase of the thickness; when the thickness of the coal seam >15 m, the AVO intercept becomes stable and does not change with the change of the thickness. The amplitude gradient (AVO gradient) also shows similar variation with thickness. From the crossplots of coal seam thickness and AVO attributes (Figs. 8.38 and 8.39), the above rules are clear. Therefore, in AVO analysis, the range and law of the thickness variation of the target layer should be determined based on other data, such as boreholes, in order to fully consider the influence of coal seam thickness on AVO inversion.

8.3.4 AVO Responses of CBM Enrichment Zones

We selected three 2D seismic lines (G2, G5 and L6B) near gas outburst points as targets (gas outburst occurred after seismic exploration). The coordinates of the gas outburst point in Panji No.3 Coalmine on March 25, 1994 is at (77,841, 33,535), the nearest line is G2, and the CDP number of the nearest point is 174. The coordinates of the gas outburst accident point on May 14, 1995 are (77,916, 32,926), the nearest line is L6B, and the CDP number of the nearest point is 306. The coordinates of the gas outburst point on November 25, 2002 is at (79,941, 32,345), the nearest line is G5, and the CDP number of the nearest point is 431 (Fig. 8.41). The distances between the locations of the gas outburst and the nearest CDP points corresponding to the two-dimensional lines are shown in Table 8.4.

According to the locations of coal seam 13-1 and gas outburst, offset-reflection amplitude curves on some CDPs on G2, G5 and L6B lines were extracted. Figure 8.42 is the AVO response near the gas outburst point on Line G2. Figure 8.43 is the AVO response near the non-gas outburst point on Line G2. Figure 8.44 is the AVO response near the gas outburst point on Line G5. Figure 8.45 is the AVO response near the non-gas outburst point on Line G5. Figure 8.46 is the AVO response near the gas outburst point on Line L6B. Figure 8.47 is the AVO response near the non-gas outburst point on Line L6B.

The analysis results of the above curves show that the AVO responses of the gas outburst points near CDP 174 on Line G2, CDP 431 on Line G5 and CDP 306 on Line L6B are similar. The AVO attributes of the six points in Fig. 8.42 can be divided into two categories. One includes CDP 171, CDP 174 and CDP 175, and the other are CDP 178, CDP 179 and CDP 185. Their corresponding three curves are similar in shape and can be regarded as the same response. Points CDP 171, CDP 174 and CDP 175 are located near the gas outburst points, with higher CBM content. The curve shape can be regarded as the AVO response model of CBM.

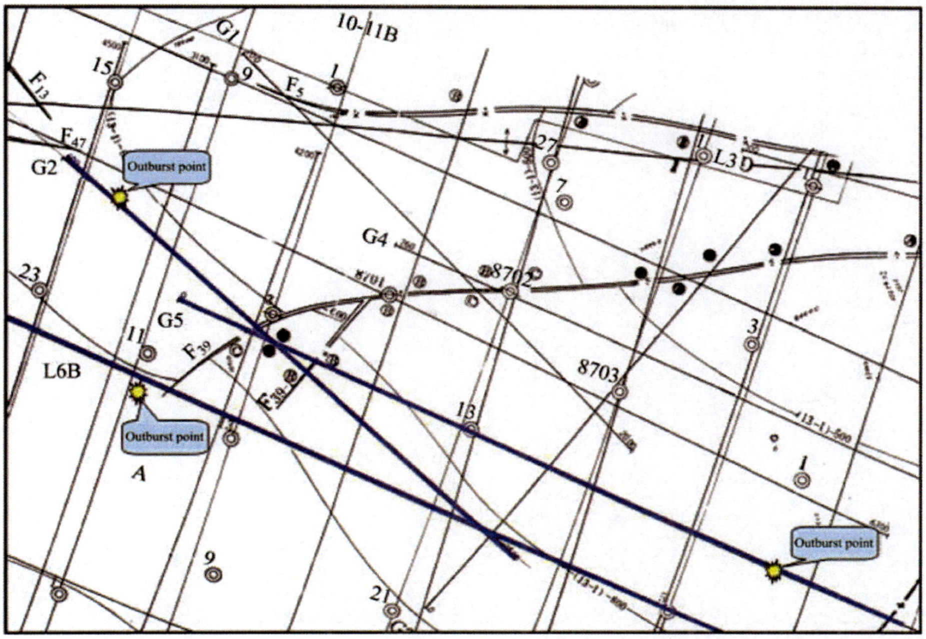

Fig. 8.41 Locations of 2D Line G2, Line G5 and Line L6B

Table 8.4 Data of gas outburst points

Type	Coordinates (*x*, *y*)	Date	Outburst volume	Nearest line number to outburst point/CDP/Coordinates/ Distance to outburst point/m
Run off	(77,841,33,535)	1994–03-25	37.4	G2/174/(77,842, 33,527.5)/7.57
Forced out	(77,916,32,926)	1995–05-14	18.0	L6B/306/(77,926, 32,944)/20.59
Forced out	(79,941,32,345)	2002–11-25	10.0	G5/431/(79,945, 32,344)/4.12

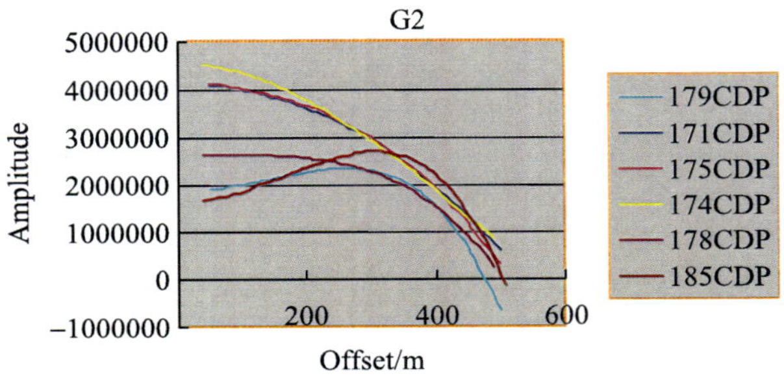

Fig. 8.42 AVO response of the gas outburst point on Line G2

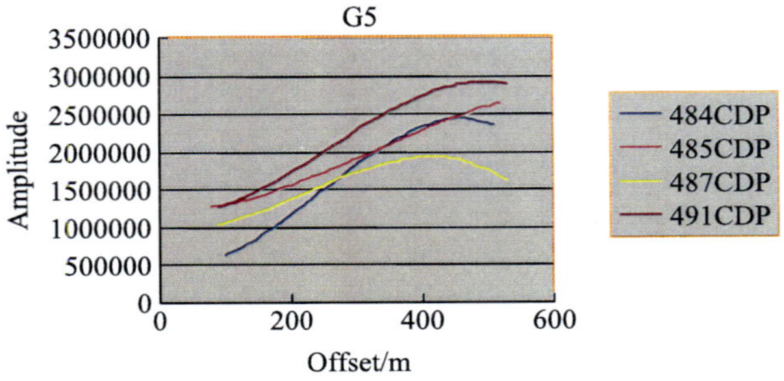

Fig. 8.45 AVO response of the non-CBM enriching zone near Line G5

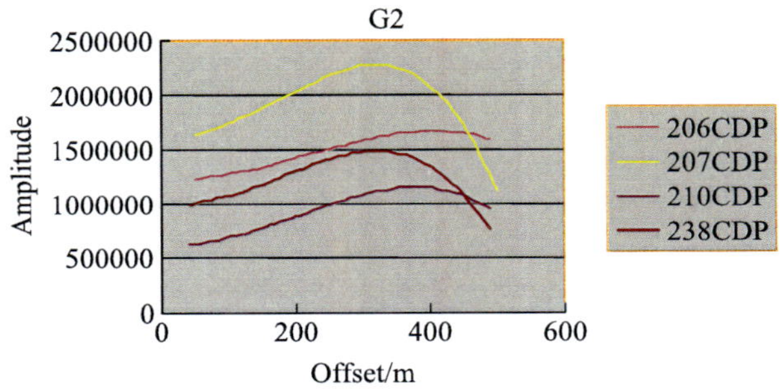

Fig. 8.43 AVO response of the non-CBM enrichment zone near Line G2

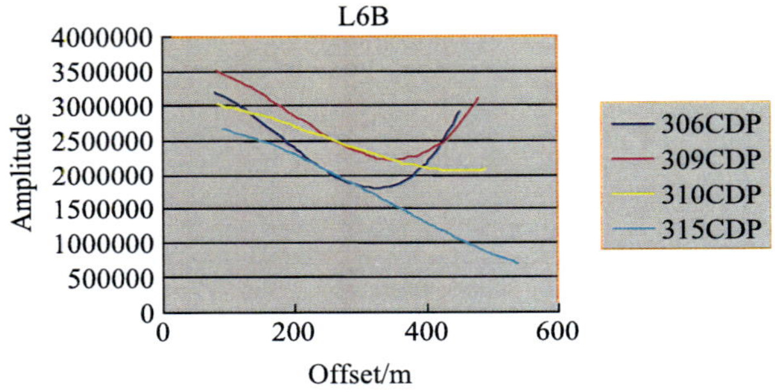

Fig. 8.46 AVO response of the gas outburst point on Line L6B

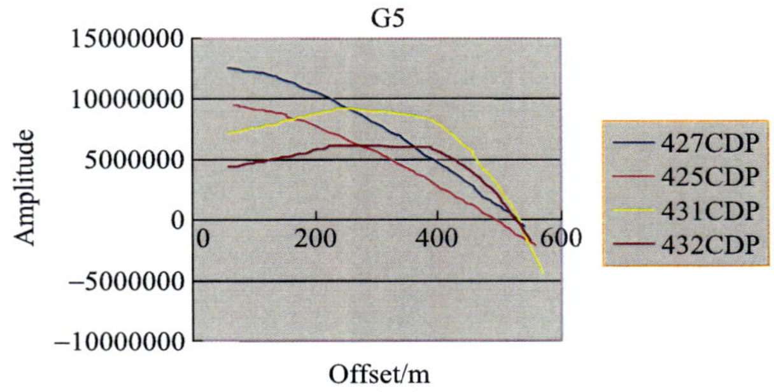

Fig. 8.44 AVO response of the gas outburst point on Line G5

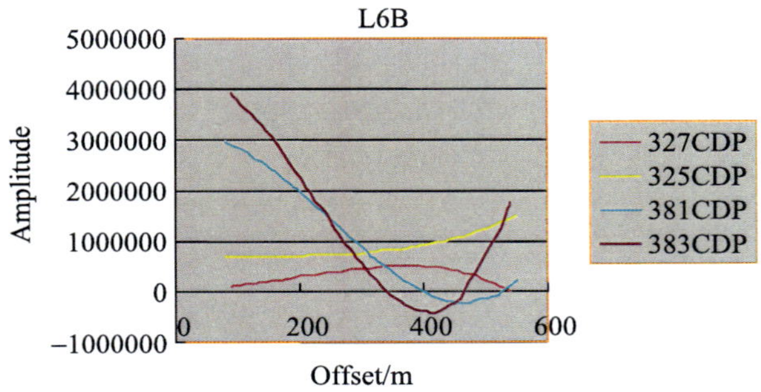

Fig. 8.47 AVO response of the non-CBM enrichment zone near Line L6B

However, points CDP 178, CDP 179 and CDP 185 are relatively far away, with lower CBM content. Compared with the AVO responses of four CDP points in the non-CBM enrichment zone near Line G2 in Fig. 8.43, it is in the middle of the two response modes, and corresponds to the CBM content. In Fig. 8.44, the four curves corresponding to the four CDP points on Line G5 can also be divided into two categories, one are CDP 425 and CDP 427, the other are CDP 431 and CDP 432. The gas contents at CDP 425 and CDP 427 are higher than those at CDP 431 and CDP 432. The CDPs on Line L6B are far away from the outburst points (the nearest is CDP 306, more than 20 m far). Its reflecting mode is difficult to distinguish.

On the other hand, in the relative density maps of Lines G2 and G5, it is easy to find that the density of the coal seam near the gas outburst is relatively low, and fractures or faults are well developed. It shows that there are better CBM reservoirs near the gas outburst points. During mining, a large amount of desorbed gas will accumulate in these reservoirs due to stress release.

Based on the above analysis of AVO seismic attributes of G2 and GS lines, we can get the following understandings: (i) CBM content near gas outburst points is slightly higher than that near non-outburst points; (ii) There are better CBM reservoirs near gas outburst points.

8.3.5 Three-Parameter AVO Method for CBM Zones

At present, the conventional AVO analysis method extracts amplitude from seismic data, and connects the AVO with rock physical properties through intercept and slope attributes. As the intercept is a function of v_P and ρ, and the slope is the function of v_S and v_P, any two of the three main elastic parameters (v_S, v_P, and ρ) are tied together and cannot be separated. In many cases, the inversion of two parameters will fail, mainly in the following two common cases: (i) drastic lithological changes; (ii) low gas saturation. The change of amplitude depends on the comprehensive influence of rock physical properties. In some cases, the amplitude of sandstone varies greatly after it contains gas; in other cases, the change of amplitude may be caused by lithology change, and fluid change may be very small. A small amount of gas in a brine-filled reservoir often results in a significant decrease in v_P, and a small change in density and v_S. When the reservoir is fully filled with gas, the density decreases and the S-wave velocity increases compared with that filled with brine. In two-parameter AVO, v_P is always related to other parameters. Therefore, on one hand, in the case of containing gas partly, the reflected wave becomes stronger; on the other hand, in the case of fully containing gas, the amplitude section also shows a bright spot. At the same time, the amplitude increases with the increase of offset in both cases. Therefore, conventional amplitude analysis method is difficult to identify the gas content.

Kelly et al. (2001) utilized:

$$\mathrm{AMP}(\vartheta) = A + B\sin^2\vartheta + C\sin^2(\vartheta)\tan^2\vartheta \tag{8.32}$$

to obtain:

$$A = \frac{\Delta v_P}{2v_P} + \frac{\Delta\rho}{2\rho} \tag{8.33}$$

$$B = \frac{\Delta v_P}{2v_P} - 2\left[\frac{v_S}{v_P}\right]^2\left[\frac{2\Delta v_S}{v_S} + \frac{\Delta\rho}{\rho}\right] \tag{8.34}$$

$$C = \frac{\Delta v_P}{2v_P} \tag{8.35}$$

AVO inversion results are actually $\Delta\rho/\rho$, $\Delta v_P/v_P$, and $\Delta v_S/v_S$. Obviously, in some special cases, these three properties cannot directly reflect the characteristics of three elastic parameters (density, bulk modulus and shear modulus). Therefore, we proposed a new AVO inversion method with three parameters. This method directly inverts three elastic parameters (density, bulk modulus and shear modulus) and uses these three parameters to discuss lithological problems.

The Eq. (8.32) can be rewritten by trigonometric function relation:

$$R_P = \frac{\Delta\rho}{2\rho} - 2\left[\frac{v_S}{v_P}\right]^2\left[\frac{\Delta\rho}{\rho} + \frac{2\Delta v_S}{v_S}\right]\sin^2\phi + \frac{\Delta v_P}{2v_P}\sec^2\phi \tag{8.36}$$

Let $A = \frac{\Delta\rho}{2\rho}; B = -2\left[\frac{v_S}{v_P}\right]^2\left[\frac{\Delta\rho}{\rho} + \frac{2\Delta v_S}{v_S}\right]; C = \frac{\Delta v_P}{2v_P}$. For n points (t_1, t_2, ..., t_i, ..., t_n, $i = 1, \ldots, n$) on a seismic trace in time domain. Δt is the sampling interval, then:

$$2A(t_i) = \frac{\Delta\rho(t_i)}{\rho(t_i)} \tag{8.37}$$

$$2A(t_i) = \Delta\ln\rho(t_i) \tag{8.38}$$

To summarize the two ends of the above equations:

$$2\sum_{i=1}^{n} A(t_i) = \sum_{i=1}^{n}\Delta\ln\rho(t_i) \tag{8.39}$$

$$\frac{\rho(t_n)}{\rho(t_1)} = \exp[\ln\rho(t_n) - \ln\rho(t_1)] = \exp\int_{t_1}^{t_n} d\ln\rho(t) \approx \exp\sum_{i}^{n}\Delta\ln\rho(t_i) \tag{8.40}$$

To substitute Eq. (8.39) into Eq. (8.40), then:

$$\frac{\rho(t_n)}{\rho(t_1)} = \exp 2\sum_i^n A(t_i) \tag{8.41}$$

$$\rho(t_n) = \rho(t_1)\exp 2\sum_i^n A(t_i) \tag{8.42}$$

where $\rho(t_n)$ is medium density.

Similarly,

$$2\sum_{i=1}^n C(t_i) = \sum_{i=1}^n \Delta \ln v_P(t_i) \tag{8.43}$$

$$\begin{aligned}\frac{v_P(t_n)}{v_P(t_1)} &= \exp[\ln v_P(t_n) - \ln v_P(t_1)] \\ &= \exp\int_{t_1}^{t_n} d\ln v_P(t) \approx \exp\sum_i^n \Delta \ln v_P(t_i)\end{aligned} \tag{8.44}$$

To substitute Eq. (8.43) into Eq. (8.44), then:

$$\frac{v_P(t_n)}{v_P(t_1)} = \exp 2\sum_i^n C(t_i) \tag{8.45}$$

$$v_P(t_n) = v_P(t_1)\exp 2\sum_i^n C(t_i) \tag{8.46}$$

$$\begin{aligned}K(t_n) &= [v_P(t_n)]^2\rho(t_n) \\ &= [v_P(t_1)]^2\rho(t_1)\exp\left[4\sum_i^n C(t_i) + \exp 2\sum_i^n A(t_i)\right]\end{aligned} \tag{8.47}$$

where $K(t_n)$ is bulk modulus.

From Eq. (8.45), we can obtain:

$$B(t_i) = -2\left[\frac{v_S(t_i)}{v_P(t_i)}\right]^2\left[\frac{\Delta\rho(t_i)}{\rho(t_i)} + 2\frac{\Delta v_S(t_i)}{v_S(t_i)}\right] \tag{8.48}$$

$$-\frac{1}{2}B(t_i)v_P^2(t_i)\rho(t_i) = v_S^2(t_i)\rho(t_i)\left[\frac{\Delta\rho(t_i)}{\rho(t_i)} + 2\frac{\Delta v_S(t_i)}{v_S(t_i)}\right] \tag{8.49}$$

$$-\frac{1}{2}B(t_i)v_P^2(t_i)\rho(t_i) = \Delta v_S^2(t_i)\rho(t_i) \tag{8.50}$$

$$-\frac{1}{2}\sum_{i=1}^n B(t_i)v_P^2(t_i)\rho(t_i) = \sum_{i=1}^n \Delta v_S^2(t_i)\rho(t_i) \tag{8.51}$$

$$\begin{aligned}v_S^2(t_n)\rho(t_n) - v_S^2(t_1)\rho(t_1) &= \int_{t_n}^{t_1} d\left[v_S^2(t)\rho(t)\right] \\ &\approx \sum_{i=1}^n \Delta v_S^2(t_i)\rho(t_i)\end{aligned} \tag{8.52}$$

To substitute Eq. (8.46) into Eq. (8.47), then:

$$v_S^2(t_n)\rho(t_n) - v_S^2(t_1)\rho(t_1) = -\frac{1}{2}\sum_{i=1}^n B(t_i)v_P^2(t_i)\rho(t_i) \tag{8.53}$$

$$\mu(t_n) = v_S^2(t_n)\rho(t_n) = v_S^2(t_1)\rho(t_1) - \frac{1}{2}\sum_{i=1}^n B(t_i)v_p^2(t_i)\rho(t_i) \tag{8.54}$$

where $\mu(t_n)$ is shear modulus.

Based on the above understanding of AVO phenomenon, we have developed a new three-parameter AVO software. The new software gives seven attributes (Table 8.5).

Table 8.5 AVO attributes

AVO attribute	Influencing factors and significance
Density	Influencing factors: lithology, mineral composition, pore, fracture and compaction degree, fluid content
Shear modulus	Used to mark lithological changes, distinguish porous from fractured reservoirs, and distinguish solid from fluid
P-wave velocity	Influencing factors: lithology, mineral composition, pore, fracture, compaction degree, content, density
Velocity ratio of S-wave/P-wave	Influencing factors: lithology, fluid
Poisson's ratio	Influencing factors: lithology, fluid
Bulk modulus	Influencing factors: lithology, mineral composition, pore, fracture, compaction degree, fluid content
Density × shear modulus × bulk modulus	Density, shear modulus, bulk modulus

The advantage of the new three-parameter AVO analysis method is that it can directly reflect the seismic response of a medium, while the P and G parameters obtained by the two-parameter AVO analysis method are functions of density, bulk modulus and shear modulus, and cannot directly reflect the seismic response of a medium. The three-parameter AVO analysis method by Kelly and Skidmore et al. (2001) is influenced by stratigraphic sequence. The three-parameter AVO analysis method proposed in this book can revert three elastic parameters representing geological bodies, so it can more accurately reflect the geological differences of geological bodies (in the sense of seismic response).

8.3.6 Three-Parameter AVO Prediction of CBM Enrichment Zones

According to the AVO response model obtained by 2D seismic lines, we conducted fidelity processing, calculated three-parameter AVO attributes to the seismic data recorded

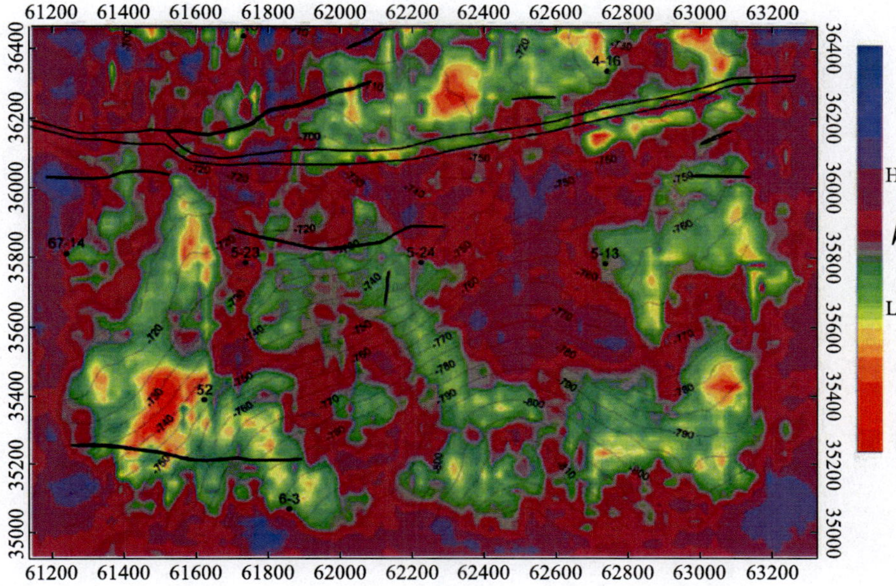

Fig. 8.48 Relative density distribution of No. 13-1 coal beam in Guqiao Coalmine

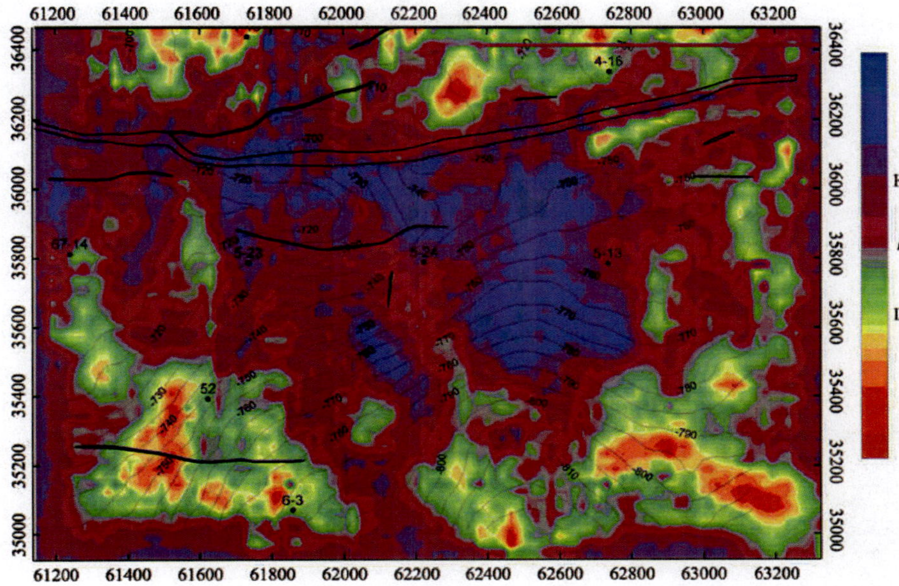

Fig. 8.49 Shear modulus distribution of No. 13-1 coal beam in Guqiao Coalmine

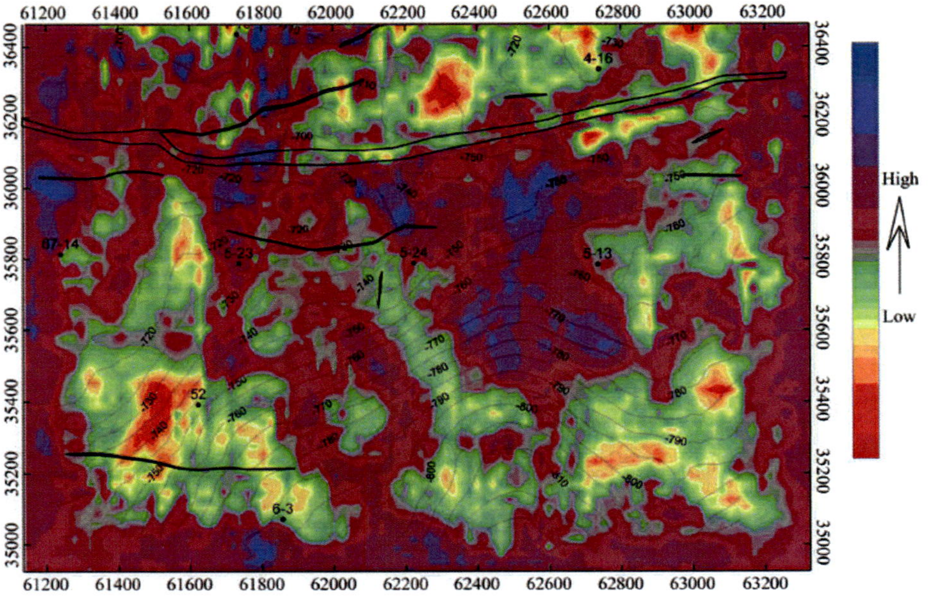

Fig. 8.50 Bulk modulus distribution of No. 13-1 coal beam in Guqiao Coalmine

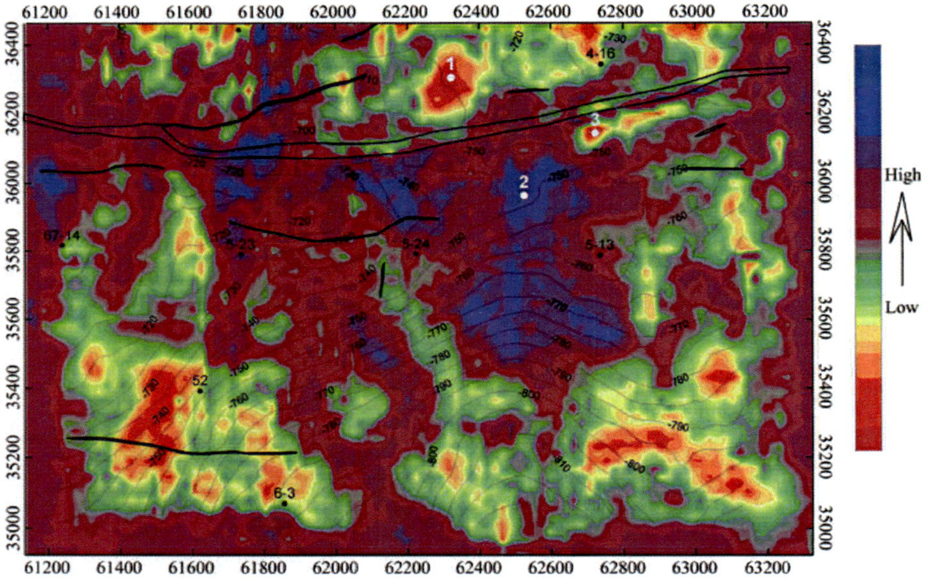

Fig. 8.51 Prediction of CBM enrichment zones of No. 13-1 coal beam in Guqiao Coalmine

in the Guqiao Coalmine in Huainan Coalfield in Anhui Province, and obtained the relative density (Fig. 8.48), shear modulus (Fig. 8.49) and bulk modulus (Fig. 8.50) of No. 13-1 coal seam. As the density, shear modulus and bulk modulus of a CBM enrichment zone are relatively low, the product of density, shear modulus and bulk modulus are taken as the basis for predicting CBM enrichment zones (Fig. 8.51). The zones showing low values are CBM enrichment zones.

References

Aki K, Richards PG (1980) Quantitative seismology, theory and methods, Vol 1. W H Freeman and Co Vambridgy Press, pp 144–154

Kelly M, Skidmore C, Ford D (2001) AVO inversion, Part 1: isolating rock property contrasts. Lead Edge 20(3):320–323

Knott CG (1899) Reflection and refraction of elastic waves with seismological applications. Philos Mag Lett 48:64–97

Skidmore C, Kelly M, Cotton R (2001) AVO inversion, Part 2: isolating rock property contrasts. Lead Edge 20(4):425–428

Zoeppritz K (1919) Über reflexion and durchang seismischer wellen durch unstetigkeits-flächen. Gott Nachr Math Phys K1:66–84

9 Comprehensive Evaluation of Geological Conditions for Coalbed Methane Development

9.1 Favorable Zones for Coalbed Methane Development in China

9.1.1 Overview of Coalbed Methane Resources in China

Coalbed methane (CBM) is a mineral resource associated with coal, and the two are closely related. China's coal resources exploration, prediction and evaluation are very thorough, and a lot of data have been accumulated. The prediction of CBM resources in China is based on a large number of coalfield geological data and reliable prediction of coal resources. According to the needs and conditions, CBM resource prediction is carried out at different levels.

Since the 1980s, in order to find out China's CBM resources at the national level, many units and individuals in China have made many predictions of China's CBM resources in different periods, and obtained corresponding results, as shown in Table 9.1.

At the regional level, in order to meet the needs of the development of CBM exploration and development, some enterprises have carried out CBM resource evaluation in some important coal basins and regions, and the amount of CBM resources have been predicted, as shown in Table 9.2. In addition, in the mid-1990s, in the project of "CBM Resources Development in China" funded by the United Nations Development Programme (UNDP), 17 coal mining areas in China, such as Songzao and Huainan, were evaluated for CBM resources and development prospects.

In order to deepen the understanding of CBM development prospects in coal mine areas and improve control on gas disaster, the development potential of CBM resources in 17 key coalmine areas in China was evaluated in the project of "China's CBM resources development" funded by the United Nations Development Programme (UNDP).

The CBM resources in China are basically characterized by huge amount, various types, wide distribution but obvious differences, complex and diversified geological conditions. All the above evaluation works reveal the basic characteristics, and lays a foundation for the exploration and development of CBM resources.

Although China has made many prediction of CBM resources, there is an obvious gap between the predictions of CBM resources and conventional natural gas. And there are many shortcomings manifested in the following aspects: only predicting the in-situ volume without the recoverable, limited scope, single method and lack of standardization, etc. These cases can not meet the needs of the development of CBM industry.

Firstly, as a mineral resource, the exploitability of CBM is an important attribute. Although it is known that the amount of CBM resources in China is huge, the problem concerned by relevant government departments and industry is how many of the CBM resources can be exploited in China. It is impossible to answer this sharp question without quantitative evaluation of the exploitability of CBM resources.

Secondly, in terms of the CBM resources prediction, only bituminous coal and anthracite coalfields shallower than 2000 m were evaluated without lignite and its distribution areas, which is the conventional practice. The new round of evaluation organized by the Ministry of Land and Resources and the National Development and Reform Commission in 2006 is different. However, according to the practice of CBM resources development in the Fenhe Basin in the United States, a large amount of CBM exists in lignite, and the thickness of the lignite seam is large, permeability is good, CBM resources have great development potential.

According to the latest prediction results of coal resources, the lignite resources shallower than 2000 m in China are 3191×10^8 t, accounting for 5.7% of the total coal resources in China. The forming age of the lignite resources is mainly K_1, E, N and the resources mainly distributed in northeast and southwest China. The CBM resources in lignite in China can not be ignored. In addition, in many lignite distribution areas (such as many coal basins west of Daxing'an Mountains), the shallow part is lignite, while the deep part often turns into bituminous coal of different ranks owing

Y. Song et al., *Coalbed Methane in China*,
https://doi.org/10.1007/978-981-33-4725-0_9

Table. 9.1 Estimated results of CBM resources in China

Research unit or researcher	Resource/10^{12} m^3		Computational range
	In-situ	Technical recoverable	
Jiaozuo Institute of Mining Technology (1987)	31.92	–	All mineable coal seams countrywide
Li et al. (1990)	32.15	–	All mineable coal seams countrywide
Institute of Petroleum Geology, Ministry of Geology and Mineral Resources(1990)	10.60–25.20	–	–
Zhang et al. (1991)	30.00–35.00	–	Lignite and coal seams in Tibet, Guangdong, Fujian, Taiwan and South China excluded
China National Coal Mine Corporation (1992)	24.75	–	Recoverable CBM amount in all mining seams countrywide
Duan (1992)	36.30	–	–
Guan (1992)	25.00–50.00	–	–
Zhang et al. (1995)	32.68	–	Lignite and coal seams in Tibet, Guangdong, Fujian, Taiwan and South China excluded
Ye et al. (1998)	14.34	–	Areas with shallow CBM content less than 4 m^3/t excluded
Zhang et al. (2000)	31.46	–	Lignite and rotten coal seams in Tibet, Guangdong, Fujian, Taiwan and southern China excluded
Ministry of Land and Resources, National Development and Reform Commission (2006)	36.81	10.87	No official information
The Book	32.86	13.90	Lignite and rotten coal seams in Tibet, Guangdong, Fujian, Taiwan and southern China excluded

Table. 9.2 Prediction of CBM resources

Basin	Year	Company and contractor	Results/10^8 m^3	Goal
Qinshui Basin	1997	Contractor: North China Geological Bureau; Company: China United Coalbed Methane Co., Ltd	53,000	Favorable blocks
Ordos Basin (C–P)	2002	Contractor: Xi'an Research Institute; Company PetroChina	126,101	Preliminary preparation
Tuha Basin	2002	Contractor: Xi'an Research Institute; Company: Tu-Ha Oilfield	74,429	Analysis of prospects
Panguan and Gemudi syncline	2004	Contractor: Xi'an Research Institute; Company: PetroChina	3520	Exploration target evaluation
Hailaer Basin	2003	Lu et al.	10,792	Preliminary assessment around Daqing oilfield

to the metamorphism of deep-seated coal. In the past, when these areas were regarded as lignite distribution areas without calculating CBM resources, the CBM resources in the bituminous coal of this area were also eliminated, which is obviously unreasonable.

The volume method (also known as CBM content method) was used to predict the national CBM resources. The estimated results of 32.86×10^{12} m^3 are considered to be underestimated and conservative. According to our definition, the CBM resources are “estimated amount of methane (including heavy hydrocarbons) in underground coal reservoirs and their surrounding rocks, which is expected to be technically feasible and economically reasonable for final exploitation on the basis of their occurrence, quantity and quality”. But in fact, CBM in roof and floor and non-recoverable coal seams was not counted in the 32.86×10^{12} m^3.

In addition to roof and floor and non-recoverable coal seams, there are also a large amount of CBM below 2000 m and in sea areas, which are potential areas of CBM resources in China.

It is an indisputable fact that roof and floor have certain gas. A considerable part of CBM will be released in the process of coal mining and CBM exploitation, which has realistic resource significance.

In the Nantong Coalfield of Chongqing, the gas content of the coal seam and its roof and floor was measured by borehole core desorption method. The results are shown in Table 9.3.

In the Jiaoping Coalfield in Shaanxi Province, the 4–2 coal seam of Yan’an Formation in Early–Middle Jurassic is a gas seam with gas content of 4–6 m^3/t, and its direct roof is Xiaojie sandstone, which is a gas-bearing seam. Coal mining practice shows that about 50–70% of the gas emission comes from Xiaojie sandstone roof. According to the estimation of gas resources in Chenjiashan Coalmine, Jiaoping Coalfield, the gas resources in the 4–2 coal seam are 3.6×10^8 m^3 and Xiaojie sandstone 34.0×10^8 m^3. According to the analysis, the great destructive effect of the huge gas exploding accident in the Chenjiashan Coalmine on November 28, 2004 was related to the gas emission from the roof sandstone.

In the Yangquan Coalmine, Shanxi Province, the gas content is high in the adjacent strata above and below the coal seam. The results of underground ventilation gas monitoring and gas drainage research show that 50% of the gas emission and drainage come from the surrounding rock. In the process of coalfield geological exploration and construction, the phenomenon of gas emission from limestone and sandstone near the coal seam happened many times, and the maximum gas emission could reached 9.6 m^3/min and the concentra- tion of methane was about 95%.

Non-mining coal seam refers to the coal seam whose thickness is less than 0.7 m (or 0.5 m). Because such coal seam is not calculated in the calculation of coal resources (reserves), there is no coal seam in the calculation of CBM resources (reserves). In fact, it is an indisputable fact that there are a lot of CBM in non-mineable coal seams.

The process of calculating mine gas (CBM) emission includes all the recoverable and non-recoverable seams within the influenced range of mining rock movement, when analyzing the gas source. Due to the lack of measured data, the proportion is usually determined to be 15% of the gas emission from the coal seam.

In the process of CBM exploitation, with the continuous drainage and pressure reduction of the completed coalbed, the radius of formation pressure drop funnel is expanding, and the desorption range of CBM is expanding correspondingly. As a result, pressure reduction and desorption occur in uncompleted non-recoverable coalbed seams, contributing to the production of CBM wells. The gas production practice of the oak forest CBM development demonstration project in Black Warrior Basin, USA, provides a good example.

From the above analysis, it can be seen that the amount of CBM resources in China shallower than 2000 m should be as follows, including non-mining thin seams and surrounding rock.

Table. 9.3 Methane content in the coal seams, roof and floor in the extended exploration area of Yutianbao-Donglin Coalmine in Nantong Coalfield (adapted from Zhang et al. 1991)

Borehole	Depth/m	Formations	$CH_4/(m^3/t)$
903	858.63	Argillaceous siltstone, K_3 roof	1.49
	862.53	K_3	8.62
	863.83	Sandy mudstone, K_3 floor	0.02
	905.44	Mudstone, K_1 roof	0.63
	909.54	K_1	6.25
	909.89	Aluminous mudstone, K_1 floor	0.87
905	762.22	Argillaceous siltstone, K_3 roof	0.02
	766.39	K_3	17.38
	769.99	Mudstone, K_3 floor	0.06

$$32.86 \times (1 + 0.20) = 39.43 \quad (9.1)$$

The total is 39 × 10^{12} m^3. This quantity can objectively reflect the situation of CBM resources shallower than 2000 m (1500 m in Northeast China).

In the past, people thought that the theoretical burial depth of CBM to be exploited was 2000 m, which became the lower limit of the depth of calculating CBM resources. This artificial boundary has been broken through with further development.

In the Baiheduomu Coalfield of Piceance Basin, the burial depth of Tom Brown's CBM production wells is 5300–8400 ft (1615–2560 m). Due to the improvement of drilling, completion and hydraulic fracturing technology, the maximum output of 65 wells was 33 × 10^6 ft^3/d by October 2002, and the average output per well was 50.8 × 10^4 ft^3/d (1.4 × 10^4 m^3/d) (Fig. 9.1), half of which came from coal seams, and the other half came from tight sandstone beds.

Based on the method of geological inference and analogy, a rough calculation of 15 coal-bearing basins (areas) in China, such as Ordos, was made. Within the range from 2000 to 5000 m, the CBM resources amount to 68 × 10^{12} m^3 (Table 9.4), which is considerable.

Of course, our understanding of the potential of CBM resources is only limited to those shallow than 2000 m, which has not yet been included in the field of exploration and development.

9.1.2 Comprehensive Evaluation and Optimum Selection of Favorable Areas for CBM Development

9.1.2.1 Distribution of CBM

The area of 115 CBM enrichment zones each in China ranges from 10 to 19,070 km^2, with an average of 109.5 km^2, the resource abundance ranges from 0.06 to 877 × 10^8 m^3/km^2, with weighted average of 116 × 10^8 m^3/km^2. According to the classification criteria of resource scale of CBM enrichment areas in China (Ye et al. 1998), the above weighted averages were used as the national resource abundance of CBM, and the boundaries of 200 and 900 km^2 CBM enrichment areas were obtained. Combined with the two resource abundance boundaries of 0.5 × 10^8 and 150 × 10^8 m^3/km^2, 115 CBM enrichment areas are classified into 9 categories (Table 9.5).

The regional distribution of CBM enrichment zones in China is extremely uneven (Table 9.6). The gas enrichment zones are mainly distributed in North China and South China, accounting for 81.74% of the total. In addition, the large gas enrichment zones are almost distributed in these two areas, especially in North China. The Northeast China is dominated by large and medium gas-bearing zones, and the number of gas enrichment areas is also large, accounting for one third in the whole region. The Qinshui gas accumulation belt and Ordos gas accumulation belt are all composed of

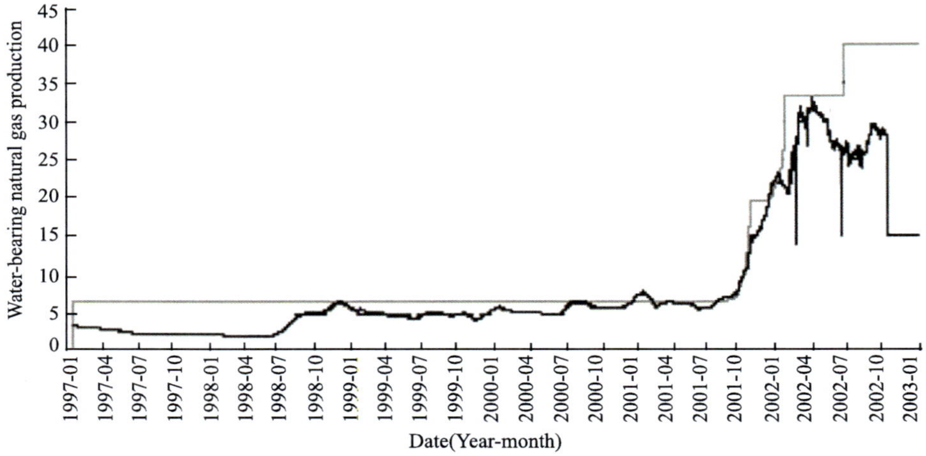

Fig. 9.1 Water-bearing natural gas content in Baiheduom Coalfield

Table. 9.4 Estimation of CBM resources in coal bearing basins with buried depth below 2000 m in China

Region	Era	Burial depth/m	Area/m^2	Coal seam thickness/m	Bulk density/ (t/m^3)	Coal resources/10^8t	Coal rank	Methane content in coal seam/ (m^3/t)	Conceptual resources of CBM/10^8 m^3	CBM resources / 10^8 m^3
Deep Songliao Basin	K_1	–	No data, not calculated	–	–	–	–	–	–	–
Hailar Basin group	K_1	≥ 1500	–	–	–	–	Long-flame coal	–	–	–
Duolun Basin group	K_1	≥ 1500	–	–	–	–	Long-flame coal	–	–	–
Ordos Basin	C–P	2000–5000	125,000.0	7.0	1.55	13,562.500	Anthracite	18.89	256,195.625	128,097.813
Ningwu Basin	C–P	2000 to inclined bottom	640.0	30.0	1.36	261.120	Coking coal	11.17	2916.710	1458.355
Coal bearing area of east Henan	C–P	2000–5000	3500.0	24.5	1.55	1329.125	Anthracite	18.89	25,107.171	12,553.586
Coal bearing area of east Taihang Mountain	C–P	2000–5000	4600.0	10.0	1.55	713.000	Anthracite	18.89	13,468.570	6734.285
Northeast area of central Hebei	C-P	2000–5000	1825.1	17.0	1.36	421.968	Fat coal, coking coal	10.79	4553.032	2276.516
Junggar Basin	J_{1-2}	2000 to inclined bottom	58,167.5	92.0	1.32	70,638.600	Gas coal, coking coal	11.91	841,305.724	420,652.862
Yili Basin	J_{1-2}	2000 to inclined bottom	3090.0	60.0	1.32	2447.280	Gas coal	6.40	15,662.592	7831.296
Tu-Ha Basin	J_{1-2}	2000 to inclined bottom	14,600.4	50.0	1.32	9636.284	Gas coal, coking coal	11.91	114,768.140	57,384.070
Youerdusi Basin	J_{1-2}	2000 to inclined bottom	2000.0	6.0	1.32	158.400	Gas coal	6.40	1013.760	506.880

(continued)

Table. 9.4 (continued)

Region	Era	Burial depth/m	Area/m^2	Coal seam thickness/m	Bulk density/ (t/m^3)	Coal resources/10^8t	Coal rank	Methane content in coal seam/ (m^3/t)	Conceptual resources of CBM/10^8 m^3	CBM resources / 10^8 m^3
Coal bearing area of Yanqi	J_{1-2}	2000 to inclined bottom	1000.0	44.0	1.32	580.800	Gas coal	6.40	3717.120	1858.560
Coal bearing area of north Tarim Basin	J_{1-2}	2000–5000	2100.0	30.0	1.32	831.600	Gas coal	6.40	5322.240	2661.120
Sichuan Basin	T_3	1000–5000	96,984.0	1.4	1.45	1968.776	Coking coal-lean coal	14.72	28,980.383	14,490.191
Deep coal bearing area of Huaying Mountain	P_2	1500–5000	3028.0	9.5	1.50	431.490	Coking coal-anthracite	18.89	8150.846	4075.423
Deep part of Hong County-Xuyong-Gulin-Xishui-Nanchuan	P_2	1500–5000	25,056.0	5.5	1.50	2067.120	Anthracite	18.89	39,047.897	19,523.948
Total	–	–	–	–	–	105,048.060	–	–	1,360,209.811	680,104.905

Note CBM resource amount = conceptual resource amount of CBM × 50%, 50% is the discount coefficient selected in consideration of such as geological structure damage

Table. 9.5 Classification of CBM enrichment zones in China

Resource abundance/(10^8 m^3/km^2)	Assess zones		
	Large zone > 900 km^2	Medium zone = 200–900 km^2	Small zone < 200 km^2
Rich gas > 1.5	12	10	11
Bearing gas = 0.5–1.5	15	15	19
Lean gas < 0.5	15	14	15

large gas-accumulation areas. The gas accumulation areas in Northeast China are mainly small to medium gas-bearing and enrichment zones, and also have large gas-bearing areas. The gas accumulation areas in Northwest China are all composed of small gas-accumulation areas, but there are relatively more gas enrichment areas. The types of gas-accumulation areas in South China are relatively balanced. There are a lot of small-medium gas-accumulation areas and poor gas-accumulation areas. The large gas-accumulation areas are located in southern Sichuan, northern Guizhou, eastern Yunnan and western Guizhou, and the large gas enrichment zones are only distributed in the eastern Yunnan- western Guizhou gas-accumulating belt.

9.1.2.2 Factors Considered in Selecting Gas Enrichment Zones

The above-mentioned methods are used to optimize the gas-accumulation areas of CBM.

Among them, the second step of the algorithm considers eight main factors, including area, resource abundance, gas content, gas saturation, coal rank, critical desorbing pressure, sand-mud ratio of caprock and development basis.

The third step considers nine main factors, including area, resource abundance, gas content, gas saturation, coal rank, critical desorbing pressure, sand-mud ratio of caprock, permeability and development basis.

The fourth step considers nine main factors, including area, resource abundance, gas content, gas saturation, coal rank, sand-mud ratio of caprock, permeability, imminent reservoir pressure and development basis.

9.1.2.3 Gas Accumulation Areas Selected

Table 9.7 lists the CBM accumulation areas shallower than 1500 m. Table 9.8 lists the CBM accumulation areas at 1500–2000 m.

9.1.2.4 Analysis, Validation, Application

The results are classified into 4 categories: A, B, C and D.

Class A less than 1500 m includes 3 areas, which are Hancheng, Yangquan-Shouyang and Fengfeng-Handan. Class B includes 5 areas, which are Huaibei, Pingdingshan, Liliu-Sanjiao, Jincheng and Kailuan. Class C includes 5 areas, which are Huainan, Wubu, Anyang-Hebi, Jiaozuo and Hongyang. Fushun is the only gas accumulation area in Class D.

Class A at 1500–2000 m includes Hancheng. Class B includes 7 areas, which are Huaibei, Huainan, Liliu-Sanjiao, Jincheng, Yangquan-Shouyang, Anyang- Hebi and Kailuan. Class C includes 3 areas, which are Wubu, Jiaozuo and Hongyang. Class D only includes Fushun.

Class A includes 2 areas, which are Hancheng, Yangquan-Shouyang; Class B includes 4 areas, which are Huaibei, Liliu-Sanjiao, Jincheng, Fengfeng-Handan. Class C includes 5 areas, which are Huainan, Pingdingshan, Wubu, Anyang-Hebi, Kailuan. Class D includes 3 areas, which are Jiaozuo, Hongyang and Fushun.

At present, the exploration and development of CBM at home and abroad generally does not exceed 1500 m depth. From the above analysis results, Classes A and B gas accumulation areas are favorable for CBM exploration at present, Classes C and D gas accumulation areas are prospective ones. The upsurge of CBM exploration and development in China is also concentrated in the first two classes, such as Hancheng, Yangquan-Shouyang, Liliu-Sanjiao and Jincheng, which shows the consistency between the results and exploration and development practice. At the same time, it also shows that the results have practical guiding significance for CBM exploration and development.

9.2 Evaluation and Optimization of CBM Enrichment Zone (Target Area) in Key Basins

9.2.1 Evaluation Methods, Parameter and Criteria

9.2.1.1 Theoretical Basis and General Idea

Overview of Analytic Hierarchy Process

Analytic Hierarchy Process (AHP) is a simple, flexible and practical multi-criteria decision-making method proposed by Professor Saaty in the early 1970s.

Table. 9.6 Regional distribution types of CBM accumulation areas in China

Type of gas-accumulation areas		Name of gas-accumulation areas			
		Northeast China	North China	Northwest China	South China
Large (31)	Gas enrichment areas (12)	–	Yangquan-Shouyang, Lu'an, Jincheng, to the north of Sanjiao, Liliu-Sanjiao, Wubu, Hancheng, Huaibei, Huainan	–	Guishan, Liupanshui, Zhina
	Gas bearing areas (15)	Jixi	Dacheng, Heshun-Zuoquan, Huodong, Xishan of Taiyuan, Huozhou, Fugu, Xiangning, Chenghe, Qingyang, Yuzhou	–	Guxu, Furong, Junlian, Northwest Guizhou
	Poor gas-accumulation areas (4)	–	Tongchuan	–	Zhenxiong, Xingyi, Guiyang
Medium (39)	Gas enriched areas (10)	Hegang, Hongyang	Kailuan, Anyang-Hebi, Jiaozuo, Yinggong, Xinmi	–	Zhongliang Mountain, Libi canyon, Songzao
	Gas bearing areas (15)	Jixian-Suibin, Shuangyashan, Boli	Fengfeng, Ningwu, Pubai, Weizhou, Yanlong, Xin'an, Dengfeng, Pingdengshan	–	Tianfu, Nanwu, Nantong, Enhong
	Poor gas-accumulation areas (14)	–	Xuanxia, Shanmian, Linru, Yongxia	–	Xuanjing, Fengcheng, Yangqiao-Yuancun, Lianshao, middle segment of Huaying Mountain, Luoguan Mountain, Qingshanling, Dong Mountain- Gufuo Mountain, Northwest Guizhou, Xuanwei
Small (45)	Gas enriched areas (11)	Fushun	Daqingshan, Shizuishan, Hulusitai	Laojun Temple in Urumqi, Baiyanghe in Urumqi, Fukang-Dahuang Mountain, Aiweiergou, Ehuobulake	Ping Xiang Panzhihua
	Gas bearing areas (19)	Tiefa, Fuxin, Hunjiang	Xinglong, Liujiang, Jiyu, Lincheng, Zhuozishan, Rujigou, Maliantan, Yiluo, Jiulishan of Xuzhou	Baoji Mountain of Jingyuan, Yaojiehanshi, Muli, Xuka	Leping, Chenlei, Xishan
	Poor gas-accumulation areas (15)	Shenbei	Lingshan, to the north of the Yellow River	Xining	Sunan, Changxing-Guangde, Quren, Guangwang, Emei Mountain, Zhongshan, North segment of Huayingshan, Hongmao, Luocheng, Heshan

Table. 9.7 The CBM accumulation areas shallower than 1500 m

Gas accumulation areas	Queue coefficient	Resource coefficient	Insurance coefficient	Class	Risk coefficient
Hancheng	1.01	0.29	0.31	A	0.61
Yangquan-Shouyang	1.10	0.37	0.35	A	0.72
Fengfeng-Handan	0.86	0.19	0.18	A	0.37
Huaibei	0.87	0.19	0.19	B	0.38
Pingdingshan	0.85	0.22	0.13	B	0.34
Liliu-Sanjiao	0.99	0.31	0.26	B	0.57
Jincheng	0.98	0.30	0.25	B	0.55
Kailuan	0.91	0.27	0.17	B	0.44
Huainan	0.83	0.21	0.09	C	0.29
Wubu	0.85	0.28	0.07	C	0.35
Anyang-Hebi	0.91	0.31	0.15	C	0.45
Jiaozuo	0.96	0.39	0.15	C	0.54
Hongyang	0.90	0.34	0.10	C	0.44
Fushun	0.96	0.48	0.10	D	0.58

Table. 9.8 The CBM accumulation areas at 1500–2000 m

Gas accumulation areas	Queue coefficient	Resource coefficient	Insurance coefficient	Class	Risk coefficient
Hancheng	1.00	0.28	0.31	A	0.59
Huaibei	0.89	0.22	0.19	B	0.42
Huainan	0.82	0.20	0.09	B	0.29
Liliu-Sanjiao	0.97	0.29	0.26	B	0.54
Jincheng	1.01	0.35	0.25	B	0.61
Yangquan-Shouyang	1.10	0.37	0.35	B	0.72
Anyang-Hebi	0.91	0.31	0.15	B	0.46
Kailuan	0.91	0.29	0.17	B	0.46
Wubu	0.84	0.25	0.07	C	0.32
Jiaozuo	0.96	0.41	0.151	C	0.55
Hongyang	0.90	0.34	0.10	C	0.44
Fushun	0.96	0.48	0.10	D	0.58

In the systematic analysis of social, economic and scientific management problems, people often face a complex system consisting of many interrelated and mutually restrictive factors, which often lacks quantitative data. In such a system, one of the interesting questions is how to assign the degree (ranking weight) of the nature of a thing in terms of a property shared by n different things, so that these values can objectively reflect the difference in the nature of different things. Analytic hierarchy process (AHP) provides a new, concise and practical modeling method for decision-making and optimization of such problems. It decomposes complex problems into constituent factors and forms a hierarchical structure according to the dominant relationship. Then the relative importance of decision-making schemes is determined by comparing the two schemes.

AHP is widely used in management decision-making in economy, science and technology, culture, military, environment and even social development. It is often used to solve such problems as comprehensive evaluation, selection of decision-making schemes, estimation and prediction, and allocation of input.

The basic principle of AHP is the principle of ranking, that is, ultimately sorting the advantages and disadvantages of various methods (or measures) as the basis for decision-making. That is, the problems of decision-making are regarded as a large system affected by many factors. These interrelated and interdependent factors can be arranged into several levels from high to low according to their membership relationship, called hierarchical structure construction. Then, experts, scholars and authoritative persons are invited to compare the importance of each factor, and then use the mathematical method to rank each factor layer by layer. Finally, the ranking results are analyzed to assist decision-making.

The main characteristic of AHP is to combine qualitative analysis with quantitative analysis, express human subjective judgment in quantitative form and deal with it scientifically. Therefore, it is more suitable for the circumstances of complex social sciences and more accurately reflects the problems. At the same time, although this method has a profound theoretical basis, but the form is very simple, easy to understand and accept, therefore, this method has been widely used.

The application of AHP to solve problems can be divided into four steps:

The first step is to establish the hierarchical structure of a problem.

The second step is to construct a pairwise comparison and judgment matrix.

The third step is to calculate the relative weights of the elements to be compared by the judgment matrix.

The fourth step is to calculate the combined weights of elements at all levels.

(1) Establish hierarchical structure
Firstly, complex problems are decomposed into components called elements, and these elements are divided into several groups according to their attributes to form different levels. Elements of the same level as criteria play a dominant role in some elements of the next level, while it is also dominated by elements of the upper level. This top-down dominant relationship forms a hierarchy, with only one element at the top level, which is generally the pre-enriched or ideal result of the analysis of the problem. The middle level is generally the criteria and sub-criteria, and the lowest level includes the decision-making scheme. The dominant relationship of elements between levels is not necessarily complete, that is, there can be such elements, it does not dominate all elements of the next level.

Secondly, the number of levels is related to the complexity of the problem and the degree of detail that needs to be analyzed. In general, there are not too many elements in each level, because the excessive number of elements in one layer will make it difficult for pairwise comparisons.

Thirdly, a good hierarchy is very important for solving problems. Hierarchical structure is based on the decision makers' comprehensive and in-depth understanding of the problems. If they are uncertain in the division of levels and the determination of the dominant relationship between levels, it is better to re-analyze the problems and clarify the relationship between the various parts of the problem in order to ensure a reasonable hierarchical structure.

A hierarchical structure should have the following characteristics:

① There are dominant relationships from top to bottom, and they are expressed by straight line segments. In addition to the first layer, each element is dominated by at least one element in the upper layer. Except for the last layer, each element dominates at least one element in the next layer. The relationship between the upper and lower elements is much stronger than that of the elements in the same level, so it is considered that there is no dominant relationship between the same level and the non-adjacent elements.
② The number of hierarchies in the whole structure is not limited.
③ There is only one element at the highest level, and the elements dominated by each element generally do not exceed 9 elements and can be further grouped.
④ Virtual elements can be introduced into some sub-hierarchical structures to make them hierarchical structures.

(2) Construct a pairwise comparison and judgment matrix

After the hierarchical structure is established, the subordinate relationship of elements between the upper and lower levels is determined. Suppose that the element C_k of the upper level is taken as the criterion, and $A_1, \ldots, A_n$ of the next level is dominated by the criterion C_k, our aim is to assign $A_1, \ldots, A_n$ corresponding weight according to their relative importance under the criterion C_k.

For most social and economic problems, especially those where human judgment plays an important role, it is not easy to get the weights of these elements directly, and it is often necessary to derive their weights by appropriate methods. The AHP uses the method of pairwise comparison.

Firstly, in the process of comparing the two, the decision maker has to answer the question repeatedly: for criterion C_k, two elements A_i and A_j, which one is more important and how important. It is necessary to assign a certain amount of value to the importance. The scale of 1–9 is used here, and their significance is shown in Table 9.9. For example, criteria are social and economic benefits, and sub-criteria can be divided into economic, social and environmental benefits. If we think that economic benefit is obviously more important than social benefit, their scale is 5, and the scale of social benefit to economic benefit is 1/5.

The scale method of levels 1–9 is a good way to quantify the judgment of thinking. First, people always use the same,

Table. 9.9 Significance of scale

1	Indicates two elements has the same importance
3	Indicates that one element is slightly more important than the other
5	Indicates that one element is obviously more important than the other
7	Indicates that one element is strongly more important than the other
9	Indicates that one element is extremely important more important than the other
Reciprocal	If the importance ratio of element i to element j is a_{ij}, then $a_{ij} = 1/a_{ij}$
2, 4, 6, 8 are the median of the above adjacent judgements mentioned above	

stronger, slightly stronger, very stronger and extremely stronger language when distinguishing the difference between things. Further subdivision can insert a compromise between two adjacent levels, so for most decision-making judgments, the scale of levels 1–9 is applicable. Secondly, psychological experiments show that most people's ability to distinguish the difference in the attributes at the same degree is between 5 and 9 levels, and the scales from levels 1–9 is used to reflect the judgment ability of most people. Thirdly, when the attributes of the elements being compared are in different order of magnitude, it is generally necessary to further decompose the elements of higher order of magnitude, which can ensure that the elements being compared have the same order of magnitude or close to the attributes being considered, so that they can be applied to the scales of levels 1–9.

Secondly, for n elements $A_1,\ldots,A_n$, by pairwise comparison, we can get the judgment matrix $\boldsymbol{A}$.

$$\boldsymbol{A} = \left(a_{ij}\right)_{n\times n} \quad (9.2)$$

The judgment matrix has the following properties:

① $a_{ij} > 0$;
② $a_{ij} = 1/a_{ij}$;
③ $a_{ii} = 1$.

We call $\boldsymbol{A}$ a positive reciprocal matrix.

According to the properties ② and ③, in fact, only $n(n-1)/2$ triangular elements of the upper (lower) triangle can be judged for the n-order judgment matrix.

(3) Compute the relative weights of the elements to be compared from the judgment matrix

(1) Calculate the relative weights of elements under a single criterion

This step is to solve the problem of computation of ranking weight of n elements $A_1,\ldots,A_n$ under criterion C_k.

For n elements $A_1,\ldots,A_n$, we can get the judgment matrix $\boldsymbol{A}$ and solve the eigenvalue problem by pairwise comparison.

$$\mathbf{A}\boldsymbol{\omega} = \lambda_{\max}\boldsymbol{\omega}$$

The obtained $\boldsymbol{\omega}$ is normalized as the ranking weight of element $A_1,\ldots,A_n$ under the criterion C_k. It is called the eigenvalue method for calculating the ranking vector.

The theoretical basis of the eigenvalue method is the Perron theorem of the following positive matrices, which guarantees the validity and uniqueness of the ranking vectors obtained.

In the theorem, if n-order square matrix $\boldsymbol{A} > 0$ and $\lambda_{\max}$ is the largest eigenvalue of modulus $\boldsymbol{A}$, then there are.

① $\lambda_{\max}$ must be a positive eigenvalue, and its corresponding eigenvector is a positive vector;
② Any other eigenvalue λ of $\boldsymbol{A}$ always has $|\lambda| < \lambda_{\max}$
③ $\lambda_{\max}$ is the single eigenvalue of $\boldsymbol{A}$, so its corresponding eigenvector is unique except for a constant factor. The maximum eigenvalue $\lambda_{\max}$ and eigenvector $\boldsymbol{\omega}$ in the eigenvalue method can be calculated directly by the Matlab software.

In addition, the maximum eigenvalue $\lambda_{\max}$ and eigenvector $\boldsymbol{\omega}$ can also be calculated by power method. Its steps are.

① Let the initial vector be $\boldsymbol{x}^{(0)} = (x_{01}, x_{02},\ldots,x_{0n})^{\mathrm{T}}$, for example, $\boldsymbol{x}^{(0)} = (1/n,\ldots,1/n)^{\mathrm{T}}$, calculate $\boldsymbol{\omega}^{(0)} = \frac{\boldsymbol{x}^{(0)}}{\sum_{i=1}^{n} x_{0i}}$.

② For $k = 1, 2, \cdots$, iterative computation

$$\boldsymbol{x}^{(k)} = \boldsymbol{A}\omega^{(k-1)}, \boldsymbol{x}^{(k)} = (x_{k1}, x_{k2},\ldots,x_{kn})^{\mathrm{T}},$$

$$\omega^{(k)} = \frac{\boldsymbol{x}^{(k)}}{\sum_{i=1}^{n} x_{ki}} = (\omega_{k1}, \omega_{k2},\ldots,\omega_{kn})^{\mathrm{T}};$$

③ For a given precision ε in advance, if $\max\limits_{1\le i\le n}\left\{\left|\omega_{ki} - \omega_{k-1,i}\right|\right\} < \varepsilon$, then $\omega^{(k)}$ is the result.

④ Calculate $\lambda_{\max} = \frac{1}{n}\sum_{i=1}^{n}\frac{x_{ki}}{\omega_{k-1,i}}$

In the case of low precision requirement, the maximum eigenvalue $\lambda_{\max}$ and eigenvector ω can also be calculated by approximation method. The commonly used methods are "sum method" and "root method".

(2) Consistency test of judgment matrix

The basic properties of judging matrix **A** are given. In special cases, the elements of judgment matrix **A** have transitions, that is to say, satisfies the equation

$$a_{ij} \cdot a_{jk} = a_{ik} \tag{9.3}$$

For example, when the importance ratio of A_i and A_j is 3, and the importance ratio of A_j and A_k is 2, the importance ratio of A_i and A_k should be 6 according to a transitive judgment. When the above formula holds true for all elements of matrix **A**, the judgment matrix **A** is called a consistency matrix.

Generally speaking, we do not require judgment to have such transitivity and consistency, which is determined by the complexity of objective things and the diversity of people's understanding. However, when constructing a pairwise judgment matrix, it is necessary to make judgments generally consistent. It is against common sense to judge that A is more important than B, B is more important than C, and C is more important than A. A confused judgment matrix is unable withstand deliberates, which may lead to decision-making errors, and when the judgment matrix deviates too much from the consistency, the reliability of the ranking weight calculated by the above-mentioned methods is also doubtful. Therefore, the consistency of judgement matrix must be checked.

The steps of consistency test of judgment matrix are as follows:

① Compute the consistency index CI : $\mathrm{CI} = \frac{\lambda_{\max} - n}{n-1}$, where n is the order of the judgment matrix;
② Average the random consistency index RI. The average random consistency index is obtained by calculating the eigenvalue of the random judgment matrix repeatedly (more than 500 times). The average random consistency index of 1000 times of repeated calculation of 1–15 order judgment matrices obtained by Gong Musen and Xu Shubai in 1986 is shown in Table 9.10.

③ Calculate Consistency Ratio CR : $\mathrm{CR} = \frac{\mathrm{CI}}{\mathrm{RI}}$. When $\mathrm{CR} < 0.1$, it is generally considered that the consistency of judgement matrix is acceptable.

(4) Calculate the combination weight of elements in each layer

In order to obtain the relative weights of all elements in each level of the hierarchical structure to total enrichment, it is necessary to properly combine the above calculation results and check the overall consistency. This step is carried out from top to bottom layer by layer. Finally, the lowest level elements, i. e. the relative weight of the priority of the decision scheme and the consistency test of the judgement of the whole hierarchical model, are obtained.

Assuming that there are m layers in the hierarchical structure and $n_k(k = 1, 2, \ldots, m)$ elements.

The combination ordering weight vector of k–1 elements $A_1, A_2, \ldots A_{nk-1}$, at layer $n_{k-1}$to total enrichment,

$$\omega^{(k-1)} = (\omega_1^{(k-1)}, \omega_2^{(k-1)}, \ldots, \omega_{nk-1}^{(k-1)})^{\mathrm{T}} \tag{9.4}$$

and single ordering weight vector of n_k elements at layer $k, B_1, B_2, \ldots, B_{nk}$ to layer n_{k-1} element (k–1) $A_1, A_2, \ldots, A_{nk-1}$ are calculated.

$$p_i^{(k)} = \left(p_{1j}^{(k-1)}, p_{2j}^{(k-1)}, \ldots, p_{nkj}^{(k-1)}\right)^{\mathrm{T}}, \quad i = 1,\ 2, \ldots, n_k \tag{9.5}$$

The weight of elements not dominated by A_j is 0.
Make $n_k \times n_{k-1}$ order matrix

$$P^{(k)} = \left(p_1^{(k)}, p_2^{(k)}, \ldots, p_{nk-1}^{(k)}\right) \tag{9.6}$$

Then the combination ordering weight vector of n_k elements at layer $k, B_1, B_2, \ldots, B_{nk}$ to total enrichment is

$$\omega^{(k)} = \left(\omega_1^{(k)}, \omega_2^{(k)}, \ldots, \omega_{nk}^{(k)}\right)^{\mathrm{T}} = P^{(k)}\omega^{(k-1)} \tag{9.7}$$

And the general formula is

$$\omega^{(k)} = P^{(k)}P^{(k-1)} \ldots P^{(3)}\omega^{(k-1)} \tag{9.8}$$

Table. 9.10 RI of 1–15 order judgment matrices

Orders	1	2	3	4	5	6	7	8
RI	0	0	0.52	0.89	1.12	1.26	1.36	1.41
Orders	9	10	11	12	13	14	15	
RI	1.46	1.49	1.52	1.54	1.56	1.58	1.59	

For the consistency test of hierarchical models, it needs to be calculated layer by layer similarly. If CI_{k-1}, RI_{k-1}, CR_{k-1} of $k-1$ level are calculated respectively, then the corresponding index at the k level is

$$CI_k = \left(CI_k^1, \ldots, CI_k^{n_{k-1}}\right)\omega^{(k-1)} \quad (9.9)$$

$$RI_k = \left(RI_k^1, \ldots, RI_k^{n_{k-1}}\right)\omega^{(k-1)} \quad (9.10)$$

$$CR_k = CR_{k-1} + \frac{CI_k}{RI_k} \quad (9.11)$$

where CI_k^j and RI_k^j are consistency index and mean random consistency index of judgment matrix of n_k element, $B_1, B_2, \ldots, B_{nk}$, at layer k under the criterion of $A_j(j = 1, 2, \ldots, n_{k-1})$.When $CR_k < 0.1$, hierarchy at level k has satisfactory consistency in the whole judgment.

Ideas for Optimizing CBM Accumulation Areas

(1) Characteristics of CBM accumulation area

CBM accumulation areas in China have the following characteristics:

① There are many gas accumulation areas, including 5 gas accumulation area, 30 gas accumulation belt and 115 CBM accumulation areas.
② The geographic locations of the gas accumulation areas are dispersed, and they are distributed all over the country except Tibet, Taiwan and Hainan.
③ There are great differences in the scale, geological conditions and the basis of CBM development in the gas accumulation areas. According to the existing knowledge, the development prospects of the gas accumulation areas are also quite different.
④ Researches on the gas accumulation areas are uneven. Some gas accumulation areas have been studied deeply, and development has been carried out in an all-round way. Some have few data. Only some gas accumulation areas have complete factors that can be discussed, but many enrichment areas have incomplete factors.

(2) Ideas for optimizing CBM accumulation areas in China

According to the above characteristics, the optimum ranking of CBM accumulation areas should be multi-level, that is, it is impossible to carry out the optimum ranking of all CBM accumulation areas according to the unified standard. For all accumulation areas, the factors available should be adopted, and for the areas with a higher degree of research, more factors can be used. Therefore, the optimization work is progressive, that is, with the rise of the optimization level, the optimization results are more and more close to the actual situation. The optimization method used here can also be called "multi-level comprehensive progressive optimization method", according to the specific situation, the following four levels of optimization can be used.

At the first level, gas content as the unique factor to screen, i.e. by "one vote veto".

At the second level, area and resource abundance are conditions to screen. The accumulation area is screened mainly considering the size of the accumulation area and the amount of resources. And the result is further optimized from the term of CBM resource. The factors considered include the evaluation area, resource abundance, gas content, gas saturation, coal rank, critical desorbing pressure, sand-mud ratio of caprock and development basis, etc.

The third level depends on permeability, which is used as a key factor. Only the accumulation area that has been tested can be selected. Other factors considered include the accumulation area, resource abundance, gas content, gas saturation, coal rank, critical desorbing pressure, sand-mud ratio of caprock, permeability and basic factors of development.

The fourth level takes reservoir pressure as a key factor. Only the accumulation area with coal reservoir pressure tested can be selected. Other factors considered include the accumulation area, resource abundance, gas content, gas saturation, sand-mud ratio of caprock, permeability, pressure ratio of immediate reservoir and basic factors of development.

As can be seen from the above, with the improvement of the optimal ranking level, the more comprehensive and representative the key factors are considered, the closer the optimal results are to the actual situation. The optimal ranking framework is shown in Fig. 9.2.

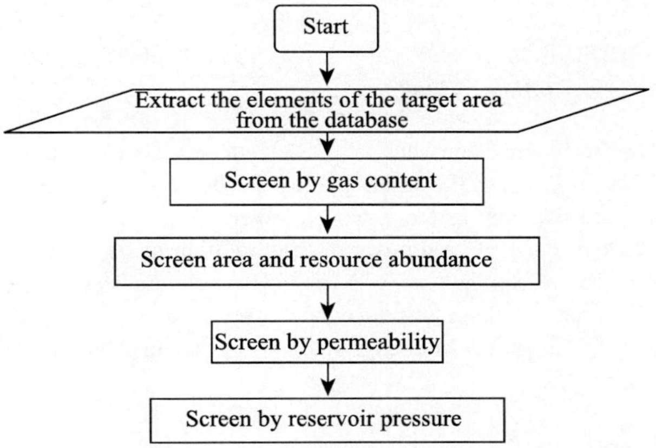

Fig. 9.2 Idea diagram for optimum selection of CBM accumulation areas

9.2.1.2 Key Parameters for Selecting Favorable CBM Areas

There are many factors affecting the quality of CBM accumulation area, and the relationship is complex. Through detailed analysis, the factors of CBM accumulation area are divided into five categories, including gas-bearing factors, coal reservoir factors, caprock factors, gas-controlling geological factors and basic development factors. Detailed analysis of all factors will be helpful to ascertain the degree of control and influence on CBM accumulation areas, and establish a unified, scientific, reasonable and feasible evaluation index system for optimizing CBM enrichment areas.

Among the many factors of CBM exploration and development, not all factors play the same important role, some factors are necessary and sufficient conditions, and some factors are only necessary conditions. In other words, there are still some "key factors" which play a decisive role in the prospects of CBM exploration and development.

Among the five types of factors mentioned above, gas-bearing factors and coal reservoir factors play a key or even decisive role in the optimization of favorable gas accumulation areas for CBM.

Gas-Bearing Factors

Gas-bearing factors are composed of gas content, resource content, resource abundance, gas gradient and gas composition. Among them, resources are determined by coal reserves and gas content. According to the conversion relationship, gas content is an independent factor, which plays a decisive role in the optimization of accumulation areas (McKee et al. 1986; Harris and Gayer 1996). There are many factors affecting gas content; other factors can be converted according to basic factors and reservoir geometric factors. However, under the weathering zone, CBM is basically composed of alkane gases, and the gas components lose their comparative significance in the range of evaluation depth, so they are not considered in comprehensive evaluation.

(1) Distribution characteristics of gas content in coal reservoirs in China

The basic characteristics of CBM content distribution in China are high in the south and low in the north and high in the east and low in the west. The average gas content (dry ash-free base) of coalmines in South China is 10–15 m^3/t, and that of Songzao is 27.1 m^3/t. The coalmines in North China, except Shandong Province, are mostly 5–10 m^3/t, of which Yangquan, Jincheng, Jiaozuo, Hebi and Anyang are higher. The coalmines in Northeast China, except Hongyang and Fushun, are generally lower, which is 4–6 m^3/t (Ye et al. 1998) (Fig. 9.3).

(2) Gas content increases with the increase of coal rank.

Kim studied the adsorption capacity of different ranks of coal. It was concluded that the adsorption capacity of high rank coal was significantly higher than that of low rank coal at the same temperature and pressure (depth).

The study of main coal seams of Carboniferous–Permian in North China found that when R^o = 0.79–0.88%, VL (Langmuir volume) = 20–33.33 $m^3/t_{combustible\ base}$, with an average of 26 m^3/t; when R^o = 1.4–1.72%, it is 32–47 m^3/t, with an average of 38 m^3/t. This indicates that the adsorption capacity of coal increases with the increase of its metamorphism.

Based on the summary of the gas content measurements of more than 300 coal samples in the United States and the data of anthracite gas content in the Jiaozuo area of China, the basic law of gas content increase with the increase of coal rank was also obtained (Table 9.11).

With the increase of coal rank, the adsorption capacity of coal is enhanced and the gas content is generally high. The main reasons are as follows: (i) with the thermal evolution (coal rank increasing), the volatile components are further discharged to form a large number of micro-pores, which improves the adsorption capacity; (ii) the cleavage in coking coal is the most developed, and in lean and anthracite coal, cleavages gradually close, which can partly inhibit the gas escaping from the coal seam and improve the gas storage capacity of the coal seam.

(3) Gas content decreases with the increase of mineral content

The inorganic minerals in coal are mainly silica-alumina, carbonate ore and pyrite. The ash yield in industrial analysis reflects the mineral content. The adsorption capacity of minerals to methane is weak. Therefore, with high mineral content, the adsorption capacity of coal seams reduces and the gas content reduces accordingly.

Samples taken from the same boreholes and same depth in the Raton Mesa Basin and Piceance Basin of the United States were tested by the same test method (desorption method). The results show that with the increase of mineral content, the gas content of coal, carbonaceous shale, sandstone and shale decreases in turn.

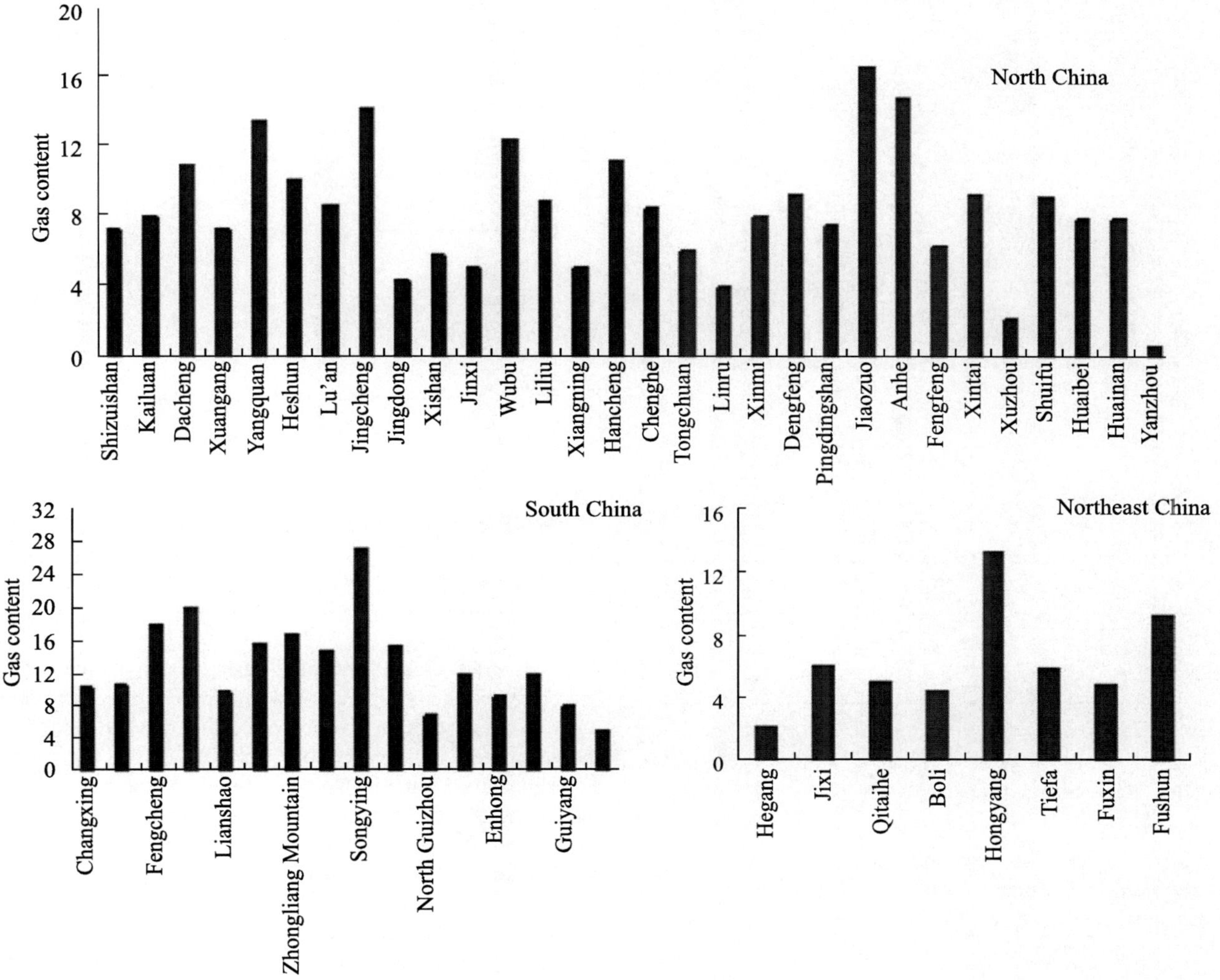

Fig. 9.3 Distribution characteristics of gas content in coal reservoirs in China

Table. 9.11 Statistics of gas content in different ranks of coal seams

Coal rank	Average gas content/(m^3/t)	Number of samples
Anthracite	15.60	57
Bituminous coal with low volatility	13.43	21
Bituminous coal with medium volatility	7.96	22
Bituminous coal with high volatility A	4.36	217
Bituminous coal with high volatility A	2.79	86
Bituminous coal with high volatility B	1.28	42

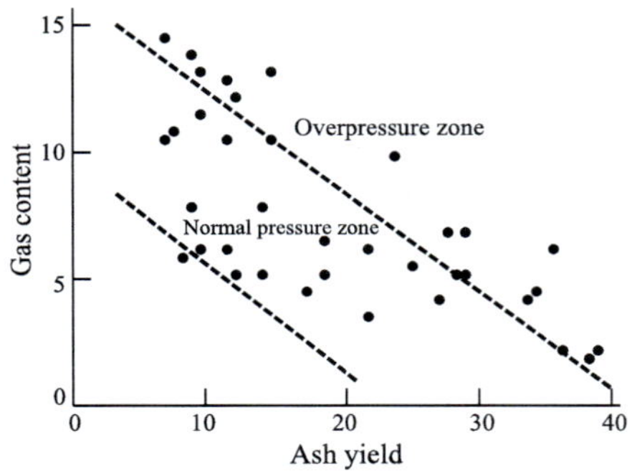

Fig. 9.4 Relationship between measured gas content and ash content in C–P main coal seams in North China (Quan 1995)

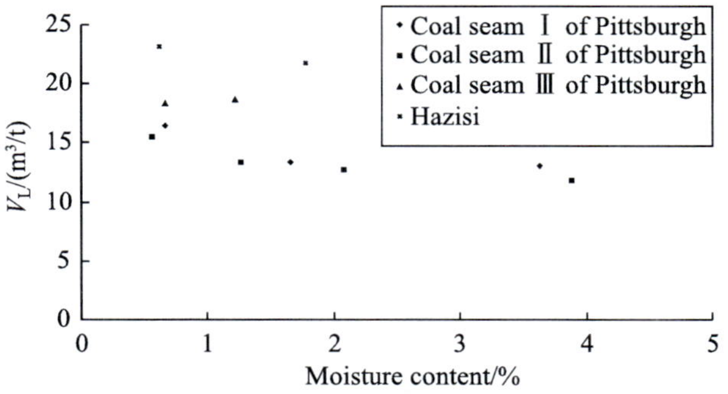

Fig. 9.5 Relationship between moisture and adsorption capacity

Gas content in several main coal seams of Carboniferous-Permian in some areas of North China has been measured. The results also show that gas content is inversely proportional to ash yield (Fig. 9.4).

(4) Gas content decreases with the increase of moisture

Coal moisture refers to the intrinsic moisture in coal samples, that is, water adsorbed or condensed in the capillary pores inside coal particles. Many scholars believe that when moisture content is between 1.5 and 2.0%, methane adsorption capacity at atmospheric pressure decreases by 1/3–1/2 compared with that of dry coal.

Isothermal adsorption experiments (Joubert and Grein 1973) were carried out on coal samples from four coal seams in the United States under different water conditions (Fig. 9.5). The results show that the methane adsorption capacity of the same coal sample decreases with the increase of moisture content, which is manifested by the decrease of Langmuir volume (V_L). Moreover, when the moisture

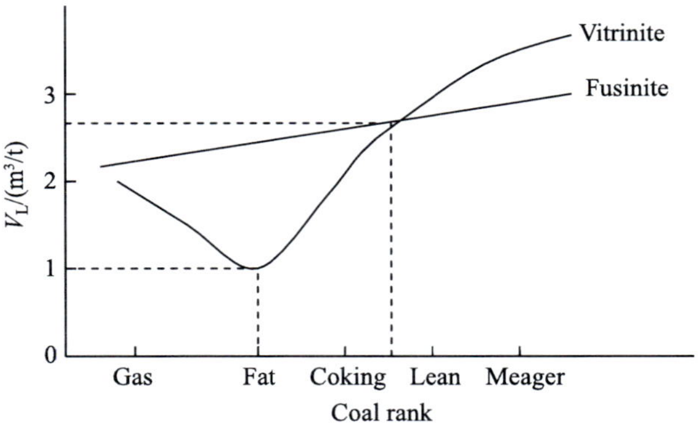

Fig. 9.6 Relationship between macerals and adsorption capacity of coal rock

content increases to a critical value, the influence of water on the adsorption capacity of coal seam is basically stable.

(5) The relationship between adsorption capacity and macerals of coal varies with coal rank

Coal macerals include vitrinite, fusinite and exinite. The content of exinite is generally low, and it does not contribute much to the adsorption capacity of coal. A detailed study on the adsorption capacity of coal macerals shows that when experimental pressure is 0.1 MPa, the adsorption capacity of fusinite macerals increases with the increase of coal rank, but the increase rate is low, showing a smooth straight line. The relationship between the adsorption capacity of vitrinite and coal rank is a spoon curve (Fig. 9.6). In the fat coal stage, vitrinite has the lowest adsorption capacity. If the coal rank is between coking coal and lean coal, vitrinite has similar adsorption capacity to fusinite coal. If the coal rank is lower than coking coal, vitrinite has lower adsorption capacity than fusinite coal, but higher than lean coal, on the contrary. When the degree of metamorphism of a coal seam is higher than that of coking coal, the higher vitrinite content, the stronger the adsorption capacity. When the degree of metamorphism is lower than that of fat coal, the high content of fusinite can increase the adsorption capacity of coal. If the experimental pressure is increased (above 0.1 MPa), the above law will become more obvious. Generally, the microscopic components in coal rock are mainly vitrinite groups (including semi-vitrinite groups). Therefore, the adsorption capacity of the vitrinite group basically reflects the adsorption capacity of the entire coal rock.

The relationship between the vitrinite content and the adsorption capacity of several major coal seams in the

Table. 9.12 Relation between vitrinite content and adsorption capacity

Vitrinite content	V_L/(m^3/t)	Average V_L/(m^3/t)
50–70	20–34	25
70–80	32–42	37
80–100	32–60	47

Carboniferous–Permian in some areas of North China was found to be proportional to the Langmuir volume (V_L) (Table 9.12).

(6) Adsorption capacity increases with pressure and decreases with temperature.

Adsorption kinetics studies show that (Debor 1964), the relationship between adsorption capacity and temperature (T) and pressure (P) is as follows.

$$V = CN\tau_c P e^{Q/(RT)} / (2\pi MRT)^{1/2} \quad (9.12)$$

where V is adsorption capacity, m^{-2}; N is Avogadro constant, 6.0226×10^{22} mol^{-1}; τ_c is the vibration time of adsorbed molecule, s; M is gas molecular mass, g; R is gas constant, 8.314 J/(mol·K); P is pressure, MPa; T is temperature, K; Q is the heat of adsorption, J/mol; C is unit conversion coefficient.

For specific adsorbents (coal) and adsorbates (methane), when τ_c and M are constants, the above formula becomes

$$V = k_0 p \quad (9.13)$$

where $k_0 = CN\tau/(2\pi MR)^{1/2}$ is a constant. If the pressure is constant, then there is $V = k_1 e^{Q/(RT)}/T^{1/2}$, that is, the adsorption capacity decreases exponentially with increasing temperature. If the temperature is constant (i.e., isothermal), there is $V = k_2 p$ where k_2 is constant, indicating that the adsorption amount and the pressure increase linearly. Theoretical derivation and specific experiments have shown that this formula is only applicable at medium pressure. When the pressure is large, the Langmuir equation is followed. When the temperature is constant, the adsorption capacity and pressure has a positive relationship. Both temperature and pressure are important factors that restrict coal adsorption capacity, and underground temperature and pressure are mainly related to depth. As depth increases, formation pressure and temperature increase. However, in general, especially in shallow layers, the formation pressure increases rapidly, while the formation temperature (absolute temperature) increases relatively small. Therefore, the general trend is that the ability of coal to adsorb methane is directly proportional to formation pressure (buried depth), as evidenced by the measured gas content (Table 9.13). However, this increase is not endless. This is because formation temperature is also increasing with burial depth and formation pressure. Although adsorption capacity can increase with the increase of pressure (When pressure is low, it increases linearly), it decreases exponentially with the increase of formation temperature. Experts believe that when temperature increases from 0 to 42 °C, the methane adsorbed by coal will reduce by 68%.

(7) Coal seam thickness and gas content

Because of the complexity of gas-controlled geological factors, there is no causal relationship between the thickness of coal reservoirs and their gas-bearing properties in many areas. But there are also many examples of positive correlation between the two, such as Huainan, Xingtai, Lincheng, Shizuishan, Baoji Mountain, Pingxiang, Fengcheng, Heshan, Luocheng, Yuanjia, Hongshandian, Laochang, Guishan and Tiefa mining areas or minefields (Qin et al. 2000).

According to provincial data from China National Administration of Coal Geology, the thick Upper Jurassic No.7 coal seam belt (primary coal seam) in the Tiefa accumulation zone in eastern Liaoning is located in the southwestern Daxing Minefield and Dalong Minefield where the gas content is generally greater than 6 m^3/t, and the coal seam gradually becomes thinner towards north and east, and

Table. 9.13 Relationship between coal seam depth and average gas content in some basins in the United States

Basin	Depth/m	Average gas content/(m^3/t)	Basin	Depth/m	Average gas content/(m^3/t)
Black Warrior Basin	0–150	0.25	Arkoma Basin	0–150	6.60
	150–300	2.00		150–300	13.50
	300–450	7.50		300–450	16.70
	450–600	13.10		450–600	18.80
	600–750	18.60		600–900	21.00

Adapted from Rieke, Hewitt et al.

the gas content decreases accordingly. The average thickness of the Upper Permian No.5 + 2, No.6, No.7, No.9, No.11 and No.15 coal seams in the Guishan Coalmine in eastern Yunnan is 1.32 m, 0.90 m, 1.23 m, 1.68 m, 1.97 m and 1.87 m, the average gas content is 7.99 m^3/t, 5.31 m^3/t, 6.97 m^3/t, 7.48 m^3/t, 11.30 m^3/t and 9.57 m^3/t, respectively, and the thicker the coal seams is, the higher the gas content is. This positive correlation is not directly related to the burial depth. The average thickness of the Upper Permian No.1, No.3 and No.4 coal seams in the Hongshandian Minefield in central Hunan is 0.92 m, 1.11 m and 1.95 m, and the average gas content is 18.24 m^3/t, 19.50 m^3/t and 21.70 m^3/t respectively. The average thickness of the C coal seam of Longtan Formation in Xuanjing Coalfield in southern Anhui is 1.16 m, 1.10 m, 1.27 m and 1.43 m, the average gas content is 0.73 m^3/t, 4.15 m^3/t, 5.70 m^3/t, 4.51 m^3/t and 6.11 m^3/t, respectively. The coal seam of Heshan Formation in Maolan Coalmine of northern Guangxi is thick in the east and thin in the west, and the gas content also shows the trend of high in the east and low in the west. The positive correlation between coal thickness and gas content is more typical in wells in the Nantong accumulation area and its surrounding (Qin et al. 2000).

For some specific areas, the correlation between coal seam thickness and gas content can be found by mathematical statistics. For example, in the Baojishan Coalmine of Jingyuan Coalfield in northeastern Gansu Province, the following empirical equation was obtained based on borehole data and interpolation of coal seam thickness and gas content contour map.

$$G = 2.94H^{-1/3} + 0.71, \ (r = 9.76) \tag{9.14}$$

where G is CBM content, m^3/t; H is coal seam thickness, m.

The main mode of CBM emission is diffusion, and the concentration difference between two points is the main driving force of CBM diffusion. According to Fick's law and mass balance, the mathematic model of methane diffusion from coal seams is established. Under similar initial conditions, the thicker the coal reservoir is, the longer it takes to reach median concentration or terminate diffusion (Wei 1998). Further analysis shows that coal reservoir itself is a highly compact and low permeability rock. The upper and lower layers play a strongly sealing role on the middle layer. And the thicker the coal reservoir is, the longer the path of CBM diffusion to the roof and the floor from the middle layer, the greater the diffusion resistance, and the better the preservation of CBM. This may be the fundamental reason why there is a positive correlation between the thickness and gas content in some coalmines or minefields.

Coal Reservoir Factors

Coal reservoir factors have a wide range of meanings, including geometric and physical elements, as well as material elements. Geometric elements and material elements can be regarded as independent elements, while physical elements are the result of their comprehensive effects. Geometric factors, including the area, thickness, burial depth, shape and stability of coal reservoirs, play a decisive role in the economic evaluation of CBM. Physical properties include pore-cut rationality, permeability, adsorbability, reservoir pressure, critical desorption pressure, displacement pressure, gas saturation, coal structure and so on, and are combined into some derivative parameters with major significance, such as reservoir pressure gradient, near-reservoir pressure ratio, etc. Among them, there is a causal relationship between permeability and other physical properties of the reservoir, which plays a decisive role in the success of CBM development. At the same time, reservoir pressure also plays an important role in the development of CBM.

(1) Permeability

The factors of coal seam permeability are very complex. Geological structure, stress state, coal seam depth, coal structure, coal rock and coal quality, coal rank and natural fissures affect coal seam permeability to varying degrees, sometimes comprehensive action of multiple factors affect the result. Sometimes a factor plays a major role. Generally speaking, internal factors such as cleavage and fracture play a leading role. On the complex geological background in China, external factors such as in-situ stress, burial depth and natural cracks have a particularly significant impact on coal seam permeability (Ye et al. 1999).

(1) Permeability of coal reservoirs in China

The analysis of available data shows that the coal reservoir permeability from well test in China is from 0.002 × 10^{-9} to 16.17 × 10^{-9} m^2, with an average of 1.27 × 10^{-3} m^2. 35% of the permeability is less than 0.1 × 10^{-9} m^2, 37% from 0.1 × 10^{-9} to 1 × 10^{-9} m^2, 28% greater than 1 × 10^{-9} m^2, and seldom less than 0.01 × 10^{-9} m^2 or greater than 10 × 10^{-9} m^2. Coal reservoirs with permeability above 5 × 10^{-9} m^2 are only distributed in Hancheng, Liulin and Shouyang enrichment areas in North China. Coal reservoirs with permeability of 1 × 10^{-9}–5 × 10^{-9} m^2 are relatively widely distributed, including 10 enrichment areas such as Hancheng, Liulin, Shouyang, Jincheng, Huaibei, Huainan, Jiaozuo, Doufeng, Tiefa and Pingdingshan. The permeability in other enrichment areas is lower than 1 × 10^{-9} m^2. The average permeability in 7 of 23 enrichment areas is larger than 1×10^{-9} m^2, accounting for 30.4%, that in 11 of them is 0.1×10^{-9}–1×10^{-9} m^2, accounting for 47.8%, and 21.7% is less than 0.1×10^{-9} m^2, indicating that the permeability of coal seams in China is mainly 0.1×10^{-9}–1×10^{-9} m^2. The coal reservoirs in the Black warrior Basin in the United States are usually at negative to normal pressure, and the absolute permeability is between 1×10^{-9} m^2 and 25×10^{-9} m^2. The absolute permeability of coal seams in some high-pressure areas of the San Juan Basin is 5×10^{-9}–15×10^{-9} m^2. In contrast, the overall permeability of coal reservoirs in China is relatively low, but there are high-permeability coal reservoirs in relatively low-permeability coal reservoir areas. Therefore, predicting and searching for high-permeability coal reservoirs in low-permeability coal reservoir areas is important to the strategy of CBM exploration and development in China.

(2) Coal seam thickness affects tested permeability

According to well test data of 47 coal seams of the Carboniferous-Permian in Qinshui, Ordos and Tiefa Basins in North China and coal-bearing areas in Huainan and Huaibei, western Henan and eastern Hebei, the relationship between coal seam thickness and permeability is in two opposite trends: the coal thickness is negatively correlated to the tested permeability in coal reservoirs with developed structural coal, but in coal reservoirs with undeveloped structural coal, when the permeability is less than 0.5 mD, the coal seam is thicker, and the permeability generally increases, when the permeability is greater than 0.5 mD, the permeability decreases with the increase of the coal thickness.

(3) In-situ stress

Coal seam permeability is most sensitive to stress. The permeability of coal seam decreases with the increase of effective stress, that is

$$K_a = K_{ai} e^{3c\Delta\delta} \tag{9.15}$$

where K_a is the absolute permeability at certain stress; K_{ai} is the absolute permeability at no pressure, c is the pore compression coefficient of coal, $\Delta\delta$ is the variation of effective stress from the initial to a certain stress state.

The tested permeability were plotted as in-situ burial depth vs. permeability according to three-level in-situ stresses (Fig. 9.7). If the in-situ stress is above 20 MPa, and the coal seam permeability is mainly less than 0.1×10^{-9} m^2. If the in-situ stress is less than 10 MPa, the coal seam permeability is more than 0.1×10^{-9} m^2. If the stress state is 10–20 MPa, the variation of coal seam permeability is large, and the law is not very obvious.

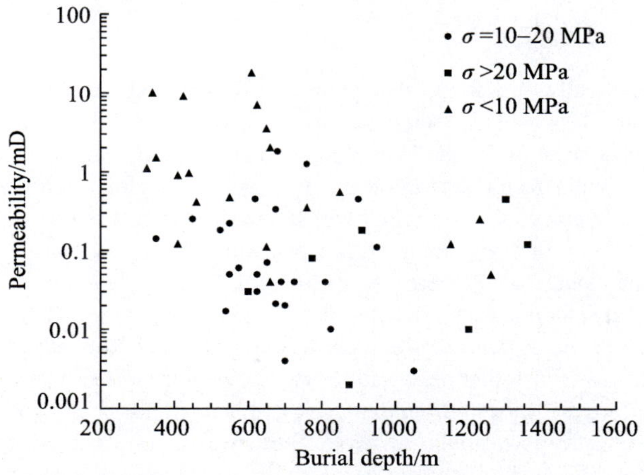

Fig. 9.7 Permeability distribution in different in-situ stress environments

So how does permeability change in a gas accumulation or enrichment zone? Fig. 9.8 shows the relationship between permeability and in-situ stress from the samples selected from the eastern margin of Ordos, Hancheng, Qinshui, east Taihang Mountain, Huainan, Huaibei, Hongyang, Shenbei, Tiefa and Dacheng gas accumulation belts or enrichment zones. Generally speaking, the permeability decreases significantly with the increase of stress.

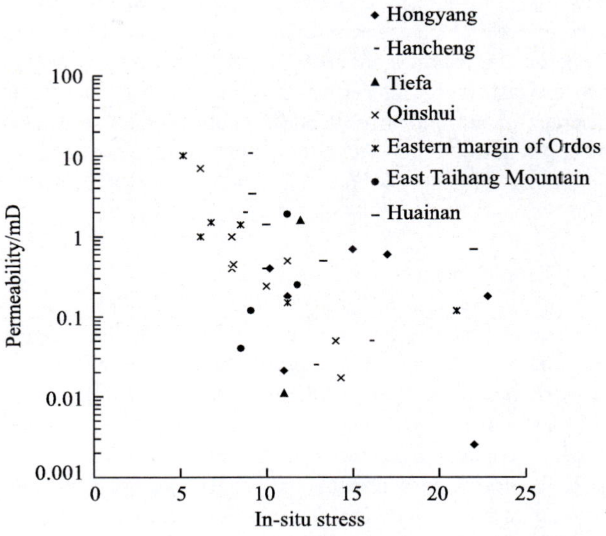

Fig. 9.8 Relationships between permeability and in-situ stress of coal reservoirs in some gas accumulation areas or enrichment belts in China

(4) Burial depth

Generally speaking, the permeability of coal reservoirs decreases with the increase of burial depth, but there are preconditions for the existence of such regularity. In China, the main factor determining permeability is tectonic stress. The change of permeability with depth is a function of stress. Only by distinguishing stress environments can we analyze the change of permeability with depth. The permeability of coal reservoir decreases with the increase of stress in China. The general trend between permeability and buried depth in China is also that permeability decreases with buried depth. However, in some depth interval, the distribution of permeability is relatively discrete. Further analysis shows that the relationship between permeability and depth is different in different areas (Fig. 9.9). There are three types in China: (i) Permeability decreases with the increase of depth, showing a negative correlation, which is the mainstream, such as Hongyang, Huainan and Huaibei, Qinshui, east Taihang Mountain, Hancheng and other gas accumulation or enrichment zones. (ii) Permeability increases with the increase of depth, which often occurs in different coal seams in a well, The reason may be related to the development of natural fissures and roof fissures in the coal seam itself, resulting in higher permeability in the lower coal seam than in the upper coal seam, i.e. the coal seam permeability of Taiyuan Formation in Shouyang gas accumulation area in Qinshui gas enrichment belt is often higher than that of Shanxi Formation. (iii) The permeability of coal reservoirs does not change with depth, which often occurs in a deep section, such as Liulin gas accumulation area in the depth range of 350–420 m, and the permeability distribution is very discrete. At present, the burial depth of coal seams in CBM accumulation areas in China is basically 500–800 m and the distribution of permeability in this depth interval is discrete, or the influence of burial depth on coal seam permeability is not important. With the further increase of burial depth, its influence on permeability is gradually obvious (Fig. 9.9).

(5) Natural fractures in coal seams

The natural fracture system in coal seams is to some extent an important factor on the permeability. Permeability is proportional to the square of the fracture width (Mckee et al. 1986). In theory, the development of natural fractures in coal seams is conducive to increasing the permeability.

Laboratory measurements of coal rock permeability shows that once the natural fracture in the sample develops, the permeability is good, and other factors such as coal rock type, coal quality and coal rank play a secondary role.

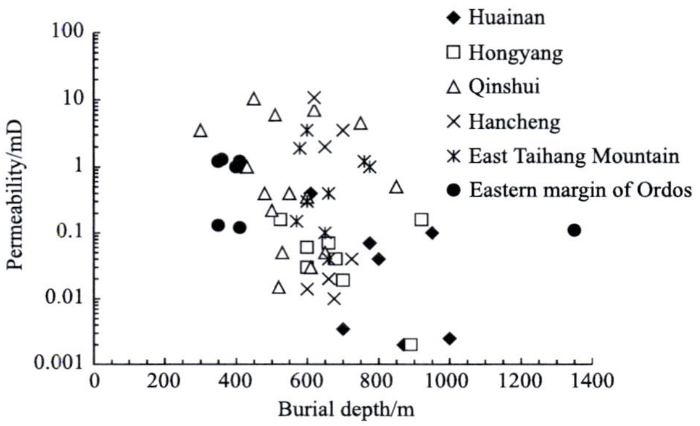

Fig. 9.9 Relationships between permeability and burial depth of coal reservoirs in some gas accumulation areas or enrichment belts in China

In laboratory, a coal core column is used to measure the permeability at confining pressure, which reflects that the coal permeability is extremely sensitive to stress. The permeability of coal rock decreases sharply with the increase of confining pressure. If the pressure increases by 10 times, the permeability will decrease by 2–3 orders of magnitude. For the coal with undeveloped fissures such as dull coal, when the pressure is above 4 MPa, the permeability is reduced by more than 95% from the initial value. For the coal samples with fissures developed, the permeability decreases less. When the confining pressure is 6.6–10 MPa, the permeability drops to over 90% of the initial value, which also indicates that the permeability decreases slowly and keeps better with the increase of confining pressure in the presence of certain fractures. Therefore underground, when formation is folded and uplifted, effective stress decreases and fractures open, which is beneficial to increase the permeability of coal seams. One view is that water and gas filling into reservoir may support the fracture system and reduce the effective stress. Therefore, when the coal reservoir is overpressured, it is beneficial to increase the permeability of the coal seam.

(2) Coal reservoir pressure

Coal reservoir pressure refers to the fluid pressure in coal pores fractures, so it is also called pore fluid pressure, which is equivalent to the oil pressure or gas pressure in conventional oil and gas reservoirs. The coal reservoir pressure is generally measured by well test, that is, the extrapolation method is used to obtain the initial pressure of the relative equilibrium state under the original formation conditions. The coal reservoir pressure is closely related to the gas content of the coal seam. The relative relationship between it and the adsorbability (especially the critical desorption pressure) directly affects the ease of drainage and pressure drop during gas production. Therefore, the study of coal reservoir pressure is not only important for the evaluation of gas-bearing property and geological conditions, but also provides important parameters for well completion process.

China's coal reservoir pressure data is derived from CBM parameter wells and test wells constructed in recent years. A total of 64 layers/times of coal reservoir pressure from 21 gas accumulation areas and 42 layers/times of geostress from 19 gas accumulation areas were collected. They provide an important basis for understanding the overall distribution of coal reservoir pressure in China.

In general, the coal reservoir pressure is closely related to the buried depth of the coal seam. The buried depth of the coal seam increases, the reservoir pressure increases accordingly. There is a significant linear relationship between the two. When the buried depth is less than 500 m, the average pressure of coal reservoirs is less than 5 MPa, such as Jincheng, Hancheng, Qinyuan, Liulin, etc. When the buried depth is less than 1000 m, the average pressure of coal reservoirs in most of the accumulation areas is less than 10 MPa, while the reservoir pressure in Hancheng, Taiyuan Xishan, Kailuan and other gas accumulation areas is close to 11 MPa. When the burial depth is more than 1000 m, the average pressure of coal reservoirs is greater than 10 MPa, such as Dacheng, Pingdingshan, Huainan and Wubu and so on, except Kaili accumulation area.

The pressure gradient of coal reservoirs in China is at least 2.24 kPa/m and the highest is 17.28 kPa/m. From 64 layers in 21 accumulation areas in China, 29 layers are at underpressure (the pressure gradient <9.30 kPa/m), accounting for 45.3%;14 layers are at normal pressure (the pressure gradient is 9.30–10.30 kPa/m), accounting for 21.9%; 17 layers at abnormally high pressure (the pressure gradient = 10.30–14.70 kPa/m), accounting for 26.6; and 4 layers at extremely high pressure (the pressure gradient > 14.70 kPa/m), accounting for 6.2% (Fig. 9.10; Table 9.14).

9.2.1.3 Evaluation System of CBM Reservoir

Methods and Models

In order to achieve the above optimization ideas, we must select appropriate calculation methods and models to rationalize the evaluation results. For this reason, three evaluation methods and hierarchical structure model were introduced, including risk probability, comprehensive queuing coefficient and fuzzy comprehensive evaluation (interval number).

(1) Risk probability method

This method is a basic method for quantitative sequencing of conventional hydrocarbon traps in the world. Based on the analysis of geological risk factors, main geological factors on gas control are computerized by probability addition method, and the geological risk coefficients reflecting the comprehensive risk of each evaluation unit are obtained, and then the risk coefficients are ranked according to the magnitude. If an evaluation unit (i) contains n main risk factors and the relative risk probability of a factor (j) is P_i

$$P_i = \sum_{j=1}^{n} p_{ij} = \sum_{j=1}^{n} \frac{f_{ij} Q_j}{f_{j,\max}} \quad i = 1, 2, \ldots, n \tag{9.16}$$

where f_{ij} is the absolute value of jth risk element in the i-th evaluation unit, Q_j is the weight value of jth risk element; $f_{j,\max}$ is the maximum value of jth risk element in all evaluation units.

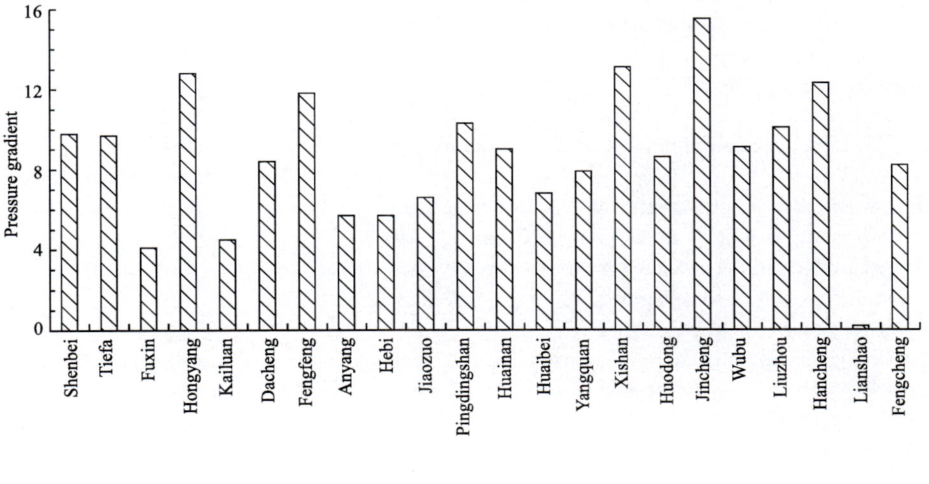

Fig. 9.10 Histogram of pressure gradient distribution in some accumulation areas in China

Table. 9.14 Coal reservoir pressure in some accumulation areas in China

Gas accumulation belt	Gas accumulation area	Reservoir pressure	Gas accumulation belt	Gas accumulation area	Reservoir pressure
Songliao-Liaoxi	Tiefa	Underpressure-high pressure	Eastern margin of Ordos	Liliu-Sanjiao	Underpressure-high pressure
	Shenbei	Normal pressure		Wubu	Normal pressure
	Fuxin	Underpressure	Weibei	Hancheng	Normal pressure-overpressure
Hunjiang-Hongyang	Hongyang	Normal pressure-overpressure	Qinshui	YangquanShouyang	Underpressure-normal pressure
Middle Hunan-Middle Jiangxi	Fengcheng	Underpressure		Huodong	Underpressure
West Henan	Pingdingshan	High pressure		Jincheng	Underpressure
Xuhuai	Huainan	Underpressure-high pressure		Tiyuan Xishan	Underpressure-overpressure
	Huaibei	Underpressure		Fengfeng	Underpressure-high pressure
Central Hebei Plain	Dacheng	Underpressure-normal pressure	East Taihang Mountain	Anyang-Hebi	Underpressure
Jingtang	Kailuan	Underpressure		Jiaozuo	Underpressure

The risk probability is the risk coefficient, and its numerical value is between 0 and 1. The normalization is introduced in the algorithm, so the risk coefficient is only a measure or ranking basis of the relative probability between the evaluation units, and cannot be regarded as absolute probability. Obviously, the higher the risk coefficient is, the worse the prospects for CBM exploration and development in the evaluation unit will be, and vice versa.

The final results can be obtained by ranking the risk coefficients of all the evaluation units according to their sizes, and the optimal segmentation method is used to process the ranking results. According to the similarity of risk probability, the ranking results are divided into several groups to facilitate the further evaluation of exploration risk level and the compare with the results of "step-by-step analysis of key factors".

(2) Comprehensive queuing coefficient method

This method was put forward by Wu Shoucheng, an expert of petroleum resource evaluation in China, in 1994. He further integrated the risk factors screened out by geological risk analysis into two categories of geological risk assessment (R_i) and resource quantity (Q_i), and assigned quantitative significance of x and y axes in the rectangular coordinate system. Y is the resource quantity and X is the average probability of other factors.

According to the above two types of coefficients, the comprehensive queuing coefficient (R_a) is calculated, and then the evaluation units are comprehensively ranked by their sizes. Mathematically, the distance of evaluation unit A (1, 1), which represents that evaluation unit $P(x, y)$ has the greatest theoretical potential. Therefore, the smaller R_a, the greater the resource potential. During the processing, the maximum resource coefficient is defined as 1, so R_a is distributed between 0 and 1.

According to the characteristics of CBM resources and its controlling factors are different from conventional oil and gas resources, the comprehensive optimization coefficient method was modified in this paper. The x-axis was redefined as resource coefficient, which is the probability sum of gas content, resources, resource abundance and theoretical saturation. The y-axis is the conservative factor G_i, whose value is equal to $1–R_i$, where R_i is the probability sum of other major risk factors.

The expression of comprehensive optimization coefficient R_a was obtained as follows:

$$R_i = \left[(1-x)^2 + (1-y)^2\right]^{-\frac{1}{2}} \tag{9.17}$$

There are several factors in the resource coefficient and conservative factor. The calculating principle and method of these two coefficients are the same as that of the risk probability value mentioned above.

(3) Fuzzy comprehensive evaluation—interval number

Fuzzy comprehensive evaluation method (FCE) is one of the multi-factor comprehensive evaluation methods widely used. It has a good processing ability for the uncertain evaluation factor system expressed by fuzzy numbers. However, for the evaluation factors expressed by interval numbers (i.e. a bounded closed interval), FCE has been powerless, and the key is how to solve the ranking problem of interval numbers. This paper constructs a mathematical model by ranking interval numbers for fuzzy comprehensive evaluation:

Let $X = \{x_1, x_2, \ldots, x_m\}$ be a factor set, in which x_i is evaluation index, such as "buried depth" and "coal thickness", and some of the factors are expressed by interval numbers; $Y = \{y_1, y_2, \ldots, y_n\}$ is a comment set, in which y_i is a fuzzy language, such as "A" and "B". Let A is the object of evaluation, such as an interval of a coalfield. The evaluation steps are as follows.

(1) Single factor evaluation

Because of some uncertainties of the evaluation object A itself, for a given factor x_i of A, if A is an accurate value, the degree of A belonging to y_i is expressed by a fuzzy value, while if A is uncertain, it can be expressed by an interval value. In addition, according to the fact that an ordinary real number is a special interval number, the evaluation index expressed by a fuzzy value is also expressed by an interval number. Therefore, for a given evaluation factor x_i, the degree of A belonging to y_i can be expressed as an interval number $[r_{ij}^-, r_{ij}^+]$ [0,1], $i = 1, 2, \ldots n; j = 1, 2, \ldots m$.

Then a fuzzy mapping f: $x \to$ IF(Y) of an interval number is obtained.

$$x_i \to f(x_i) = \left(\left[r_{i1}^-, r_{i1}^+\right], \left[r_{i2}^-, r_{i2}^+\right], \ldots, \left[r_{im}^-, r_{im}^+\right]\right)$$

where, IF(Y) is a fuzzy set of all interval numbers on Y. The fuzzy comprehensive evaluation matrix of the interval numbers is obtained as follows.

$$\boldsymbol{R}_{\mathrm{A}} = \begin{bmatrix} \left[r_{11}^-, r_{11}^+\right] & \left[r_{12}^-, r_{12}^+\right] & \cdots & \left[r_{1m}^-, r_{1m}^+\right] \\ \left[r_{21}^-, r_{21}^+\right] & \left[r_{21}^-, r_{21}^+\right] & \cdots & \left[r_{2m}^-, r_{2m}^+\right] \\ \vdots & \vdots & \vdots & \vdots \\ \left[r_{n1}^-, r_{n1}^+\right] & \left[r_{n2}^-, r_{n2}^+\right] & \cdots & \left[r_{nm}^-, r_{nm}^+\right] \end{bmatrix} \quad (9.18)$$

(2) Determining the weight of the evaluation index

Let $W = (\omega_1, \omega_2, \ldots, \omega_n) \in F(X)$, where $F(X)$ is a total fuzzy set on X. W_i is the weight of each factor. This paper uses grey correlation method to calculate the weight of each factor, and satisfy $\omega_1 + \omega_2 + \cdots + \omega_n = 1$.

$\boldsymbol{WR}_A = \left(\left[d_1^-, d_1^+\right], \left[d_2^-, d_2 +\right], \ldots, \left[d_m^-, d_m^+\right]\right)$ is obtained by matrix multiplication.

Where,

$$\left[d_j^-, d_j^+\right] = \sum_{i=1}^{n} \omega_i\left[r_{ij}^-, r_{ij}^+\right] = \sum_{i=1}^{n}\left[\omega_i r_{ij}^-, \omega_i r_{ij}^+\right] = \left[\sum_{i=1}^{n} \omega_i r_{ij}^-, \sum_{i=1}^{n} \omega_i r_{ij}^+\right] \quad j = 1, 2, \ldots, m \quad (9.19)$$

(3) Rank

$\left[d_j^-, d_j^+\right]$, $(j = 1, 2, \ldots, m)$ is ranked by the interval number ranking method. If $[d_k^-, d_k^+] = \max\{[d_j^-, d_j^+] \mid 1 \le j \le m\}$, and the evaluated object A finally belongs to the comment y_k.

(4) Establishment of hierarchical structure model

There is an important problem in any method, that is, how to determine the weight of each factor. In order to avoid the effect of human factors, the analytic hierarchy method is used to determine the weight.

The principle of determining factor weight by analytic hierarchy method is: for a given factor at a given level, the two-way judgment matrix of the next-level element is established, and the relative weight of the factor at this level to the upper level is calculated at one time.

Selection Method of CBM Accumulation Zones in China

Based on the above ideas, methods and models, the optimal selection method of CBM accumulation zone is formed.

Step 1. Screen by gas content as a key factor. The threshold is set according to the experts' opinions at present.
Low coal rank ($R^{\circ}_{max} < 0.65\%$): 2m^3/t;
Medium coal rank ($R^{\circ}_{max} \le 0.65 - 2.0\%$): 4 m^3/t;
High coal rank ($R^{\circ}_{max} > 2.0\%$): 8 m^3/t。

Step 2. Screen by evaluated area and resource abundance as key factors, mainly considering the size and resources of a CBM accumulation area. The threshold is set according to experts' opinions as follows:
Only large, medium and small gas enrichment zones and large and medium gas-bearing zones are selected, but large, medium and small poor gas zones and small gas-bearing zones are screened out.

The scale of an accumulation zone is determined according to its area: large > 900 km^2, medium is 900–200 km^2, small < 200 km^2.

The gas content of an accumulation zone is determined by the resource abundance: a gas accumulation zone > 1.0×10^8 m^3/km^2, a gas-bearing zone is 0.5×10^8 m^3/km^2, and a poor gas zone < 0.5×10^8 m^3/km^2.

After screening, rank the CBM accumulation zones quantitatively by the factors on CBM resource, which are put forward by experts' opinions, including area, resource abundance, gas content, gas saturation, coal rank, critical desorption pressure, sand-mud ratio of caprock and development basis, etc. Risk probability method is used to compare and verify the results, and combined with the comprehensive queuing coefficient method.

The calculation process of risk probability value is as follows: first, establish the judgment matrix of the eight factors; then, solve the weight of each factor according to experts' opinions; next, calculate the weight of each factor in the accumulation zones; finally, using the above formulas to calculate the risk probability of each accumulation zone.

The calculation procedure of the comprehensive queuing coefficient have been discussed before.

Step 3. Screen by permeability as a key factor, and other factors include the area, resource abundance, gas content, gas saturation, coal rank, critical desorption pressure, sand-mud ratio of caprock, permeability and basic factors of development, etc.

The risk probability method and the comprehensive queuing coefficient method are used to optimize the ranking, in which the factors considered and the number of factors may not be exactly the same.

Step 4. Screen by reservoir pressure as a key factor. Only the accumulation zones with coal reservoir pressure tested can be optimized. Other factors include the area, resource abundance, gas content, gas saturation, sand-mud ratio of caprock, permeability, critical desorption pressure and basic development factors, etc. Similarly, the risk probability method and the comprehensive queuing coefficient method are used to optimize the ranking, in which the factors considered and the number of factors may not be exactly the same.

Principle of Optimal Segmentation Method

In order to make the optimal results clearer and more reasonable, the optimal segmentation method is used to classify the results.

There are n samples arranged in some order. Each sample measures k indexes and arranges them into the following matrix.

$$\boldsymbol{X} = \begin{bmatrix} x_{11} & x_{12} & \cdots & x_{1k} \\ x_{21} & x_{22} & \cdots & x_{2k} \\ \cdots & \cdots & \cdots & \cdots \\ x_{n1} & x_{n2} & \cdots & x_{nk} \end{bmatrix} \tag{9.20}$$

where n rows denote n samples; k columns denote k indexes; and matrix element x_{ij} denotes the jth index of the ith sample.

It is required to divide the n samples into g segments in order (Table 9.15), so as to minimize the sample differences within each segment, while the sample differences among the segments are as large as possible and it need to divide them best.

In order to simplify the problem, the optimal segmentation method is introduced in the case of $k = 1$, where the matrix $\boldsymbol{X}$ is reduced to

$$\boldsymbol{X} = \begin{bmatrix} x_1 \\ x_2 \\ \cdots \\ x_n \end{bmatrix} \tag{9.21}$$

$\{i, \ldots, j\}$ is used to represent the segment from sample ith to sample jth, where $1 \le i \le j \le n$ is introduced in this segment

$$d(i,j) = \sum_{\alpha=i}^{j} [x_\alpha - \bar{x}(i,j)]^2 \tag{9.22}$$

where,

$$\bar{x}(i,j) = \sum_{\alpha=i}^{j} \frac{x_\alpha}{j-i+1} \tag{9.23}$$

$d(i, j)$ represents the sample difference within a segment $\{i, \ldots, j\}$. The smaller the $d(i, j)$ is, the smaller the difference between samples is, that is, there is consistency or concentration. On the contrary, the larger the $d(i, j)$ is, the greater the difference between the samples is, and it is discrete or non-uniform. $d(i, j)$ is the diameter of segment $\{i, \ldots, j\}$.

The segmentation method divides n samples into g segments, as shown in the above table. The diameter of each segment is determined as follows

$$d(1,p), d(p+1,q), \ldots, d(v+1,n), \tag{9.24}$$

Table. 9.15 n samples divided into g intervals

Interval 1	Interval 2	…	Interval g
$\{1,\ldots,p\}$	$\{p+1,\ldots,q\}$	…	$\{v+1,\ldots,n\}$

For this method, the sum of diameters S of all segments can be obtained.

$$S = d(1,p) + d(p+1,q) + \cdots + d(v+1,n) \quad (9.25)$$

An optimal segmentation means to find the method that makes S reach the minimum, called the optimal g-segment method.

(1) Optimal two-segment method

For n ordered samples, when, $j(1 \leq i \leq j \leq n)$, two segments can be determined.

$$\{1,\ldots,j\}, \{j+1,\ldots,n\} \quad (9.26)$$

The sum of the diameters of the two segments is

$$S_j^{(2)}(n) = d(1,j) + d(j+1,n) \quad (9.27)$$

where n in $S_j^{(2)}(n)$ denotes the number of segmented samples, the upper corner (2) denotes the number of segments, and j denotes the jth sample, i.e. the last point.

It's easy to understand that $S_j^{(2)}(n)$ is the sum of the square of the intra-segment difference (S_e) of segments $\{1,\ldots,j\}$ and $\{j+1,\ldots,n\}$. Let S_T be the total sum of squares of difference, and S_A be the sum of squares of intersegment, then

$$S_T = S_A + S_e \quad (9.28)$$

Since $S_T = d(1, n)$ is a fixed quantity for given n samples, S_e is minimized, that is, S_A is maximized.

$$S_A = S_T - S_e = d(1,n) - S_j^{(2)}(n) \quad (9.29)$$

So only the appropriate j can make $S_j^{(2)}(n)$ to minimize, we can determine the optimal segmentation method.

For all $1 \leq j \leq n–1$, find the minimum value of $S_j^{(2)}(n)$. Suppose that when $j = I_2$, $S_j^{(2)}(n)$ reaches the minimum value, i.e.

$$S_j^{(2)}(n) = \min_{1 \leqslant j \leqslant n-1} S_j^{(2)}(n) \quad (9.30)$$

Then optimal two segments are

$$\{1,\ldots,I_2\}, \quad \{I_2+1,\ldots,n\} \quad (9.31)$$

That is, the first I_2 samples are in segment 1, and the last ($n–I_2$) samples are in segment 2. The corresponding S_e and S_A are as follows

$$S_e^{(2)} = S_{I_2}^{(2)}(n), \quad S_A^{(2)} = d(1,n) - S_{I_2}^{(2)}(n) \quad (9.32)$$

(2) Optimal three segments

For any $j(2 \leq j \leq n–1)$, the first j samples are calculated by two-segment method in the first segment.

$$S_i^{(2)}(j) = d(1,i) + d(i+1,j) \quad (9.33)$$

Calculate the minimum value of the ith sample

$$S_{I_2(j)}^{(2)}(j) = \min_{1 \leqslant i \leqslant j-1} S_i^{(2)}(j) \quad (9.34)$$

So the optimal two segments of the first j samples are

$$\{1,\ldots,I_2(j)\}, \quad \{I_2(j)+1,\ldots,j\} \quad (9.35)$$

$S_{I_2(j)}^{(2)}(j)$ represents the S_e of the two intervals.

Now the optimal segments of the first j samples together with $\{j + 1,\ldots, n\}$ to form three segments, and then finds an appropriate sample number j.

$$S_j^{(3)}(n) = S_{I_2(j)}^{(2)}(j) + d(j+1,n), \quad (j = 2,3,\ldots,n-1) \quad (9.36)$$

where, we need to keep the $S_j^{(3)}(n)$ as small as possible, and find the minimum value for all $j(2 \leq j \leq n–1)$. Suppose that when $j = I_3$, $S_j^{(3)}(n)$ reaches the minimum value, i.e.

$$S_j^{(3)}(n) = \min_{2 \leqslant j \leqslant n-1} S_j^{(3)}(n) \quad (9.37)$$

Then optimal three segments are determined.

$$\{1,\ldots,I_2(I_3)\}, \quad \{I_2(I_3)+1,\ldots,I_3\}\{I_3+1,\ldots,n\} \quad (9.38)$$

Find out the best three segments that make the sum of the diameters of all segments reach the minimum value, the corresponding S_e and S_A are as follows

$$S_e^{(3)} = S_{I_3}^{(3)}(n), \quad S_A^{(3)} = d(1,n) - S_{I_3}^{(3)}(n). \quad (9.39)$$

(3) Optimal g segments

Assuming that the optimal (g–1) segments are known, for any $j \leq (g–1)$,

$$\{1,\ldots,I_2\}, \{I_2+1,\ldots,I_3\},\ldots,\{I_{g-1}+1,\ldots,j\} \quad (9.40)$$

The sum of the diameters of all segments is

$$S_{I_{g-1}}^{(g-1)}(j) = \min_{g-2 \leqslant i \leqslant j-1} S_i^{(g-1)}(j) \; (g-1 \leqslant j \leqslant n-1) \quad (9.41)$$

where,

$$S_i^{(g-1)}(j) = S_{I_{g-1}(i)}^{(g-2)}(i) + d(i+1,j) \tag{9.42}$$

Now let take $j \le (g\text{–}1)$, the optimal segments for the (g–1) segments of first j samples and together with $\{j+1,\ldots,j\}$ to form g segments of n samples (but not necessarily optimal)

$$\{1,\ldots,I_2\},\{I_{g-1}+1,\ldots,j\},\ldots,\{j+1,\ldots,n\} \tag{9.43}$$

Then from all possible g segments obtained by the above method, optimal segments are obtained, i.e. $j(g\text{–}1 \le j \le n\text{–}1)$ is selected to minimize the total diameter.

$$S_i^{(g)}(j) = S_{I_{g-1}(j)}^{(g-1)}(j) + d(j+1,n) \tag{9.44}$$

Finding the minimum value of $S_j^{(g)}(n)$ for j

$$S_j^{(g)}(n) = \min_{g-1\leqslant j\leqslant n-1} S_j^{(g)}(n) \tag{9.45}$$

The optimal g segments are determined to get the last point I_g, and the former I_g samples can be divided into (g–1) segments, so

$$\{1,\ldots,I_2\},\{I_2+1,\ldots,I_3\},\ldots,\{I_g+1,\ldots,n\} \tag{9.46}$$

It can be proved that the g segments obtained by above method make the sum of the diameters reach the minimum and S_e is

$$S_e^{(g)} = S_{I_g}^{(g)}(n) \tag{9.47}$$

And the S_A is $S_A^{(g)} = d(1,n) - S_e^{(g)}$.

(4) The number of segments g

The number of segments can be determined until g segments or a positive number δ given in advance, so that the sum of the diameters of all segments, $S_{I_g}^{(g)}(n) < \delta$, and g is the final number. It is clear that $S_{I_g}^{(g)}(n)$ decreases monotonously with the increase of g, and such figure can be drawn. When $g = 5$, the curve is relatively flat, so it is more appropriate to divide it into five segments.

(5) In the case of $k \le 2$.

When a sample has more than two indexes, the data are shown in the matrix.

$$X = \begin{bmatrix} x_{11} & x_{12} & \cdots & x_{1k} \\ x_{21} & x_{22} & \cdots & x_{2k} \\ \cdots & \cdots & \cdots & \cdots \\ x_{n1} & x_{n2} & \cdots & x_{nk} \end{bmatrix} \tag{9.48}$$

Define the diameter of the segment $\{i,\ldots,j\}$ is as follows

$$d(i,j) = \sum_{\alpha=i}^{j}\sum_{\beta=1}^{k}\left[x_{\alpha\beta} - \bar{x}_\beta(i,j)\right]^2 \quad 1\leqslant i\leqslant j\leqslant n \tag{9.49}$$

where,

$$\bar{x}_\beta(i,j) = \sum_{\alpha=i}^{j}\frac{x_{\alpha\beta}}{j-1+1} \quad (\beta = 1,2,\ldots,k) \tag{9.50}$$

Which is the average value of the βth index in $\{i,\ldots,j\}$. The steps of optimal g segments are exactly the same as that of one index.

(6) Two explanations

① In the case of more indexes, in order to eliminate the difference in magnitude, of these indexes, they can be normalized or standardized beforehand, so that each index is at a unified metric standard. This can avoid the disadvantages of highlighting or weakening the role of some indexes.
② The diameter of the sample segment can be defined by other methods, such as

$$d(i,j) = \sum_{\alpha=i}^{j}\sum_{\beta=1}^{k}\left|x_{\alpha\beta} - \bar{x}_\beta(i,j)\right| \tag{9.51}$$

It can also be defined by the distance in another metric space.

(7) Specific procedures

① First normalize indexes

$$z_{ij} = \frac{(x_{ij} - \min\limits_{1\leqslant t\leqslant n} x_{ij})}{(\max\limits_{1\leqslant t\leqslant n} xi_j - \min\limits_{1\leqslant t\leqslant n} x_{ij})} \quad (i = 1,\ldots,n; j = 1,\ldots,k) \tag{9.52}$$

② Calculated from z_{ij}

$$d(i,j) = \sum_{\alpha=i}^{j}\sum_{\beta=1}^{k}\left[z_{\alpha\beta} - \bar{z}_\beta(i,j)\right] \tag{9.53}$$

where,

$$\bar{z}_\beta(i,j) = \sum_{\alpha=i}^{j} \frac{z_{\alpha\beta}}{j-1+1} \quad (9.54)$$

D matrix is

$$D = \begin{bmatrix} d(1,1) & d(1,2) & d(1,3) & \cdots & d(1,n) \\ & d(2,2) & d(2,3) & \cdots & d(2,n) \\ & & \cdots & \cdots & \cdots \\ & & & \cdots & \cdots \\ & & & & d(n,n) \end{bmatrix} \quad (9.55)$$

where $d(i,i) = 0,\ (i = 1,2,\ldots,n)$, the lower triangle is redundant and can be evaluated by a symmetric matrix. Let $S_{I_1(j)}^{(1)}(j) = d(1,j), I_1(j) = 0, (j = 1,2,\ldots,n)$. According to formula

$$S_{I_g(j)}^{(g)}(j) = \min_{g-1\leqslant i\leqslant j-1} S_i^{(g)}(j) \quad (9.56)$$

and

$$S_{I_g(j)}^{(g)}(j) = S_{I_{g-1}(i)}^{(g-i)}(i) + d(i+1,j)$$
$$(j = g, g+1,\ldots,j;\ g = 2,3,\ldots,n-1)$$

The $S_{I_g(j)}^{(g)}(j)$ and point $I_e(j)(g = 2,3,\ldots,n-1; i = g, g+1,\ldots,n)$ corresponding to optimal g segments of any j samples are obtained step by step, and save the calculating results. Thus, the points for g segments $(g = 2,3,\ldots,n-1)$ of n samples are obtained and arranged from large to small as follows

$$I_g(n), I_{g-1}(I_g), I_{g-2}(I_{g-1}),\ldots,I_2(I_3) \quad (9.57)$$

In addition, the corresponding diameter sum $S_{I_g(n)}^{(g)}(n)$ can be determined. The optimal g segments are

$$\{1,\ldots,I_2\}, \{I_{2+1},\ldots,I_3\},\ldots,\{I_{g+1},\ldots,n\} \quad (9.58)$$

where, take the appropriate number of segment or select the smallest g that satisfies the relationship based on a given positive number δ, and then establish the segmenting points.

9.2.2 Evaluation of CBM Enrichment Zones (Target) in Key Coal Basins

CBM target zones in Ordos Basin, Qinshui Basin and Junggar Basin were evaluated and optimized.

9.2.2.1 Geological Survey of Key Coal Basins

Ordos Basin

The Ordos Basin rises from the Lüliang Mountain in the east, the Yinchuan Graben-Liupan Mountains in the west, the Ulam Geer uplift in the north and the Weibei uplift in the south, with an exploration area of about 25 × 10^4 km^2. It includes seven geological units: the western margin thrust belt, the Tianhuan syncline, the Yishan slope, the central paleo-uplift, the west Shanxi fold belt, the Ulam Gerr uplift and the Weibei uplift. Because the exploration of CBM is restricted by the distribution and burial depth of coal seams, Paleozoic CBM exploration in the basin is mainly distributed in the west Shanxi fold belt and Weibei uplift belt, and then sporadically in the western margin thrust belt. The exploration area of Mesozoic CBM is mainly located in the Yishan slope, Tianhuan depression, central paleo-uplift and Weibei uplift area. The coal-bearing strata are mainly Carboniferous-Permian and Jurassic.

The vitrinite content of the Yan'an Formation coal in the basin is 19.4–95.2%, with an average of 58.5%, which is related to the sedimentary environment during the coal-accumulating period in the basin. During the evolution from river facies in the edge of the basin to lake-marsh facies in the middle and east of the basin, the vitrinite content gradually increased and transformed from oxidizing to reducing environment. The vitrinite content of the Shanxi Formation and Taiyuan Formation coal is 71–90%, with an average of 79%.

The hydrogeological characteristics of the Jurassic coal-bearing strata are as follows: according to total salinity and Cl concentration, the Jurassic Yan 8, 9 and 10 Formations in the Heshui-Ningxian area are divided into 3 water types—$CaCl_2$, $NaHCO_3$ and Na_2SO_4. The total salinity of $CaCl_2$ water is the highest, ranging from 5 to 100 g/L, in which Ca^{2+} and Cl^- account for more than 70% of the particle concentration. Na_2SO_4 water has a total salinity of 0.5–60 g/L, and Na^+ and $SO_4{}^{2-}$ account for more than 60%. Na_2CO_3 water has a total salinity of 5–80 g/L, and $CO_3{}^{2-}$ accounts for 40–50%. According to the regional hydrogeological conditions, Na_2SO_4 water in the Yan 10 Formation is in the Xunyi-Binxian area and the ancient channels; $NaHCO_3$ water is in the Changwu-Zhengning area; $CaCl_2$ water is in the Yanwu and a large part in northern Gucheng. Atmospheric precipitation on the west and south margins of the Heshui-Ningxian infiltrates into the water supply area along coal outcrops and faults, while paleo-high surface water alternates slowly or in a viscous flow state, and the water type is represented by $CaCl_2$ or $NaHCO_3$. In the low of the paleo-channels, water alternates actively, forming a drainage area, and the water type is dominated by Na_2SO_4. The distribution of the water type in the Yan 9 Formation basically maintains the characteristics of the Yan 10 Formation, but the area with $NaHCO_3$ water expands along the ancient channels or the ancient highlands. The area with $CaCl_2$ water decreases and the area with Na_2SO_4 water expands northward. The horizontal distribution of the Yan 8 Formation is basically the same as that of the Yan 9 Formation. The water supply is still from the west and the south, the drainage area is located in the ancient channel and the confined area is in the ancient highland. In summary, groundwater flows from southwest to northeast in the accumulation area of Heshui-Ningxian. With the change of stratigraphic age, the area with $CaCl_2$ water decreases and that with Na_2SO_4 water increases, but $CaCl_2$ water area is always distributed in ancient the highlands and forming confined area. If CBM is well preserved, Na_2SO_4 water area located in the ancient channels or southern area is forming the drainage area or the water supply area respectively. If CBM is not well preserved, the water type is $NaHCO_3$ at the junction of the ancient channel and the ancient highland, and the storage condition of CBM is between the two. At the same time, it can be seen that the distribution of the ancient channel in the Yan 10 Formation has a great influence on later hydrogeological conditions. The ancient channel controls the water types around the accumulation zone and has a certain geological significance for the preservation of CBM.

The Carboniferous–Permian hydrogeological characteristics are as follows: vertically, thick mudstones of the Shihezi Formation and the Shiqianfeng Formation prevent vertical flow of top formation water. Thicker mudstone and bauxite mudstone of the Benxi Formation prevent Ordovician limestone water from channeling upward, making the coal-bearing strata an independent and closed hydrological system. The sealing is mainly manifested in the different hydrological characteristics at the top and bottom of the coal seam. Through the analysis of the vertical aquifers in 17 wells in the Wubu area, the artesian discharge of the upper Triassic aquifer is 0.5–4.19 L/S and the salinity is 20–60 g/L, where the hydrochemical type is $CaCl_2$ and the local is Na_2SO_4. The artesian discharge of the aquifer in the Shanxi to Taiyuan Formations is 0.9–8.7 L/s and the salinity is 10–25 L/s, where the water type is dominated by transitional $NaHCO_3$. The artesian discharge of limestone aquifer in the Majiagou Formation is 28.5–61.05 L/s and the salinity is 1–100 g/L, where the water type is $CaCl_2$ and $MgCl_2$. From the water-bearing characteristics of these three aquifers, the vertical hydrogeological characteristics of the coal-bearing formations are independent and closed, which is beneficial to the preservation of CBM.

The geological CBM resources are $98{,}196.88 \times 10^8$ m^3 from the new round evaluation in this basin.

Qinshui Basin

Qinshui Basin is located in the southeast of Shanxi Province, with the north latitude of 35°–38° and the east longitude of 112°00′–113°50′. It extends along the NNE direction with an elliptical middle contraction. It is about 120 km wide from east to west and 330 km long from south to north, with a total area of about 30,000 km^2. The overall structure is a compound syncline, whose axis is roughly located in the line of Yushe–Qinxian–Qinshui. The structure is relatively simple and the faults are not well developed. Generally speaking, the western part is characterized by the superimposition of Mesozoic folds and Cenozoic normal faults, while the northeastern and southern parts are dominated by Mesozoic EW and NE-trending folds, while the NE-NE folds in the central basin are well developed. Coal-bearing strata are mainly in the Upper Carboniferous Taiyuan Formation and the Lower Permian Shanxi Formation.

The coal seams are composed of humic coal, and the macroscopic coal composition is mainly bright coal, followed by dull coal, and less vitrain and fusain. On the coal composition, bright component is relatively riche, mostly densely distributed in strips and lines, with shell-like or step-like faults, and endogenous fractures are developed. The content of dull component is relatively low, and distributed in broad strips or lenticular shape, with stepped or heterogeneous faults, dense and uniform. Coal types are relatively complete, including gas coal and anthracite, but mainly metamorphic bituminous coal and anthracite.

The coal ranks of the Shanxi Formation are mainly dry coal, lean coal and anthracite, followed by coking coal, fat coal and gas coal. The central basin contains mainly dry coal and anthracite, and the surrounding area contains lean coal, while the southern basin contains basically anthracite, and the western basin covers mainly gas coal and coking coal. The coal rank of the Taiyuan Formation is basically the same as that of the Shanxi Formation, but its distribution of gas coal is smaller than that of the Shanxi Formation. From the distribution of coal ranks in the whole basin, the west part is mainly coking coal and gas coal, the east part is mainly lean coal and dry coal, and the north part is mainly lean coal, dry coal and anthracite, and the south part is basically anthracite.

Analysis of the Shanxi Formation and the Taiyuan Formation show that the ash and volatile matter in the coal are relatively low. The volatile matter in the Shanxi Formation is 7–38.92% (higher in some areas), with an average of 17.23%. The volatile matter of the main coal seams in the Taiyuan Formation is generally 8–21%, with an average of 14.36%. The ash content in the Shanxi Formation is generally 2.6–24.15%, with an average of 11.11%, while that in the Taiyuan Formation is 4.8–25.49%, with an average of 13.26%. The moisture of the raw coal ranges from 0.83 to 2.26%, generally 1% in Xishan, Huoxi and Lu'an, >1% in

Yangquan and >2% in Jincheng. The sulphur content of the Shanxi Formation is generally less than 1%, which is low sulphur coal; it is generally more than 1% in the Taiyuan Formation, which is medium and high sulphur coal.

The contents of vitrinite and inertinite in the Shanxi Formation are 45%–70% and 20%–36%, and those in the Taiyuan Formation are 65%–80% and 16%–30% respectively. It is clear that the vitrinite content in the Taiyuan Formation is higher than that in the Shanxi Formation, while the inertinite content is lower than that in the Shanxi Formation.

The porosity is measured by nitrogen method, the lowest porosity is 1.5%, the highest is 12.2%, and generally less than 5%. Statistical research shows that the porosity is related to the degree of metamorphism. Generally, the porosity of fat coal and coking coal is the lowest, and that of lean coal is high.

Using the burial depth, thickness and gas content to divide the calculation units, and calculating the area of each calculation unit and the amount of CBM resources, the total geological resources of CBM in the Qinshui Basin are $39{,}668.70 \times 10^8$ m^3.

Junggar Basin

The Junggar Basin is located in the northern part of Xinjiang Uygur Autonomous region, which is one of the three major basins in Xinjiang. The basin is triangular in shape and surrounded by boundary mountains. The western boundary is the Zaire Mountain and the Hala'alate Mountain, the northeastern boundary is the Altai Mountain, the Qinggelidi Mountain and the Kelamai Mountain, the southern boundary is the Yilin Heibilgen Mountain and the Bogda Mountain. Most of the hinterland is covered by the Gurbantunggut desert. The basin covers an area of 13×10^4 km^2. Since the formation of late Carboniferous, the basin has undergone the transformation of Hercynian, Indosinian, Yanshanian and Himalayan tectonic movements, and has become a typical large composite gas-bearing basin with 85.6×10^8 t oil resources and 2.1×10^{12} m^3 natural gas resources.

Coal-bearing strata are widely developed, with large thickness and huge coal resources. Also the CBM resources are abundant. While petroleum and natural gas exploration, the evaluation and exploration of CBM resources has been paid more and more attention, which is an important part of the energy strategy of the western development. The shallow coalfield geological data, deep seismic data and special geological data in the Junggar Basin are abundant, which lay a good foundation for studying the geological background of CBM in the basin.

The thickness of the coal seam in the Badaowan Formation is 0–50 m. There are four thick coal areas more than 20 m thick. The southern coal area is centered on Well Ka1 in Western Urumqi, with the maximum coal thickness of more than 70 m and an EW-trending distribution. The western coal area is located in east of Karamay–Urhe, with the maximum coal thickness of more than 50 m, where the southern coal accumulation center near Daguai is the thickest, the central coal accumulation center is located near Well Can 1, and the northern coal accumulation center is located near Well Xia 13, with the maximum thickness of more than 30 m and a NE-SW-trending distribution. The eastern coal area is located in the Cainan area and centered on Well Shanan 1, with the maximum coal thickness of nearly 40 m and a NE-SW-trending distribution. The central and western thick coal area is located in the line of Well Lunan 1–Well Lu 3, with the maximum coal thickness of 30 m, and the coal accumulation center is located at both sides, which is a saddle shape and extends in NE-SW direction. There is no coal belt in the northern basin. Most coal seams are less than 10 m thick, and less than 5 m in the central part. The variation of the coal thickness in the Badaowan Formation is basically consistent with that of coal seam accumulation. The No.2 and No.4 coal seams of the Badaowan Formation are developed continuously, and the thickness of coal seams is relatively large, which are the main coal reservoirs in the basin. The thickness of No.2 coal seam is 0–14 m, mainly distributed in the southeastern, middle-eastern and southern margins of the basin. There are a few areas distributed and not thick in the northeast margin. The coal seam is undeveloped in the west and northwest. There are two 5-m-thick coal areas. One is located between Fukang and Cainan, largely distributed and thicker, up to 14 m thick in a shape of flower near Fukang, and thick in east and thin in west. The other is located in the southern area of Well Qing 1 and the distribution is relatively large, more than 5 m thick and in a shape of a wide fan. The No.4 coal seam is the most widely distributed, and 0–12 m thick. Thick coal belts are basically distributed on the basin margin, and there are three thick coal belts more than 4 m thick. The eastern thick coal belt includes two coal-accumulating centers in the south and north, where the northern one is near Well Cai 2, with the maximum coal thickness of more than 7 m and in a shape of tongue, and the southern one is near Well Fu 1, with the maximum coal thickness of 12 m, and like a long lens thinning in all directions. The thick coal belt in the western margin includes three coal-accumulating centers in the south, middle and north, whose scale decreases from south to north. The southern coal-accumulating center is located around Well Ke 76 with the maximum coal thickness of more than 10 m. The middle coal-accumulating center is located around Well Ma 2 with the maximum coal-seam thickness of more than 6 m. The northern coal-accumulating center is located around Well Qi 2 with the maximum coal-accumulating thickness of more than

4 m. The coal-accumulating center is like a long lens with thick middle and thin edge. The thick coal belt in the central and western part is a narrow strip with two small coal-accumulating centers in the east and west. One is located in the north side of Well Lunan 1, and the other is located near Well Lu 1, with the maximum thickness of coal seam of more than 6 m. A thick coal area is developed in the southern margin, which is in the west of Urumqi. The maximum coal seam thickness is more than 10 m, and the coal body is a long lens extending from east to west. Coal-free areas are developed in the northern margin and central part of the basin, while the coal seams in other areas of the basin are not more than 4 m thick.

The thickness of the coal seam in the Xishanyao Formation is 0–40 m, and there are two thick coal belts more than 20 m thick. The primary coal-accumulating center of the southern thick coal belt is located in its west and the south of Well Qing 1, more than 40 m thick. The secondary coal-accumulating center is located near Haojiagou in the western part of Urumqi and up to 20 m thick, and the extension of the coal-accumulating center is consistent with that of the coal-accumulating belt. The eastern thick coal belt is composed of Fukang and Cainan coal-accumulating centers. The Fukang coal-accumulating center is the thickest in the basin, up to nearly 50 m, and EW distributed. The Cainan coal-accumulating center is near the line of Well Caitu 2 to Well Fu 1, more than 20 m thick. There are three thick coal areas greater than 20 m thick: the northeastern thick coal area is centered on Well Luncan 1, with the maximum coalbed thickness of 24 m, the western thick coal area is located near Daguai, east of Karamay, with the maximum thickness of 25 m, and the central and western thick coal area is centered on Well Lu 3, with the maximum thickness of more than 20 m. The maximum coalbed thickness in the most area of the basin generally does not exceed 10 m, and the thickness of the coal seam in the center, north and southwest of the basin is the smallest or even missing. The No.6 coal seam of the Xishanyao Formation is the main coal seam in this area, but its distribution is limited.

The southern coal seam of the Badaowan Formation is characterized by strip structure. Macroscopic coal types include bright coal, followed by semi-bright type. Macroscopic coal composition is mainly bright coal, vitrain, fusinized lens and a few durain strips, with rare minerals. The vitrinite content is 38.8–100%, with an average of 77.2%, inertinite content is 0–55%, with an average of 9.8%, liptinite content is 0–40%, with an average of 8.7%, raw coal ash is 6.00–25.00%, total sulfur content is 0.30%–0.58% and volatile matter is 15.86–49.78%, indicating low-medium ash and low sulfur coal. Coal rank ranges from flame coal to fat coal and vitrinite reflectance is 0.5–1.0%. Coal ranks increase from west to east—flame coal in Sikeshu-Changji, gas coal and flame coal in Changji-Urumqi, and gas coal and fat coal in Urumqi-Baiyanghe. Coal seams in the eastern basin have linear, banded and lenticular structures. Macroscopic coal lithotypes are bright type and semi-bright type. The microcomponents are mainly vitrinite of 57–100% and an average of 79.5%. The second liptinite is 0–47%, with an average of 16.6%, which contains a certain amount of sapropelic components. The inertinite content is low, ranging from 0 to 6%, with an average of 3.8%. From the edge to the basin center, the vitrinite content decreases and the liptinite content increases. The ash content is 4.36–33.82%, total sulfur content is 0.19–0.39%, and volatile matter is 48.07–54.50%, indicating medium–low ash coal. The shallow are old lignite and flame coal, and the deep are gas coal. The vitrinite reflectance is 0.45–0.68% and increases toward the basin center. Coal seams in the western and northwestern parts have stripped or linear structures. Macroscopic coal lithotype is bright type, and macroscopic coal composition is mainly vitrain and bright coal. The vitrinite content is 73–100% with an average of 82.7%, the liptinite content is 0–7% with an average of 4.3%, but it is as high as 63% for the coal samples from Well Xia 6, and the stratified distribution of sporophyte and cutinite was observed. The inertinite content is 0–21% with an average of 13%. The ash content of raw coal is 11.58–23.90%, the total sulfur content is 1.08%, and the volatile matter is 51.98–59.49%, indicating medium-ash, low-sulfur and highly volatile coal. The vitrinite reflectance ranges from 0.4 to 0.68%, with an average of 0.54%. The coal rank is lignite to gas coal, mainly flame coal, and the coal rank from the basin edge to the basin center is increasing. There are a few coal core samples and coal data in the hinterland of the basin. The vitrinite content is 64–100% with an average of 81.3%, the liptinite content is 0–23% with an average of 14.3%, and the inertinite content is 0–13% with an average of 4.3%. The vitrinite reflectance ranges from 0.66 to 0.73%, with an average of 0.69%. The coal rank changes from gas coal to fat coal, and the vitrinite reflectance reaches 1.3% near the southern margin of the basin,

The coalbed of the Xishanyao Formation in the southern basin shows stripped and homogeneous structures, and a few of them are granular. Macroscopic coal lithotypes are mainly bright type and semi-bright type, followed by semi-dull coal. The vitrinite content is 20–100% with an average of 75.2%, and that of the liptinite content is 0–85%, with an average of 6.7%. Only a few coal seams are rich in liptinite and some coal seams contain sapropel components, while that of inertinite content is 0–55% with an average of 15.5%. The maceral differentiation of upper coal seams is obvious, where the ash content is 6–15%, low to only about 3% and high to 27%; the sulfur content is 0.12–1.0% and low to 1.01–1.68%; the volatile content is generally 30–40%, low to less than 20% and high to about 50%, indicating low-ash, low-sulfur and highly volatile bituminous coal. The vitrinite reflectance ranges from 0.47 to 1.0%, with an average of 0.68%. The coals are mainly flame coal, followed by gas coal and fat coal. The coal rank increases from west to east, and from shallow to deep. Coal seams in the eastern basin have linear and stripped structures. Macroscopic coal lithotypes are mainly dull coal and semi-dull coal, followed by semi-bright coal and less bright coal. The vitrinite content is 30–70% with an average of 46.3%, inertinite content is 0–55% with an average of 23.3%, liptinite content is 0–70% with an average of 30.5%. Maceral composition is characterized by high liptinite, inertinite and low vitrinite. The ash content is 6.12–13.59%, total sulfur content is 0.13–0.78%, volatile matter is 28.72–38.09%, indicating bituminous coal with low ash, low sulfur and medium volatile content. The vitrinite reflectance ranges from 0.48 to 0.65%, with an average of 0.57%. It is dominated by flame coal and a small amount of lignite and gas coal. From coal macerals in the western-northwestern basin, the vitrinite content is 50–95% with an average of 77.8%, inertinite content is 0–42% with an average of 9.6%, liptinite content is 5–21% with an average of 12.6%; the vitrinite reflectance ranges from 0.53 to 0.59%, with an average of 0.56%; the coal rank is flame coal. From coal macerals in the hinterland of the basin, the vitrinite content is 73–95% with an average of 87.2%, liptinite content is 3–24% with an average of 10.6%, inertinite content is 0–6% with an average of 2.2%; the vitrinite reflectance ranges from 0.5 to 0.95%; and the coal ranks are mainly gas coal, a little fat coal and flame coal.

The new round evaluation on CBM resource shows that the total amount of shallow CBM resources at 2000 m in the Junggar Basin is 38,698.37 × 10^8 m^3, including 14,349.65 × 10^8 m^3 in the Badaowan Formation, and 24,348.72 × 10^8 m^3 in the Xishanyao Formation.

9.2.2.2 Case Study on Ordos Basin

CBM Enrichment Zones Divided

To divide CBM enrichment zones, the following aspects should be considered: basic geological factors such as the age and burial depth of coal-bearing strata; main geological faults, overall structural characteristics and hydrogeological conditions; previous evaluation of CBM resources and calculation units, and traditional block names; and differences of all factors in different blocks. According to the above principles, the Ordos Basin was divided into 19 CBM enrichment zones: 3 on the eastern margin, including Fugu, Wubu and Daning-Jixian; 3 on the southern margin, including Hancheng, Chengcheng and Tongchuan; 9 on the western margin, including southern Xifeng, Xifeng, Huanxian, Weizhou, western Pingluo, Rujigou and Shitanjing, Shizuishan and Wuhai; and 4 in the central part, including Etuoke Banner, Wushen Banner, Hangjin Banner and Dongsheng Banner.

Screened by Area and Resource Abundance

Table 9.16 lists the basic data of the area and resource abundance of the 19 zones. Screening by area and resource abundance, Rujigou, Shitanjing and Shizuishan were screened out, and the other 16 are retained.

Screened by Gas Content

Table 9.17 lists the basic gas contents of 19 zones.

According to gas content, Wuhai, Etuoke Banner and Dongsheng Banner were selected, and the other 13 zones are retained.

Ordered by Key Factors

Used the risk probability method and the comprehensive queuing coefficient method to rank the remaining 13 enrichment zones (Table 9.18).

Selected by Fuzzy Comprehensive Evaluation Method and Optimal Segment Method

Because of some uncertainties of the evaluated object A itself, for a factor x_i of A, if A is an accurate value, the degree of A belonging to y_i is expressed by a fuzzy value; if A is uncertain, then the degree of A belonging to y_i is expressed by an interval. In addition, according to the fact that an ordinary real number is a special interval, the evaluation index expressed by a fuzzy value is also expressed by an interval. Therefore, for an evaluation factor x_i, the degree of A belonging to y_i can be expressed as intervals $[r_{ij}^-, r_{ij}^+]$ [0, 1], $i = 1, 2, \ldots, n$; $j = 1, 2, \ldots, m$.

Then an interval fuzzy mapping $f{:}x \to \mathrm{IF}(Y)$ is obtained.

$$x_i \to f(x_i) = \left(\left[r_{i1}^- - r_{i1}^+\right], \left[r_{i2}^-, r_{i2}^+\right], \ldots, \left[r_{im}^-, r_{im}^+\right]\right)$$

Where, IF(Y) is a fuzzy set for all intervals on Y. The fuzzy comprehensive evaluation matrix is as follows

$$\boldsymbol{R}_{\mathrm{A}} \begin{bmatrix} [r_{11}^-, r_{11}^+] & [r_{12}^-, r_{12}^+] & \cdots & [r_{1m}^-, r_{1m}^+] \\ [r_{21}^-, r_{21}^+] & [r_{21}^-, r_{21}^+] & \cdots & [r_{2m}^-, r_{2m}^+] \\ & & \vdots & \\ [r_{n1}^-, r_{n1}^+] & [r_{n2}^-, r_{n2}^+] & \cdots & [r_{mm}^-, r_{mm}^+] \end{bmatrix} \tag{9.59}$$

First selected CBM enrichment zones using the fuzzy comprehensive evaluation method and then classified them by the optimal segment method (Table 9.19).

Table. 9.16 Basic data of the area and resource abundance of 19 CBM enrichment zones

CBM enrichment zone	Daning-Jixian	Wubu	Fugu	Hancheng	Chengcheng	Tongchuan	Southern Xifeng	Xifeng	Huanxian
Resource abundance/(10^8 m^3/km^2)	1.4	1.1	0.8	1.5	1.1	0.9	0.8	0.7	0.8

CBM enrichment zone	Weizhou	Pingluoxi	Rujigou	Shitanjing	Shizuishan	Wuhai	Etuoke Banner	Dongsheng Banner	Wushen Banner	Hangjin Banner
Resource abundance/(10^8 m^3/km^2)	0.6	0.8	1.1	1.0	1.1	0.8	0.9	0.6	0.8	0.7
Area/km^2	2200	3100	180	163	185	3600	2600	4400	3500	1700

Table. 9.17 Basic gas contents and coal ranks

CBM enrichment zone	Daning-Jixian	Wubu	Fugu	Hancheng	Chengcheng	Tongchuan	Southern Xifeng	Xifeng	Huanxian
Average gas content/(m^3/t)	13.2	8.5	7.3	10.1	6.1	6.2	4.3	5.0	6.5
Average R^{o}_{max}	1.2	1.1	0.82	1.3	1.2	1	0.5	0.6	0.6

CBM enrichment zone	Weizhou	Pingluoxi	Rujigou	Shitanjing	Shizuishan	Wuhai	Etuoke Banner	Dongsheng Banner	Wushen Banner	Hangjin Banner
Average gas content/(m^3/t)	6.6	12.1	11.3	9.1	4.2	3.8	2.5	3.0	4.3	3.2
Average R^{o}_{max}	1.5	2.6	1.6	1.8	0.8	0.7	0.7	0.6	0.55	0.5

Table. 9.18 Ranked result

CBM enrichment zone	Comprehensive queuing coefficient	Risk probability
Daning-Jixian	0.8543547	0.3510743
Hancheng	0.873957	0.3835036
Wubu	0.8652104	0.3861968
Etuoke Banner	0.8757177	0.3863557
Wushen Banner	0.8915156	0.4158366
Hangjin Banner	0.9009197	0.4336644
Fugu	0.9049333	0.4542109
Chengcheng	0.9195208	0.4655008
Tongchuan	0.9230684	0.4679414
Southern Xifeng	0.9243153	0.4732202
Xifeng	0.9305209	0.4870079
Huanxian	0.9452734	0.5112469
Dongsheng Banner	0.9545026	0.5227991

Table. 9.19 Optimized enrichment zones in Ordos Basin

Category	CBM enrichment zone
A	Daning-Jixian,Daning-Jixian,Wubu
B	Wushen Banner,Hangjin Banner,Fugu
C	Chengcheng,Southern Xifeng,Dongsheng Banner

Table. 9.20 Optimized CBM zones

Basin / Class	A	B	C
Qinshui	Jincheng, Yangcheng, Anze, Yangquan	Tunliu, West Tunliu, Qinyuan, Xishan	Wuxiang, Huoxi, West Qinyuan
Ordos	Daning-Jixian, Hancheng, Wubu	Wushen Banner, Hangjin Banner, Fugu	Chengcheng, South Xifeng, Dongsheng Banner
Junggar	Fukang, Huainan	Jimusar, Shaqiu River	Xiazijie, Shixi

9.2.2.3 Optimized CBM Enrichment Zones in Typical Basins

The CBM enrichment zones in typical basins such as Ordos Basin, Qinshui Basin and Junggar Basin were optimized and ranked by using this software. Nine favorable CBM zones were selected, namely Jincheng, Yangcheng, Anze and Yangquan in Qinshui Basin, Daning-Jixian, Hancheng and Wubu in Ordos Basin, Junnan and Fukang in Junggar Basin (Table 9.20).

Taking Qinshui Basin, Ordos Basin and Junggar Basin, which have rich resources and have been explored, as examples, the evaluation method, parameter system and evaluation criteria have been built, and then 9 CBM exploration targets were optimized, which can be exploited in the near future.

References

Debor JH (1964) Adsorption dynamic characteristics. Liu ZH, et al (trans). Science Press, Beijing (in Chinese)

Hrris I, Gayer R (1996) Coalbed methane and coal geology. London Geological Society, London

Joubert JI, Grein CT (1973) Sorption of methane in moist coal. Fuel 52:181–185

McKee CR, Bumb AC, Way SC et al. (1986) Application of the relationship of permeability and depth to evaluation of the potential of coalbed methane. In: North China Petroleum Geological Bureau (ed) Coalbed methane translation collection. Henan Science and Technology Press, Zhengzhou

Qin Y, Fu XH, Yue W et al (2000) Relationship between sedimentary system and coalbed methane reservoir and caprock characteristics. J Palaeogeogr 2(1):77–83 (in Chinese)

Quan YK (1995) Preliminary study on several factors affecting CBM content. Nat Gas Ind 15(5):1–5 (in Chinese)

Wei ZT (1998) Numerical simulation on geological evolution history of coalbed methane. China University of Mining and Technology Press, Xuzhou (in Chinese)

Ye JP, Qin Y, Lin DY, et al. (1998) China's coalbed methane resources. China University of Mining and Technology Press, Xuzhou, pp 38–39 (in Chinese)

Ye JP, Shi BS, Zhang CC (1999) Coal reservoir permeability and influencing factors. J China Coal Soc 24(2):118–122 (in Chinese)

Zhang XM, Zhang SA, Zhong LW, et al. (1991) China's coalbed methane. Shaanxi Science and Technology Press, Xi'an (in Chinese)

10 Desorption-Seepage Mechanism and Development Schemes

10.1 Elastic Mechanics of Coal Rock in Multiphase Medium

In the process of CBM drainage and pressure relief, on the one hand, as fluid pressure decreases, CBM will be desorbed from coal matrix, resulting in shrinkage of coal matrix and enlargement of pores and fractures, and accordingly increasing permeability; on the other hand, as fluid pressure decrease, effective stress will increases, pore and fractures will close and permeability will reduce. The comprehensive effect of the above two aspects determines the change of permeability, and to some extent determines the continuous production of CBM wells. In the process of CBM drainage and pressure relief, either positive or negative effects of permeability are closely related to the elastic strain characteristics of coal rock, and rock mechanic and physical properties and the changing trend of coal rock are the decisive factors on the change of permeability, which is the most sensitive parameter in CBM production.

Rock mechanical properties mainly refer to rock deformation and strength. The former is usually expressed by stress–strain, and the latter is reflected by compressive strength, tensile strength, shear strength, or Young's modulus, Poisson's ratio, etc. Generally, the whole stress–strain curve of rock is composed of several segments such as compaction, linear elastics, non-linear deformation and residual strength (unstable development of fractures). In the process of CBM development, the first two stages are mainly involved in coal rock.

Coal reservoir is a three-phase medium, i.e. solid, liquid and gas. Triaxial mechanical experiment on natural coal samples, water-saturated coal samples and gas and water saturated coal samples is one of the important methods to study the elastic deformation of coal matrix and the deformation of coal reservoir in the process of CBM drainage and pressure relief.

10.1.1 Triaxial Mechanical Experiment

10.1.1.1 Experimental Equipment

Mechanical experiments on coal rock was carried out in the laboratory of Fracturing Center of Langfang Branch, Research of Institute of Petroleum Exploration & Development, using the Rock Mechanics Testing System produced by TerraTek Company, USA (Fig. 10.1).

The mechanical behaviors of coal rock were simulated at reservoir temperature, stress and pressure, i.e., the simulated overlying stress is 0–800 MPa, horizontal stress 0–140 MPa, reservoir pressure 0–100 MPa, and reservoir temperature 0–200 °C. Data acquisition was controlled by computer. Besides the Rock Mechanics Testing System, supporting equipment includes tools for sample preparation and conventional core analysis. All sensors used are calibrated and checked annually by the National Bureau of Standards and Metrology.

10.1.1.2 Experimental Samples

The experimental coal samples are anthracite collected from the Changping Coalmine and Sihe Coalmine in the Jincheng mining area in the south-central Qinshui Basin, which are the most promising blocks for CBM exploration and development in China.

10.1.1.3 Preparation of Samples

On a new working face, two large and relatively regular coal blocks with side length more than 30 cm each were taken and then transported to the transport tunnel, where they were cut into cubes of 20 cm × 20 cm × 20 cm each by a steel saw. After wrapped by several layers of toilet paper, and put into black plastic bags, and tightly wrapped with wide tape, they were carried out to the ground (Fig. 10.2). In laboratory, according to the experimental requirements, four cylindrical samples with diameter of 25 mm and height of 50 mm were drilled on every coal cube along the plane

Y. Song et al., *Coalbed Methane in China*,
https://doi.org/10.1007/978-981-33-4725-0_10

Fig. 10.1 Rock mechanics testing system

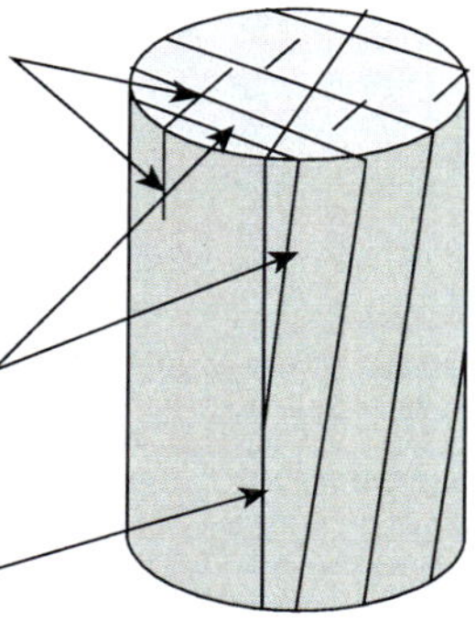

Fig. 10.2 Coal sample

direction, and the end of the coal cylinder was cut flat. The processing accuracy is in accordance with the requirements of mercury experiment and isothermal adsorption experiment defined by the International Institute of Rock Mechanics.

A group of natural coal samples were left before mechanical experiments, and the other three groups were immersed in 5% KCl solution and saturated by vacuumizing for 24–48 h. The diameter and length of the coal sample were measured on the bottom. Sample W-4 has a diameter of 23.72 mm and a length of 49.76 mm, and it weighs 32.5625 g before water saturation and 33.0685 g after water saturation. Sample WW-4 has a diameter of 23.58 mm and a length of 50.58 mm, and it weighs 30.9021 g before water saturation and 31.3835 g after water saturation.

10.1.1.4 Experiment Item

The mechanical experiment items are listed in Table 10.1. Under the conditions of air drying, water saturation and gas and water saturation, triaxial compression experiments and adsorption experiments were carried out on each coal sample. During the adsorption experiment, the deformation and permeability of the coal rock samples were measured.

10.1.1.5 Experimental Method

The experiments were carried out in strict accordance with the standards recommended by the International Society of Rock Mechanics (ISRM). The designed confining pressure is 8 MPa and the loading rate is 0.035 MPa/s. Data were recorded every 5 s, including time (s), axial stress difference (MPa), confining pressure (MPa), axial strain (mm/mm), radial strain 1 (mm/mm) (vertical interface fractures), radial strain 2 (vertical end fractures) (mm/mm), average radial strain (mm/mm) and volumetric strain (mm/mm).

10.1.2 Experimental Principle

The experimental data obtained from each coal sample are very large. The Graph-tool software prepared for the computer converted the experimental data into Excel files for subsequent data processing.

10.1.2.1 Static Elastic Modulus and Poisson's Ratio

In the literatures and experimental regulations on the mechanical properties of coal rock, the elastic modulus of coal rock generally refers to secant modulus, i.e. in uniaxial compression experiments, the slope of the straight line on the stress–strain curve is called Young's modulus, and the ratio of transverse strain to longitudinal strain is called Poisson's ratio. Since coalbed methane is produced

Table 10.1 Triaxial compression experiment

Coal sample	Sample No.	Triaxial compression		
		Natural coal sample	Water saturated coal sample	Gas and water saturated coal sample
No. 1	W-1	√		
	W-2		√	
	W-3			√
No. 2	WW-1	√		
	WW-2		√	
	WW-3			√

underground, what are the most concerned about are the mechanical properties of coal underground, that is the water and gas saturated coal reservoir how to express the mechanical behavior at confining pressure and stress–strain deformation, which are very different from the elastic modulus shown in uniaxial compression experiments in a general sense. In order to distinguish them, the modulus measured at confining pressure is called engineering elastic modulus, or elastic modulus for short, because it is the measured deformation of a coal sample after applied a static load. It is also called static elastic modulus and Poisson's ratio.

The secant modulus is the slope of the straight line from triaxial stress–strain relationship. In order to investigate the change of elastic modulus and Poisson's ratio in the stress–strain process, the formula of triaxial tangential elastic modulus and Poisson's ratio can be used to calculate the elastic modulus and Poisson's ratio point by point in a linearly elastic deformation stage.

The formula for triaxial tangent modulus of elasticity and Poisson's ratio are as follows.

$$E = \frac{\sigma_1(\sigma_1 + \sigma_3) - \sigma_2}{(\sigma_1 + \sigma_3)\varepsilon_1 - (\sigma_2 + \sigma_3)\varepsilon_2} \tag{10.1}$$

$$v = \frac{\sigma_2\varepsilon_1 - \sigma_1\varepsilon_2}{(\sigma_1 + \sigma_3)\varepsilon_1 - (\sigma_2 + \sigma_3)\varepsilon_2} \tag{10.2}$$

In pseudo-triaxial mechanics experiments, because $\sigma_2 = \sigma_3$, the above formula can be simplified as follows.

$$E = \frac{\sigma_1(\sigma_1 + \sigma_2) - \sigma_2}{(\sigma_1 + \sigma_2)\varepsilon_1 - 2\sigma_2\varepsilon_2} \tag{10.3}$$

$$v = \frac{\sigma_2\varepsilon_1 - \sigma_1\varepsilon_2}{(\sigma_1 + \sigma_2)\varepsilon_1 - 2\sigma_2\varepsilon_2} \tag{10.4}$$

where E is elastic modulus; v is Poisson's ratio; σ_1, σ_2, σ_3 are triaxial pressures in MPa, σ_1 respectively; represents vertical pressure, i.e. the axial pressure in experiment; σ_2 and σ_3 represents horizontal pressure, i.e. the confining pressure in experiment. In the experiment of pseudo-triaxial mechanics, σ_2 is equal to σ_3; ε_1 represents vertical strain, i.e. the axial strain in the experiment; ε_2 represents the transverse strain and refers to the average radial strain in the experiment.

By using the axial strain, average radial strain, axial pressure and confining pressure obtained in experiments in the Eqs. (10.3) and (10.4), the static elastic modulus (E) and the static Poisson's ratio (v) at each point can be calculated.

10.1.2.2 Volume Compressibility and Bulk Modulus

Volume compressibility is a measure of the relative volume change caused by confining pressure rising by 1 MPa at given temperature. The volume compressibility C_v is expressed as follows.

$$C_v = -\frac{1}{V}\frac{dV}{dP} \tag{10.5}$$

where V represents the volume of a coal rock sample, cm^3; dP represents pressure variation, MPa; dV represents volume variation, cm^3.

Obviously, pressure and volume change in the opposite direction, that is, as the pressure increases, the volume is compressed. The dimension of volume compressibility and pressure is the inverse of each other. The reciprocal of volume compressibility is bulk modulus (K_v).

$$K_v = \frac{1}{C_v} \tag{10.6}$$

The ratio of the volumetric strain rate to the differential of confining pressure obtained from experiments is the volume compressibility, which varies with the variation of confining pressure.

10.1.3 Triaxial Mechanical Characteristics

The triaxial compression experiments on natural coal samples, water-saturated coal samples, gas-water-saturated coal samples were carried out in the Changping Coalmine and Sihe Coalmine in Jincheng mining area. The stress–strain

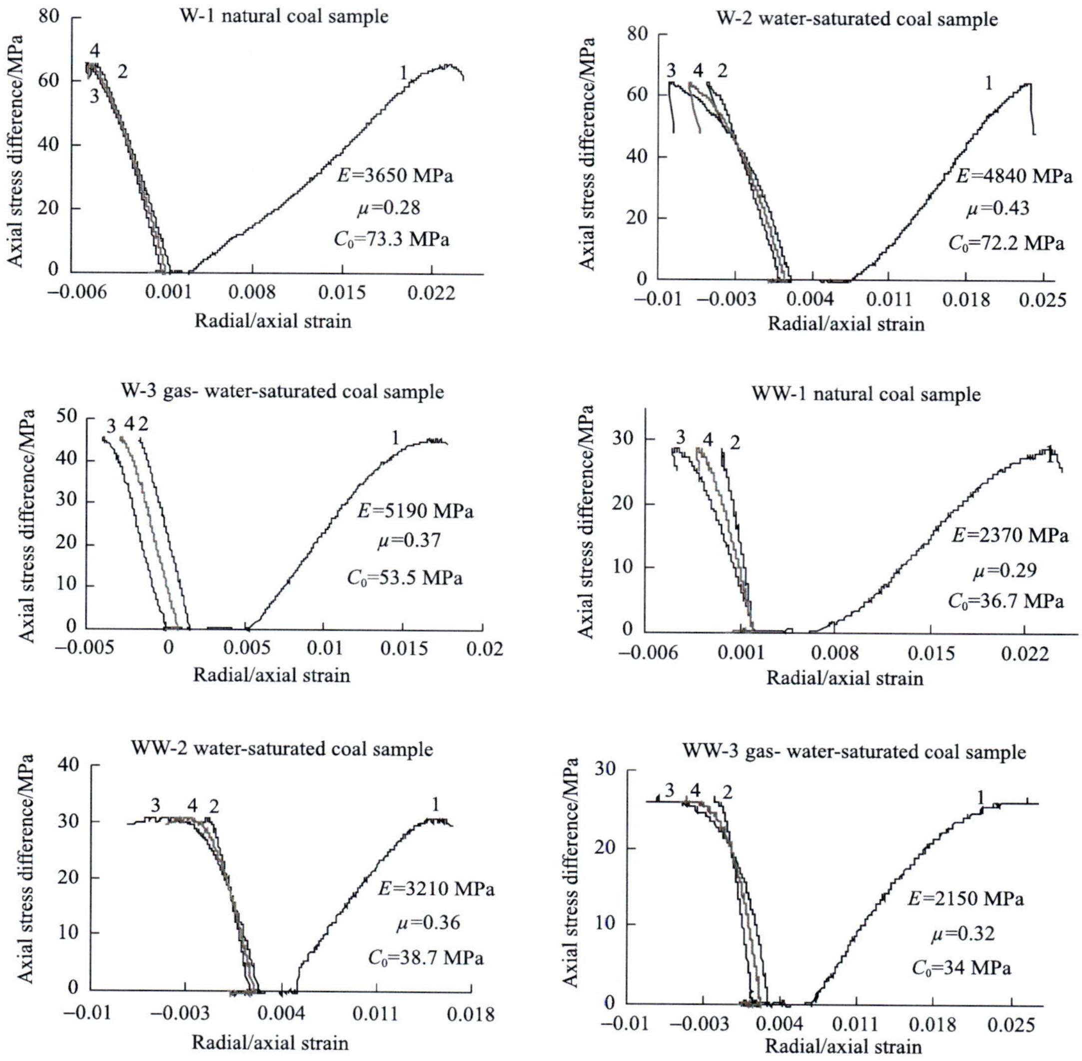

Fig. 10.3 Stress–strain curve from triaxial experiment. 1. Axial strain; 2. Radial strain 1; 3. Radial strain 2; 4. Average radial strain

curve of triaxial compression (Fig. 10.3) shows relatively homogeneous coal samples W-1, H-2 and Q-1, and radial strain 1 and radial strain 2 almost coincide; the anisotropy of gas- water-saturated coal samples is the greatest, that is, the difference between radial strain 1 and radial strain 2 is the greatest, followed by water-saturated coal samples.

The elastic modulus, Poisson's ratio and volume compressibility on each point can be obtained by substituting the experimental data into the above formulas, as shown in Table 10.2.

Based on the experimental data, the column contrast diagrams of compressive strength, elastic modulus and Poisson's ratio of natural coal samples, water-saturated coal samples and gas- water-saturated coal samples were obtained (Figs. 10.4, 10.5 and 10.6).

From the experimental results, it can be seen that the elastic modulus and compressive strength of natural coal samples are larger than that of water-saturated coal samples, and the elastic modulus and compressive strength of water-saturated coal samples are larger than that of gas-water-saturated coal samples (Figs. 10.4 and 10.5). The elastic modulus and compressive strength of coal samples have good consistency—large elastic modulus means high compressive strength. The Poisson's ratio of natural coal samples is less than that of water-saturated coal samples, and that of water- saturated coal samples is less than that of gas-water-saturated coal samples (Fig. 10.6), which indicates that the radial strain of gas-water-saturated.

coal samples increases. It can be seen that saturated water not only affects the deformation and strength of coal

Table 10.2 Triaxial mechanical test results of coal samples

Coal sample	Saturation	Lithology	Confining pressure/MPa	Young's modulus/MPa	Poisson's ratio	Volume compressibility/(1/MPa)	Compressive strength/MPa
W-1	Natural coal sample	Anthracite	8	3650	0.28	4.88×10^{-4}	73.3
W-2	Water-saturated coal sample	Anthracite	8	4840	0.43	8.24×10^{-4}	72.2
W-3	Gaswater-saturated coal sample	Anthracite	8	5190	0.37	3.17×10^{-4}	53.5
WW-1	Natural coal sample	Anthracite	8	2370	0.29	8.79×10^{-4}	36.7
WW-2	Water-saturated coal sample	Anthracite	8	3210	0.36	5.97×10^{-4}	38.7
WW-3	Gaswater-saturated coal sample	Anthracite	8	2150	0.32	8.03×10^{-4}	34.0

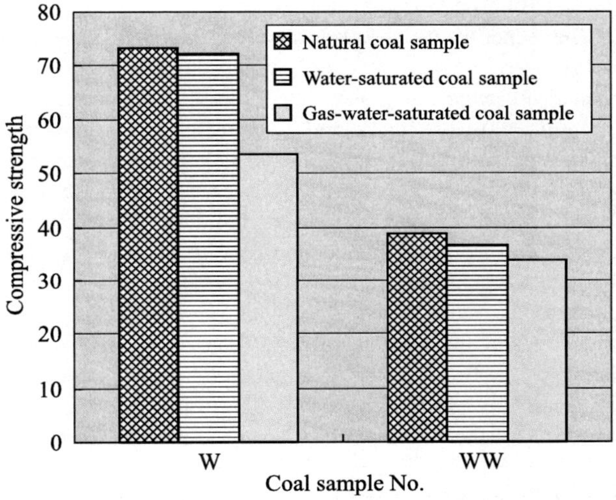

Fig. 10.4 The column contrast of compressive strength of natural coal sample, water-saturated coal sample and gas- water-saturated coal sample

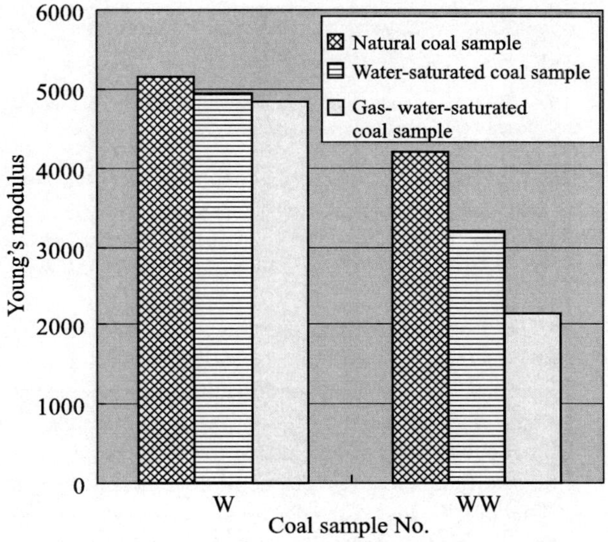

Fig. 10.5 The column contrast of Young's modulus of natural coal sample, water-saturated coal sample and gas- water-saturated coal sample

samples, but also has an important impact on the deformation mechanism of coal samples. The failure of natural coal samples is typical brittle failure. The relationship between static elastic modulus, compressive strength and Poisson's ratio, and coal rank is not clear in different media.

In summary, after gas and water saturation, the elastic modulus and compressive strength of coal sample decrease, while Poisson's ratio increases.

10.1.4 Volume Compressibility and Bulk Modulus

Under 3D isobaric condition, volumetric strain increases with confining pressure, and volume compressibility decreases logarithmically with confining pressure.

$$C_{\mathrm{v}} = -a \ln P_{\mathrm{c}} + b \tag{10.7}$$

where a and b are fitting coefficients (Table 10.3). On the contrary, bulk modulus increases exponentially with the increase of confining pressure. The maximum confining pressure of our experiment was 8 MPa. The volume compressibility and bulk modulus was obtained directly at 5 and 8 MPa, as shown in Table 10.3. The least square method was used to determine the coefficients a and b, that is, to minimize the sum of squares of residual errors.

$$a = \frac{n \sum_{i=1}^{n} C_{\mathrm{v}_i} \ln P_{\mathrm{c}_i} - \sum_{i=1}^{n} C_{\mathrm{v}_i} \sum_{i=1}^{n} \ln P_{\mathrm{c}_i}}{n \sum_{i=1}^{n} (\ln P_{\mathrm{c}_i})^2 - \left(\sum_{i=1}^{n} \ln P_{\mathrm{c}_i}\right)^2} \tag{10.8}$$

$$b = \frac{\sum_{i=1}^{n} C_{\mathrm{v}_i} - a \sum_{i=1}^{n} \ln P_{\mathrm{c}_i}}{n} \tag{10.9}$$

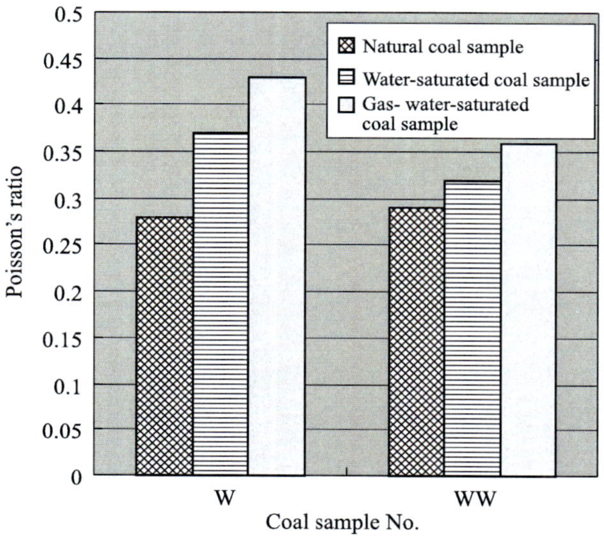

Fig. 10.6 The column contrast of Poisson's ratio of natural coal sample, water-saturated coal sample and gas- water-saturated coal sample

The triaxial mechanical experiments on natural coal, water-saturated coal and gas- water-saturated coal samples were carried out, and the experimental data were analyzed and processed (Fig. 10.7). The conclusions are as follows:

① At 3D isobaric pressure, the volumetric strain of coal samples increases with the increase of confining pressure, and the volume compressibility decreases logarithmically with the increase of confining pressure.

② The Young's modulus is related to the composition of the coal sample and the degree of coal metamorphism. Generally, the higher the degree of coal metamorphism, the greater the Young's modulus. The relationship between Poisson's ratio and coal rank is not clear.

③ The Young's modulus is positively correlated with the burial depth of the coal sample. The deeper the depth, the greater the Young's modulus.

10.2 Permeability Change of Coal Reservoir While Mining

10.2.1 Permeability Experiment

10.2.1.1 Equipment and Samples

The experiments on adsorption expansion and permeability of coal rock samples were carried out in the Development Research Institute of Langfang Branch, Research Institute of Petroleum Exploration & Development. The equipment and coal samples are the same as the triaxial mechanical experiment. The samples were polished with coarse sandpaper, and measured with a vernier caliper until the two bottoms are parallel, and finally polished with fine sandpaper. Two coal samples (lean and anthracite coals) were tested in adsorption expansion and permeability experiments.

The experiments of adsorption expansion and permeability were carried out by keeping the effective stress unchanged, which eliminates the effect of pressure on the compressibility of matrix and fracture, and the measured strain is the result of volume change, and the volume change is only affected by the change of gas composition. Because the experiment temperature and the humidity of the sample affect gas adsorption, the experiment was carried out at constant temperature and controlled humidity.

Since the maximum permeability is in the facial cleat direction and the cleat is perpendicular or nearly perpendicular to the bedding, the quantitative relationship between fluid pressure and radial expansion has more practical significance for the change of coal reservoir permeability.

10.2.1.2 Experiment Scheme and Procedures

The diameter and length of the bottom of the coal sample were measured. Sample W-4 has a diameter of 23.72 mm and a length of 49.76 mm, and it weighs 32.5625 g before water saturation and 33.0685 g after water saturation. Sample WW-4 has a diameter of 23.58 mm and a length of

Table 10.3 Volume compressibility and bulk modulus of coal rock

Coal sample No.	Saturation	Fitting coefficients		$P_c = 5$ MPa		$P_c = 8$ MPa	
		a	b	C_v	K_v	C_v	K_v
W-1	Natural coal sample	5.065	15.412	9.87	1.20	4.88	2.04
W-2	Water-saturated coal sample	3.647	15.823	15.48	0.71	8.24	1.20
W-3	Gas-water-saturated coal sample	4.532	12.594	6.32	1.52	3.17	3.15
WW-1	Natural coal sample	3.241	9.530	17.31	0.26	8.79	1.13
WW-2	Water-saturated coal sample	3.817	13.907	10.94	0.94	5.97	1.67
WW-3	Gas-water-saturated coal sample	4.213	16.790	15.92	0.87	8.03	1.24

Note P_c is confining pressure; C_v is volume compressibility, $\times 10^{-4}$; K_v is volume compression modulus, $\times 10^{3}$

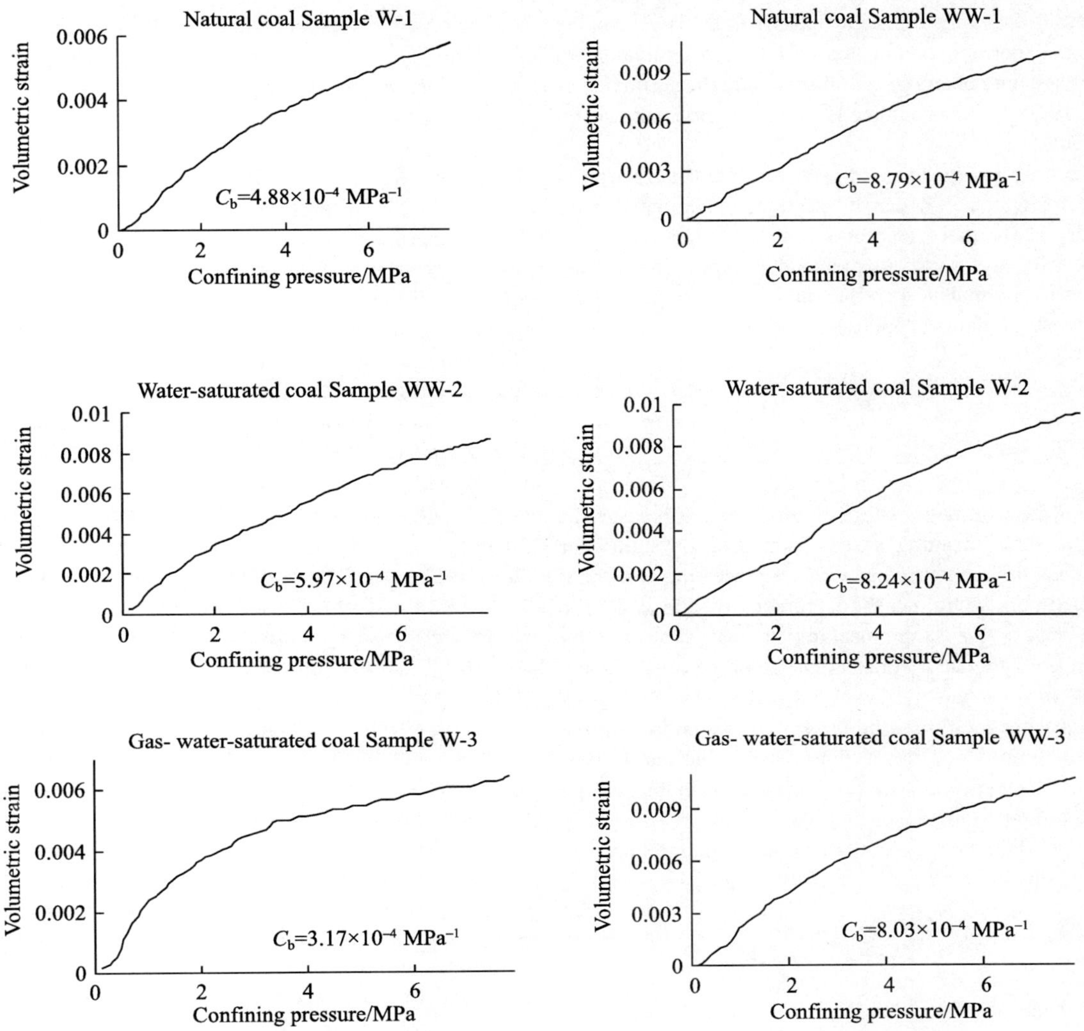

Fig. 10.7 Volumetric strain versus confining pressure

50.58 mm, and it weighs 30.9021 g before water saturation and 31.3835 g after water saturation.

At constant effective stress, the water-saturated coal sample was injected with CO_2 at 99.99% purity. Longitudinal and radial volumetric expansion, average CO_2 flow rate and permeability of the coal sample were measured at each CO_2 pressure (1.0, 2.0, 3.0 and 4.0 MPa) and each confining pressure (2.0, 3.0, 4.0 and 5.0 MPa). The experiment was stable for 6 h at each pressure, and then the outlet flow rate was measured, based on which the permeability was calculated (using the soapsuds method). Data were recorded every 60 s. Only the deformation and permeability caused by CO_2 adsorption were observed.

10.2.1.3 Experiment Principles

In the process of CBM drainage and pressure relief, with the desorption of gas and the drainage of water, the coal matrix shrinks and the permeability of coal reservoir increases, which is commonly called the positive effect of shrinkage of coal matrix. At the same time, the decrease of pressure and the increase of effective stress result in shrinkage of coal matrix and decrease of fractures, thus decreasing permeability, which is the negative effect of shrinkage of coal matrix.

Adsorption expansion and desorption of coal are reciprocal reversible processes. In the process of surface drainage and depressurization to develop CBM, the deformation of coal is within the range of elastic deformation. Therefore, the adsorption expansion parameters of coal are equivalent to the shrinkage parameters of coal matrix, and the shrinkage of coal matrix can be simulated by adsorption expansion experiments.

Research by Sharpalani et al. showed that the adsorption strain of CH_4 and CO_2 can be accurately simulated by a Langmuir isothermal adsorption model. Therefore, an equation with the same mathematical expression as the Langmuir isothermal adsorption model is suitable for the adsorption strain data, as follows.

$$\varepsilon_v = \frac{\varepsilon_{max} P}{P + P_{50}} \tag{10.10}$$

where ε_v is the volumetric strain of adsorption at pressure P; ε_{max} is the same meaning as that expressed by Langmuir volume data in Langmuir equation, representing the theoretical maximum strain, i.e. the incremental value at infinite pressure; P_{50} is the same meaning as that expressed by Langmuir pressure data, representing the pressure when coal rock mass reaches half the maximum strain.

The experiments by Levine (1996) further show that the adsorption strain is not linear with pressure, but curvilinear. The curve is steeper at low pressure and smooth at high pressure, similar to the adsorption isotherm.

In Eq. (10.10), the differential of volumetric strain at any pressure is the strain rate (M_s), as follows.

$$M_s = \frac{d\varepsilon_v}{dP} = \frac{\varepsilon_{max} P_{50}}{(P + P_{50})^2} \tag{10.11}$$

The calculation of ε_{max} and P_{50} is carried out in two steps. The first step is to transform the Eq. (10.10) into a straight line type, as follows.

$$\frac{\varepsilon_v}{P} = -\frac{1}{P_{50}}\varepsilon_v + \frac{\varepsilon_{max}}{P_{50}} \tag{10.12}$$

The intercepts $\frac{\varepsilon_{max}}{P_{50}}$ and $\frac{1}{P_{50}}$ are obtained by linear fitting of Eq. (10.12).

The values of ε_{max} and $\frac{1}{P_{50}}$ can be obtained by calculating $\frac{\varepsilon_{max}}{P_{50}}$ and $\frac{1}{P_{50}}$ in the second step. The Langmuir type adsorption expansion equation can be obtained by substituting ε_{max} and P_{50} into Eq. (10.11). In practical experiments, their values are obtained by the least square method [i.e. Eqs. (11.8) and (11.9)].

10.2.1.4 Adsorption Expansion and Desorption Shrinkage

In the experiment on adsorption expansion of a coal sample, because coal has a stronger ability to adsorb CO_2 than CBM, and it is relatively fast to reach balance, so CO_2 is used in the experiment of adsorption expansion. At constant effective stress and temperature, the volume deformation (ε_v) at different fluid pressures (P) was observed (Table 10.4). The relationship between ε_v and ε_v/P is close to a straight line, and the relation between ε_v and P is the same as that of Langmuir equation (Fig. 10.8). The adsorption capacity of coal to gaseous medium increases with the increase of pressure, so it can be said that the larger the adsorption capacity, the larger the volume deformation of a coal sample.

Through experiments, at constant effective stress and temperature, the relationship between fluid pressure and volume deformation was built as the following graph (Fig. 10.8).

On the curve, intuitively, the volume deformation increases rapidly at lower fluid pressure, and increases slowly with the increase of fluid pressure. At variable unit pressure, the adsorption deformation rate decreases as a straight line until it reaches the maximum and is stable. On the contrary, with the decrease of fluid pressure and variable unit pressure, the shrinkage increases as a straight line. That is to say, the relationship between adsorption expansion and desorption shrinkage, and fluid pressure indicates that the shrinkage of coal matrix will increase gradually with the gradual discharge of CBM and the decrease of reservoir pressure in the actual development process, which has very important practical significance for improving reservoir permeability.

The results of adsorption expansion and permeability experiment were summarized in Tables 10.5 and 10.6.

The relationship between fluid pressure and permeability was obtained (Fig. 10.9). With the increase of fluid pressure and CO_2 adsorption, the permeability decreases rapidly at the beginning, but gradually tends to be stable. At constant

Table 10.4 Volume deformation of coal after adsorbing CO_2

Sample W-4			Sample WW-4		
P	ε_v	ε_v/P	P	ε_v	ε_v/P
0.986784	0.001541	0.001562	0.987137	0.00094	0.000952
1.976773	0.001969	0.000996	1.974267	0.002853	0.001436
2.990174	0.002293	0.000767	2.989413	0.003918	0.001311
3.990411	0.002419	0.000606	3.989508	0.004762	0.001194

Note P is fluid pressure, MPa; ε_v is volume deformation, mm/mm

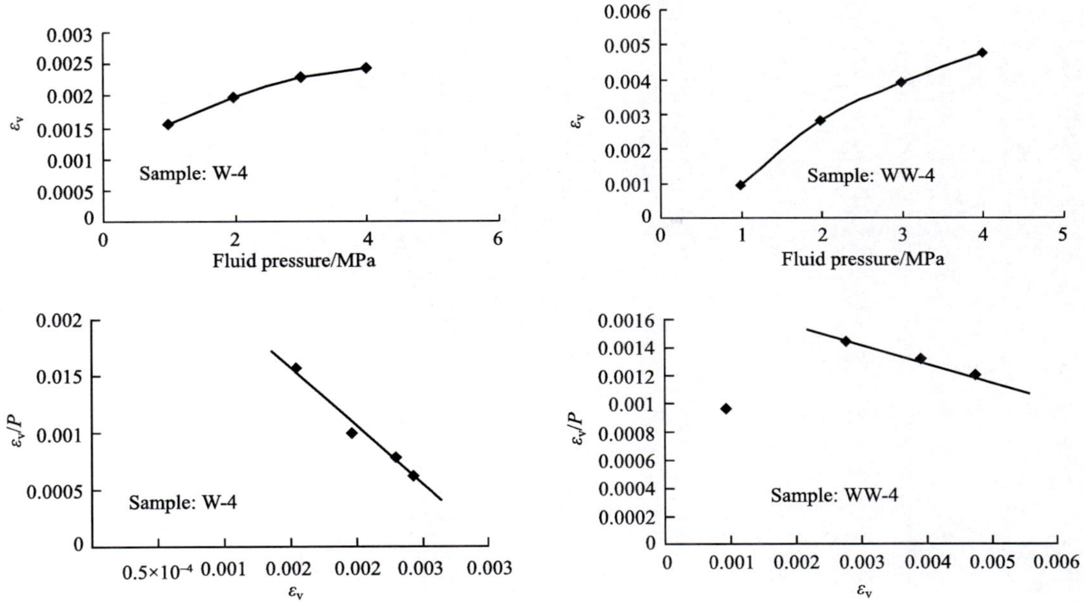

Fig. 10.8 Volume deformation (ε_v) versus fluid pressure (P)

Table 10.5 Experiments result on adsorption expansion and permeability of sample W-4

Sample No.	Diameter/mm	Length/mm	Confining pressure/MPa	Inlet pressure/MPa	Outlet pressure/MPa	Flow/mL	Time/s	Average flow rate/(mL/s)	Permeability/μm^2	Permeability/mD	Volumetric strain/(mm/mm)
W-4	23.72	49.76	2	1	0.1	4	86.89	0.046035	4.03000×10^{-4}	0.002618	0.402856
	23.72	49.76	3	2	0.1	6	57.31	0.104694	9.99567×10^{-5}	0.002073	0.099957
	23.72	49.76	4	3	0.1	5	33.18	0.150693	4.43634×10^{-5}	0.001742	0.044363
	23.72	49.76	5	4	0.1	4	33.07	0.120956	2.49423×10^{-5}	0.001600	0.024942

Table 10.6 Experiments result on adsorption expansion and permeability of sample WW-4

Sample No..	Diameter/mm	Length/mm	Confining pressure/MPa	Inlet pressure/MPa	Outlet pressure/MPa	Flow/mL	Time/s	Average flow rate/(mL/s)	Permeability/μm^2	Permeability/mD	Volumetric strain/(mm/mm)
WW-4	23.58	50.58	2	1	0.1	6	23.13	0.259403	0.000414	0.414371	0.005069
	23.58	50.58	3	2	0.1	4	11.79	0.33927	0.000103	0.102814	0.003777
	23.58	50.58	4	3	0.1	5	16.16	0.309406	0.000046	0.045632	0.002298
	23.58	50.58	5	4	0.1	5	16.30	0.306748	0.000026	0.025655	0.000262

effective stress, the permeability decreases linearly with the decrease of average gas pressure.

From Fig. 10.10, at constant effective stress, the permeability of the coal sample has a linear negative correlation with pressure reciprocal, and the permeability decreases with the increase of average fluid pressure.

There are two reasons for this change. On the one hand, when the effective stress is unchanged, the average fluid pressure increases and the confining pressure increases, which leads to the increase of the stress on the coal framework, the denser pore structure and the smaller permeability. On the other hand, the average pressure is the collision force from the gas in coal rock pores to unit area of the pore wall, which depends on the energy and the density of gas molecules. With the increase of pressure, the momentum of gas molecule increases, and the gas density increases, that is, the gas concentration increases. As a result, the mean free path decreases, the collision opportunity between gas molecules increases, the fluidity of molecules decreases, the slippage effect is not obvious, and the permeability decreases.

At constant effective stress, with the increase of fluid pressure, the volumetric strain increases gradually. The

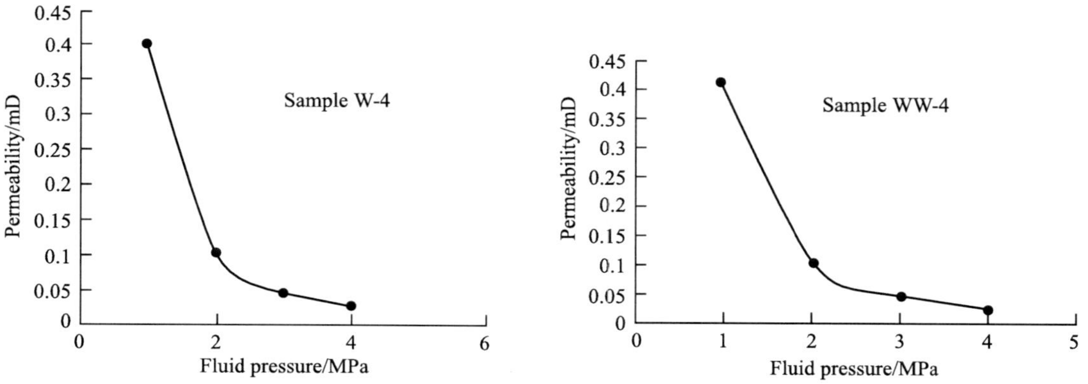

Fig. 10.9 Permeability versus fluid pressure

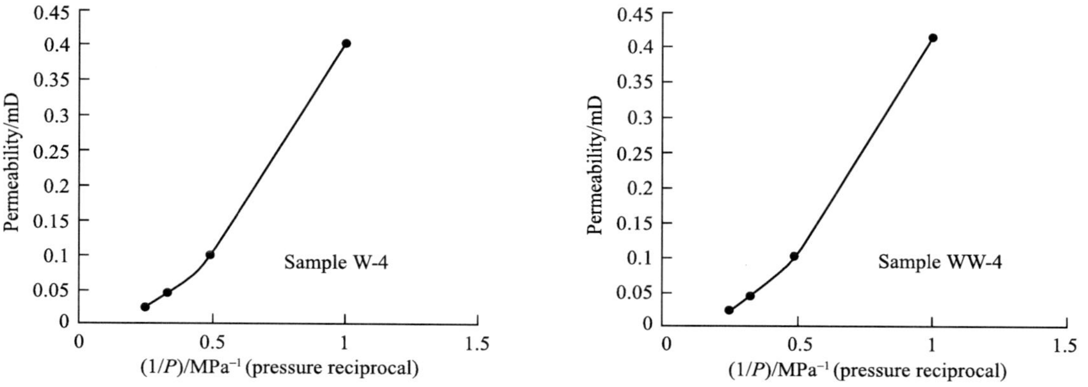

Fig. 10.10 Permeability versus pressure reciprocal

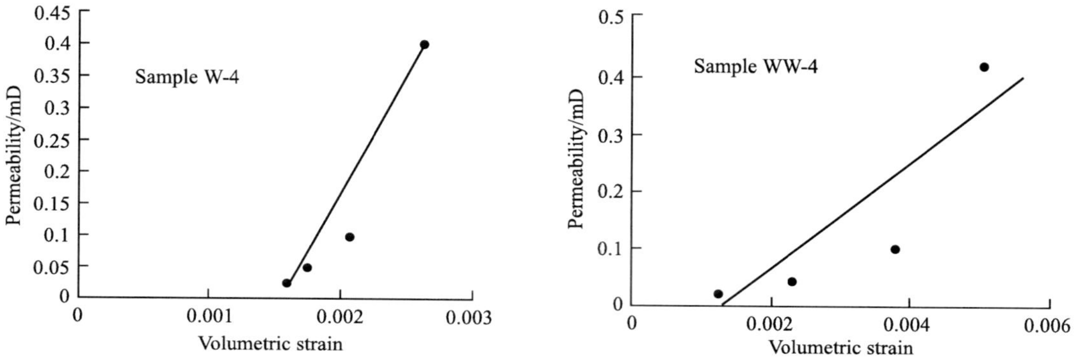

Fig. 10.11 Volumetric strain versus permeability

permeability of the sample was measured, and the variation of volumetric strain and permeability was obtained (Fig. 10.11).

The relationship between volume deformation and permeability can be fitted into a linear relationship. At the beginning of injecting CO_2, the permeability decreases rapidly and the deformation is large, and then the permeability decreases slowly and the deformation tends to be stable.

10.2.1.5 Field Measurement

Well TL-003 is the first CBM well of China United CBM Company in the south Qinshui Basin. The primary targets are coal seams No. 3 and No. 15, which were fractured separately. At different times, the well was produced at different modes, such as separate fracturing and commingled production, separate fracturing and production. The production history of the well is summarized as follows:

① From March 16, 1998 to March 16, 1999, commingled drainage was carried out, then workover and off production for 265 days.
② From Dec. 4, 2000 to Oct. 23, 2003, commingled production resumed, and the daily gas production was much worse than before.
③ From October 28, 2003 to March 9, 2004, No. 15 coal was plugged and only No. 3 coal produced. But the result showed that the daily gas production changed little. Then the well was shut again for 105 days.
④ From June 22, 2004 to October 1, 2005, with the cooperation between China and Canada, CO_2 injection was used to improve CBM recovery. The well experienced several short workovers, but little effect was seen on the production. Then after injecting CO_2, the daily production continued to increase.

Until October 1, 2005, after 1966 days (i.e. 5.4 years), Well TL-003 produced cumulative water of 5.5×10^4 tons, gas of 188×10^4 m^3, and the highest gas production is 7000 m^3/d (956 m^3/d on average). Figure 10.12 shows how the gas production changes with engineering work.

According to the production history, the production of a well depends on what measures were done to it.

① After commingled production for 365 days, the average gas production was 2323 m^3/d, the average water production was 38 m^3/d from the day when it began to work. Then it was shut in for 265 days.
② After opened again, the gas production decreased to 580.9 m^3/d and the water production was 37 m^3/d by commingled production, indicating the shut-in operation affected the production significantly.
③ After plugging coal 15 and only producing coal 13, the gas production was 557 m^3/d, and the water production was 0.7 m^3/d for 134 days. The gas production changed little, but seemed better on the production performance curve, indicating coal 15 contributed little.
④ After injecting CO_2 in June 2004, the gas production increased to 987.8 m^3/d, and the water production was 0.44 m^3/d.

According to incomplete understanding, only 2 of the 1400 CBM wells in China have been measured permeability before and during production. The static permeability measured in Well TL-003 before production (or before fracturing) in 1998 was 0.95 mD, and 13.6 mD after 6 years in May 2004. The permeability increased fourteen times after six years of production, which is basically consistent with

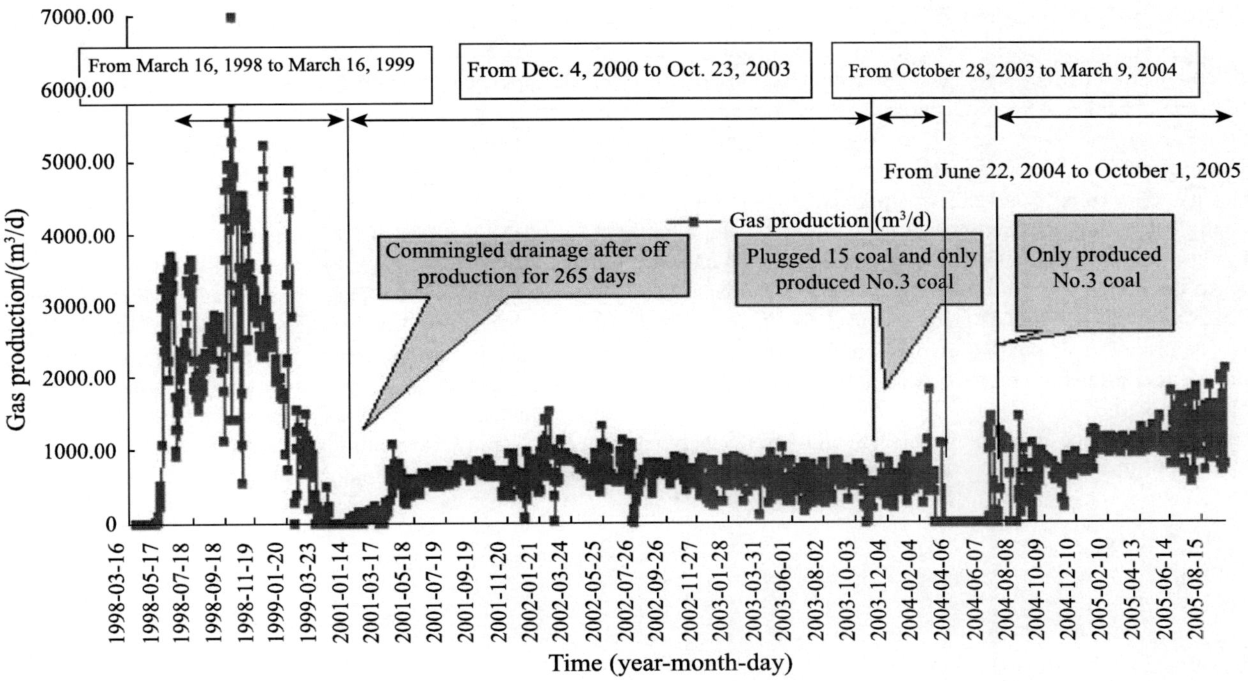

Fig. 10.12 Gas production history

the theory of permeability change, that is, the permeability of high-rank coal reservoirs in the Qinnan Basin increases with production and drainage.

In summary, when coal adsorbs methane, the volume expands (CO_2 is used in laboratory experiments, and the principle is the same). When desorbing, the shrinkage, adsorption expansion and desorption shrinkage of coal matrix are reversible. Therefore, when discussing the characteristics of reservoir pressure relief, the properties of coal matrix desorption shrinkage can be reflected by adsorption expansion experiment. Therefore, in laboratory, by injecting CO_2 into a coal sample, the coal sample expands, and then the change of permeability is investigated to reflect the changes of coal reservoir characteristics in the process of pressure relief.

The experiment shows that the volume of the coal sample increases with the increase of fluid pressure at constant effective stress and temperature. On the contrary, in the production process, with the drainage of CBM, the reservoir pressure decreases, and the shrinkage of coal matrix increases, which is conducive to the increase of permeability. It is of great practical significance. Experiments show that the permeability decreases with the increase of fluid pressure when injecting CO_2. At the same time, the experimental data of adsorption expansion show that the volume deformation increases rapidly at low fluid pressure. With the increase of fluid pressure, the volume deformation increases slowly until it reaches the extreme value and tends to be stable.

10.2.2 Self-regulating Effect Model

Although the permeability of conventional oil and gas reservoirs changes with production, it is much smaller than that of coal reservoirs, because the response of coal reservoirs to stress is more sensitive than that conventional oil and gas reservoirs. Coal reservoir permeability is not only controlled by internal factors such as fracture development, but also strongly influenced by external factors which may make the permeability increase by two orders of magnitude (Su et al. 2001). External factors, especially stress, induce the deformation of coal reservoir, which is related to the mechanical properties of coal, such as soft, brittle and low elastic modulus.

There are three external factors affecting coal reservoir permeability as follows.

(1) Effective stress

Effective stress is the result of total stress (perpendicular to fractures) minus fluid pressure (in fractures), called effective normal stress. It is a controlling factor on fracture width. With the increase of effective stress, fractures decrease in width or even closes, which leads to the sharp decreases of the permeability.

(2) Klinkenberg effect

When gas flows in porous media, the laminar velocity near the solid surface is nearly zero due to the viscous property of the fluid. But for some gases, there is no such phenomenon, but molecular slip phenomenon. In porous media, because the average free path of gas molecule is on an order of magnitude with the fluid passage, the gas molecule interacts with the passage wall (collision), which results in the slip of gas molecule along the pore surface and increases the molecular velocity. This phenomenon is called molecular slip. This effect due to the interaction between gas molecules and solids is called Klinkenberg effect, which was proposed by Klinkenberg (1941).

Klinkenberg (1941) found that the permeability measured with air is different from that measured with liquid and the permeability measured with air is always higher than that measured with liquid. On the basis of experiments, Klinkenberg proposed that the velocity of liquid on the surface of sand grains is zero, while gas on the surface of sand grains has a certain velocity. That is to say, the gas has slippage phenomenon on the surface of sand grains. Because of the slippage effect, the gas has a larger velocity at a given pressure drop. Klinkenberg also found that for a given porous medium, the calculated permeability decreases with the increase of average fluid pressure.

(3) Shrinkage effect of coal matrix

Experiments show that coal mass expands while adsorbing and shrinks when desorbing gas. In the process of CBM production, when the reservoir pressure drops below the critical pressure, CBM begins to desorb. With the desorption of methane, the coal matrix shrinks. Because the coal mass is affected laterally by confining pressure, the shrinkage of coal matrix can not cause the horizontal strain of the whole coal seam, but can only cause local lateral adjustment and strain along fractures. The shrinkage of the matrix along fractures causes the increase of fracture width and permeability.

Because of different properties, the shrinkage rate is also different, some coals hardly shrink, and others shrink greatly. There are few experimental data on strain caused by desorption or adsorption of coal, due to difficult experiments and less researches. The shrinkage test of coal matrix is carried out by linear or volumetric strain of coal matrix at different gas pressures.

The influences of the above three factors on the permeability of coal reservoir varies with the properties of coal itself. Coal seams not easy to shrink or not to shrink at all are mainly affected by effective stress, and permeability decreases with the increase of effective stress. Coal seams easy to shrink primarily show matrix shrinkage, and the permeability increases with the increase of shrinkage.

10.2.2.1 Study on Effective Stress

Many scholars studied the effects on permeability of coal samples by increased triaxial stress (Somerton et al. 1975). The experimental data show that the permeability of coal sample decreases exponentially with the increase of effective stress. Later, McKee et al. (1987) and Seidle et al. (1992) carried out theoretical research on the exponential relationship in this respect, but the premise of their research is that solid grains are incompressible.

$$k = k_0 e^{-3C_p \Delta\sigma} \tag{10.13}$$

In the formula above, k is permeability, mD; k_0 is initial permeability, mD; C_p is average volume compressibility; and σ is effective stress.

10.2.2.2 Coal Matrix Shrinkage

The effect of shrinkage of coal matrix on fracture permeability was first quantified by Gray (1987). Since then, a large number of scholars have studied the permeability depending on pore pressure, and put forward many models. Foreign research is mainly limited to low-rank coal, but less to high-rank coal.

Any change in the volume of coal matrix will cause the change of fracture width and gas flow in coal. The shrinkage of coal matrix caused by gas desorption will increase fractures, which will lead to the increase of coal permeability. When CBM is discharged and produced, the liquid pressure decreases and the effective stress increases, which closes the fractures and reduces the permeability. Therefore, when the shrinkage of coal matrix is less than the effective stress, the permeability decreases, while when the shrinkage is greater than the effective stress, the permeability increases.

Palmer and Mansouri (1996) have developed a new theoretical formula for stress-dependent permeability, in which stress and shrinkage of coal matrix are taken into account.

$$\frac{\varphi}{\varphi_0} = 1 + \frac{C_m}{\varphi_0}(P - P_0) + \frac{C_0}{\varphi_0}\left(\frac{K}{M} - 1\right)\left(\frac{bP}{1+bP} - \frac{bP_0}{1+bP_0}\right)$$

$$\frac{k}{k_0} = \left(\frac{\varphi}{\varphi_0}\right)^3 \tag{10.14}$$

where k is permeability, mD; φ is fracture porosity; C_m is compressibility of coal matrix, psi[1]; C_0 is Langmuir volume constant; b is Langmuir pressure constant, psi^{-1}; P is pressure, psi; and subscript 0 indicates initial state.

Production and experimental results in the San Juan Basin show that when pore pressure decreases, CBM desorption and permeability increase. In the process of drainage and depressurization, the change of fracture permeability is affected by the initial effective horizontal stress, and the volume of matrix shrinkage is proportional to the volume of desorbed gas.

Sawyer et al. (1990) developed a 3D dual-medium reservoir model to simulate the changes of internal stress and shrinkage versus permeability of coal matrix during gas production.

$$\varphi = \varphi_i[1 + C_p(P - P_i)] - C_m(1 - \varphi_i)\frac{\Delta P_i}{\Delta C_i}(C - C_i)$$

$$k = k_i\left(\frac{\varphi}{\varphi_i}\right)^3 \tag{10.15}$$

where φ is fracture porosity; φ_i is initial fracture porosity; C_p is pore volume compressibility, psi^{-1}; P_i is pore pressure of initial reservoir, psi; P is pore pressure of reservoir, psi; C_m is compressibility of coal matrix, psi^{-1}; C_i is gas concentration of initial reservoir; c is gas concentration of reservoir; ΔP_i is maximum pressure; ΔC_i is the gas concentration change based on initial desorption pressure.

The most common use of the model is in a CBM reservoir numerical simulator, such as COMET 3D simulator. Geometrically, the change of permeability is caused by shrinkage of coal matrix. The following formula is used.

[1] 1 psi = 1 1 bf/in^2 = 6.89 × 10^3 Pa.

$$\frac{k_{f2}}{k_{f1}} = \frac{(1 + 2C_x\Delta P/\varphi_{f1})^3}{(1 - C_x\Delta P)} \quad (10.16)$$

where k_f is fracture permeability, mD; C_x is compressibility, psi^{-1}; ΔP is pressure change, psi; φ_f is fracture porosity; subscripts 1 and 2 represent initial and final values respectively.

10.2.2.3 Self-regulating Effect Model of Coal Matrix

In the process of CBM drainage, on the one hand, fluid pressure decreases and CBM is desorbed from coal matrix, which causes the shrinkage of coal matrix and enlarges fractures, and results in the increase of permeability. On the other hand, as fluid pressure decreases, the effective stress of coal reservoir increases, pores and fractures close, and permeability reduces. The comprehensive effect of the two factors determines the changing law of permeability in the continuous production process of CBM wells, and thus to some extent determines the continuous production of CBM wells.

The model of coal matrix self-regulating effect mainly considers the negative effect of effective stress and the positive effect of coal matrix shrinkage due to the decrease of fluid pressure in the process of CBM development.

(1) Effective stress

Previous experiments have proved that the change of effective stress leads to the exponential change of permeability. Volume compressibility is directly related to rock mechanics such as Young's modulus and Poisson's ratio.

(2) Shrinkage of coal matrix

The shrinkage of coal matrix is related to the amount of desorbed CBM and the volume deformation caused by the desorption of CBM.

(3) Self-regulating effect model

Permeability is one of the most important influencing factors in CBM production. In the process of CBM depressurization and drainage, the increase of effective stress causes the compression of coal matrix, the decrease of pores and fractures, the decrease of permeability and negative effect. At the same time, due to the desorption of CBM, coal matrix shrinks, fractures expand, permeability increases and produces positive effects. Two effects act simultaneously. When the negative effect is greater than the positive effect, permeability decreases and productivity decreases. When the positive effect is greater than the negative effect, permeability increases and productivity increases. The influences of the two effects on the permeability of coal reservoir is influenced by many factors. Although predecessors have done some work in this respect, they all built models through factors they payed attention to. So although there are many models, they are far from fitting the actual situation correctly. There is still much work to be done in this respect. Most foreign scholars paid more attention to medium- and low-rank coal. In China high-rank coal is more important, but less researches were carried out on it. It can be confirmed that the self-regulating effect of coal matrix is related to coal rank, and is closely related to coal rock mechanical properties.

It is concluded that the two aspects of the self-regulating effect of coal matrix are related to the elastic mechanical properties of coal matrix, and there are many factors affecting the self-regulating effect. In a specific mine field, the elastic properties of coal matrix are relatively fixed, so it should be feasible to describe the elastic properties related to the self-regulating effect of coal matrix using some comprehensive indicators. Based on the results of experiments and previous studies, we propose a preliminary model based on comprehensive indicators.

$$\frac{k_j}{k_i} = \left(\frac{\varphi_j}{\varphi_i}\right)^3 = \left(\frac{\varphi_i + C_v(P_j - P_i) + \left(\frac{\varepsilon_{max}P_j}{P_j + P_{50}}\right)}{\varphi_i}\right)^3 \quad (10.17)$$

where C_v is volume compressibility; ε_{max} has the same meaning as Langmuir's volume data in Langmuir equation, representing the theoretical maximum strain, i.e. the asymptotic value at infinite pressure; P_{50} has the same meaning as Langmuir's pressure data, representing the pressure when coal rock mass reaches half the maximum strain, psi; φ_j is the porosity of fractures; φ_i is the porosity of initial fractures; P_i is the pore pressure of initial reservoir, psi. P_j is the pore pressure of reservoir, psi; k_j is permeability, mD; k_i is initial permeability, mD; C_v, ε_{max} and P_{50} can be obtained by experiments.

10.3 Desorption-Seepage Mechanism of CBM in Production Process

10.3.1 Characteristics of CBM Production

The methane in coal seam exists in three phases, adsorbed, free and dissolved. The methane in these three phases is in a dynamic equilibrium process and can be expressed as follows.

$$\text{Adsorption phase: dissolved phase: free phase} \quad \begin{array}{c}\xrightarrow{\text{Pressure decreases}} \\ \xleftarrow[\text{Pressure increases}]{}\end{array}$$

A large number of isothermal adsorption experiments and CBM development practices have proved that CBM is mainly adsorbed in reservoirs. There are three evidences.

① By comparing measured CBM content with theoretical adsorption capacity from isothermal adsorption experiments, it is found that most samples are under-saturated or near-saturated, and very few are over-saturated. This fact fully proves that the occurrence of CBM is dominated by adsorption phase.
② CBM development practice further proves that almost all CBM production wells begin to produce gas after drainage and depressurization, and do not have the characteristics of free gas production.
③ Although pores and fractures in coal seams are full of water, the amount of methane dissolved in water is relatively insignificant compared with the measured CBM content. The experiment of methane dissolution in water shows that the methane dissolved per liter of water is no less than 0.05–3.11 L at normal temperature, pressure and salinity. If assuming the porosity at 30% (the assumed value is much larger than the actual, the maximum water content per ton of coal is only 0.25 m^3. If the maximum solubility of 3 m^3/t_{coal} is used, the dissolved methane per ton of coal is only 0.75 m^3.

The non-linear relationship between methane adsorption capacity and reservoir pressure is described by Langmuir equation or extended Langmuir equation by most scholars.

$$C(P) = \frac{V_L P}{P_L + P} \tag{10.18}$$

where V_L is Langmuir volume, m^3/t; P_L is Langmuir pressure, psi; P is reservoir pressure, psi.

When the fluid pressure in coal reservoir decreases, the CBM adsorbed on the inner surface of coal matrix pores in the form of physical adsorption is desorbed into free methane. After that, methane in free state diffuses or seeps into natural fractures through coal matrix. Finally, free methane reaches wellbore and produces.

10.3.1.1 Desorption Mechanism

When the pressure in coal reservoir decreases, the adsorbed methane molecule separates from the inner surface of coal and desorbs to free phase. Most scholars use the Langmuir equation or extended Langmuir equation to describe the desorption characteristics. See Eq. (10.18) for details. Of course, in the case of under-saturated adsorption, methane molecules can be desorbed from the inner surface of coal matrix pores only after the pressure in coal reservoir is lower than the critical desorption pressure.

10.3.1.2 Diffusion Mechanism

The diffusion of CBM is a process in which methane molecules move from a high concentration area to a low concentration area. The process can be described by the Fick's first theorem.

$$\overline{q}_m = D\sigma V_m \rho_g [C(t) - C(P)] \tag{10.19}$$

where $\overline{q}_m$ is the amount of methane diffused from coal matrix, lb/d; D is diffusion coefficient, ft^2/d; σ is shape coefficient, ft^{-2}; ρ_g is the density of methane, lb/ft^3; V_m is the volume of coal matrix, ft^3; $C(t)$ is the average concentration of methane in coal matrix, scf/ft^3; and $C(P)$ is the equilibrium concentration of methane on the matrix cleat boundary, scf/ft^3.

10.3.1.3 Seepage Mechanism

The flow of CBM in a fracture system conforms to the Darcy's theorem. In a fracture system, methane and water miscibe flow in separate phases. The Darcy's theorem needs to consider the phase permeability of each fluid, i.e. effective permeability. The ratio of effective permeability to absolute permeability is called relative permeability. Relative permeability is usually used in practical research, and it is usually considered as the function of saturation. According to the Darcy's theorem, the seepage theorem of each phase can be written as follows.

$$V_l = \frac{K_l}{u_l}\frac{\Delta P_l}{L} \tag{10.20}$$

$$K_l = KK_{rl} \tag{10.21}$$

where V_l is the seepage velocity of l phase, cm/min; u_l is the viscosity coefficient of l phase, mPa s; ΔP_l is the pressure difference of l phase, MPa; L is the length of seepage path, cm; K_l is the effective permeability of l phase, mD; K is the absolute permeability of porous media, mD; K_{rl} is the relative permeability of l phase, mD.

10.3.2 Kinetic Characteristics and Desorption Behavior of CBM Desorption

10.3.2.1 Kinetic Characteristics

Essence and Difference of Adsorption and Desorption of CBM

Modern adsorption theory holds that solid adsorbs gas to a certain extent due to the mobility, roughness, incompleteness and surface energy of surface atoms on the solid. According to the properties of interaction force between solid (adsorbent) surface and adsorbed gas (adsorbate) molecules, the adsorption of solid to gas is usually divided into physical adsorption and chemical adsorption.

From the essential difference between physical adsorption and chemical adsorption at the solid–gas interface, it can be

Table 10.7 Essential differences between physical adsorption and physical desorption of CBM

Condition	Physical adsorption	Physical desorption
Acting process	Adsorption with coal thermal evolution and hydrocarbon generation and expulsion (a "spontaneous process")	Water drainage-decompression-desorption process (a passive process)
Acting time	Adsorption is a long process for millions of years	Desorption is a relatively fast process, for days or hours
Acting types	Physical and chemical adsorption	Physical desorption, decreasing pressure, rising temperature, displacing and diffusing
Acting conditions	Coal has a strong adsorption capacity. The gas generated with thermal evolution of coal is first adsorbed by coal, in the course of evolution, coal seams gradually dehydrate, warm up and pressurize	Coal has a stronger adsorption capacity; limited depressurization and very limited pore space; almost constant temperature
Influence factor	Coal quality, inner surface area of coal matrix pores, etc.	Escaping velocity of CBM when desorbed to free state

seen that the gas adsorbed on the solid surface by physical adsorption is easier to be desorbed, as long as the pressure decreases or the temperature rises. While the desorption of the chemically adsorbed gas is more difficult, it is necessary to surmount the energy barrier from the chemical adsorption state to the physical adsorption state, and the energy barrier is the energy required to form chemical bonds between adsorbate and adsorbent surfaces.

It is found that there are essential differences in the adsorption conditions, adsorption process and desorption process of CBM in natural coal seams (Table 10.7). These differences are mainly manifested in many aspects, such as the process of action, the time of action, the type of action, the conditions of action and the influencing factors.

(1) Differences in adsorption and desorption processes and acting types

CBM adsorption occurs in the process of hydrocarbon generation and expulsion during coal thermal evolution. It is a very complex process. Coal adsorbing CBM is a physical or chemical process.

In fact, desorption process is much simpler than adsorption process. In the production process of CBM, only the CBM adsorbed on the inner surface of the pore of coal matrix is desorbed by draining water and decreasing pressure. There is no obvious temperature rise and related chemical changes in the process of CBM production. Therefore, in the process of CBM development, the desorption of CBM can only be physical, but not chemical.

(2) Differences in adsorption and desorption conditions

The adsorption process of CBM is long, which takes place in the process of hydrocarbon generation and expulsion during the thermal evolution of coal for millions of years. Therefore, with the increase of thermal evolution, the amount of generated hydrocarbon gradually increases, and at the same time, the adsorption capacity of coal to gas gradually increases. Because coal itself has such a strong adsorption capacity, it is bound to cause that the hydrocarbon gas generated with thermal evolution first meets its own adsorption, and then it is possible to discharge during the process of hydrocarbon expulsion from inside to outside. The desorption process of CBM is relatively short, and the maximum time unit can only be measured in days. The desorption of CBM in the process of production is restricted by many factors, such as pressure drop, dropping rate and pore space of coal matrix.

Dynamic Characteristics and Desorption Types

Based on the theoretical understanding that physical desorption is primary in the process of CBM development. Therefore, our research focuses on the mechanism of physical desorption of CBM. As far as the physical adsorption and desorption of CBM are concerned, there are also great differences between the two processes, of which the most significant difference is the process of action and the constraints.

According to desorption conditions and characteristics, physical desorption of CBM can be divided into four sub-categories: depressurizing desorption, temperature-rising desorption, displacing desorption and diffusing desorption. Depressurizing desorption is the most important and contributes the most to CBM production.

(1) Depressurizing desorption

Depressurizing desorption is the most characteristic physical desorption process, and also the most important desorption in the process of CBM development. The basic characteristic of depressurizing desorption is that the CBM molecules adsorbed on the inner surface of the pore of coal matrix become more active due to the decrease of "external pressure", so that they are free from the bondage of van der Waals force and change from adsorbed to free state. According to the basic understanding of depressurizing desorption at present, its desorption behavior basically obeys the Langmuir equation. In view of the fact that depressurizing desorption is a very familiar mechanism, here we won't describe it in details.

(2) Temperature-rising desorption

According to modern physical and chemical studies, the adsorption capacity of adsorbent to adsorbate is a function of adsorbate, adsorbent properties and their interaction, pressure and temperature at adsorption equilibrium. Temperature is negatively correlated with adsorption and positively correlated with desorption. The increase of temperature accelerates the thermal movement of gas molecules, which makes them more capable of escaping from the bondage of Van der Waals force and being desorbed. Some people attribute the influence of temperature on desorption rate and desorption amount to a influencing factor. We believe that temperature, like pressure, is a driving force for desorption and should be defined as a type of desorption. In the process of CBM development, temperature is almost constant.

This desorption type has been confirmed in the experiment for determining CBM content. We can find that when determining CBM content, after the desorption tank is put into a constant-temperature water tank, even if the pressure in the desorption tank is increasing, the desorption of CBM will accelerate.

(3) Displacing desorption

The essence of displacing desorption is that unabsorbed water molecules or other gas molecules displace the methane molecules in adsorption state in order to achieve dynamic equilibrium, so that the original adsorbed methane molecule becomes free. This is a typical competitive adsorption process of different components, and it is also a desorption type commonly existing in the process of CBM development.

The essence of displacing desorption can be defined as the natural law of "survival of the fittest". On the one hand, water molecules and other gas molecules that are not adsorbed constantly break away from the chance of adsorption under the Van der Waals force, which exists widely among atoms and molecules. On the other hand, the thermodynamic properties of gas molecules determine that these adsorbed gas molecules are constantly breaking away from the Van der Waals force bond and changing from the adsorbed state to free state.

(4) Diffusing desorption

According to the theory of molecular diffusion, as long as there is a concentration difference, there will be molecular diffusion movement, which is determined by the thermodynamic properties of gas molecules. In the coal seam being mined, methane molecules are highly enriched on the inner surface of the coal pore, which constitutes concentration difference at a high gradient with the fluid in the pore and fracture, that forces methane molecules to diffuse and then causes the desorption in reality. Due to the ubiquitous existence of diffusion, diffusing desorption is also an important desorption in the process of CBM development. The essence of diffusing desorption is the desorption caused by gas diffusion due to concentration difference. This diffusion itself is coupled with desorption and is the coupling of desorption and diffusion. Therefore, we call it diffusing desorption from the perspective of desorption.

10.3.2.2 Experiments on Desorption Behavior of CBM

Almost all experts, scholars and engineers in the field of CBM at home and abroad generally believe that the adsorption and desorption of CBM is a reversible process. Whether CBM adsorption or desorption, the same Langmuir equation has been used to describe it in the field of CBM. Therefore, it is one of the main directions for the CBM industry to deeply and systematically study the mechanism of adsorption and desorption of CBM.

Experimental Methods and Equipment

Since 1990s, China has introduced several CBM isotherm adsorbents from the United States. These instruments are manufactured by TerraTek and RavenRideg Resource in the United States. Over the past decade, it has been found that the adsorbents supplied by the two companies have the same defects. Firstly, the size of experimental samples is too small (20–30 g or 60–70 g), which can not satisfy the desorption experiment on CBM production process. Secondly, the reproducibility of the experiment is not good. When a group of samples are tested in parallel, the error is large. At the same time, the experiment is only used to understand the adsorption characteristics of coal. The desorption behavior of CBM (the initial stage of production) can not be described in details and can not guide the production practice of CBM.

In order to solve the problem of experimental conditions, aiming at the experimental research on the desorption mechanism of CBM and the shortcomings of the adsorption and desorption experiments of CBM at home and abroad at present, AST-1000 and AST-2000 CBM adsorption and

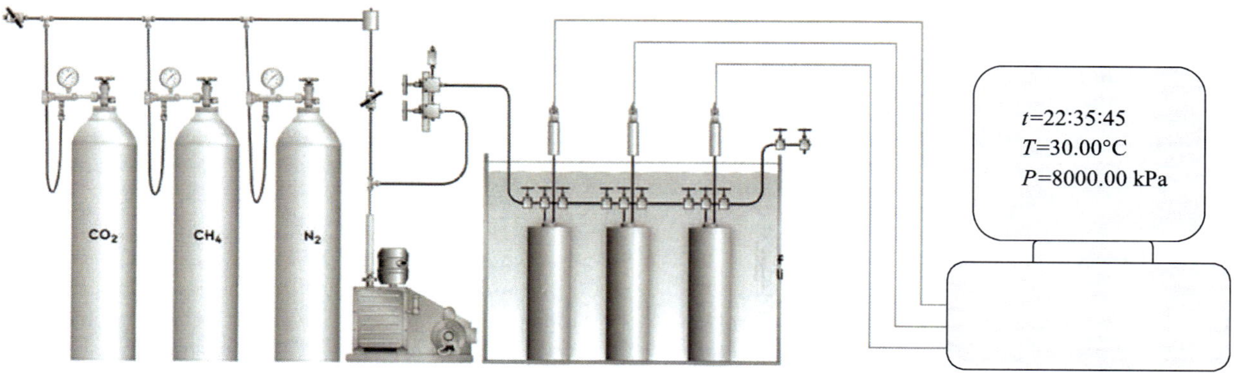

Fig. 10.13 The schematic diagram of AST-2000

desorption simulators for large samples have been developed jointly by China University of Petroleum, Beijing and Xi'an University of Science and Technology. Parallel and repetitive experiments prove that the equipment runs normally and the experimental data are reliable. The structure and working principle of the experimental equipment for isothermal adsorption and desorption (AST-2000) are shown in Fig. 10.13.

Absorption and Desorption Reversibility Experiments

For a long time, most scholars and experts have demonstrated the desorption characteristics of CBM mainly through isothermal adsorption experiments of single-component, multi-components and different- rank coal samples. At the end of the twentieth century, some scholars at home and abroad began to carry out the reversibility experiments on CBM adsorption and desorption. At present, this kind of experiment is not standardized, which leads to a great difference in the experimental results, so that different conclusions are drawn. In order to further study the desorption mechanism in the process of CBM development, the physical and chemical characteristics and reversibility of methane adsorption and desorption are experimentally and theoretically discussed in this paper.

Reversibility experimental results show that the adsorption and desorption of methane by coal under equilibrium water conditions is very reversible and the desorption lags to a certain degree (Figs. 10.14, 10.15 and 10.16). The moisture of coal samples not only significantly affects the adsorption capacity, but also significantly affects the reversibility in low-rank coal (Figs. 10.14 and 10.15).

Experiments on Desorption Characteristics of CBM

The isothermal desorption experiment began at the highest adsorption pressure. Then the pressure dropped gradually

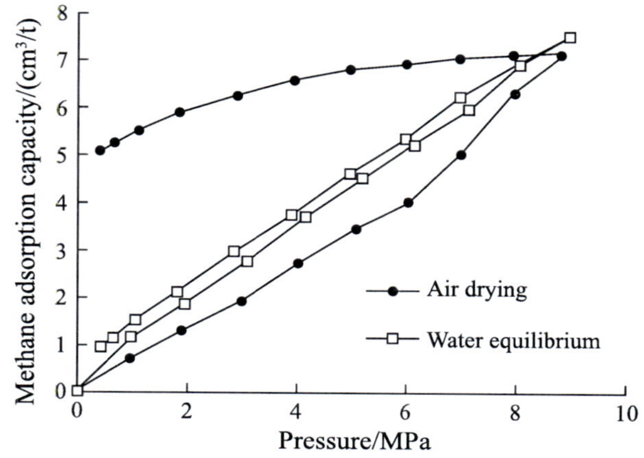

Fig. 10.14 Simulated isothermal adsorption and desorption experiment on coal sample BYH-01 (lignite)

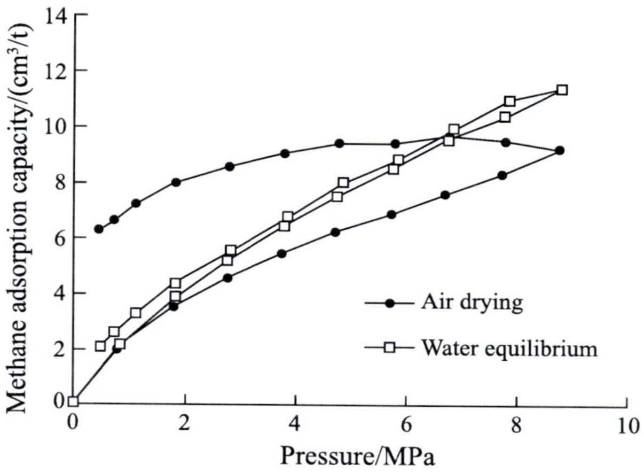

Fig. 10.15 Simulated isothermal adsorption and desorption experiment on coal sample MNT-02 (long-flame coal)

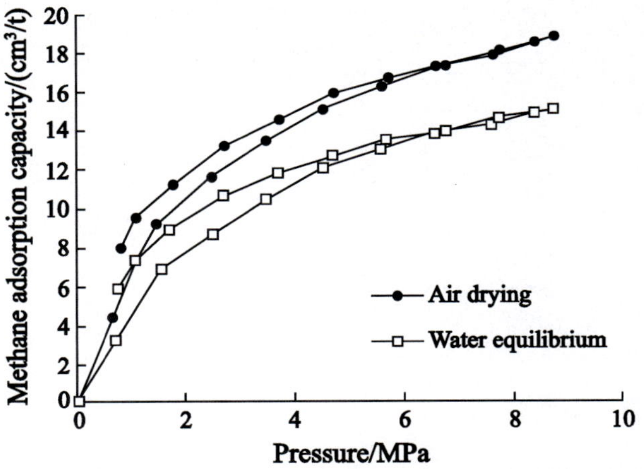

Fig. 10.16 Simulated isothermal adsorption and desorption experiment on coal sample XSH-03 (lean coal)

from high to low. The cumulative desorbed CBM was measured on 5–8 pressure points (Tables 10.8, 10.9, and 10.10). The isothermal desorption curves are shown in Figs. 10.14, 10.15 and 10.16.

10.3.2.3 Desorption Mechanism of CBM

It is found that the adsorption or desorption behavior of gas molecules adsorbed on the inner surface of coal matrix is influenced not only by the physical and chemical characteristics of coal itself, but also by the physical conditions such as pressure, temperature and water saturation of pore fluid, as well as by the pore structure of coal, especially the CBM desorption from low-rank coal. The results of isothermal adsorption and desorption are very similar to the curves of mercury intrusion and evacuation of coal samples in mercury penetration experiment, as shown in Fig. 10.17.

Table 10.8 Experimental data of air dried coal sample BYH-01 (lignite)

Pressure/MPa	Gas content in adsorption process/(m^3/t)	Pressure/MPa	Gas content in desorption process/(m^3/t)
0	0	8.815	7.127433
0.955	0.68124480	7.905	7.120050
1.910	1.27340259	6.930	7.035544
3.015	1.91833269	5.960	6.928000
4.040	2.71960791	4.950	6.792492
5.070	3.46651040	3.920	6.579182
6.015	4.01977811	2.895	6.263527
6.990	5.03629408	1.845	5.880091
7.935	6.34540876	1.105	5.527147
8.815	7.12728594	0.680	5.216873
		0.430	5.095655

Table 10.9 Experimental data of water balanced coal sample XSH-03 (lean coal, 60–80 meshes)

Pressure/MPa	Gas content in adsorption process/(m^3/t)	Pressure/MPa	Gas content in desorption process/(m^3/t)
0	0	8.992	18.52702
0.992	2.29398576	8.407	18.78301
2.127	6.05582160	7.627	18.44896
3.467	9.18377048	6.772	17.72997
5.057	12.85976200	5.987	17.36804
6.622	15.0508013	5.077	16.82267
8.097	17.1825770	4.137	15.93046
8.992	18.5270214	3.232	14.50600
		2.297	13.13543
		1.427	12.63987
		1.022	10.35537

Table 10.10 Experimental data of water balanced coal sample SHK-05 (anthracite, 60–80 meshes)

Pressure/MPa	Gas content in adsorption process/(m^3/t)	Pressure/MPa	Gas content in desorption process/(m^3/t)
0	0	9.130	25.536700
0.475	5.17640584	8.510	26.176630
1.210	9.16161886	7.635	26.569180
2.210	12.53203800	6.360	25.866230
3.320	15.48194280	4.990	24.797410
4.595	18.09466680	3.965	23.588420
5.840	20.1170980	3.115	21.893890
7.100	22.06955370	2.220	19.967080
8.140	24.2039950	1.425	17.428730
9.130	25.53670210	0.940	15.275110
		0.625	13.785610
		0.435	12.442770

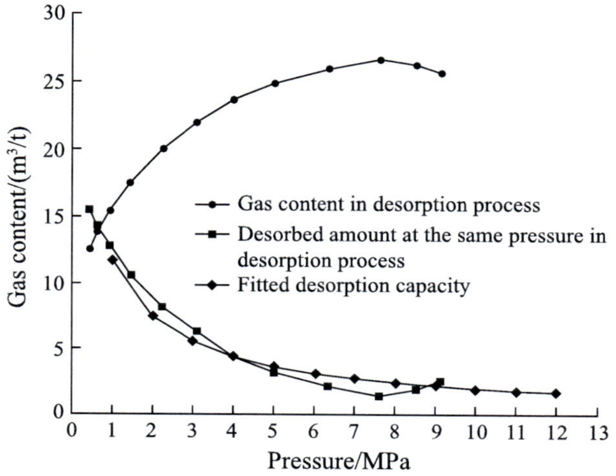

Fig. 10.17 Isothermal desorption curve and fitted Weibull function of Sihe anthracite

As a typical porous solid, coal usually has irregular pore structures. It is difficult to accurately describe the geometrical characteristics of the pore. The pore size can be characterized by half pore width r, and the differential distribution function of half pore width r can be expressed by the Weibull function of asymmetric distribution.

When $r \geq r_a$,

$$f(r) = \frac{q}{a}(r - r_a)^{q-1}\exp\left[-\frac{(r - r_a)^q}{a}\right] \quad (10.22)$$

When $r < r_a$,

$$f(r) = 0 \quad (10.23)$$

where r_a is the molecular diameter of adsorbate (methane).

According to the analysis of surface potential function distribution on a Lenard-Jones medium, the energy distribution on the surface of coal is asymmetric, which also obeys the Weibull function. According to the Weibull function, the mathematical model of adsorption/desorption of CBM can be deduced.

$$V = V_0 \cdot [1 - \exp(-bP^q)] \quad (10.24)$$

The above formula is the Weibull function of CBM adsorption and desorption, in which parameters V_0 and b have clear meanings. At supercritical temperature, the saturated absolute adsorption V_0 is a constant, while the b value is the heat of adsorption, which can be obtained by physical and chemical experiments. Usually, the maximum adsorbed gas V_d in the desorption process is less than the maximum adsorption capacity V_a in the adsorption process, because the adsorption heat b is larger than the adsorption heat b in the desorption process. The factors affecting b are temperature, pressure and surface activation energy.

The amount of desorbed CBM can be calculated by the Weibull function in desorption process, that is $V = V_0–V_d$. In the desorption process, with the decrease of pressure, CBM is desorbed, and the maximum desorption amount of coal with different metamorphic degree is distributed in different pressure sections.

The isothermal desorption experimental data of three coal samples at different metamorphic stages (Sihe Coalmine No. 3 coal, Hancheng No. 5 coal and Liudaowan Coalmine No. 43 coal) were fitted. The results are shown in Table 10.11, Figs. 10.17, 10.18 and 10.19.

Table 10.11 Fitted results of Weibull function

Sample	Adsorption V_a	Desorption V_d
Sihe WYM	$V_a = 97.629 \cdot (1 - e^{-0.087 \cdot p^{0.561}})$	$V_d = 29.121 \cdot (1 - e^{-0.799\, p^{0.504}})$
Xiangshan SM	$V_a = 26.701 \cdot (1 - e^{-0.335 \cdot p^{0.882}})$	$V_d = 18.871 \cdot (1 - e^{-0.469\, p^{0.562}})$
Liudaowan CYM	$V_a = 13.235 \cdot (1 - e^{-0.313 \cdot p^{0.759}})$	$V_d = 11.756 \cdot (1 - e^{-0.593\, p^{0.647}})$

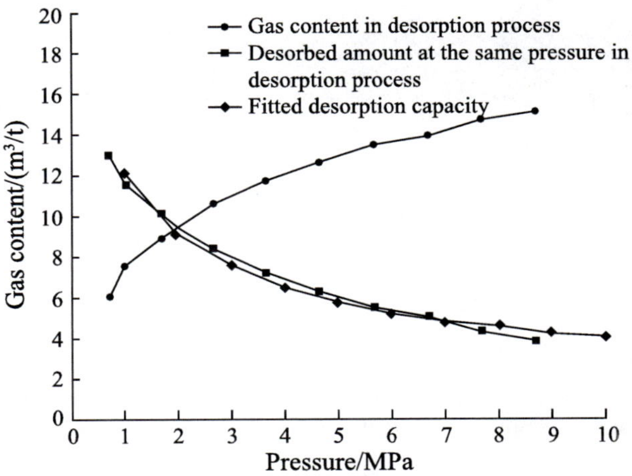

Fig. 10.18 Isothermal desorption curve and fitted Weibull function of Xiangshan lean coal

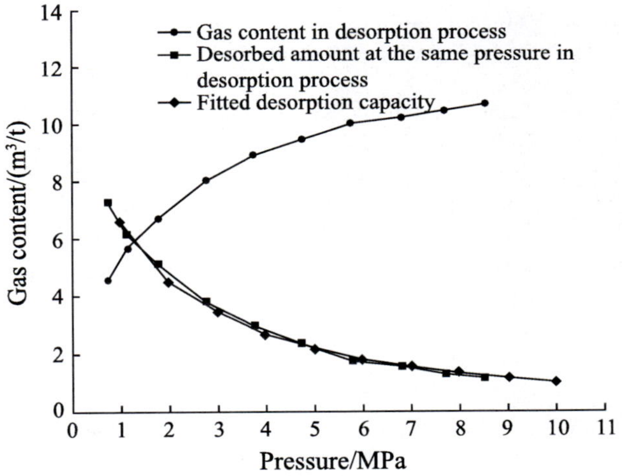

Fig. 10.19 Isothermal desorption curve and fitted Weibull function of Liudaowan flame coal

10.3.3 CBM Seepage Mechanism

10.3.3.1 Experiments on Influencing Factors and Degree

Because of the extremely low permeability of coal reservoir, the seepage of CBM in coal seam is difficult, so that the production of CBM wells is generally low. Therefore, before studying the mechanism of CBM seepage in depth, it is necessary to study the controlling factors on CBM seepage, especially effective overburden, gas pressure, fluid and temperature.

Basic Experimental Methods

(1) Preparation of pulverized coal samples. Take the same coal sample and screen 60–80 meshes after crushing, dry for 24 h, weigh and load into a compactible holder, and finally compact the pulverized coal until its density is the same as the coal sample taken.
(2) Preparation of coal seam samples. Drill out coal cylinders of 25 mm (diameter) by 4–6 cm (length) each, from the coal samples taken in the primary coal seam in the Jincheng block. The coal cylinders sampled horizontally are used to study horizontal permeability. Beddings and natural microfractures can be seen on the surface of the sample, and the sample is compact.
(3) Experimental device. The equipment includes an OPP-1 (permeation tool) provided by the Seepage Mechanism Laboratory of China University of Petroleum (Beijing), and a sample holder (quasi-triaxial pressure). Figure 10.20 shows the experimental principle. The experimental procedures are in accordance with to applicable petroleum industry standards.

Experiments on Seepage Conditions

(1) How effective overburden influences the permeability of coal seam samples

The experiment conditions include fixed gas pressure and varying effective confining pressure. The experiment procedures are as follows:

Dry a sample and then put it into the sample holder of the OPP-1. Turn on N_2 and keep the inlet pressure at about 0.2 MPa. SET effective the confining pressure at 0.6 MPa, 1.0 MPa, 1.6 MPa, 2.4 MPa, 3.2 MPa, 5 MPa, 7.5 MPa and 10 MPa, respectively, and measure the permeability of the sample (Fig. 10.21).

The results show that the permeability of the samples decreases with the increase of the effective confining pressure, which is the result of the shrinkage of pores and the closure of microfractures in the sample which is compressed

Fig. 10.20 Flow chart of gas seepage experiment

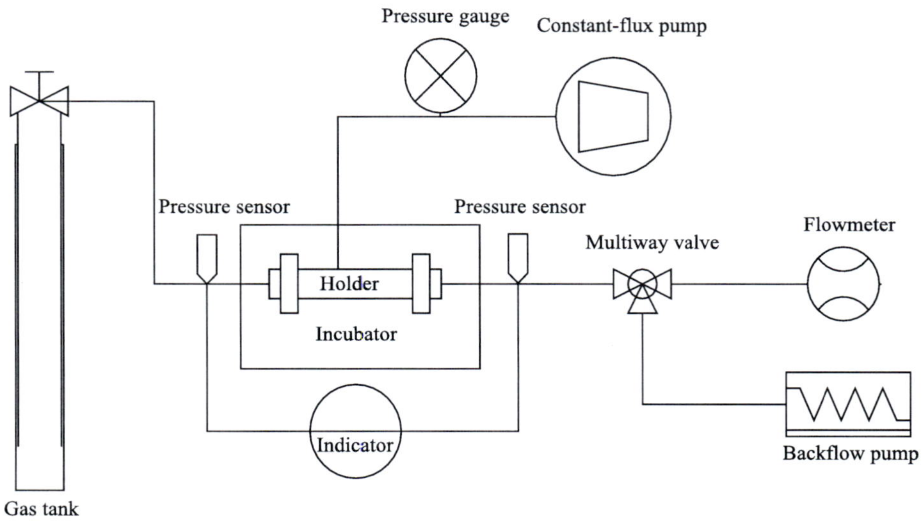

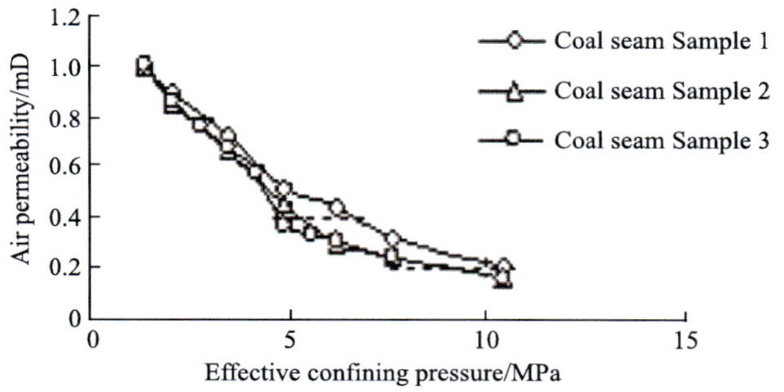

Fig. 10.21 Effect of effective confining pressure on permeability of coal seam samples

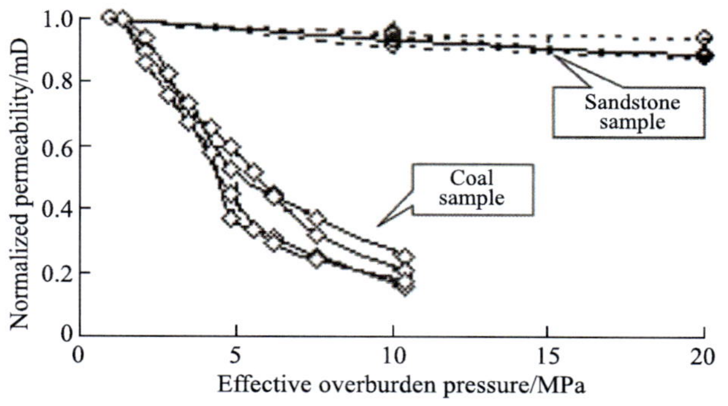

Fig. 10.22 Effect of effective confining pressure on the permeability of coal seam samples and conventional sandstone reservoir samples

at higher effective confining pressure. The effective confining pressure changes rapidly at 2–5 MPa and slowly after exceeding 5 MPa.

Compared with a conventional sandstone sample, the stress sensitivity of the coal sample is stronger (Fig. 10.22).

At 10 MPa effective confining pressure, the permeability of a sandstone sample is 90–95% of the initial value (at 1 MPa confining pressure), but that of a coal sample is only 18–25% of the original value.

(2) How gas pressure influences the permeability of coal seam samples.

The experiment conditions include fixed effective confining pressure and varying gas pressure. The experiment procedures are as follows:

Dry a sample and then put it into the sample holder of the OPP-1. Turn on N_2 and keep the inlet pressure at about 2 MPa. SET gas pressure at 0.6 MPa, 1.0 MPa, 1.6 MPa, 2.4 MPa, 3.2 MPa, 5.0 MPa 0.6 MPa, 1.0 MPa, 1.6 MPa, 2.4 MPa, 3.2 MPa, 5 MPa, 7.5 MPa and 10 MPa, respectively, and measure the permeability of the sample (Fig. 10.23). Then measure how the permeability changes

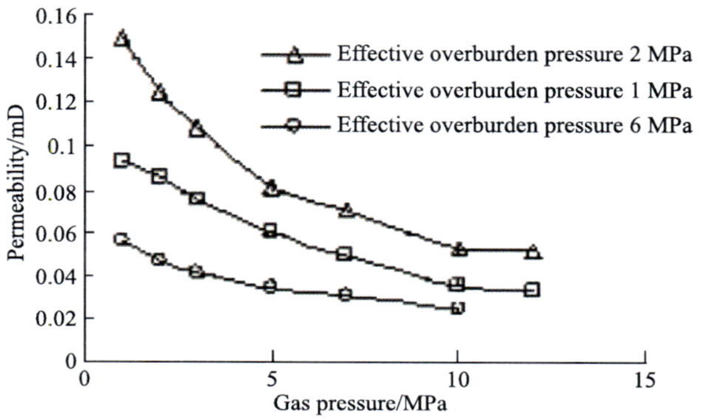

Fig. 10.23 Effect of gas pressure on permeability of coal seam samples

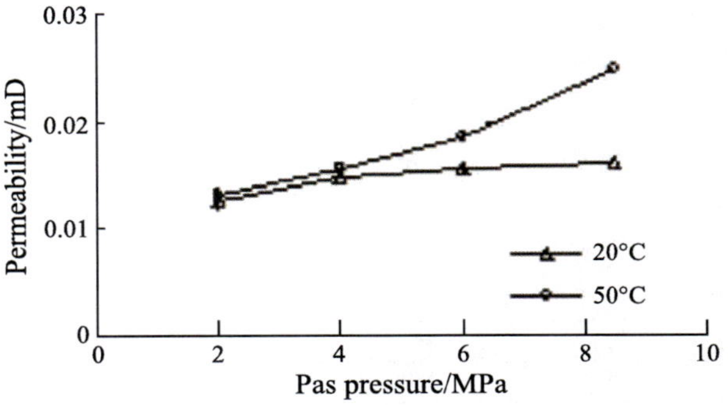

Fig. 10.24 Gas permeability at different pressures

with the gas pressure at confining pressure 4.0 MPa and 6.0 MPa, respectively.

The experiment result shows that with the increase of gas pressure, the effective permeability of the coal seam sample decreases. When $1/P$ tends to zero, the intercept on the longitudinal axis is Klinkenberg permeability, which is in line with the Klinkenberg effect.

(3) How temperature influences the permeability of coal seam samples

At 14 MPa confining pressure, the permeability of the same sample was measured at room temperature (20 °C) and 50 ° C. The experimental results (Fig. 10.24) show that temperature has a great influence on the permeability of the coal seam sample. The permeability at 50 °C is higher than that at normal temperature. That rising temperature changes the permeability of the coal seam sample may be caused by the following aspects.

① At high temperature, gas molecule movement is intensified and the free path of the gas molecules is shortened, so the permeability measured increases.
② At high temperature, adsorbed gas on the pore surface decreases, the adsorption layer becomes thin and the effective flow radius increases. From the adsorption isotherms at different temperatures, it can be seen that the adsorption capacity varies greatly at high pressure, so the permeability varies greatly. The adsorption capacity varies slightly at low pressure, so the permeability varies slightly.
③ The saturation of coal seam water decreases at high temperature, so the permeability increases.
④ Permeability at different gas pressure.

The experimental results (Fig. 10.25) show that, gas permeability is related to the type of gas, i.e. molecular weight, that is, the lower the molecular weight of the gas, the

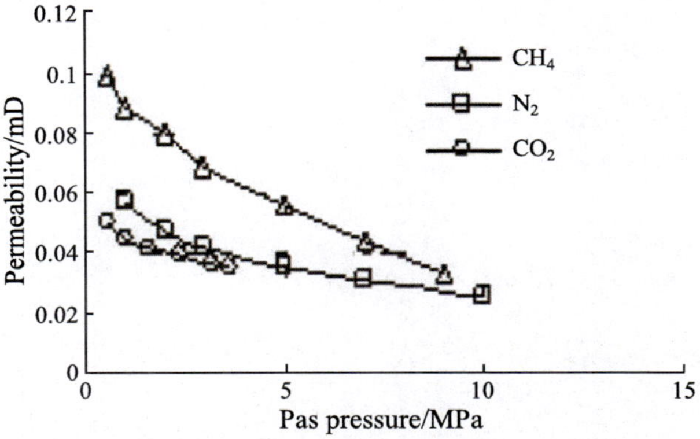

Fig. 10.25 Permeability of different gases

higher the gas permeability. Methane permeability is the highest, CO_2 permeability is the lowest, and N_2 permeability is between them. However, there must be some influence of coal on the adsorbability difference of different gases such kind of experiment.

10.3.3.2 Experiments on Seepage Mechanism of CBM

Experiment Method

In order to further explore the seepage mechanism of CBM, corresponding laboratory experiments have been carried out. According to the seepage experiment, coal samples were collected from the coal face of the Xishan mine field by channel method. A sample is 400 mm by 400 mm by 400 mm. The coal rank is coking coal (R^o = 1.23%, moisture content is 0.85%, ash content is 10.44%, volatile content is 21.59%). In laboratory, the coal samples were manually cut to cuboids, 11.336 cm (length) by 6.034 cm (width) by 3.524 cm (height) each.

Experiment Procedures

① At 5 MPa confining pressure and 0.10 MPa outlet pressure, recorded experimental data (Table 10.12 and Fig. 10.26).
② At 10 MPa confining pressure and 0.10 MPa outlet pressure, recorded experimental data (Table 10.13 and Fig. 10.27).
③ At 15 MPa confining pressure and 0.10 MPa outlet pressure, the experimental data are shown in Table 10.14 and Fig. 10.28.

At 15 MPa confining pressure and 0.10 MPa outlet pressure.

Table 10.12 Experiment result at 5 MPa confining pressure and 0.10 MPa outlet pressure

Pressure difference/MPa	Gas flow rate/ (cm^3/s)	Pressure gradient/ (MPa/m)
0.3972	0.0142	3.504
0.8931	0.0790	7.878
1.3865	0.2882	12.231
1.8774	1.0582	16.562
2.4160	3.1888	21.330
2.8990	6.5963	25.573

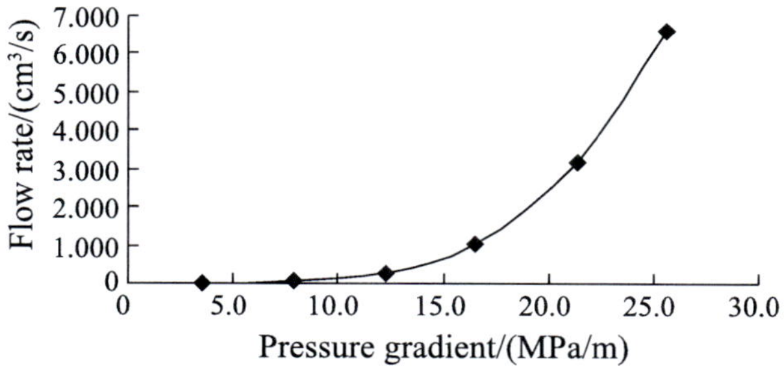

Fig. 10.26 Flow rate versus pressure gradient

Table 10.13 Experiment result at 10 MPa confining pressure and 0.10 MPa outlet pressure

Pressure difference/MPa	Gas flow rate/ (cm^3/s)	Pressure gradient/ (MPa/m)
0.42	0.0070	3.701
0.91	0.0080	8.031
1.92	0.0100	16.910
2.38	0.0170	21.020
3.89	0.0612	34.320
4.87	0.1230	42.980
5.85	0.4790	51.600
6.84	1.8330	60.350

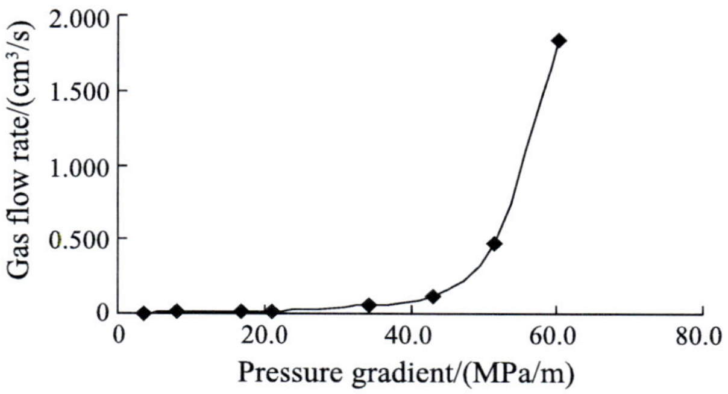

Fig. 10.27 Gas flow rate versus pressure gradient

Table 10.14 Experiment result at 15 MPa confining pressure and 0.10 MPa outlet pressure

Pressure difference/MPa	Gas flow rate/ (cm^3/s)	Pressure gradient/ (MPa/m)
0.385	0.0140	3.394
0.915	0.0137	8.075
1.932	0.0127	17.043
2.874	0.0160	25.360
3.866	0.0230	34.100
4.868	0.0340	42.940
5.840	0.0490	51.510
6.866	0.0760	60.570

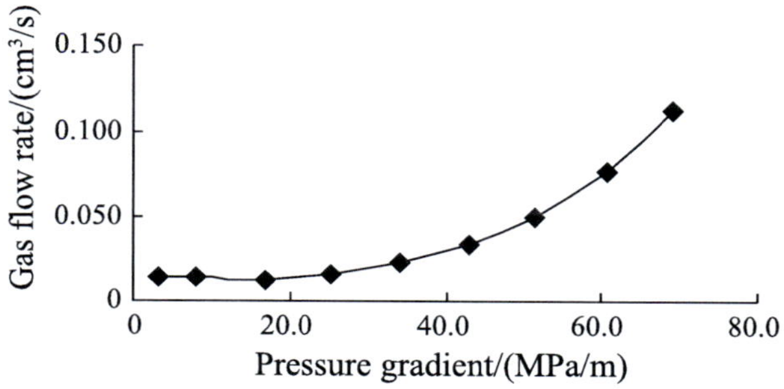

Fig. 10.28 Gas flow rate versus pressure gradient

10.3.3.3 CBM Seepage Mechanism

Gas Flow Rate Versus Pressure Gradient

The experimental results of CBM seepage mechanism (Figs. 10.25, 10.26 and 10.27) show that the CBM seepage does not meet the Darcy seepage characteristics. The experimental results show that no matter how high the confining pressure is, the relationship between seepage velocity (gas flow rate) and pressure gradient is not linear, but a typical downward bending non-linear relationship, and there is a significant starting pressure gradient.

From Figs. 10.26, 10.27 and 10.28, it can be clearly seen that the smaller the confining pressure, the larger the starting pressure gradient. When the pressure gradient is greater than the starting pressure gradient, the seepage velocity increases rapidly with the increasing pressure gradient, which is an approximately increasing quadratic curve. With the further increase of pressure gradient, the seepage velocity increases rapidly. There are many factors causing non-Darcy seepage of CBM, but there are two key factors. Firstly, coal reservoir belongs to typical ultra-low permeability reservoir (Fig. 10.29), so it has the unique seepage characteristics of

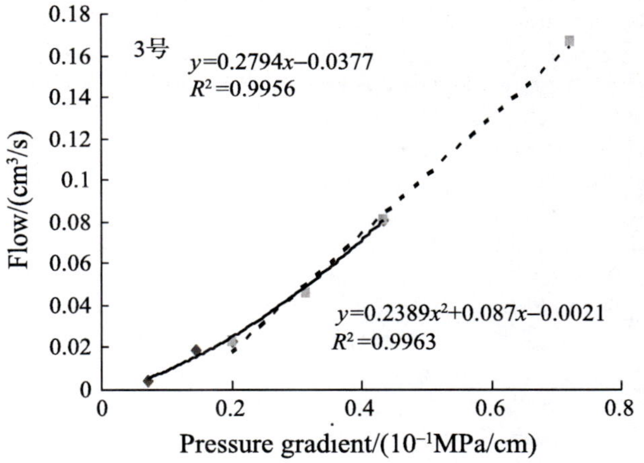

Fig. 10.29 Seepage velocity versus pressure gradient in ultra-low-permeability sandstone core ($K\infty$ = 0.2 mD)

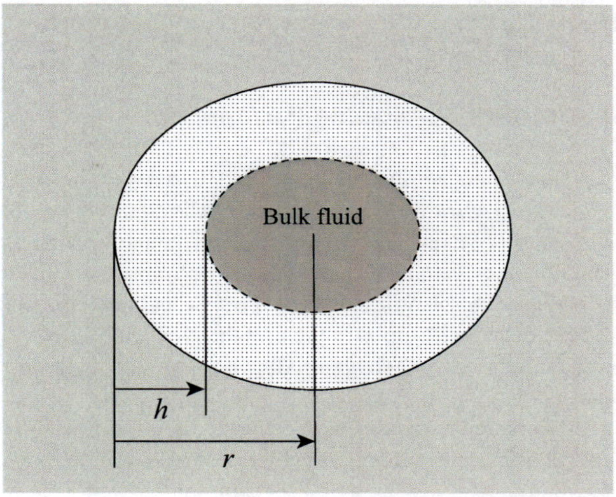

Fig. 10.30 Schematic diagram of bulk and boundary fluids in pores of coal reservoir

low permeability reservoir, i.e. low-rate non-linear seepage, affected by starting pressure gradient. Secondly, coal has a strong adsorption capacity. In the process of CBM seepage, an adsorption boundary exists in the pores and fractures in coal due to the influence of CBM re-adsorption.

How Adsorbing Boundary Layer Influence CBM Seepage Law

The adsorbed boundary layer on solid surface is caused by the interaction force between the solid surface and its neighboring molecules or atoms. The influence of the thickness of the fluid–solid boundary layer on seepage characteristics is as follows.

① The influence of capillary radius. At the same driving pressure gradient, the smaller the capillary radius, the thicker the boundary layer.
② The influence of driving pressure gradient. The thickness of the effective boundary layer decreases with the increase of driving pressure gradient. When the driving pressure gradient reaches a certain level, the boundary layer is a solidified one, and the thickness will reduce less.
③ The influence of fluid viscosity. If other conditions are fixed, and the pressure gradient is the same, the greater the viscosity is, the thicker the boundary layer is.
④ The effect of fluid components. The more polar components are in the fluid, the thicker the adsorbed boundary layer is.

Fluids in the seepage environment include bulk and boundary fluids. Bulk fluid is the fluid whose properties are not affected by interfacial phenomena, and boundary fluid is the fluid whose properties are affected by interfacial phenomena, as shown in Fig. 10.30.

The relationship between the influence of adsorbed boundary layer and permeability is that the smaller the permeability, the greater the influence of boundary layer; on the contrary, the larger the permeability, the smaller the influence of boundary layer.

The influence of adsorbed boundary layer on permeability is that the larger the adsorbed boundary layer, the lower the permeability.

Analysis of CBM Seepage Mechanism

According to the low-rate non-linear seepage theory, the starting pressure gradient theory, the adsorption boundary layer theory and the experimental results of CBM seepage mechanism in ultra-low-permeability reservoirs, the seepage characteristics of CBM can be described by the following formula.

$$v = A\left(\frac{\Delta P}{L}\right)^2 + B\left(\frac{\Delta P}{L}\right) + C \tag{10.25}$$

where v is seepage velocity, cm/s; $\Delta P/L$ is pressure gradient, 10^{-1} MPa/cm; A is adsorption boundary factor; B is the parameter related to fluid viscosity and seepage area; and C is starting pressure gradient, 10^{-1} MPa/cm.

10.4 Optimal CBM Reservoir Development Schemes

10.4.1 CBM Production Performance

At present, there are two well types of CBM development in the southern Qinshui Basin: vertical wells, accounting for more than 95%, and horizontal wells, accounting for less than 5%. Next we only describe how to optimize the vertical well pattern. A scientific, rational, economical and effective well pattern should aim at improving the producing reserves, production, recovery, stable production term and economic benefits.

Under certain geological conditions, it is necessary to fully consider the scientificity of the lateral well pattern, including layout, density and orientation of the well pattern, and also to fully consider the combination of production layers, gas production, stable production term, decline period, stable production measures, production cycle, etc.

Successful experience of CBM development at home and abroad tells us that in order to realize commercial production of CBM, efforts must be made to promote multi-well linkage and areal pressure drop, and the basic premise is to achieve the inter-well interference at a reasonable time. A CBM well pattern may be rectangular, rhombic, irregular, etc. According to the NNE structure, a five-point rhombic well pattern was selected in the southern Qinshui Basin. Well density involves the evaluation of gas field development index and economic benefit. Its size depends on the influence of reservoir properties and production scale on economy and the requirement to recovery. A too dense or sparse well pattern will cause a high or low recovery and a short or long production cycle, and it is not economic. Therefore, in view of the geological conditions of coal reservoirs in the southern Qinshui Basin, we have carried out in-depth summary and analysis of well production and test data or well group for many years. Under the current technical and economic conditions, numerical simulation was used to simulate and predict three well spacings (300 m × 300 m, 300 m × 400 m, 400 m × 500 m). According to the principle of maximum benefit, the 300 m × 300 m well scheme was selected. It can not only effectively utilize inter-well interference to ensure continuous production increase, but also meet the basic requirements of economic operation (Table 10.15, Fig. 10.31).

According to the well pattern selected, it is assumed that the average production of a well is 2000 m^3/d and the service life is 15 years, then the average annual production of a well can reach 70 × 10^4 m^3, the cumulative production can reach 1050 × 10^4 m^3, the average yearly recovery can reach 6.67%, and the final recovery can reach 54.3%. The production performance is predicted to be stable for 9 years and then declines 6 years from the tenth to the fifteenth year. The production decline is compensated by drilling infill wells.

In 2005, the first phase of the Qinnan development project entered the production test stage with 40 gas wells in total, of which 38 wells were put into operation after hydraulic sand fracturing (Fig. 10.32). In the initial gas production stage, because the liquid level was shallow and high liquid column had high backpressure on bottom hole, the production was low, even a well might produce gas several hundreds of m^3 a day. As water drainage lasted long,

Table 10.15 Well production prediction in a demonstration area

Production life/year	CBM production/ (m^3/d)	Average CBM production/ (10^4 m^3/a)	Cumulative CBM production/10^4 m^3
1	2164.68	71.43	71.43
2	2184.51	72.09	143.52
3	2360.79	77.91	221.43
4	2548.21	84.09	305.52
5	2687.48	88.69	394.21
6	2613.21	86.24	480.44
7	2434.88	80.35	560.79
8	2227.15	73.50	634.29
9	2069.06	68.28	702.57
10	1871.48	61.76	764.33
11	1701.55	56.15	820.48
12	1555.91	51.35	871.82
13	1432.91	47.29	919.11
14	1324.85	43.72	962.83
15	1229.39	40.57	1003.40

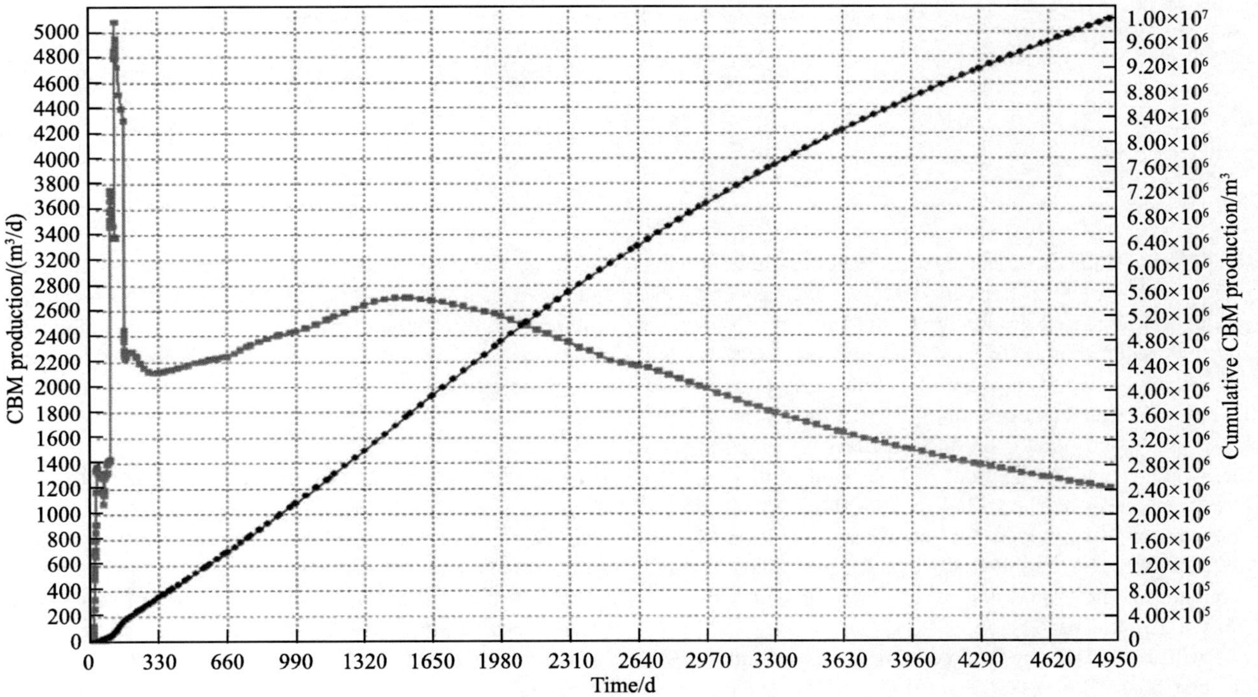

Fig. 10.31 Well numerical simulation

Fig. 10.32 Number of phase I producing wells and their initial producing date in the Qinnan development project

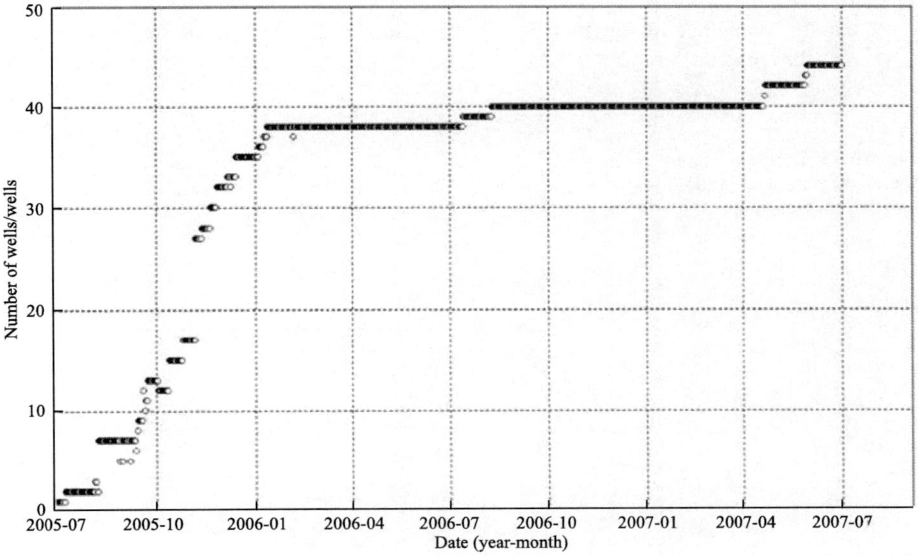

Table 10.16 Average gas and water production per well in January 2006

Coal seam	Gas production		Water production	
	Average gas production/(m^3/d)	Cumulative gas production/10^4 m^3	Average water production/(m^3/d)	Cumulative water production/10^4 m^3
No. 3 coal seam	1364	406.7864	5.0	2.3445

the dynamic liquid level decreased, the bottom hole flowing pressure decreased, the pressure difference increased, and the production increased gradually. At the same time, after the spacing reduced to 300 m × 300 m, gas breakthrough advanced greatly, from more than one year to more than 10 days and less than one month. By the first phase of 2006, the average daily water production decreased rapidly, and the average gas production of the well drilled in the No. 3 coal seam reached 1364 m^3/d (Table 10.16).

As water was pumped out, the average liquid level continued to decrease, and the daily average gas production and cumulative gas production also continued to rise (Figs. 10.33 and 10.34).

By 30 June 2007, the liquid column in the second stage of the first phase decreased (Fig. 10.35), the wellhead casing pressure increased (Fig. 10.36), the average bottom hole flowing pressure was 1.4 MPa, the gas production continued to rise, and the average gas production was 2630 m^3/(d·well) (Fig. 10.37).

It can be seen that in the first few years (usually 2–3 years), the production of a CBM well is rising. The 300 m × 300 m well pattern made the expected increase and change of the well production basically meet the requirements of economic production, which proved the well pattern reasonable and it has played an important guiding role in the rational development in the Qinnan gas field.

10.4.2 Comparison of Development Schemes

Well TL003 drilled through the Quaternary, the upper Shihezi Formation of the Upper Permian, the lower Shihezi Formation and Shanxi Formation of the Lower Permian, the Taiyuan Formation of the Upper Carboniferous, the Benxi Formation of the Middle Carboniferous and the Fengfeng Formation of the Middle Ordovician from top to bottom. Among them, No. 3 coal seam and No. 15 coal seam are primary gas pay layers, 6.33 m and 0.90 m thick, burial depth of 472.37 m and 583.26 m deep, respectively, and K_2 limestone of 9.14 m thick.

According to the coal seams and aquifers and fluid flow regimes, four schemes were proposed.

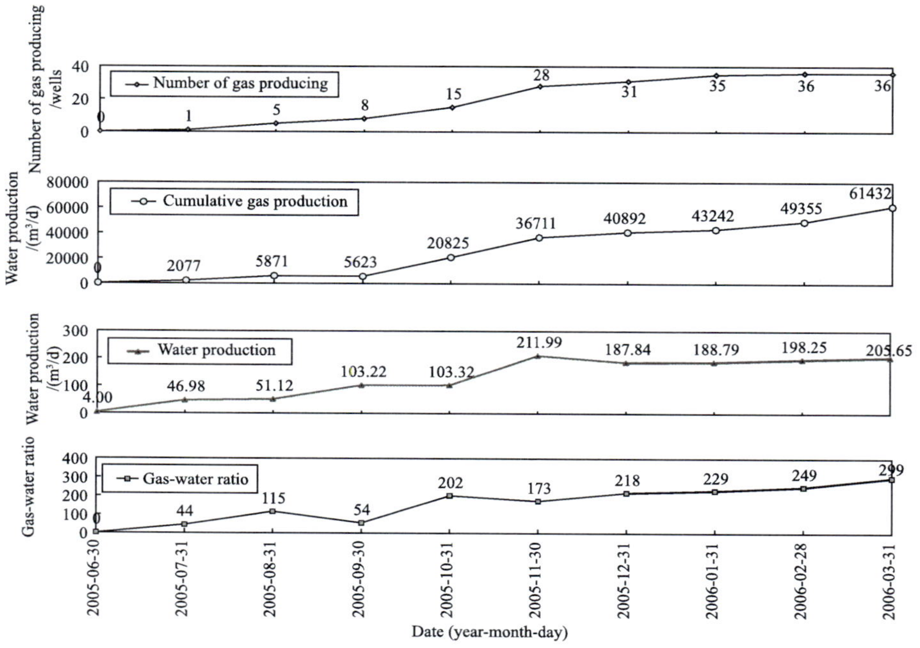

Fig. 10.33 Daily production in the Panhe pilot area in the first stage of phase I

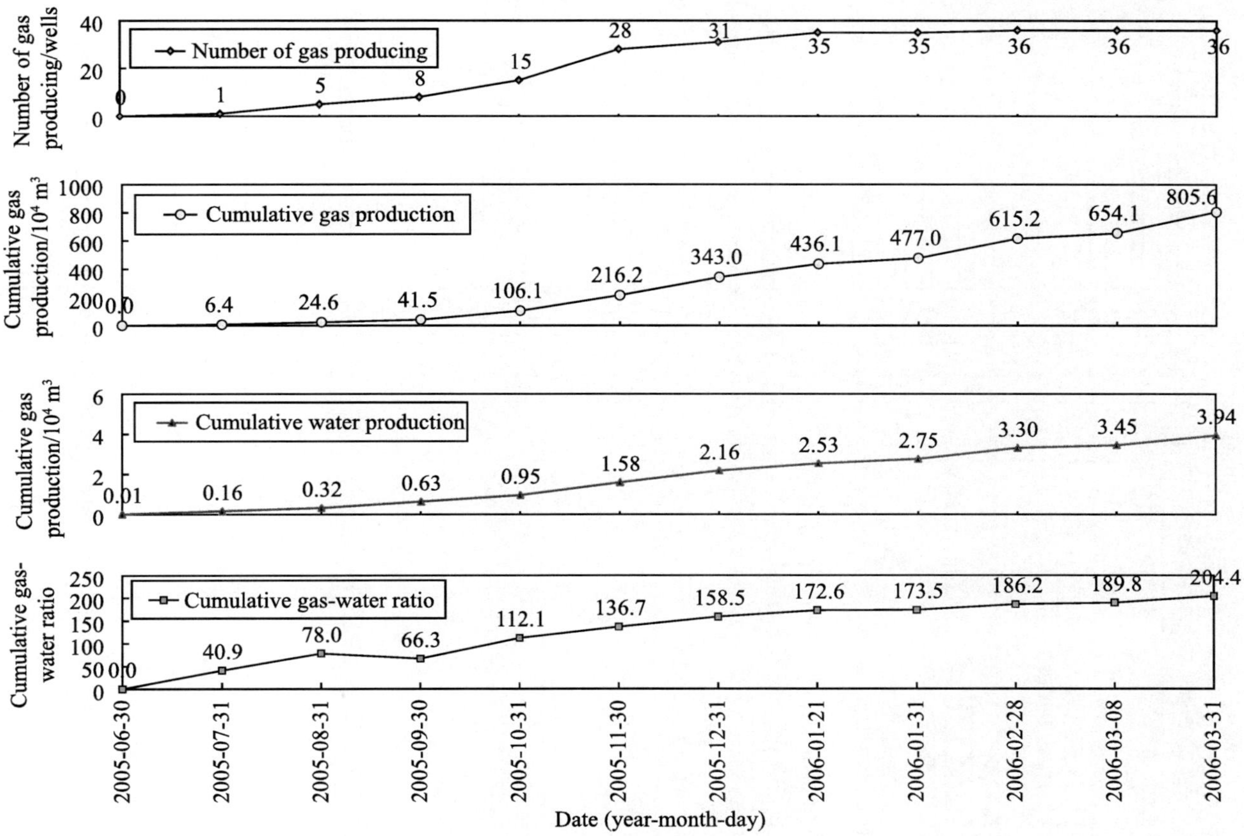

Fig. 10.34 Cumulative production in the Panhe pilot area in the first stage of phase I

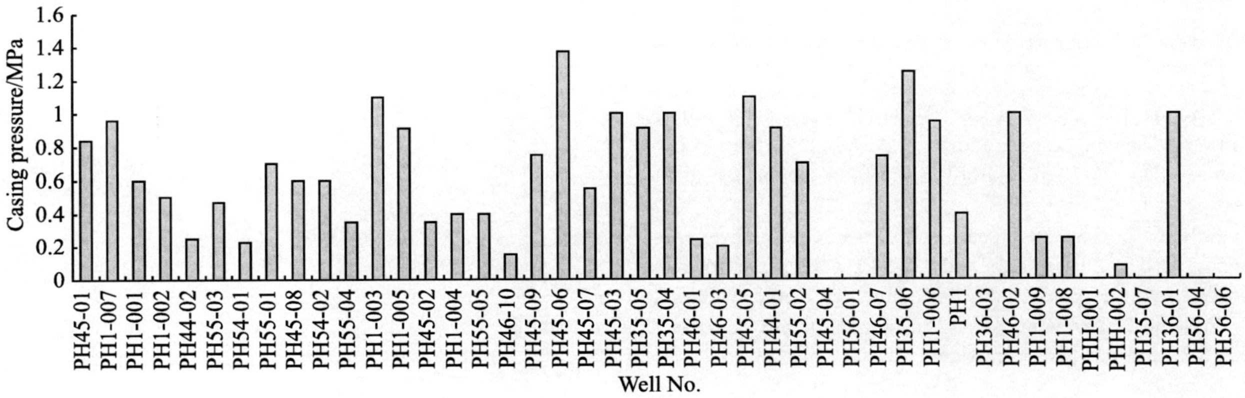

Fig. 10.35 Liquid level in the second stage of phase I

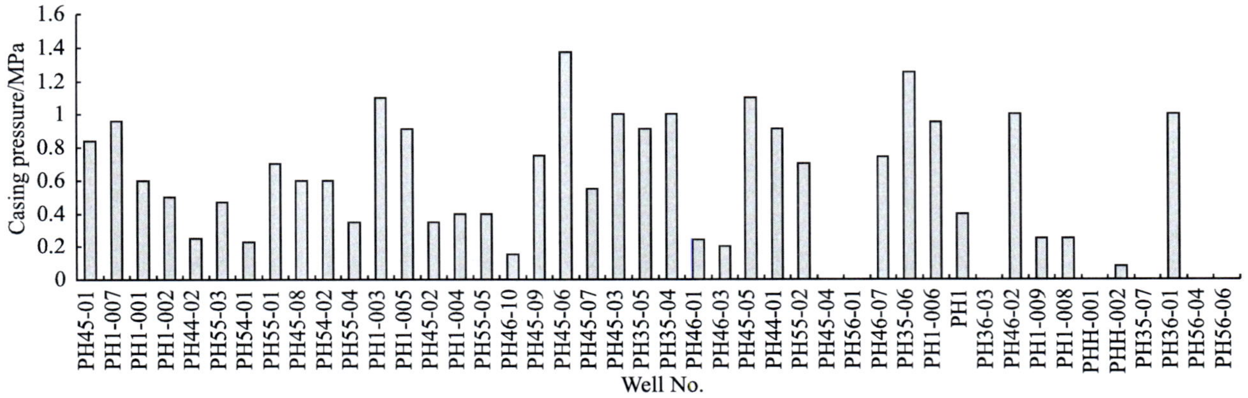

Fig. 10.36 Average wellhead casing pressure in the second stage of phase I

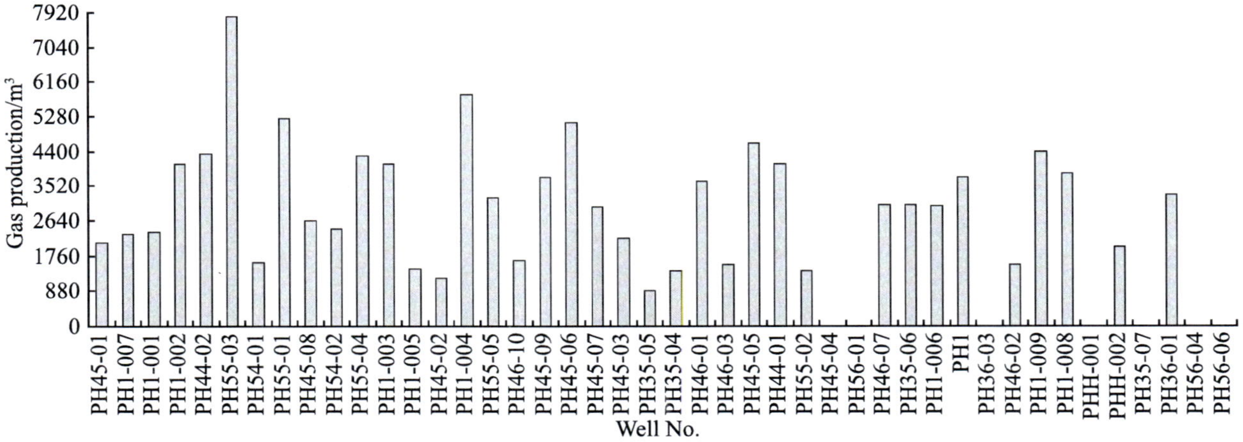

Fig. 10.37 Average well gas production in the second stage of phase I

Scheme 1: No. 3 coal seam, No. 15 coal seam and K_2 limestone are pay zones, and K_2 limestone and No. 15 coal seam builds a unified hydrodynamic field, connected to the wellbore.

Scheme 2: No. 3 coal seam and No. 15 coal seam are pay zones, and K_2 limestone and No. 15 coal seam forms a unified hydrodynamic field, not connected to the wellbore.

Scheme 3: No. 3 and No. 15 coal seams are pay zones, but K2 limestone not considered.

Scheme 4: only No. 3 coal seam as pay zone.

Using a mathematical model, we predicted the production of every scheme by developing a numerical simulation program.

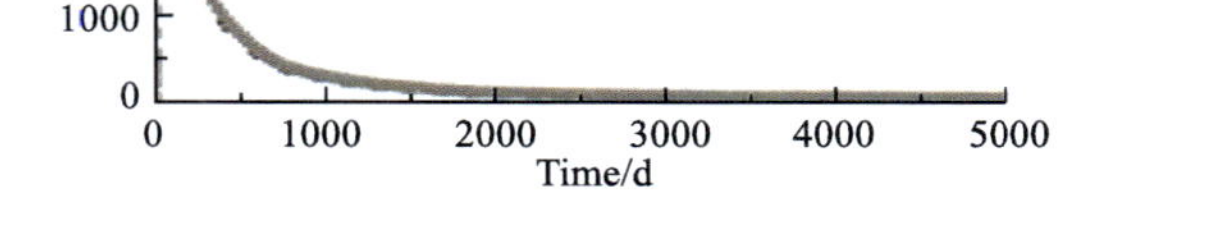

Fig. 10.38 Gas production curve of different schemes

From Figs. 10.38 and 10.39, it can be seen that, the scheme that No. 3 and No. 15 coal seams are pay zones and production or recharge of K_2 limestone is not considered, has the highest daily and cumulative gas production. The scheme that No. 3 coal seam, No. 15 coal seam and K_2 limestone are pay zones, also has high daily and cumulative gas production. However, if we consider that K_2 limestone and No. 15 coal seam constitute a unified hydrodynamic field and are not connected to the wellbore, this scheme is not much different from Scheme 4 that only No. 3 coal seam

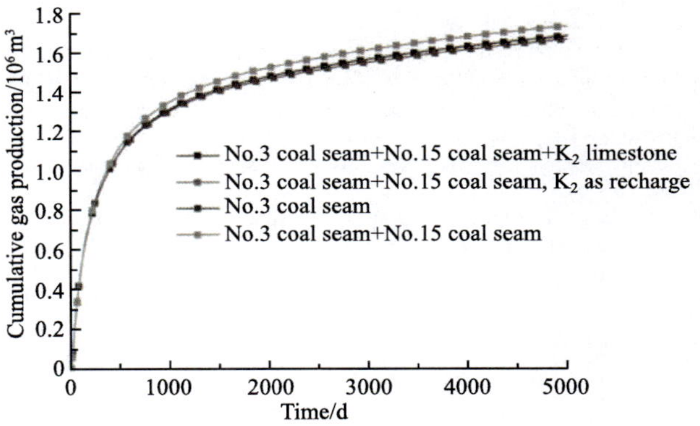

Fig. 10.39 Cumulative gas production of different schemes

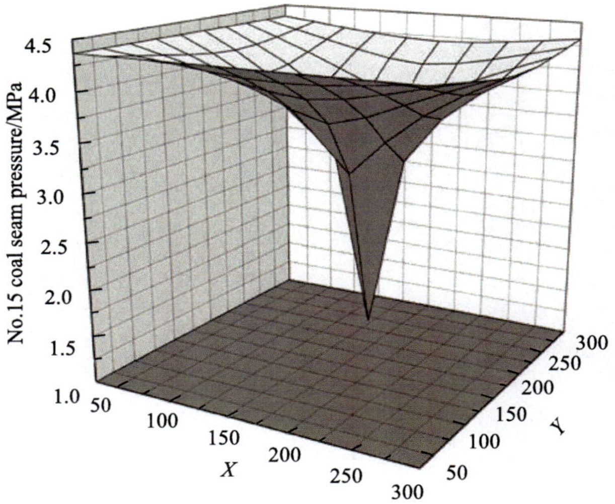

Fig. 10.40 The pressure drop funnel formed in No. 15 coal seam after producing 5000 days (Scheme 1)

is a pay zone in terms of daily or cumulative gas production, indicating that No. 15 coal seam has little or no contribution to gas production.

From Figs. 10.40, 10.41 and 10.42, it can be seen that the influence range of the pressure drop funnel formed in Scheme 3 is obviously larger than those in Scheme 1 and Scheme 2. The influence range of the pressure drop funnel formed by Scheme 1 is larger than that in Scheme 2. In actual production, if water production remains high, recharge from K_2 limestone should be considered. Therefore, to separate No. 3 and No. 15 coal seams from K_2 limestone and to take No. 3 and No. 15 coal seams as pay zones is the best scheme to improve the productivity of a CBM well.

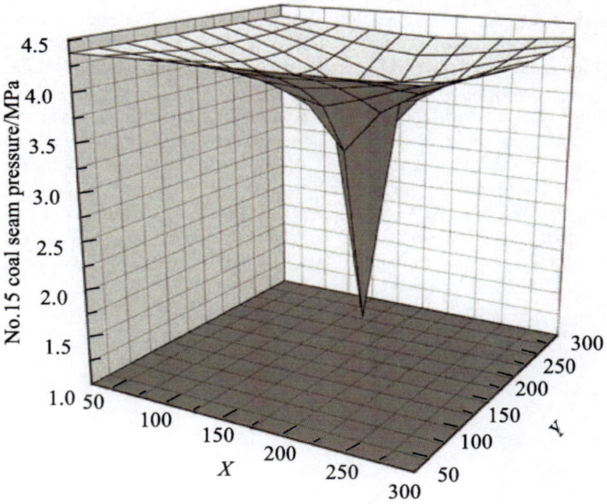

Fig. 10.41 The pressure drop funnel formed in No. 15 coal seam after producing 5000 days (Scheme 2)

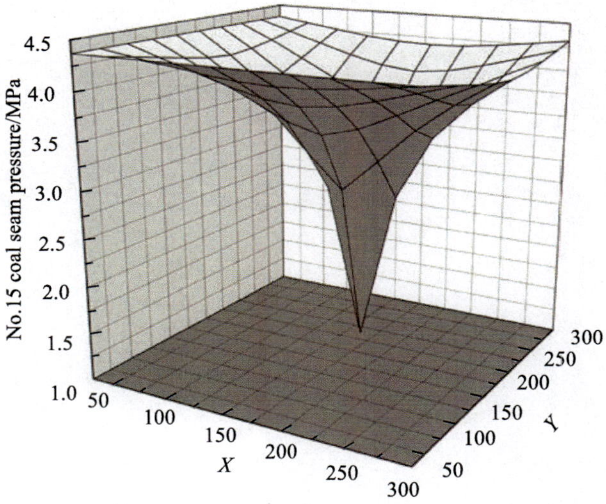

Fig. 10.42 The pressure drop funnel formed in No. 15 coal seam after producing 5000 days (Scheme 3)

References

Gray L (1987) Reservoir engineering in coal seams: part 1, the physical process of gas storage and movement in coal seams. SPE Reserv Eng 228–234

Klinkenberg LJ (1941) The permeability of porous media to liquid and gases. In: API drilling and production practices, pp 200–213

Levine JR (1996) Model study of the influence of matrix shrinkage on absolute permeability of coal bed reservoirs. Geol Soc Publ 199:197–212

McKee CR, Bumb AC, Koening RA (1987) Stress-dependent permeability and porosity of coal. In: Proceedings of the 1987 coalbed methane symposium. University of Alabama, Tuscaloosa, Alabama, pp 183–193

Palmer I, Mansouri J (1996) How permeability depends on stress and pore pressure in coalbeds: a new model. In: Proceedings of the 71st annual technical conference. Society of Petroleum Engineers, Denver, Colorado, pp 557–564

Sawyer WK, Paul GW, Schraufnagel RA (1990) Development and application of a 3D coalbed simulator. In: Proceedings of international technical meeting of Petroleum Society of CIM and Society of Petroleum Engineers, Calgary, pp 1191–1199

Seidle JP, Jeansonne MW, Erickson DJ (1992) Application of matchstick geometry to stress dependent permeability in coals. In: Proceedings of the SPE rocky mountain regional meeting. Society of Petroleum Engineers, Casper, Wyoming, pp 433–445

Somerton WH, Soylemezoglu IM, Dudley RC (1975) Effect of stress on permeability of coal. Int J Rock Mech Min Sci Geol Abstr 12:129–145

Su XB, Feng YL, Chen JF et al (2001) The characteristics and origin of cleat in coal from western North China. Int J Coal Geol 47:51–62 (in Chinese)

11 Stimulation Mechanism and Application

11.1 Hydraulic Fracturing Simulation

11.1.1 Experimental Study

11.1.1.1 Mechanical Properties of Coal Rock

Before a CBM well is put into production, fracturing is often required to induce fractures of a certain height and width in order to improve productivity. In the design of fracture height and width, it is necessary to have an accurate understanding of in-situ stress and coal rock mechanics. In addition, in the study of fracture pressure and wellbore stability in fracturing construction, it is also necessary to test and study the rock mechanical parameters.

According to vitrinite reflectance, China's coal resources can be classified into 10 coal ranks: lignite, long-flame coal, gas coal, fat coal, coking coal, lean coal, meagre coal, anthracite 3#, anthracite 2# and anthracite 1#. The pore structures in coal seams, a dual porous medium with pores and cleats, have been studied as early as the stage of mining coal. Figure 11.1 shows a CT picture of coal rock. It confirms that coal rock is different from sandstone.

Therefore, measuring the mechanical parameters and fractures in coal rock by experiments can not only provide the necessary rock mechanical parameters for coal seam fracturing simulation, but also understand the unique mechanical characteristics of coal seam.

Triaxial Mechanical Properties

The mechanical properties of coal rock mainly include elastic modulus E, Poisson's ratio v, compressive strength P_o, tensile strength C_t and internal friction angle φ.

(1) Experimental equipment

The mechanics experiments were conducted in the Laboratory of Fracturing Center of Langfang Branch, Research Institute of Petroleum Exploration & Development; and the Geological Research Institute, Sinopec Shengli Oilfield. The experimental equipment includes a Rock Mechanics Testing System produced by TerraTek Company, USA, which is used to simulate rock mechanical behaviors at reservoir temperature, stress and pressure. The simulated overlying stress is between 0 and 800 MPa, reservoir pressure is between 0 and 100 MPa, reservoir temperature is between 0 and 200 °C. Data were automatically collected by computer.

(2) Preparation of samples

Figure 11.2 shows the photograph of coal rock samples. Original coal samples are divided into two types: square coal samples of about 15 cm × 15 cm × 15 cm each, which were taken from the new working face at 200 m deep in the Datong area and the Qinshui Basin; and coal rock cores with a diameter of 65 mm each, drilled at 800–1200 m deep in the Ordos Basin. In laboratory, coal cylinders (samples) were drilled on large coal samples according to ASTMD2938, USA. A cylinder has a diameter of 25 mm, and the ratio of length to diameter is about 2:1, to avoid end effect. In addition, the surface of the coal sample should be smooth, the parallelism of two end faces must be kept within 0.02 mm, and the perpendicularity of end face to axis should be within 0.05 mm.

(3) The triaxial compression experiment

Coal matrix contains pores and fractures which contain water, methane and other gases. These fluids make the mechanical properties of coal rock more complex. In the coal seam at tens of meters to kilometers underground, hydrostatic pressure acts on the inner surface of coal pores, and rocks around the coal seam exert pressure on the coal seam, which make the coal seam in a three-dimensional stress state. In the coal seam, the coupling of fluid and solid leads to the deformation of the coal matrix skeleton, the fluid flows in pores, and the pores and fractures constantly change. In the process of fracturing the coal seam, the solid and fluid regions are mutually contained and intertwined, so it is difficult to distinguish them clearly. Therefore, the fluid phase and solid phase must be regarded as quasi-continuum

Y. Song et al., *Coalbed Methane in China*,
https://doi.org/10.1007/978-981-33-4725-0_11

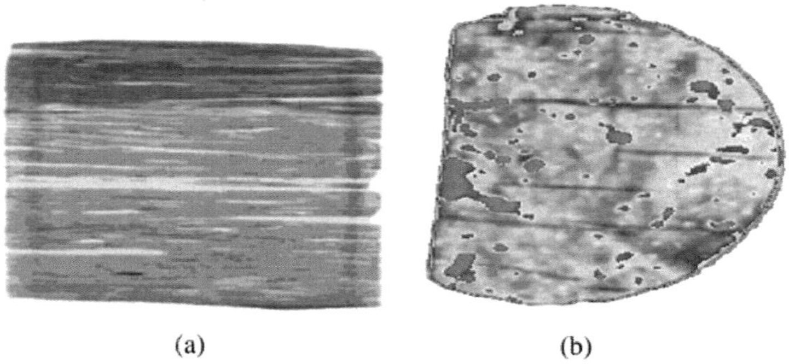

Fig. 11.1 CT picture of coal rock. **a** Longitudinal section; **b** axial section

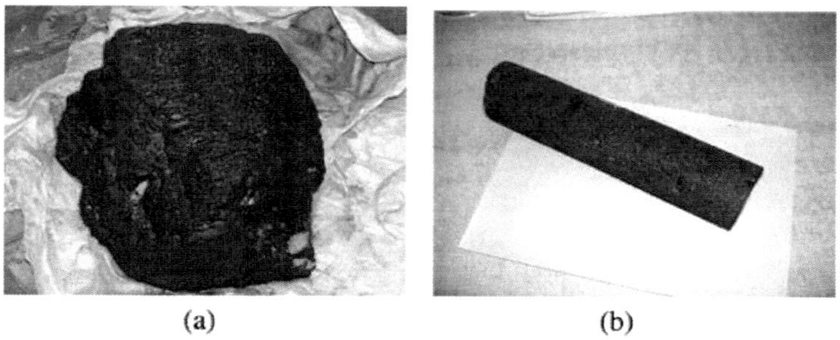

Fig. 11.2 The photographs of coal rock for experiment. **a** Coal block from Datong mine; **b** coal core from Ordos Basin

overlapping by each other. Conventional experiments on coal rock mechanics (uniaxial compression experiments) usually use dry coal samples, and measure the elastic modulus, Poisson's ratio and compressive strength, but no confining pressure. Such results can not reflect the mechanical properties of underground coal rock.

At present, the development of CBM is carried out underground, so the mechanical properties of coal rock at underground stress are more concerned, that is, the mechanical behavior of coal rock saturated with water and gas at confining pressure. Under underground conditions, CBM is mostly adsorbed onto pore surface or dissolved in formation water in coal seams, so the influence of formation gas on the properties of coal seam is not more obvious than that of formation water. In our experiment, dry coal samples and formation water saturated samples were used. By comparing the experimental results, we can understand the real mechanical properties of formation rock.

The formation sample with diameter of 25 mm and length of 50 mm was used in the experiment. The experiments were carried out in parallel with the cleat direction and perpendicular to the cleat direction.

The confining pressures applied to the sample were 10, 9, 8, 7 and 4 MPa. While keeping the confining pressures unchanged, pore pressure was applied, and then axial pressure, at a rate of 0.035 MPa/s. Data were collected by a computer every 10 s, including confining pressure, axial pressure, fluid pressure, axial strain, radial strain and volumetric strain. Before the experiment, the ready core was immersed in 5% KCl solution and saturated with fluid for 48 h.

The elastic modulus of coal rock generally refers to secant modulus, that is, the slope of the straight line of the stress–strain relationship from uniaxial compression experiments, also known as Young's modulus, the ratio of transverse strain to longitudinal strain is called Poisson's ratio. The experiment was carried out at simulated formation confining pressure and pore fluid pressure. The secant modulus was determined by the slope of the straight line through the triaxial stress–strain relationship. At the same time, in order to investigate the changes of elastic modulus and Poisson's ratio at confining pressure and axial pressure, the elastic modulus and Poisson's ratio at the stage of linear elastic deformation were calculated point by point using the formula of triaxial tangent elastic modulus and Poisson's ratio.

The constitutive equation of an elastomer at triaxial stress is as follows.

$$\varepsilon_x = \frac{1}{E}\left[\sigma_x - v\left(\sigma_y + \sigma_z\right)\right] \tag{11.1}$$

$$\varepsilon_y = \frac{1}{E}\left[\sigma_y - v\left(\sigma_x + \sigma_z\right)\right] \tag{11.2}$$

$$\varepsilon_z = \frac{1}{E}\left[\sigma_z - v\left(\sigma_x + \sigma_y\right)\right] \tag{11.3}$$

To transform the above equation, the results of elastic modulus and Poisson's ratio are as follows.

$$E = \frac{\sigma_z(\sigma_z + \sigma_y) - \sigma_x(\sigma_y + \sigma_x)}{(\sigma_z + \sigma_y)\varepsilon_z - (\sigma_x + \sigma_y)\varepsilon_x} \quad (11.4)$$

$$v = \frac{\sigma_x\varepsilon_z - \sigma_z\varepsilon_x}{(\sigma_z + \sigma_y)\varepsilon_z - (\sigma_x + \sigma_y)\varepsilon_x} \quad (11.5)$$

Because the core used in the experiment is cylindrical and the experiment is pseudo-triaxial, then $\sigma_x = \sigma_y$, and the above formula can be simplified as follows.

$$E = \frac{\sigma_z(\sigma_z + \sigma_x) - 2\sigma_x^2}{(\sigma_z + \sigma_x)\varepsilon_z - 2\sigma_x\varepsilon_x} \quad (11.6)$$

$$v = \frac{\sigma_x\varepsilon_z - \sigma_z\varepsilon_x}{(\sigma_z + \sigma_x)\varepsilon_z - 2\sigma_x\varepsilon_x} \quad (11.7)$$

where E is elastic modulus, Pa; v is Poisson's ratio, dimensionless; σ_x, σ_y and σ_z are horizontal and vertical stresses, Pa; ε_x, ε_y and ε_z are horizontal and vertical strains, dimensionless.

(4) Experimental results

Twelve coal samples were measured in the triaxial and uniaxial compression experiments. The results of some coal samples are shown in Figs. 11.3, 11.4, 11.5, 11.6, 11.7 and 11.8. Generally, the stress and strain at near formation pressure are linearly elastic.

Table 11.1 shows the experimental results of triaxial mechanics of coal rock. The results show that.

① For dry coal samples, the stress–strain curve is almost a straight line, and there is no obvious segment. No. 10 and No. 12 coal samples were taken from 1100 m deep coal seams in the Ordos Basin. They were compacted very seriously at high stress. Even without confining pressure, their stress–strain curves are approximately straight, showing good linear elastic deformation. There was no obvious non-linear compaction in the uniaxial compression experiment on coal samples taken from shallow mines. At confining pressure, the stress–strain curves of No. 9 and No. 11 cores taken in the Datong Coalmine also show a good linear

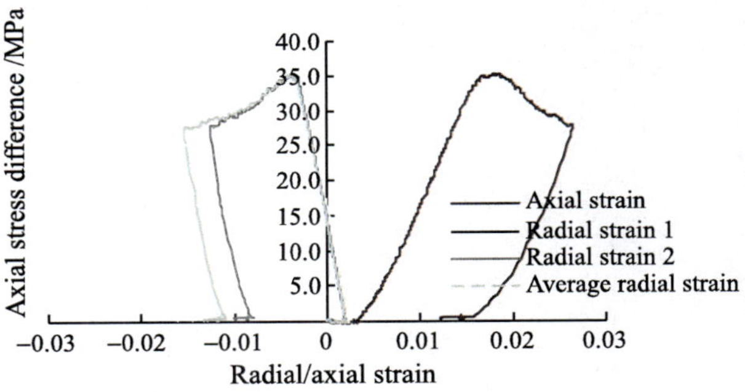

Fig. 11.3 Stress–strain curve of no. 1 coal sample. 10 MPa confining pressure, 4 MPa pore pressure

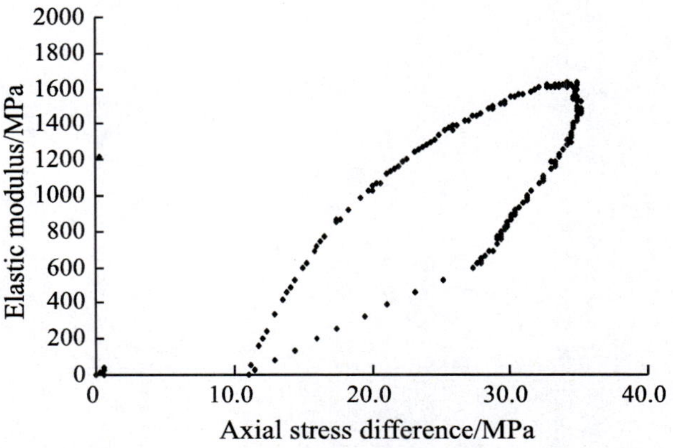

Fig. 11.4 Elastic modulus versus axial stress difference of no. 1 coal sample. 10 MPa confining pressure, 4 MPa pore pressure

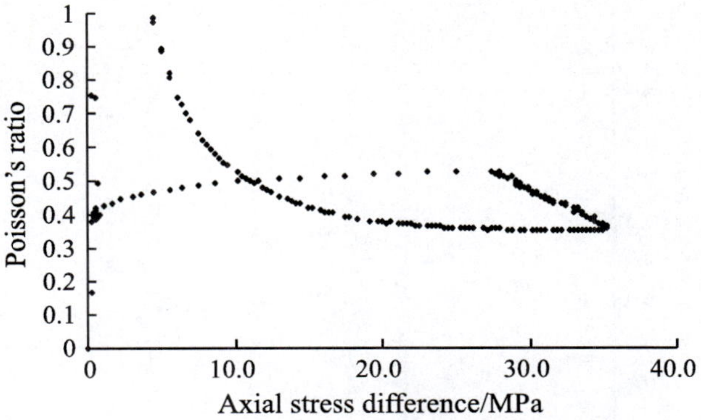

Fig. 11.5 Poisson's ratio versus axial stress difference of no. 1 coal sample. 10 MPa confining pressure, 4 MPa pore pressure

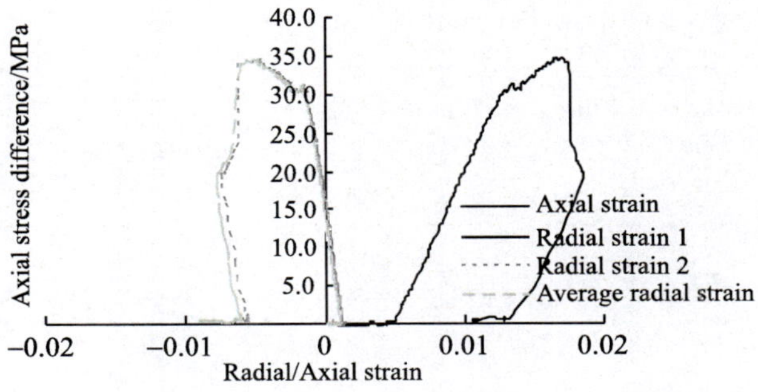

Fig. 11.6 Stress–strain curve of no. 2 coal sample. 9 MPa confining pressure, 4 MPa pore pressure

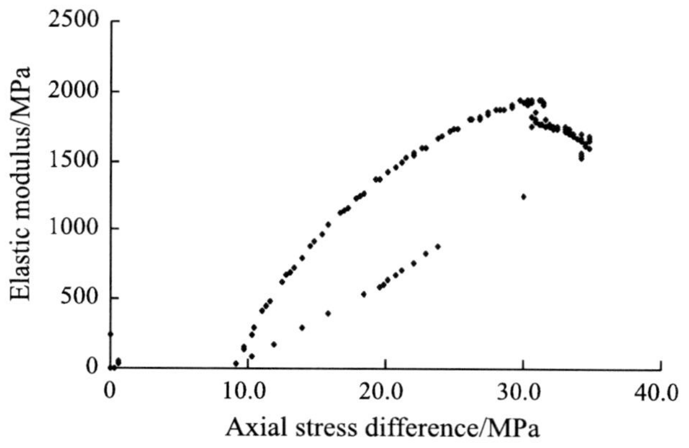

Fig. 11.7 Elastic modulus versus axial stress difference of no. 2 coal sample. 9 MPa confining pressure, 4 MPa pore pressure

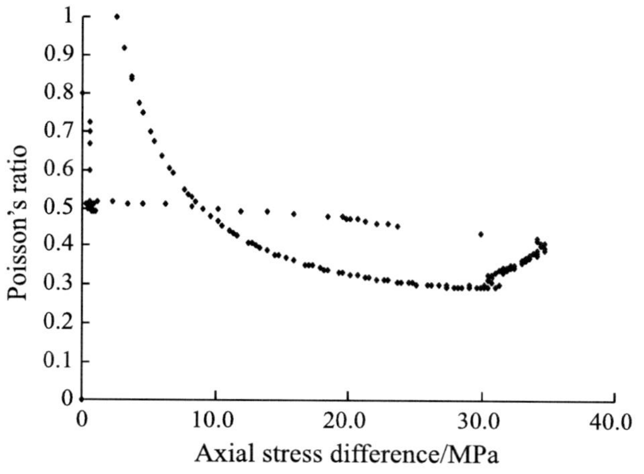

Fig. 11.8 Poisson's ratio versus axial stress difference of no. 2 coal sample. 9 MPa confining pressure, 4 MPa pore pressure

relationship, which indicates that confining pressure has a great influence on the deformation of coal rock, even in cores drilled in the shallow seam, there was no compaction. For dry cores, the changes of the elastic modulus and Poisson's ratio obtained point by point are also small. For example, the elastic modulus and Poisson's ratio of core No. 12 are approximately constant, except that the instability of the instrument at the beginning of the experiment and the failure of the rock at the later stage led to unreasonable results. From the above experiment results, it can be seen that for highly compacted dry coal rock and dry coal samples at confining pressure, they can be approximately regarded as linear elastomers before fracturing.

② For coal rock saturated with KCl solution, the stress–strain curve bended upward at the beginning stage, indicating that the coal rock swelled after saturated with brine. The initial axial stress compacted the coal rock skeleton and applied a force on the fluid. It is a continuous compaction process. The linear section of the stress–strain curve is shorter, indicating that the fluid has changed the mechanical properties of the dry coal rock which is approximately linear. The elastic modulus and Poisson's ratio of the coal rock saturated with brine vary with the axial stress. The elastic modulus increases with the increase of axial stress, while the Poisson's ratio decreases gradually. The mechanical properties show a gradual hardening process. The change of elastic modulus is more obvious than that of Poisson's ratio in the whole process of increasing axial stress.

Generally speaking, the static elastic modulus and compressive strength of a natural coal sample are larger than that

Table 11.1 Triaxial mechanical experiment results of coal rock

Core no	Elastic modulus/MPa	Poisson's ratio	Shear modulus/MPa	Bulk modulus	Compressive strength
1	2854	0.38	1034	3961	42.3
2	4043	0.31	1545	3513	39.4
3	4215	0.34	1572	4412	43.1
4	4226	0.33	1592	4082	44.4
5	2914	0.32	1101	2747	30.3
6	3937	0.36	1448	4671	27.7
7	2604	0.41	926	4645	32.7
8	4213	0.31	1609	3687	45.6
9	3340	0.25	1336	2227	38.1
10	3980	0.29	1543	3159	44.3
11	4100	0.32	1553	3796	39.2
12	4190	0.33	1575	4108	45.1

of a coal sample saturated with brine. However, due to the serious heterogeneity of coal, the coincidence of the two radial strain curves of the coal sample is not as good as that of sandstone. The experiment results at 4 and 8 MPa show that the elastic modulus and compressive strength of the coal sample increase with the increase of confining pressure, which indicates that the elastic modulus and compressive strength are affected by confining pressure.

Tensile Strength

In the tensile strength (Brazilian) experiment, initial coal samples were processed into samples with diameters of about 50 mm and thickness of about 25 mm each, and the ratio of the diameter to the thickness is almost 2–1. Pressure was applied on the Brazilian experiment machine until tensile fractures were induced. The force applied at that time was recorded as P. The tensile strength (C_t) of the rock samples was calculated by Eq. (11.8). The calculation results are shown in Table 11.2.

$$C_t = \frac{2P}{\pi d L} \quad (11.8)$$

where P is the pressure when the coal sample is destroyed, N; L is the thickness of the coal sample, mm; d is the diameter of the coal sample, mm.

Internal Friction Angle

In order to determine the internal friction angle of the coal samples from different coal seams, the coal sample was placed in the experimental chamber, and firstly, confining pressure σ_3 was applied to the set value, and then vertical pressure σ_1 was applied until the coal sample was destroyed. According to the σ_1 and σ_3, a failure stress circle can be drawn in the $\sigma - \tau$ coordinate system. After experimenting on multiple samples from the original same sample, a series of σ_1 and σ_3 can be obtained, and a group of failure stress circles can be drawn. The envelope of the stress circle is the shear strength curve of the coal. The σ coordinate of any point on the envelope represent the shear strength τ along the shear failure surface at some confining pressure and vertical pressure, the angle φ between the tangent of any point and the τ coordinate axis represents the internal friction angle on the corresponding shear failure surface, and the intercept between the tangent and the σ coordinate axis represents the cohesion τ_0 of the shear failure surface. Experiments show that when the confining pressure is large, the envelope is generally a quadratic curve.

Figure 11.9 is a simplified linear envelope. The internal friction angle φ and cohesion τ_0 vary with the shear surface at some confining pressure. In other words, the internal friction angle and cohesion change with the confining pressure. When the confining pressure is high and shear failure occurs, the internal friction angle decreases and cohesion increases. On the contrary, at lower confining pressure, the internal friction angle increases, and the cohesion decreases. Table 11.3 is calculated internal friction angle.

Dynamic and Static Mechanical Parameters

According to the interval velocities of P-wave and S-wave, the dynamic Young's modulus and Poisson's ratio can be calculated by using the following formulas.

$$E_D = \rho_B v_S^2 \frac{3v_P^2 - 4v_S^2}{v_P^2 - v_S^2} \times 10^{-3} \quad (11.9)$$

Table 11.2 Tensile strength

Coal sample	Coal rank	Diameter/mm	Thickness/mm	Pressure applied	Fracturing/ (load/N)	Tensile strength/MPa	Average/MPa
Jin 1	Anthracite 2#	50.20	25.30	Parallel with bedding	2891	1.45	1.820
Jin 2		50.00	25.00	Parallel with bedding	4604	2.35	
Jin 3		50.00	24.96	Vertical along bedding	3247	1.66	
Yang 1	Anthracite 3#	50.00	25.10	Parallel with bedding	3203	1.63	1.160
Yang 2		50.10	24.60	Parallel with bedding	1379	0.71	
Yang 3		50.10	25.00	Vertical along bedding	2224	1.13	
Feng1	Coking coal	49.00	25.10	Parallel with bedding	111	0.06	0.245
Feng 2		49.90	25.00	Vertical along bedding	845	0.43	

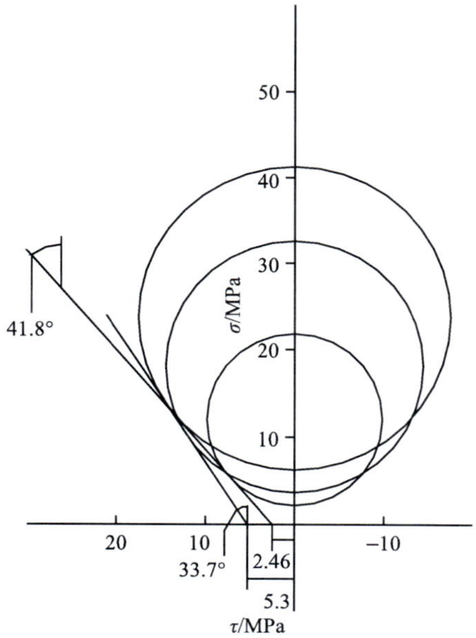

Fig. 11.9 Simplified linear envelope

$$\upsilon_{D} = \frac{\frac{1}{2}v_{P}^{2} - v_{S}^{2}}{v_{P}^{2} - v_{S}^{2}} \tag{11.10}$$

where v_P represents P-wave velocity, m/s; v_S is S-wave velocity, m/s; E_D is dynamic Young's modulus, MPa; υ_D is dynamic Poisson's ratio; ρ_B is coal density, t/m^3.

In the laboratory, two ultrasonic wave transmitter and receiver caps were placed at both ends of a coal sample, so that dynamic and static mechanical parameters were obtained simultaneously. Table 11.4 is the experimental result. From Table 11.4, for the same coal sample, the dynamic Young's modulus is larger than the static Young's modulus, and the Poisson's ratio changes little. The ratio of dynamic Young's modulus to static Young's modulus of four kinds of coal samples is 1.68 (Yangquan Coalmine), 1.36 (Jincheng Coalmine), 1.89 (Fengfeng Coalmine) and 1.32 (core drilled in Well Jin Test 1). According to the logging curves of Well Jin Test 1, the velocity ratio of P-wave to S-wave is 1.62, the dynamic Young's modulus is 7500 MPa, and the Poisson's ratio is 0.36. According to laboratory data, the velocity ratio of P-wave to S-wave is 2.13, the dynamic Young's modulus value is 5235 MPa, and the Poisson's ratio is 0.19. The ratio of the measured to the experimented dynamic Young's modulus is 1.43. With less data available, it is impossible to regress. Table 11.5 lists the experimental data of dynamic and static parameters of sandstone and mudstone at different depths in Well Da Test 1 and Well Jin Test 1, from which we can get the following understandings: the experimented dynamic Young's modulus is different from the logged in Well Jin Test 1; the former is larger than the latter; the experimented dynamic Poisson's ratio is larger than the experimented static Poisson's ratio. Using the following Eq. (11.11), logged Young's modulus can be regressed to static Young's modulus.

$$E_{\text{static state}} = 0.61E_{\text{longging}} + 1059 \tag{11.11}$$

There are only three set of sandstone and mudstone data from Well Da Test 1, so that it is impossible to build an relationship like Eq. (11.11). Based on the above two points, we can conclude that the dynamic Young's modulus from laboratory is different from the logging value, the dynamic Poisson's ratio is larger than the static Poisson's ratio from laboratory. Using Eq. (11.11), the logging value (dynamic) can be converted into static. Equation (11.11) is only applicable to the Jincheng area. To get Equations from other areas, rock samples from those areas must be selected for indoor dynamic and static experiments.

Laboratory Hydraulic Fracturing Experiment

CBM exists in coal seams in adsorbed state, and is usually exploited by pressure-lowering desorption method. In most

Table 11.3 Internal friction angle

Coalmine	Simulated well depth/m	Coal sample	Confining pressure/MPa	Compressive strength/MPa	Shear strength/MPa	Internal friction angle/(°), cohesion/MPa		
						Group I ①–②	Group II ②–③	Group III ①–③
Yangquan	600	Yang 1	2	43	20.5	$\varphi = 11.5°$	$\varphi = 30°$	$\varphi = 22.6°$
		Yang 2	4	46	21	$\tau = 16.3$	$\tau = 9.8$	$\tau = 12.8$
		Yang 3	6	52	23			
Jincheng	600	Jin 1	2	75	36.5	$\varphi = 11.5°$	$\varphi = 65.1°$	$\varphi = 56.4°$
		Jin 2	4	78	37	$\tau = 29.4$	$\tau = 0$	$\tau = 8$
		Jin 3	6	119	56.5			

Table 11.4 Experimental results of static and dynamic mechanical parameters

No	Coal sample	Simulated well depth/m	Confining pressure/MPa	Axial stress difference/MPa	Static		Dynamic		Volume compressibility/ (l/MPa)	Compressive strength/MPa
					Young's modulus/MPa	Poisson's ratio	Young's modulus/MPa	Poisson's ratio		
1	Yang 5	600.0	4	10	3380	0.36	5681	0.34	5.48×10^{-4}	45
2	Jin 5	600.0	4	20	4575	0.36	6272	0.34	2.27×10^{-4}	109
3	Feng 5	600.0	4	9	2590	0.28	4904	0.36	9.79×10^{-4}	46
4	No. 3 coal seam in Well Jin Test 1	524.5	3	10	3971	0.30	5235	0.36	5.27×10^{-4}	43

Table 11.5 Static and dynamic mechanical parameters

Well no	Depth/m	Lithology	Logging value		Confining pressure/MPa	Experimental value					
			Young's modulus/MPa	Poisson's ratio		Static		Dynamic		Volume compressibility/(10^{-4}/MPa)	Compr-essive/MPa
						Young's modulus/MPa	Poisson's ratio	Young's modulus/MPa	Poisson's ratio		
Well Jin Test 1	501.8–502	Mudstone	17,500	0.35	3	9925	0.15	37,963	0.19	2.77	94
Well Jin Test1	527.9–530	Mudstone	22,300	0.35	3	9431	0.19	32,130	0.23	3.34	89
Well Jin Test 1	532.01	Mudstone	33,500	0.33	3	15,992	0.07	44,535	0.22	2.12	125
Well Jin Test 1	604–606.1	Mudstone	50,600	0.25	4	31,579	0.19	76,678	0.29	0.29	191
Well Jin Test 1	604.5	Limestone	47,300	0.25	4	32,343	0.14	83,859	0.29	0.29	>213
Well Jin Test 1	606.75	Limestone	23,400	0.25	4	28,503	0.16	66,197	0.32	0.32	168
Well Jin Test 1	610–612	Mudstone	27,000	0.33	4	14,791	0.28	31,762	0.24	0.24	81
Well Da 1-1	1152–1161	Mudstone	28,300	0.30	6	9570	0.23	32,796	0.32	0.32	91
Well Da 1-1	1152–1161	Limestone	28,300	0.30	6	14,100	0.12	43,648	0.21	0.21	108
Well Da 1-1	1174–1178	Mudstone	25,200	0.32	6	21,352	0.24	38,632	0.29	0.29	124
Well Da 1-1	1235–1250	Limestone	22,800	0.31	7	20,761	0.22	41,206	0.24	0.24	175
Well Da 1-1	1251–1263	Mudstone	30,200	0.29	7	20,723	0.20	45,575	0.25	0.25	184
Well Da 1-1	1266–1269	Mudstone	22,300	0.35	7	23,064	0.21	49,164	0.26	0.26	128

areas, the permeability of coal seams is low, usually less than 1 mD. In China, as the permeability of coal seams is generally 0.001–0.1 mD, according to the criterion of reservoir classification based on oil and gas reservoir permeability, coal seams belong to ultra-low-permeability reservoirs or tight reservoirs, so CBM wells need to be stimulated by hydraulic fracturing.

The reservoir characteristics of coal seams are quite different from those of conventional oil and gas reservoirs. Coal rock is a dual medium with pores and fractures. Fracturing to coal seams is different from to conventional oil and gas reservoir because of the great difference in the microstructures of two mediums.

Because there was no true equipment for triaxial experiment before, the triaxial experiment on rock is actually false triaxial. In the experiment, a cylinder core is used, and a rubber cylinder is installed outside the core. Equal radial pressure is applied by hydraulic pressure, and another pressure is applied in the axial direction. Because this kind of equipment can not realize true triaxial and cylindrical core is used, it is very difficult to simulate hydraulic fracturing because the core is small. In order to accurately obtain the stress–strain relationship under formation conditions, true triaxial experiment on core must be carried out (The equipment is shown in Fig. 11.10). The maximum axial pressure is 1500 kN and the maximum confining pressure is 600 kN.

Although uniaxial experiments on coal seam can reflect some mechanical properties of coal rock, the simulation to hydraulic fracturing is the most accurate. Therefore, we introduced the physical simulation of hydraulic fracturing into the experiment, and defined a horizontal stress difference coefficient Kh. The experiment was carried out at Kh = 0.25, 1.0 and 2.5, respectively. The experimental results (Fig. 11.11) show that when the horizontal stress difference is small, induced fractures would propagate along natural fractures, and when the horizontal stress difference is large, induced fractures would be perpendicular to the minimum principal stress. The results of coal rock experiments also confirm that the orientation of induced fractures in fractured rock is the result of both the horizontal stress and the natural fractures. The analysis shows that natural fractures are mostly open at low confining pressure, and very permeable, so in weak rock, induced fractures would propagate along natural fractures; when confining pressure is high, natural fractures are mostly close, and the bonding ability between fracture surfaces is strengthened at the confining pressure, that is the fractures are bonded, so induced fractures are nearly perpendicular to the minimum principal stress.

Many representative results have been obtained. The surface of induced fractures is extremely irregular, not smooth or stepped; induced fractures either parallel to or cross natural fractures (Fig. 11.12). The propagating pressure of the fractures is very high, sometimes greater than the fracturing pressure. The flow resistance increases due to the formation of a large amount of coal powders during fracturing and the irregular fractures and their unsmooth surface.

The fracture direction is no longer vertical to the direction of the minimum principal stress, as is considered in conventional reservoirs. The direction of the fracture is affected by the horizontal stress, and complex fracture shapes appear. Different from conventional sandstone gas reservoir, natural fractures in coal seam are developed, and its elastic modulus is small and Poisson's ratio is large, so fracture expansion in coal seam is extremely complex, showing a large number of irregular fractures. Generally, induced fractures in coal seams are radially distributed around wellbore. The fractures

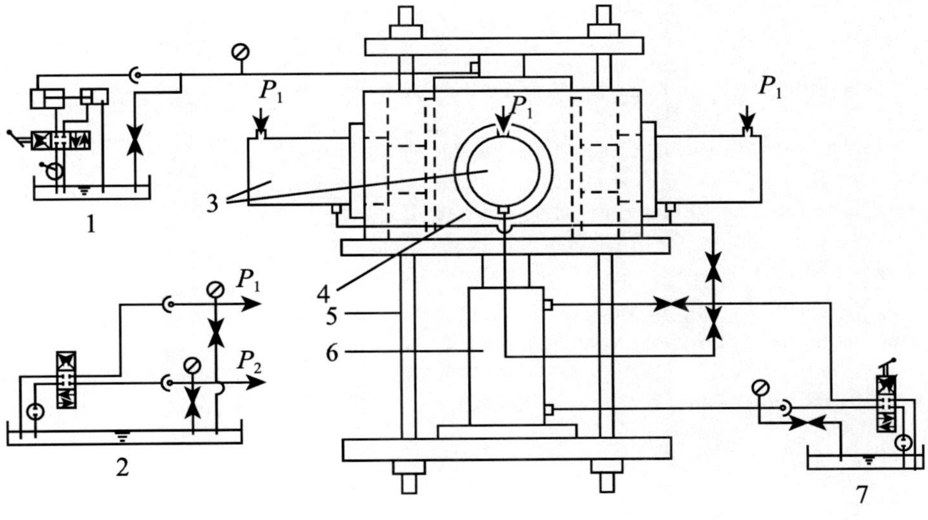

Fig. 11.10 True triaxial experiment machine. 1. hydraulic power system; 2. confining pressure power system; 3. confining pressure cylinders (14); 4. confining pressure loading frame; 5. loading rack; 6. axial pressure cylinder; 7. axial pressure power system

Fig. 11.11 Experimental results at different confining pressures. **a** $Kh = 0.25$; **b** $Kh = 2.5$

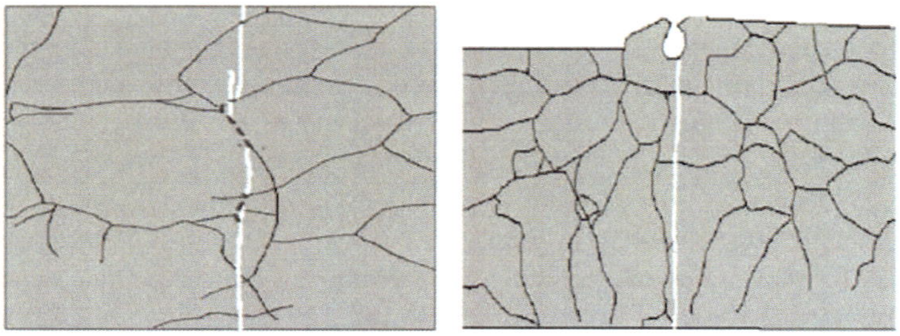

(a) (b)

Fig. 11.12 Experimental results at different confining pressures

(a) (b)

are stepped, cornered and asymmetric, and may be "T" and "I" fractures on the boundary of the coal seam. According to the measured results of the Qinshui Basin, it is confirmed that the induced fractures are very irregular. Unlike in conventional reservoirs, induced fractures are horizontal in shallow layers and vertical in deep layers. Induced fracture in coal seams are very random in shapes. Their directionality is not obvious in the same formation of the same basin, but they are possible to appear in a direction at the same formation, which is not consistent with natural cleat direction.

The Influence of Mechanical Properties of Coal Rock on Hydraulic Fracturing Design

(1) The influence of Young's modulus on fracture height

In a 2D model with constant fracture height, for Newtonian fluid, the fracture width (W) is inversely proportional to the one-fourth power of Young's modulus, i.e. $W \propto E^{-1/4}$. For non-Newtonian fluids, the fracture width is inversely proportional to the $\frac{1}{2n+2}$ power of Young's modulus, i.e. $W \infty \frac{1}{E^{1/(2n+2)}}$. If $n = 0.5$, the fracture width is inversely proportional to the 1/3 power of Young's modulus. This shows that the fracture width will be reduced by 16–20% if the Young's modulus of the formation is doubled.

(2) The influence of Young's modulus on fracturing pressure

According to the 2D fracture simulation theory, for Newtonian fluid, the net pressure in fracturing operation (fluid pressure in fracture minus fracture closing pressure) is inversely proportional to the 3/4 power of Young's modulus, i.e. $\Delta P_{\rm f} \propto E^{-3/4}$. For non-Newtonian fluids, the net pressure is proportional to the $\frac{2n+1}{2n+2}$ ower of Young's modulus, i.e. $\Delta P_{\rm f} \propto E^{-\frac{2n+1}{2n+2}}$. Therefore, when the Young's modulus is doubled, the net pressure will increase by 1.6–1.7 times.

(3) The influence of interlayer Young's modulus difference on fracture height

At little different interlayer stress, Young's modulus may become an important factor to control fracture extension. Following is an example of the effect of Young's modulus difference on fracture geometry simulated by TerraFrac 3D fracturing design software. In the simulation example, the values of the Young's modulus of the reservoir and the overlying and underlying layers are different. Table 11.6 lists the Young's modulus and simulated fracture length and height. Case A: the Young's modulus of the boundary layer is 5 times that of the reservoir. Case B: the Young's modulus of the boundary layer is 2.5 times that of reservoir. Case C:

Table 11.6 Young's modulus, simulated fracture length and height

Case	Young's modulus/MPa			Full fracture length/m	Fracture height/m		
	Top layer	Reservoir	Bottom layer		Extend upward	Extend downward	Total fracture height
A	34,483	6879	34,483	556	12	17	29
B	34,483	13,739	34,483	181	13	151	164
C	34,483	20,690	34,483	161	55	127	182
D	34,483	34,483	34,483	157	55	138	193
E	3448.3	34,483	3448.3	98	61	208	269

the Young's modulus of the boundary layer is 1.67 times that of the reservoir. Case D: the Young's modulus of the boundary layer is the same as that of the reservoir. Case E: the Young's modulus of the boundary layer is 1/10 of that of the reservoir.

Other input parameters are the same, including in-situ stress, fracture toughness, filtration, thickness, rheological property of fracturing fluid and displacement of each layer.

(4) Fracturing particularity of coal rock

According to a large number of measured rock mechanic parameters, fracture initiating simulation and fracture conductivity experiments, coal rock shows different characteristics from conventional sandstone reservoir in structure, rock mechanics and physical properties, which makes induced fractures in coal rock different from conventional hydraulic fracturing to sandstone reservoir.

Fracture extension is controlled by both in-situ stress and natural fractures. When the difference of horizontal stress is small, induced fractures will propagate along natural fractures. When the difference of horizontal stress is large, induced fractures will be perpendicular to the minimum principal stress. The surface of induced fractures is extremely irregular, not smooth or stepped. The fractures are either parallel to or across natural fractures.

Coal rock is characterized by low strength, low elastic modulus and high Poisson's ratio. These mechanical properties have important impacts on the development and geometric size of induced fractures. On the one hand, coal rock is easy to fracture because of low strength, especially low tensile strength. On the other hand, because of the high Poisson ratio, the lateral formation pressure increases, which makes it difficult to fracture. Therefore, the difficulty in fracturing coal rock depends on the actual calculation at specific condition. According to practical experience, the pressure gradient of bottom hole of coal seam is much higher than that of conventional sandstone at the same depth (the fracturing pressure gradient of coal rock bottom hole is 0.031–0.056 MPa/m). However, it is certain that the low elastic modulus and high Poisson's ratio of coal rock will lead to the decrease of fracture length and increase of fracture width. This is because the Young's modulus reflects the mechanical properties of rock itself under the conditions of stress and deformation, and has an important influence on the geometric sizes of induced fractures. According to the Lamb equation theory, the width of the hydraulic fracture is inversely proportional to the Young's modulus of the rock. The smaller Young's modulus, the wider the fracture width, which is the main reason for the wider fracture in coal seams. With the increase of fracture width, the increase of fracture length is limited at the same construction scale. Therefore, compared with conventional sandstone, short and wide fractures are more likely to form in coal rock.

Induced fractures in coal seams are irregular. Coal seam is different from conventional sandstone gas reservoir. Natural fractures are developed in coal seams, and there are a lot of cleats in the matrix. The elastic modulus of coal seam is small, Poisson's ratio is large, so fracture expansion in coal seam is extremely complex, showing a large number of irregular fractures. In hydraulically fracturing coal seams, vertical fractures and horizontal fractures coexist, or more vertical (horizontal) fractures exist, that is, the so-called complex fracture system. This phenomenon is the result of combined influence of factors such as natural cleats, different mechanical properties between coal rock and overlying and underlying layers, and blocked fracture ends caused by coal powders. After hydraulic fracturing, "T" shaped fractures (horizontal in top layer and vertical in coal seams) can be observed on the new roadway.

11.1.1.2 Experiments on Conductivity of Coal Rock

Experimental Equipment

FCS-100 diversion instrument was used, and the diversion chamber was designed according to API standard. Figure 11.13 shows the physical diagram of the FCS-100. Figure 11.14 is the working principle of the FCS-100. The instrument can simulate formation conditions and evaluate

Fig. 11.13 The physical diagram of FCS-100 diversion instrument

short-term or long-term conductivities of different types of proppants. The maximum experimental temperature is 150 °C and the maximum closure pressure is 200 MPa.

Principle of Experiment

The experimental procedure is shown in Fig. 11.14. The experimental principle can be expressed by Darcy's law.

$$k = \frac{Q\mu L}{A\Delta P} \tag{11.12}$$

where k is the permeability of propped fracture, cm^2; Q is the flow rate in fractures, cm^3/s; μ is fluid viscosity, MPa s; L is the length of sample section, cm; A is the cross-sectional area of propped fractures, cm^2; ΔP is the pressure difference between the two ends of the sample section, atm.

The FCS-100 diversion instrument uses API diversion chamber and operates strictly according to API standard. The formula for calculating proppant permeability and conductivity can be expressed as follows.

The permeability of propped fracture,

$$k = \frac{5.411 \times 10^{-4}\mu Q}{\Delta P W_{\mathrm{f}}} \tag{11.13}$$

The conductivity of propped fractures,

$$k = \frac{5.411 \times 10^{-4}\mu Q}{\Delta P} \tag{11.14}$$

where W_{f} is the width of propped fracture, cm; Q is the flow rate in the fracture, cm^3/min, and other parameters are the same as above.

Evaluation Methods of Short-Term Conductivity and Their Limitations

The short-term evaluation method is generally used to evaluate the conductivity of proppants in laboratory, but the short-term conductivity can not show the real conductivity of proppants underground, which should be expressed as instantaneous conductivity.

1. API short-term conductivity evaluation method

The short-term conductivity evaluation method recommended by API is China's industry standard for petroleum and natural gas (SY/T6302–1997). The standard involves experimental methods and instruments. Its purpose is to establish standard steps and conditions for evaluating short-term conductivity of various proppants in laboratory. By comparing the conductivities of different proppants under certain conditions, it is possible to build criterion for selecting proppants.

(1) Experimental conditions

The experimental liquid is NaCl solution, the ambient temperature is 24 °C, and the experimental instrument is recommended by API.

(2) Experimental procedures

According to the procedures, it is necessary to add enough closure pressure on the sample for a long time to make proppants enter fractures until to a semi-steady state. When liquid flows through the propped layer at some closure pressure, measure the width, pressure difference and flow rate, and then calculate the conductivity and permeability of the propped layer. Three flow rates are used at each closure pressure, and the result is the average of the results of three flow rates.

(3) Experimentation requirements

At given flow rate and room temperature, there is no non-Darcy flow or inertia effect. After increasing one closure pressure to another, it must take a certain time to make proppants reach semi-stable state.

(4) Limitation

The evaluation method of short-term conductivity can be used to compare the performance of different proppants and select proppants in laboratory. However, this method can not

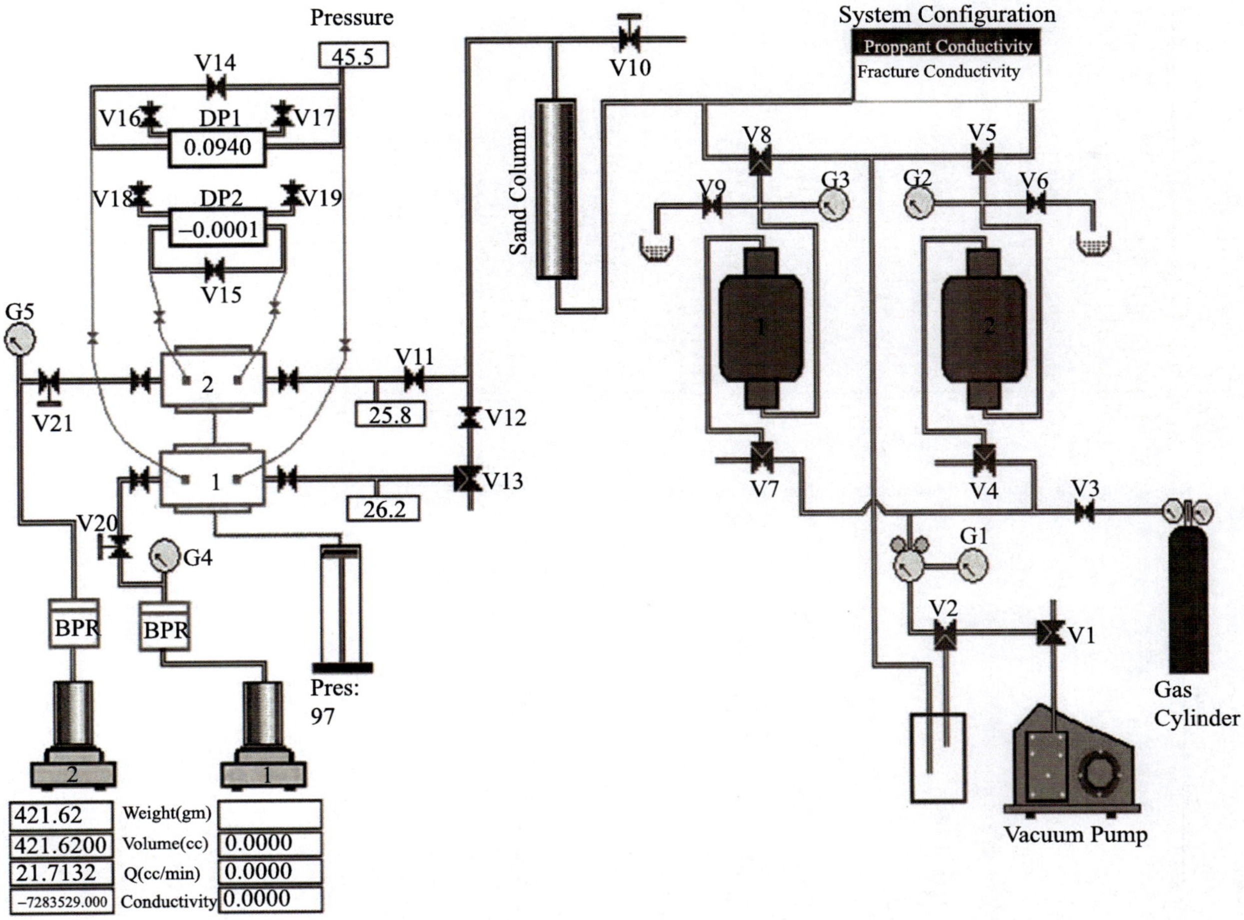

Fig. 11.14 The working principle of FCS-100 diversion instrument

obtain the absolute value of the conductivity of propped fractures in reservoir, nor can it provide a reference to long-term conductivity.

2. Comparison of short-term and long-term conductivities by experiments

There has been no quantitative comparison between long-term and short-term conductivities. In order to determine whether long-term or short-term conductivity is used in the experiment on coal rock conductivity, common sandstone samples were used to study the difference between long-term and short-term flow conductivities.

Figures 11.15 and 11.16 are experimental curves of short-term and long-term flow conductivity of quartz sand and walnut shell at closure pressure of 30 MPa and 40 MPa, respectively. The result shows that the conductivity decreases rapidly in a short period when the closure pressure increases, and then the conductivity tends to stabilize. At 40 MPa, after 25 h, the conductivity has been very small, more than 80% lower than before. Figures 11.17 and 11.18 are the experimental curves of the short-term and long-term conductivity of quartz sand and walnut shell after changing their composition, at closed pressure of 30 and 40 MPa. After 20 h, the conductivity tends to be stable. From the above experiments, it can be seen that the short-term conductivity measured after about 5 h can not represent the conductivity of proppants after compacting in fractures, and the short-term conductivity can only be used to compare the relative difference of the conductivity at different proppant concentration. If we want to truly reflect the conductivity in fractures, the experiment result of long-term conductivity are more reasonable. Therefore, for coal rock, the experiment period at the same closure pressure should be 24 h for the long-term flow conductivity, so as to determine the stable conductivity of fractures in coal rock.

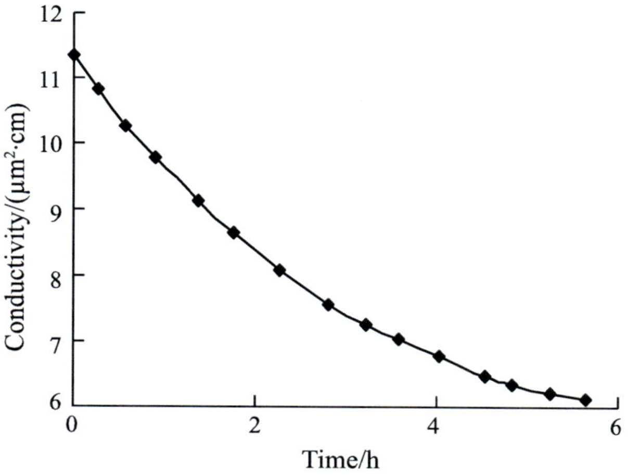

Fig. 11.15 Short-term conductivity of quartz sand and walnut shell at 30 MPa

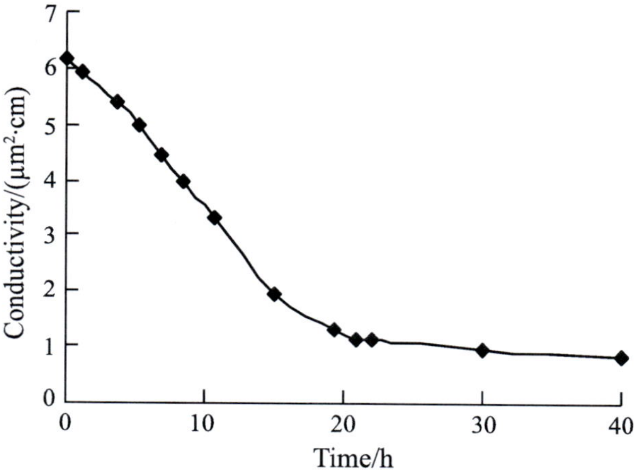

Fig. 11.16 Long-term flow conductivity of quartz sand and walnut shell at 40 MPa

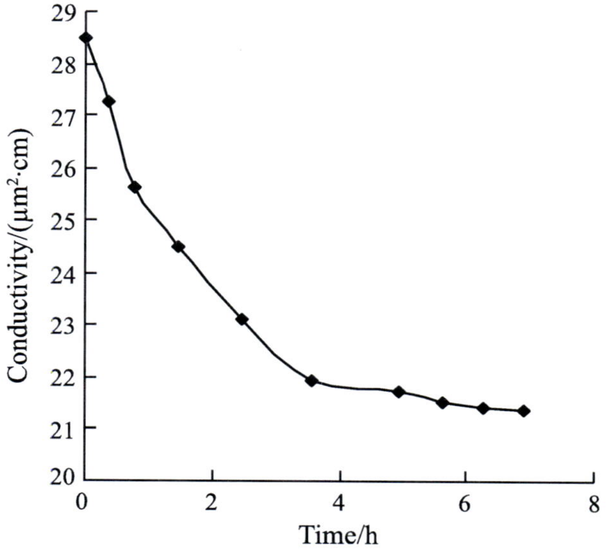

Fig. 11.17 Short-term flow conductivity of quartz sand and walnut shell at 30 MPa

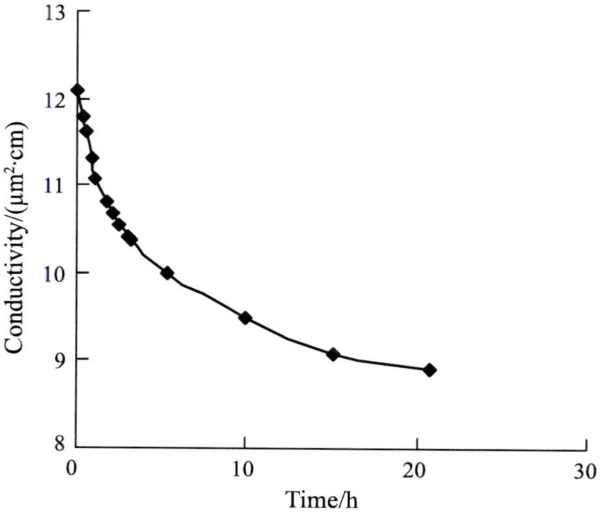

Fig. 11.18 Long-term flow conductivity of quartz sand and walnut shell at 40 MPa

Experimental Conditions and Samples

In order to truly reflect the actual state of proppants in underground fractures, the simulated temperature was set 50 °C, and the experiment on long-term flow conductivity was carried out. Data were measured on an experiment pressure point for 24 h. The CBM well in China may be 1200 m deep, so the maximum closure pressure was set 40 MPa, and the proppants were 20/40 mesh quartz sand which is widely used now. The fluid was 4% NaCl brine and the flow rate was 2–5 mL/min. Cores taken from 200 m underground in the Datong Coalmine were used to make slice samples, 17.7 cm long, 3.8 cm wide and 1–2 cm thick each, and the end of the sample is semi-circular.

Because natural fractures are developed in coal seams, it is very difficult to obtain intact and non-destructive cores. Cores which were not very seriously damaged were bonded the fractures with chemical glue, and then used in experiments after smoothing the bond.

Five groups of coal samples were used in the conductivity experiments. The experimental results are shown in Table 11.7.

Experimental Results

Figures 11.19, 11.20, 11.21, 11.22, 11.23 and 11.24 are experiment results. The curve was regressed using the experiment results. The experiment results at 5 kg/m^2 proppant concentration show that the conductivity is

Table 11.7 Conductivity experiments

No	Sample	Proppant concentration/(kg/cm^2)	Fluid
1	Steel plate	5	Fresh water
2	Coal	5	Fresh water
3	Coal	10	Fresh water
4	Coal	5	0.4% fracturing fluid
5	Coal	5	Fresh water

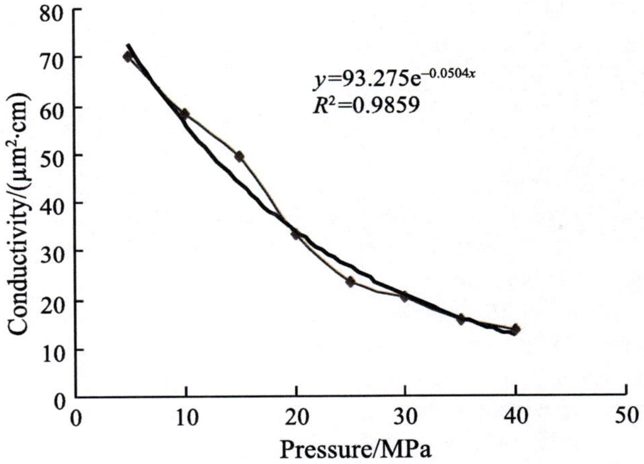

Fig. 11.19 Long-term conductivity of steel plate ($5\ kg/m^2$)

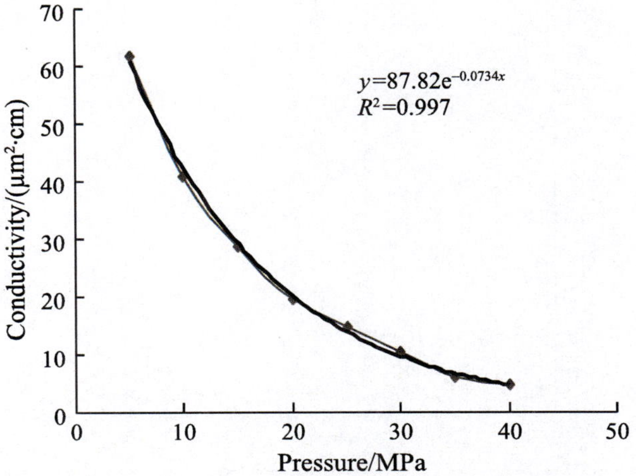

Fig. 11.20 Long-term conductivity of coal rock ($5\ kg/m^2$)

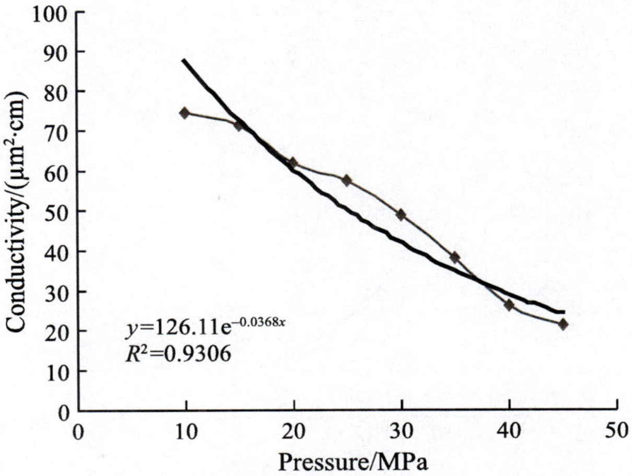

Fig. 11.21 Long-term conductivity of coal rock ($10\ kg/m^2$)

regressed with the closure pressure according to the exponential equation, and correlated with the curve by above 0.98, indicating that the conductivity decreases exponentially with the increase of closure pressure at $5\ kg/m^2$ proppant concentration. But at $10\ kg/m^2$, the correlation reduced to 0.93.

Figure 11.24 shows the conductivity of coal rock fractures at two proppant concentrations: the conductivity at $10\ kg/m^2$ is obviously greater than that at $5\ kg/m^2$, and before 20 MPa, the attenuation of the conductivity at $10\ kg/m^2$ is less than that at $5\ kg/m^2$. At low proppant concentration and low closure pressure, a large amount of proppants embedded fractures, making the conductivity of the fractures decrease rapidly. At high proppant concentration, embedding proppants gave less influence on the conductivity of fractures, so the conductivity reduced less. However, when the closure pressure was higher than 20 MPa, more proppants were broken, so the conductivity at high proppant concentration reduced. When the proppant concentration was low, proppants embedded fractures, but less were broken, so the conductivity of the fractures decreased slightly, and decreased slowly.

Figure 11.25 shows the conductivity of a steel plate and a coal sample at $5\ kg/m^2$ proppant concentration. The results show that the conductivity of the steel plate is larger than that of the coal sample, indicating proppants embedded.

Figure 11.26 shows the original coal sample. Figure 11.27 shows the coal sample with embedded proppants after the

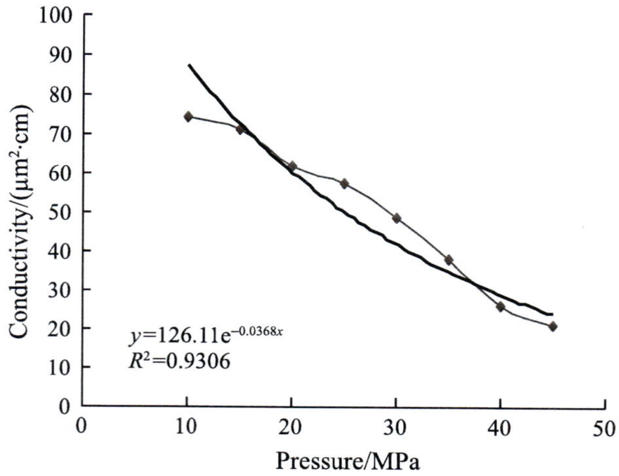

Fig. 11.22 Long-term conductivity of coal rock (5 kg/m^2)

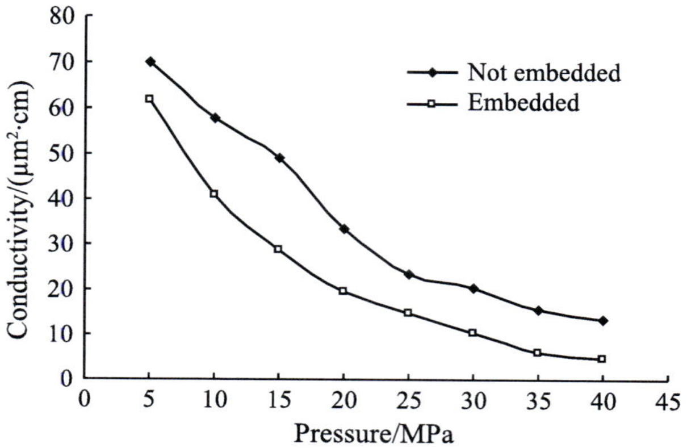

Fig. 11.25 Conductivity versus proppant concentration: 20/40 mesh, 5 kg/m^2

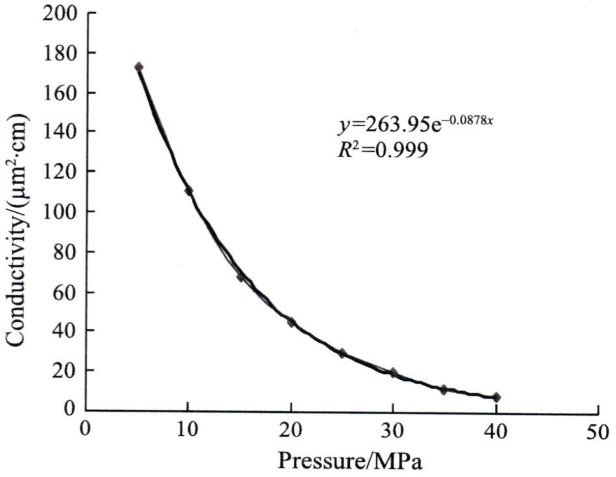

Fig. 11.23 Long-term conductivity of coal rock (5 kg/m^2)

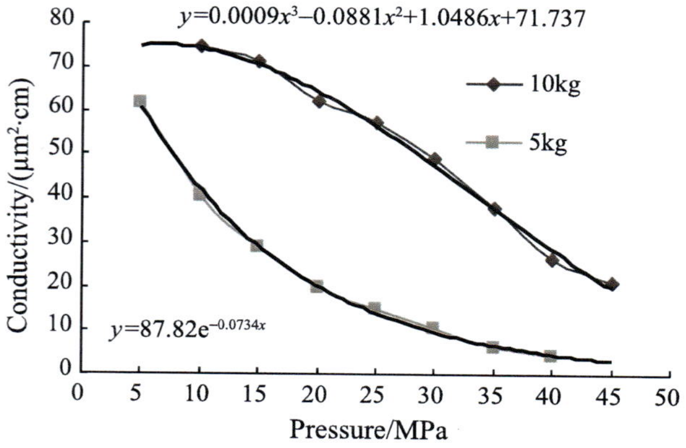

Fig. 11.24 Long-term conductivity of coal rock

experiment. Proppants embedded very serious, and obvious pits appeared on the surface of the sample, which reduces the actual width of the fracture. The result of the conductivity in Fig. 11.25 is reasonable.

Figure 11.28 shows the original sandstone sample. Figure 11.29 shows the sandstone sample with embedded proppants. It can be concluded that more proppants embedded into the coal sample than into the sandstone sample.

Experiments show that embedded proppants and coal powder are harmful to the conductivity of the fractures in coal rock. Low compressive (tensile) strength, serious embedded proppants and blocked fractures by coal powder induced by fracturing and propagating pressures may seriously damage the conductivity of fractures.

11.1.2 Characteristics of Induced Fractures

Recent data of fractured CBM wells constructed by PetroChina were collected from the Daning-Jixian area, the Jincheng area of the Qinshui Basin, the Wubu area of Shaanxi Province and the Dacheng area of Hebei Province. Data from the Daning-Jixian area and the Jincheng area are complete, and the fracturing results are good. They are the pilot CBM production areas. Therefore, the hydraulic fracturing parameters of the two areas were analyzed statistically, and the parameters of other areas were used as supporting data.

A total of 26 layers in 15 wells in the Daning-Jixian area and the Jincheng area of the Qinshui Basin were analyzed, of which 11 layers were fractured with fresh water, 5 by active water and 10 by gel fluid. The fracturing fluid injected is 285.4–564 m^3, 375.1 m^3 on average. The pad fluid is 12.8–52%, 36.7% on average. The volume of proppants is 14.7–

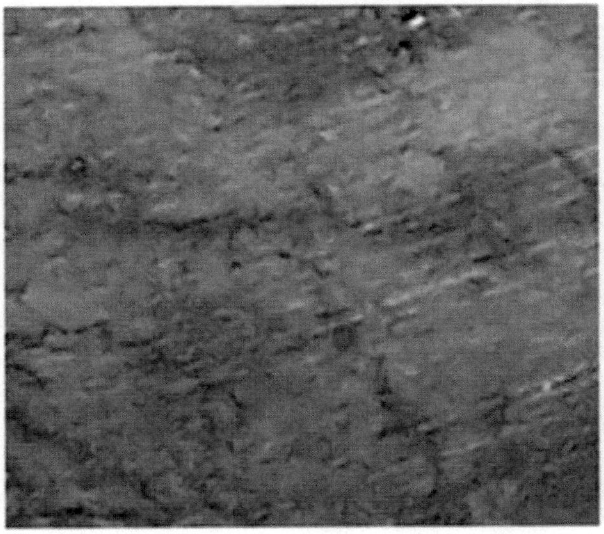

Fig. 11.26 Original coal sample

Fig. 11.28 Original sandstone sample

Fig. 11.27 Coal sample with embedded proppants

Fig. 11.29 Sandstone sample with embedded proppants

51.1 m^3, 36.5 m^3 on average. The proppant intensity is 2.3–17 m^3/m, 7.8 m^3/m on average. The displacement is 4.0–7.2 m^3/min, 5.9 m^3/min on average.

Generally speaking, hydraulic fracturing to coal seams is complex. To improve the reliability of evaluation, stress, well temperature, surface potential, cross-well seismic data, dynamic observation, pressure curve and well test interpretation should be used to diagnose hydraulic fractures.

11.1.2.1 Fracture Azimuth and Length

Fracture Azimuth

Statistical results of fracture azimuths in 26 wells/layers in the Daning-Jixian area, Jincheng area, Wubu area and Dacheng area of Shaanxi Province show that although the fracture azimuths vary with local tectonic stress and cleat development in coal seams, they are mainly controlled by stress. The fracture azimuth in these three areas is basically NE (Fig. 11.30).

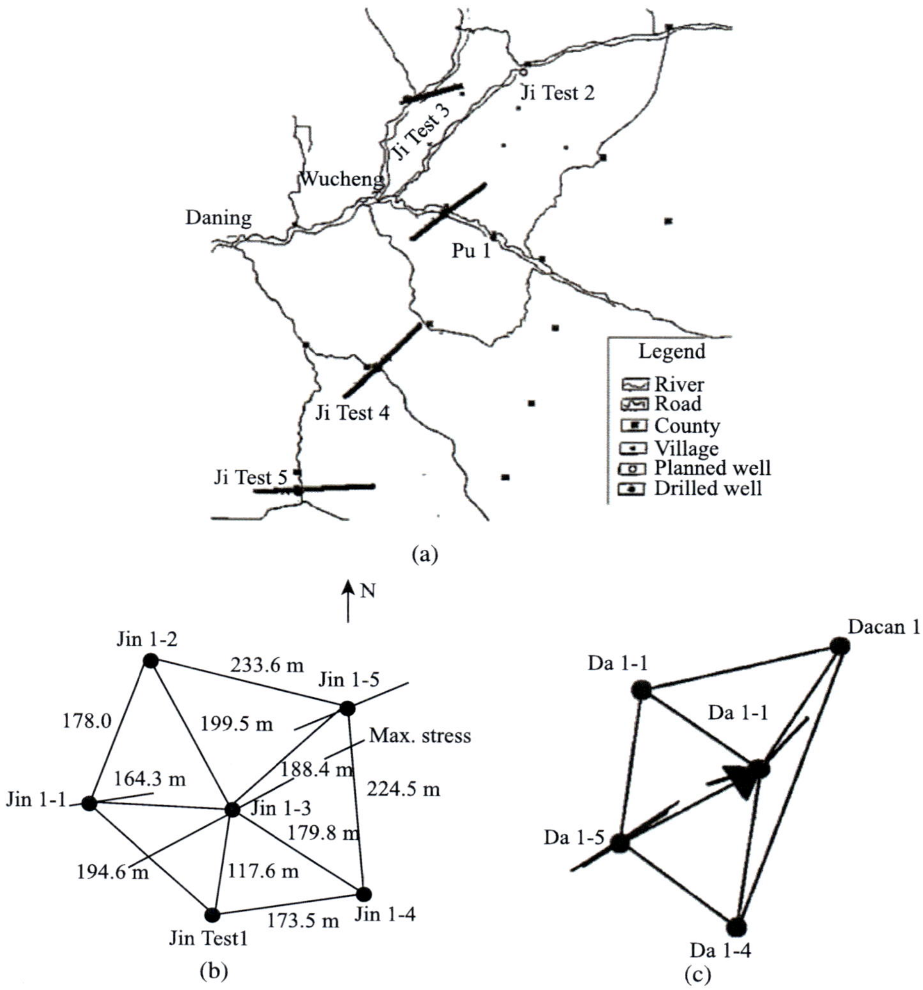

Fig. 11.30 Fracture azimuth monitored. **a** Fracture azimuth in Daning-Jixian; **b** schematic fracture azimuth in Well Jin Test 1; **c** schematic fracture azimuth in CBM wells in Dacheng

Fracture Length

From the ground potential interpretation of 16 layers in 13 wells in the Daning-Jixian area and Jincheng area (Table 11.8), the fracture shape is generally symmetrical and unequal or asymmetrical and unequal. The fracture half-length is 44–106.6 m, and the average half-length is 65.8 m.

From the results of fracture monitoring and fracturing curve, the azimuth and shape of the hydraulic fractures are irregular in coal seams, and the fracture pressure gradient is higher, generally greater than 0.02 MPa/m, i.e. 0.02–0.03 MPa/m, compared with those in sandstone (Table 11.9). The fracturing curve shows that multiple fractures were induced at the beginning and during fracture propagation, and it is more possible to induce fractures in the coal seam with relatively developed cleats.

Prediction of Effective Fracture Length Contributing to Gas Production

The fracture lengths based on ground potential interpretation and well test data were analyzed in 10 layers in 8 wells in the Daning-Jixian area and the Jincheng area, and the following understandings were concluded.

① The fracture length from ground potential data is dynamic. The half-length of the dynamic fracture is 50–80 m. According to conventional hydraulic fracturing calculation, the propped fracture length is generally 80% of the dynamic fracture length, and the half-length of the propped fracture measured by potential is 40–60 m.

② For sandstone whose compressive strength is higher than that of coal rock, the effective fracture length is

Table 11.8 Fracture azimuth and length measured by surface potential

Well no	Coal seam	Perforation top/m	Thickness/m	Fracture				
				Fracture height/m	Fracture azimuth	Fracture half-length/m	Fracture azimuth	Fracture half-length/m
1	No. 5	977.8	5.4	14.0	NE45°	69.0	SW45°	53
2	No. 8	1051.6	8.8	13.0	NE45°	89.0	SW45°	66
3	No. 8	1267.7	6.7	20.5	NE79°	72.0	–	–
4	No. 8	1134.9	9.3	–	NE37°	68.0	SW37°	44
5	No. 5	904.4	7.0	16.0	NE86°	65.0	SW86°	54
6	No. 5	1128.2	5.6	–	NE75°	48.2	SW75°	83.9
7	No. 5	1011.2	7.2	35.0	NE75°	58.7	SW75°	93.5
8	No. 5	967.0	6.6	24.0	NE75°	Short	SW75°	106.6
J1	No. 3	521.6	5.8	14.0	NE55°	82.0	SW55°	65
J2	No. 3	514.2	6.4	17.0	NE105°	58.0	NW85°	75
J3	No. 3	837.8	5.4	20.0	NE105°	55.0	NW55°	67
J3	No. 15	941.8	5.2	–	NE80°	50.0	NW70°	76
J4	No. 3	525.6	6.4	–	NE80°	54.0	SW80°	73
J4	No. 15	612.8	3.2	8.0	NE80°	57.0	SW80°	62
J5	No. 3	539.0	5.4	–	NE65°	51.0	SW65°	60
J5	No. 15	626.4	3.0	10.0	NE65°	53.0	SW65°	65

Table 11.9 Fracturing pressure gradient in Daning-Jixian area

Well no	Coal seam	Depth/m	Fracturing pressure/MPa	Fracturing pressure gradient/ (MPa/m)
3	No. 5	1166.6–1201.5	34.78	0.029
3	No. 8	1267.7–1277.3	29.63	0.023
5	No. 5	904.4–15.5	19.76	0.022

only about 2/3 of the propped fracture length due to the influence of proppant embedding and liquid damage.

③ The fracture half-length from well test data is 20–50 m, which basically represents the effective fracture length.

④ Multiple fractures cause the fracture length and effectiveness to decrease. Considering the mechanical properties, fracture tortuosity, influences of multiple fractures and coal power, it is estimated that the effective fracture length, which is about 20–30 m, is far less than the fracture length from ground potential data.

11.1.2.2 Diagnosis of Fracture Height

The upward and downward extension of fracture height influences on fracture height is significant. The difference in in-situ stress between top and bottom and coal seam is the primary factor controlling the growth of fracture height.

In-situ Stress

According to the acoustic time difference of P-wave and S-wave of long-distance acoustic logging curve, the in-situ stress in six CBM wells in the Jincheng area were studied. The results show that the maximum difference of in-situ stress between coal seam and bottom and top interlayers is not more than 4 MPa, generally between 2 and 4 MPa, so it is difficult to control hydraulic fractures in coal seam (Fig. 11.31).

Fracture Height

Temperature logging is a better method to determine fracture height in shallow wells. According to the temperature logging interpretation results of four areas—Daning-Jixian, Jincheng, Wubu and Dacheng in Shaanxi province, some fracture heights were obtained. Statistics show that, due to the small difference of in-situ stress, the fracture height is not completely controlled in the coal seam, but extends greatly upward and downward. The whole fracture height is about 1.7–4.8 times the thickness of the coal seam. The fracture extension upward and downward is relatively small in Daning-Jixian, and is larger in Wubu (Table 11.10).

After analyzing a large amount of fractured well data, we got the understandings of fracture distribution law as follows.

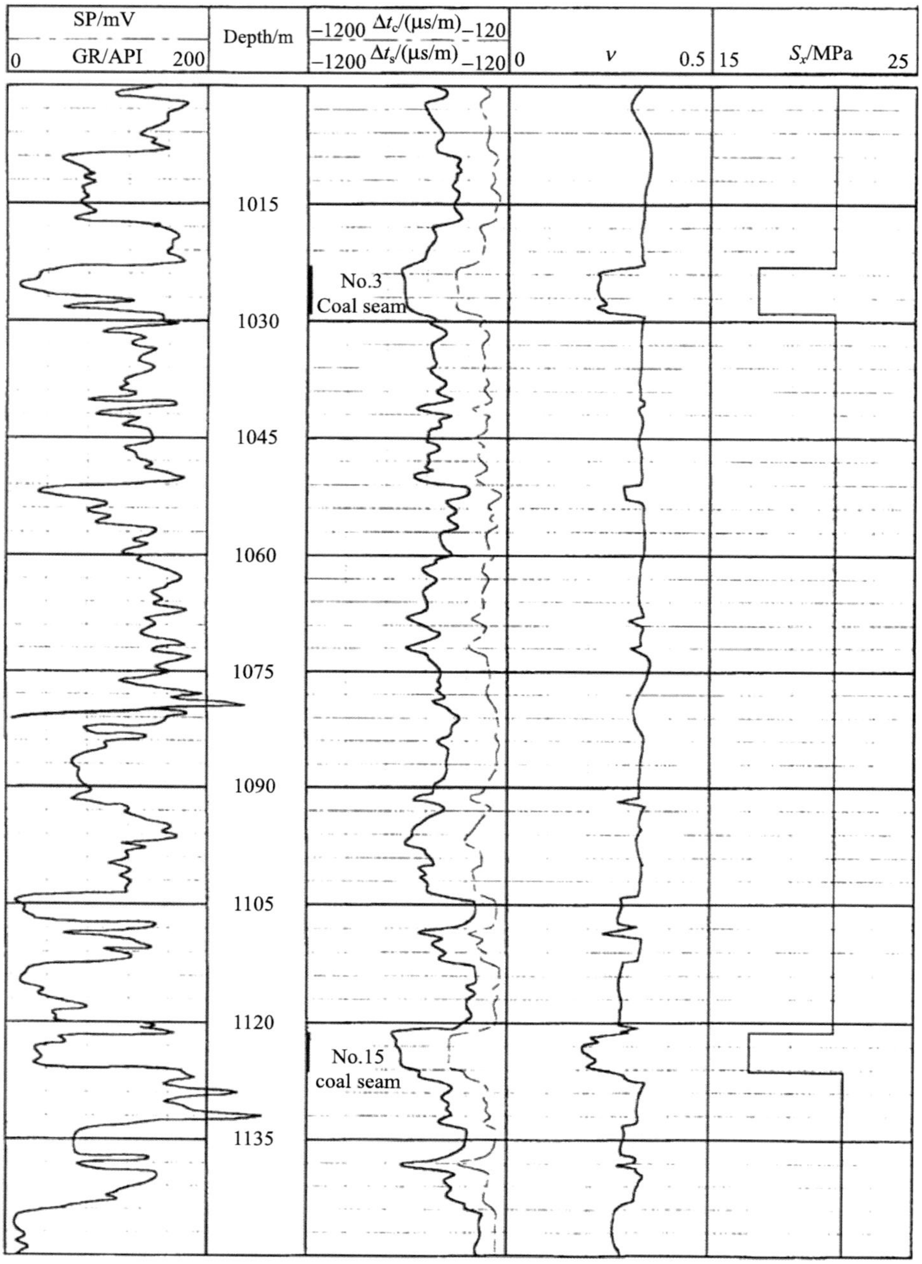

Fig. 11.31 Interpreted in-situ stress in a well in Jincheng

① The direction of fracture extension is mainly controlled by tectonic stress. Statistical results of 26 wells/layers in Daning-Jixian, Jincheng, Wubu and Dacheng show that, although the azimuth of the fracture varies with local tectonic stress and cleats development, the azimuth of the hydraulic fracture in most wells is mainly controlled by in-situ stress, and is basically north-east oriented. However, the fracture propagation is not completely controlled by in-situ stress, but also by primary natural fractures. The law of fracture propagation is complex.

② There are obvious multi-fracture phenomena. Fracturing curve fitting analysis shows that multiple fractures and fracture bending occur in the process of initiation and extension of fractures, which reduces the effective length of the fracture.

③ The fracture length is short, and the fracture height extends beyond the coal seam. Because of the low

Table 11.10 Temperature logging data in Daning-Jixian and Qingcheng

Well no	Coal seam	Perforated interval/m	Abnormal temperature/m	Interpreted fracture height/m	Height/Perforated thickness
1	No. 5	977.8–987.4	975–989	14	1.46
	No. 8	1050.6–1059.4	1048–1061	13	1.67
3	No. 5	1166.6–1201.5	1165–1170	5	–
	No. 8	1267.7–1277.3	1263.5–1284.0	20.5	3.06
4	No. 5	1061.6–1070.6	1034.0–1073.0	39	4.81
	No. 8	1134.9–1144.2	1119	–	–
5	No. 5	904.4–915.5	903–919	16	2.29
J1	No. 15	612.8–616.0	610–623	13	4.06
	No. 3	523.0–525.6	523	–	–
J2	No. 15	606.2–609.4	604–615	11	3.44
	No. 3	519.6–525.0	517–530	13	2.41
J3	No. 15	604.0–607.0	–	–	–
	No. 3	519.0–524.4	517–535	18	3.33
J4	No. 15	603.0–605.6	–	–	–
	No. 3	518.6–524.0	515–530	15	2.78
J5	No. 15	626.4–629.4	624–633	9	3
	No. 3	539–544.4	537–560	23	4.26
J6	No. 15	606.6–609.6	602–620	18	6
J7	No. 3	514.2–520.6	511–528	17	2.66
J8	No. 3	509.2–515.2	500–516	16	2.67
J9	No. 3	521.6–527.4	No abnormal temperature observed due to low temperature		
J9	No. 15	613.8–619.7	612–630	18	3.46
J10	No. 3	837.8–843.2	831.5–851.5	20	3.7
J10	No. 15	941.8–947.0	No sandstone coal seam observed, but abnormal temperature in no. 3 coal seam observed		
J11	No. 3	1023.0–1029.0	1015–1035	20	3.33

Young's modulus of the coal seam, the width of the fracture is large and the length of the fracture is short, and the half-length of the propped fracture is generally 50–80 m. The difference of in-situ stress between coal seam and top and bottom layers is small, generally 2–4 MPa, and the fracture height extends upward and downward to a certain extent.

④ The induced fractures are irregular, symmetrical and unequal or asymmetrical and unequal. Fractures either parallel to the cleats in the coal seam or cross the cleats vertically, and the fracture surface is unsmooth and stepped.

11.1.3 Fracture Distribution Model

There are some limitations in applicable commercial software for fracture propagation on multiple fields. For example, fracture propagation path should be pre-defined. In this section, based on a mechanical model, finite element numerical simulation using a proprietary software to hydraulic fracturing process is described under the jointing action of fluid, solid and heat. At present, 2D simulation has been done. 3D simulation is feasible in theory, the simulation process is time-consuming, with higher requirements for computer hardware. Commercial software can't provide what algorithms are used, but in our proprietary software, they are known, so it is helpful to study the influences of various parameters on fracture propagation, and to explore deeply the inherent mechanical mechanism of hydraulic fracturing technology.

11.1.3.1 Models and Control Equations

Figure 11.32 is a horizontal strain model of vertical induced fractures. It shows the fractures propagate under the combined action of temperature field, seepage field and stress field. The engineering hydraulic fracturing is regarded as a quasi-static

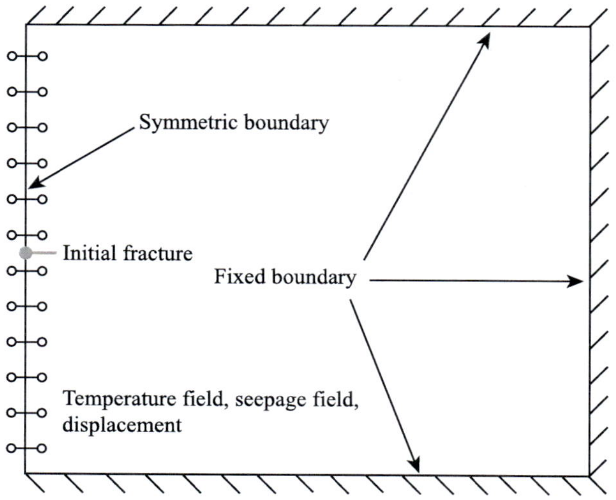

Fig. 11.32 Plane strain model of vertical induced fracture propagation (horizontal section)

progressive process, which has given the control equations and boundary conditions of the physical fields of heat transfer, seepage, displacement, and stress, the coupling equations between different physical fields and the fracture propagation criterion equation. The model takes into account the heterogeneous distribution of reservoir physical parameters, and can realize fracture propagation in any direction. It is assumed that the permeable medium is always completely saturated. The corresponding finite element mesh is shown in Fig. 11.33.

Heat Transfer Equation

$$K\nabla^2 T = 0 \tag{11.15}$$

The temperature on the fixed boundary is constant at 60 °C, the injected liquid temperature at the perforation position is 10 °C, and heat can not flow through the symmetrical boundary.

Seepage Equation

$$K\nabla^2 P = 0 \tag{11.16}$$

The seepage pressure on the fixed boundary is constant 10 MPa, the seepage pressure at the perforation position is working load, the pressure on the symmetrical boundary is balanced, and the seepage velocity is 0.

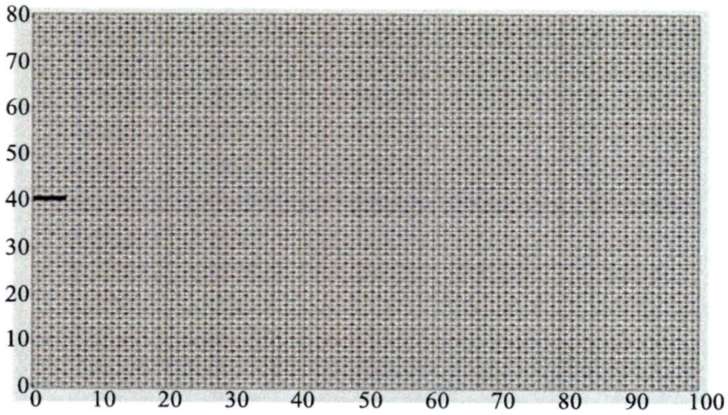

Fig. 11.33 Finite element mesh of plane strain model for vertical fractures

Stress Field Equation

Equilibrium equation,

$$\frac{\partial \sigma_{ij}}{\partial x_j} + b_i = 0 \tag{11.17}$$

Geometric equation,

$$\varepsilon_{ij} = \frac{1}{2}(u_{i,j} + u_{j,i}) \tag{11.18}$$

Constitutive equation,

$$\sigma'_{ij} = \sigma_{ij} - p\delta_{ij} = [D \cdot (\varepsilon_{ij} - \varepsilon_{\mathrm{T}}) + \sigma_0] - P\delta_{ij} \tag{11.19}$$

The displacement on the fixed boundary is 0, and the normal displacement on the symmetrical boundary is 0. Based on the general stress–strain equation, the initial in-situ stress term, boundary seepage pressure term and thermal strain term are added to the constitutive equation above to characterize the coupling between them.

Fracture Propagation Criteria

Maximum tensile strength criterion:

$$\sigma_1 > \sigma_{\mathrm{t}} \tag{11.20}$$

Mohr–Coulomb (M-C) criterion:

$$\sigma_3 - \sigma_1 \frac{1 + \sin\varphi}{1 - \sin\varphi} < -\sigma_{\mathrm{c}} \tag{11.21}$$

See Fig. 11.34 for M-C criterion. The stress symbol in this graph is positive. In the above formulas, x_j is the space coordinate of a node, s is the incremental step of load, T is temperature, P is seepage pressure, u_i is displacement, σ_{ij} is stress, K is thermal conductivity coefficient, k is permeability, ε_{ij} is strain, ε_{T} is thermal expansion strain, σ_0 is prestress, σ_{ij} is effective stress, σ_i is principal stress ($\sigma_1 < \sigma_3$), and b_i is bulk force.

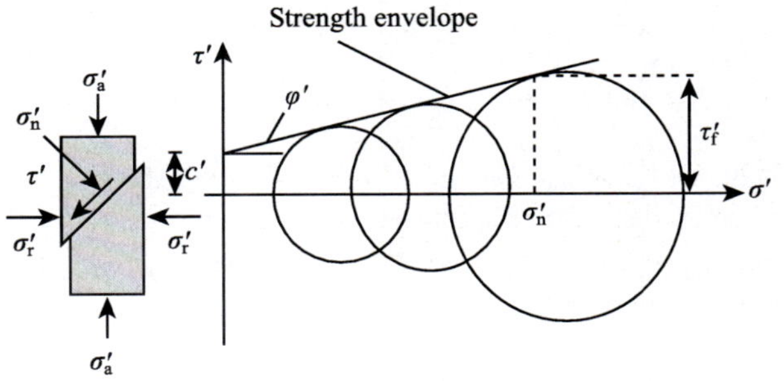

Fig. 11.34 Schematic diagram of Mohr–Coulomb failure criterion

Disposal of Damage

When unit stress reaches the above limit, failure will occur. The elastic modulus is set to 1/50 of the initial average elastic modulus, the permeability is set to 1000 times of the initial average permeability, and the thermal conductivity coefficient is set to 50 times of the initial average thermal conductivity coefficient. The unit itself is not removed from the model. It is calculated iteratively according to the physical parameters of new material until the system reaches equilibrium and no new unit is destroyed, and then the next incremental load step begins.

Interaction of Multiple Fields

① Permeability is related to the distribution of temperature field.
② Permeability is related to the failure of stress field units.
③ Thermal conductivity coefficient is related to the failure of stress field unit.
④ Stress constitutive equation includes the thermal expansion pre-strain term caused by temperature field.
⑤ Stress constitutive equation includes the pre-strain term caused by seepage field.

Figure 11.35 shows the direct influence relationship between the three fields, where the arrows indicate the influencing factors, and the three fields are completely decoupled.

Non-uniform Medium Distribution

The medium with weak strength has low elastic modulus, high permeability and high thermal conductivity coefficient. The elastic modulus obeys the Weibull random distribution law. The larger the m, the stronger the uniformity. The density function of random distribution is as follows.

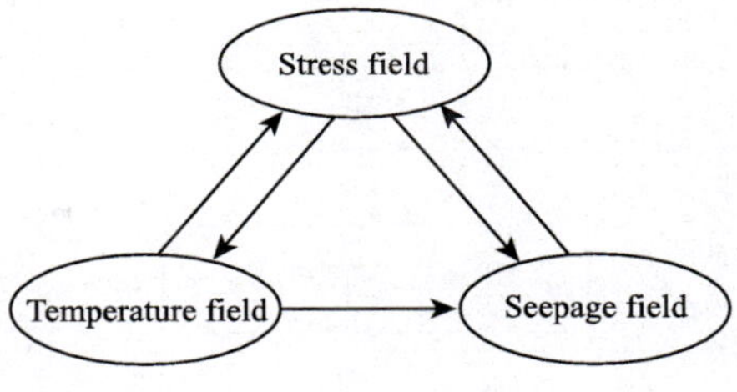

Fig. 11.35 Interaction diagram of three physical fields

$$f(E,m) = \frac{m}{E_0}\left(\frac{E}{E_0}\right)^{m-1}\exp\left[-\left(\frac{E}{E_0}\right)^{m}\right] \tag{11.22}$$

11.1.3.2 Numerical Simulation Algorithm

Based on the above boundary conditions of the control equation, the finite element formula was derived, and finite element solution program and post-processing program were developed by MATLAB.

The solution strategy of multi-field coupling is decoupling and iterative solution. The calculation flow chart is shown in Fig. 11.36. The equation of heat transfer and permeation is generally the basic form. Its finite element formulation is relatively easy, which can refer to the general textbooks. Because of the complexity of the stress field equation, coupled with seepage pressure, initial stress and thermal strain, its finite element formulation is deduced as follows.

The stress (total stress) strain relationship can be written as follows.

$$\sigma = [D\cdot(\varepsilon - \varepsilon_{\mathrm{T}}) + \sigma_0] \tag{11.23}$$

Thermal strain term:

$$\varepsilon_T = \alpha(T - T_0)[1, 1, 0]^{\mathrm{T}} \tag{11.24}$$

By using the principle of minimum potential energy, without considering the body force and boundary force, the functional expression of the problem is as follows.

$$\prod(\boldsymbol{u}) = \int_{\Omega}\left(\frac{1}{2}\boldsymbol{\varepsilon}^{\mathrm{T}}\boldsymbol{D}\varepsilon - \boldsymbol{\varepsilon}^{\mathrm{T}}\boldsymbol{D}\varepsilon_{\mathrm{T}} + \boldsymbol{\varepsilon}^{\mathrm{T}}\boldsymbol{\sigma}_0\right)\mathrm{d}\Omega + \text{External Force Term} \tag{11.25}$$

The stiffness equation can be obtained from functional variation equal to 0. The stiffness coefficient matrix in the stiffness equation is identical to the general plane strain. The difference is that the load vectors are different, and the initial stress and thermal stress will produce additional load items. The additional load term is as follows.

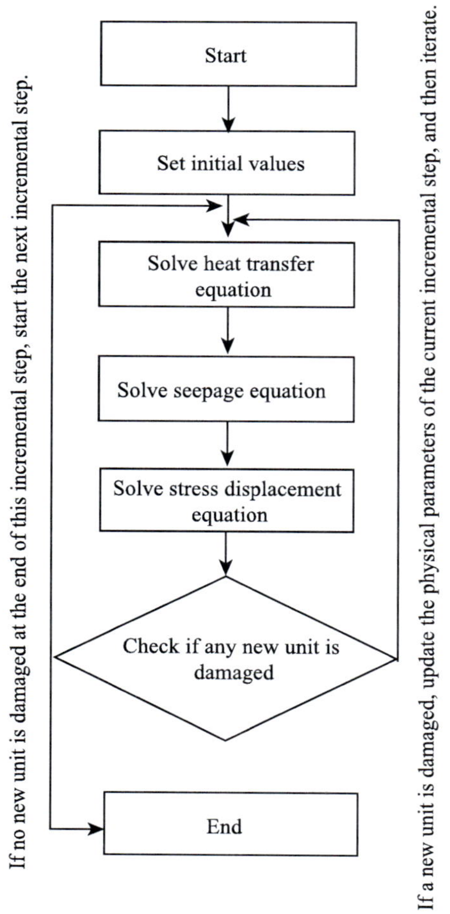

Fig. 11.36 Flow chart of estimating fracture propagation on multiple physical fields

$$P_{\varepsilon_T+\sigma_0} = \int_{\Omega} B^T(D\varepsilon_T - \sigma_0)d\Omega \quad (11.26)$$

11.1.3.3 Numerical Simulation Analysis of Vertical and Horizontal Fractures

The Planar Model of Vertical Fractures

The plane strain model for vertical fractures has 8181 nodes and 16,000 triangular units. The in-situ stress is 25 MPa in the *X* direction and 25 MPa in the *Y* direction. The initial formation pressure is 10 MPa and the total hydraulic load is 60 MPa. The increment step of the load is 1 MPa from 10 MPa. Without considering the effect of temperature field here, the initial fracture length is 5 m, which is 1/20 of the simulated size. The purpose of the study is to numerically simulate the process of hydraulic fracture propagation, and obtain the relationship between fracture length and width and load.

Due to the limitation of calculated scale, the unit size is much larger than the actual fracture width. The fracture space is simulated by element failure, but the fracture width is not equal to the unit thickness. To calculate the fracture width is to calculate the displacement of the failed node. The different displacements of the nodes, along the normal direction of the fracture surface, on two sides of the failed unit is defined as the fracture width.

Numerical Simulation Results and Analysis

Figure 11.37 shows how fracture propagation varies with hydraulic load. It can be concluded that the initiating pressure is 31 MPa under the current working condition. With the increase of hydraulic pressure, fractures continue to expand, fracture length increases nonlinearly, and the growth rate slows down with the increase of fracture length, which is due to the attenuation of hydraulic load with the fracture length. The maximum length of fracture propagation is 46 m.

Figure 11.38 shows how fracture width varies with hydraulic load at the top (the symmetrical plane). Whether before or after initiating fracture, the maximum fracture width increases approximately linearly with the increase of hydraulic load. Near the initiating load, the fracture width increases rapidly with the increase of hydraulic load. The maximum fracture width is 31 mm under all loads.

Figure 11.39a–c shows fracture shapes under three different load steps. At the current in-situ stress, the fracture propagates along the straight fracture surface. The elastic modulus in the Figure is shown by color. The elastic modulus of the damaged unit is very low in black.

Figure 11.40 shows the distribution of hydraulic load (seepage pressure) on fracture surface in fluid–solid coupling simulation at the incremental step of load 45 (hydraulic load 45 MPa). The hydraulic load is 45 MPa at the top and 10 MPa of the seepage pressure at the far field near the boundary as the fracture surface attenuates. Because the permeability of the damaged and undamaged units is

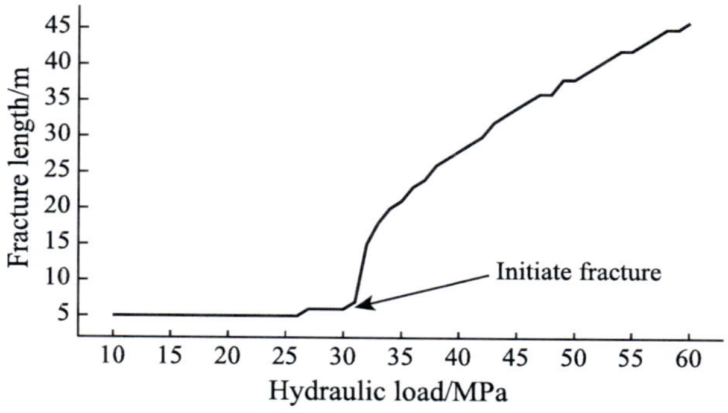

Fig. 11.37 Fracture length versus hydraulic load

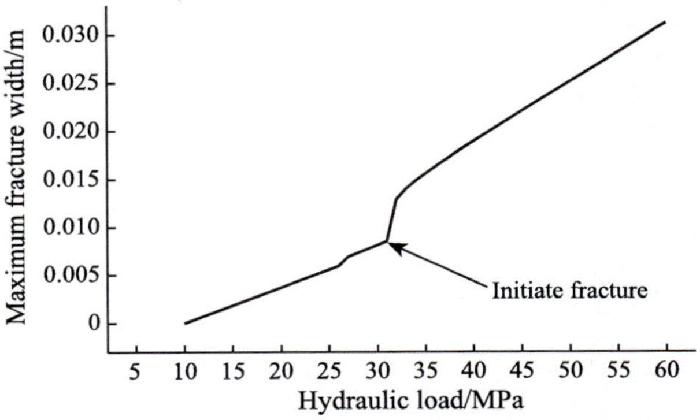

Fig. 11.38 Maximum fracture width varies with hydraulic load

different, the attenuating velocity is different on both sides of the fracture tip. The specific values depend on the proportional relationship between the permeability of the damaged unit and that of the undamaged unit. In this paper, the permeability of the damaged unit is 50 times that of the undamaged unit. The fracture tip is located at $X = 33$ m at the incremental step of load 45.

As for the calculation of fracture width, the width of a failed unit is not taken as the fracture width, but as the normal displacement of the failed unit on the fracture surface. As shown in Fig. 11.41, the fracture width is the sum of the normal displacement of the nodes on the upper and lower fracture surfaces. The left shadow is before the deformation of the unit and the dashed line is after the deformation.

Figure 11.42 shows how the normal displacements of the upper and lower surfaces of a fracture vary with the fracture length when the incremental step of load is 45. Based on the symmetry of the model, the normal displacement is also symmetrical. The Y displacement of the upper surface is upward, and the Y displacement of the lower surface is

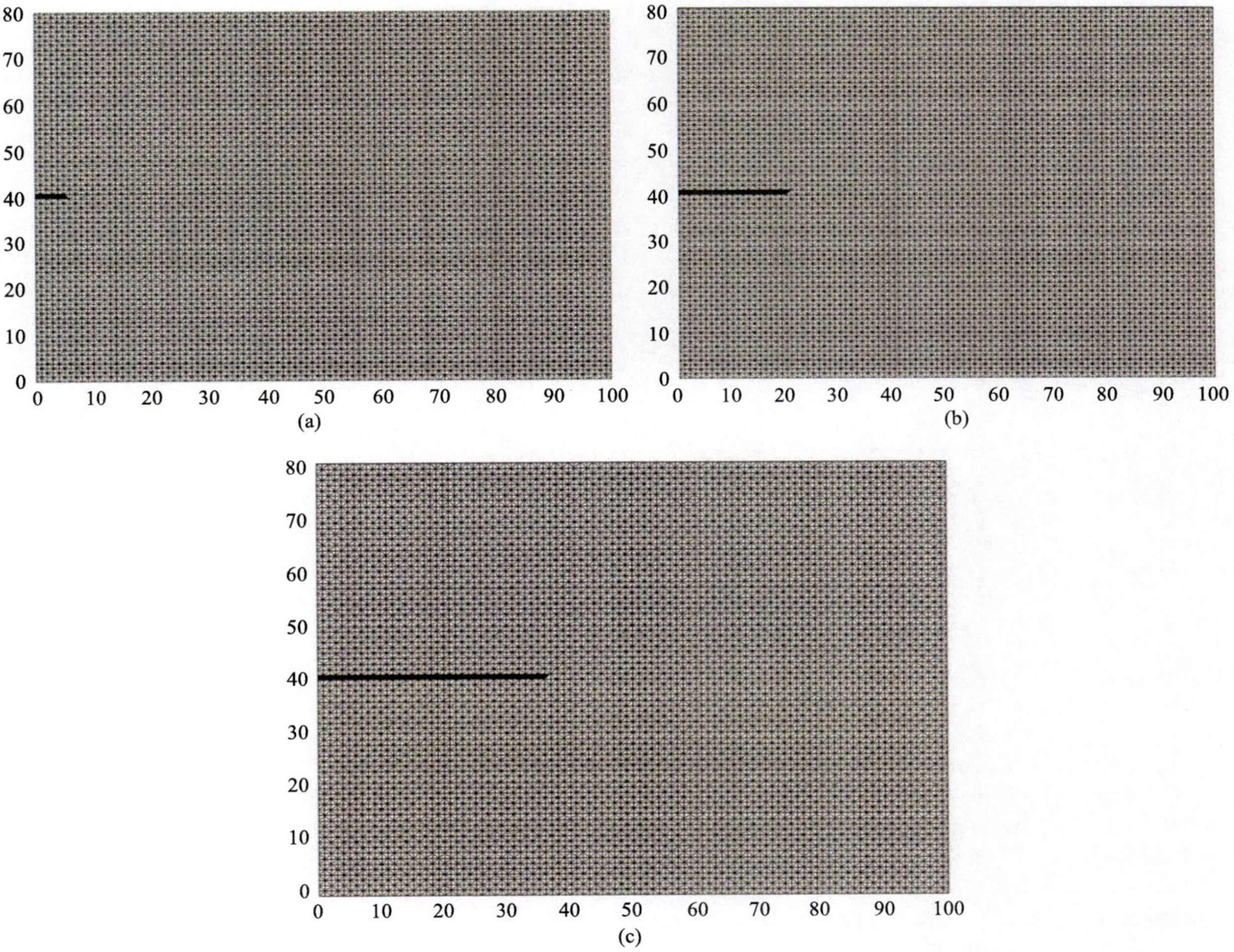

Fig. 11.39 Fracture shapes. **a** Load step 31; **b** load step 35; **c** load step 50

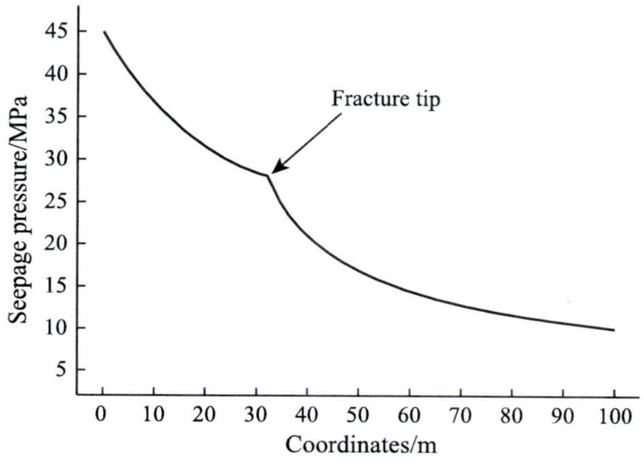

Fig. 11.40 Distribution of hydraulic load on fracture surface

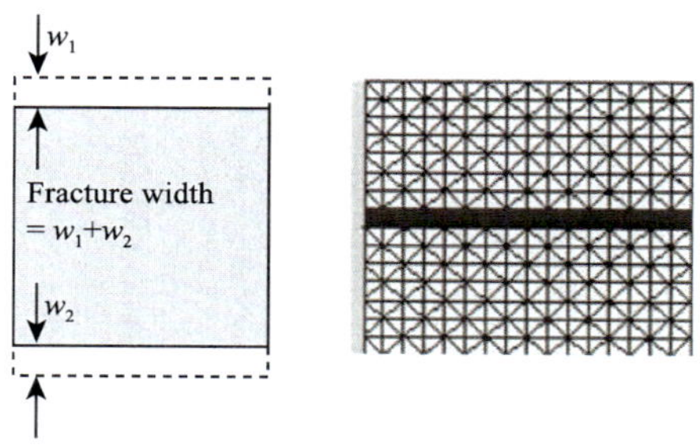

Fig. 11.41 Schematic diagram for calculating fracture width

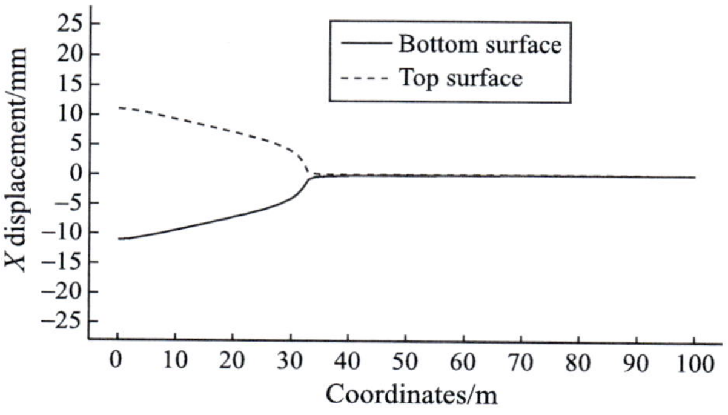

Fig. 11.42 Normal displacements of upper and lower surfaces of a fracture

downward, that is, a fracture is propped up by water pressure. The fracture tip is at 33 m, and the normal displacement of the undamaged unit behind the fracture tip is also found due to the normal tensile stress, but it is much smaller than that of the failed unit. Figure 11.43 shows the fracture width calculated. The maximum fracture width is 22 mm, corresponding to the maximum fracture width at the load of 45 MPa in Fig. 11.40.

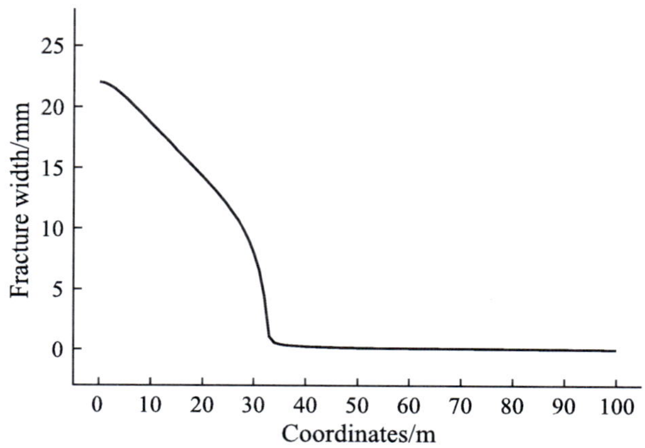

Fig. 11.43 Change of fracture width (load increment step 45)

Numerical Simulation and Analysis of a Plane Model of Horizontal Fractures

The surface of a horizontal fracture is horizontal. The vertical symmetrical cross section was taken for plane strain model analysis. The model has 4141 nodes and 8000 triangular units. The in-situ stress is 30 MPa in the X direction and 25 MPa in the Z direction. The initial seepage pressure is 10 MPa and the total hydraulic load is 60 MPa. The increment step of the load is 1 MPa from 10 MPa. The initial fracture length is 5 m, which is 1/20 of the simulated size.

Figures 11.44 and 11.45 show how fracture length and maximum width vary with hydraulic load, respectively. The initiation pressure is 34 MPa and the fracture length is 26 m when loaded to 60 MPa. At that time, the maximum fracture length is 22 mm.

11.1.3.4 Numerical Simulation to How In-situ Stress Influences Fracture Propagation

Effect on Fracture Length

In the above analysis of vertical fracture propagation, the simulated plane model is the horizontal cross section of

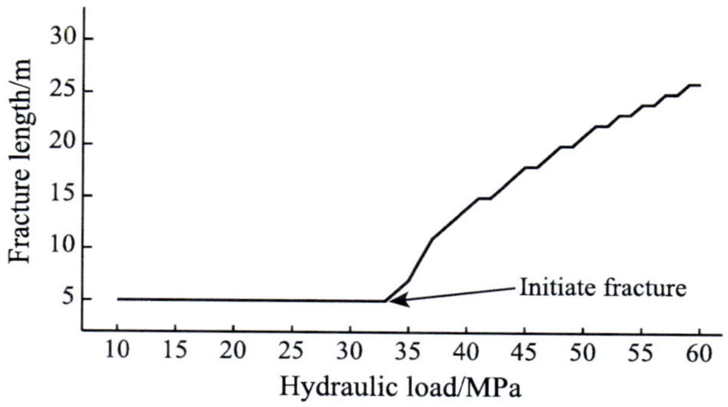

Fig. 11.44 Horizontal fracture length varies with hydraulic load

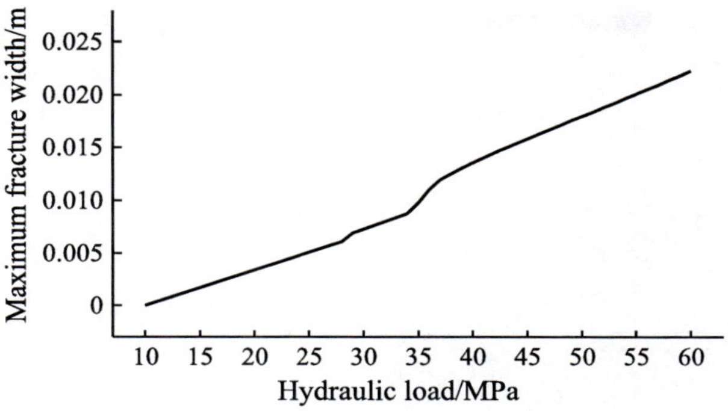

Fig. 11.45 Maximum horizontal fracture width varies with hydraulic load

vertical fractures, and the in-situ stresses in *X* and *Y* directions are both horizontal stresses at 25 MPa each. In some geological strata, the in-situ stresses in horizontal plane are not necessarily isotropic. Next, the influence of in-situ stress on fracture propagation is described.

Keeping other parameters are the same as those in the previous section, we investigated the changes of fracture propagation and fracture shapes at *X* and *Y* in-situ stresses of 15–40 MPa, respectively.

The circular symbols in Fig. 11.46 represent the changes of fracture length at 25 MPa *X* in-situ stress and 15 MPa to 40 MPa *Y* in-situ stress. The *Y* in-situ stress hinders fracture propagation. With the increase of the *Y* in-situ stress, the fracture length decreases. The triangle symbols in Fig. 11.46 represent the change of fracture length at 25 MPa *Y* in-situ stress and 20 MPa to 40 MPa *X* in-situ stress. The change of the *X* in-situ stress has no significant effect on fracture length. From the theoretical analysis, it is known that the fracture extends along the *X* direction, mainly in a "I" shape under the action of *Y* tensile stress, so the *Y* in-situ stress has great influence. The *X* in-situ stress also has some influence, because the failure criterion is based on the principal stress rather than the *Y* in-situ stress.

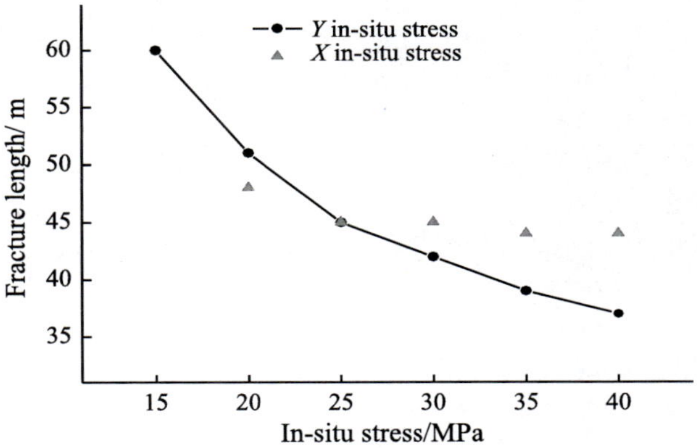

Fig. 11.46 Influence of *Y* in-situ stress on fracture length

Simulation to Fracture Bifurcating

Further simulation shows that the effect of in-situ stress on fracture propagation is not only the change of fracture length, but also the great influence on fracture shape. When the in-situ stresses in *X* and *Y* directions are quite different, fractures may not propagate along the original straight direction, but turn and bifurcate.

Figure 11.47 is the fracture propagation diagram when *X* in-situ stress is 15 MPa, *Y* in-situ stress is 30 MPa and hydraulic load is 60 MPa. Because of the large compressive stress in *Y* direction, it is difficult for the fracture to continue expanding along the straight fracture (that is, *X* direction), and the straight fracture bifurcates, forming two fractures and continuing expanding. Figure 11.48 is the fracture propagation diagram when the in-situ stress in *X* direction is 15 MPa, the in-situ stress in *Y* direction is 40 MPa and the hydraulic load is 60 MPa. At that time, the difference of in-situ stress between the two directions is greater, and the shape of the bifurcating fracture has changed to some extent.

When the difference of in-situ stress between the two directions is small or the in-situ stress in *Y* direction is smaller than that in *X* direction, the straight fracture continues expanding along the original direction. When the in-situ stress in *Y* direction is larger than that in *X* direction to a certain extent, the fractures begin to bifurcate, and with the increase of in-situ stress difference, the angle away from the original extension direction becomes larger. The deviated angle in Fig. 11.47 is about 45°, while that in Fig. 11.48 it is close to 90°.

The advantage of using unit failure to study fracture propagation is that it does not need to specify fracture path or possible path in advance, and it can fully realize fracture propagation in any direction. This requires a very dense units, which results in a very large amount of calculation. At the same time, the process of fracture propagation is a strong non-linear process, which requires iterative solution. In this book, we tend to develop this method and carry out exploratory research to discover the law of fracture propagation, and the solution scale is limited to 20,000 units. What needs further study is how to conduct large-scale parallel computation to get more accurate results by simulating more than one million units.

The analysis of fracture bifurcating angle is also affected by the limitation of unit division and calculation scale. In theory, with the change of the in-situ stress in two directions, the fracture bifurcating angle can change continuously from 0° to 90°.

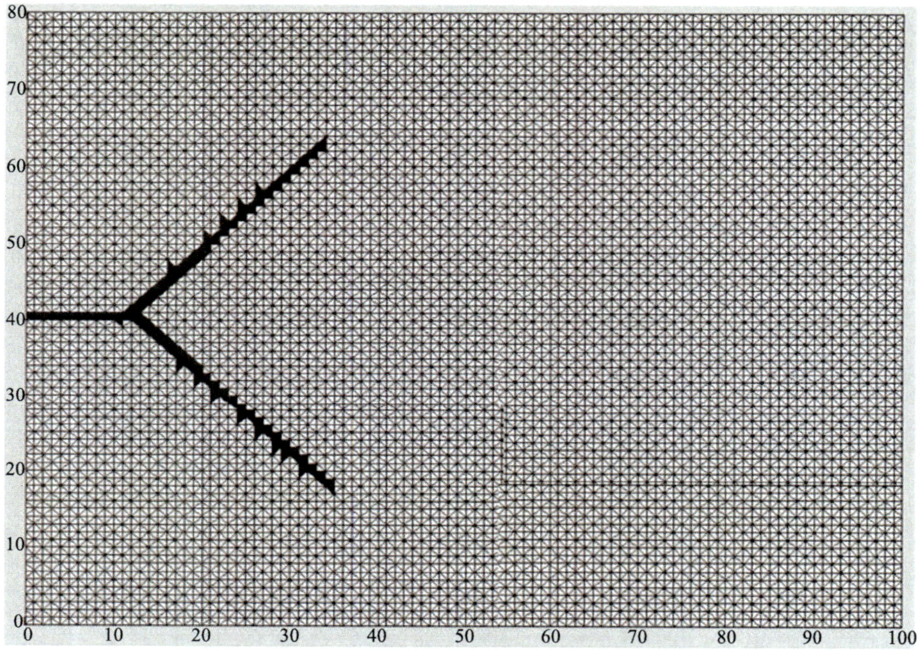

Fig. 11.47 Fractures bifurcated. 15 MPa *X* in-situ stress, 30 MPa *Y* in-situ stress

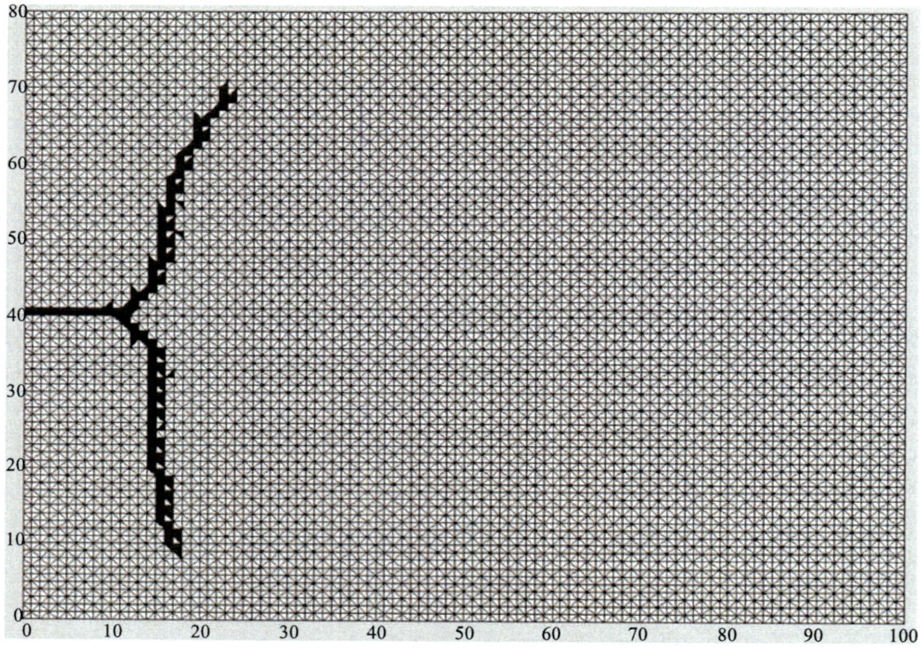

Fig. 11.48 Fracture bifurcated. 15 MPa *X* in-situ stress, 40 MPa *Y* in-situ stress

11.1.3.5 Numerical Simulation to How Natural Fractures Influence Fracture Propagation

If the natural fracture is parallel to the primary fracture and located on the primary fracture surface, it is obvious that the natural fracture will converge with the primary fracture, and the primary fracture will continue expanding according to the original law. Two kinds of natural fractures were simulated, which are orthogonal and oblique to the primary fracture respectively. The propagation of initial primary fractures after encountering natural fractures was simulated.

Figure 11.49a–c shows the effect of vertical natural fractures on the propagation of initial primary fractures at different in-situ stresses. When the in-situ stress in *X* direction is large and the in-situ stress in *Y* direction is small, induced fractures are easy to propagate along the *X* direction, so the primary fracture will turn along the natural fracture to the boundary after encountering the natural fracture, and continue to propagate along the *X* direction, reaching 53 m. When there is no natural fracture, the fracture length at the in-situ stress is 57 m. The natural fracture is not conducive to the propagation of induced fractures, partly because of the hydraulic load is lost on the non-primary fracture path, as shown in Fig. 11.49a. When the in-situ stress in the *X* direction is small and the in-situ stress in the *Y* direction is large, the fracture may bifurcate, as shown in Figs. 11.47 and 11.48. After the induced primary fracture meets the natural fracture, the propagation law is the same as that of the upper section. When the difference of in-situ stress between the two directions is greater, the angle between the bifurcating fracture and the primary fracture is larger, as shown in Figs. 11.49b, c. In the two models, the induced primary fracture did not bifurcate until reached the natural fracture, but became zigzagged. The main reason is that the natural fracture affected the distribution of seepage pressure, so that the induced primary fracture continued moving along a straight line.

Figure 11.50a–c shows the effect of vertical natural fractures on the propagation of induced initial primary fractures at different in-situ stresses. Figure 11.50a shows that when the initial primary fracture first converges with the natural fracture at 25 MPa *X* and 25 MPa *Y* in-situ stresses, it turns and advances along the initial propagating direction (*X* direction) when it reaches the end of the natural fracture. Figure 11.50b shows at 15 MPa *X* and 35 MPa *Y* in-situ stresses, the initial primary fracture converges with the natural fracture until reaching the end of the natural fracture along two directions. The fracture shapes become complex —one end continues expanding along the direction of natural fractures, while the other end extending from 135° to 90°, then to 45°, in the same direction as that without natural fractures in the upper section.

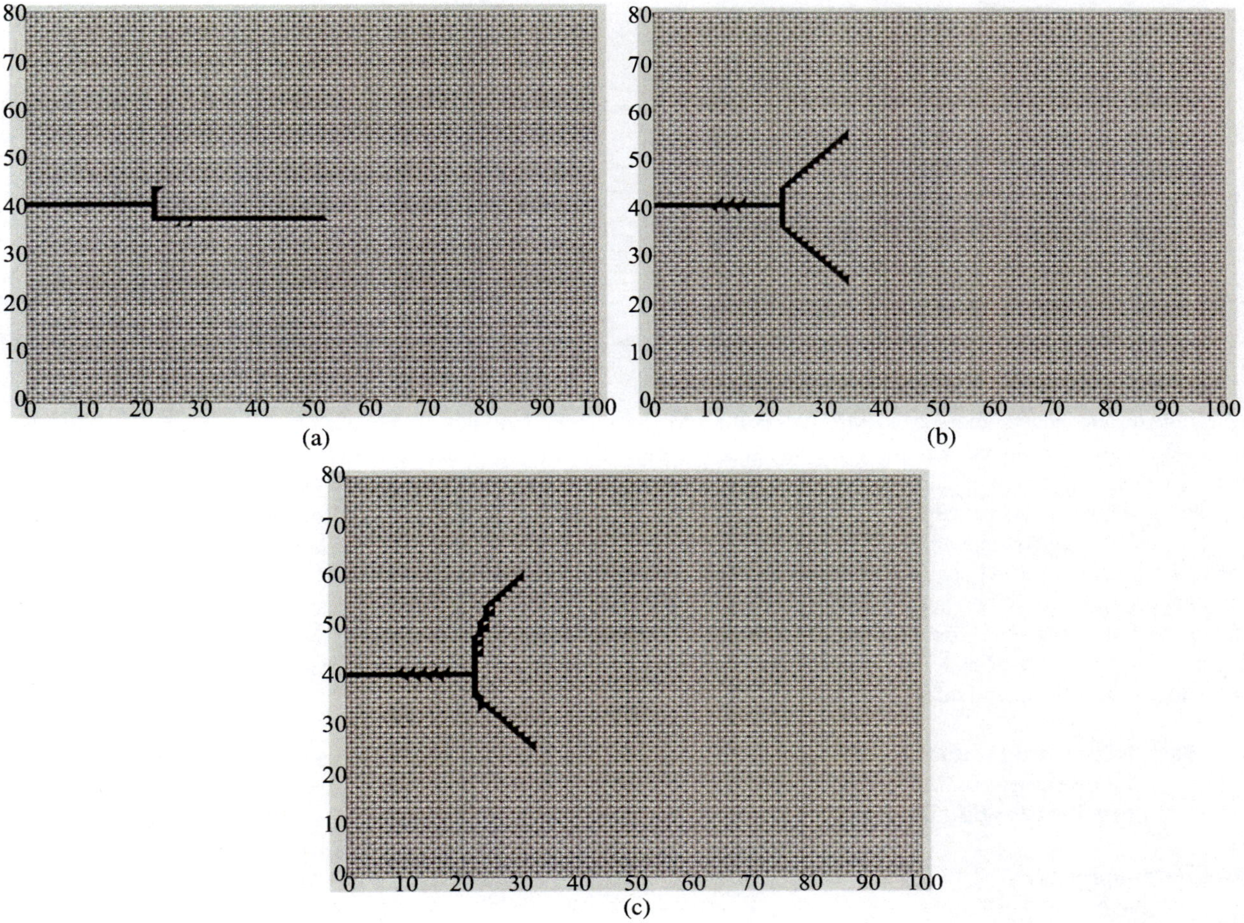

Fig. 11.49 Propagation of induced primary fracture influenced by vertical natural fractures. **a** 40 MPa *X* and 15 MPa *Y* in-situ stresses; **b** 15 MPa *X* and 30 MPa *Y* in-situ stresses; **c** 15 MPa *X* and 40 MPa *Y* in-situ stresses

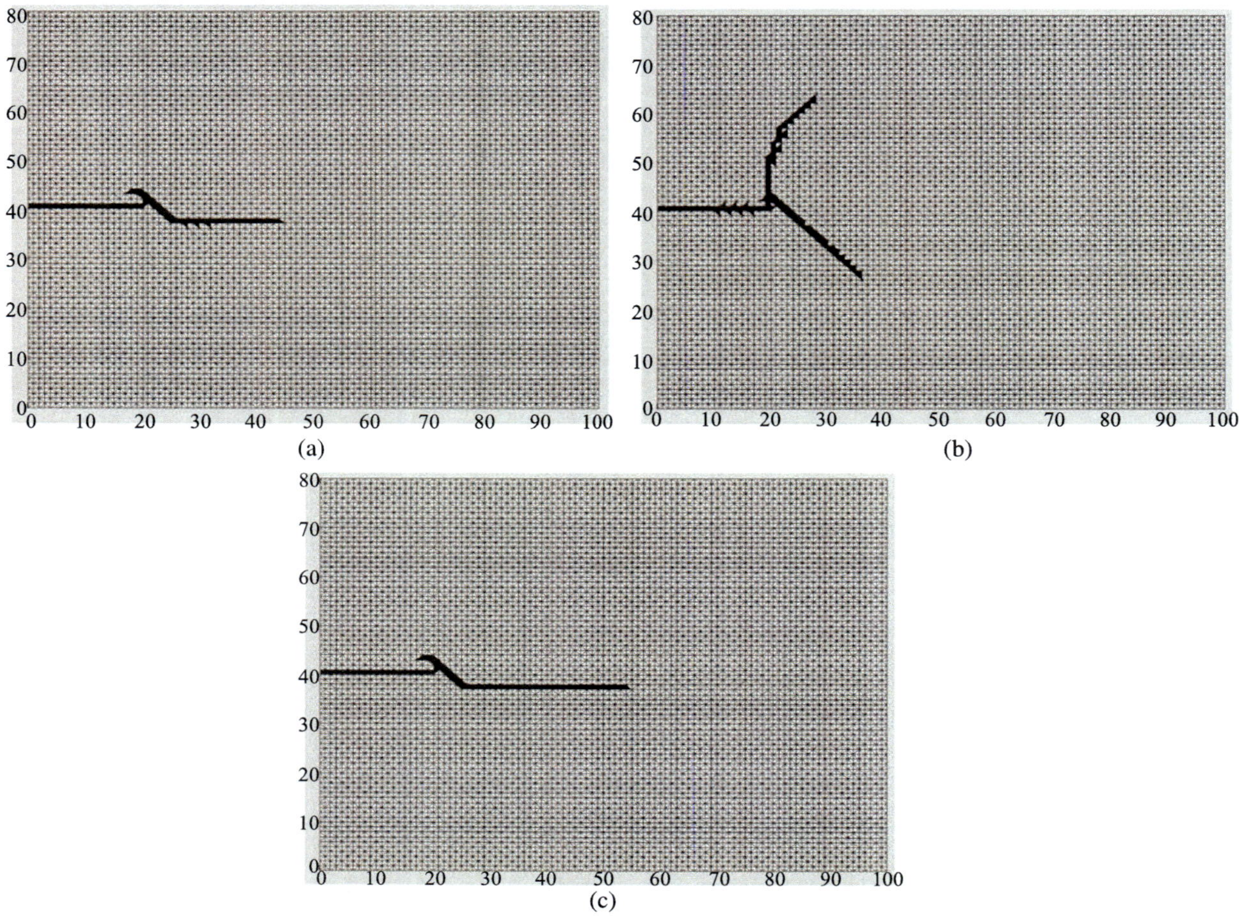

Fig. 11.50 Propagation of induced primary fracture influenced by vertical natural fractures. **a** 25 MPa *X* and 25 MPa *Y* direction in-situ stresses; **b** 15 MPa *X* and 35 MPa *Y* in-situ stresses; **c** 40 MPa *X* and 15 MPa *Y* in-situ stresses

Figure 11.50c shows that at 40 MPa *X* and 15 MPa *Y* in-situ stresses, the induced primary fracture reaches the end of the natural fracture, and then turns and continues expanding along the horizontal direction.

The above analysis shows that natural fractures have complex influences on the shape of induced fractures, and the influences are is closely related to the location, angle and in-situ stress state of natural fractures. In many cases, natural fractures cause the loss of hydraulic load, reduce the length of fracture propagation, and affect fracturing effect.

11.1.3.6 Numerical Simulation to How Heterogeneous Media Influence Fracture Propagation

Rock Heterogeneity

The strength of rock determines the physical parameters such as elastic modulus, permeability and thermal conductivity. Assuming that the elastic modulus obeys Weibull distribution, and the permeability and thermal conductivity are directly related to the elastic modulus, Weibull random elastic modulus is applied to all elements to ensure that its average is the average elastic modulus of the rock. The elastic modulus is ω times unit of the average elastic modulus, and permeability and thermal conductivity coefficient are $1/\omega$ of their respective average values. The degree of heterogeneity is determined by the m in the Weibull random distribution function. The larger the m, the more heterogeneous the medium is.

Simulation Results and Analysis of How Heterogeneous Parameters Influence Fracture Propagation

When the heterogeneity of rock is strong, and some local weak parts are on the fracture propagation path, the fracture may easily deviate from the original path, and move toward weak part, leading to complex fractures. Non-straight fractures may lead to loss of hydraulic load and decrease fracture length, thus affecting fracturing effect.

11.1.3.7 Numerical Simulation to How Uneven Temperature Field Influences Fracture Propagation

Temperature Field Model

The influence of uneven temperature field on fracture propagation is various. This book mainly considers two aspects. One is that temperature field distribution affects permeability, and accordingly hydraulic load and fracture propagation. The other is that uneven temperature field produces thermal stress which affects fracture propagation.

The initial temperature is set at 60 °C, the temperature of the fracturing fluid injected into the reservoir is set at 10 °C, and the calculation of thermal stress is based on 60 °C. The thermal expansion coefficient is set at 10^{-5} °C. It is assumed that the temperature difference at 50 °C will lead to 20% decrease in rock permeability and linearly change with temperature.

Considering vertical fractures, the in-situ stress state is 25 MPa in both directions, the total load is 60 MPa, and the straight fractures propagate. The study is divided into four working conditions: (i) permeability varying with temperature, and thermal stress; (ii) thermal stress, but permeability not varying with temperature; (iii) permeability varying with temperature, no thermal stress; (iv) thermal stress, but permeability not varying with temperature. The influence of temperature field is discussed in two respects.

Numerical Simulation Results

It can be seen that both permeability and thermal stress increased the length and width of fracture propagation, and the effect of thermal stress is particularly significant. Considering the temperature sensitivities of thermal stress and permeability, the initiating pressure dropped to 21 MPa, but the fracture shape did not change, and it is still straight fracture, but the fracture length has changed significantly. Before initiating fracture, because the temperature of fracturing fluid in the initial primary fracture was only 10 °C, lower than that in the far field, and the temperature in the area near the initial fracture decreased due to heat transfer, the rock shrank when it was cooled, but because the boundary condition of the far field was fixed, tensile stress was produced in the rock. The thermal stress is equivalent to the initial tension stress, which is opposite to the initial in-situ stress. The in-situ stress hinders fracture propagation, while the thermal stress is beneficial to fracture propagation.

The sensitivity of permeability to temperature has no significant effect on fracture growth, but it has some. With corresponding data of physical parameters varying with temperature, it is assumed that the permeability decreases when the temperature decreases, and that the permeability decreases by 20% when the temperature decreases from 60 to 10 °C, the seepage equation is rewritten as: $\nabla \cdot [k(x,y) \cdot (\nabla \cdot p)] = 0$. It can be seen that

$$k(x,y) \cdot (\nabla \cdot p) = \text{const} \tag{11.27}$$

Therefore, the decrease of permeability leads to the increase of pressure gradient. Near the fracture tip, the rock temperature decreases and the seepage pressure gradient increases, so the local payload increases, which promotes the fracture propagation.

11.1.4 Hydraulic Fracturing Measures and Echnology

For low permeability and ultra-low permeability CBM wells, hydraulic fracturing is one of the main measures to improve their productivity. For such reservoirs, it is necessary to obtain enough propped fractures and certain fracture conductivity to effectively expand fluid seepage area and improve productivity. On the basis of experimental research and summary of actual wells, two ways to improve the hydraulic fracturing effect on CBM wells are put forward: increase effective fracture length by controlling fracture height and preventing excessive extension of fracture height and controlling multiple fractures; reduce secondary damage to reservoir and improve fracture conductivity.

11.1.4.1 Increase Effective Fracture Length

For CBM wells with low porosity and low permeability, only through hydraulic fracturing technology can reservoirs be deeply reformed, so as to expand flowing area, improve flowing channels, and improve well productivity. Current hydraulic fracturing monitoring and testing results show that the dynamic fracture length is less than 100m, generally 50–80 m, and the propped fracture length is 40–60 m. After considering the influence of various damage factors, the actual effective length contributing to production is estimated to be about 20–40 m, so how to improve the effective propped fracture length becomes more and more urgent. The primary way to increase the effective propped fracture length is to control the excessive extension of fracture height and prevent multiple fractures.

Control Fracture Height

The difference of in-situ stress between producing layer and upper and lower interlayers is the primary factor controlling fracture height, but this parameter can not be changed and controlled. At certain in-situ stress difference, the control on fracture height can be achieved by controlling fracturing fluid viscosity and displacement, adopting the combination of pad fluid and mixed-gas pad fluid and other technical means.

(1) Control the viscosity of fracturing fluid

Fracture fluid viscosity also affects the vertical extension of fractures. The greater the viscosity, the more serious the vertical extension is. Therefore, reducing fracturing fluid viscosity properly can control fracture height. This is also the mechanism of controlling fracture height by low-viscosity clean fracturing fluid, and the mechanism of controlling fracture height by lowering the polymer concentration of fracturing fluid. Therefore, on the premise of guaranteeing the proppant carrying capacity of fracturing fluid, in order to control fracture height, the fracturing fluid viscosity should be reduced as much as possible.

(2) Mixed pad fluid

Pad fluid is the carrier of initiating fractures. Pad fluid should be viscous, but too high viscosity will lead to excessive extension of fracture height. It is necessary to deal with the contradiction between these two aspects and play its best role. Gel pad fluid has strong ability to initiate fractures, but can cause excessive extension of fracture height. Clean water pad fluid has low viscosity and can control the extension of fracture height, but it has inadequate ability to initiate fractures. Through the simulation with 3D software, it is proved that if gel and water are combined half to half, they can fully play their advantages and make up for their shortcomings. The results show that the extension of fracture height with gel as pad fluid decreases by 18%, the length of fracture decreases by only 8%. However, the height of fracture with clean water as pad fluid decreases by 30%, and the length of fracture decreases by 21% (Table 11.11).

(3) Mixed pad fluid with gas

In fracturing construction, gas is pumped while pad fluid is injected, but gas preferentially blocked natural fractures, and reduce filtration and control fracture height to a certain extent. In addition, after the completion of construction, gas can facilitate drainage.

Table 11.11 Effects of pad fluids on fracture size

Pad fluid	Ratio of fracture height to pay layer thickness	Fracture length/m
Gel	2.04	85
Gel to water: half to half	1.70	79
Clean water	1.40	65

Control Multiple Fractures and Increase Effective Fracture Length

Due to the well development of coal seam cleats, it is easy to induced multiple fractures in early fracturing construction and result in sand plugging. At the same time, at constant construction scale, multiple fractures enlarge liquid filtration, shorten fracture length, narrow fracture width, reduce low conductivity and stimulated effect. Therefore, controlling multi-fractures is an important means to increase effective fracture length in fracturing coal seams. The main methods to control the formation of multiple fractures are as follows.

(1) Inject proppant slug to grind fracture tortuousness

Inject proppant slugs with pad fluid to grind fracture tortuousness and reduce the bending of fracture. It is conducive to the extension of fractures along primary fractures and reduce the probability of sand plugging.

(2) Inject 400 mesh resin, 100 mesh silty sand or silty ceramsite to control solid particles to reduce filtration

Into coal seams with cleats, 400 mesh resin, inject 100 mesh silt or silty ceramsite with pad fluid to control solid particles, and build bridge plugs in natural micro fractures, reduce filtration, prevent sand plugging caused by over-filtration and ensure the length of primary fractures.

(3) Sand out in fracture tips

Sand out in fracture tips means that pad fluid is used up and fracture length is frozen at the late stage by controlling the percentage of pad fluid. Then after continuing adding sand, the fracture length will not increase, but the fracture width will increased, in this way to improve fracture conductivity.

Based on the data from the Daning area, 3D simulation proved the effect of sand out technique. Taking gel fracturing fluid as an example, the proportion of pad fluid (30 m^3) in non-sand out design is 40% and that in sand out design is 25%, we got the fracture length, propped fractures and conductivity in Fig. 11.51a, b and Table 11.12.

The above research results show that the design of sand out fracturing can achieve the purpose of improving flow conductivity, but the fracture length is shortened to a certain extent. In the two designs, the fracture conductivity is increased by 19.6% by end sand out technology, but the fracture length is shortened by 20%. The results are shown in Table 11.12.

Sand out design can improve fracture conductivity, but the application should depend on CBM wells. In the area

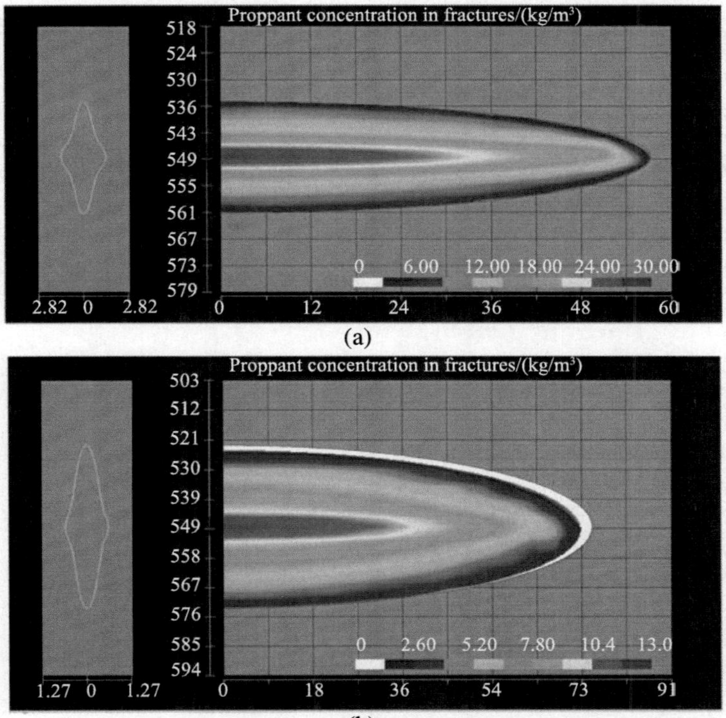

Fig. 11.51 Propped fracture. **a** Sand out design; **b** non-sand out design

Table 11.12 Fracture parameters of sand out design and non-sand out design

Fracture parameters	Sand out	Non-sand out
Propped fracture length/m	58	73
Maximum fracture width at the opening/cm	2.8	1.2
Fracture conductivity at the opening/d cm	56	44

with more development of multiple fractures, it is not recommended to adopt the sand out technique when getting longer propped fracture is the dominant target. If the permeability of the target layer is good and the higher fracture conductivity is the dominant target, it is suggested that the sand out technology should be considered, but the initial test should be based on the understanding of formation filtration through small-scale fracturing test.

11.1.4.2 Reduce Secondary Reservoir Damage and Improve Fracture Conductivity

Improve Fracturing Fluid System to Reduce Reservoir Damage

① Continue developing and improving the fracturing fluid system of breaking gel at low temperature to improve flowback and reduce damages to reservoirs and fracture system.

② Use VES clean fracturing fluid system. The fluid is characterized by no solid residue, little damage to reservoirs and fractures, low viscosity and high back-flow rate. As the viscosity is relatively low, it is advantageous to control fracture height, and at the same volume of proppants, it is easier to obtain long fractures, and improve reservoir contact.

③ Develop less viscous fracturing fluid which has good proppant-carrying ability, lefts less solid residue and induces less damage to reservoir.

Develop and Apply Low-Density and High-Strength Proppants

Proppants with low density and high strength require low proppant carrying capacity of fracturing fluid, thus it is possible to reduce the use of thickener, and even clear water can carry proppants. Damage to reservoirs and fractures can be reduced.

Clean Fractures to Improve Fracture Conductivity
Cleaning fractures means to use fracture cleaning agent to dissolve the residues in fractures, so as to improve fracture surface permeability and propped fracture conductivity. In the late stage of flowback, put fracture cleaning agent through wellhead and then shut in the well for 5–8 h. To reduce proppant settlement, don't add cleaning agent with replacement fluid.

11.2 Stimulation Mechanism of Multi-branch Horizontal Wells

11.2.1 Mathematical and Numerical Models

In order to improve the production of CBM wells, some measures need to be taken. At present, the latest and proven effective stimulation technologies abroad include hydraulic fracturing stimulation, oriented pinnate horizontal wells and multi-component gas displacement.

Multi-branch horizontal well technology is a patent technology developed by CDX International Company in the late 1990s. It has been used in CBM development in recent years with remarkable results. Practice has proved that multi-branch horizontal CBM wells are generally suitable for reservoirs with thick coal and continuous distribution of coal seams, and they are most effective when the horizontal wellbore is perpendicular to the direction of the maximum permeability. But under what conditions, to what extent and how to improve CBM recovery of multi-branch horizontal wells still need to be quantitatively analyzed. At present, there is no report in this field in China and no public report in foreign countries.

In this section, mathematical models and numerical simulation of multi-branch horizontal CBM well are established, the solution of the numerical model is given, and the numerical program for simulating the production of a multi-branch horizontal well is worked out. The calculation results can give satisfactory answers to the above questions.

11.2.1.1 Mathematical Model

Whether vertical or multi-branch horizontal wells are used to exploit CBM, the gas flow in coal seams will follow the same law. From the point of view of mathematics, the partial differential equations, initial conditions and external boundary conditions for flow control are the same, but the internal boundary conditions, i.e., well treatment methods, are different. However, for the sake of the integrity of this section, a complete mathematical model for CBM production by multi-branch horizontal wells will be listed below.

Flow Equation of CBM and Water in Dual Media
Coal seam is regarded as a dual medium with pores and fractures. In the original state, fractures are filled with water and contain a small amount of free gas. A large amount of gas exists in the matrix in the form of adsorption, considering gravity and ignoring capillary force. The gas in fractures satisfies the real gas state equation. The velocity of gas motion is regarded as the sum of the macroscopic seepage velocity and the gas diffusion velocity following the Fick's law. That is,

$$v_{\mathrm{g}} = -\left(\frac{K_{\mathrm{g}}}{\mu_{\mathrm{g}}}\beta_{\mathrm{gs}}\nabla\Phi_{\mathrm{g}} + \frac{D_{\mathrm{f}}}{C_{\mathrm{f}}}\nabla C_{\mathrm{f}}\right) \tag{11.28}$$

For water phase in fractures,

$$v_{\mathrm{w}} = -\frac{K_{\mathrm{w}}}{\mu_{\mathrm{w}}}\beta_{\mathrm{ws}}\nabla\Phi_{\mathrm{w}}$$

where β_{ls} is a modified coefficient of the Darcy's law. When $\left|\frac{\partial\Phi_{\mathrm{l}}}{\partial_{\mathrm{s}}}\right| > \lambda$, $\beta_{\mathrm{ls}} = 1 - \frac{\lambda}{\left|\frac{\partial\Phi_{\mathrm{l}}}{\partial_{\mathrm{s}}}\right|}$. When $\left|\frac{\partial\Phi_{\mathrm{l}}}{\partial_{\mathrm{s}}}\right| < \lambda$, $\beta_{\mathrm{ls}} = 0$. Among them, λ is the critical pressure gradient, subscript 1 denotes gas and water (g, w), and subscript s denotes X, Y and Z directions. The gas desorption process in the matrix is regarded as quasi-steady diffusion, which satisfies the Fick's first law.

$$\begin{aligned}\frac{\mathrm{d}C(t)}{\mathrm{d}t} &= D_{\mathrm{m}}F_{\mathrm{S}}[V_{\mathrm{E}}(P_{\mathrm{g}}) - C(t)] \\ &= 1/\tau[V_{\mathrm{E}}(P_{\mathrm{g}}) - C(t)]\end{aligned} \tag{11.29}$$

where $V_{\mathrm{E}}(P_{\mathrm{g}})$ is the concentration in equilibrium with the gas pressure in fractures.

$$\text{When } P_{\mathrm{g}} \geq P_{\mathrm{d}}, \quad V_{\mathrm{E}}(P_{\mathrm{g}}) = V_{\mathrm{E}}(P_{\mathrm{d}}) \tag{11.30}$$

When $P_{\mathrm{g}} < P_{\mathrm{d}}$, satisfy the Langmuir equation

$$V_{\mathrm{E}}(P_{\mathrm{g}}) = V_{\mathrm{L}}\frac{P_{\mathrm{g}}}{P_{\mathrm{L}} + P_{\mathrm{g}}} \tag{11.31}$$

The flow rate from matrix to fractures is the source of the gas flow equation in the fractures.

$$q_{\mathrm{m}} = -F_{\mathrm{G}}\frac{\mathrm{d}C(t)}{\mathrm{d}t} \tag{11.32}$$

Because the whole well section is open, the production flowing into each branch and section is taken as the convergence term of the gas and water phase flow equations, the gas-phase flow equation and the water-phase seepage equation in the fracture are respectively as follows.

$$\frac{\partial}{\partial t}\left(\frac{\phi_f S_g \Phi_g}{Z}\right) = \nabla \cdot \left[\frac{p_g}{Z}\beta_{gs}\frac{K_g}{\mu_g}\nabla\Phi_g + \frac{D_f}{S_g}\nabla\left(\frac{S_g p_g}{Z}\right)\right] + \frac{RT}{M}q_m - q_g'\frac{P_g}{Z} \tag{11.33}$$

$$\frac{\partial}{\partial t}\left(\frac{\phi_f S_w}{Z}\right) = \nabla \cdot \left(\frac{K_w}{B_w \mu_w}\beta_{ws}\nabla\Phi_w\right) - \frac{q_w'}{B_w} \tag{11.34}$$

Where Φ_g is gas flow potential, Φ_w is water flow potential; q_g' is gas production and q_w' is water production from unit volume reservoir in unit time, 1/s.

$$q_g' = \frac{q_g}{\Delta x_i \Delta y_j \Delta z_k}, q_w' = \frac{q_w}{\Delta x_i \Delta y_j \Delta z_k} \tag{11.35}$$

Where q_g is gas production and q_w is water production from a grid volume per unit time, m^3/s.

$$q_g = \text{PID}\ \beta_{gs}\frac{k_{rg}(P_g - P_{wf})}{\mu_g}, q_w = \text{PID}\beta_{ws}\frac{k_{rw}(P_w - P_{wf})}{\mu_w}$$

Where P_{wf} is the flow pressure in the wellbore of each micro-section, and PID is the well index.

$$\text{PID} = 2\pi\frac{K_e L_p}{\ln\frac{r_b}{r_w} + S} \tag{11.36}$$

Where K_e is the equivalent isotropic permeability of anisotropic medium, L_p is the length of the interval in the transformed space grid, r_b is the equivalent radius of the well grid, r_w is the equivalent well diameter, and S is the skin factor.

Saturation equation:

$$S_w + S_g = 1 \tag{11.37}$$

Pressure-Sensitive Models of Porosity and Permeability

The compressibility of pores in fractured coal seams is much larger than that of clastic rock and carbonate rock, so the porosity and permeability of coal seams are more sensitive to pressure. In addition, besides the effect of effective stress, gas desorption can cause coal matrix to shrink, and the latter item of Eq. (11.38) reflects the effect of shrunk matrix.

$$\begin{aligned}\phi_f &= \phi_i[1 + C_p(P - P_i)] - C_m(1 - \phi_i)\\ &\frac{P_d - P_{sc}}{C(P_d) - C(P_{sc})}[C(P) - C(P_d)]\end{aligned} \tag{11.38}$$

In general pressure-sensitive reservoirs, the relationship between permeability and pressure is approximately exponential $k = k_i e^{-\alpha_{k1}(p_i - p)}$. In coal seams, the effect of effective stress and gas desorption lead to the decrease and increase of permeability, and dominate the early and late stages of pressure reduction. Therefore, the relationship between pressure difference and permeability can be approximated by Eq. (11.39). Among them, α_{k1} and α_{k2} are pressure-sensitive coefficients, and P_k is the pressure corresponding to the lowest permeability on the pressure-sensitive curve, which can be measured by experiments.

$$K = \begin{cases} K_i e^{-\alpha_{k1}(P_i - P)} & P \geq P_k \\ K_i e^{-\alpha_{k1}(P_i - P_k)} e^{\alpha_{k2}(P_k - P)} & P < P_k \end{cases} \tag{11.39}$$

Simplified Model of Multi-branch Horizontal Wells

Whether a vertical well is used or an electric submersible pump is directly put into a horizontal wellbore, there is a common feature, that is, an outlet controls a horizontal branch. To be simple, all outlets are regarded as a point on the target layer profile in the simplified model of a multi-branch horizontal well, so that it is convenient to set and treat the boundary conditions inside the outlets. Since the distance between the outlets is shorter than that of the whole well area in the target layer profile, this simplification is reasonable.

Wellbore Pressure Drop Model

There are many differences between horizontal well production and conventional vertical well production. Because the horizontal section in the production layer is longer, the flow condition of the horizontal section will have a certain impact on the production performance of the horizontal well. But for convenience, the initial treatment of A horizontal wellbore assumes that the pressure along the whole horizontal section is constant. In the case of low-permeability formation or a short horizontal section, neglecting the flow condition in the wellbore will not cause great deviation, but it does not accord with the actual flow condition, because in fact, the flow from the end of the horizontal section to the production section can only be realized at a certain pressure difference. For high-permeability formation or long horizontal section, the pressure difference of the horizontal section is relatively large (Zhou et al. 2002). In 1989, Dikken first proposed that wellbore pressure drop should not be neglected in order to reliably predict the production performance of horizontal wells. At present, people mostly analyze the mechanism of complex flow in wellbore on the basis of experimental research, and establish corresponding empirical or theoretical relation formula. Generally speaking, the pressure drop in wellbore is divided into frictional pressure drop due to the roughness of wellbore wall and accelerated pressure drop due to the continuous convergence of fluid and the increase in quality. In a perforated wellbore, there is another drop called injection pressure drop caused by damaged flow pattern and boundary layer in the casing at high perforation velocity. At present, the former two have more mature calculation formulas.

In a multi-branch horizontal well, friction pressure drop and accelerated pressure drop are considered because of the open hole completion. However, the roughness of an open-hole horizontal wellbore is greater than that of a conventional cased wellbore, and the inflow of reservoir fluid into the wellbore can cause the change of momentum, which will change the pressure distribution along the wellbore. Therefore, a conventional cased-hole friction coefficient correlation formula can not reasonably predict a horizontal well. In a specific coal seam, the friction coefficient between coal seam fluid and open-hole horizontal wellbore needs to be measured by experiments. In this model, the conventional friction coefficient formula of Haaland equation is used for the time being.

In a multi-branch horizontal well, formation fluid can flow from the branch to the primary branch, then to the outlet and to surface, or directly from the primary branch to the outlet and to surface. The primary branch and the branches are regarded as consisting of several micro-sections. In the wellbore, considering friction pressure drop and accelerated pressure drop, the pressure relationship of the upstream and downstream adjacent micro-sections in the primary branch or the branch is expressed as follows.

$$P_{\mathrm{wf}_{i-1}} = P_{\mathrm{wf}_i} + 0.5(\Delta P_{\mathrm{wf}_{i-1}} + \Delta P_{\mathrm{wf}_i}) \tag{11.40}$$

where $P_{\mathrm{wf}_{i-1}}$ and P_{wf_i} represent the pressure at the center of the adjacent upper and lower sections respectively, $\Delta P_{\mathrm{wf}_{i-1}}$ and ΔP_{wf_i} represent the pressure drop through adjacent upper and lower well sections respectively.

When producing at constant pressure, the outlet pressure is set as P_{wf_c}, and the central pressure of the micro-section adjacent to the outlet is as follows.

$$P_{\mathrm{wf}_i} = P_{\mathrm{wf}_c} + 0.5\Delta P_{\mathrm{wf}_i} \tag{11.41}$$

When producing at constant gas production, it is necessary to add an equation and the gas production should be the sum of the productions of all branches and primary branches.

$$q_{\mathrm{sum}} = \sum q_{g_i} \tag{11.42}$$

The pressure drop in the wellbore is regarded as the sum of friction pressure drop and accelerated pressure drop, that is

$$\Delta P_{\mathrm{wf}_i} = \Delta P_{\mathrm{fric}_i} + \Delta P_{\mathrm{acc}_i} \tag{11.43}$$

Following is a detailed derivation of the wellbore pressure drop model. For the control body (a micro-section in Fig. 11.52), the flow in the wellbore is assumed to be one-dimensional, stable and incompressible, and the surface forces acting on the fluid in the extended direction of the wellbore are upstream end pressure $P_{\mathrm{wf}_{i-1}}$, downstream end pressure P_{wf_i}, and casing wall friction resistance τ_{w}.

According to the momentum theorem,

$$\begin{aligned} &P_{\mathrm{wf}_{i-1}}A - P_{\mathrm{wf}_i}A - \tau_{\mathrm{w}}\pi D\Delta x \\ &= \Delta(m_i V_i) = (\rho_i A V_i)V_i \\ &- (\rho_{i-1}AV_{i-1})V_{i-1} \end{aligned} \tag{11.44}$$

Among them, m_i is mass flow.

$$m_i = \rho_i A V_i \tag{11.45}$$

The formula considering the friction resistance,

$$\tau_{\mathrm{w}} = \rho_i f_i \overline{V}_i^2/8 \tag{11.46}$$

Then,

$$\begin{aligned} P_{\mathrm{wf}_{i-1}} - P_{\mathrm{wf}_i} &= [(\rho_i V_i)V_i - (\rho_i V_{i-1})V_{i-1}] \\ &- \tau_{\mathrm{w}}\pi D\Delta x/(\pi D^2/4) \end{aligned} \tag{11.47}$$

The first term on the right side is regarded as accelerated pressure drop and the second term as frictional pressure drop, represented as follows.

$$\Delta P_{\mathrm{fric}_i} = \frac{1}{2}\frac{\rho_i f_i}{D}\overline{V_i^2}\Delta x_i \tag{11.48}$$

$$\Delta P_{\mathrm{fric}_i} = \rho_i(V_i + V_{i-1})(V_i - V_{i-1}) \tag{11.49}$$

where ρ_i is the density of the fluid in well section i, Δx_i is the length of well section i, D is borehole diameter, V_{i-1} is the velocity of fluid flowing in,and V_i is the velocity of fluid flowing out of well section i, $\overline{V}_i$ is the average velocity of the fluid in well section i.

$$\overline{V_i} = \frac{V_i + V_{i-1}}{2} \tag{11.50}$$

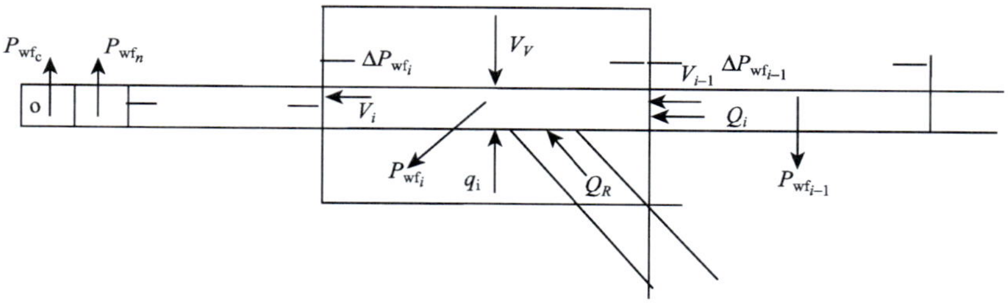

Fig. 11.52 Schematic diagram of pinnate horizontal sections

f_i is the friction coefficient between fluid and wellbore in well section i.

For laminar flow,

$$f_i = \frac{64}{\mathrm{Re}_i} \tag{11.51}$$

For turbulent flow,

$$f_i = \left\{ -1.8\lg\left[\frac{6.9}{\mathrm{Re}_i} + \left(\frac{e}{3.7}\right)^{\frac{10}{9}}\right]\right\}^{-2} \tag{11.52}$$

where e is the relative roughness of wellbore wall, Re_i is Reynolds number, $\mathrm{Re}_i = \frac{\overline{V}_i \rho_i}{\mu_i}$ is the fluid viscosity of well section i, and ρ_i is the fluid density of well section i. In the case of gas–water two-phase flow,

$$\mu_i = \mu_{\mathrm{w}}^{f_{\mathrm{w}}} \mu_{\mathrm{g}}^{1-f_{\mathrm{w}}} \tag{11.53}$$

Among them,

$$f_{\mathrm{w}} = \frac{\rho_{\mathrm{w}} q_{\mathrm{w}i}}{\rho_{\mathrm{w}} q_{\mathrm{w}i} + \rho_{\mathrm{g}} q_{\mathrm{g}i}} \tag{11.54}$$

$$\rho_i = \frac{\rho_{\mathrm{g}} q_{\mathrm{g}i} + \rho_{\mathrm{w}} q_{\mathrm{w}i}}{q_{\mathrm{g}i} + q_{\mathrm{w}i}} \tag{11.55}$$

From the conservation of mass,

$$\rho_{i-1} V_{i-1} \frac{\pi D^2}{4} + \rho_i V_{\mathrm{V}} \pi D \Delta x - \rho_i V_i \frac{\pi D^2}{4} + \rho_i \frac{\pi D^2}{4} V_{\mathrm{R}} = 0 \tag{11.56}$$

it can be get that:

$$V_i - V_{i-1} = \frac{4V_{\mathrm{V}}}{D} \Delta x + V_{\mathrm{R}} \tag{11.57}$$

$$\overline{V_i} = \frac{V_i + V_{i-1}}{2} = V_{i-1} + \frac{(V_i - V_{i-1})}{2} \tag{11.58}$$

where V_{R} is the velocity from a branch flowing into the primary branch; and V_{V} is the velocity of seepage flow directly from formation to the primary branch. $Q_i = \frac{V_{i-1}\pi D^2}{4}$, $q_i = V_{\mathrm{V}} \Delta x \pi D$, $Q_{\mathrm{R}i} = \frac{V_{\mathrm{R}i}\pi D^2}{4}$, thus, the pressure drop on each micro-section is expressed in the following three cases, respectively.

Section i is in a branch,

$$\Delta P_{\mathrm{wf}i} = \left[\frac{2f_i\rho_i}{\pi^2 D^5}(2Q_i + q_i)^2 \Delta x_i + \frac{16\rho_i(g_i + Q_{\mathrm{R}i})}{\pi^2 D^4}(2Q_i + q_i)\right] \tag{11.59}$$

where subscript i denotes the ith section of the branch, counted from the upstream end, q_i is the flow flowing from formation into the branch, $q_i = q_{\mathrm{g}i} + q_{\mathrm{w}i}$, Q_i is the flow from the upstream section to section i, $Q_i = \sum_{k=1}^{i-1} q_{k\mathrm{b}}$.

Section i is in the primary branch, and there are branches in section i:

$$\Delta P_{\mathrm{wf}i} = \left[\frac{2f_i\rho_i}{\pi^2 D^5}(2Q_i + q_i + Q_{\mathrm{R}i})^2 \Delta x_i + \frac{16\rho_i(q_i + Q_{\mathrm{R}i})}{\pi^2 D^4}(2Q_i + q_i)\right] \tag{11.60}$$

where q_i is the flow from formation to the primary branch of section i, $q_i = q_{\mathrm{g}i} + q_{\mathrm{w}i}$, Q_i is the flow from the upstream section to section i, $Q_i = \sum_{k=1}^{i-1} q_{k\mathrm{pb}} + \sum q_{\mathrm{ub}}$, $Q_{\mathrm{R}i}$ is the flow flowing from the branch of well section i into section i, $Q_{\mathrm{R}i} = \sum_{k=1}^{n} q_{ik\mathrm{b}}$, and the two subscripts of $q_{ik\mathrm{b}}$ denote the numbers of the primary branch and the branch respectively.

Section i is in the primary branch, and there aren't branches in section i:

$$\Delta P_{\mathrm{wf}i} = \left[\frac{2f_i\rho_i}{\pi^2 D^5}(2Q_i + q_i)^2 \Delta x_i + \frac{16\rho_i q_i}{\pi^2 D^4}(2Q_i + q_i)\right] \tag{11.61}$$

In the formula, subscript i denotes section i of the branch counted from the upstream end, Q_i is the flow from the upstream section to section i, and it is the sum of flows from all primary branches and branches. $Q_i = \sum_{k=1}^{i-1} q_{k\mathrm{pb}} + \sum q_{\mathrm{ub}}$, and q_i is the flow from formation to the primary branch of section i, $q_i = q_{\mathrm{g}i} + q_{\mathrm{w}i}$.

In the above equation, basic SI unit system is used to all terms.

The above gas–water two-phase flow equation and wellbore pressure drop equation can finally be transformed into a set of non-linear equations containing only formation pressure, formation gas saturation and bottom hole pressure, which constitutes the mathematical model of the entire problem. The gas and water production of the whole multi-branch horizontal well group should be the sum of outlet productions of all primary branches and branches.

11.2.1.2 The Numerical Model of CBM Production in Multi-branch Horizontal Wells

Differential Discretization of Gas–Water Two-Phase Flow Equation in Coal Seams

As shown in Fig. 11.53, locate a group of branch wellbores along four symmetrical directions and use center grids on a 3D rectangular coordinate system, and define the thickness direction of the coal seam is Z direction (positive downward), and multi-branch horizontal wells located on the XY plane with vertical coordinates $Z = 0$. The grid number in the directions of X, Y and Z is expressed by i, j and

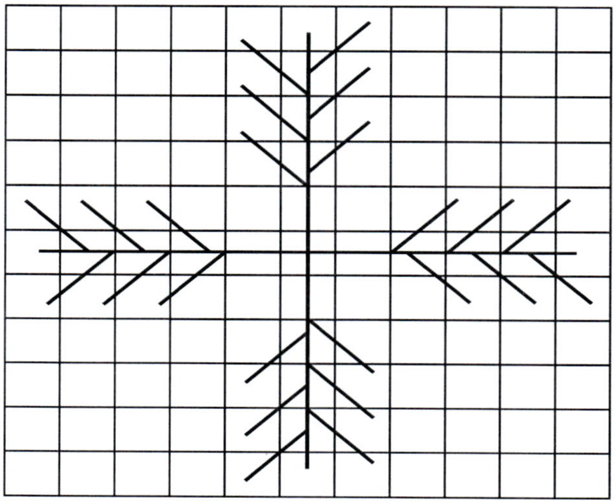

Fig. 11.53 Sketch map of grid section of directional pinnate horizontal wells

k respectively. The serial number of X is from left to right, Y from inside to outside, and Z from top to bottom, and the gravity direction is downward.

The grids on the XY plane divide the multi-branch horizontal wells into micro-sections. Because each section is relatively short, it can be considered that the flow rate from reservoir into a section is uniformly distributed. For arbitrary grids, q_g and q_w in Eqs. (11.62) and (11.63) denote the production from a grid volume per unit time,

$$q_{g_{i,j,k}} = 86.4\mathrm{PID}_{i,j,k}\frac{k_{rg}(p_{g_{i,j,k}} - P_{wf_{i,j,k}})}{\mu_{g_{i,j,k}}} \tag{11.62}$$

$$q_{w_{i,j,k}} = 86.4\mathrm{PID}_{i,j,k}\frac{k_{rw}(P_{g_{i,j,k}} - P_{wf_{i,j,k}})}{\mu_{w_{i,j,k}}} \tag{11.63}$$

In the formula above, $P_{wf_{i,j,k}}$ donates the wellbore flow pressure of section i contained in $\mathrm{grid}_{i,j,k}$, and $\mathrm{PID}_{i,j,k}$ is well index. In the grid without a well section passing through, PID = 0. In the grid with well sections passing through,

$$\mathrm{PID} = 2\pi\frac{k_e L_P}{\ln\frac{r_b}{r_w} + S} \tag{11.64}$$

In the formula above, k_e is the equivalent isotropic permeability of an anisotropic medium, $k_e = (k_x k_y k_z)^{1/3}$, L_P is the length of the interval in the transformed space grid.

$$L_P = \sqrt{L_x^2 + L_y^2}, L_x = L\sqrt{\frac{k_e}{k_x}}\cos\omega, L_y = L\sqrt{\frac{k_e}{k_y}}\sin\omega \tag{11.65}$$

In the formula, L is the length of the section located in the grid; ω is the angle with X axis and r_b is the equivalent radius of well grid.

$$r_b = \sqrt{r_{bx}^2 + r_{bx}^2}, r_{bx} = R_{bx}\cos\omega, r_{by} = R_{by}\sin\omega \tag{11.66}$$

$$\begin{aligned} R_{bx} &= 0.14\sqrt{(k_e/k_y)\Delta y^2 + (k_e/k_z)\Delta z^2}, \\ R_{by} &= 0.14\sqrt{(k_e/k_x)\Delta x^2 + (k_e/k_z)\Delta z^2} \end{aligned} \tag{11.67}$$

r_w is equivalent borehole diameter.

$$r_w = \sqrt{r_{wx}^2 + r_{wy}^2}, r_{wx} = R_{wx}\cos\omega, r_{wy} = R_{wy}\sin\omega \tag{11.68}$$

$$R_{wx} = \frac{R_w}{2}\left(\sqrt{\frac{k_e}{k_y}} + \sqrt{\frac{k_e}{k_z}}\right), R_{wy} = \frac{R_w}{2}\left(\sqrt{\frac{k_e}{k_x}} + \sqrt{\frac{k_e}{k_z}}\right) \tag{11.69}$$

R_w is real borehole diameter.

According to the mathematical model established, taking the inner boundary as the fixed bottom hole pressure condition and the outer boundary as the closed formation boundary condition, the mathematical model is discretized directly by using the finite difference method through the conservation of matter in the differential grid.

The gas phase flow equation is discretized as follows:

$$\begin{aligned} & 86.4\Delta y_j\Delta z_k\Bigg\{\left[\frac{P_g}{Z(P_g)}\frac{k(P_g)k_{rg}(S_g)}{\mu_g}\right]_{i+\frac{1}{2}}\frac{P_{gi+1}-P_{gi}}{\Delta x_{i+\frac{1}{2}}} + \left[\frac{D_f\phi_f(P_g)}{s_g}\right]_{i+\frac{1}{2}}\frac{\left[\frac{S_gP_g}{Z(P_g)}_{i+1} - \frac{S_gP_g}{Z(P_g)}_i\right]}{\Delta x_{i+\frac{1}{2}}} \\ & + \left[\frac{P_g}{Z(P_g)}\frac{k(P_g)k_{rg}(S_g)}{\mu_g}\right]_{i-\frac{1}{2}}\frac{P_{gi-1}-P_{gi}}{\Delta x_{i-\frac{1}{2}}} + \left[\frac{D_f\phi_f(P_g)}{s_g}\right]_{i-\frac{1}{2}}\frac{\left[\frac{S_gP_g}{Z(P_g)}_{i-1} - \frac{S_gP_g}{Z(P_g)}_i\right]}{\Delta x_{i-\frac{1}{2}}}\Bigg\} \\ & + 86.4\Delta x_i\Delta z_k\Bigg\{\left[\frac{P_g}{Z(P_g)}\frac{k(P_g)k_{rg}(S_g)}{\mu_g}\right]_{j+\frac{1}{2}}\frac{P_{gi-1}-P_{gi}}{\Delta y_{j+\frac{1}{2}}} + \left[\frac{D_f\phi_f(P_g)}{s_g}\right]_{j+\frac{1}{2}}\frac{\left[\frac{S_gP_g}{Z(P_g)}_{j+1} - \frac{S_gP_g}{Z(P_g)}_j\right]}{\Delta y_{j+\frac{1}{2}}} \\ & + \left[\frac{P_g}{Z(P_g)}\frac{k(P_g)k_{rg}(S_g)}{\mu_g}\right]_{j-\frac{1}{2}}\frac{P_{gj-1}-P_{gj}}{\Delta y_{j-\frac{1}{2}}} + \left[\frac{D_f\phi_f(P_g)}{s_g}\right]_{j-\frac{1}{2}}\frac{\left[\frac{S_gP_g}{Z(P_g)}_{j-1} - \frac{S_gP_g}{Z(P_g)}_j\right]}{\Delta y_{j-\frac{1}{2}}}\Bigg\} \\ & + 86.4\Delta x_i\Delta y_j\Bigg\{\left[\frac{P_g}{Z(P_g)}\frac{k(P_g)k_{rg}(S_g)}{\mu_g}\right]_{k+\frac{1}{2}}\left(\frac{P_{gk+1}-P_{gk}}{\Delta z_{k+\frac{1}{2}}} - \frac{\gamma_g(P_g)_{k+\frac{1}{2}}(z_{k+1}-z_k)}{10^6\Delta z_{k+\frac{1}{2}}}\right) \\ & + \left[\frac{D_f\phi_f(P_g)}{s_g}\right]_{k+\frac{1}{2}}\frac{\left[\frac{S_gP_g}{Z(P_g)}_{k+1} - \frac{S_gP_g}{Z(P_g)}_k\right]}{\Delta z_{k+\frac{1}{2}}} + \left[\frac{P_g}{Z(P_g)}\frac{k(P_g)k_{rg}(S_g)}{\mu_g}\right]_{k-\frac{1}{2}} \\ & \left(\frac{P_{gk-1}-P_{gk}}{\Delta z_{k-\frac{1}{2}}} - \frac{\gamma_g(P_g)_{k-\frac{1}{2}}(z_{k-1}-z_k)}{10^6\Delta z_{k-\frac{1}{2}}}\right) + \left[\frac{D_f\phi_f(P_g)}{s_g}\right]_{k-\frac{1}{2}}\frac{\left[\frac{S_gP_g}{Z(P_g)}_{k+1} - \frac{S_gP_g}{Z(P_g)}_k\right]}{\Delta z_{k-\frac{1}{2}}}\Bigg\} \\ & - \frac{F_G}{\Delta t}24\frac{RT}{M}\Delta x_i\Delta y_j\Delta z_k(C^{n+1} - C^n) - \mathrm{PID}_{i,j,k}\frac{k_{rg}(P_{g_{i,j,k}} - P_{wf_{i,j,k}})}{\mu_g(P_g)}86.4\left(\frac{P_g}{Z}\right)_{i,j,k} \\ & = 24\frac{\Delta x_i\Delta y_j\Delta z_k}{\Delta t}\left[\left(\frac{\phi_f(P_g)S_gP_g}{Z(P_g)}\right)_{i,j,k}^{n+1} - \left(\frac{\phi_f(P_g)S_gP_g}{Z(P_g)}\right)_{i,j,k}^{n}\right] \end{aligned} \tag{11.70}$$

The water phase flow equation is discretized as follows:

$$\begin{aligned}
&86.4\Delta y_j\Delta z_k\left\{\left[\frac{k(P_g)k_{rw}(S_g)}{\mu_w B_w}\right]_{i+\frac{1}{2}}\frac{P_{gi+1}-P_{gi}}{\Delta x_{i+\frac{1}{2}}}+\left[\frac{k(P_g)k_{rw}(S_g)}{\mu_w B_w}\right]_{i-\frac{1}{2}}\frac{P_{gi-1}-P_{gi}}{\Delta x_{i-\frac{1}{2}}}\right\}\\
&+86.4\Delta x_i\Delta z_k\left\{\left[\frac{k(P_g)k_{rw}(S_g)}{\mu_w B_w}\right]_{j+\frac{1}{2}}\frac{P_{gj+1}-P_{gj}}{\Delta y_{j+\frac{1}{2}}}+\left[\frac{k(P_g)k_{rw}(S_g)}{\mu_w B_w}\right]_{j-\frac{1}{2}}\frac{P_{gj-1}-P_{gj}}{\Delta y_{j-\frac{1}{2}}}\right\}\\
&+86.4\Delta x_i\Delta y_j\left\{\left[\frac{k(P_g)k_{rw}(S_g)}{\mu_w B_w}\right]_{k+\frac{1}{2}}\left(\frac{P_{gk+1}-P_{gk}}{\Delta z_{k+\frac{1}{2}}}-\frac{\gamma_{wk+\frac{1}{2}}(z_{k+1}-z_k)}{10^6\Delta z_{k+\frac{1}{2}}}\right)\right.\\
&\left.+\left[\frac{k(P_g)k_{rw}(S_g)}{\mu_w B_w}\right]_{k-\frac{1}{2}}\left(\frac{P_{gk-1}-P_{gk}}{\Delta z_{k-\frac{1}{2}}}-\frac{\gamma_{wk-\frac{1}{2}}(z_{k-1}-z_k)}{10^6\Delta z_{k-\frac{1}{2}}}\right)\right\}-\mathrm{PID}_{i,j,k}86.4\frac{k_{rw}(S_g)(P_{gi,j,k}-P_{wf\,i,j,k})}{\mu_w B_w}\\
&=24\frac{\Delta x_i\Delta y_j\Delta z_k}{\Delta t}\left[\left(\frac{\phi_f(P_g)(1-S_g)}{B_w}\right)_{i,j,k}^{n+1}-\left(\frac{\phi_f(P_g)(1-S_g)}{B_w}\right)_{i,j,k}^{n}\right]
\end{aligned}\tag{11.71}$$

For the boundary, when $i = 1$, $P_{gi-1} = P_{gi}$, $S_{gi-1} = s_{gi}$;
when $i = ix$, $P_{gi+1} = P_{gi}$, $S_{gi+1} = s_{gi}$;
when $j = 1$, $P_{gj-1} = P_{gj}$, $S_{gj-1} = s_{gj}$;
when $j = jy$, $P_{gj+1} = P_{gj}$, $S_{gj+1} = s_{gj}$;
when $k = 1$,

$$P_{gk-1} = P_{gk} + \frac{\gamma_g(P_g)_{k-\frac{1}{2}}}{10^6}(z_{k-1} - z_k), S_{gk-1} = S_{gk};$$

When $k = kz$,

$$p_{gk+1} = p_{gk} + \frac{\gamma_g(P_g)_{k+\frac{1}{2}}}{10^6}(z_{k+1} - z_k), S_{gk+1} = S_{gk}.$$

In the formula, $\Delta x_{i+\frac{1}{2}} = x_{i+1} - x_i$, $\Delta x_{i-\frac{1}{2}} = x_i + x_{i-1}$, $\Delta y_{j+\frac{1}{2}} = y_{j+1} - y_j$, $\Delta y_{j-\frac{1}{2}} = y_j - y_{j-1}$, $\Delta z_{k+\frac{1}{2}} = z_{k+1} - z_k$, $\Delta z_{k-\frac{1}{2}} = z_k - z_{k-1}$.

The spatial subscript in the equation, such as $i \pm \frac{1}{2}, i \pm 1, i, j \pm \frac{1}{2}, j \pm 1, j, k \pm \frac{1}{2}, k \pm 1, k$ omit the other two indices, for example, $i - \frac{1}{2}$ donates $i - \frac{1}{2}, j, k$, $j - \frac{1}{2}$ donates $i, j - \frac{1}{2}, k$, $k - \frac{1}{2}$ donates $i, j, k - \frac{1}{2}$, etc.

The spatial subscript in the difference equation is the coefficient of $i - \frac{1}{2}$, $i + \frac{1}{2}$, $j - \frac{1}{2}$, $j + \frac{1}{2}$, $k - \frac{1}{2}$, $k + \frac{1}{2}$. The absolute permeability and fracture porosity are taken as harmonic average values, that is:

$$\begin{aligned}
k_{i+\frac{1}{2}} &= \frac{2k(P_{gi,j,k})k(P_{gi+1,j,k})}{k(P_{gi,j,k}) + k(P_{gi+1,j,k})},\\
\phi_{fk+\frac{1}{2}} &= \frac{2\phi_f(P_{gi,j,k})\phi_f(P_{gi,j,k+1})}{\phi_f(P_{gi,j,k}) + \phi_f(P_{gi,j,k+1})}
\end{aligned}\tag{11.72}$$

For relative permeability, viscosity and the compression factor of natural gas is determined by upstream weight, that is:

$$\begin{aligned}
\mu_{i\pm\frac{1}{2}} &= \begin{cases}\mu(P_{gi,j,k}), P_{gi,j,k} > P_{gi\pm1,j,k}\\ \mu(P_{gi\pm1,j,k}), P_{gi\pm1,j,k} > P_{gi,j,k}\end{cases}\\
k_{ri\pm\frac{1}{2}} &= \begin{cases}k_r(S_{g_{i,j,k}}), P_{gi,j,k} > P_{gi\pm1,j,k}\\ k_r(S_{g_{i\pm1}\,j,k}), P_{gi\pm1,j,k} > P_{gi,j,k}\end{cases}\\
Z_{i\pm\frac{1}{2}} &= \begin{cases}Z(P_{gi,j,k}), P_{gi,j,k} > P_{gi\pm1,j,k}\\ Z(P_{gi\pm1,j,k}), P_{gi\pm1,j,k} > P_{gi,j,k}\end{cases}
\end{aligned}\tag{11.73}$$

For gas phase pressure and gas phase saturation, the arithmetic mean is obtained directly, that is:

$$\begin{aligned}
S_{gi\pm\frac{1}{2}} &= \frac{(S_{gi,j,k} + S_{gi\pm1,j,k})}{2},\\
P_{gi\pm\frac{1}{2}} &= \frac{(P_{gi,j,k} + P_{gi\pm1,j,k})}{2}
\end{aligned}\tag{11.74}$$

For gas phase gravity, the arithmetic mean is taken, that is:

$$\begin{aligned}
\gamma_g(P_g)k \pm \frac{1}{2} &= \frac{1}{2}(\gamma_g(P_{gi,j,k\pm1}) + \gamma_g(P_{gi,j,k}))\\
&= \frac{1}{2}\left(\frac{P_g M_g}{ZRT_{i,j,k\pm1}} + \frac{P_g M_g}{ZRT_{i,j,k}}\right)
\end{aligned}\tag{11.75}$$

Discrete Form of Wellbore Pressure Drop Model

In the whole formation area, multi-branch horizontal wells are naturally divided into several small wellbore sections by difference grids. Therefore, for each grid with wellbore sections, the discrete form of the wellbore pressure drop model can be written as follows.

$$P_{wfi-1} = P_{wfi} + 0.5(\Delta P_{wfi-1} + \Delta P_{wfi})\tag{11.76}$$

If the outlet is the constant pressure boundary condition, then

$$P_{wfi} = P_{wfc} + 0.5\Delta P_{wfi}\tag{11.77}$$

If the outlet is the constant gas production boundary condition, then

$$q_{sum} = \sum q_i\tag{11.78}$$

If the grid has only branches or only primary branches passing through it, then

$$\Delta P_{wfi} = \frac{1}{10^6}\left[\frac{2f_i\rho_i}{\pi^2D^5}(2Q_i + q_i)^2\Delta x_i + \frac{16\rho_i q_i}{\pi^2D^4}(2Q_i + q_i)\right]\tag{11.79}$$

If the grid has branches or primary branches passing through it, then

$$\begin{aligned}
\Delta P_{wfi} = \frac{1}{10^6}&\left[\frac{2f_i\rho_i}{\pi^2D^5}(2Q_i + q_i + Q_{Ri})^2\Delta x_i\right.\\
&\left.+\frac{16\rho_i(q_i + Q_{Ri})}{\pi^2D^4}(2Q_i + q_i + Q_{Ri})\right]
\end{aligned}\tag{11.80}$$

In this case, the micro-sections are divided naturally by difference grids. The difference Eqs. (11.62)–(11.80) all adopt SI practical units, including k (μm^2), r_w (m), h (m), P_g (MPa), P_d (MPa), P_L (MPa), V_L (mPa s), μ_w (mPa•s),

$\Delta x(\text{m})$, $\Delta y(\text{m})$, $\Delta z(\text{m})$, $\Delta t(\text{h})$, D_{f} (m^2/d), T (k), M (kg/mol), $R\left(\frac{\text{MPa m}^3}{\text{mol K}}\right)$, C (kg/m^3), P_{wf} (MPa), q (m^3/d), γ_{g} (N/m^3).

Because the numerical model of wellbore pressure drop represents the pressure relationship between two adjacent sections upstream and downstream of a primary branch or a branch, in multi-branch horizontal wells, the positions of several wellbore sections divided by difference grids in the whole formation area must be determined first in the whole multi-branch horizontal well system, then the pressure relationship of the adjacent two wellbore sections can be determined. In this way, we need to number the wellbore sections of a primary branch and a branch separated naturally by difference grids. As shown in Fig. 11.55, the numbering principle is as follows. For the wellbore section located on a primary branch, it is necessary to determine the serial number of the primary branch to which the section belongs and which section of the primary branch this wellbore section is located on, that is, the serial number of the section, which consists of two elements. For the section located on a branch, it is necessary to determine the primary branch number where the branch is located, which side of the primary branch the branch is located on, the serial number of the branch, and which section of the branch this wellbore section is located on, i.e. the serial number of the wellbore section, which consists of four elements. The serial number of the primary branch can be given clockwise or counterclockwise at will. The relative position of the branch and the primary branch is according to the left or right side seen from the beginning of the primary branch (outlet). The number of the branch is determined according to the distance from the outlet. The nearer to the outlet, the smaller the number of the branch is. When numbering well sections, the upstream number is small and the downstream number is large. For example, the numbers of the two wellbore sections in black bold in Fig. 11.54 are primary branch [1] Sect. 11.2 and primary branch [1] left branch (1) Sect. 11.2, respectively.

The specific numbering steps are as follows.

① Define the physical coordinates of every difference grid.
② Known the number, arrangement and length of primary branches and branches, define the physical coordinates of the primary branches and branches and their ends.
③ According to the arrangement of the difference grid, determine whether each grid is passed by a primary branch or a branch.
The judgment method we adopt is to list the linear equations of each primary branch, each branch and the four edges of the grid at first, then determine the primary branch or branch that may pass through the grid according to the position of the beginning and ending points of the primary branch and branch (that is, the range of the beginning and ending points contains this grid), and then find the intersection point between the primary branch or branch and the four edges of the grid. The number in Fig. 11.55 is 4 intersection points. The Y coordinates of the four intersection points are arranged in the order of size, and the X and Y coordinates of the midpoint of the two intersection points in the second and third positions are solved. If the point is located in the grid, or the beginning and end of the primary branch and branch are located in the grid, it means that the primary branch and branch pass through the grid.
④ If the grid is passed by primary branch and branch, record which primary branch and branch they are, and then calculate the length of the well section in the grid.
⑤ Determine the number of the section on the primary branch and the branch. The method is as follows: firstly, without considering the upstream and downstream relationship, the well section in every primary branch or branch is separately numbered (and at the same time, the total number of well sections in the primary branch or branch is determined); secondly, according to the distance from the center of the section to the beginning of the primary branch (outlet) or branch (the position of the branch on the primary branch), the section is re-numbered. The farthest number is the smallest, and the nearest number is the largest.

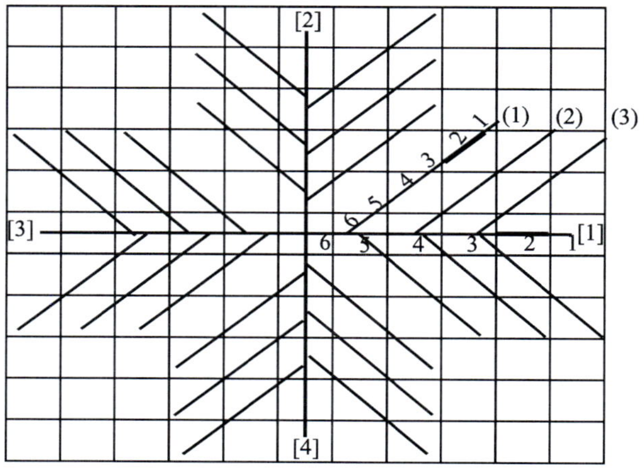

Fig. 11.54 Number directional pinnate horizontal well sections

After numbering, we can list the wellbore pressure equation for each grid containing the wellbore section. However, some grids may contain both primary branch sections and branch sections. In order to list only one wellbore pressure equation for each grid, and to ensure that the number of wellbore pressure equations is the same as the number of unknown wellbore pressure, for the kind of grid

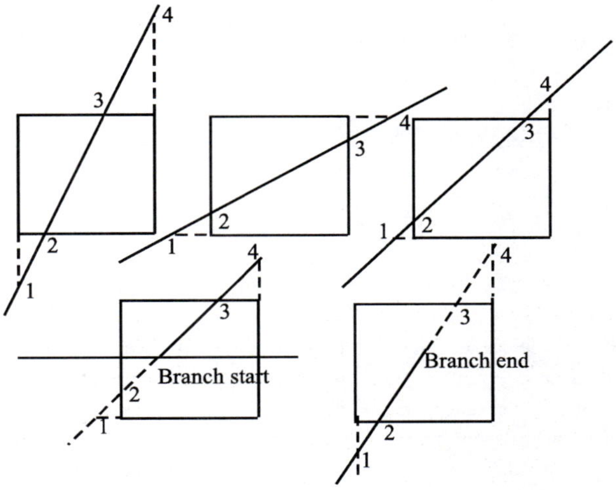

Fig. 11.55 Schematic diagram of intersection between well section and grid

containing more than two wellbore sections, only the equation for the primary branch need to be listed.

It should be pointed out that, since the unknowns in the gas–water two-phase flow equation are numbered by the difference grid system and the unknowns in the wellbore pressure drop equation are numbered by the primary branch, branch and section systems mentioned above, in order to couple the wellbore pressure drop equation with the gas–water two-phase flow equation, the unknowns (i.e. formation pressure, gas saturation, and wellbore pressure of each section) in the two groups of equations must be converted to the same numbering system. With the above method of numbering well section, we first scan the difference grid line by line and row by row. If there is a section passing through the grid, then we number the section according to the above method, and then assign the values in the corresponding difference grid to the section.

In this way, the wellbore pressure drop equation is also expressed by the values of formation pressure, wellbore section pressure and saturation in each grid. After solving the wellbore pressure drop equation, the wellbore pressure in each difference grid should be re-assigned by the pressure of each section.

So far, the differential discretization of the whole mathematical model has been completed, and a complete numerical model has been established.

Solution of Numerical Model

In this section, the solution of the numerical model is discussed firstly, and then the linearization process of the wellbore pressure drop equation is described in details.

(1) Solving method

In the last section, a numerical model in the form of difference is given. The numerical model is three groups of a non-linear equation system with formation pressure, wellbore pressure and gas saturation as unknown variables. The number of unknowns is the same as the number of equations, and the three groups of equations are coupled with each other. For this system of non-linear equations, we will solve it by transforming Newton–Raphson method into a system of linear equations.

The saturation of the left side of Eqs. (11.70) and (11.71) is taken as explicit, while formation pressure and wellbore pressure are taken as implicit. After merging, the unknown variable of saturation is eliminated and moved to the right side, and make it 0, that is,

$$\begin{aligned} f(&P_{gi+1,j,k}, P_{gi-1,j,k}, P_{gi,j+1,k}, P_{gi,j-1,k},\\ &P_{gi,j,k+1}, P_{gi,j,k-1}, P_{gi,j,k}, P_{wfi,j,k}) = 0 \end{aligned} \tag{11.81}$$

Equation (11.81) is written in a simpler form and expressed as a function of formation pressure and wellbore pressure.

$$f(P_{wf}, P_g) = 0 \tag{11.82}$$

Considering the wellbore pressure Eqs. (11.81) and (11.80), move it to the other side and make it 0, that is,

$$g(P_{wf}, P_g) = 0 \tag{11.83}$$

In order to understand the above two sets of non-linear equations, in Eq. (11.81), it is assumed that the wellbore pressure is known, i.e. the outlet pressure, and the fixed production condition can be set at any reasonable wellbore pressure, the Eq. (11.81) becomes a system of equations about formation pressure, and it is linearized as follows.

$$\begin{aligned} &\left(\frac{\partial f}{\partial P_{gi+1,j,k}}\right)^k \delta P^{k+1}_{gi+1,j,k} + \left(\frac{\partial f}{\partial P_{gi-1,j,k}}\right)^k \delta P^{k+1}_{gi-1,j,k}\\ &+ \left(\frac{\partial f}{\partial P_{gi,j+1,k}}\right)^k \delta P^{k+1}_{gi,j+1,k} + \left(\frac{\partial f}{\partial P_{gi,j-1,k}}\right)^k \delta P^{k+1}_{gi,j-1,k}\\ &+ \left(\frac{\partial f}{\partial P_{gi,j,k+1}}\right)^k \delta P^{k+1}_{gi,j,k+1} + \left(\frac{\partial f}{\partial P_{gi,j,k-1}}\right)^k \delta P^{k+1}_{gi,j,k-1}\\ &+ \left(\frac{\partial f}{\partial P_{gi,j,k}}\right)^k \delta P^{k+1}_{gi,j,k} = -f^k \end{aligned} \tag{11.84}$$

The formed seven-diagonal equations system is solved by conjugate gradient method treated with preconditions. First $\delta P^{k+1}_{gi,j,k}$, then $P^{k+1}_{gi,j,k} = P^k_{gi,j,k} + \delta P^{k+1}_{gi,j,k}$ and finally $P^{n+1}_{gi,j,k} = P^{k+1}_{gi,j,k}$ until $\left|\delta P^{k+1}_{gi,j,k}\right| < \varepsilon_P$ and $\left|f^k\right| < \varepsilon_f$.

Then the formation pressure is substituted into the wellbore pressure equation system, and solved by the Gauss elimination method after linearization, the same iteration is performed to satisfy the linearization convergence. The obtained wellbore pressure is then substituted into the formation pressure equation, and iterated until the calculated unknown number satisfies certain convergence conditions, $\left|P^{k+1}_{gi,j,k} - P^{k}_{gi,j,k}\right| < \varepsilon_{P_g}$ and $\left|P^{k+1}_{wfi,j,k} - \right|P^{k}_{wfi,j,k} < \varepsilon_{P_{wf}}$, and then the formation pressure and wellbore pressure of each section at this time can be obtained.

The above two sets of equations can be regarded as a non-linear system consisting of two unknown vectors. The unknown vectors are wellbore pressure and formation pressure, each of which has multiple unknown components. In fact, the solving process is similar to the Seidel iteration method for solving linear equations. Only here, the new values of the two vectors are obtained by solving a non-linear equation system (The system of non-linear equations is transformed into a system of linear equations by Newton Raphson's method). The convergence conditions of Newton iteration for solving two systems of nonlinear equations are that the increments and residuals approach a smaller value.

The formation pressure and wellbore pressure calculated are substituted into the water phase difference equations, and

$$h(S_{gi+1,j,k}, S_{gi-1,j,k}, S_{gi,j+1,k}, S_{gi,j-1,k}, S_{gi,j,k+1}, S_{gi,j,k-1}, S_{gi,j,k}) = 0 \tag{11.85}$$

Since the formation pressure and wellbore pressure have been solved by the above-mentioned pressure equation and wellbore pressure equation, and now they are known, the saturation is an unknown variable. Take the saturation on the left side of the equation as implicit, and calculate it by Newton Raphson's method. After linearized, Eq. (11.85) is

$$\begin{aligned}&\left(\frac{\partial h}{\partial s_{gi+1,j,k}}\right)^k \delta S^{k+1}_{gi+1,j,k} + \left(\frac{\partial h}{\partial s_{gi-1,j,k}}\right)^k \delta S^{k+1}_{gi-1,j,k}\\ &+ \left(\frac{\partial h}{\partial s_{gi,j+1,k}}\right)^k \delta S^{k+1}_{gi,j+1,k} + \left(\frac{\partial h}{\partial s_{gi,j-1,k}}\right)^k \delta S^{k+1}_{gi,j-1,k}\\ &+ \left(\frac{\partial h}{\partial s_{gi,j,k+1}}\right)^k \delta S^{k+1}_{gi,j,k+1} + \left(\frac{\partial h}{\partial s_{gi,j,k-1}}\right)^k \delta S^{k+1}_{gi,j,k-1}\\ &+ \left(\frac{\partial h}{\partial s_{gi,j,k}}\right)^k \delta S^{k+1}_{gi,j,k} = -h^k\end{aligned} \tag{11.86}$$

Since Eq. (11.86) is also a system of seven-diagonal equations, it is still solved by conjugate gradient method. First $\delta S^{k+1}_{gi,j,k}$, then $S^{k+1}_{gi,j,k} = S^{k}_{gi,j,k} + \delta S^{k+1}_{gi,j,k}$, and finally $S^{n+1}_{gi,j,k} = S^{k+1}_{gi,j,k}$ until $\left|\delta S^{k+1}_{gi,j,k}\right| < \varepsilon_s$ and $\left|h^k\right| < \varepsilon_h$.

According to the practice of reservoir numerical simulation, the whole process is equivalent to alternative solutions of implicit pressure and implicit saturation.

After calculating the pressure at each point, wellbore pressure and gas saturation of each section at every moment, gas and water production can be obtained with the following two formulas.

$$q_w = \sum \mathrm{PID}_{i,j,k} 86.4 \frac{k_{rw}(S_g)(P_{gi,j,k} - P_{wfi,j,k})}{\mu_w B_w} \tag{11.87}$$

$$q_g = \sum \mathrm{PID}_{i,j,k} 86.4 \frac{k_{rg}(S_g)(P_{gi,j,k} - P_{wfi,j,k})}{\mu_w} \cdot \frac{P_{wfi,j,k} M}{Z(P_{wfi,j,k}) R T \rho_{sc}} \tag{11.88}$$

(2) Linearization of wellbore pressure drop equations

Since the equation of pressure drop in wellbore is a non-linear equation system about formation pressure and wellbore pressure, Eq. (11.83) can be expressed as the function of formation pressure and wellbore pressure on the relevant difference grid. After calculating the formation pressure, the Eq. (11.83) is only the function of wellbore pressure. After linearized by Newton Raphson method, it is

$$\sum \left(\frac{\partial g}{\partial P_{wf}}\right)^k \Delta P^{k+1}_{wfi} = -g^k \tag{11.89}$$

Next, we discuss the expressions of the coefficients of the linearized equations in three cases.

① The section is on a branch and counted the ith section from the end of the branch, then the derivative of the kth section is,

when $k = 1, 2, \ldots, i-1$, (Subscript h denotes the kth section.)

$$\begin{aligned}\frac{\partial g_i}{\partial P_{wfk}} = &-\frac{0.5}{10^6}\left[\frac{2 f_i \rho_i \Delta x_i}{\pi^2 D^5} \cdot 2(2Q_i + q_i) + \frac{16 \rho_i q_i}{\pi^2 D^4}\right.\\ &+ \frac{2 f_{i+1} \rho_{i+1} \Delta x_{i+1}}{\pi^2 D^5} \cdot 2(2Q_{i+1} + q_{i+1})\\ &\left.+ \frac{16 \rho_{i+1} q_{i+1}}{\pi^2 D^4}\right] \cdot 2 \frac{\partial q_k}{\partial P_{wfk}}\end{aligned} \tag{11.90}$$

When $k = i$,

$$\begin{aligned}\frac{\partial g_i}{\partial P_{wfk}} = 1 &- \frac{0.5}{10^6}\left[\frac{2 f_i \rho_i \Delta x_i}{\pi^2 D^5} \cdot 2(2Q_i + q_i) \cdot \frac{\partial q_k}{\partial P_{wfk}}\right.\\ &+ \frac{16 \rho_i (2Q_i + 2q_i)}{\pi^2 D^4} \cdot \frac{\partial q_k}{\partial P_{wfk}}\\ &+ \frac{2 f_i \frac{\partial \rho_i}{\partial P_{wfk}} + 2 \rho_i \frac{\partial f_i}{\partial P_{wfk}}}{\pi^2 D^5} (2Q_i + q_i)^2 \Delta x_i\\ &+ \frac{16 \frac{\partial \rho_i}{\partial P_{wfk}}}{\pi^2 D^4} q_i (2Q_i + q_i)\\ &+ \frac{2 f_{i+1} \rho_{i+1} \Delta x_{i+1}}{\pi^2 D^5} \cdot 2(2Q_{i+1} + q_{i+1})\\ &\left.\cdot 2 \frac{\partial q_k}{\partial P_{wfk}} + \frac{16 \rho_{i+1} \cdot 2 q_{i+1}}{\pi^2 D^4} \cdot \frac{\partial q_k}{\partial P_{wfk}}\right]\end{aligned} \tag{11.91}$$

When $k = i + 1$,

$$\frac{\partial g_i}{\partial P_{\text{wf}k}} = -1 - \frac{0.5}{10^6}\left[\frac{2f_{i+1}\rho_{i+1}\Delta x_{i+1}}{\pi^2 D^5}\cdot 2(2Q_{i+1}+q_{i+1})\cdot\frac{\partial q_k}{\partial P_{\text{wf}k}} + \frac{16\rho_{i+1}(2Q_{i+1}+q_{i+1})}{\pi^2 D^4}\cdot\frac{\partial q_k}{\partial P_{\text{wf}k}} + \frac{2f_{i+1}\frac{\partial \rho_{i+1}}{\partial P_{\text{wf}k}} + 2\rho_{i+1}\frac{\partial f_{i+1}}{\partial P_{\text{wf}k}}}{\pi^2 D^5}(2Q_{i+1}+q_{i+1})^2\Delta x_{i+1} + \frac{16\frac{\partial \rho_{i+1}}{\partial P_{\text{wf}k}}q_{i+1}}{\pi^2 D^4}(2Q_{i+1}+q_{i+1})\right] \tag{11.92}$$

As for $\rho_i = \frac{\gamma_g/(g\cdot q_{gi}) + \rho_w q_{wi}}{q_{gi}+q_{wi}}$,

$$\frac{\partial \rho}{\partial P_{\text{wf}i}} = \frac{\left[\left[\left(\frac{\partial \gamma_g}{\partial P_{\text{wf}i}}\right)\Big/(g\cdot q_{gi}) + \gamma_g\Big/\left(g\cdot\frac{\partial q_{gi}}{\partial P_{\text{wf}i}}\right) + \rho_w\frac{\partial q_{wi}}{\partial P_{\text{wf}i}}\right](q_{gi}+q_{wi}) - [\gamma_g/(g\cdot q_{gi}) + \rho_w q_{wi}]\cdot\left(\frac{\partial q_{gi}}{\partial P_{\text{wf}i}} + \frac{\partial q_{wi}}{\partial P_{\text{wf}i}}\right)\right]}{(q_{gi}+q_{wi})^2} \tag{11.93}$$

It is approximated that the derivative of friction coefficient to bottom hole pressure is zero, that is

$$\frac{\partial f}{\partial P_{\text{wf}i}} \approx 0 \tag{11.94}$$

when $k = i + 2, \ldots, N$,

$$\frac{\partial g_i}{\partial P_{\text{wf}k}} \approx 0 \tag{11.95}$$

② The well section i is on a primary branch and no branch in the grid.

If $i = N$, that is to say, the section i is located in the grid at the outlet.

$$\frac{\partial g_i}{\partial P_{\text{wf}k}} = -\frac{0.5}{10^6}\left[\frac{2f_i\rho_i\Delta x_i}{\pi^2 D^5}\cdot 2(2Q_i+q_i) + \frac{16\rho_i q_i}{\pi^2 D^4}\right].2\frac{\partial q_k}{\partial P_{\text{wf}k}} \tag{11.96}$$

When $k = i$,

$$\frac{\partial g_i}{\partial P_{\text{wf}k}} = 1 - \frac{0.5}{10^6}\left[\frac{2f_i\rho_i\Delta x_i}{\pi^2 D^5}\cdot 2(2Q_i+q_i)\cdot\frac{\partial q_k}{\partial P_{\text{wf}k}} + \frac{16\rho_i(2Q_i+2q_i)}{\pi^2 D^4}\cdot\frac{\partial q_k}{\partial P_{\text{wf}k}} + \frac{2f_i\frac{\partial \rho_i}{\partial P_{\text{wf}k}} + 2\rho_i\frac{\partial f_i}{\partial P_{\text{wf}k}}}{\pi^2 D^5}(2Q_i+q_i)^2\Delta x_i + \frac{16\frac{\partial \rho_i}{\partial P_{\text{wf}k}}}{\pi^2 D^4}q_i(2Q_i+q_i)\right] \tag{11.97}$$

When $k = i + 1, \ldots, N$,

$$\frac{\partial g_i}{\partial P_{\text{wf}k}} = 0 \tag{11.98}$$

For all sections kf on a primary branch when $k < i$ (upstream), there is

$$\frac{\partial g_i}{\partial P_{\text{wf}k}} = -\frac{0.5}{10^6}\left[\frac{2f_i\rho_i\Delta x_i}{\pi^2 D^5}\cdot 2(2Q_i+q_i) + \frac{16\rho_i q_i}{\pi^2 D^4}\right]\cdot 2\frac{\partial q_{kf}}{\partial P_{\text{wf}kf}} \tag{11.99}$$

If $i \neq N$, when $k = 1, 2, \ldots, N$, $\frac{\partial g_i}{\partial P_{\text{wf}k}}$ is the same as that of the first case, like Eqs. (11.90)–(11.95), and the subscripts k, i, and N represent the sections on a primary branch.

For all branch sections kf on a primary branch when $k < i$ (upstream), there is.

$$\frac{\partial g_i}{\partial P_{\text{wf}k}} = -\frac{0.5}{10^6}\left[\frac{2f_i\rho_i\Delta x_i}{\pi^2 D^5}\cdot 2(2Q_i+q_i) + \frac{16\rho_i q_i}{\pi^2 D^4} + \frac{2f_{i+1}\rho_{i+1}\Delta x_{i+1}}{\pi^2 D^5}\cdot 2(2Q_{i+1}+q_{i+1}) + \frac{16\rho_{i+1}q_{i+1}}{\pi^2 D^4}\right]\cdot 2\frac{\partial q_{kf}}{\partial P_{\text{wf}kf}} \tag{11.100}$$

③ The well section i is on a primary branch and there are branches in the grid.

If $i = N$, that is to say, the section i is located in the grid at the outlet.

For sections on a primary branch, $k = i + 2, \ldots, N$, $\frac{\partial q_i}{\partial P_{\text{wf}k}}$ is similar to the second case, and only the q_i in Eqs. (11.96)–(11.98) changed to $q_i + Q_{\text{R}_i}$.

For all branch sections kf on a primary branch when $k < i$ (upstream), there is.

$$\frac{\partial g_i}{\partial P_{\text{wf}kf}} = -\frac{0.5}{10^6}\left[\frac{2f_i\rho_i\Delta x_i}{\pi^2 D^5}\cdot 2(2Q_i+q_i+Q_{\text{R}i}) + \frac{16\rho_i(q_i+Q_{\text{R}i})}{\pi^2 D^4}\right]\cdot 2\frac{\partial q_{kf}}{\partial P_{\text{wf}kf}} \tag{11.101}$$

If $i \neq N$, for all branch sections on a primary branch, $k = 1, 2, \ldots, N$, $\frac{\partial g_i}{\partial P_{\text{wf}k}}$ is similar to the first case, and only the q_i changed to $q_i + Q_{\text{R}_i}$, and the subscripts k, i, and N represent the sections on the primary branch.

For all branch sections kf on a primary branch when $k < i$ (upstream), there is.

$$\frac{\partial g_i}{\partial P_{\text{wf}kf}} = -\frac{0.5}{10^6}\left[\frac{2f_i\rho_i\Delta x_i}{\pi^2 D^5}\cdot 2(2Q_i+q_i+Q_{\text{R}i})\frac{\partial q_{kf}}{\partial P_{\text{wf}kf}} + \frac{16\rho_i\cdot(2Q_i+2q_i+2Q_{\text{R}i})}{\pi^2 D^4}\frac{\partial q_{kf}}{\partial P_{\text{wf}kf}} + \frac{2f_{i+1}\rho_{i+1}\Delta x_{i+1}}{\pi^2 D^5}\cdot 2(2Q_{i+1}+q_{i+1})\cdot 2\frac{\partial q_{kf}}{\partial P_{\text{wf}kf}} + \frac{16\rho_{i+1}\cdot q_{i+1}}{\pi^2 D^4}\cdot 2\frac{\partial q_{kf}}{\partial P_{\text{wf}kf}}\right] \tag{11.102}$$

Ideas of Programming Design

Based on the solving method of the above numerical model, a numerical simulation program for CBM production with multi-branch horizontal wells was developed on the Compaq visual FORTRAN 6.5 platform. The program is based on FORTRAN 95 language, modular design and global variables, global arrays and dynamic arrays. It saves computer resources, and is easy to use and maintain.

For data entry, original data are input by data files independent of the source program. By inputting and adjusting formation parameters, geological models can be established, and numerical simulation study on the CBM production of multi-branch horizontal wells can be carried out under different geological conditions. By adjusting the parameters of the casing program, the casing program can be optimized and a reasonable development scheme can be provided. By adjusting the control parameters used in the calculation, the calculation program itself can be further optimized and improved, which is conducive to the maintenance of the program.

The original data files mainly include the following categories. The data files of formation parameters mainly include fracture porosity, *x*, *y* and *z* absolute permeability (three vertical directions), water volume factor, fracture and pore volume compressibility, matrix skeleton shrinkage coefficient, formation temperature, formation original pressure, bottom hole pressure (fixed), gas production (fixed), gas diffusion coefficient, gas desorption time constant, critical desorption pressure, Langmuir pressure constant, Langmuir volume constant, wellbore radius, outer boundary radius, matrix geometric factor, CBM molar mass, density of CBM at standard state, density of coal seam water at standard state, the length and width of a coal seam area and the thickness of the coal seam.

The data file of the casing program mainly includes relative roughness of wellbore, outlet coordinates, the number, azimuth and length of primary branches, the number of branches, the root location of every branch on the primary branch (branch spacing, the location of the branch the nearest to the outlet), the length of the branch and the angle between branch and primary branch.

The data file used in simulation mainly include start time, simulation period, time step, convergence error of iteration and the maximum iterations.

For data output, in addition to CBM production, specific saturation and formation pressure will be output, which can be used to understand the conditions of CBM production and production mechanism.

Because both initial and iterated wellbore pressures are expressed by the differential grid system, the formation pressure, wellbore pressure and saturation expressed by the primary branch, branch and section system should first be assigned to the values expressed by the grid variables when solving the linearized coefficient of wellbore pressure.

11.2.2 Stimulation Mechanism of Multi-branch Horizontal CBM Wells

We utilize the numerical simulation software developed by ourselves to predict the productivity of multi-branch horizontal CBM wells, and verify the reasonableness and reliability of the numerical simulation software by comparing with the results of similar simulation software abroad. By comparing with vertical wells, this paper illustrates the stimulation effect of multi-branch horizontal wells, further reveals the stimulation mechanism of multi-branch horizontal wells of CBM production, and carries out the analysis of the influencing factors on the production of multi-branch horizontal wells.

11.2.2.1 Analysis of Simulation Mechanism

In order to analyze the stimulation mechanism of multi-branch horizontal CBM wells, we utilize the same formation parameters to simulate and compare the vertical wells and multi-branch horizontal wells.

Because the permeability is isotropic in the vertical well model, here the absolute permeability is set to 0.3 mD, the radius of the outer boundary of the simulated area is 1551.9 m, and the thickness of the coal seam is 8 m. In this way, for any of the two types of wells, the control area is 7.56 km^2 and the control volume is 6088×10^4 m^3. According to the Langmuir constant and critical desorption pressure, the gas content of the coal seam is 31.1 m^3/m^3 (22.08 kg/m^3), and the total gas content of the well area is 189336.8×10^4 m^3.

Figure 11.56 is a comparison result of gas production between a vertical well and a multi-branch horizontal well group under the same formation conditions. It can be seen that the daily gas production of the horizontal well group can reach tens of thousands of m^3, while the production of a vertical well in the center of the well area is only several m^3 or even lower (uneconomically) at the same formation conditions and working system, without any stimulation measures.

When CBM is developed by multi-branch horizontal wells, the gas and water productions tend to change similarly to vertical wells, that is, a large amount of water is produced at the beginning of production, and gas production begins when the formation pressure falls below the critical desorption pressure (16 days in our case). Gas production increases first and then decreases, while water production decreases monotonously, but the production is much higher than that of vertical wells. Figure 11.57 shows the profiles of formation pressure and gas saturation around a vertical CBM

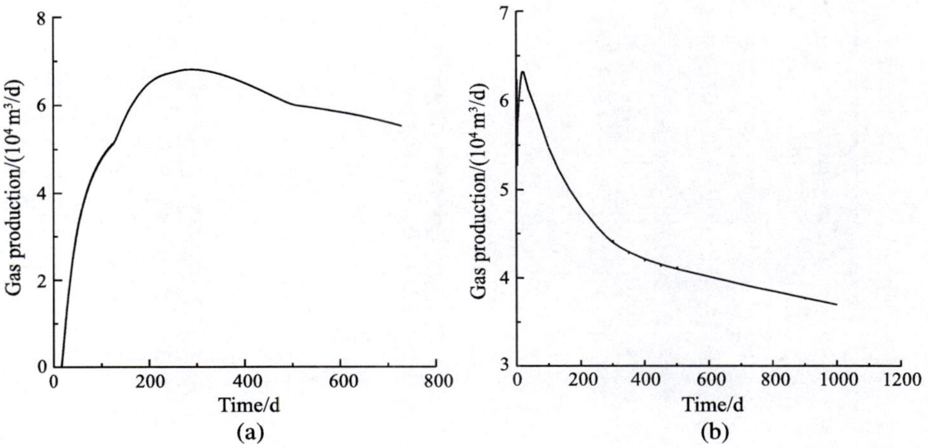

Fig. 11.56 Gas production of a vertical well (**b**) and a multi-branch horizontal well group (**a**)

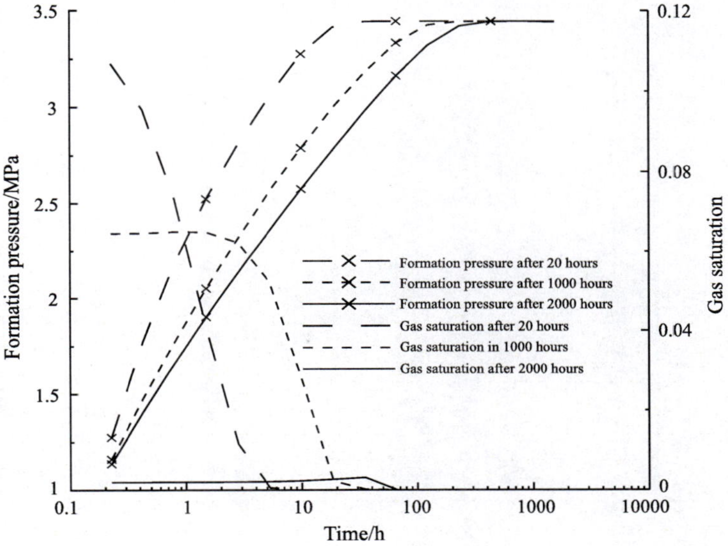

Fig. 11.57 Formation pressure and saturation in a vertical CBM well

well. The variation of saturation is the result of gas desorption (increasing saturation), diffusion (evenly distributed saturation) and production (decreasing saturation). As can be seen from Fig. 11.57, in early production, the pressure near the well decreases rapidly and the gas saturation rises rapidly, while in the area far from the well, the gas in the matrix is not fully desorbed due to small pressure drop, and the gas saturation rises slowly. With more gas recovered and diffusion, the pressure drop funnel gradually expands, but the pressure drop tends to be gentle, and the gas saturation decreases near the vertical well; on the contrary, the pressure drop in the area far from the vertical well becomes larger, and the gas saturation gradually rises from zero; and finally, the gas saturation in the whole well area increases in waves. In this production mode, the formation pressure is always distributed in the way of a pressure drop funnel. Without effective pressure drop, the gas in the matrix can not be effectively desorbed, and the production potential can not be fully activated, resulting in low production.

Figures 11.58 and 11.59 show the isoline maps of formation gas saturation and formation pressure during two years of production. It can be seen from Fig. 11.59 that the formation pressure drop was gradually diffused outward from the multi-branch horizontal well. From the plane point of view, the horizontal wellbore with the whole length of 16800 m can be regarded as the "source" of formation pressure drop. In contrast, the vertical well has only one point as the source of pressure drop. Obviously, the pressure drop in the multi-branch horizontal well is more effective, and there isn't a so-called pressure drop funnel which appears in a vertical well, so almost the whole coal seam is producing, and the production potential is fully developed.

From the above analysis, it can be seen that the branches of a multi-branch horizontal well extensively and evenly extend

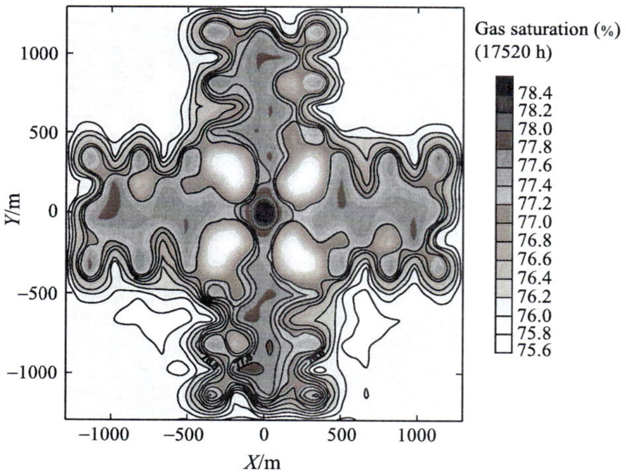

Fig. 11.58 Gas saturation isolines in a pinnate horizontal CBM well

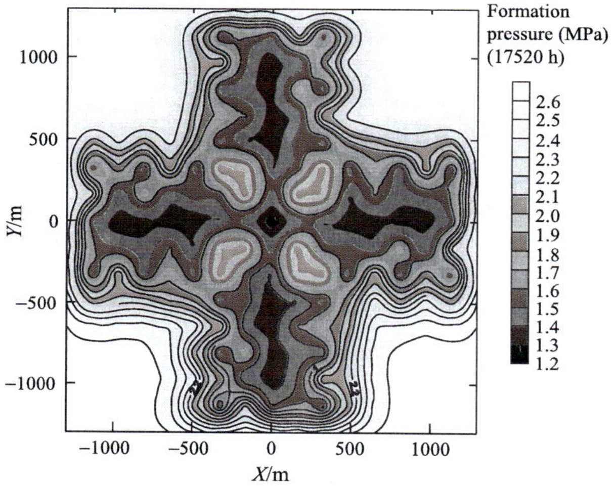

Fig. 11.59 Formation pressure isolines in a pinnate horizontal CBM well

in formation, which makes formation pressure uniformly and rapidly decline, increases the chance of gas desorption and diffusion, and greatly increases the producing area. This is the fundamental factor on promoting the increase of CBM production of a multi-branch horizontal well.

11.2.2.2 Influencing Production Factors

CBM production is a complex process related to desorption, diffusion and seepage mechanisms, and the change characteristics of productivity are affected by many parameters. In a particular coal seam, all parameters is interrelated, and different geological conditions create different combination of coal seam parameters. These coal seam parameters can be classified into four categories, including gas abundance index, production value index, gas production performance index and coal thermal evolution index. Generally speaking, the lower the degree of thermal evolution and the rank of coal, the more advantageous the physical characteristics of coal seams. Table 11.13 lists the classification of some parameters related to the simulation of development indexes.

In order to better understand the impact of various factors on the production of multi-branch horizontal wells, the sensitivity of each parameter is analyzed separately below.

Our software can quantitatively analyze the sensitivity of each parameter. Sensitivity analysis of parameters can get the general law of causality between coal seam properties and gas production characteristics, further evaluate the production value of coal seams, and then guide production practice.

Next the influences of some parameters will be studied, including absolute permeability, relative permeability, Langmuir constant, adsorption time constant and gas saturation (determined by formation pressure and critical desorption pressure on the same isothermal adsorption line), on the CBM production of multi-branch horizontal wells.

The parameters used in the following analysis and calculation are changed in turn when studying the influence of a factor.

(1) The influence of relative permeability

Figure 11.60 shows two typical relative permeability curves with different characteristics. Figure 11.61 shows the different responses of these two relative permeability curves. As can be seen from Fig. 11.61, the steeper the relative permeability curve, the smaller the residual gas saturation, and the more right the intersection point of the relative permeability curve is, the more advantageous it is to increase gas production. The reasons are as follows. The production of CBM is the result of gas diffusion and seepage in the coal seam. When the formation pressure drops to the critical desorption pressure, gas is desorbed and becomes free. However, only when the gas saturation is greater than the residual gas saturation can gas participate in seepage and be produced in a large quantity, otherwise gas just diffuses. The steeper the relative permeability curve is, the faster the relative permeability of gas phase rises and the larger the seepage flow rate is.

(2) The influence of adsorption time constant

Adsorption time constant mainly affects the time when gas production reaches its peak. The shorter the adsorption time, the earlier the peak period reaches, but the faster the production decreases after the peak. However, as shown in Fig. 11.62, for a multi-branch horizontal well, the effect of adsorption time constant on production is not as obvious as that of a vertical well.

Table 11.13 Classification of important parameters of CBM target evaluation

Gas abundance			Production value						Production performance		Evolution	
Reserves billion/m^3	Coal seam thickness/m		Gas content/(m^3/t)		Gas saturation/%		Formation pressure and desorption pressure/(P_d/P_i)		Primary permeability/mD		Coal rank	
Largest >3000	Thickest	>50	High	>20	High	>80	High	>0.6	High	>5	Lignite, long flame coal	Low
Larger 3000–300	Thicker	10–50	Higher	15–20	Large	60–80	Large	0.2–0.6	Higher	0.5–5	Gas coal, fat coal	↓
Large 30–300	thick	5–10	High	8–15	Low	<60	Low	<0.2	High	0.1–0.5	Coking coal, lean coal	↓
Small 3–30	Thin	<5	低	<8	–	–	–	–	Low	<0.1	Lean coal, anthracite	High

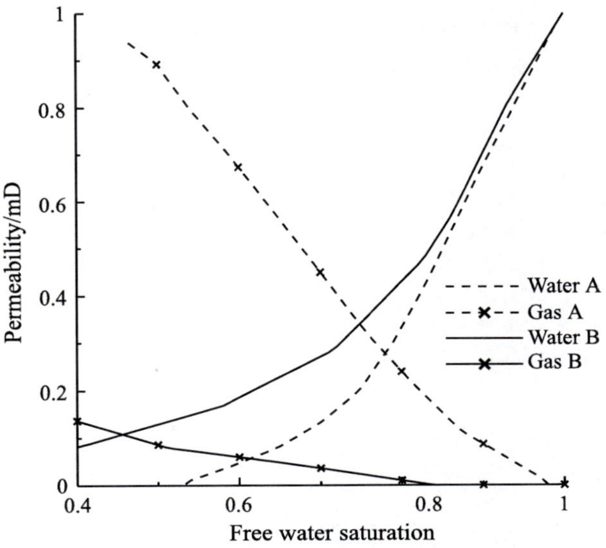

Fig. 11.60 Gas–water two-phase relative permeability curve

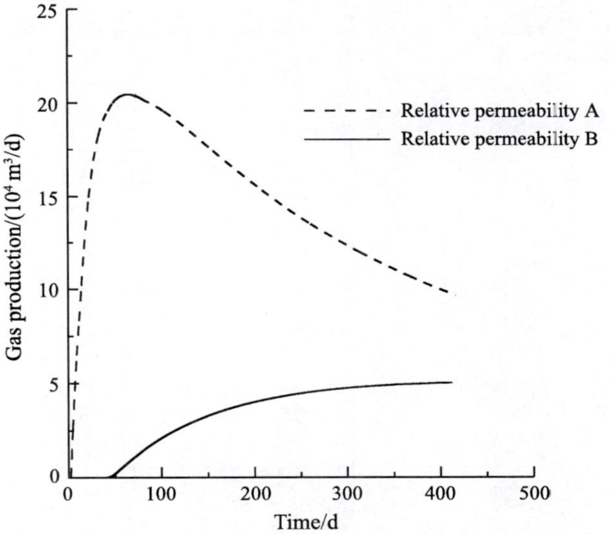

Fig. 11.61 The effect of relative permeability curve on CBM production

On the other hand, adsorption time constant mainly affects the properties of gas sources. For a vertical well, a long adsorption time constant makes desorbed gas unable to meet the need of increasing production, so the production increases relatively slowly. However, for a multi-branch horizontal well, the uniform decrease of formation pressure makes gas desorption easier and alleviates the contradiction between supply and demand. Therefore, a multi-branch horizontal well is actually eliminating the adverse effects of a long adsorption time on gas production.

Adsorption time constant is expressed as $\tau = S^2/(8\pi D_m)$, which indicates that adsorption time constant is related to diffusion distance and is proportional to the square of fracture spacing. In a sense, a multi-branch horizontal well is just like creating many large artificial fractures in coal seams. These "fractures" densify natural fractures, greatly reduce the fracture spacing, and thus reduce the equivalent adsorption time constant. Because the influence of these "fractures" on the adsorption time is much greater than that of the natural fractures, the effect of the adsorption time constant itself on the production of the multi-branch horizontal well is almost concealed, which makes the difference of several curves in Fig. 11.62 very small.

(3) The influence of Langmuir constant

As ever pointed out for a vertical well, Langmuir constant determines the change of adsorption capacity by changing the steep degree of the isothermal adsorption line in the range of formation pressure drop, thus affecting the change

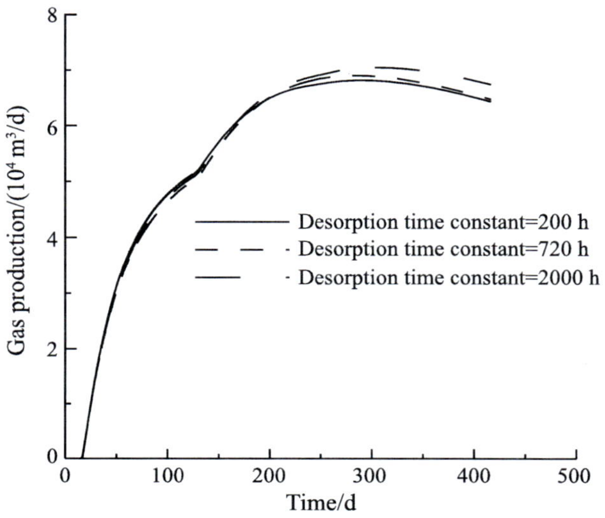

Fig. 11.62 The influence of adsorption time constant on CBM production

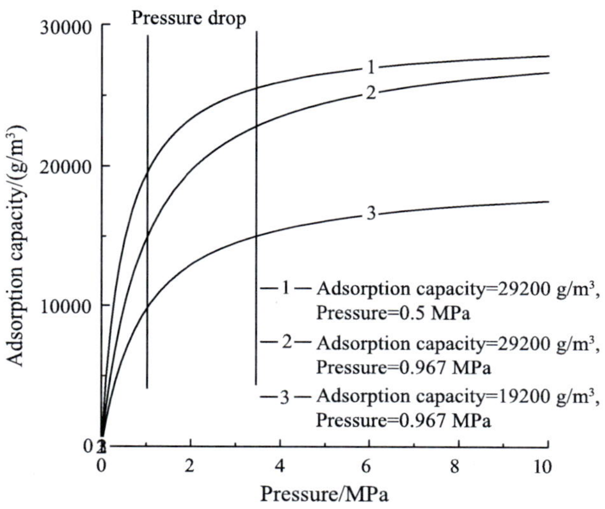

Fig. 11.63 Isothermal adsorption lines at different Langmuir constants

of gas production. Figure 11.64 shows the same effect of Langmuir constant on the production of a multi-branch horizontal well. From the three isotherm adsorption lines in Fig. 11.63, in the range of pressure drop, the change of line 2 is the steepest, line 1 is steeper, and line 3 is gentle, therefore, in the corresponding production curves, line 2 is the highest, line 1 is higher, and line 3 is low.

(4) The influence of absolute permeability

In the areas with drilled CBM wells in China, the permeability is above 1 mD only in a few areas, but in the most less than 1 mD. According to Table 11.13, we selected two

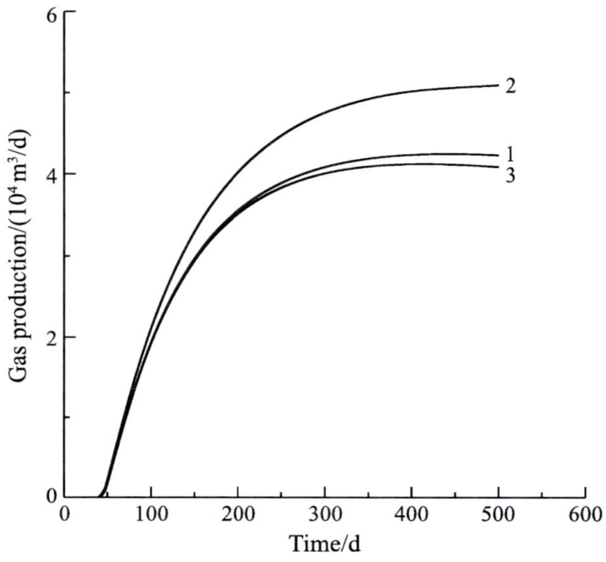

Fig. 11.64 The influence of Langmuir constant on CBM production

groups: medium and high absolute permeability. Two groups of beddings perpendicular to each other are developed in the coal seam, and sedimentation and compaction resulted in the vertical permeability smaller than the horizontal permeability. From the two groups of permeability in Fig. 11.65, the increase of permeability leads to a significant increase in gas production. It shows that in the two-phase flow of CBM and water dominated by seepage, coal seam permeability is one of the important factors affecting CBM

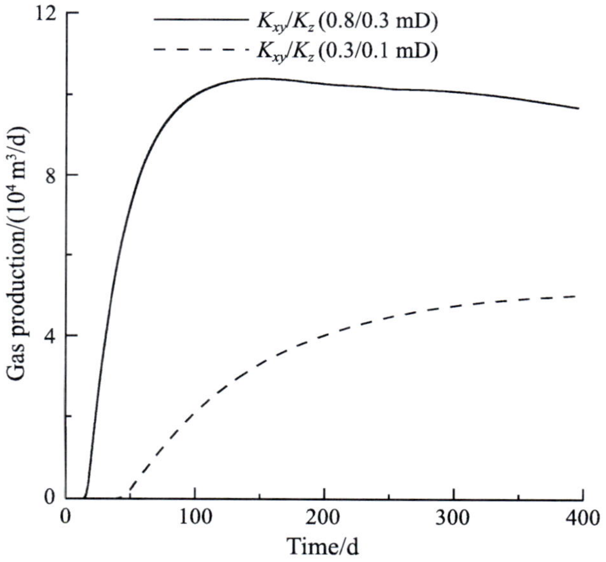

Fig. 11.65 The influence of absolute permeability on CBM production

production, which is consistent with the characteristics of general oil and gas reservoir. In actual production, in order to increase productivity, an effective way is to increase absolute permeability by fracturing stimulation. The primary branches and branches of a multi-branch horizontal well are just like many artificial fractures, and equivalent to increasing the effective permeability, thus greatly increasing the productivity.

(5) The influence of adsorbed gas saturation

From Table 11.13, it can be seen that high formation pressure to critical desorption pressure ratio and high adsorbed gas saturation are important parameters to measure whether a coal seam has a great production value. Under the same original formation pressure, different critical desorption pressures lead to different adsorbed gas saturation, which actually reflects the difference of gas content in coal seams.

On Fig. 11.66, the original formation pressure is 3.445 MPa on both curves, but the critical desorption pressure is 3.345 MPa and 3 MPa respectively. At the given Langmuir constant, the adsorbed gas saturation of the two curves is calculated to be 99.5% and 96.9% respectively. It is obvious from Fig. 11.66 that the higher the adsorbed gas saturation, the higher the production of CBM.

(6) The influence of formation pressure

If adsorbed gas saturation reflects CBM content, that is, whether the source of CBM is sufficient or not, formation pressure affects the difficulty of CBM production from the perspective of energy. With the same saturation and content

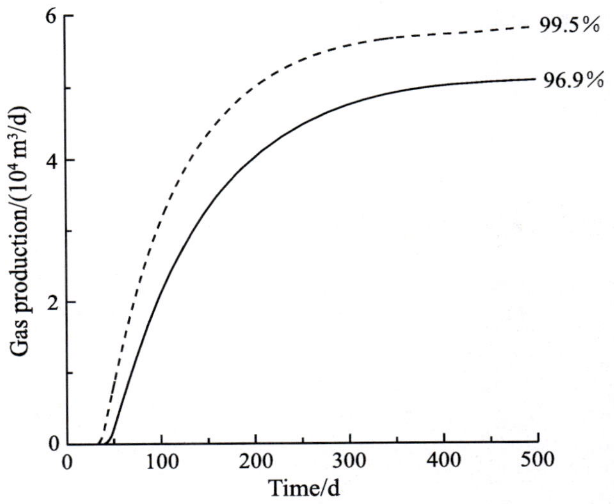

Fig. 11.66 The influence of adsorbed gas saturation on CBM production

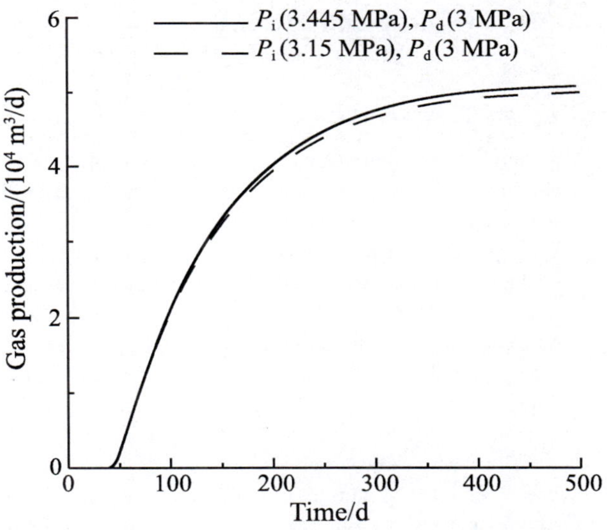

Fig. 11.67 The influence of formation pressure on CBM production

of adsorbed gas, high formation pressure obviously makes it easier to exploit CBM. The Langmuir constants and critical desorption pressures of the two curves in Fig. 11.67 are the same, both of which are 3 MPa, i.e. the original gas content is the same, but the original formation pressure is 3.15 MPa and 3.445 MPa, respectively. It shows that the higher the formation pressure is, the more favorable the production of CBM will be.

11.2.3 Application Conditions and Economic Analysis

11.2.3.1 Technical Introduction

A multi-branch horizontal well refers to drilling a primary horizontal well with multiple branches on both sides as gas flow channels. A multi-branch wellbore can penetrate more coal seams, maximize the contact with natural fracture systems, increase gas flowing channels and formation conductivity, thereby improving well production. In the field where underground reservoir permeability is very low or aboveground conditions are very poor, vertical wells are not suitable for developing CBM, but multi-branch horizontal wells have unique advantages. In order to reduce costs and consider specific geological conditions, sometimes a group of horizontal wells are drilled in three or four symmetrical directions, or two sets of branch wellbores are drilled to develop two coal seams on a well site at the same time.

Using such technology to develop CBM, it is advantageous to require less surface equipment and less land use, reduce ground obstacles and impact on the environment, and accordingly, drilling less wells can reduce investment and improve the comprehensive economic benefits.

11.2.3.2 Production Characteristics and Applicable Geological Conditions

Production Characteristics

① A multi-branch horizontal well creates multiple branches in coal seams, which increase contact with coal seams and water production.
② Fast pressure drop in a multi-branch horizontal well facilitates quicker peak gas production, and the peak gas production is high. A conventional fractured vertical well may spend a longer time to reach peak gas production. In the San Juan Basin, CDX multi-branch horizontal wells are drilled in coal seams with low permeability. The gas production is 2.83×10^4–$1^4 \times 10^4$ m^3/d, which is 10–20 times that of a vertical well.
③ The effective production life of a multi-branch horizontal well is short, but its cumulative gas production is high. A conventional vertical well takes longer to achieve the same cumulative production.

Suitable Reservoir Conditions

The advantages of CBM development with pinnate horizontal wells include high gas productivity, small land use, and strong adaptability to topographic conditions. For coal seams in mountainous areas or with low permeability, which are not suitable for vertical wells, pinnate horizontal wells are very suitable.

Multi-branch horizontal wells are suitable for coal seams with simple structures, stable and thick distribution and good coal, including high-rank and harder coal, but not suitable for coal seams of high permeability and high porosity and thinner coal seams with limestone or sandstone.

Economic Analysis

(1) Development scheme

In the Fanzhuang block, pinnate horizontal wells will be drilled to develop CBM—38 wells will be drilled and 2.5×10^8 m^3 of CBM production capacity will be built in 3 years. The scheme design is shown in Table 11.14.

Table 11.14 Technical scheme and development index prediction of pinnate horizontal wells

Year	The first year	The second year	The third year	The fourth to eighth years
Wells to be drilled per year	9	9	11	9
Annual production/10^8 m^3	0.6	1.2	1.8	2.5

(2) Investment estimation

According to applicable provisions, the total investment includes construction investment, interest during the construction period and circulating fund. The construction investment includes development project investment (well drilling and completion and ground facilities), deferred assets, intangible assets and basic reserve funds. According to engineering design, the investment of a well to drilling and completion is RMB 12 million, and the investment of surface construction is RMB 150 million.

The deferred assets account for 5% of the ground construction investment, involving project management and supervision, environmental impact assessment, production preparation, office and living equipment and furniture, joint commissioning, on-site foreign technicians, expatriate personnel, translation and duplication of drawing and files, etc.

The intangible assets take 8% of the ground construction investment, involving mining right, proprietary technology, goodwill, research and test, transfer of land use right, technology transfer, feasibility report, survey and design, power and water resources and so on.

The reserve fee refers to unpredictable fee, which accounts for 18% of the sum of the project investment, deferred assets and intangible assets. The total investment of the scheme is RMB 70,506 million (Table 11.15).

(3) Economic evaluation

The basic parameters selected for economic evaluation are listed in Table 11.16. CBM production is a state-supported industry, which is exempted from value added tax and mineral resources compensation, and the subsidy of selling 1 m^3 CBM is 0.2 yuan.

The results of the main evaluation indicators are shown in Table 11.17. It can be seen that, when wellhead gas price is 0.6 yuan/m^3, the internal rate of return is 12.85%; and when wellhead gas price is 0.8 yuan/m^3, the internal rate of return is 21.1%. Good economic benefits can be obtained.

11.3 Application Cases

11.3.1 Fracturing Operation and Effect

In the research above, after experimental study on fracture propagation and analysis of the fracture distribution law in fractured wells, hydraulic fracturing measures were put forward, and a hydraulic fracture propagation model was established and applied to the Daning-Jixian CBM field in the eastern margin of the Ordos Basin and the Fanzhuang block in the Qinshui Basin.

Table 11.15 Estimated investment of multi-branch horizontal wells in Fanzhuang block

No	Description	Investment/10^4 yuan
I	Construction	67,554
一	Development project	60,600
(1)	Development project	45,600
(2)	Ground facilities	15,000
二	Deferred assets	750
三	Intangible assets	1200
四	Basic reserves	5004
II	Interest during the construction period	2198
III	Working capital	754
	Total investment	70,506

Table 11.16 Basic indicators of economic evaluation

Basic indicators	Value	Unit	Basic indicators	Value	Unit
Project evaluation period	8	Year	Value added tax	0	%
Project construction period	3	Year	Subsidy/m^3	0.2	yuan
Commodity rate of CBM	95	%	Mineral resources compensation	0	%
Wellhead gas price	0.6–0.8	yuan/m^3	Resource tax	2	10^3 yuan/m^3
Discount rate	8	%	Income tax	33	%

Table 11.17 Economic evaluation results

No	Major economic indicators	Gas price 0.6 yuan/m^3	Gas price 0.7 yuan/m^3	Gas price 0.8 yuan/m^3
1	Internal rate of return/%	12.85	16.8	21.1
2	Net present value/10^4 yuan	7931	14,280	20,813
3	Payback period/year	7.1	6.8	6.2

Table 11.18 Geological parameters of Well J3

Well no	Coal seam	Coal seam depth/m	Coal seam thickness/m	Permeability/mD	Gas content/m^3/t
J3	No. 5	1128.2–1136.8	5.6	–	15.31–19.36

11.3.1.1 Application to Daning-Jixian CBM Field

(1) Brief geological introduction

The No. 5 coal in Well J3 in the Daning-Jixian CBM field is at 1128.4–1136.8 m and 5.6 m thick, the gas content is 15.31–19.36 m^3/t, 18.19 m^3/t on average, and the gas saturation is 71–89%. The well was not experienced injection and pressure drop tests, so that no permeability data is available (Table 11.18).

(2) Fracturing operation

Well J3 was fractured with gel fracturing fluid. The total volume of fracturing fluid is 262.9 m^3, including 121 m^3 pad fluid, 132.7 m^3 proppant-carrying fluid, 9.2 m^3 replacement fluid and 34.5 m^3 sand. The proppant concentration is 31%, and the proppant-adding strength is 5.5 m^3/m.

(3) Fracture diagnosis

On the fracturing curve of Well J3 (Fig. 11.68), there are obvious fractured points. The fracturing pressure is 25.3 MPa. The gradient of fracturing pressure is 0.0224 MPa/m, which is between the critical value for fracturing horizontal fractures and that for fracturing vertical fractures. The fractures may form a very complex fracture system with dominant vertical fractures.

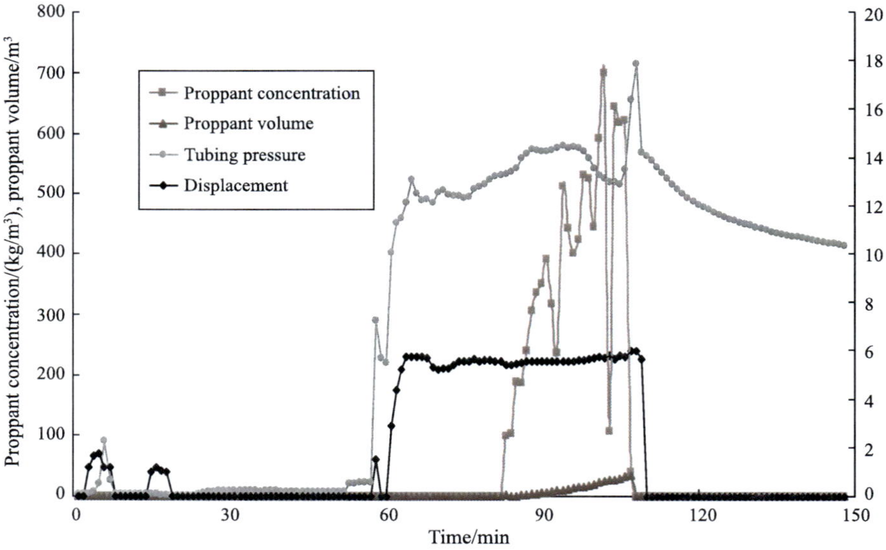

Fig. 11.68 Fracturing curve of Well J3

According to the sand addition at the later stage, the pressure rose at the constant displacement, indicating that natural fractures may be close, and then open with sand addition. A lot of fractures were induced.

Ground potential survey shows that the fracture orientation is N75°E and S75°W, and the fracture lengths are 48.2 m and 83.9 m, respectively.

Temperature logging data shows that the fracture height is 24.4 m, extending 3.2 m upward and 13.2 m downward. The fracture height is 3.7 times the coal seam thickness. Ground potential survey shows single-wing fractures. The length of the fracture at N75°E is very short, and that at S75°W is 106.6 m.

(4) Fracturing effect

Production test had been carried out for 28 days after fracturing (Fig. 11.69). Gas produced when the liquid level dropped to 461 m, and then increased with the steady decrease of the dynamic liquid level. When the dynamic liquid level dropped to 1050 m, gas production rose to 1 319 m^3/d, and water production was 21.03 m^3/d. The maximum gas production was 2446 m^3/d. When the dynamic liquid level was at 1126 m, gas production was 2006 m^3, and water production was 19.26 m^3.

11.3.1.2 Application to No. 3 Coal Seam in Xizhuang Block

(1) Geological characteristics

The No. 3 coal seam in the Fanzhuang block is generally 5–6 m thick at 350–750 m. The reservoir pressure tested is 2.77–4.20 MPa, with an average of 3.48 MPa, and the reservoir pressure coefficient is 0.67 MPa/hm, indicating low-pressure reservoir. Most of the coal is semi-bright, followed by bright coal. The macerals are mainly vitrinite. The

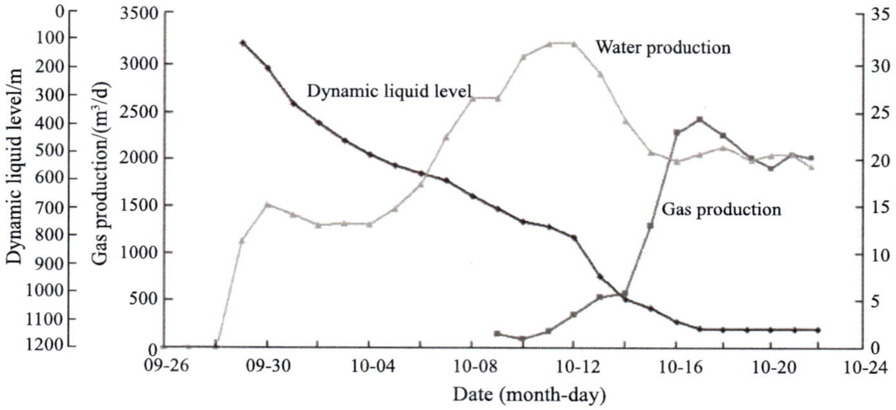

Fig. 11.69 Production test curve of Well J3

density is 1.39–1.51 t/m^3, with an average of 1.47 t/m^3. The metamorphism degree is high. The volatile yield is low, ranging from 5.20 to 9.0%, with an average of 7.48%. The moisture content is 0.95–2.19%, with an average of 1.27%. The ash yield is 14.26–17.89%, with an average of 15.36%.

The coal is very metamorphic and has a strong adsorption capacity. According to the isothermal adsorption experiment, the air drying Langmuir volume is 33.98–42.71 m^3/t, with an average of 37.88 m^3/t; the dry ash-free Langmuir volume is 42.44–53.53 m^3/t, with an average of 47.32 m^3/t; the Langmuir pressure is 3.35–3.83 MPa, with an average of 3.51 MPa; and the gas content is high at 17–25 m^3/t (Table 11.19).

According to injection and pressure drop test and numerical simulation results, the permeability is low, 0.025–0.51 mD (Table 11.20). Most wells have no natural productivity, or water production less than 0.3 m^3/d before fracturing.

(2) Fracturing difficulties and countermeasures

Through the analysis of reservoir characteristics, it is considered that the coal seam is continuous and suitable for draining water and depressurizing for gas production by fracturing stimulation. The coal seam has low porosity and permeability, but it has better sealing conditions and gas adsorbing and generating capacities. It is potential for fracturing stimulation. The critical desorption pressure is high and more conducive to coal methane desorption. Integral fracturing stimulation can play the role in reducing the reservoir pressure and realize the effective CMB development in the Fanzhuang block.

However, there are some difficulties in fracturing operation. The pressure coefficient is low, which is not conducive to the flowback of fracturing fluid. The temperature is very low, making it difficult for fracturing fluid to break completely. The low porosity, low permeability, low diffusion coefficient and strong heterogeneity of the coal seam require moderately large-scale fracturing, and the fractured target should be optimized. Moreover, the developed natural fractures and strong absorbability lead to serious filtration and being susceptible to damage the coal seam. The stress difference between the coal seam and upper and lower barriers is small, so that the control of fracture height is difficult. The Young's modulus of the coal rock is low, and the proppant embedding is serious. The stress sensitivity of the coal rock is strong, so that it is easy to cause pressure-sensitive damage to coal pores and permeability.

According to the reservoir characteristics and the fracturing difficulties, fracturing measures were put forward (Table 11.21).

(3) Fracturing effect

Eighteen wells have been fractured in the Fanzhuang block, and most of them have an initial peak in the early production (Fig. 11.70). This phenomenon shows that the induced fracture improves the flow conductivity around the wellbore, accelerates the water drainage and pressure drop near the fractured fractures, and desorbs CBM, therefore resulting in the increase of gas production. However, the induced fractures are not enough to extend far to the area with low permeability, so that it is difficult to drain water and induce pressure drop, CBM can not be desorbed in a large extent, and finally insufficient gas supply results in declining gas production.

The degree and time of production decline after initial peak depend on the seepage condition of coal seam itself. The lower

Table 11.19 Statistical gas content

Well no	No. 3 coal seam		No. 15 coal seam	
	Depth/m)	Gas content/(m^3/t)	Depth/m	Gas content/(m^3/t)
1	521.6	25.29	606.6	22.39
2	514.2	22.80	610.8	21.50
3	509.2	17.10	601.0	12.70
4	521.6	25.25	613.8	23.64

Table 11.20 Permeability of no. 3 coal seam

Well no	Interval/m	Skin factor	Permeability from injection and pressure drop test/mD
1	519.0–524.4	−3.66	0.22
2	539.0–544.4	−3.4	0.08
3	521.6–527.4	−0.53	0.51
4	514.2–520.6	−1.29	0.08
5	509.2–515.2	−1.14	0.038
6	521.6–527.4	−0.64	0.025

Table 11.21 Reservoir characteristics and fracturing measures

Reservoir characteristics	Fracturing difficulties	Fracturing measures
Low temperature, strong absorbability, and difficult to break gel	Susceptible to damage the coal seam; fracturing fluid with low damage and good gel breaking performance required	Less active water system, and filtration control
Low Young's modulus	Strong plasticity and easy proppant embedding	High sand ratio or late large sands; low initial pressure, small step and multi-stage sand addition
Developed natural fractures	Complex fracture initiation and propagation; damage to the coal seam; lost fracturing fluid; inadequate fractures; and early sand plugging	Reasonable displacement; optimized pad liquid percent; proppant slug; limited pressure and unlimited displacement in later stage of sand adding
Low porosity and permeability, poor connectivity	Large-scale fracturing operation may increase the risk of sand plugging	On-site filtration test when injecting pad liquid; optimized pad liquid volume; proppant slug; and low sand ratio and multi-stage sand addition
Low stress difference	Difficult to control fracture height	Combined pad liquid; variable displacement; downward fracture extension: after injecting the last proppant slug, shut down for 4–5 min if fluid loss is not too serious; upward fracture extension: shut down as required after fracturing
Strong stress sensitivity	Easy to induce compaction effect near wellbore during flowback	Suitable the choke and control the lowest dynamic fluid level

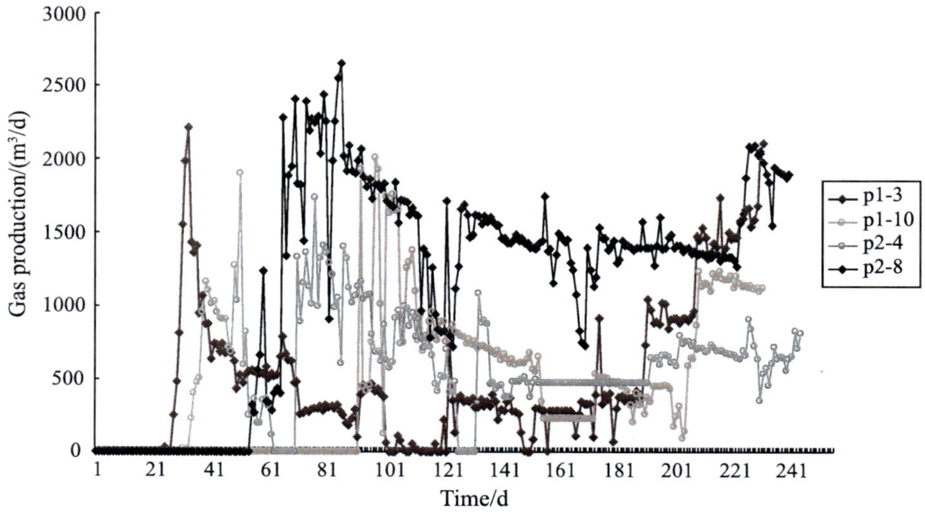

Fig. 11.70 Initial peaks of CBM wells

the permeability, the more obvious the decline of gas production will be, and the longer the decline time will be.

After the initial peak, the gas production of most wells declined to a some extent and then remained stable. However, some wells showed a steady increasing production (Figs. 11.71 and 11.72). It indicates that the desorbing area was gradually enlarged after effective water drainage and pressure drop, and the gas production increased gradually.

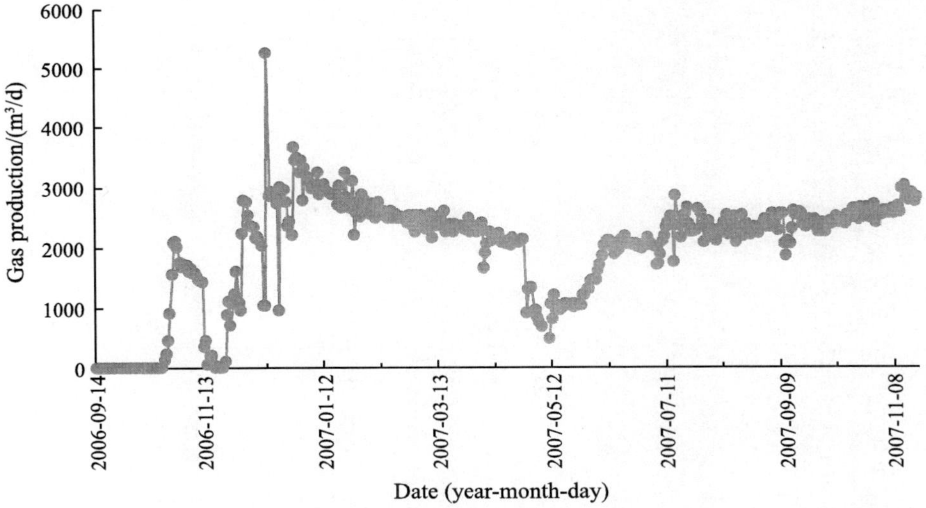

Fig. 11.71 Gas production curve of Well p1

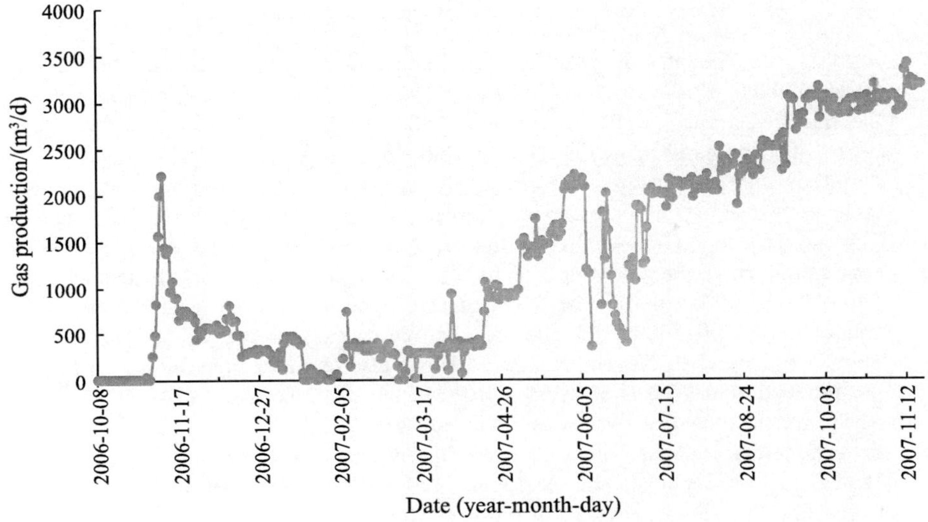

Fig. 11.72 Gas production curve of Well p4

11.3.2 Application of Multi-branch Horizontal Wells and Result Analysis

Based on the mathematical model and numerical model of CBM development by multi-branch horizontal wells, we developed a computer simulation program which has been used to optimize the design of multi-branch horizontal wells in the Fanzhuang block. The well gas production has been increased greatly.

11.3.2.1 Application Cases

In 2006, PetroChina drilled a multi-branch horizontal Well f1 in the Fanzhuang block. The horizontal section in coal seams is 5158.5 m long. After six months of production, the gas production reached over 10,000 m^3/d. This well has a large control area, effectively communicates with the natural fractures in the coal seam, and has a wide range of pressure drop and desorption. In the initial stage of gas production, there is no pressure drop funnel like that in a vertical well, and the production showed a steady increasing trend (Fig. 11.73).

11.3.2.2 Production Increase

By comparing the production of the multi-branch horizontal Well f1 with a fractured vertical well, the following understandings are obtained.

① The water drainage of the multi-branch horizontal Well f1 is large. The water production of Well f1 is 10–20 m^3/d, and that of a fractured vertical well is only 1–3

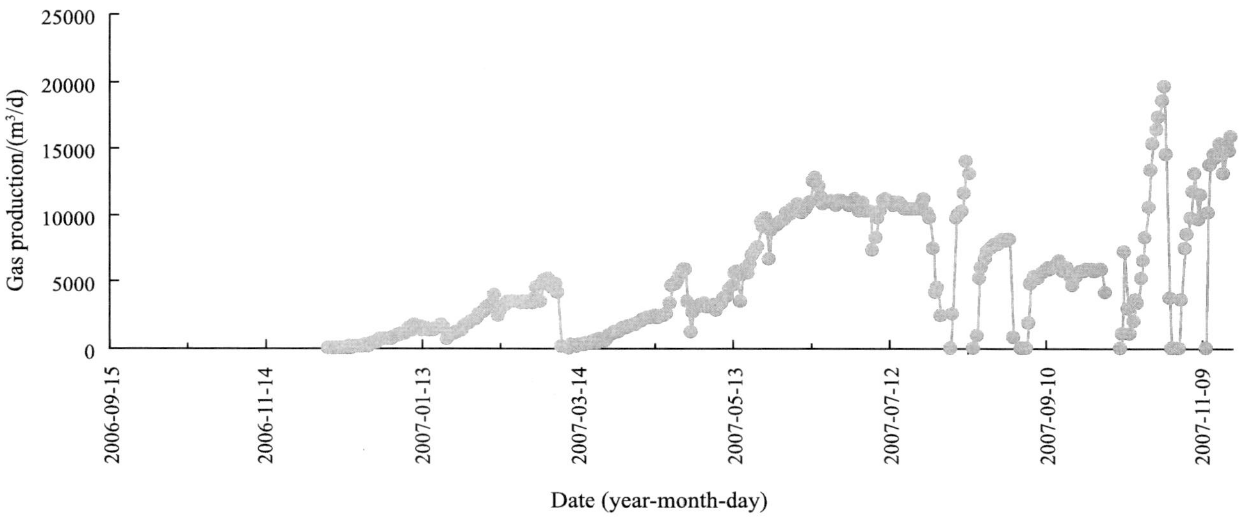

Fig. 11.73 Gas production curve of Well f1

m^3/d. By the end of November 2007, the cumulative water production of Well f1 was 4800 m^3, and that of the fractured vertical well was averaged 1500 m^3. It proves that the multi-branch horizontal well can communicate with coal seams more effectively and has a large water drainage and pressure drop area.

② The gas production of Well f1 is higher. After one year's production, the gas production of Well f1 rose to 1.6×10^4 m^3/d, but that of the fractured vertical well was generally 500–3000 m^3/d, 5000 m^3/d at the maximum. By the end of November 2007, the cumulative gas production of Well f1 was 179×10^4 m^3, but that of the fractured vertical well was averaged 24×10^4 m^3—the former is about 7 times the latter. It proves that Well f1 can communicate with coal seams more effectively and has a large area of pressure drop and desorption.

③ The bottom hole pressure of Well f1 is higher than that of the vertical well by about 0.5 MPa, indicating less loss of fluid flowing energy.

In summary, the multi-branch horizontal well has the advantages of fast water drainage and pressure drop, high productivity and low loss of fluid energy. However, there are some problems that can not be ignored. First, it is strict with geological conditions–the best coal seam should have simple structures, stable and thick distribution and good coal. Second, high drilling cost is required with applicable technology conditions.

Reference

Zhou ST, Zhang Q, Li MZ (2002) Research progress of variable mass flow in horizontal wells. Prog Mech 32(1):119–127 (in Chinese)

Epilogue

The project "**Basic Theory of CBM Geology and Development in China**" is on the way of the development of China's CBM industry. At the beginning of the project, CBM wells drilled were not more than 300, and the annual output of CBM was less than 1×10^8 m^3 in China. Commercial CBM development had not yet begun. After more than five years, CBM wells drilled have reached more than 2000, and the output has increased to 4.5×10^8 m^3 per year, and the production capacity has expanded to 15×10^8 m^3 per year. The development history of Canada's CBM industry shows that the annual production of CBM was 5×10^8 m^3, and the wells drilled were only 700 in 2003; by 2007, the production of CBM increased to 103×10^8 m^3, and the wells drilled increased to more than 16,000. China's CBM industry is at the beginning of rapid development. China's CBM resources have a great potential, and the technically recoverable resources of CBM are up to 13.9×10^{12} m^3. At present, the proven rate of CBM is less than 0.4%. The development of China's CBM industry should start from basic researches on the forming mechanism, distribution law and control factors of CBM, and finally specific technology.

The basic research plays a very positive role in the development of CBM industry. We have achieved four innovations in basic theoretical research—understood CBM genesis, occurrence, accumulation and permeability, systematically established China's CBM geological theory system, enriched and perfected natural gas geological theory, and built the theoretical basis for gas resource evaluation, enrichment prediction and economic development; and five innovations in method and technology—built an experimental platform for basic research of CBM, and developed a series of technologies for CBM recoverable resource prediction, comprehensive geological evaluation, geophysical exploration and production optimization design which have been applied in field. The application of the innovations has achieved good economic and social benefits.

However, there are challenges that restrict the development of CBM in China.

According to the experience of foreign CBM development, CBM blocks with high abundance play important roles in CBM reserves and production increase. The Fruitland CBM enrichment belt in the San Juan Basin, USA is 14 km wide and 64 km long; the abundance of the CBM resources is 1.6×10^8–3.3×10^8 km^2; and the output of a CBM well can reach 2.8×10^4–16.8×10^4 m^3/d, and the annual output accounts for 50% of the total annual output of CBM in the United States. According to the formation and distribution characteristics of CBM in the Qinshui Basin with high degree of CBM exploration and development in China, the CBM geological characteristics and drilling conditions prove that the southern part of the basin is conducive to accumulation of CBM. But researches have not yet targeted the formation, distribution and prediction of CBM enrichment zones with high abundance, and the control factors and enrichment mechanism of high abundance CBM zones are unclear. These restrict the selection and development of the high-abundance CBM enrichment zones with the similar characteristics to the Fruitland CBM enrichment belt.

The geological conditions of CBM in China are very diverse. At present, CBM production mainly depends on vertical wells and pinnate multi-branch horizontal wells; production increase depends on hydraulic fracturing stimulation; both natural and stimulated productions are low and well production is low. The primary cause is that the mechanism, adaptability and control factors of the producing methods are unclear. For example, the structural deformation of the coal seams in China is quite serious, resulting in complex coal structure, more fragile rock mechanical properties, low permeability, greatly varying production performance, and difficult and ineffective fracturing stimulation. Present economic benefits of CBM development are very unsatisfactory. It is necessary to study interwell interference, improve the theory and technology of coal reservoirs, and understand fracturing mechanism, fracture propagation and control factors, reservoir damage, pinnate horizontal wells, etc., and finally propose optimized CBM development plan to increase CBM production.

Y. Song et al., *Coalbed Methane in China*,
https://doi.org/10.1007/978-981-33-4725-0

In summary, present researches mainly focus on the microscopic mechanism of CBM occurence, the law of CBM accumulation, the forming mechanism of CBM reservoirs and the distribution of CBM resources in large regions and basins, but less on the distribution law, prediction and evaluation of CBM enrichment zones. For production technology, researches have been done only involving numerical simulation to producing CBM while draining water from single wells and horizontal wells, CBM desorption and seepage laws and seepage mechanism, but less on well group and stimulation mechanisms, even little on interwell interference, reservoir damage and protection. It is recommended to consider the following two important points in the future 973 Program: the mechanism and evaluation of abundant CBM enrichment zones; technical approaches to increase CBM production. All these will promote the basic theoretical researches on CBM geology and development, and make greater contributions to the rapid development of China's CBM industry.